**Functional Neuroimaging
of Visual Cognition**

Attention and Performance

Attention and Performance XIV: Synergies in Experimental Psychology, Artificial Intelligence, and Cognitive Neuroscience. Edited by David E. Meyer and Sylvan Kornblum, 1993

Attention and Performance XV: Conscious and Nonconscious Information Processing. Edited by Carlo Umiltà and Morris Moscovitch, 1994

Attention and Performance XVI: Information Integration in Perception and Action. Edited by Toshio Inui and James L. McClelland, 1996

Attention and Performance XVII: Cognitive Regulation of Performance: Interaction of Theory and Application. Edited by Daniel Gopher and Asher Koriat, 1998

Attention and Performance XVIII: Control of Cognitive Processes. Edited by Stephen Monsell and Jon Driver, 2000

Attention and Performance XIX: Common Mechanisms in Perception and Action. Edited by Wolfgang Prinz and Bernhard Hommel, 2002

Attention and Performance XX: Functional Neuroimaging of Visual Cognition. Edited by Nancy Kanwisher and John Duncan, 2003

Functional Neuroimaging of Visual Cognition
Attention and Performance XX

Edited by
Nancy Kanwisher
John Duncan

This book is based on the papers presented at the **Twentieth International Symposium on Attention and Performance** held at the Ettore Majorana Centre for Scientific Culture in Erice, Sicily, July 1-7, 2002

OXFORD
UNIVERSITY PRESS

Great Clarendon Street, Oxford OX2 6DP

Oxford University Press is a department of the University of Oxford.
It furthers the University's objective of excellence in research, scholarship,
and education by publishing worldwide in

Oxford New York

Auckland Bangkok Buenos Aires Cape Town Chennai
Dar es Salaam Delhi Hong Kong Istanbul Karachi Kolkata
Kuala Lumpur Madrid Melbourne Mexico City Mumbai Nairobi
Sao Paulo Shanghai Taipei Tokyo Toronto

Oxford is a registered trade mark of Oxford University Press
in the UK and in certain other countries

Published in the United States
by Oxford University Press Inc., New York

© The International Association for the Study of Attention and Performance, 2003

The moral rights of the author have been asserted

Database right Oxford University Press (maker)

First published 2004

All rights reserved. No part of this publication may be reproduced,
stored in a retrieval system, or transmitted, in any form or by any means,
without the prior permission in writing of Oxford University Press,
or as expressly permitted by law, or under terms agreed with the appropriate
reprographics rights organization. Enquiries concerning reproduction
outside the scope of the above should be sent to the Rights Department,
Oxford University Press, at the address above

You must not circulate this book in any other binding or cover
and you must impose this same condition on any acquirer

A catalogue record for this title is available from the British Library

ISBN 0 19 852845 0

10 9 8 7 6 5 4 3 2 1

Typeset by Cepha Imaging Pvt Ltd
Printed in China

Preface

The twentieth meeting of the International Association for the Study of Attention and Performance took place on July 2-6, 2002 in the beguiling surroundings of the Ettore Majorana Centre for Scientific Culture, Erice, Sicily. The meeting brought together scientists addressing visual cognition at many levels–phylogeny, neurophysiology, brain imaging, behaviour, and formal modelling. Over five days, we discussed what each perspective brings to our understanding of visual function, and the prospects for effective convergence; this book is the result.

The plan for a symposium on functional brain imaging began with a meeting of the Association's Executive Committee in July 1998. At that time, the dominant theme in imaging work still seemed to be "human brain mapping"–localization of cognitive functions in the brain. For those of us working on functional models of cognition, this theme was unsatisfying–the complaint was frequently made that knowing *where* told us little of *how*, and that in some way neuroimaging was failing to deliver on its evident promise. For Attention and Performance XX, we planned a meeting to ask whether and how imaging methods could be adapted to contribute to a real functional understanding of cognition.

By the 2002 meeting, the field had already flown by the Executive Committee's question. Instead of discussing the abstract idea of theoretical convergence, in visual cognition we had a blossoming existence proof that it could work. In paper after paper, we heard not of "human brain mapping" but of function–the sub-processes of object recognition, the mechanisms of selective attention, how vision links to action, and the plasticity of learning and development. Whether the method was human neuroimaging, single cell physiology in the behaving monkey, cognitive psychology, or formal modelling, the central question throughout was no longer where vision happens, but how it works.

Of course, this change of perspective for imaging work has happened more quickly in some places than in others. As the central topic of the meeting we chose vision exactly because we think it is the best current example. Important factors in this success include clear functional questions, direct links to animal experiments, and grounding in known sensorimotor physiology. Together, these provide a strong basis for the cognitive neuroscience enterprise and for models of how cognitive function is implemented at the neural level.

This book has four major sections. In the first, we address the key question of modularity–the extent to which vision can be analysed in terms of distinct neural modules with separate visual functions. On the one hand, much neuroimaging work in grounded on modular assumptions on the other, neuroimaging, with its simultaneous measurement across the whole brain, is perhaps the method most clearly suited to pose the question of how well these assumptions hold. In the second section we turn to the specific topic of object recognition, and its implementation in the occipitotemporal "ventral stream". In computational and empirical studies, the papers ask how everyday objects, faces, and words are represented at successive levels of the visual system, how these representations develop and work, and why they take the form they do. The final two sections address visual attention and sensorimotor integration respectively. In both fields, well-specified psychological models (for example, the quantitative models developed over 50 years' study of evidence accumulation and action choice) are now under close investigation in physiological studies of the behaving animal. The papers present both this detailed convergence of psychological and physiological levels, and how it can now be extended with the methods of fMRI, ERP, and MEG.

The book also contains two special chapters. If our general theme is convergence of cognitive, neuroimaging, and physiological work, then Giacomo Rizzolatti's chapter, based on his Association Lecture at the meeting, is a beautiful case study. Rizzolatti's discovery of "mirror neurons" in the monkey premotor cortex–neurons responding either when a particular action is carried out, or when it is observed in the behaviour of another–has led to an explosion of studies addressing the physiological basis for both action imitation and action understanding. As Rizzolatti discusses, these ideas now reach far beyond the context of simple movements like grasping the tearing, informing domains from social perception to language production. For the final paper at the meeting–and the final chapter in the book–we turn to the scientist who, more than any other, first pointed the way for a cognitive neuroscience based on imaging methods. In this final chapter, Mike Posner uses the papers of the meeting to draw out major questions both for current research and for the future. These include neural networks and cognitive modularity, combination of anatomical and time-course studies, questions of attention and volition, the central importance of development and plasticity, and the prospect from the work now beginning to use functional imaging to study genetic variation.

The organization of this meeting owes much to the advice and support of the organization's Executive Committee and Advisory Council, and especially to former secretary of the Association Stephen Monsell whose guidance, wisdom, and detailed knowledge of A&P tradition were critical to the success of the meeting. We owe an enormous debt to the Ettore Majorana Centre for Scientific Culture, who provided their spectacular meeting facilities and setting; to Danilo Mainardi, Director of the School of Ethology of the Centre, who acted as host of the conference; and in particular to our co-organizer Carlo Umiltà, whose expert liaison with the Centre on all scientific and practical matters took so much of the organizational burden off our shoulders. Superb assistance with meeting logistics and with construction of the meeting program came from Yuhong Jiang, Rebecca Saxe, and Ellen Goodman. Finally we are grateful for generous financial support from the National Institute of Mental Health, the Office of Naval Research, the Air Force Office of Scientific Research, and the National Science Foundation.

Contents

The Attention and Performance Symposia *xi*

Participants *xiii*

Association lecture

1 Understanding the actions of others *3*
  Giacomo Rizzolatti

Part 1 Modularity: Splitting the atom of visual cognition

2 Core knowledge *29*
  Elizabeth Spelke

3 The evolution of human neocortex: Is the human brain fundamentally different than that of other mammals? *57*
  Leah Krubitzer and Dianna M. Kahn

4 Analysis of topographically organized patterns of response in fMRI data: Distributed representations of objects in the ventral temporal cortex *83*
  James V. Haxby

5 Time as coding space in the cerebral cortex *99*
  Wolf Singer

6 Separate modifiability and the search for processing modules *125*
  Saul Sternberg

Part 2 Investigations of the ventral visual pathway: Extracting and representing visual objects

7 Approaches to visual recognition *143*
  Shimon Ullman

8 The functional organization of the ventral visual pathway and its relationship to object recognition *169*
  Kalanit Grill-Spector

9 The cartography of human visual object areas *195*
  R. Malach, G. Avidan, Y. Lerner, U. Hasson, and I. Levy

10 The neural bases of subliminal priming *205*
   Stanislas Dehaene

11 Audio-visual associative learning enhances responses to auditory stimuli in visual cortex *225*
   Désirée Gonzalo and Christian Büchel

12 Neural representation of object images in the macaque inferotemporal cortex *241*
   Manabu Tanifuji, Kazushige Tsunoda, and Yukako Yamane

13 Plasticity and functional brain development: The case of face processing *257*
   Mark H. Johnson

Part 3 **Mechanisms of visual attention**

14 Neurobiology of human spatial attention: Modulation, generation, and integration *267*
   Jon Driver, Martin Eimer, Emiliano Macaluso, and José van Velzen

15 Towards a neural basis of human visual attention: Evidence from functional brain imaging *301*
   Sabine Kastner

16 The influence of scene organization on attention: Psychophysics and electrophysiology *321*
   Mitchell Valdes-Sosa, Maria A. Bobes, Valia Rodríguez, Yanelis Acosta, Alejandro Pérez, Jorge Iglesias, and Mayelin Borrego

17 Endogenous and stimulus-driven mechanisms of task control *345*
   Gordon L. Shulman and Maurizio Corbetta

18 The role of competitive circuits in extrastriate cortex in selecting spatially superimposed stimuli *363*
   John H. Reynolds and Mazyar Fallah

19 The imaging of visual attention *381*
   Steven A. Hillyard, Francesco Di Russo, and Antigona Martinez

Part 4 **Sensorimotor integration**

20 Neural selection and control of action *391*
   Jeffrey D. Schall

21 Human brain imaging reveals a parietal area specialized for grasping *417*
   Jody Culham

22 A framework for the investigation of directed cortical interactions: Theoretical background and application to dynamic sensorimotor mapping *439*
   Rainer Goebel, Alard Roebroeck, Dae-Shik Kim, and Elia Formisano

23 From viewing of movements to understanding and imitation of other persons' acts: MEG studies of the human mirror-neuron system *463*
Riitta Hari and Nobuyuki Nishitani

24 Neuropsychological perspectives on sensorimotor integration: Eye-hand coordination and visually guided reaching *481*
David P. Carey

Finale

25 The achievement of brain imaging: Past and future *503*
Michael I. Posner

Author index *529*

Subject index *549*

The Attention and Performance Symposia

Since the first was held in The Netherlands in 1966, the Attention and Performance Symposia have become an established and highly successful institution. They are now held every two years, in a different country. The original purpose remains: to promote communication among researchers in experimental cognitive psychology and related areas working at the frontiers of research on 'attention, performance, and information processing'. The format is an invited workshop-style meeting, with plenty of time for papers and discussion, leading to the publication of an edited volume of the proceedings. The International Association for the Study of Attention and Performance exists solely to run the meetings and publish the volume. Its Executive Committee selects the organizers of the next meeting, and develops the program in collaboration with them, with advice on potential participants from an Advisory Council of up to 100 members. Participation is by invitation only, and the rules of the Association[1] are constructed to ensure participation from a wide range of countries, with a high proportion of young researchers, and a substantial injection of new participants from meeting to meeting.

Held usually in a relatively isolated location, each meeting has four and a half days of papers presented by a maximum of 26 speakers, plus an invited Association Lecture from a leading figure in the field. There is a maximum of 65 participants (incl. the current members of the executive committee and the organizers). There are no parallel sessions, and all participants commit themselves to attending all the sessions. There is thus time for substantial papers followed by extended discussion, both organized and informal, and opportunities for issued and ideas introduced at one point in the meeting to be returned to and developed later. Speakers are encouraged to be provocative and speculative, and participants who do not present formal papers are encouraged to contribute actively to discussion in various ways; for example, as formal discussants, by presenting a poster, or as contributors to scheduled discussion sessions. This intensive workshop atmosphere has been one of the major strengths and attractions of these meetings.

Manuscript versions of the papers are refereed anonymously by other participants and external referees and published in a high-quality volume edited by the organizers, with a publication lag similar to many journals. Unlike many edited volumes, the Attention and Performance series reaches a wide audience and has considerable prestige. Although not a journal, it is listed in journal citation indices with the top dozen journals in experimental psychology. According to the Constitution, 'Papers presented at meetings are expected to describe work not previously published, and to represent a substantial contribution...' Over the years, contributors have been willing to publish original experimental and theoretical research of high quality in the volume, and this tradition continues. A&P review papers have also been much cited. The series has attracted widespread praise in terms such as 'unfailingly presented the best work in the field' (S. Kosslyn, Harvard), 'most distinguished series in the field of cognitive psychology' (C. Bundesen, Copenhagen), 'held in high esteem throughout the field because of its attention to rigor, quality and scope...indispensable to anyone who is serious about understanding the current state of the science' (M. Jordan MIT), 'the books are an up-to-the-minute tutorial on topics fundamental to understanding mental processes' (M. Posner, Oregon).

In the early days of the Symposium, when the scientific analysis of attention and performance was in its infancy, thematic coherence could be generated merely by gathering together the most active researchers in the field. More recently, experimental psychology has ramified, 'cognitive science' has been born, and converging approaches to the issues we study have developed in neuroscience. Participation has therefore become interdisciplinary, with neuroscientists, neuropsychologists and computational modelers joining the experimental psychologists. Each meeting now focuses on a restricted theme under the general heading of 'attention and performance'. Recent themes include *Synergies in Experimental Psychology, Artificial Intelligence and Cognitive Neuroscience* (USA, 1990), *Conscious and Unconscious Processes* (Italy, 1992), *Integration of Information* (Japan, 1994), *Cognitive Regulation of Performance: Interaction of Theory and Application* (Israel, 1996), and *Control of Cognitive Processes* (UK, 1998), and *Common Processes in Perception and Action* (Germany, 2000).

---

[1] For more information about the Association and previous symposia, visit the webpage http://go.to/A&P

Participants

Alan Allport
Department of Experimental Psychology
University of Oxford
South Parks Road
Oxford OX1 3UD
UK
alan.allport@psy.ox.ac.uk

Marlene Behrmann
Carnegie Mellon University
Department of Psychology
Pittsburgh, PA 15213-3890
USA
behrmann@condor.cnbc.cmu.edu

Christian Büchel
Universitätskrankenhaus Hamburg-Eppendorf
Neurologische Klinik, Haus S10
Martinistr. 52
20246 Hamburg
Germany
buechel@uke.uni-hamburg.de

Claus Bundesen
Department of Psychology
University of Copenhagen
Njalsgade 90
DK-2300 Copenhagen S
Denmark
claus.bundesen@psy.ku.dk

David Carey
Neuropsychology Research Group
Department of Psychology
University of Aberdeen
Old Aberdeen, AB24 2UB
Scotland, UK
d.carey@abdn.ac.uk

Leo Chelazzi
Associate Professor of Neurophysiology
Department of Neurological and Vision Sciences
Section of Physiology, University of Verona
Strada Le Grazie 8
I-37134 Verona,
Italy
leonardo.chelazzi@univr.it

Maurizio Corbetta
Washington University School of Medicine
660 S. Euclid
Campus Box 8111
St. Louis, MO 63110
USA
mau@npg.wustl.edu

Laila Craighero
University of Ferrara
Dept. of Biomedical Sciences and Advanced Therapies-
Section of Human Physiology
via Fossato di Mortara 17/19
44100 Ferrara
Italy
crh@unife.it

Jody Culham
University of Western Ontario
Psychology Department
London, Ontario N6A 5C2
Canada
culham@imaging.robarts.ca

Stanislas Dehaene
Unité INSERM 562 "Neuroimagerie Cognitive"
Cognitive Neuroimaging Unit
Service Hospitalier Frederic Joliot
CEA/DRM/DSV
4 Place du general Leclerc
91401 Orsay cedex
France
dehaene@shfj.cea.fr

Paul Downing
Brigantia Building
School of Psychology
University of Wales
Bangor LL5 72AS
Wales, UK
p.downing@bangor.ac.uk

Jon Driver
Institute of Cognitive Neuroscience
University College London
17 Queen Square
London WC1N 3A
UK
j.driver@ucl.ac.uk

John Duncan
MRC-CBU Cambridge
15 Chaucer Road
Cambridge CB2 2EF
UK
john.duncan@mrc-cbu.cam.ac.uk

Martin Eimer
Dept. of Psychology
Birkbeck College
University of London
Malet Street
London WC1E 7HX
UK
m.eimer@bbk.ac.uk

Steve Engel
1285 Franz Hall
Box 951563
Los Angeles, CA 90095-1563
USA
engel@psych.ucla.edu

Jan Willem de Fockert
Room 213 Whitehead Building
Psychology Department
Goldsmiths College
University of London
New Cross, London SE14 6NW
UK
j.de-fockert@ucl.ac.uk

Isabel Gauthier
Department of Psychology
Vanderbilt University
530 Wilson Hall
Nashville, TN 37203
USA
isabel.gauthier@vanderbilt.edu

Rainer Goebel
Department of Neurocognition
Faculty of Psychology
Universiteit Maastricht
Postbus 616
6200MD Maastricht
The Netherlands
R.Goebel@psychology.unimaas.nl

Daniel Gopher
Industrial Engineering and Management
Technion
Haifa, 32000
Israel
ierbw05@ie.technion.ac.il

Gabriele Gratton
Dept. of Psychology and Beckman Institute
University of Illinois
2161 Beckman Institute

405 N. Mathews Avenue
Urbana, IL 61801
USA
grattong@uiuc.edu

Kalanit Grill-Spector
Stanford University
Psychology Department
Jordan Hall, Bldg. 420
Stanford, CA 94305-2130
USA
kalanit@psych.stanford.edu

Riitta Hari
Brain Research Unit
Low Temperature Laboratory
Helsinki University of Technology
Otakaari 3 A
02015 HUT, Espoo
Finland
hari@neuro.hut.fi

James Haxby
3-S-13 Green Hall
Department of Psychology
Princeton University
Princeton, New Jersey 08544
USA
haxby@princeton.edu

Steven Hillyard
University of California San Diego
School of Medicine
9500 Gilman Drive Room 3054
La Jolla, CA 92093
USA
shillyard@ucsd.edu

Yuhong Jiang
Harvard University
820 William James Hall
33 Kirkland St.
Cambridge, MA 02138
USA
jiang5@fas.harvard.edu

Mark Johnson
Centre for Brain and Cognitive
Development
School of Psychology
Birkbeck College
32 Torrington Square
London WC1E 7JL
UK
mark.johnson@
psyc.bbk.ac.uk

Nancy Kanwisher
MIT
NE20-454
77 Massachusetts Avenue
Cambridge, MA 02139
USA
ngk@mit.edu

Sabine Kastner
Department of Psychology
Center for the Study of Brain,
Mind & Behavior
Princeton University
Green Hall
Princeton, NJ 08544
USA
skastner@princeton.edu

Mitsuo Kawato
Hikaridai 2-2-2 "Keihanna
Science City"
Seikacyo, Sorakuguń
Kyoto 619-0288
Japan
kawato@atr.co.jp

Roberta Klatzky
Carnegie Mellon University
Department of Psychology
Baker Hall 342c
Pittsburgh, PA 15213
USA
klatzky@andrew.cmu.edu

Leila Montaser Kouhsari
Researcher in School of Cognitive Sciences
Institute for Theoretical Studies in
Physics and Mathematics(IPM)
Niavaran PO.Box:19395-5746
Tehran, Iran
montaser@ipm.ir

Leah Krubitzer
Universit of California, Davis
Center for Neuroscience
1544 Newton Court
Davis, CA 95616
USA
lakrubitzer@ucdavis.edu

Nikos Logothetis
Max-Planck Institute for Biological
Cybernetics
Spemannstr. 38
72076 Tuebingen
Germany
nikos.logothetis@tuebingen.mpg.de

Rafael Malach
Department of Neurobiology
Weizmann Institute of Science
Rehovot 76100
Israel
Bnmalach@wisemail.weizmann.ac.il

Rene Marois
Department of Psychology
Vanderbilt University
301 Wilson Hall
Nashville, TN 37203
USA
rene.marois@vanderbilt.edu

Jason Mattingley
Department of Psychology
School of Behavioural Science
University of Melbourne
Victoria 3010
Australia
j.mattingley@psych.unimelb.edu.au

Gregory McCarthy
Brain Imaging and Analysis Center
Duke University Medical Center
PO Box 3918
Durham, NC 27710, USA
gregory.mccarthy@duke.edu

Janine Mendola
Center for Advanced Imaging
HSC-S, Radiology, PO Box 9236
West Virginia University School of
Medicine
Morgantown, WV 26506-9236
USA
jmendola@hsc.wvu.edu

Stephen Monsell
University of Exeter
School of Psychology
Washington Singer Laboratories
Exeter EX4 4QG, UK
S.Monsell@exeter.ac.uk

Cathleen Moore
Department of Psychology
Pennsylvania State University
Moore Building
University Park, PA 16802
USA
cmm15@psu.edu

Helen Neville
Department of Psychology
1227 University of Oregon
Eugene, OR 97403-1227
USA
neville@uoregon.edu

Alvaro Pascual-Leone
Department of Neurology
Beth Israel Hospital
Harvard Medical School
330 Brookline Ave
Boston, MA 02215
USA
apleone@caregroup.harvard.edu

Eraldo Paulesu
Psicobiologia e Psicologia Fisiologica
Facoltà di Psicologia
Università degli Studi di Milano Bicocca
Piazza dell'Ateneo Nuovo 1
20126 Milano, Italy
eraldo.paulesu@unimib.it

Michael Posner
Department of Psychology
1227 University of Oregon
Eugene, OR 97403-1227
USA
mposner@darkwing.uoregon.edu

Mary C. Potter
MIT
NE20-453
77 Massachusetts Ave.
Cambridge, MA 02139
USA
mpotter@mit.edu

Wolfgang Prinz
Max Planck Institute for Psychological Research
Amalienstr. 33
D-80799 Munich, Germany
prinz@psy.mpg.de

Robert Rafal
School of Psychology
University of Wales, Bangor
Brigantia Building
Penrallt Road
Bangor, Gwynedd LL57 2AS
Wales, UK
r.rafal@bangor.ac.uk

Geraint Rees
Institute of Cognitive Neuroscience
University College London
Alexandra House
17 Queen Square
London WC1N 3AR, UK
g.rees@fil.ion.ucl.ac.uk

John H. Reynolds
The Salk Institute for Biological Studies
10010 North Torrey Pines Road
La Jolla, CA 92037-1099
USA
reynolds@salk.edu

Giacomo Rizzolatti
Dipartimento di Neuroscienze
Sezione di Fisiologia
Università di Parma
Via Volturno, 39
43100 Parma
Italy
giacomo.rizzolatti@unipr.it

David Rosenbaum
Pennsylvania State University
Department of Psychology
642 Moore Building
University Park, PA 16802-3104
USA
dar12@psu.edu

Rebecca Saxe
Harvard University
1168 William James Hall
33 Kirkland St.
Cambridge, MA 02138
USA
saxe@mit.edu

Jeffrey Schall
Vanderbilt University
Department of Psychology
Wilson Hall
Nashville, TN 37203, USA
jeffrey.d.schall@vanderbilt.edu

Gordon Shulman
Dept. of Neurology, Box 8111
Washington University School of Medicine
4525 Scott Ave.
St. Louis, MO 63110
USA
gordon@npg.wustl.edu

Wolf Singer
Max-Planck-Institut für Hirnforschung
Deutschordenstrasse 46
D-60528 Frankfurt am Main
Germany
prinz@mpipf-muenchen.mpg.de

Elizabeth Spelke
Harvard University
1130 William James Hall
33 Kirkland Street
Cambridge, MA 02138
USA
spelke@wjh.harvard.edu

Saul Sternberg
Department of Psychology
University of Pennsylvania
3815 Walnut Street
Philadelphia, PA 19104-6196
USA
saul@psych.upenn.edu

Catherine Tallon-Baudry
CNRS UPR640 LENA
47 Bd de l'Hôpital
75651 Paris cedex 13
France
catherine.tallon-baudry@chups.jussieu.fr

Manabu Tanifuji
Brain Science Institute, RIKEN
2-1 Hirosawa, Wako-shi, Saitama
351-0198
Japan
tanifuji@postman.riken.go.jp

Shimon Ullman
Faculty of Mathematics And Computer Science
Weizmann Institute of Science
Rehovot, 76100
Israel
shimon@wisdom.weizmann.ac.il

Carlo Umiltà
Dipartimento di Psicologia Generale
Università di Padova
Via Venezia, 8
35131 Padova
Italy
carlo.umilta@unipd.it

Mitchell Valdes-Sosa
Cognitive Neuroscience Dept
Cuban Center for Neuroscience
Apartado 6880
La Habana
Cuba
mitchell@cneuro.edu.cu

Wim Vanduffel
MGH-NMR Center
Bldg. 149-2301
13th St.
Charlestown, MA 02129
USA
wim@nmr.mgh.harvard.edu

Patrik Vuilleumier
Laboratory for Neurology and Imaging of Cognition
Department of Neurosciences &
Clinic of Neurology
University Medical Center
1 rue Michel-Servet
1211 GENEVA 4
Switzerland
patrik.vuilleumier@medecine.unige.ch

Noriko Yamagishi
Department of Cognitive Neuroscience
ATR Computational Neuroscience Laboratories
2-2-2 Hikaridai Seika-cho Soraku-gun
Kyoto 619-0288
Japan
n.yamagishi@atr.co.jp

Association lecture

Chapter 1

Understanding the actions of others

Giacomo Rizzolatti

Abstract

There is a set of stimuli that play a fundamental role in our life: the actions of our conspecifics. How do we understand their actions? What are the neurophysiological bases of this ability? It is possible that the capacity to understand the actions of others is based exclusively on the visual analysis of the observed action. However, there is growing evidence that a mechanism underlying this capacity is based on mapping visual information of the observed action on to an internal motor representation of the same action. Here I will discuss evidence for such a mechanism, review its properties, and explore its implication for other cognitive capacities, such as action imitation, understanding intentionality and language.

1.1 Introduction

Most primates have big brains. This fact seems to be easy to explain. Big brains evolved because primates had to overcome a series of challenges posed by the physical environment. They had to find food, defend themselves against other animals, and make or find a shelter in which to keep their offspring. In order to overcome these difficulties, and in the absence of specific abilities of attack or defense, they needed to acquire intellectual capacities. They developed these capacities, and this explains why they have big brains.

At first glance this explanation seems to be satisfactory. In fact, it is only partially so. Several years ago Nick Humphrey (1976), working among the mountain gorillas in Rwanda, was struck by the fact that gorillas, primates undoubtedly endowed with a large brain and great intelligence, have a rather perplexing behavior in the wild. He wrote: 'There was no obvious sign of the gorillas using their intelligence to any practical advantage. Hard as I looked I never saw them do anything that struck me as being clever, let alone any sign that they were having to solve difficult conceptual problems' (Humphrey 2002).

Other ethologists have made similar observations (e.g. Jolly 1966; Byrne and Whiten 1988). There should, therefore, be something else, beside the challenge of the physical world and the presence of animals of other species, predators or competitors, which led to the evolution of big brains. The hypothesis that Humphrey and others put forward (see Byrne 1995) was that the principal driving force leading to the evolution of big brains and intelligence was not the interaction with the physical world in which an individual lives, but with their social world. To deal with conspecifics presents an intellectual challenge. Primates must adapt to this challenge. They therefore developed a specific intelligence that primarily takes care of what their conspecifics are doing and are going to do.

The recognition that the social world shaped the brain evolution of primates had a very limited impact on cognitive neuroscience. Yet, if the social hypothesis is true, a considerable part of the brain should be devoted to mechanisms involved in understanding actions. The aim of this chapter is to discuss these mechanisms and to examine the organization of the cortical circuits involved in action understanding.

1.2 Understanding action: theoretical considerations

How do we understand others' actions? Broadly speaking there are two main possibilities. The first one, which I will refer to as the 'visual hypothesis', maintains that we understand actions on the basis of a visual description of the elements that form the action. No motor involvement is required. For example, when we observe a hand grasping an apple, the action will be recognized because the visual system describes the hand, the apple and the movement of the hand toward the apple. The association of these elements will be sufficient to give the observer the notion of 'grasping'.

The second hypothesis, which I will call the 'direct-matching hypothesis', holds that we understand actions by mapping the visual representation of the observed action on to our motor representation of the same action. According to this view, an action is understood when its observation causes the motor system to 'resonate'. Thus, in the example mentioned above, the individual understands the meaning of a grasping action because the visual representation of the action activates those same neurons that control the execution of grasping movements. It is this activation that transforms a description that has no meaning for the individual into something that belongs to his personal knowledge and therefore has a well-defined meaning.

The two hypotheses are not necessarily mutually exclusive. It may well be that for some actions we use the direct-matching mechanism, for others we rely on visual description of the action and their consequences. Furthermore, as will be discussed later, visual descriptions and motor descriptions of an action interact strictly in the nervous system. In addition, actions with emotional content—'hot actions'—are most likely recognized using a mechanism similar to that postulated by the direct-matching hypothesis but subserved by circuits where motor activity is strictly linked with vegetative activity. In the next sections I will limit my discussion to 'cold actions' only, and, in particular, I will review the evidence in favor of the existence of a direct matching mechanism.

1.3 Evidence in favor of a direct matching hypothesis: the monkey mirror neuron system

The idea that we understand actions performed by others through an 'internal act' that somehow replicates the observed action is one of the fundamental tenets of phenomenology (e.g. Merleau-Ponty 1962). However, until recently this fascinating idea lacked neurophysiological support. Some years ago evidence was provided that a mechanism exists in the monkey that matches visual representation of an action made by others on the motor representation of that action. The neural system endowed with these properties was called the 'mirror neuron system' (Di Pellegrino *et al.* 1992; Gallese *et al.* 1996; Rizzolatti *et al.* 1996*a*). In this section I will review the neurophysiological properties of this system and discuss its possible role in cognitive functions.

Mirror neurons were originally found in the monkey ventral premotor cortex, and precisely in area F5. This area contains a rich representation of mouth and hand movements. It is adjacent to another premotor area, F4, where arm and head movements are represented (see Rizzolatti and Luppino 2001).

An important characteristic of F5 is that many of its neurons discharge during specific motor acts; that is, movements with a specific goal. These neurons become active regardless of whether the movement goal (e.g. grasping an object) is obtained using the right hand, the left hand, or, frequently, the mouth. Conversely, they are not active when movements similar to those effective in triggering them are used for another motor purpose (e.g. pushing away). Using the motor act effective in triggering them as a classification criterion, F5 neurons were subdivided into various classes. Among them, the most represented are grasping, holding, tearing, and manipulating neurons.

A second fundamental motor characteristic of F5 is that many of its neurons specify the way in which a goal can be achieved. For example, the majority of grasping neurons discharge only if the grasping is made using a particular type of prehension, such as, for example, precision grip, finger prehension, and, much more rarely, whole-hand prehension. Therefore these neurons code specific motor prototypes or, looking at their properties from a dynamic point of view, motor schemas on how an action can be achieved.

The properties of F5, and more generally of the ventral premotor cortex (F4 and F5), were conceptualized as storage ('vocabulary') of motor acts (Rizzolatti *et al.* 1988). The motor vocabulary is constituted of 'words' represented by populations of specific F4 and F5 neurons. Some words code the general goal of the action (e.g. grasping, holding, reaching, bringing to the mouth), others select specific motor prototypes (e.g. precision grip), and others specify the temporal aspect of the action to be executed. Among the various consequences of this conceptualization of premotor cortex, one is particularly important: premotor cortex neurons code motor acts. This high-order motor representation may be activated internally and, provided that other contingencies allow it, may be transformed into movements or, as will be shown below, addressed by sensory stimuli. Thus, through an associative mechanism, sensory descriptions of the external events can be transformed into actions (see Jeannerod *et al.* 1995; Rizzolatti and Luppino 2001).

Studies in which F5 neurons were tested with visual stimuli showed that about 20% of them respond to these stimuli (Murata *et al.* 1997). Some of these visuomotor neurons discharge when the monkey sees a three-dimensional object that is congruent with the type of grip coded by the studied neurons. These neurons have been called 'canonical neurons' (Rizzolatti *et al.* 2000). Other neurons discharge when the monkey observes an action, made by another individual, that corresponds to that coded motorically by the studied neuron. These neurons have been dubbed 'mirror neurons' (Gallese *et al.* 1996; Rizzolatti *et al.* 1996a).

An example of a mirror neuron is shown in Figure 1.1. This neuron discharges both when the monkey grasps an object and when the experimenter makes the same action. The neuron does not discharge when the object is grasped with a tool.

Typically, in order to be triggered, F5 mirror neurons require an interaction between hand and object (as the neuron shown in Fig. 1.1). The sight of the agent alone mimicking an action, or of the object alone, is ineffective. Object significance has no influence on the mirror neuron response. Grasping a piece of food or a geometric solid produces responses of the same intensity. Similarly, the action subsequent to the effective action does not influence the mirror neuron response. Grasping mirror neurons are active both when the experimenter grasps a piece of food and then gives it to the monkey and when another monkey grasps a piece of food and then eats it.

A particularly important feature of mirror neurons is the relation between their visual and motor properties. In this respect, the most striking finding is the close relationship, observed in most neurons (93%), between the visual action they respond to and the motor action they code. According to the type of congruence between motor and visual responses, we partitioned mirror neurons into two broad classes: 'broadly congruent' and 'strictly congruent' neurons.

'Broadly congruent' mirror neurons are defined as those minor neurons that do not require the observation of exactly the same action as they code motorically in order to be triggered. Some of them, for example, have very strict motor requirements, discharging only during the execution of a particular type of action (e.g. grasping) when executed in a specific way (e.g. precision grip). However, they respond to the observation of grasping made by another individual, regardless of the type of grip used. Other broadly congruent neurons discharge in association with a single specific motor action made by the monkey (e.g. holding), but respond to the observation of two actions made by others (e.g. grasping and holding). Mirror neurons of the 'broadly congruent' class occur most frequently (61% of mirror neurons).

'Strictly congruent' mirror neurons are defined as those minor neurons for which the effective observed and effective executed actions correspond, both in terms of goal (e.g. grasping) and means; that is, how the action is executed (e.g. precision grip). They correspond to 32% of F5 mirror neurons.

UNDERSTANDING THE ACTIONS OF OTHERS

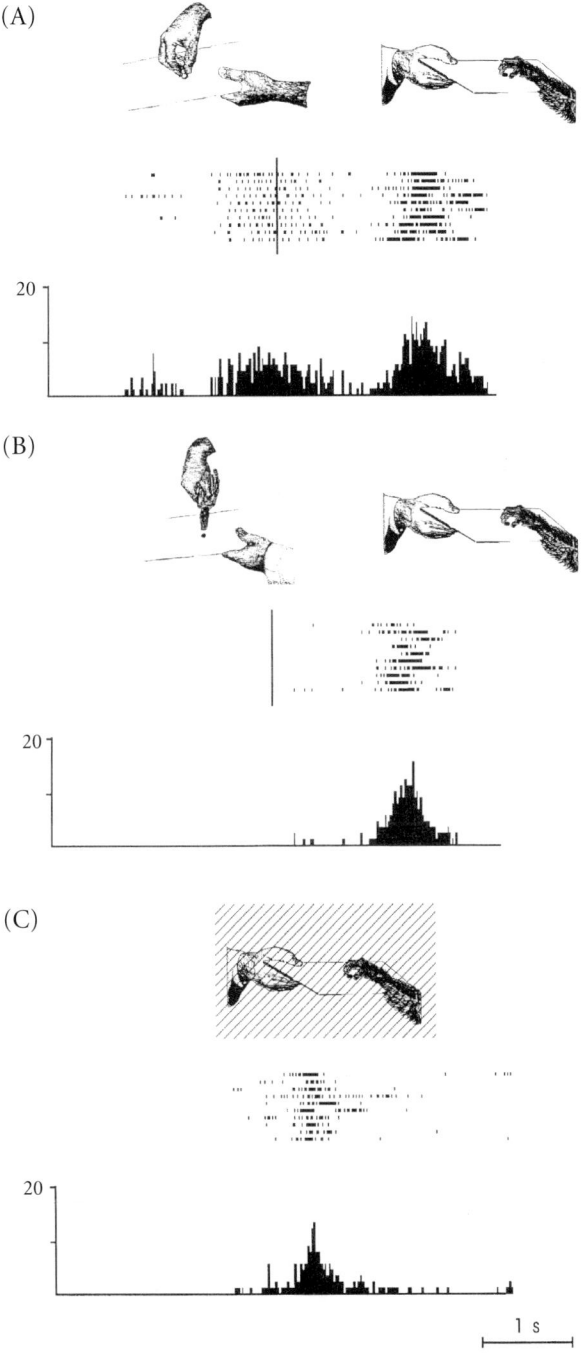

Fig. 1.1 Example of mirror neurons. Testing conditions are illustrated schematically above the rasters. (A) The experimenter grasps a piece of food placed on a tray. Subsequently, the tray, with a piece of food on it, is moved toward the monkey, which grasps it. Histograms below the rasters are the sum of eight trials. The events area synchronized with the moment when the experimenter grasped the food. (B) As above. The experimenter grasps the food with a pair of pliers. (C) The monkey grasps the food in the dark. The discharge in the latter situation demonstrates that the neuron discharge was not due to seeing the hand. (From Rizzolatti *et al.* 1996*a*. Reprinted with permission.)

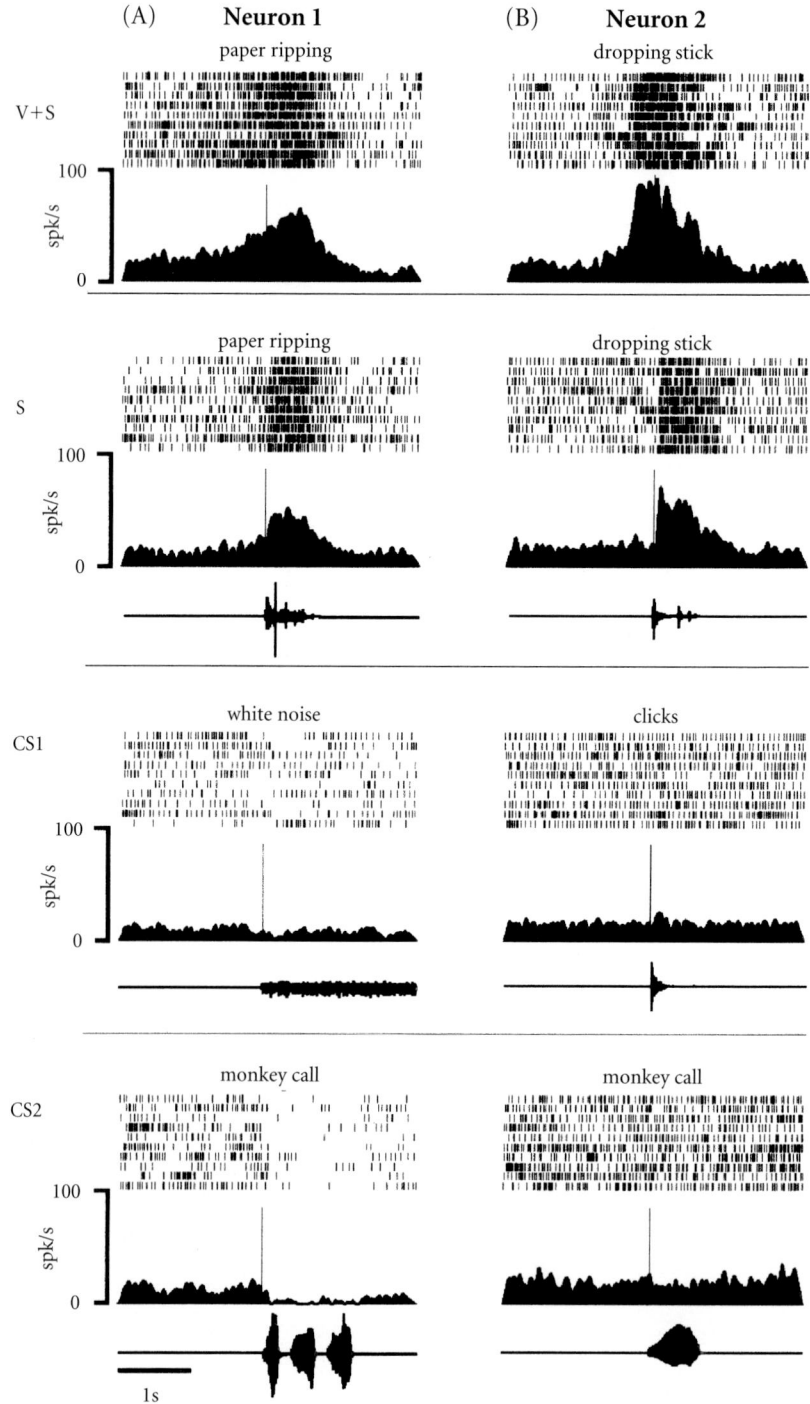

1.3.1 New evidence for a role of F5 mirror neurons in understanding action

From the very discovery of mirror neurons, it was proposed that they are involved in understanding action. The core of this notion is that an observed action acquires meaning for the observer when it activates, in the observer's brain, motor schemas of known outcomes. Before discussing this proposal in more detail, let me examine some new evidence in favor of this interpretation.

A way to test whether mirror neurons are involved in action understanding is to place the monkey in situations in which it is able to understand the meaning of an action, but in which the sensory situation is different from the one that typically triggers mirror neurons. If mirror neurons are involved in action understanding, their activity should reflect the action meaning and not the sensory contingencies leading to understanding action.

In everyday life many actions are accompanied by characteristic sounds. For example, ripping a sheet of paper has a particular sound, breaking a peanut another. When individuals hear one of these sounds, they are able to recognize the actions that caused them, even without seeing them. Recently, we tested F5 neurons with both visual and auditory stimuli (Kohler *et al.* 2002). In an initial group of neurons, we used sounds produced by actions made by the experimenter in front of the monkey and non-action-related sounds generated by a computer; in a subsequent group, digitized, pre-recorded, action-related sounds were used.

The results showed that about 15% of the tested neurons discharge, both when the monkey performs a hand action and when it hears the action-related sound. An example of these 'audio-visual' neurons is shown in Fig. 1.2A (Neuron 1). This neuron responded to the vision (and sound) of a tearing action (paper ripping, 'V+S'). The sound of the same action performed out of the monkey's sight was equally effective ('S'). In order to control for unspecific factors, such as arousal or fear, non-action-related sounds (white noise, chimpanzee calls) were also presented. These stimuli did not evoke any excitatory response ('CS1' and 'CS2'). Another example is presented in Fig. 1.2B (Neuron 2).

Fig. 1.2 Two examples of audio-visual mirror neurons. Rastergrams are shown together with spike density functions. (A) Neuron 1 responded to the vision accompanied by sound ('V+S') and to the sound alone ('S') of a tearing action (paper ripping). (B) Neuron 2 responded to the vision accompanied by sound ('V+S') of a stick dropped to the ground and to the sound alone ('S') of the stick hitting the ground. Non-action-related sounds—white noise ('CS1') and chimpanzee calls ('CS2') for Neuron 1, clicks ('CS1') and chimpanzee calls ('CS2') for Neuron 2—did not produce any significant excitation. The vertical lines indicate the time at which the sound occurred. The traces under the spike density functions in 'S' and in 'CS' conditions are oscillograms of the sounds used to test the neurons. (From Kohler *et al.* 2002. Reprinted with permission.)

In another experiment of the same study (Kohler *et al.* 2002), the neurons' capacity to differentiate actions on the basis of auditory and visual characteristics of the action was tested. The experimental design included two hand actions presented randomly in vision-and-sound, sound-only, vision-only, and motor conditions (monkeys performing object-directed actions). Twenty-nine out of 33 audio-visual mirror neurons tested showed auditory selectivity. Taken together, these findings indicate that F5 neurons discharge when there are sufficient sensory cues to evoke the meaning of an action in the monkey. What these cues are is irrelevant. They could be visual or auditory. Once the action meaning is specified, the neuron fires.

Another way to test whether action understanding triggers F5 mirror neurons is to eliminate both vision and sound and to give the monkey cues to the action meaning. Human beings recognize an action even when it is hidden. If I see a girl near a basket and I know that the basket is full of apples, I will have no doubts that when she moves her hand toward the basket, she is going to pick up an apple. Even if I do not see the apple and the hand grasping it, I know that the girl is grasping it.

An experiment to test whether mirror neurons fire when a crucial aspect of the action, the hand–object interaction, is hidden from the monkey's view, was carried out by Umiltà *et al.* (2001). Only those mirror neurons were studied that responded selectively to hand–object interaction (e.g. the late phase of reaching and grasping) or to holding. The experimental paradigm consisted of two basic conditions. In one, the monkey was shown a fully visible action directed toward an object ('full vision' condition). In the other, the same action was presented but with its final critical part hidden behind a screen ('hidden' condition). Before each trial, the monkey was shown that there was an object behind the screen.

The results showed that about 50% of mirror neurons studied also discharged in the hidden condition ($n = 19$). Of these, seven did not show any difference between hidden and full vision conditions, while nine responded more strongly in full vision. Of the remaining three, the response was either more pronounced in the hidden condition than in full vision, or showed a temporal shift (anticipation) in the hidden condition.

In conclusion, the experiments in which the stimulus conditions were altered clearly indicate that F5 mirror neuron activation correlates with conditions in which actions are understood, thus strongly supporting the notion that F5 activity is involved in action understanding.

1.3.2 Action coding in other brain centers of the monkey

Neurons responding to the observation of actions are also present in other monkey brain areas. An important series of studies carried out by Perrett and his coworkers (Perrett *et al.* 1989, 1990; Jellema and Perrett 2002; Jellema *et al.* 2002) showed that a variety of neurons responding to the observation of biological actions are present in the rostral part of the superior temporal sulcus (STS). Movements effective in eliciting neuronal responses were walking, turning the head, bending the torso, and moving the arms.

A small set of neurons was found to discharge during the observation of goal-directed hand movements (Perrett *et al.* 1990).

By comparing these data with those concerning area F5, two differences emerge. First, the STS appears to code a larger number of movements than F5. Secondly, although the issue of STS motor properties was not specifically studied, STS neurons appear to be visual neurons not endowed with motor properties. Finally, although any conclusion on the possible motor properties of the STS should await more experiments, the anatomical characteristics of the STS region and the available functional data suggest that motor properties, if present, should be much less prominent in the STS than in F5.

The STS does not send direct connections to F5. Information on the observed actions cannot, therefore, reach F5 directly from the STS. The intermediate station between the STS and F5 appears to be represented by the inferior parietal lobule and, in particular, by area PF (area 7b of Brodmann), which receives afferents from the STS and is connected with F5 (Seltzer and Pandya 1994).

Evidence in favor of this possibility was provided recently by experiments in which we investigated the functional properties of area PF neurons, with the specific aim of examining their responses to biological actions (Fogassi *et al.* 1998; Gallese *et al.* 2002). In agreement with previous studies (Leinonen and Nyman 1979; Leinonen *et al.* 1979, Hyvarinen 1982), we found that the majority of PF neurons respond to somatosensory stimuli, visual stimuli, or both. A set of visually responsive neurons requires specific biological actions to be triggered, as do STS neurons. Most interestingly, some of these neurons also fire during active movements, and their motor properties are congruent with their visual properties. They have, therefore, the functional properties that define F5 mirror neurons: they discharge during the execution of a given action and during the observation of a similar action made by another individual.

The STS region, besides being connected with PF, is also a part of a circuit that includes the amygdala and the orbitofrontal cortex (Amaral *et al.* 1992). Several years ago Brothers and coworkers (Brothers *et al.* 1990; Brothers and Ring 1992) examined the functional properties of amygdala neurons, using complex social stimuli. They found that a population of amygdala neurons is triggered specifically by social stimuli. It is likely that visual information on motor action is transmitted by the STS to this subcortical center. The STS–amygdala circuit may be the basis of a direct-matching system for emotionally charged stimuli.

1.3.3 Machinery for understanding action

From these findings, and with the caveats that the STS data, on the one hand, and those for F5 and PF, on the other, were obtained from different groups of researchers, one may draw the following tentative conclusions.

An observed action is described visually in the STS, where it is defined from both a viewer-centered and an object-centered perspective (Perrett *et al.* 1989; see

Jellema and Perrett 2002). The STS therefore provides the organism with a basic description of a variety of actions. The computational logic of the STS is not different from that of F5, discussed above. As F5 is a store of motor actions, the STS is a store of pictorially described actions. Accepting the Shepard perception theory (1984) and translating it into neurophysiological terms, STS neurons would be the basic perceptual elements that 'resonate' when the appropriate stimulus is present in the environment. According to the Shepard theory, the resonating elements respond maximally to one category of stimuli, but are also able to respond to incomplete or corrupted stimuli. These properties are particularly useful for extracting the correct information from complex stimuli, such as motor actions made by other individuals.

Action understanding occurs when an association is formed between the visual prototypes coded in the STS and the motor prototypes coded in the premotor cortex. When this association is formed, the premotor neurons resonate every time a visual description of an action matches its motor description. This motor activation is the basis for understanding action.

This visuomotor matching does not occur directly. As described above, PF forms an intermediate stage between the STS and premotor areas. It is outside the scope of the present article to discuss how this association may occur and to examine the advantage that an intermediate layer between sensory and motor resonators gives to the network. However, it is important to stress that, beside forward connections from STS to PF and from PF to F5, the system is endowed with rich backward connections. These connections form the anatomical substrate that renders possible the association between visual and motor prototypes. Furthermore, the presence of an intermediate layer, represented by the inferior parietal lobe, gives the system the flexibility necessary to categorize the pictorially described actions in terms of their meaning without changing their basic properties. Through these dynamic relations, the capacity to understand action originally related to the activity of premotor neurons may extend to the parietal cortex. The motor cortex 'teaches' the parietal lobe the meaning of action that can then be recognized by neurons located in this structure.

1.4 The mirror neuron system in humans

A mirror neuron system similar to that discovered in the monkey also exists in humans. Evidence for this derives from electroencephalographic (EEG), magnetoencephalographic (MEG), transcranial magnetic stimulation (TMS), and brain imaging studies (for a review, see Rizzolatti et al. 2001). Here I will focus on the latter studies.

Earlier brain imaging studies aimed at assessing whether a mirror neuron system does exist in humans were positron emission tomography (PET) studies, where subjects were required to observe hand/arm actions made by another individual. These experiments demonstrated that observation of action activates, besides occipital visual areas, area STS, the inferior parietal lobe, and a ventral frontal region including Broca's area (Grafton *et al.* 1996; Rizzolatti *et al.* 1996*b*; Decety *et al.* 1997; Grèzes *et al.* 1998). The activated areas correspond to those areas where neurons responding to hand actions are located in the monkey.

Grèzes *et al.* (1998) investigated whether the areas that become active when subjects observe meaningful actions—pantomimes of bimanual transitive actions—are the same as those that become active when subjects observe meaningless actions—actions derived from American Sign Language. The results confirmed that the observation of meaningful hand actions activates the left premotor cortex (area 44 and area 6), the left inferior parietal lobe, and various occipital and inferotemporal areas. They also showed that, during the observation of meaningless gestures, there was an increase in activation of the right superior parietal lobule and a marked decrease of activation in Broca's area. It is likely that the activation shift toward the right parietal lobe was due to the fact that, in this condition, the participants paid greater attention to the movements that formed the actions than to action *per se*. In contrast, attention was immediately allocated to the goal of the action in the case of meaningful actions. The same explanation probably holds true for the decrease in area 44 activation. Movements of the arms, rather than the goal of the action, were the focus of attention.

Since the time of von Bonin and Bailey (1947) area 44 has been considered to be the human homologue of monkey F5 (see also Petrides and Pandya 1997). In spite of this, the activation of Broca's area during observation of action was an unexpected finding, because of the classical neurological notion that Broca's area is a speech motor area. This fact led some authors (e.g. Grèzes and Decety 2001) to raise doubts whether activation of Broca's area during observation action was due to its involvement in an action recognition system or to an internal verbalization. There is no doubt that this interpretation might be true for some complex experimental tasks. However, it is a very unlikely explanation for the cases when the subject observes the same simple action several times (e.g. grasping or finger lifting). No subject repeats over and over: grasping, grasping, grasping when he/she observes this action, or lifting, lifting, lifting when he/she observes a finger movement.

New data on the functional organization of the human Broca's area, coming mostly from functional magnetic resonance imaging (fMRI) studies, fully confirmed the homology between the '*pars opercularis*' of this area (basically corresponding to area 44) and monkey area F5. A series of experiments using different experimental procedures, all showed that the dorsal part of Broca's area opercular sector becomes active during hand movements (Krams *et al.* 1998; Binkofski *et al.* 1999, Iacoboni *et al.* 1999; Ehrsson *et al.* 2000; Gerardin *et al.* 2000; Simon *et al.* 2002).

Particularly important in this respect is the study by Binkofski *et al.* (1999), who correlated the functional activation of area 44 obtained during object manipulation with the cytoarchitectonic maps of Broca's area of Amunts *et al.* (1999). The data showed that the premotor activation was clearly inside area 44, as rigidly defined by those authors. In addition, Ehrsson et al. (2000) showed that, in perfect analogy with F5, prehension grip, but not power grip, is richly represented in area 44. Thus, the organization of F5 and area 44 appear to be very similar in their basic structure. In both areas, hand actions are represented dorsally and mouth movements ventrally, with a large overlap between the two representations.

Earlier brain imaging experiments studied activations related to the observation of hand and arm movements. Recently, experiments have been carried out to find the cortical areas active during observation of other types of actions (Buccino et al. 2001). These experiments also tested the validity of the explanation of frontal lobe activation as due to verbalization. In fact, if the interpretation based on verbal mediation is correct, Broca's area should be active regardless of the type of observed action and effector used. But, if activation of Broca's area reflects a specific anatomical and functional localization for hand actions, this activation should be absent during the observation of actions made with other effectors, such as the foot.

Participants were instructed to observe video-clips showing mouth (biting an apple), hand/arm (grasping a cup or an apple), and foot (kicking a ball or pushing a brake) actions. Action observation was contrasted with observation of a static picture of a face, a hand, and a foot, respectively. In addition to actions directed toward an object, the participants were shown actions not involving an object: chewing, mimicking a reaching to grasp movement, and pretending to kick a ball or push a brake.

The results showed that the observation of both object- and non-object-related actions led to somatotopic activation of the premotor cortex. Mouth action produced activations centered in areas 44 and 45. Observation of reaching to grasp movements produced two activation foci: one located in area 44, and precisely in that part of this area where grasping and manipulation movements are represented, and a second one more dorsally in area 6. This sector of area 6 is related mostly to reaching movements (see Rizzolatti et al. 2002). Finally, observation of foot movements produced dorsal premotor activation, not including Broca's area. During the observation of object-related actions, a roughly somatotopic activation was also found in the inferior parietal lobule. The results are shown in Fig. 1.3.

These data are important for two reasons. First, they indicate that the activation of Broca's area is a genuine mirror phenomenon related to some particular types of actions. Secondly, and most importantly, they show that in humans the mirror neuron system is a large system, comprising a considerable part of the premotor cortex and of the inferior parietal lobule. The mirror mechanism appears to be a fundamental aspect of human cortical organization.

Recently, in a new fMRI study (Buccino et al. 2003), we tested whether the observation of actions performed by animals activates the same areas as those activated when observing human actions. Normal volunteers were presented with video-clips showing a man or an animal (monkey or dog) either biting a piece of food or performing communicative actions (silent speech, silent lip-smacking, and silent

UNDERSTANDING THE ACTIONS OF OTHERS | 15

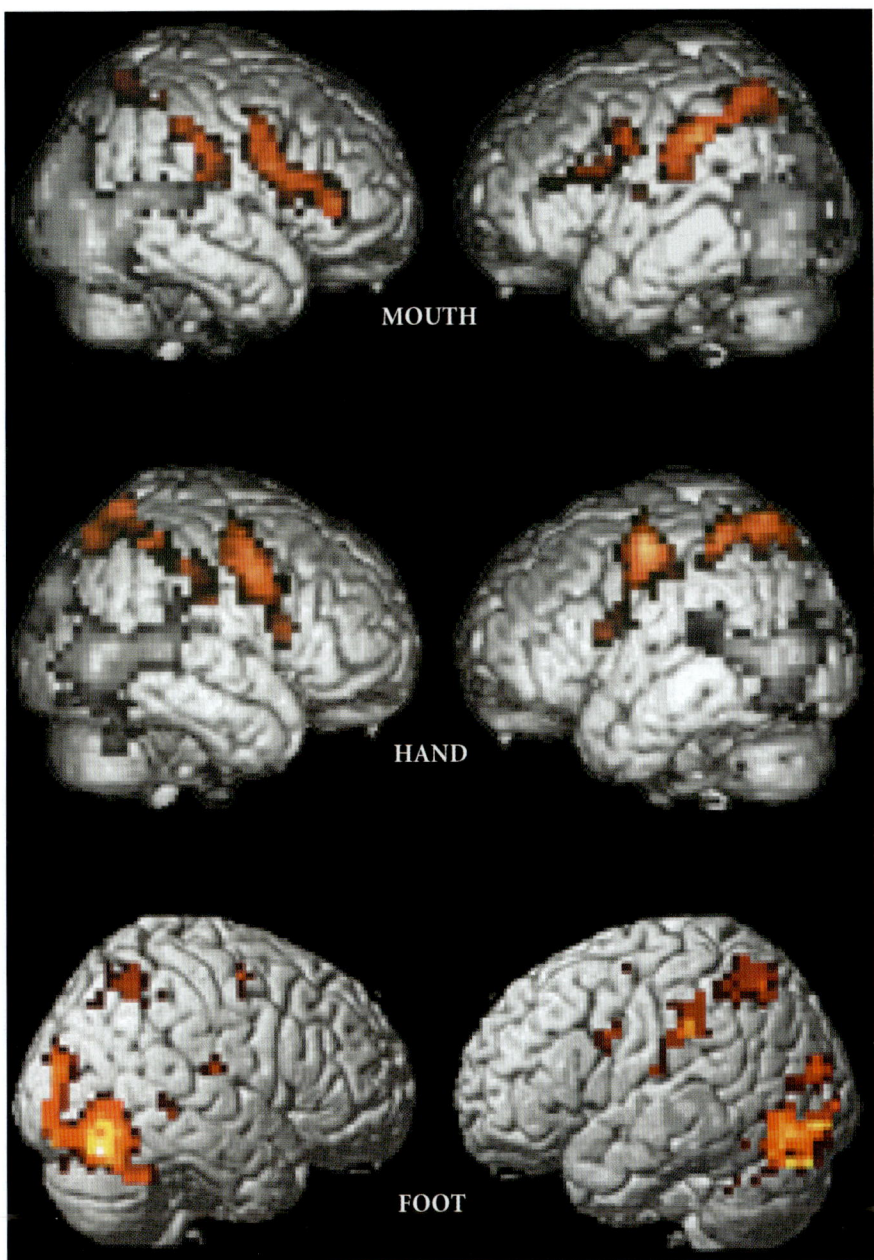

Fig. 1.3 Brain activation in frontal and parietal areas during the observation of mouth-, hand- and foot object-directed actions. For explanation, see text. (From Buccino *et al.* 2001. Reprinted with permission.)

barking, respectively). As a control, for each sequence showing a mouth action, the subjects had to observe a static frame of the same mouth action.

Observation of a man biting food confirmed the result of the study just reviewed. Besides activation of the visual areas, activations were found in area 44, ventral area 6, and inferior parietal lobule (Fig. 1.4). Observation of the same actions performed by a monkey or a dog produced activations in the same sites as those found during observation of the human action, but less broadly distributed.

Observation of lip-reading led to a strong activation of left Broca's area, posterior area 22 and the STS region bilaterally (Fig. 1.5). Lip-smacking, a very common communicative action used by monkeys, produced a small activation of the left Broca's area and a stronger activation of the right area 44. Activations were also found in the inferior parietal lobule and STS, in locations similar to those found during biting observation. Finally, observation of a dog barking gave no activation in frontal or parietal lobe.

It therefore appears that there are two ways of coding actions made by non-conspecifics. When the observed action has, unambiguously, the same meaning as a human action, the same parieto-premotor circuits are activated as during the observation of the homologous human action. Note that, in the case of a dog biting, the visual stimulus has, obviously, very little resemblance to the stimulus showing human biting. The similarity is in the action meaning.

In the case of actions that are not in the human natural motor repertoire, such as barking, the understanding of action appears to be based on visual description of the observed action. Although no subject had doubt about what he/she observed, the interpretation of the action was not based on a 'personal' motor experience. It was recognized as if made by an inanimate agent, without any mirror involvement.

Finally, in the case of communicative stimuli that are ambiguous for the human observer, such as lip-smacking, the mirror system areas related to communication only weakly resonate, and without the left hemisphere dominance observed during lip-reading. Thus, lip-smacking appears to have been interpreted in 'personal' motor terms, but with ambiguity about its meaning. Interestingly enough, the debriefing of the subjects after the experiment reflected the neural ambiguity. Some said that the observed action was related to food, others that the monkey wanted to 'speak', and others told the experimenter that they were uncertain about these two alternatives.

1.5 The mirror neuron system and higher cognitive functions

The presence in humans of a large mirror neuron system comprising considerable parts of the inferior parietal lobe and premotor cortex is hardly justified from an evolutionary perspective, unless some cognitive functions evolved in addition to it. What is the advantage of having a system that describes in detail actions made by others if this knowledge has no consequences for the owner of this system? A possible answer to this is that some high-order cognitive functions evolved with the mirror neuron system.

UNDERSTANDING THE ACTIONS OF OTHERS | 17

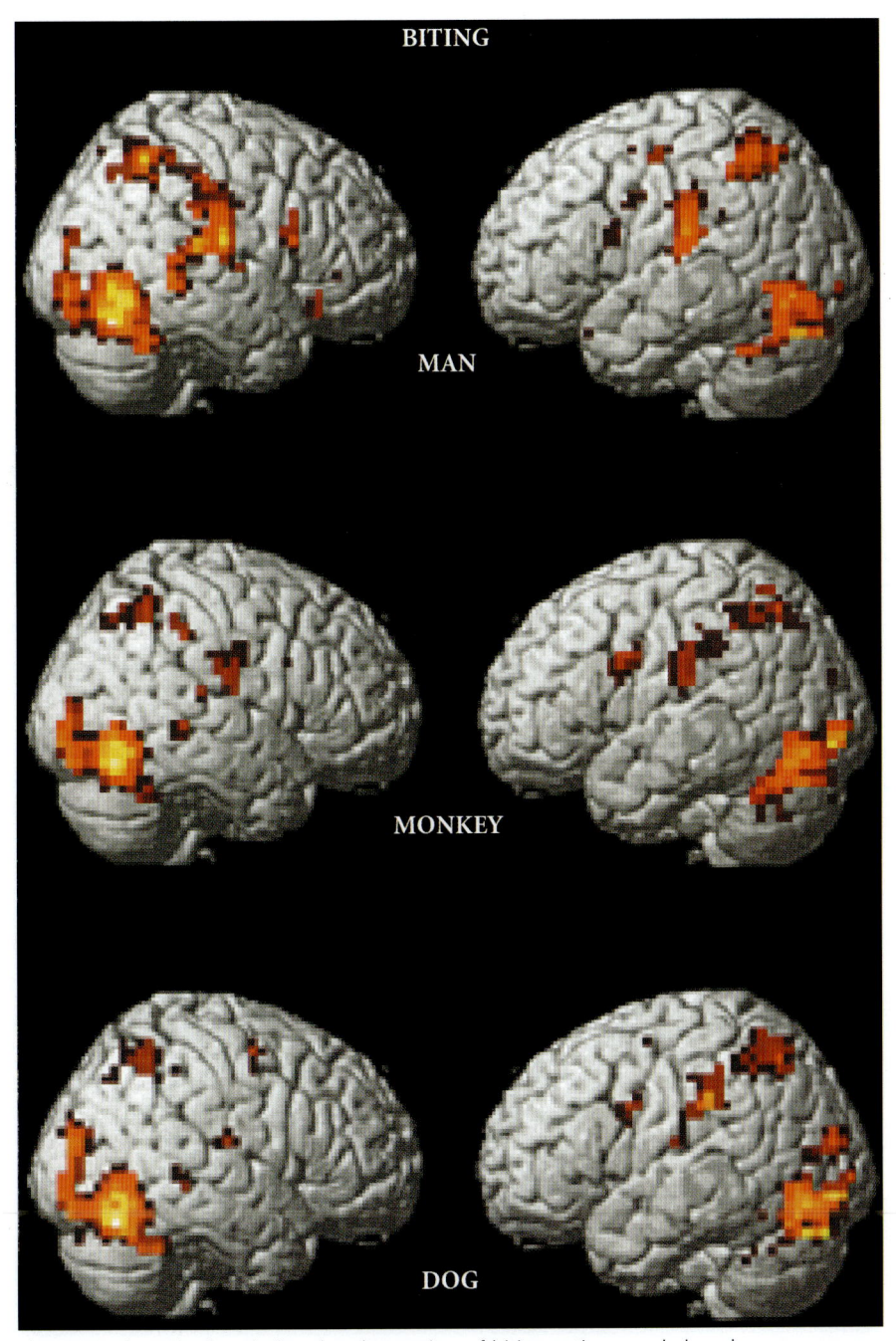

Fig. 1.4 Brain activation during the observation of biting actions made by a human, a monkey and a dog. For explanation, see text.

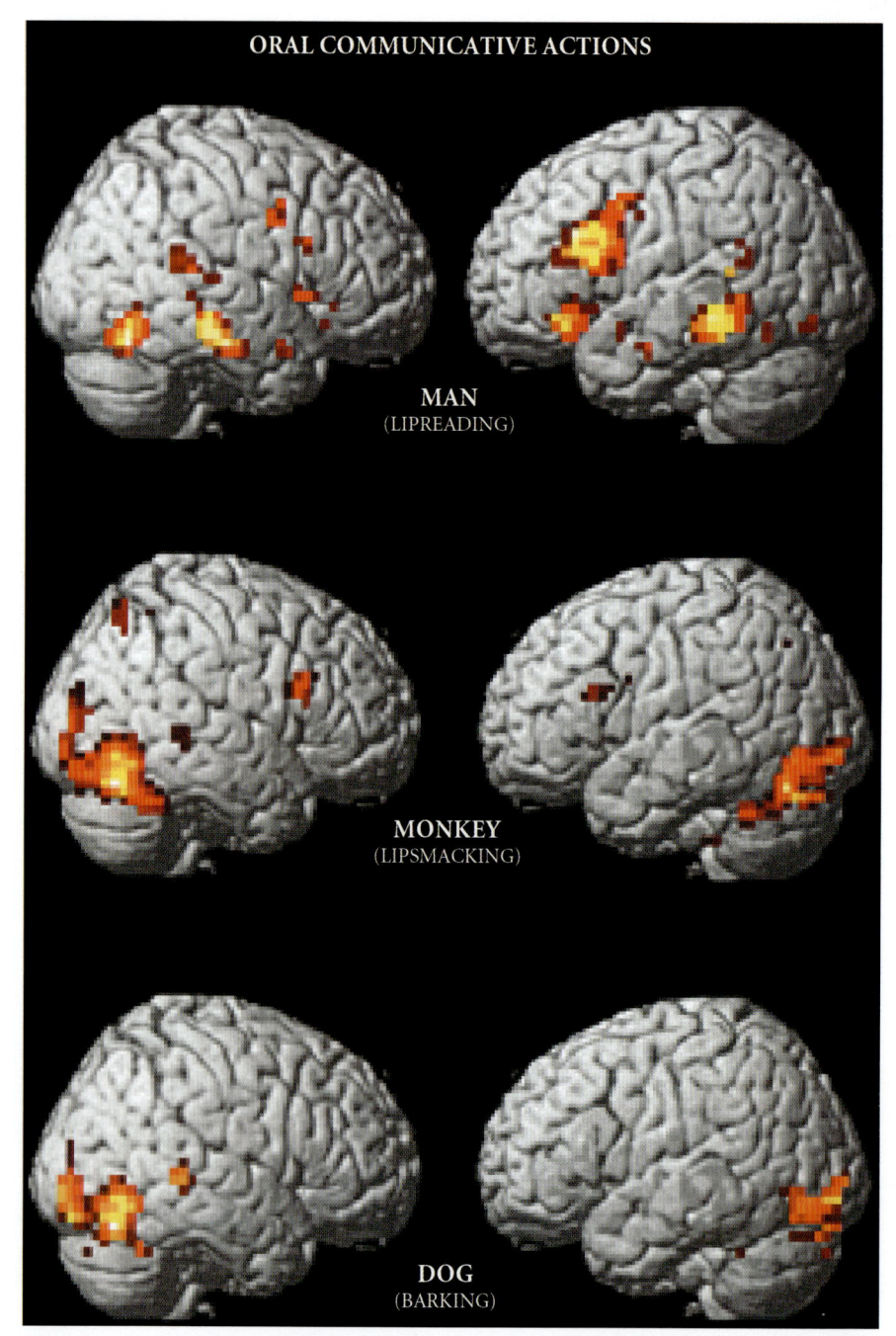

Fig. 1.5 Brain activation during the observation of communicative actions. For explanation, see text.

Cognitive functions that are possibly related to the mirror neuron system are: intention understanding, imitation, and language.

In order to interact effectively with others one must understand what the other is doing. Human beings do not just understand *what* another person is doing, but also, in most cases, *why* this person is doing something. The problem of how we understand the intention of others is a matter of debate. Two different accounts of this problem have been proposed. According to one account, known as 'theory theory', mental states are represented as inferred posits of a naive psychological theory. The other account is known as 'simulation theory'. According to it, other people's intentions are understood by adopting their perspective, by simulating their mental states (for references see Gallese and Goldman 1998; Blakemore and Decety 2001).

During action observation, the mirror neuron system is activated by the observed stimuli. It automatically describes the 'what' of the observed, but not the 'why' of it. Thus, the existence of a mirror system is not, *per se*, an empirical proof of simulation theory. This said, one must also recognize that is very tempting to admit that a neuronal mechanism that describes the observed present action (action understanding) is also involved in describing what the future actions of the acting individual will be (intention understanding). A simple hypothesis that may account for intention understanding is the following. While an individual observes an action made by another person, in addition to activation of mirror neurons related it, there is a sub-threshold activation of many other mirror neurons describing actions probabilistically related to the observed one. According to the context in which the observed action is performed (and possibly other factors), a specific set of neurons, previously activated sub-threshold, prevails on the others and fires. The activity of these mirror neurons represents the most likely action that the observer thinks the acting individual is going to do. In other words it represents his/her intentions.

Another fundamental cognitive property very likely related to the mirror neuron system is the capacity to imitate actions made by others. Imitation is a highly developed capacity in the human species (Byrne 1995; Tomasello and Call 1997; Visalberghi and Fragaszy 2001), forming the basis of human culture and civilization (see Donald 1991). Imitation has been defined in many ways. For sake of simplicity, I will distinguish here two senses in which this word is used: 'imitation' as used by psychologists and 'imitation' as used by ethologists. Imitation for psychologists defines actions that individuals perform in response to observed actions. The rules of actions evoked by observing others are studied and defined (Bekkering and Wohlschlaeger 2002; Prinz 2002). Imitation for ethologists is the capacity to acquire a motor behavior not previously present in the observer motor repertoire (see Byrne 1995; Tomasello and Call 1997). The stress here is on motor learning and specifically on learning based on a translation of actions coded in visual terms into actions coded in motor terms.

Experiments on imitation in psychology have received a strong impetus in the past few years from the work of Meltzoff (see Meltzoff 2002) and Prinz and his colleagues (see Prinz 2002, Bekkering and Wohlschlaeger 2002). In particular, the latter authors re-elaborated the concept of ideomotor compatibility put forward originally by Lotze (1852), James (1890), and Greenwald (1970). On the basis of a large number of experimental observations, they concluded that there are 'common representational resources for perception and action: perceptual cognition shares representational resources with action planning' (Prinz 2002, p. 160).

It is evident that there is a close relationship between these conclusions and the basic property of the mirror neuron system. The existence of this link was tested empirically by Iacoboni *et al.* (1999) in an fMRI experiment. An experimental paradigm conceptually similar to that employed by Brass and his colleagues (Brass *et al.* 2000) was used.

Normal human volunteers were tested in two basic conditions: 'observation-only' and 'observation-execution'. In the observation-only condition, they were asked to pay attention to the following stimuli: a moving finger, a cross that appeared on a stationary finger, a cross on an empty background. In the 'observation-execution' condition, they were required to lift a finger in response to the same stimuli. The main result of the experiment was that the movement of the finger triggered by the observation of finger movement produced activations stronger than the same movements triggered by the other stimuli, in four regions. These regions were area 44, superior parietal lobule, parietal operculum (SII region), and STS (this activation, just under the statistical threshold, was found to be significant in a subsequent experiment, see below).

Similar results were obtained by Nishitani and Hari (2000) using event-related neuromagnetic recordings. In their experiments, normal human participants were requested, in different conditions, to grasp a manipulandum, to observe the same movement performed by an experimenter, and to observe and simultaneously replicate the observed action. The results showed that during execution, there was an early activation in area 44 with a response peak appearing approximately 250 ms before the touch of the target. This activation was followed within 100–200 ms by activation of the left precentral motor area and 150–250 ms later by activation of the right precentral motor area. During observation imitation, the pattern and sequence of frontal activations were similar to those found during execution, but the frontal activations were preceded by an occipital activation due to visual stimulation occurring in the former conditions.

Thus, during imitation of simple finger movements, there is first a pictorial description of the movement in the STS, after which this information is transmitted to Broca's area, most likely via the inferior parietal lobe (PF). As mentioned above, PF activation is strong in the case of transitive action, while it is often sub-threshold during intransitive actions. In addition, during imitation there are three other activations that are not present during action observation. They are located in the superior parietal lobule (see Grèzes *et al.* 1998 for similar observations), parietal operculum and STS.

On the basis of the properties of area PE (Mountcastle *et al.* 1975), the most plausible interpretation of the superior parietal lobule activation during imitation is that, during movement observation for imitation, a kinesthetic copy of the movement to be performed is formed in the right parietal lobe. This copy, indicating the final and possibly some intermediate joint positions, is then used during action execution.

A similar interpretation is valid for the parieto-opercular activation. It is known from monkey studies that several somatosensory areas are located in this sector of the parietal lobe (Robinson and Burton 1980*a*, *b*; Krubitzer *et al.*, 1995). Recent brain imaging data indicate that a similar organization is also present in humans (Disbrow *et al.*, 2000). On the basis of these findings, it is possible to hypothesize that parieto-opercular activation represents a tactile copy of the intended action. It is interesting to note that the pure observation of a hand manipulation action decreases signals evoked in the SII region by median nerve stimulation (Avikainen *et al.* 2002). This finding further supports the sensory copy hypothesis.

The activation found in the STS is particularly interesting. This activation was specifically investigated in an experiment by Iacoboni *et al.* (2001), in which volunteers either observed ('observation-only') or executed ('observation-execution') a finger movement. The observed stimuli involved the right hand or the left hand. In half of the trials, the stimulus was a finger movement, in half it was a small black cross, presented on the finger. The results showed that during observation-only, the best stimulus was the movement of the hand *anatomically* corresponding to that used by the subjects in the experiment (right hand), whereas during the observation-execution condition, the best stimulus was the hand *spatially* corresponding to that of the subject. Thus, when imitation is required, what is privileged is the space common to the acting hand and to the observed hand (right hand of the subject/ left hand of the actor).

The reversal of activation intensity observed in the STS when the observed movement had to be imitated is consistent with behavioral data showing that humans tend preferentially to imitate mirror image movements (Shofield 1976*a*, *b*; Bekkering *et al.* 2000; see Gattis *et al.* 2002). Furthermore, this finding, together with the data on activation of the parietal operculum and superior parietal lobe, suggests that a re-afferent mechanism based on mirror activity underlies imitation. When imitation is requested, a re-afferent discharge coming from mirror neurons activates the STS hand prototypes that are pictorially *and* spatially congruent with the action to be imitated. This activation overcomes that caused by hand observation, and renders the spatially congruent stimuli more effective.

In conclusion, these experiments clearly demonstrate that the mirror neuron system is involved in imitation as defined by psychologists. It is hard to believe that the mirror neuron system is not also involved in imitation as defined by ethologists. An empirical demonstration of this is lacking. In particular, it is not clear which mechanism, in addition to the mirror neuron system, intervenes to allow learning of new motor behavior. In this respect, it is of interest to mention a hypothesis advanced recently by Byrne (2002). His proposal is that a behavior to be imitated is dissected by the observer into a string of simpler sequential components that are already present in the observer's motor repertoire. Then an additional mechanism assembles these acts into a new complex motor action. Mirror neurons appear to be the ideal neural substrate for the purpose of the first of these operations. They code elementary motor acts and are activated both by motor act observation and during motor action production.

A third fundamental human capacity that appears to be linked to the mirror neuron system is language. As suggested by Rizzolatti and Arbib (1998), the mirror neuron system provides a bridge between 'doing' and 'communicating'. The gestures of others 'resonate' in my brain and are understood by me. Thus, thanks to the mirror neuron system, action becomes information, a necessary prerequisite for the evolution of language.

There is no space here to discuss the possible relations between the mirror neuron system and the origin of language. However, a new finding, relevant for this issue, is worth mentioning: the discovery in area F5 of audio-visual mirror neurons (Kohler *et al.* 2002, see above). These neurons, like all mirror neurons, represent action content but, in addition, they respond to specific *auditory* stimuli. The sound of an action represents a specific action, and the vision of the action is not required to understand it. These neurons therefore create a semantic link between the speaker and the *listener*. This indicates that in the macaque monkey there is already a neural substrate from which speech might potentially have evolved. It is likely that a similar substrate also existed in primates of the evolutionary line from which humans descended.

Acknowledgement

This work was supported by Centro Nazionale delle Ricerche (CNR) and by EU Contract number IST-2000–29689.

References

Amaral, D.G., Price, J.L., Pytkanen, A., and Carmichael, S.T. (1992). Anatomical organization of the primate amigdaloid complex. In J.P. Aggleton (Ed.), *The amygdala: Neurobiological aspects of emotion, memory, and mental disfunction*, pp. 1–66. Wiley-Liss, New York NY.

Amunts, K., Schleicher, A., Buergel, U., Mohlberg, H., Uylings, H.B.M., and Zilles, K. (1999). Broca's region re-visited: cytoarchitecture and intersubject variability. *J. Comp. Neurol.*, **412**, 319–341.

Avikainen, S., Forss, N., and Hari, R. (2002). Modulated activation of human SI and SII cortices during observation of hand actions. *NeuroImage*, **15**, 640–6.

Bekkering, H. and Wohlschlaeger, A. (2002). Action perception and imitation: a tutorial. In W. Prinz and B. Hommel (Eds), *Attention and performance XIX. Common mechanisms in perception and action*, pp. 294–333. Oxford University Press, Oxford.

Bekkering, H., Wohlschlaeger, A., and Gattis, M. (2000). Imitation of gestures in children is goal directed. *Q. J. Exp. Psychol.*, **53A**, 153–64.

Binkofski, F., Buccino, G., Posse, S., Seitz, R.J., Rizzolatti, G., and Freund, H. (1999). A fronto-parietal circuit for object manipulation in man: evidence from an fMRI-study. *Eur. J. Neurosci.*, **11**, 3276–86.

Blakemore, S.-J., and Decety, J. (2001). From the perception of action to the understanding of intention. *Nature Rev. Neurosci.*, **2**, 561–7.

Brass, M., Bekkering, H., Wohlschlaeger, A., and Prinz, W. (2000). Compatibility between observed and executed finger movements: comparing symbolic, spatial and imitative cues. *Brain Cognit.*, **44**, 124–43.

Brothers, L. and Ring, B. (1992). A neuroethological framework for the representation of minds. *J. Cogn. Neurosci.*, **4**, 107–18.

Brothers, L., Ring, B., and Kling, A. (1990). Response of neurons in the macaque amygdala to complex social stimuli. *Behav. Brain Res.*, **41**, 199–213.

Buccino, G., Binkofski, F., Fink, G.R., Fadiga, L., Fogassi, L., Gallese, V., Seitz, R.J., Zilles, K., Rizzolatti, G., and Freund, H.-J. (2001). Action observation activates premotor and parietal areas in a somatotopic manner: an fMRI study. *Eur. J. Neurosci.*, **13**, 400–4.

Buccino, G., Lui, F., Canessa, N., Patteri, I., Lagravineze, G., Benuzzi, F., Porro, C. A., and Rizzolatti, G. (2003). Neural circuits involved in the recognition of actions performed by non-conspecifics: a fMRI study. *J. Cognitive Neurosci*, in press.

Byrne, R.W. (1995). *The thinking ape. Evolutionary origins of intelligence.* Oxford University Press, Oxford.

Byrne, R.W. (2002). Seeing actions as hierarchically organized structures: Great ape manual skills. In Melzoff, A.N. and Prinz, W. (Eds), *The imitative mind. Development, evolution and brain bases*, pp. 122–40. Cambridge University Press, Cambridge.

Byrne, R.W. and Whiten, A. (1988). *Machiavellian intelligence: Social expertise and the evolution of intellect in monkeys, apes and humans.* Clarendon Press, Oxford.

Decety, J., Grezes, J., Costes, N., Perani, D., Jeannerod, M., Procyk, E., Grassi, F., and Fazio, F. (1997). Brain activity during observation of actions. Influence of action content and subject's strategy. *Brain*, **120**, 1763–77.

Di Pellegrino, G., Fadiga, L., Fogassi, L., Gallese, V., and Rizzolatti, G. (1992). Understanding motor events: A neurophysiological study. *Exp. Brain Res.*, **91**, 176–80.

Disbrow, E., Roberts, T., and Krubitzer, L. (2000). Somatotopic organization of cortical fields in the lateral sulcus of homo sapiens: Evidence for SII and PV. *J. Comp. Neurol.*, **418**, 1–21.

Donald, M. (1991). *Origins of the modern mind.* Harvard University Press, Cambridge MA.

Ehrsson, H.H., Fagergren, A., Jonsson, T., Westling, G., Johansson, R.S., and Forssberg, H. (2000). Cortical activity in precision- versus power-grip tasks: an fMRI study. *J. Neurophysiol.*, **83**, 528–36.

Fogassi, L., Gallese, V., Fadiga, L., and Rizzolatti, G. (1998). Neurons responding to the sight of goal directed hand/arm actions in the parietal area PF (7b) of the macaque monkey. *Society of Neuroscience Abstracts*, **24**, 257–7.

Gallese, V. and Goldman, A. (1998). Mirror neurons and the simulation theory of mind-reading. *Trends in Cognitive Sciences*, **12**, 493–501.

Gallese, V., Fadiga, L., Fogassi, L., and Rizzolatti, G. (1996). Action recognition in the premotor cortex. *Brain*, **119**, 593–609.

Gallese, V., Fogassi, L., Fadiga, L., and Rizzolatti, G. (2002). Action representation and the inferior parietal lobule. In Prinz, W. and Hommel, B. (Eds), *Attention and performance XIX. Common mechanisms in perception and action*, pp. 334–55. Oxford University Press, Oxford.

Gattis, M., Bekkering, H., and Wohlschlaeger, A. (2002). Goal-directed imitation. In Melzoff, A.N. and Prinz, W. (Eds.), *The imitative mind. Development, evolution and brain bases*, pp. 143–62. Cambridge University Press, Cambridge.

Gerardin, E., Sirigu, A., Lehericy, S., Poline, J.B., Gaymard, B., Marsault, C., Agid, Y., and Le Bihan, D. (2000). Partially overlapping neural networks for real and imagined hand movements. *Cereb. Cortex*, **10**, 1093–1104.

Grafton, S.T., Arbib, M.A., Fadiga, L., and Rizzolatti, G. (1996). Localization of grasp representations in humans by PET: 2. Observation compared with imagination. *Exp. Brain Res.,* **112**, 103–11.

Greenwald, A.G. (1970). Sensory feedback mechanisms in performance control: With special reference to the ideo-motor mechanism. *Psychol. Rev.,* **77**, 73–99.

Grèzes, J. and Decety, J. (2001). Functional anatomy of execution, mental simulation, observation, and verb generation of actions: A meta-analysis. *Human Brain Mapping,* **12**, 1–19.

Grèzes, J., Costes, N., and Decety, J. (1998). Top-down effect of strategy on the perception of human biological motion: a PET investigation. *Cognitive Neuropsychol.,* **15**, 553–82.

Humphrey, N. (1976). The social function of intellect. In Bateson, P.G. and Hinde, R.A. (Eds), *Growing points in ethology.* Cambridge University Press, Cambridge.

Humphrey, N. (2002). *The inner eye.* Oxford University Press, Oxford.

Hyvarinen, J. (1982). Posterior parietal lobe of the primate brain. *Physiol. Rev.,* **62**, 1060–1129.

Iacoboni, M., Woods, R.P., Brass, M., Bekkering, H., Mazziotta, J.C., and Rizzolatti, G. (1999). Cortical mechanisms of human imitation. *Science,* **286**, 2526–8.

Iacoboni, M., Koski, L.M., Brass, M., Bekkering, H., Woods, R.P., Dubeau, M.-C., Mazziotta, J.C., and Rizzolatti G. (2001). Reafferent copies of imitated actions in the right superior temporal cortex. *Proc. Natl Acad. Sci.,* **98**, 13995–9.

James, W. (1890). *Principles of psychology,* vol. 2. Holt, Dover edition, New York NY.

Jeannerod, M., Arbib, M.A., Rizzolatti, G., and Sakata, H. (1995). Grasping objects: the cortical mechanisms of visuomotor transformation. *Trends Neurosci.,* **18**, 314–320.

Jellema, T. and Perrett, D.I. (2002). Coding of visible and hidden actions. In W. Prinz and B. Hommel (Eds), *Attention and performance XIX. Common mechanisms in perception and action,* pp. 356–80. Oxford University Press, Oxford.

Jellema, T., Baker, C.I., Oram, M.W., and Perrett, D.I. (2002). Cell populations in the banks of the superior temporal sulcus of the macaque monkey and imitation. In Melzoff, A.N. and Prinz, W. (Eds), *The imitative mind. Development, evolution and brain bases,* pp. 143–62. Cambridge University Press, Cambridge.

Jolly, A. (1966). Lemur social behavior and primate intelligence. *Science,* **153**, 501–6.

Kohler, E., Keysers, Ch., Umiltà, M.A., Fogassi, L., Gallese, V., and Rizzolatti, G. (2002). Hearing sounds, understanding actions: Action representation in mirror neurons. *Science,* **297**, 846–8.

Krams, M., Rushworth, M.F., Deiber, M.P., Frackowiak, R.S., and Passingham, R.E. (1998). The preparation, execution and suppression of copied movements in the human brain. *Exp. Brain Res.,* **120**, 386–98.

Krubitzer, L., Clarey, J., Tweedale, R., Elston, G., and Calford, M. (1995). A redefinition of somatosensory areas in the lateral sulcus of Macaque monkeys. *J. Neurosci.,* **15**, 3821–39.

Leinonen, L. and Nyman, G. (1979). II. Functional properties of cells in anterolateral part of area 7 associative face area of awake monkeys. *Exp. Brain Res.,* **34**, 321–33.

Leinonen, L., Hyvarinen, J., Nyman, G., and Linnankoski, I. (1979). I. Functional properties of neurons in the lateral part of associative area 7 in awake monkeys. *Exp. Brain Res.,* **34**, 299–320.

Lotze, R. (1852). *Medizinsche psychologie oder physiologie der seele.* Leipzig: Weidmannsche Buchandlung.

Meltzoff, A.N. (2002). Elements of a developmental theory of imitation. In Meltzoff, A.N. and Prinz, W. (Eds), *The imitative mind. Development, evolution, and brain bases,* pp. 19–41. Cambridge University Press, Cambridge.

Merleau-Ponty, M. (1962). *Phenomenology of perception* [translated from the French by C. Smith]. Routledge, London.

Mountcastle, V. B., Lynch, J. C., Georgopoulos, A. Sakata, H., and Acuna, C. (1975). Posterior parietal association cortex of the monkey: command functions for operations within extrapersonal space. *J. Neurophysiol.* **38**, 871–908.

Murata, A., Fadiga, L., Fogassi, L., Gallese, V., Raos, V., and Rizzolatti, G. (1997). Object representation in the ventral premotor cortex (area F5) of the monkey. *J. Neurophysiol.*, **78**, 2226–30.

Nishitani, N. and Hari, R. (2000). Temporal dynamics of cortical representation for action. *Proc Natl Acad Sci*, **97**, 913–18.

Perrett, D.I., Harries, M.H., Bevan, R., Thomas, S., Benson, P.J., Mistlin, A.J., Chitty, A.J., Hietanen, J.K., and Ortega, J.E. (1989). Frameworks of analysis for the neural representation of animate objects and actions. *J. Exp. Biol.*, **146**, 87–113.

Perrett, D.I., Mistlin, A.J., Harries, M.H., and Chitty, A.J. (1990). Understanding the visual appearance and consequence of hand actions. In Goodale, M.A. (Ed.), *Vision and action: The control of grasping*, pp. 163–342. Ablex, Norwood NJ.

Petrides, M. and Pandya, D.N. (1997). Comparative architectonic analysis of the human and the macaque frontal cortex. In Boller, F. and Grafman, J. (Eds), *Handbook of neuropsychology* Vol. IX, pp. 17–58. Elsevier, New York.

Prinz, W. (2002). Experimental approaches to imitation. In Melzoff, A.N. and Prinz, W. (Eds), *The imitative mind. Development, evolution and brain bases*, pp. 143–62. Cambridge University Press, Cambridge.

Rizzolatti, G. and Arbib, M.A. (1998). Language within our grasp. *Trends Neurosci.*, **21**, 188–94.

Rizzolatti, G., Fogassi, L., and Gallese, V. (2000). Cortical mechanisms subserving object grasping and action recognition: a new view on the cortical motor functions. In Gazzaniga, M.S. (Ed.), *The cognitive neurosciences*, (2nd edn), pp. 539–52. MIT Press, Cambridge MA.

Rizzolatti, G. and Luppino, G. (2001). The cortical motor system. *Neuron*, 31, 889–901.

Rizzolatti, G., Camarda, R., Fogassi, L., Gentilucci, M., Luppino, G., and Matelli, M. (1988). Functional organization of inferior area 6 in the macaque monkey: II. Area F5 and the control of distal movements. *Exp. Brain Res.*, **71**, 491–507.

Rizzolatti, G., Fadiga, L., Fogassi, L., and Gallese, V. (1996a). Premotor cortex and the recognition of motor actions. *Cogn. Brain Res.*, **3**, 131–41.

Rizzolatti, G., Fadiga, L., Matelli, M., Bettinardi, V., Paulesu, E., Perani, D., and Fazio, F. (1996b). Localization of grasp representation in humans by PET: 1. Observation versus execution. *Exp. Brain Res.*, **111**, 246–52.

Rizzolatti, G., Fogassi, L., and Gallese, V. (2001). Neurophysiological mechanisms underlying the understanding and imitation of action. *Nature Rev. Neurosci.*, **2**, 661–70.

Rizzolatti, G., Fogassi, L., and Gallese, V. (2002). Motor and cognitive functions of the ventral premotor cortex. *Curr. Opin. Neurobiol.*, **12**, 149–54.

Robinson, C.J. and Burton, H. (1980a). Somatotopographic organization in the second somatosensory area of *M. fascicularis*. *J. Comp. Neurol.*, **192**, 43–67.

Robinson, C.J. and Burton, H. (1980b). Organization of somatosensory receptive fields in cortical areas 7b, retroinsula, postauditory, and granular insula of *M. fascicularis*. *J. Comp. Neurol.*, **192**, 69–92.

Seltzer, B. and Pandya, D.N. (1994). Parietal, temporal, and occipital projections to cortex of the superior temporal sulcus in the rhesus monkey: a retrograde tracer study. *J. Comp. Neurol.*, **15**, 445–63.

Shepard, R.N. (1984). Ecological constraints on internal representation: resonant kinematics of perceiving, imaging, thinking and dreaming. *Psychol. Rev.*, **9**, 417–47.

Shofield, W.N. (1976a). Do children find movements which cross the body midline difficult? *Q. J. Exp. Psychol.*, **28**, 571–82.

Shofield, W.N. (1976b). Hand movements which cross the body. Findings relating age differences to handedness. *Pecept Mot Skills*, **42**, 643–6.

Simon, O., Mangin, J.F., Cohen, L., Le Bihan, D., and Dehaene, S. (2002). Topographical layout of hand, eye, calculation, and language-related areas in the human parietal lobe. *Neuron*, **33**, 475–87.

Tomasello, M. and Call, J. (1997). *Primate cognition.* Oxford University Press, Oxford.

Umiltà, M.A., Kohler, E., Gallese, V., Fogassi, L., Fadiga, L., Keysers, C., and Rizzolatti, G. (2001). 'I know what you are doing': A neurophysiological study. *Neuron*, **32**, 91–101.

Visalberghi, E. and Fragaszy, D. (2002). Do monkeys ape? Ten years after. In Dautenhahn, K. and Nehaniv, C. (Eds), *Imitation in animals and artifacts*, pp. 471–99. MIT Press, Boston MA.

Von Bonin, G. and Bailey, P. (1947). *The neocortex of* macaca mulatta. University of Illinois Press, Urbana IL.

Part 1

Modularity: Splitting the atom of visual cognition

Chapter 2

Core knowledge

Elizabeth Spelke

Abstract

Does human cognition depend on a set of representational systems that are domain-specific, task-specific, and encapsulated? Studies of human infants and of non-human animals suggest that representations of extended surface layouts, manipulable objects, intentional agents, and approximate numerosities depend on just such modular *core knowledge* systems. In contrast, studies of human adults, and some studies of trained non-human animals, suggest that representations of space, objects, agents, and numerosity depend on a more flexible cognitive architecture. In this review, I attempt to bridge these findings and suggest that mature human cognitive performance depends in part on the same domain-specific systems found in infants and non-human animals. However, mature performance also depends on at least two kinds of domain-general systems for combining core representations: associative learning mechanisms that are common to humans and other animals, and symbolic systems that are uniquely human. In the future, parallel behavioral and neuroimaging studies of human infants, children, and adults may play a key role, both in probing the transition from encapsulated to flexible performance and in decomposing mature cognitive processes into the core representations that form their building blocks.

2.1 Introduction

One of the most debated questions in contemporary cognitive science and cognitive neuroscience concerns the modularity of human cognitive processing. Are the human brain and mind organized into systems that are domain-specific (i.e. that represent particular kinds of entities, such as manipulable objects or persons), task-specific (i.e. that function for particular purposes, such as object manipulation or social interaction), and encapsulated (i.e. that operate on a privileged subset of the information that the perceiver detects and remembers, independently of explicit knowledge or instruction)? Alternatively, does human cognition depend on a set of general-purpose cognitive mechanisms? Powerful arguments have been advanced on both sides of this issue, in part, because human cognition shows two striking features. First, it is flexible and open-ended. Over the course of history, humans have created and transformed countless concepts and conceptual domains, from science and art to sports and cuisine, and new concepts emerge with every technological advance or cultural innovation. Secondly, human knowledge grows rapidly and spontaneously in infants and young children, with minimal shaping by the environment. Children master the rules of their native language simply by hearing fragments of it, even fragments that are distorted by non-native speakers or multilingual environments. By age 4, and perhaps younger, children conceive of people's actions as guided by beliefs and desires—intangible entities that no child has seen and few parents have attempted to explain. Even younger children are sensitive to the mechanical properties of objects and to the geometry of the surrounding layout. These two features of human cognition seem to demand quite different architectures. Because black holes, soufflés, and home runs are too diverse and recent in origin for special-purpose cognitive devices to have evolved to encompass them, the most plausible accounts of the flexibility of mature human cognition appeal to general-purpose learning mechanisms. Because children's rapidly developing knowledge seems far to outrun the information on which such general-purpose mechanisms operate, the most plausible accounts of children's cognitive development appeal to specialized learning mechanisms. Can both these accounts be true?

The debate over domain-specific cognitive mechanisms is fueled, as well, by the rich and contradictory evidence for and against such mechanisms that has emerged from research in cognitive psychology and cognitive neuroscience. Some behavioral, neuropsychological, and neuroimaging experiments offer compelling evidence that human adults have localized, domain-specific mechanisms for representing entities such as faces, bodies, goal-directed actions, and places (see Caramazza 2000; Kanwisher *et al.* 2001). However, such evidence is countered by experiments using other displays or tasks that suggest that the relevant mechanisms are distributed and domain-general (e.g. Tarr and Gauthier 2000; Haxby *et al.* 2001). Why does mature human cognitive architecture look modular through some lenses and non-modular through others?

In this chapter, I consider these questions from the perspectives of comparative and developmental psychology. Studies of human infants and of non-human animals, I suggest, favor the view that human cognition depends on a set of systems that are domain-specific, task-specific, and relatively encapsulated: what I will call *core knowledge* systems. Most of these systems have a long phylogenetic history and so are not unique to humans. Moreover, the systems emerge early in ontogeny and operate throughout life, and therefore are common to adults and infants. Nevertheless, studies comparing the performance of human adults to that of infants or non-human animals suggests that these systems, by themselves, do not account for the flexibility of mature cognition. In addition to core knowledge systems, humans have at least two domain-general learning systems that serve to construct new concepts and belief systems.

One is a system of associative learning that we share with other animals. The other is a symbolic capacity that is unique to humans and is best exemplified by natural languages.

I focus here on four core knowledge systems for representing *places, objects, agents,* and *number*. Studies of non-human animals and human children suggest that each of these systems is a cognitive module in Fodor's sense (Fodor 1983), that each is largely constant over human phylogeny and ontogeny, and that together these systems provide the building blocks of mature human intelligence. However, the same experiments that provide evidence for these core systems also provide evidence that cognition becomes increasingly flexible over development. Here I discuss evidence for the construction of new spatial representations for navigation, of new representations of objects, of new concepts of humans and animals that are both mechanical and intentional, and of a new set of number concepts—the natural numbers—that serve as the entry to formal mathematics. I end with two questions. First, how many core and combinatorial systems do humans have, and how can we discover what these systems are? Secondly, how can we tease apart the respective roles of core and combinatorial systems so as to understand complex, uniquely human cognitive capacities? I suggest that neuroimaging studies comparing adult humans and non-human primates may play a central role in revealing both the nature and the limits of core knowledge systems in mature functioning, and that neuroimaging studies of infants and children may serve to elucidate how core systems are orchestrated to permit flexible cognitive performance. Future understanding of human cognition may therefore be linked, in part, to future progress in comparative and developmental cognitive neuroscience.

2.2 Geometric representations of the spatial layout

The best evidence for core representations of the environmental layout comes from studies of animal navigation. Animals as diverse as insects, fish, birds, and rodents find their way through the environment by drawing on representations of exquisite precision. The most important navigational mechanisms subserve processes of path integration, updating representations of the egocentric distances and directions of significant places in the layout as the animal moves (e.g. Mittelstaedt and Mittelstaedt 1980; Wehner and Wehner 1990). When an oriented animal approaches a significant location, it corrects for small errors and imprecisions in its reckoned position by drawing on stored, view-specific representations of the location (e.g. Collett *et al.* 1999). When it becomes fully disoriented, moreover, it reorients itself by relying on a representation of the shape of the surrounding surface layout (Cheng 1986; Margules and Gallistel 1988; Vallortigara *et al.* 1990). Representations of the shape of the layout, which are activated spontaneously when oriented rats explore new environments (O'Keefe and Burgess, 1996), serve as cues to reorientation in the same environments.

Evidence for a purely geometric representation of the environment comes from experiments by Cheng (1986), in which rats searched for food that they had previously observed being hidden in a rectangular environment with many featural cues to the food's location, including distinctive odors and visual patterns. After disorientation, rats ignored all this featural information and searched reliably at the two locations specified by the shape of the symmetrical enclosure. This experiment and many follow-ups provided evidence that rats store and use information about the shape of the environment to reorient themselves. Because rats failed to use featural information to distinguish between possible food locations in this rectangular environment, Cheng proposed that the geometry-based reorientation system is modular. Nevertheless, both his experiments and those of others provide evidence that disoriented rats and other animals do use featural information when they are trained to do so (Cheng, 1986; Gouteux *et al.*, 2001), especially when motivation levels are high, as in an escape task (Dudchenko *et al.* 1997; Sovrano *et al.* 2002). In these cases, animals may learn to associate geometric and featural representations (for a discussion, see Wang and Spelke 2002).

Studies of young children provide further evidence for modular geometric representations. Children from 18 months to 4 years have been tested in a situation similar to that used by Cheng: they are introduced to a rectangular room with distinctive features that break its symmetry (e.g. a distinctively colored wall or asymmetrically placed landmark), observe a toy hidden in one corner of the room, and then are disoriented by slow turning with eyes covered. When children are released and encouraged to find the toy, they reliably search in the correct corner and the geometrically equivalent opposite corner, providing evidence that they are sensitive to the shape of the room. Children fail to search the correct corner more than the geometrically equivalent opposite corner, ignoring all the featural information that distinguishes the corners (Hermer and Spelke 1996; Stedron *et al.* 2000). In both respects, children's search resembles that of rats.

Children's reorientation appears to depend on a mechanism that is domain-specific, task-specific, and encapsulated. Evidence for domain-specificity comes from experiments testing 4-year-old children's reorientation in an environment that presented a rectangular arrangement of detached walls, detached corners, or objects (Gouteux and Spelke 2001). Children reoriented by the geometrical arrangement only when presented with walls, providing evidence that their geometry-based system was specific to the domain of extended surfaces. Evidence for task specificity comes from experiments comparing the search performance of children who were oriented versus disoriented in identical environments (Hermer and Spelke 1996). After watching a toy being hidden in a location with distinctive geometric and non-geometric properties, children's eyes were covered and the boxes were moved so that their featural and geometric properties were dissociated. In one condition, children stood with eyes closed but remained oriented and so presumably attempted, when their eyes were uncovered, to determine where the container with the hidden object had moved. In a second condition, children were disoriented and presumably attempted, when their eyes were uncovered, to reorient themselves. Children relied primarily on the features of the container in the oriented condition and on the geometry of the room in the disoriented condition, providing evidence that the geometry-based system is specific to the task of reorientation. Finally, evidence for encapsulation comes from experiments in which children were observed over multiple sessions in a square room with one red and three white walls (Wang *et al.* 1999). For example, one child was trained to hit the red wall so as to activate a musical sequence. Once she did this reliably, she watched a toy being hidden in one corner of the room (immediately left or right of the red wall), was disoriented, and then was asked either to 'find the toy' or 'make the music'. When asked to make the music, the child immediately searched for the red wall, providing evidence that she was sensitive to the wall's location and had learned to use the wall as a direct cue to a trained action. When asked to find the toy, however, she ignored the wall and searched the four corners equally. This finding and others (e.g. Hermer and Spelke 1996; Wang *et al.* 1999) provide evidence that children's reorientation system is encapsulated: it fails to take account of environmental information that the child detects and uses for other purposes. Studies of children therefore support Cheng's (1986) original claim for a modular process of reorientation.

Although children perform very similarly to a variety of non-human animals on these tasks, human adults perform quite differently. When adults were tested in the original experiment of Hermer and Spelke (1996), they reoriented in accord both with the shape of the room and with the room's non-geometric features (a distinctively colored wall). Asked how they found the object, many adults spontaneously noted that it was hidden, for example, 'left of the blue wall'. Developmental experiments by Hermer-Vazquez provide evidence that children shift from modular to adult-like performance at about 7 years of age, and that the best predictor of this shift is the emergence of the capacity to produce spatial expressions involving the terms 'left' and 'right' (Hermer-Vazquez *et al.* 2001). Finally, experiments with adults using a dual-task procedure revealed that under conditions of verbal interference (but not non-verbal interference), adults performed like young children and rats: they continued to reorient in accord with the shape of the environment, continued to use a distinctively colored wall as a direct cue to action, and failed to use the same wall as a landmark for reorientation (Hermer-Vazquez *et al.* 1999). These findings suggest that the core, geometric system of spatial representation continues to exist in human adults and guides adults' reorientation independently of language. Nevertheless, adults are able to go beyond the limits of this core system and navigate more flexibly, and their flexible navigation performance depends in some way on the active use of language.

Why might language in general, and spatial language in particular, allow older children and adults to reorient by combining geometric and non-geometric information, overcoming the limits of their modular systems of spatial representation? As a medium of representation, natural languages are non-modular systems in two respects. First, a language has a *domain-general lexicon* containing words for diverse kinds of entities including objects ('soufflé'), people ('Mary'), places ('hotel'), properties ('scrumptious'), abstract entities ('three'), and spatial relations ('at'). Secondly, languages have a *combinatorial syntax* and *compositional semantics*. Therefore, once we learn the meanings of individual words and their rules of combination, we know the meanings of novel expressions containing those words without further learning (e.g. 'Mary baked three scrumptious soufflés at the hotel'). These complex expressions can serve to combine information across domains.

Figure 2.1 illustrates how language learning might allow children to form new and more flexible spatial representations. Because the infant's geometric representations of the spatial layout capture sense relationships, children may learn the meaning of the word 'left' by relating expressions involving that term to purely geometric representations of the surface layout. Because the child's object representations capture kinds and properties of objects, children may learn terms such as 'blue' and 'truck' by relating expressions involving those terms to representations of objects. Having learned these words and expressions, the child can formulate further expressions that combine them (e.g. 'left of the blue truck'). Such new expressions serve to link information in different core systems and so will allow the child to represent these combinations flexibly.

However, language is not the only domain-general representational system that allows for the combination of information across different core systems. Since Pavlov's dogs learned to associate the taste of food with the sound of a bell, we have known that both humans and other animals have a domain-general learning mechanism for combining diverse kinds of information. Associative learning allows appropriately trained animals to search for food to the left of a non-geometrically defined

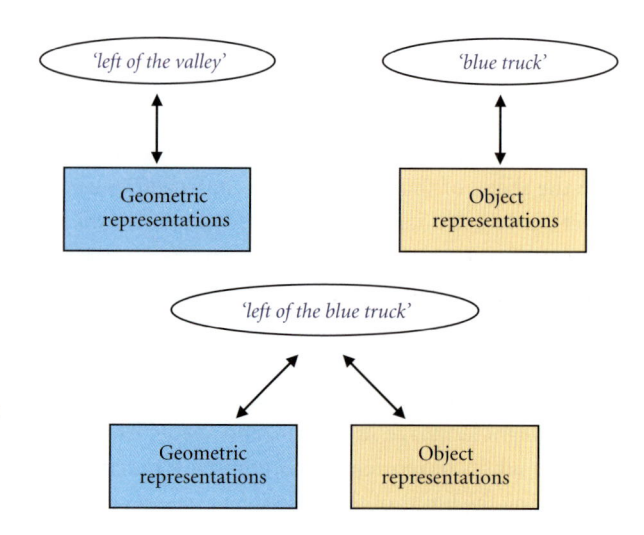

Fig. 2.1 How children might learn to use language to combine geometric and featural information.

landmark (Cheng 1986; Gouteux *et al.* 2001), and it would probably allow suitably trained children and language-impaired adults to do the same. Because each new combination of information must be learned individually, however, associative processes do not enable the learner to combine information rapidly, flexibly, and productively. That capacity may be available only to speakers of a natural language.

In summary, young human children and non-human animals have a system of core knowledge of the spatial layout. The system is domain-specific (it serves to represent extended surfaces but not bounded objects), task-specific (it serves to specify the navigating child's position within the environment but not the relative positions of objects), and encapsulated (it is sensitive to information about the shape of the surrounding layout but not to information about non-geometric features of the layout). Human adults also have this system, and it guides their navigation when they are tested with little training and with verbal interference. However, adults are also able to combine representations from different core systems. One combinatorial mechanism, shared with other animals, is associative learning. A second mechanism, unique to humans, depends on natural language with its domain-general lexicon, combinatorial syntax, and compositional semantics. The latter mechanism is highly flexible and efficient, for it allows information to be combined reliably in the absence of any specific training. Thus, the same modular systems exist and function in human adults, infants, and other animals, but they are partly obscured, and their signature limits are overcome, by other, domain-general systems. Domain-general combinatorial systems may account for the flexibility of human spatial behavior, whereas domain-specific systems account for its emergence and core functioning.

2.3 Spatio-temporal representations of objects

Evidence for a core system for representing objects first came from studies of human infants. Experiments using a variety of methods, including habituation and novelty preference, object-directed reaching and manipulation, and expectancy violation, provide evidence for an early developing ability to parse the visible surface layout into bodies that are internally connected, bounded, separately movable, solid, and continuously existing over occlusion. Infants parse visual arrays into objects by analyzing spatio-temporal properties of the arrays in accordance with three general constraints on object motion: cohesion (objects move as connected and bounded wholes), continuity (objects move on connected and unobstructed paths, and contact (objects change their motion on contact with other objects). Infants fail to parse visual arrays into objects by analyzing featural properties of the array, such as surface texture, coloring, and shape (Spelke 1990), although infants detect these properties and use them for other purposes, including texture segmentation (e.g. Quinn and Bhatt 1998) and categorization (e.g. Quinn and Eimas 1996). Infants also fail to use the same spatio-temporal relationships to perceive non-solid entities, such as piles of sand or stacks of blocks (Huntley-Fenner *et al.* 2000; Chiang and Wynn 2001; although see Wynn *et al.* 2002).

These findings provide evidence for a system of object representation that is domain-specific (it serves to represent objects but not surface markings or non-solid substances), task-specific (it serves to parse the continuous surface layout into bodies but not to parse individual surfaces into regions of homogeneous texture or to categorize bodies in accord with their similarity), and encapsulated (it operates on spatio-temporal information but not on information about the featural properties of objects or the kinds to which they belong).

Evidence for the encapsulation of the infant's processes of object parsing comes from experiments by Xu et al. (1999), using a visual preference-for-novelty procedure. Ten-month-old infants were presented with a toy duck and a toy truck, arranged so that the first object rested on top of the second. Because younger infants reliably categorize such objects as toy animals and vehicles (Mandler and McDonough 1993), and because infants in separate experiments discriminated between the objects, it is highly likely that infants detected the objects' features. Infants' ability to parse the array into units was tested by comparing their looking times to events in which a hand grasped the toy duck and lifted it into the air and either the duck moved by itself or the duck and truck moved together as a single unit. If the duck initially was slid over the surface of the truck, providing spatio-temporal evidence for a boundary between the objects, infants looked longer at the event in which the duck and truck moved as one body. This finding replicates earlier experiments (e.g. Spelke et al. 1989) and provides evidence that infants perceived the two objects as separately movable bodies. When the duck and truck were stationary, in contrast, infants showed no differential looking between the two test events. Although infants detected the distinctive features and categorical identities of the duck and truck, they failed to use this information to parse the array into two bodies. Similar findings were obtained by Xu and Carey (1996) in a different task, focusing on infants' representation of the numerical distinctness of two objects of different kinds that appeared at different places and times.

Research with newborn chicks and monkeys has yielded converging findings. Like human infants, these animals parse visual arrays into objects by analyzing the spatial arrangements and motions of surfaces (Regolin and Vallortigara 1995; Vallortigara and Zanforlin 1995a, b; Lea et al. 1996; Williams and Carey 2000). Experiments using an imprinting method, focusing on the tendency of day-old chicks to approach a familiar inanimate object that has accompanied them in the absence of any conspecifics, provide evidence that chicks not only perceive visible objects as do infants, but that they represent hidden objects and solve object search tasks whose memory demands pose problems for human infants for many months (Regolin et al. 1995a, b). These findings suggest that the system for parsing visual arrays into objects has a long phylogenetic history.

Recent research by Scholl (Scholl and Pylyshyn, 1999; Scholl et al. 2001, 2002) provides evidence that this spatio-temporal system of object representation continues to exist and function in human adults. Scholl presented adults with a multiple-object tracking task (Pylyshyn and Storm 1988), in which they viewed a set of featurally identical, independently moving objects, kept track of a subset of the objects, and reported whether a change occurred in any member of that subset. When objects are continuously visible and move on non-intersecting paths, adults can perform this task well with subsets of up to four objects. In one series of experiments, Scholl asked what kinds of entities adults are able to track in this manner. Like infants, adults could track entities that were internally connected and externally bounded (e.g. spatially separated disks or squares) and failed to track entities that lacked these properties (e.g. disks that were connected in pairs to form 'barbells' but that had to be tracked as individual object parts), providing evidence that attentive object tracking accorded with the cohesion constraint. In further experiments, Scholl asked what kinds of transformations during object motion would impair adults' attentive tracking. Like infants, adults were able to track objects that moved continuously in and out of view behind occluders, and they failed to track objects that appeared and disappeared discontinuously or that broke apart and coalesced, providing evidence that attentive object tracking accorded with the continuity and cohesion constraints. Finally, adults' attentive tracking was not influenced by featural properties of the objects or changes in those properties. All these findings and others (Kahneman and Treisman 1984; Kahneman et al. 1992; Kanwisher and Driver 1992; Mitroff et al. in press) suggest that a common system of representation underlies the object representations of adults tested in attentive tracking tasks and of infants tested in object parsing tasks.

Despite this convergence, there are striking developmental changes in object parsing, both in humans and in other animals. Twelve-month-old human infants and adult monkeys parse visual arrays into objects by combining spatio-temporal and featural information (e.g. Xu et al. 1999; Munakata et al. 2001). In humans, this developmental change correlates with the acquisition of words naming the objects (Xu and Carey 1996), and it is accelerated when objects are presented with distinctive words to label them (Xu 2002).

How does language aid infants' individuation of objects? Young infants, I suggest, have two domain- and task-specific systems for representing objects. One is the spatio-temporal system described above, which guides object parsing and object-directed reaching. The other is a featural system, which uses distinctions of color, texture, and shape to recognize and categorize objects (e.g. Quinn and Eimas 1996). Although we know little about the neural mechanisms subserving these systems in human infants, they may correspond, respectively, to the mechanisms associated with dorsal and ventral visual pathways of adult humans and monkeys (see Bertenthal 1996). Both systems arise early in human development, but initially they operate in relative independence. However, when children learn the first words for objects, the words may serve to link information from the two systems, because of the ways in which words link to objects. Early word learning has been proposed to accord with two constraints (Markman 1989; Bloom 2000): a 'whole object constraint' that leads infants to apply count nouns to the spatio-temporal

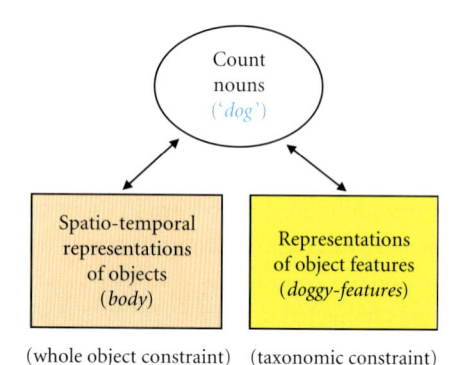

Fig. 2.2 How children might use count nouns to link representations of spatio-temporal and featural properties of objects.

[Diagram: "Count nouns ('dog')" at top, with arrows pointing down from it to two boxes: "Spatio-temporal representations of objects (body)" labeled "(whole object constraint)" and "Representations of object features (doggy-features)" labeled "(taxonomic constraint)"]

bodies picked out by the infant's object parsing system (e.g. Soja *et al.* 1991), and a 'taxonomic constraint' that leads infants to generalize count nouns across objects that share perceptual features and are members of the same kind (e.g. Smith *et al.* 1996; Gopnik and Sobel, 2000). Early count nouns therefore may serve to link representations of objects as spatio-temporal bodies to representations of objects as bearers of properties (see Fig. 2.2). Once these links are formed, then children can use property information rapidly and flexibly to enrich their spatio-temporal representations of object boundaries.

Nevertheless, language is not the only means for linking spatio-temporal and featural representations of objects, both because monkeys make the same links (Williams and Carey 2000; Munakata *et al.* 2001), and because prelinguistic human infants use featural information to parse objects when featural and spatio-temporal information are correlated and infants have the opportunity to learn the correlation (Needham and Baillargeon, 1998). These findings suggest that information from the two systems of object representation can become associatively linked when human infants or non-human animals have repeated experience with the objects. Both the acquisition of language and the accumulation of experience therefore allow perceivers to combine spatio-temporal and featural information about objects and to use both sources of information, in concert, to parse visual arrays.

2.4 Representations of goal-directed agents

The most important entities in the child's visual environment are agents: persons and other self-propelled objects that move from within, act on inanimate objects, and interact with one another and with the infant. Considerable research, using methods for studying infants' object representations, suggests that a distinct, domain-specific system underlies representations of agents.

Early studies focused on the constraint of no action at a distance. Intuitively, inanimate objects are subject to this constraint whereas agents are not: one must come into contact with a ball to set it in motion, but one can signal a person at a distance to move her to action. Woodward *et al.* (1993) tested whether 7-month-old infants are sensitive to this difference by presenting events in which either two people or two inanimate objects moved in succession. During habituation, the transition from movement of the first to movement of the second person or object took place behind a screen; during the test, the screen was removed and the motion of the second person or object was shown, alternately, to occur with or without contact with the first person or object. In the condition with two inanimate objects, infants looked longer at the no-contact event, replicating previous evidence that infants represent the motions of inanimate objects as subject to the constraint of no action at a distance (see Leslie 1994). In the condition with two persons, infants looked equally at the two test events, suggesting that they did not apply this constraint to the motions of people.

Further studies provide evidence that young infants represent human actions, but not inanimate object motions, as goal-directed. Woodward (1998) presented 6-month-old infants with events in which either a person or an inanimate object of a similar size and shape reached for one of two targets (a toy bear or a ball) in a single fixed position. After habituation, the two targets switched position and the person or object reached alternately for the same target (old goal, new direction of reach) or the other target (new goal, old direction of reach). Infants looked longer at the latter event, providing evidence that they encoded the initial reach primarily in relation to the goal object. In contrast, they showed no such preference in the condition with the inanimate object. These findings and those of Woodward *et al.* (1993) present a double dissociation: infants represent inanimate object motions as subject to a constraint of no action at a distance but not as goal-directed, and they represent human actions as goal-directed but not as subject to a constraint of no action at a distance.

Further differences between representations of inanimate objects and human agents emerge from studies of infants' capacities to determine whether an entity seen on one occasion is the same individual as an entity seen on another occasion. Studies of infants' representations of inanimate objects provide evidence that infants individuate such objects by analyzing the spatio-temporal continuity or discontinuity of object motion: an object seen at one place and time is the same individual as one seen at another place and time just in case a continuous path of motion links the two appearances (Spelke *et al.* 1995; Xu and Carey 1996). As reviewed above, infants under 12 months of age fail to individuate inanimate objects consistently when spatio-temporal information is ambiguous and featural information (differences in color, texture, shape) is available to distinguish them (Xu and Carey 1996). Parallel studies presenting people rather than inanimate objects provide evidence for just the reverse pattern of performance: infants fail to use spatio-temporal information for the continuity or discontinuity of motion in individuating people (Sorrentino *et al.* 1999; Kuhlmeier, Bloom, and Wynn 2003), and they successfully use featural information for individuation (Bonatti *et al.* 2002). This second double dissociation suggests that infants' representations of human agents depend on a system that is domain-specific (it applies to people but not to inanimate objects) and encapsulated (it is sensitive to featural information and to the goal-directedness of motion but not to spatio-temporal properties of motion, even though the latter properties are detected, are used for other purposes, and would improve performance on the person-individuation task).

Research suggests that infants' representations of agents serve at least two interrelated functions. First, infants use the behavior of other agents to direct their attention to, and learning about, relevant features of the world (Tomasello 2001). As early as 2 months, infants follow the direction of gaze of a person to fixate objects where the person is looking (Hood et al. 1998; see also Butterworth 1991). Infants aged 9 months and beyond imitate agents' object-directed actions and thereby learn about the functional properties of objects (Meltzoff 1995; see also Gergely et al. 2002). These attention and learning patterns are domain-specific, for they are not observed when infants interact with mechanical objects (Meltzoff 1995; Johnson et al. 1998). Secondly, infants use representations of agents and their actions to learn how people behave and how to communicate with them. Even 2-month-old infants engage in reciprocal social communication, adjusting their behavior to that of their social partner (see Legerstee 2001, for a review). At the start of the second year, infants use information about a speaker's attention and actions to determine the referents of her words and expressions (Baldwin 1991; Tomasello 2001). These processes of social learning also depend on the presence of featural or behavioral information that the social partner is an agent (Legerstee 2001).

In summary, research on infants provides evidence for a core system for representing agents. This system is domain-specific (it encompasses agents but not other kinds of objects), task-specific (it serves to direct attention, learning, and communication, but not object tracking or manipulation), and relatively encapsulated (it is blind to detectable information about spatio-temporal constraints on the agent's behavior). Comparative studies suggest that very similar representations of agents and their goal-directed actions exist in non-human primates. Rhesus and tamarin monkeys, for example, represent the actions of humans and animals as self-propelled and goal-directed (Hauser 1998; Santos and Hauser 1999), and they detect correspondences between the actions of other monkeys or humans and their own actions (Rizzolatti et al. 2000). When chimpanzees see a person or conspecific acting on an object in a particular way, their attention to, and actions on, the object are enhanced (Tomasello 1999). Finally, recent experiments focusing on competition and dominance relationships provide evidence that chimpanzees predict that an opponent's behavior will depend on what that opponent has seen (Hare et al. 2000, 2001). In both human infants and non-human primates, agents and inanimate objects are represented as distinct kinds of entities that are subject to distinct constraints.

Once again, a consideration of human adults suggests a more complex picture. Adults distinguish among different kinds of agents: persons, animals, and complex machines are treated alike in certain contexts and differently in other contexts. Adults also can reason about a person in multiple ways: as a *material body* subject to mechanical constraints; a *living being* that grows, reproduces, eats, and is subject to disease; an *agent* that acts in pursuit of goals; and a *sentient being* whose chosen actions are guided by beliefs and desires. These conceptions permit more differentiated and flexible reasoning than do the isolated core representations of agency found in infants. Studies of children suggest that these conceptions begin to differentiate at the end of the preschool years (Keil 1979; Carey 1985; Schult and Wellman 1997), and that differentiated concepts emerge as children link their core representations of persons with their core representations of objects (Carey and Spelke 1994). Language may facilitate this process, as words either call attention to, or serve as placeholders for, new concepts (Carey 2001). For example, when a single expression, such as 'is thirsty', is applied to people, animals, and plants, it may signal the existence of a new ontological category, *living thing*. And when diverse expressions such as 'is heavy', 'is thirsty', 'is walking', and 'wants an apple' all are applied to the same person, they may signal that a human being belongs to a number of nested ontological categories (Keil 1979) (Fig. 2.3). Nevertheless, the multiple, mature conceptions of persons are not without tensions and contradictions, as attested by the extensive philosophical literature on the mind/body problem.

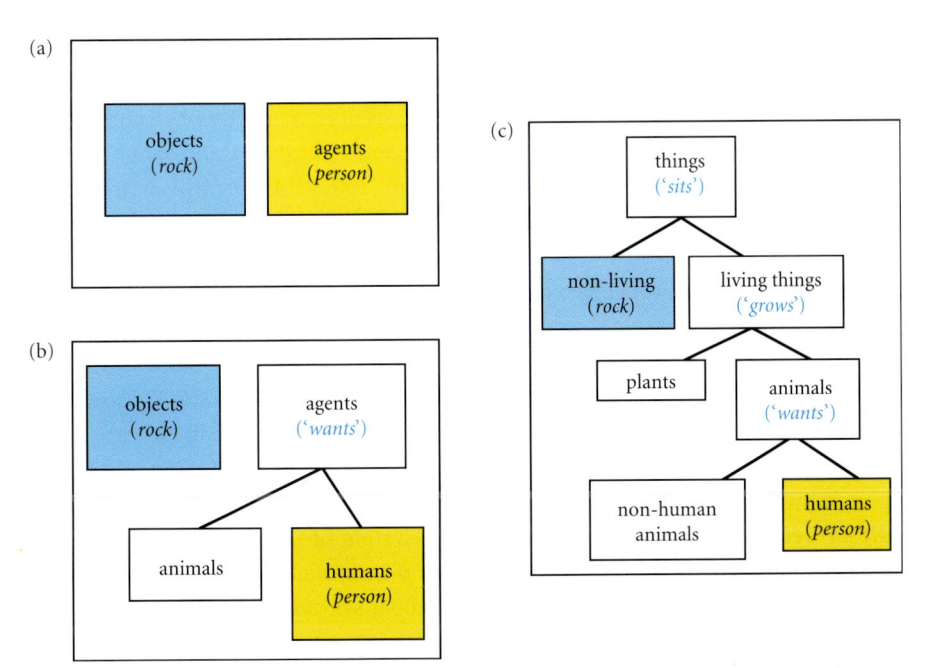

Fig. 2.3 Partial taxonomies of the natural world in (a) human infants, (b) preschool children, and (c) adults (after Keil 1979; Carey 1985).

2.5 Number sense

Although formal mathematics is unique to educated humans, a central building block of this system exists in animals and human infants: a sense of approximate numerical magnitudes or *number sense* (Dehaene 1997). Evidence for number sense comes from a variety of studies in which rats, pigeons, or other laboratory animals are trained to make a given number of responses, to discriminate between two different numbers of events, or to respond to displays that match in numerosity (e.g. Brannon and Terrace 1998; Nieder *et al.* 2002; see Gallistel 1990, for a review of the older literature). When animals learn to perform these tasks, their performance shows several key features. First, animals base their responses on numerosity, not on continuous variables that typically correlate with numerosity, such as the amount of effort required by a series of lever presses or the amount of light emitted over a series of flashes. Secondly, the error in animals' performance is proportional to the target numerosity, in accord with Weber's law. As a consequence, animals' numerical discriminations depend on the ratio of the two set sizes. Thirdly, number representations are sufficiently abstract to permit transfer across modalities and formats; for example, animals trained to discriminate numerosity with tone sequences show transfer to sequences of light flashes (Fernandes and Church 1982). The first and second findings have recently been replicated in untrained monkeys using a habituation/novelty detection paradigm, providing evidence that monkeys form representations of approximate numerical magnitudes spontaneously (Hauser *et al.* in press).

Studies of human infants and young children provide evidence for the same system of representation. When tested with preferential looking methods, human infants discriminate between arrays of visual forms on the basis of numerosity, when continuous variables such as element size, element density, image size, filled area, and contour length are controlled (Xu and Spelke 2000*a*, *b*; Brannon, 2002). Infants also discriminate between sound sequences when tested under conditions similar to those used with untrained monkeys (Lipton and Spelke in press). By 11 months, infants discriminate ascending from descending numerical sequences (Brannon 2002), providing evidence that their number representations are ordered. Like animals, infants' discrimination depends on the ratio of the two set sizes: at 6 months, infants discriminate numerosities that differ by a ratio of 2.0 (4 versus 8, 8 versus 16, and 16 versus 32), and they fail to discriminate numerosities that differ by a ratio of 1.5 (4 versus 6, 8 versus 12, 16 versus 24). The critical ratio narrows to 1.5 at 9 months of age (Lipton and Spelke in press). Infants show the same ratio limits on discrimination with visual–spatial and auditory–temporal arrays, suggesting that the system of number representation is amodal. Under certain conditions, infants also show transfer of numerical discrimination across modalities or formats (Feron *et al.* 2002; Kobayashi *et al.* 2002), although this ability has only been tested with small numbers and is not found in all experiments (Mix *et al.* 1997). These findings suggest that a similar system of representation underlies numerical discrimination by non-human animals and human infants.

What can animals and children do with this system? Gallistel (1990) has argued that animals use their representations of numerosity to perform computations: to add, subtract, multiply, and divide numerical values. However, the evidence for arithmetic operations in animals is indirect and is open to multiple interpretations (see Brannon *et al.* 2001; Dehaene 2001). More direct evidence that non-symbolic numerosity representations can enter into arithmetic computations comes from studies of 5-year-old children who have learned little or no formal arithmetic (La Mont *et al.* 2003). Children were presented with an array of blue dots that was covered by a screen and then a second array of blue dots that moved behind the screen. Finally, they were shown a single array of red dots and asked whether there were more blue or red dots. Children's responses, to both this problem and to problems involving simple comparisons of dot arrays, provided evidence that they focused on the number of dots rather than on continuous variables such as dot density or area, and that they were capable of performing these non-symbolic additions reliably, provided that the true sum and the comparison array differed by a sufficiently large ratio (1.5 or higher).

The numerosity representations found in non-human animals and in young children also exist in human adults. When adults are prevented from counting, they nevertheless discriminate between visual–spatial, visual–temporal, and auditory–temporal arrays on the basis of numerosity when other continuous variables are controlled (e.g. Whalen *et al.* 1999; Barth *et al.* 2003). Adults' discrimination depends on set size ratio, with a discrimination ratio that is considerably better than that of infants and comparable to that of trained adult animals (about 1.15: van Oeffelen and Vos 1982). Adults can discriminate and compare numerosities across changes in modality and format almost as well as within a single modality and format (Barth *et al.* 2003). Adults can also perform operations of addition, subtraction, multiplication, and division on non-symbolic numerosity representations (Barth 2002). Nevertheless, human adults have further numerical abilities that are found in no non-human animal. By using verbal counting and symbolic notation, we can represent large numerosities exactly, with no upper bound and no set size ratio limit on discrimination. This ability emerges between 3 and 4 years of age when children learn verbal counting (Wynn 1990, 1992).

Although there is no consensus concerning the process by which the natural number concepts emerge (for discussion, see Spelke 2000; Carey 2001), investigators have suggested that two systems of representation are involved in their construction. One is the system of object representation discussed earlier, which serves to represent up to three or four objects at once. This system may provide young children with one component of natural number concepts: the concept of *numerically distinct individuals*. The other is the system of number sense, which serves to represent the approximate numerosities of sets of objects or events. This system may provide children with a second component of natural number concepts: the concept of a *set with a cardinal value*. When children learn number words, the words may serve to combine these components into the concept of a *set of individuals*. The verbal counting routine then may serve to overcome

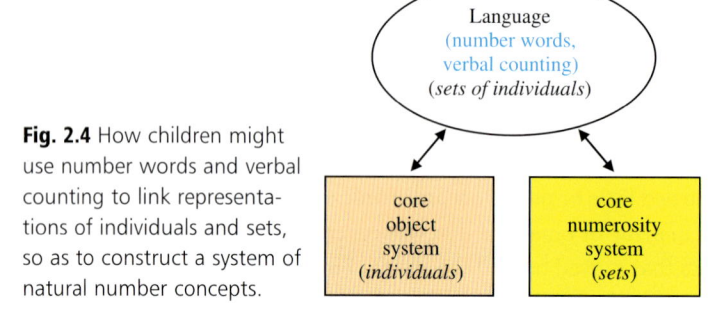

Fig. 2.4 How children might use number words and verbal counting to link representations of individuals and sets, so as to construct a system of natural number concepts.

the signature limits of the two initial systems: the set size limit of core object representations and the set size ratio limit of core number sense (Spelke 2000) (Fig. 2.4). If these suggestions are correct, then mature representations of the natural numbers depend on representations from three systems: representations of objects, representations of approximate numerosities, and the language of number words and verbal counting.

Evidence that symbolic number abilities depend, in part, on the core object system is suggested by the literature on 'subitizing'. When adults enumerate a set of four or more elements exactly, their response time rises approximately linearly with increasing set size, suggestive of a verbal counting process. When they enumerate three or fewer elements, however, responses are faster and show less rise with increasing numerosity, suggesting that 1, 2, or 3 objects can be represented by a parallel process (Mandler and Shebo 1982). Trick and Pylyshyn (1994) found a close correspondence between the conditions under which this subitizing effect occurs, and the conditions that allow parallel individuation of objects in multiple object tracking tasks with adults, and object representation tasks with infants. This finding suggests that the core system of object representation is active when adults enumerate objects, and serves as the basis of adults' response when numerosities are small.

The evidence that symbolic number abilities depend in part on number sense has been reviewed by Dehaene (1997). When adults compare two symbolic numerosities or verify simple arithmetic computations, they perform faster when the numerosities are more distant. In a comparison task, adults judge more quickly that 9 > 5 than that 6 > 5; in an arithmetic verification task they reject 19 more quickly than 13 as the sum of 7 + 5. Young children who have recently learned to represent numbers with words and Arabic symbols show similar effects (Temple and Posner, 1998). Moreover, adult neurological patients with normal numerical abilities prior to injury but subsequent impairments to number sense show impaired symbolic numerical abilities, and normal adults tested in neuroimaging experiments [both functional magnetic resonance imaging (fMRI) and event-related potentials (ERPs)] show activation, during number processing tasks, of certain areas of parietal cortex that are damaged in patients with impaired number sense (e.g. Dehaene and Cohen 1997; Pinel *et al.* 2001). These findings suggest that representations of the natural numbers depend, in part, on number sense.

Finally, four sources of evidence suggest that natural number representations depend, in part, on language. First, aphasic patients, with impaired abilities to use number words, show a characteristic deficit in symbolic number calculation: they may lose access to exact number facts but continue to be sensitive to approximate number facts (for example, a patient may judge that 7 + 5 is '13 roughly'; Warrington 1982). Secondly, when normal adults perform exact arithmetic with problems presented in Arabic notation, their response time shows a phonological length effect: they respond more slowly when the Arabic numerals correspond to words that require more syllables to pronounce. In contrast, when the same problems are presented under conditions that require only estimation of the approximate sum, no phonological length effect is observed (Lemer 2000). These findings suggest that adults spontaneously translate Arabic notation into a phonological code when performing exact, but not approximate, addition. Thirdly, when bilingual adults are taught, in one language, new arithmetic facts or verbal material containing exact numerosities (e.g. 37 + 52 = 89; the king is 43 years old), they show a cost when asked to retrieve the facts in their other language. In contrast, when they are taught material containing approximate numerosities (e.g. 37 + 52 is about 90; the king is about 40 years old), they show no such cost (Dehaene *et al.* 1999; O'Kane and Spelke 2001; Spelke and Tsivkin 2001). Finally, fMRI studies of adults performing exact versus approximate addition on symbolically presented numbers show activation of secondary language areas for the exact, but not the approximate, tasks (Dehaene *et al.* 1999; Stanescu-Cosson *et al.* 2000). All the findings of the approximate number tasks confirm the evidence presented above that a system of approximate number representation exists in adults and is independent of language. Moreover, the contrasting findings of the exact number tasks provide evidence that exact number representations depend, in part, on a specific natural language, consistent with the thesis that language links core representations of objects and approximate numerosities to form new representations of exact numerosity.

In summary, studies of non-human animals, human infants and children, and human adults provide evidence that a core system of number sense is enhanced by language to form representations of natural numbers. Studies of highly trained animals suggest that language may not be the only system for forming new number representations: after years of training, both chimpanzees and parrots have learned symbols for numbers from 1 to 9, suggesting that associative learning can serve to construct the meanings of individual number words (Matsuzawa 1985, 2000; Pepperberg 1994; Boysen 1997). However, in contrast to children, animals' constructions are not productive: after learning the meanings of the Arabic symbols 1–5, for example, the chimpanzee Ai still required extensive training to learn the meaning of the symbol 6, showing little or no 'learning set' for number (Matsuzawa 1985). When children come to understand the verbal counting system, in contrast, they require no further learning in order to represent the meanings of all the number words in their counting routine (Wynn 1990, 1992). Indeed, the system of number concepts that children form carries them beyond their finite vocabulary and allows them to conceive of numbers as increasing indefinitely (Gelman 1991) and as each designating a unique cardinal value (Lipton 2003). Therefore, as in the case of spatial representations and object representations, studies of number representations suggest that core number sense can be enriched and made precise in two different ways, by associative learning and by the acquisition of a natural language. Humans and animals share the associative mechanism, but only humans have the more flexible, rapid, and productive system that depends on natural language. This system begins to function when children learn the meanings of number words and the verbal counting routine, and it continues to function whenever human adults represent large, exact numerical values.

2.6 Other core knowledge systems?

The research reviewed above provides evidence for four systems of core knowledge encompassing places, manipulable objects, agents, and number. Stepping back from these cases, we may ask how many more core systems reside in the human mind? Does human cognition depend on hundreds or thousands of domain-specific systems, or only on a few? Although existing evidence does not provide a definitive answer to this question, I believe it suggests that humans are endowed with core systems only in a small number of domains. Within each domain, however, multiple, relatively encapsulated sub-systems may serve to identify the entities in the domain and to represent distinct aspects of their behavior.

For example, studies of infants suggest that at least four systems serve to represent agents: a face-recognition system, a biological motion system that parses agents' bodies into parts, and systems for analyzing goal-directed action and contingent social interaction (e.g. Bertenthal *et al.* 1987; Johnson and Morton 1991; Johnson *et al.*, 1998; Woodward, 1998). Similarly, studies of animal navigation provide evidence that multiple, distinct sub-systems contribute to any single act of navigation, including systems for determining compass heading, for perceiving the distances and directions of landmarks, and for apprehending the overall shape of the layout (see Gallistel 1990; Collett and Collett 2000).

Some evolutionary psychologists have proposed that humans are endowed with a much richer array of domain-specific systems, including systems for representing food objects, predators, and non-reciprocating social partners (e.g. Cosmides and Tooby 2000; cf. Fodor 2000). However, in contrast to the cases considered above, there is no evidence for such systems in infants, and little evidence in young non-human primates. For example, both human and non-human adult primates parse food objects at color/texture boundaries (Munakata *et al.* 2001), but human and monkey infants do not (Williams and Carey 2000; Condry *et al.* 2002). Moreover, adult monkeys generalize learning about individual food objects to other objects of the same color independently of shape, and they generalize learning about individual artifacts to other objects of the same shape independently of color (Santos *et al.* 2001), but juvenile monkeys do not show these differentiated generalization patterns (Santos *et al.* 2001). For omnivores and promiscuous travelers, such as humans and monkeys, the perceptible features of food objects may be too diverse and variable for an effective, innately specified recognition system. Instead, core systems for representing objects and agency may allow animals to learn what objects count as food by observing the eating patterns of conspecifics. Consistent with this suggestion, a number of animal species, including rats and monkeys, have been found to learn rapidly that a given type of object is edible by observing whether other animals eat the object (e.g. Galef 1988; Santos *et al.* 2001; see also Rozin *et al.* 2000).

Although there may be more domain-specific, core knowledge systems than the four systems discussed in this chapter, these observations suggest there are not vast numbers of them. The richness and flexibility of human cognition may stem primarily from the rich and variable ways in which domain-general combinatorial systems, especially natural languages, construct new concepts from a limited set of core systems.

2.7 Suggestions for cognitive neuroscience

If mature human cognitive performance depends both on domain-specific, core representations and on domain-general systems that combine these representations, then studies of human adults are likely to yield a complex picture of the human mind. When human adults are tested in circumstances that activate representations from one core system in relative independence of other systems, mature cognition will appear to depend on a modular architecture. When adults are tested in circumstances that activate multiple core representations that either are linked by well-established associations or are orchestrated by a central system such as language, mature cognition will appear to depend on a non-modular architecture. The portrait of cognitive architecture suggested by comparative and developmental studies therefore may help to explain why it has been so difficult to obtain conclusive evidence for or against domain-specific modules in human adults. How can cognitive psychologists cut through this complexity, study the nature and functioning both of core systems and of combinatorial systems, and tease apart their respective roles in cognitive performance?

Progress on these questions will be most enhanced, in my view, by a marriage of the fields of comparative, developmental, and cognitive psychology, and by research using parallel behavioral and neuroimaging methods. Behavioral and neuroimaging experiments, first, could serve to investigate the functional properties of core representational systems in human infants and non-human animals. These studies might provide tools for detecting the operation of core systems by revealing their distinctive signatures. The studies reviewed in this chapter already suggest a number of behavioral signatures of core systems, including the spatio-temporal signatures of object representations and the set size ratio signature of number sense. However, our ability to detect core systems and study their operation would be enhanced by the discovery of neural signatures of their functioning.

Neuroimaging studies of human adults suggest where to look for signatures of some of the core systems discussed above. The core system for representing the shape of the extended surface layout may be localized, in part, in a specific cortical region near the hippocampus that shows intense activation in fMRI experiments when subjects view photographs of scenes, regardless of whether the scenes are familiar or novel, indoor or outdoor, and furnished or bare (Epstein and Kanwisher 1998). The core system for representing spatio-temporally defined objects may be localized in part in regions of parietal cortex that are activated during multiple object tracking tasks, and activation of which rises with increases in the number of objects to be tracked (Culham *et al.*, 1998, 2001). The core system for representing agents may depend, in part, on systems for representing faces, bodies, and goal-directed actions that evoke intense activity in quite specific, localized regions of the temporal and occipital lobes, including a region in the fusiform gyrus (Kanwisher *et al.* 1997), a region in extra-striate visual cortex (Downing *et al.* 2001), and regions of the superior temporal sulcus (Allison *et al.* 2000; Gallagher *et al.* 2000). Finally, the core system of number sense may be localized, in part, in specific regions of the intraparietal sulcus (e.g. Dehaene 1997; Pinel *et al.* 2001). These suggestions are tentative, because the neuroimaging methods used with adults differ considerably from the behavioral methods used with infants and non-human animals, and because no developmental data currently link these sets of findings. To clarify both the continuities and changes in core systems over ontogeny and phylogeny, we need neuroimaging experiments that use a common set of behavioral tasks to compare systematically performance across ages and species.

As the building blocks of mature cognitive functioning are isolated, further studies become possible. In particular, conjoint behavioral and neuroimaging experiments can serve to study human adults' performance of complex, uniquely human tasks, such as mathematics or map use, using the signatures revealed by comparative and developmental studies so as to probe both for the activity of core systems and for the activity of the combinatorial systems that orchestrate them. To highlight the role of the combinatorial systems, it is useful to compare performance on tasks that are closely similar, one of which relies primarily on core representations and the other on constructed representations (examples might include studies of navigation guided by geometrically versus featurally specified landmarks, of object tracking based on spatio-temporal versus featural information, or of mental arithmetic based on exact versus approximate numerical information). Developmental versions of these experiments can focus on changes in children's performance of the complex tasks, so as to begin to reveal how core representations are linked and new concepts are constructed over the course of human development and education.

Neuroimaging studies of human children and non-human animals have lagged behind studies of human adults, which often focus on complex cognitive functions unique to mature humans. If complex human cognition depends on components that humans share with other animals, parallel studies of humans and non-human animals could be highly revealing, both of the shared components and of the mechanisms and processes that are unique to humans. If those components emerge in human infancy, and their orchestration into complex skills occurs during human development, then parallel studies of human adults, infants, and children may shed light both on humans' unchanging core capacities and on the processes by which we build new cognitive skills. Cognitive psychologists and cognitive neuroscientists widely assume that the architecture of human cognition is best investigated through studies of adults: once that architecture is revealed, then studies of children and animals can address subsidiary questions concerning its ontogeny and phylogeny. Here, I suggest a reversal of these assumptions. Mature human cognition is so complex and flexible that its underlying architecture is obscured by layers upon layers of acquired skills. Nevertheless, even the most complex skills are built from simpler components. Comparative and developmental studies may shed light on the architecture of mature human cognition by isolating those components in human infants and non-human animals.

References

Allison, T., Puce, A., and McCarthy, G. (2000). Social perception from visual cues: Role of the STS region. *Trends in Cognitive Sciences*, **4**, 267–78.

Baldwin, D. A. (1991). Infants' contribution to the achievement of joint reference. *Child Development*, **63**, 875–90.

Barth, H. C. (2002). Numerical cognition in adults: Representation and manipulation of nonsymbolic quantities. Doctoral dissertation, Department of Brain and Cognitive Sciences, Massachusetts Institute of Technology.

Barth, H., Kanwisher, N., and Spelke, E. (2003). Construction of large number representations in adults. *Cognition*, **86**, 201–21.

Bertenthal, B. I. (1996). Origins and early development of perception, action, and representation. *Annual Review of Psychology*, **47**, 431–59.

Bertenthal, B. I., Proffitt, D. R., and Kramer, S. J. (1987). Perception of biomechanical motions by infants: Implementation of various processing constraints. *Journal of Experimental Psychology: Human Perception and Performance*, **13**(4), 577–85.

Bloom, P. (2000). *How children learn the meanings of words*. MIT Press, Cambridge MA.

Bonatti, L., Frot, E., Zangl, R., and Mehler, J. (2002). The Human First Hypothesis: identification of conspecifics and individuation of objects in the young infant. *Cognitive Psychology*, **45**, 1–39.

Boysen, S. T. (1997). Representation of quantities by apes. *Advances in the Study of Behavior*, **26**, 435–62.

Brannon, E. M. (2002). The development of ordinal numerical knowledge in infancy. *Cognition*, **83**, 223–40.

Brannon, E. M. and Terrace, H. S. (1998). Ordering of the numerosities 1 to 9 by monkeys. *Science*, **282**, 746–9.

Brannon, E. M., Wusthoff, C. J., Gallistel, C. R., and Gibbon, J. (2001). Numerical subtraction in the pigeon: Evidence for a linear subjective number scale. *Psychological Science*, **12**, 238–43.

Butterworth, G. E. (1991). The ontogeny and phylogeny of joint visual attention. In A. Whiten (Ed.), *Natural theories of mind*, pp. 223–32. Basil Blackwell, Oxford.

Caramazza, A. (2000). The organization of conceptual knowledge in the brain. In M. Gazzaniga (Ed.), *The new cognitive neurosciences*, pp. 1037–46. MIT Press, Cambridge MA.

Carey, S. (1985). *Conceptual change in childhood*. MIT Press, Cambridge MA.

Carey, S. (2001). Cognitive foundations of arithmetic: Evolution and onto-genesis. *Mind and Language*, **16**, 37–55.

Carey, S. and Spelke, E. S. (1994). Domain-specific knowledge and conceptual change. In L. Hirschfeld and S. Gelman (Eds.), *Mapping the mind: Domain specificity in cognition and culture*, pp. 169–200. Cambridge University Press, Cambridge UK.

Cheng, K. (1986). A purely geometric module in the rat's spatial representation. *Cognition*, **23**, 149–78.

Chiang, W. C. and Wynn, K. (2001). Infants' tracking of objects and collections. *Cognition*, **77**, 169–95.

Collett, T. S. and Collett, M. (2000). Path integration in insects. *Current Opinion in Neurobiology*, **10**, 757–62.

Collett, M., Collett, T. S., and Wehner, R. (1999). Calibration of vector navigation in desert ants. *Current Biology*, **9**, 1031–4.

Condry, K., Santos, L., and Spelke, E. S. (2002). Perception of natural food objects by four-month-old infants. Unpublished manuscript.

Cosmides, L. and Tooby, J. (2000). The cognitive neuroscience of social reasoning. In M. S. Gazzaniga (Ed.), *The new cognitive neurosciences*, 2nd edn, pp. 1259–70. MIT Press, Cambridge MA.

Culham, J. C., Brandt, S. A., Cavanagh, P., Kanwisher, N. G., Dale, A. M., and Tootell, R. B. (1998). Cortical fMRI activation produced by attentive tracking of moving targets. *Journal of Neurophysiology*, **80**, 2657–70.

Culham, J. C., Cavanagh, P., and Kanwisher, N. G. (2001). Attention response functions: Characterizing brain areas using fMRI activation during parametric variations of attentional load. *Neuron*, **32**, 737–45.

Dehaene, S. (1997). *The number sense: How the mind creates mathematics*. Oxford University Press, Oxford.

Dehaene, S. (2001). Subtracting pigeons: Logarithmic or linear? *Psychological Science*, **12**, 244–46.

Dehaene, S. and Cohen, L. (1997). Cerebral pathways for calculation: Double dissociations between Gerstmann's acalculia and subcortical acalculia. *Cortex*, **33**, 219.

Dehaene, S., Spelke, E., Pinel, P., Stanescu, R., and Tsivkin, S. (1999). Sources of mathematical thinking: Behavioral and brain-imaging evidence. *Science*, **284**, 970–4.

Downing, P. E., Jiang, Y., Shuman, M., and Kanwisher, N. (2001). A cortical area selective for visual processing of the human body. *Science*, **293**, 2470–3.

Dudchenko, P. A., Goodridge, J. P., Seiterle, D. A., and Taube, J. S. (1997). Effects of repeated disorientation on the acquisition of spatial tasks in rats: dissociation between the appetitive radial arm maze and aversive water maze. *Journal of Experimental Psychology: Animal Behavior Processes*, **23**, 194–210.

Epstein, R. and Kanwisher, N. (1998). A cortical representation of the local visual environment. *Nature*, **392**, 598–601.

Fernandes, D. M. and Church, R. M. (1982). Discrimination of the number of sequential events by rats. *Animal Learning and Behavior*, **10**, 171–76.

Feron, J., Streri, A., and Gentaz, E. (2002). Numerical intermodal transfer from touch to vision by 5-month-old infants. Poster presented at the International Conference on Infant Studies, Toronto.

Fodor, J. A. (1983). *The modularity of mind*. Bradford/MIT Press, Cambridge MA.

Fodor, J. A. (2000). *The mind doesn't work that way*. MIT Press, Cambridge MA.

Galef, B. G. (1988). Communication of information concerning the distant diets in a central place foraging species: *Rattus norvegicus*. In T. Zentall and B. G. Galef (Eds.), *Social learning: A comparative approach*, pp. 119–40. Erlbaum, Hillsdale NJ.

Gallagher, H. L., Happe, F., Brunswick, N., Fletcher, P. C., Frith, U., and Frith, C. D. (2000). Reading the mind in cartoons and stories: An fMRI study of 'theory of mind' in verbal and nonverbal tasks. *Neuropsychologicia*, **38**, 11–21.

Gallistel, C. R. (1990). *The organization of learning*. MIT Press, Cambridge, MA.

Gelman, R. (1991). Epigenetic foundations of knowledge structures: Initial and transcendent constructions. In S. Carey and R. Gelman (Eds.), *The epigenesis of mind: Essays on biology and cognition*. Erlbaum, Hillsdale NJ.

Gergely, G., Bekkering, H., and Király, I. (2002). Rational imitation in preverbal infants. *Nature*, **415**, 755.

Gopnik, A. and Sobel, D. (2000). Detecting blickets: How young children use information about novel causal powers in categorization and induction. *Child Development*, **71**, 1205–22.

Gouteux, S. and Spelke, E. S. (2001). Children's use of geometry and landmarks to reorient in an open space. *Cognition*, **81**, 119–48.

Gouteux, S., Thinus-Blanc, C., and Vauclair, J. (2001). Rhesus monkeys use geometric and nongeometric information during a reorientation task. *Journal of Experimental Psychology: General*, **130**, 505–19.

Hare, B., Call, J., Agnetta, B., and Tomasello, M. (2000). Chimpanzees know what conspecifics do and do not see. *Animal Behavior*, **59**, 771–85.

Hare, B. Call, J., and Tomasello, M. (2001). Do chimpanzees know what conspecifics know? *Animal Behavior*, **61**, 139–51.

Hauser, M. D. (1998). A non-human primate's expectations about object motion and destination: The importance of self-propelled movement and animacy. *Developmental Science*, **1**, 31–8.

Hauser, H. D., Tsao, F., Gascia, P., and Spelke, E. S. (in press). Evolutionary foundations of number: Spontaneous representation of numerical magnitudes by cotton-top tamarins. *Proceedings of the Royal Society* (London).

Haxby, J. V., Gobbini, M. I., Furey, M. L., Ishai, A., Schouten, J. L., and Pietrini, P. (2001). Distributed and overlapping representations of faces and objects in ventral temporal cortex. *Science*, **293**, 2425–30.

Hermer, L. and Spelke, E. S. (1996). Modularity and development: the case of spatial reorientation. *Cognition*, **61**, 195–232.

Hermer-Vazquez, L., Spelke, E. S., and Katsnelson, A. (1999). Sources of flexibility in human cognition: Dual-task studies of space and language. *Cognitive Psychology*, **39**, 3–36.

Hermer-Vazquez, L., Moffett, A., and Munkholm, P. (2001). Language, space, and the development of cognitive flexibility in humans: The case of two spatial memory tasks. *Cognition*, **79**, 263–99.

Hood, B. M., Willen, J. D., and Driver, J. (1998). Adult's eyes trigger shifts of visual attention in human infants. *Psychological Science*, **9**, 131–34.

Huntley-Fenner, G., Carey, S., and Solimando, A. (2000). Objects are individuals but stuff doesn't count: Perceived rigidity and cohesiveners influence infants' representations of small groups of distinct entities. *Cognition*, **85**, 223–50.

Johnson, M. H. and Morton, J. (1991). *Biology and cognitive development. The case of face recognition*. Blackwell, Oxford.

Johnson, S., Slaughter, V., and Carey, S. (1998). Whose gaze will infants follow? The elicitation of gaze following in 12-month-olds. *Developmental Science*, **1**, 233–8.

Kahneman, D. and Treisman, A. (1984). Changing views of attention and automaticity. In R. Parasuraman and D. R. Davies (Eds.), *Varieties of attention*, pp. xvi, 554. Academic Press, Orlando.

Kahneman D, Treisman, A., and Gibbs, B. (1992). The reviewing of object files: Object specific integration of information. *Cognitive Psychology*, **24**, 175–219.

Kanwisher, N. and Driver, J. (1992). Objects, attributes, and visual attention: Which, what, and where. *Current Directions in Psychological Science*, **1**, 26–31.

Kanwisher, N., McDermott, J., and Chun, M. M. (1997). The fusiform face area: A module in human extrastriate cortex specialized for face perception. *Journal of Neuroscience*, **17**(11), 4302–11.

Kanwisher, N., Downing, P., Epstein, R., and Kourtzi, Z. (2001). Functional neuroimaging of human visual recognition. In R. Cabeza and A. Kingstone (Eds.), *Handbook on functional neuroimaging*, pp. 109–52. MIT Press, Cambridge MA.

Keil, F. (1979). *Semantic and conceptual development: An ontological perspective*. Harvard University Press, Cambridge MA.

Kobayashi, R., Hiraki, K., and Hasegawa, T. (2002). Intermodal numerical correspondences in 6-month-old infants. Poster presented at the International Conference on Infant Studies. Toronto.

Kuhlmeier, V. A., Bloom, P., and Wynn, K. (2003). Infants do not see humans as material objects. Manuscript under review.

LaMont, K., Barth, H., and Spelke, E. S. (2003). Non-symbolic addition by preschool children. Poster presented at the society for Research in Child Development, Tampa FL, April.

Lea, S. E. G., Slater, A. M., Ryan, C. M. E. (1996). Perception of object unity in chicks: A comparison with the human infant. *Infant Behavior and Development*, **19**, 501–4.

Legerstee, M. (2001). Domain specificity and the epistemic triangle: The development of the concept of animacy in infancy. In F. Lacerda, C. von Hofsten, and M. Heimann (Eds.), *Emerging cognitive abilities in early infancy*. Erlbaum, Mahwah NJ.

Lemer, C. (2000). Representations langagieres des nombres dans la resolution de calculs mentaux complexes: Une approche par la memoire a court-terms verbale. Unpublished doctoral dissertation, Universite Libre de Bruxelles, Brussels.

Leslie, A. M. (1994). ToMM, ToBy, and Agency: core architecture and domain specificity. In L. Hirschfeld and S. Gelman (Eds.), *Mapping the mind: Domain specificity in cognition and cultural*, pp. 119–48. Cambridge University Press, New York.

Lipton, J. S. (2003). Preschool children's mapping of number words to non symbolic numerosities. Poster presented at the annual meeting of the society for Research in Child Development, Minneapolis MN, April.

Lipton, J. S. and Spelke, E. S. (in press). Origins of number sense: Large number discrimination in human infants. *Psychological Science*.

Mandler, G. and Shebo, B. J. (1982). Subitizing: An analysis of its component processes. *Journal of Experimental Psychology: General*, **11**, 1–22.

Mandler, J. M. and McDonough, L. (1993). Concept formation in infancy. *Cognitive Development*, **8**(3), 291–318.

Margules, J. and Gallistel, C. R. (1988). Heading in the rat: Determination by environmental shape. *Animal Learning and Behavior*, **16**, 404–10.

Markman, E. M. (1989). *Categorization and naming in children: Problems of induction*. MIT Press, Cambridge MA.

Matsuzawa, T. (1985). Use of numbers by a chimpanzee. *Nature*, **315**, 57–9.

Matsuzawa, T. (2000). Numerical memory span in a chimpanzee. *Nature*, **402**, 39–40.

Mechner, F. (1958). Probability relations within response sequences under ratio reinforcement. *Journal of the Experimental Analysis of Behavior*, **1**, 109–21.

Meltzoff, A. N. (1995). Understanding the intentions of others: Re-enactment of intended acts by 18-month-old children. *Developmental Psychology*, **31**(5), 838–50.

Mitroff, S. R., Scholl, B. J. and Wynn, K. (in press). Divide and conquer: How object files adapt when a persisting object splits in two. *Psychological Science*.

Mittelstaedt, M. L. and Mittelstaedt, H. (1980). Homing by path integration in a mammal. *Naturwissenschaften*, **67**, 566–7.

Mix, K. S., Levine, S. C., and Huttenlocher, J. (1997). Numerical abstraction in human infants: Another look. *Developmental Psychology*, **33**, 423–8.

Munakata, Y., Santos, L. R., Spelke, E. S., Hauser, M. D., and O'Reilly, R. C. (2001). Visual representation in the wild: How rhesus monkeys parse objects. *Journal of Cognitive Neuroscience*, **13**, 44–58.

Needham, A. and Baillargeon, R. (1998). Effects of prior experience on 4.5 month old infants' object segregation. *Infant Behavior and Development*, **21**, 1–24.

Nieder, A., Freedman, D. J., and Miller, E. K. (2002). Representation of the quantity of visual items in the primate prefrontal cortex. *Science*, **297**, 1708–11.

O'Kane, G. C. and Spelke, E. S. (2001). Language dependence in representations of number, space, and time. Poster presented at the meeting of the Psychonomic Society, Orlando FL, November.

O'Keefe, J. and Burgess, N. (1996). Geometric determinants of the place fields of hippocampal neurons. *Nature*, **381**, 425–8.

Pepperberg, I. M. (1994). Numerical competence in an African grey parrot (*Psittacus erithacus*). *Journal of Comparative Psychology*, **108**, 36–44.

Pinel, P., Dehaene, S., Riviere, D., and LeBihan, D. (2001). Modulation of parietal activation by semantic distance in a number comparison task. *Neuroimage*, **14**, 1013–26.

Pylyshyn, Z. W. and Storm, R. W. (1988). Tracking multiple independent targets: Evidence for a parallel tracking mechanism. *Spatial Vision*, **3**, 179–97.

Quinn, P. C. and Bhatt, R. S. (1998). Visual pop-out in young infants: Convergent evidence and an extension. *Infant Behavior and Development*, **21**, 273–88.

Quinn, P.C. and Eimas, P. D. (1996). Perceptual organization and categorization in young infants. In C. Rovee-Collier and L. P. Lipsitt (Eds.), *Advances in infancy research*, Vol. **10**, pp. 1–36. Ablex, Norwood NJ.

Regolin, L. and Vallortigara, G. (1995). Perception of partly occluded objects by young chicks. *Perception and Psychophysics*, **57**, 971–6.

Regolin, L., Vallortigara, G., and Zanforlin, M. (1995*a*). Object and spatial representations in detour problems by chicks. *Animal Behavior*, **49**, 195–9.

Regolin, L., Vallortigara, G., and Zanforlin, M. (1995*b*). Detour behaviour in the domestic chick: Searching for a disappearing prey or a disappearing social partner. *Animal Behavior*, **50**, 203–11.

Rizzolatti, G., Fogassi, L., and Gallese, V. (2000). Cortical mechanisms subserving object grasping and action recognition: A new view on the cortical motor functions. In M. S. Gazzaniga (Ed.), *The new cognitive neurosciences*, pp. 539–52. MIT Press, Cambridge MA.

Rozin, P., Haidt, J., and McCauley, C. R. (2000). Disgust. In M. Lewis and J. M. Haviland-Jones (Eds.), *Handbook of emotions*, 2nd edn, pp. 637–65. Guilford Press, New York NY.

Santos, L. R. and Hauser, M. D. (1999). How monkeys see the eyes: Cotton-top tamarins' reaction to changes in visual attention and action. *Animal Cognition*, **2**, 131–9.

Santos, L. R., Hauser, M. D., and Spelke, E. S. (2001). Recognition and categorization of biologically significant objects by rhesus monkeys (*Macaca mulatta*): the domain of food. *Cognition*, **82** (2), 127–55.

Scholl, B. J. and Pylyshyn, Z. W. (1999). Tracking multiple items through occlusion: Clues to visual objecthood. *Cognitive Psychology*, **38**, 259–90.

Scholl, B. J., Pylyshyn, Z. W., and Feldman, J. (2001). What is a visual object? Evidence from target merging in multiple-object tracking. *Cognition*, **80**, 159–77.

Schult, C. A. and Wellman, H. M. (1997). Explaining human movements and actions: Children's understanding of the limits of psychological explanation. *Cognition*, **62**, 291–324.

Smith, L. B., Jones, S.S., and Landau, B. (1996). Naming in young children: A dumb attentional mechanism? *Cognition*, **60**, 143–71.

Soja, N., Carey, S., and Spelke, E. S. (1991). Ontological categories guide young children's inductions of word meaning: Object terms and substance terms. *Cognition*, **38**, 179–211.

Sorrentino, C., Carey, S., and Spelke, E. (1999). Infants' individuation of persons. Unpublished manuscript.

Sovrano, V. A., Bisazza, A., and Vallortigara, G. (2002). Modularity and spatial reorientation in a simple mind: Encoding of geometric and nongeometric properties of a spatial environment by fish. *Cognition*, **85**, B51–B59.

Spelke, E. S. (1990). Principles of object perception. *Cognitive Science*, **14**, 29–56.

Spelke, E. S. (2000). Core knowledge. *American Psychologist*, **55**, 1233–43.

Spelke, E. S. and Tsivkin, S. (2001). Language and number: A bilingual training study. *Cognition*, **78**, 45–88.

Spelke, E. S., von Hofsten, C., and Kestenbaum, R. (1989). Object perception and object-directed reaching in infancy: Interaction of spatial and kinetic information for object boundaries. *Developmental Psychology*, **25**, 185–96.

Spelke, E. S., Kestenbaum, R., Simons, D., and Wein, D. (1995). Spatiotemporal continuity, smoothness of motion and object identity in infancy. *British Journal of Developmental Psychology*, **13**, 113–42.

Stanescu-Cosson, R., Pinel, P., van de Moortele, P.-F., Le Bihan, D., Cohen, L., and Dehaene, S. (2000). Understanding dissociations in dyscalculia: A brain imaging study of the impact of number size on the cerebral networks for exact and approximate calculation. *Brain*, **123**, 2240–55.

Stedron, J. M., Munakata, Y., and O'Reilly, R. C. (2000). Spatial reorientation in young children: A case of modularity? Poster presented at the 2000 meeting of the International Conference on Infant Studies, Brighton, July.

Tarr, M. J. and Gauthier, I. (2000). FFA: A flexible fusiform area for subordinate-level visual processing automatized by expertise. *Nature Neuroscience*, **3**, 764–9.

Temple, E. and Posner, M. I. (1998). Brain mechanisms of quantity are similar in 5-year-old children and adults. *Proceedings of the National Academy of Sciences, USA*, **95**, 7836–7841.

Tomasello, M. (1999). *The cultural origins of human cognition*. Harvard University Press, Cambridge MA.

Tomasello, M. (2001). *The cultural origins of human cognition*. Harvard University Press, Cambridge MA.

Trick, L. and Pylyshyn, Z. W. (1994). Why are small and large numbers enumerated differently? A limited capacity preattentive stage in vision. *Psychological Review*, **101**, 80–102.

Vallortigara, G., Zanforlin, M., and Pasti, G. (1990). Geometric modules in animals' spatial representations: a test with chicks (*Gallus gallus domesticus*). *Animal Cognition*, **4**, 191–205.

van Harle, K. and Scholl, B. J. (in press). Alternative tracking of objects vs. substances. *Psychological Science*.

van Oeffelen, M. P. and Vos, P. G. (1982). A probabilistic model for the discrimination of visual number. *Perception and Psychophysics*, **32**, 163–70.

Wang, R. F., Hermer-Vazquez, L., and Spelke, E. S. (1999). Mechanisms of reorientation and object localization by human children: a comparison with rats. *Behavioral Neuroscience*, **113**, 475–85.

Wang, R. F. and Spelke, E. S. (2000). Updating egocentric representations in human navigation. *Cognition*, **77**, 215–50.

Wang, R. F. and Spelke, E. S. (2002). Human spatial representation: Insights from animals. *Trends in Cognitive Science*, **6**, 376–82.

Warrington, E. K. (1982). The fractionation of arithmetical skills: A single case study. *Quarterly Journal of Experimental Psychology*, **34A**, 31–51.

Wehner, R. and Wehner, S. (1990). Insect navigation: Use of maps or Ariadne's thread? *Ethology, Ecology and Evolution*, **2**, 27–48.

Whalen, J., Gallistel, C. R., and Gelman, R. (1999). Non-verbal counting in humans: The psychophysics of number representation. *Psychological Science*, **10**, 130–137.

Williams, T. D. and Carey, S. (2000). Development of object individuation in infant pigtail macaques. Paster presented at the International Conference on Infant Studies, Brighton, July.

Woodward, A. L. (1998). Infants selectively encode the goal object of an actor's reach. *Cognition*, **69**, 1–34.

Woodward, A. L., Phillips, A. T., and Spelke, E. S. (1993). Infants' expectations about the motions of inanimate vs. animate objects. *Proceedings of the cognitive science society*. Erlbaum, Hillsdale NJ.

Wynn, K. (1990). Children's understanding of counting. *Cognition*, **36**, 155–93.

Wynn, K. (1992). Children's acquisition of the number words and the counting system. *Cognitive Psychology*, **24**, 220–51.

Wynn, K., Bloom, P., and Chiang, W. (2002). Enumeration of collective entities by 5-month-old infants. *Cognition*, **83**, B55–B62.

Xu, F. (2002). The role of language in acquiring object kind concepts in infancy. *Cognition*, **85**, 223–50.

Xu, F. and Carey, S. (1996). Infants' metaphysics: the case of numerical identity. *Cognitive Psychology*, **30**, 111–53.

Xu, F. and Spelke, E. S. (2000*a*). Large number discrimination in 6-month-old infants. *Cognition*, **74**, B1–B11.

Xu, F. and Spelke, E. S. (2000*b*). Large number discrimination in infants: Evidence for analog magnitude representations. Paper presented at the International Conference on Infant Studies, Brighton, July.

Xu, F., Carey, S., and Welch, J. (1999). Infants' ability to use object kind information for object individuation. *Cognition*, **70**, 137–66.

Chapter 3

The evolution of human neocortex: Is the human brain fundamentally different than that of other mammals?

Leah Krubitzer and Dianna M. Kahn

Abstract

The neocortex is composed of areas that are functionally, anatomically, and histochemically distinct. In comparison to most other mammals, humans have an expanded neocortex, with a pronounced increase in the number of cortical areas. This expansion underlies many complex behaviors associated with human capabilities, including perception, cognition, language, and volitional motor responses. We consider data from comparative studies as well as from developmental studies to gain insight into the mechanisms involved in arealization, and discuss how these mechanisms may have been modified in different lineages over time to produce the remarkable degree of organizational variability observed in the neocortex of mammals. Because any phenotype is a result of the complex interactions between genotypic influences and environmental factors, we also consider environmental, or epigenetic, contributions to the organization of the neocortex.

3.1 Introduction

How did humans evolve their remarkable cognitive abilities? What makes the human brain different from that of other animals, and the behavior it generates unique? Although these questions are fundamental to psychologists and neuroscientists alike, they are difficult to answer, for several reasons. First, evolution of the mammalian brain, and neocortical evolution in particular, is difficult to study directly. Secondly, even if we could study evolution directly, considering evolutionary contributions to phenotypic variability in isolation is too restrictive. Finally, these questions are highly subjective and therefore their answers would provide limited information regarding cortical function and evolution. However, we can circumvent some of these problems by studying evolution indirectly and making inferences about the process. In addition, we can examine the non-evolutionary mechanisms that generate phenotypic variability. Finally, we can re-formulate our questions in a more objective fashion.

Like other mammals, our sensory receptor arrays are capable of sampling only a limited portion of the physical environment. Our nervous system enhances stimulus features, generates probabilities, and constructs the reality of our world in a highly biased manner that considers only those parameters that we can actually detect. Unfortunately, the concepts we generate regarding the organization and function of the nervous system, particularly the neocortex, reflect these same biases. So, how can we get out of our own skin?

For addressing questions of human brain evolution, we can talk about complexity rather than intelligence, cognition, or any other covert behavior generated by the human brain whose definition requires subjective human experience. We appreciate that some mammals, such as human and non-human primates and cetaceans, have a relatively large neocortex that is complexly organized. For our purposes, complexity can simply be defined as a large number of functionally distinct parts that are intricately interconnected. We also appreciate that mammals that have a relatively large, complexly organized neocortex appear to generate more complex behaviors. As with the nervous system, complex behavior refers to behaviors that have many parts, and includes motor behaviors such as reaching, grasping, locomotion, articulation of sounds, or behaviors such as stimulus detection, perception, learning and memory, components of which can be quantified. Although it is difficult to detect subtle differences in complex brains and behavior using this rather gross scheme of classification, one can feel fairly confident stating that mammals that have neocortices with many functionally heterogeneous parts that are specifically interconnected, generally have more complex behaviors. Thus, we can refine our questions regarding human brain evolution, and achieve at least some modicum of objectivity if we examine how the brain evolves more functional parts (cortical fields), how connections between these parts become specified, and ultimately how the addition of these parts are specifically related to the generation of complex behavior.

However, considering only the *evolution* of cortical fields is problematic because it is far too restrictive. Evolution requires the transmission of genes from one generation to the next. When we consider evolution in isolation, we only consider those characteristics of the brain that are heritable. Yet, studies of cortical plasticity in adult and developing mammals indicate that the nervous system is capable of remarkable change within the life of the individual which takes the form of functional map reorganization in adults (Recanzone et al. 1992, 1993; Recanzone, 2000), and in the developing nervous system includes large sensory domain shifts, changes in functional map organization and changes in connectivity (Kahn and Krubitzer 2001, 2002).

The final problem associated with the questions posed at the beginning of this chapter is that evolution is difficult to study directly because the time course of change is relatively slow by individual life-span standards, and subtle changes often occur over thousands of generations. However, there are two ways to circumvent this problem: (1) use of a comparative approach, and (2) examining the developmental mechanisms that give rise to particular characteristics of complex brains. Using the comparative approach we can study the products of the evolutionary process and make inferences about the process itself. This method allows us to deduce general characteristics of nervous systems and the types of brain changes that are actually possible. A comparative approach in combination with a developmental approach allows us to examine the constraints that direct the course of evolution.

For instance, electrophysiological recording studies, architectonic analysis, and studies of cortical and subcortical connections indicate that all mammals have a constellation of specifically interconnected cortical fields (Krubitzer 1995, Krubitzer and Huffman 2000). Some of these fields include the primary and secondary sensory areas such as S1, S2, V1, V2, A1 and R (Fig. 3.1). These same types of studies indicate that the types of system changes that are possible are limited and include changes in:

- the size of the cortical sheet;
- the number of cortical fields;
- the amount of cortex devoted to a particular sensory system (sensory domains);
- the amount of a cortical field devoted to a particular portion of the sensory epithelium;
- connectivity;
- modularity of existing fields.

Within these large categories, further modifications in cell size, dendritic and axonal arborization, and laminar organization have been made to the neocortex over time. Although we propose that the types of modifications with respect to all of the possible ways in which brains could change are limited, there is still a large degree of freedom for phenotypic change within these constraints.

The limited types of modifications that are observed in extant brains, particularly those that have evolved independently, indicate that there are constrained developmental mechanisms that generate nervous systems. Thus, the second way to study the evolution of the neocortex, in particular the mechanisms that give rise to current organization and constraints imposed on evolving nervous systems, is to study the development of the neocortex. This chapter will focus on some of the modifications associated with complex brains and discuss the evolutionary (inherent, genetic contributions) and activity-dependent mechanisms that give rise to these features.

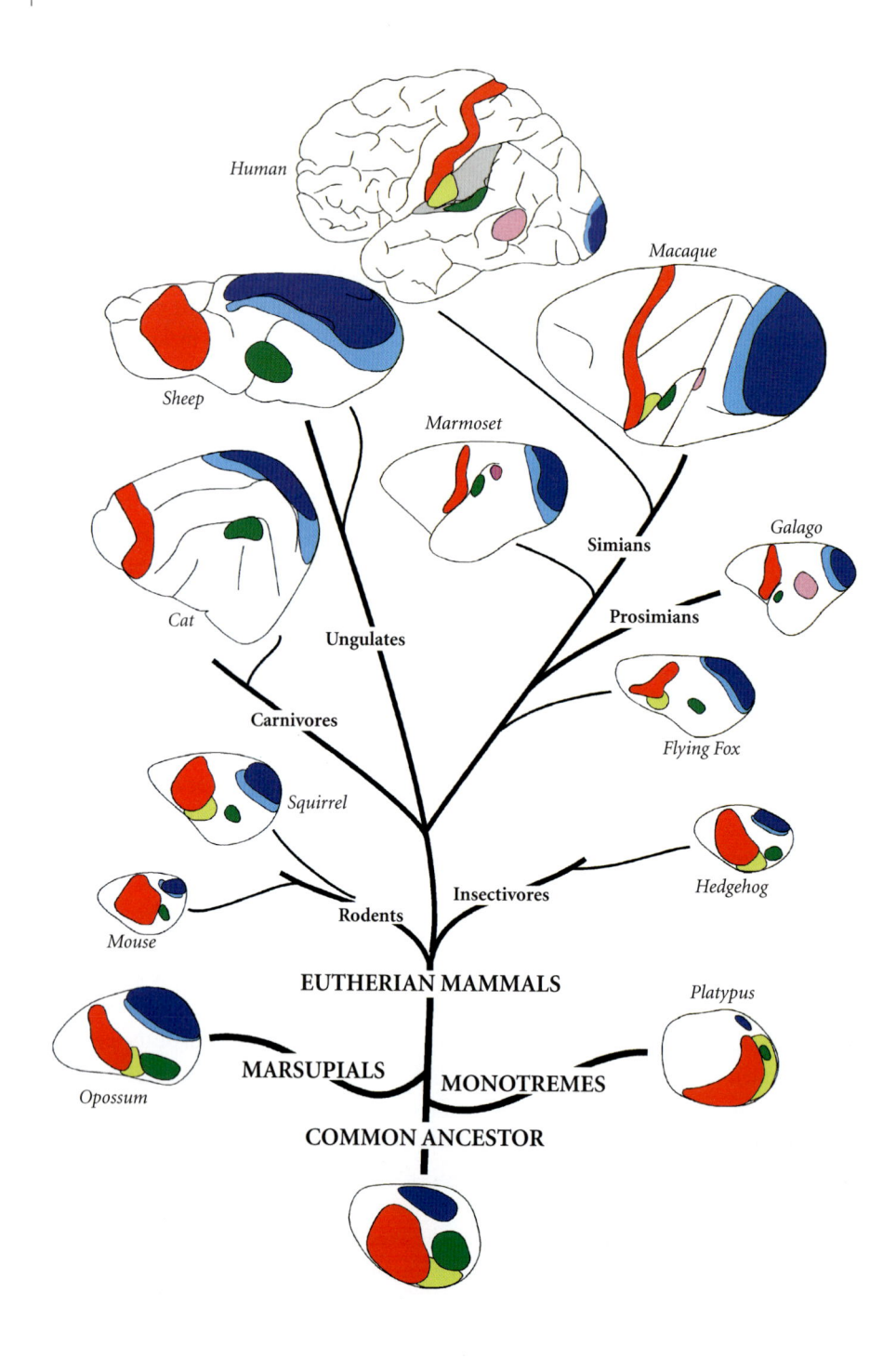

60 | FUNCTIONAL NEUROIMAGING OF VISUAL COGNITION

3.2 Increases in the size of the cortical sheet

Probably the most salient feature associated with complexity in mammalian brains, particularly the human brain, is the disproportionate increase in the size of the cortical sheet (Fig. 3.2). This feature, described several decades ago by Stephan *et al.* (1981), and termed encephalization, has recently been indexed by Finlay and Darlington (1995) and Clark *et al.* (2001). The selective pressures that led to an enlarged cortical sheet are not clearly understood, but it has been proposed that frugivory (fruit eating), longevity, and sociality may be the driving forces behind the evolution of an enlarged brain in primates (for a review, see Allman 1999). Although these conjectures are viable, they are only correlative and not necessarily causal. Even if they were causal, it is difficult to link the expansion of one portion of cortex compared to the others, with any of these global features of social organization, foraging, or longevity. Further, it is not clear exactly what a larger cortical surface area with more functional areas confers to an individual. Therefore, it is not surprising that the selective pressures that led to an expanded cerebral cortex are elusive. However, as noted earlier, it is clear that larger brains with many cortical areas that are uniquely interconnected generate more complex behavior.

Thus, an increase in the size of the neocortex is a necessary (although, perhaps not sufficient) step in the evolution of complex mammalian nervous systems. Therefore, to appreciate human brain evolution and the factors that contribute to its current phenotype, it is necessary to understand how cortical sheet size is regulated in development, the types of genetically mediated developmental mechanisms that could give rise to an expanded cortical sheet, and how and why neocortex expands at the expense of other telencephalic structures.

Two mechanisms have been proposed to explain how the cortical sheet may increase in size. One suggestion is that more cells are generated in development. Kornack and Rakic (1998) propose that a simple change in the timing of cell division cycles of progenitor cells in the ventricular zone during neurogenesis could result in an exponential

---

Fig. 3.1 An evolutionary tree depicting the phylogenetic relationship of major orders of mammals and the cortical organization of some of the sensory fields that have been described in particular species. Electrophysiological, anatomical, histochemical, and molecular analyses have revealed that certain cortical regions, such as S1, S2, A1, V1, and V2, are common to all mammals and most likely are homologous areas that arose from a common ancestor. On the other hand, some regions, such as MT (pink) have been observed in only a few orders, such as primates, and likely evolved independently in these lineages. If a number of species are compared, one can be fairly confident when assigning features of cortical organization to the unknown state, such as the common ancestor or human, even in the absence of direct data. S1 = primary somatosensory area (red), S2 = secondary somatosensory area (yellow), A1 = auditory (green), V1 = primary visual area (dark blue), V2 = secondary visual area (light blue); rostral is left, medial is up.

Mouse

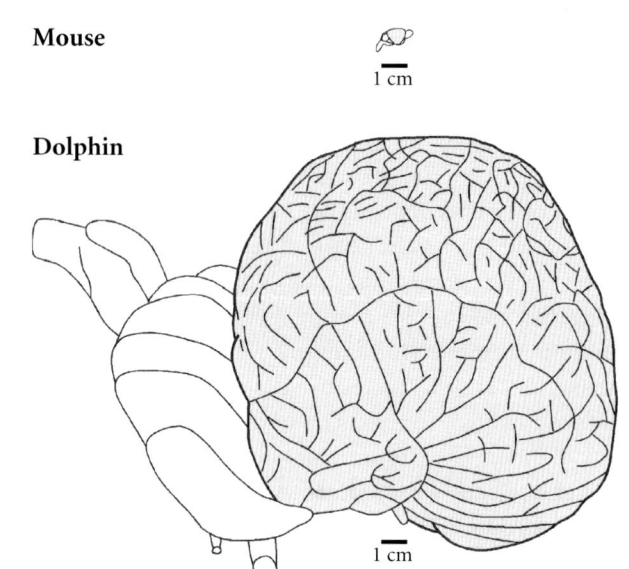

1 cm

Dolphin

Fig. 3.2 A comparison of the mouse and dolphin brain drawn to scale illustrates the dramatic differences in the size of the neocortex (gray). The difference in size is even larger in magnitude than is illustrated here, since the dolphin brain contains a number of fissures in which the neocortex is buried. Rostral is right, medial is up.

1 cm

increase in the size of the cortical sheet (Fig. 3.3). Kornack's comparative analysis on the kinetics of cell division in monkeys and rodents (Kornack and Rakic 1998; Kornack 2000) reveals that in macaque monkeys, the cell cycle duration is five times as long as in the mouse, and that there are more total cycles of cell division than in the mouse. This prolonged and accelerated cell division during cortical neurogenesis could account for the pronounced increase in the cortical sheet in some lineages, such as anthropoid primates.

Another proposition of how the cortical sheet increases in size is that there is a decrease in naturally occurring cell death (apoptosis) during corticogenesis. Several genes and their products (proteins) have been demonstrated to decrease the rate of apoptosis. For example, in mutant mice in which a gene associated with cell death (caspase 9; *Casp9*) is deleted, a larger proliferative zone is observed in the forebrain, along with an increase in the size of the neocortex (Kuida et al. 1998). Additionally, there is evidence that the apoptotic process may be further regulated by certain genes in the Bcl-2 family, which function to inhibit or facilitate apoptosis by acting upon caspases (Boise et al. 1993; Motoyama et al. 1995; Roth et al. 2000). Like the former mechanism proposed by Kornack, a small change in the timing of the expression of a gene or genes involved in apoptosis could change the size of the cortical sheet dramatically.

While the genes responsible for the kinetics of cell division of progenitor cells and rates of apoptosis during development are not well known, there is evidence indicating that the protein β-catenin, and the genes that regulate its production, may be involved in the determination of cortical sheet size in different lineages. β-Catenin, which activates signaling molecules involved in cell growth and cell fate (Peifer and Polakis 2000), is expressed in neuroepithelial precursor cells (which will become the neocortex) in the

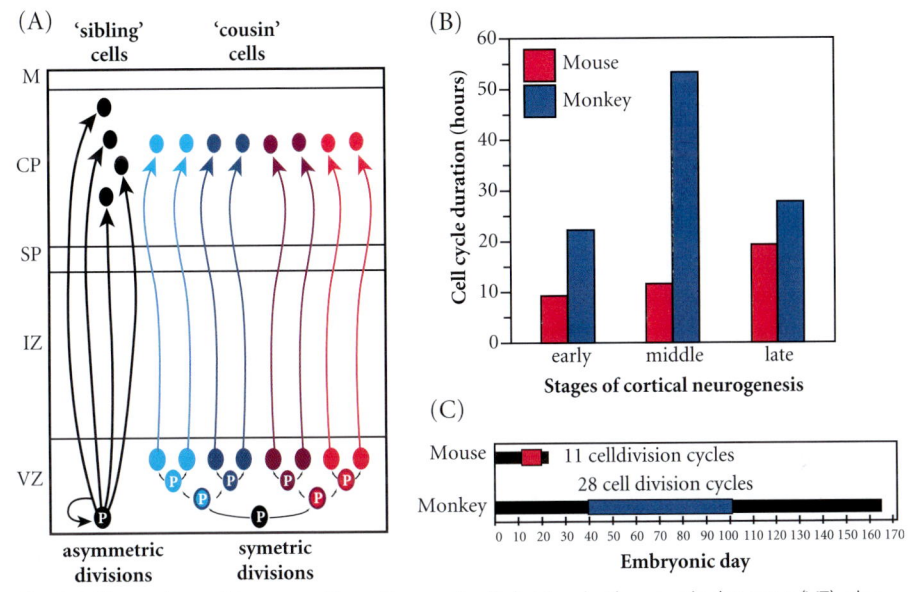

Fig. 3.3 Illustrations of how specific patterns of cell division in the ventricular zone (VZ) give rise to the patterns of clonally related neurons in the neocortex. In (A), asymmetric divisions from a single progenitor cell (P) generate 'sibling' cells that migrate sequentially to different layers of the cortical plate (CP). This type of cell division determines cortical thickness. Symmetric divisions from a single progenitor cell generate several progenitor cells that, in turn, simultaneously generate 'cousin' cells which then migrate, in parallel, to the same cortical layer. This type of division determines cortical sheet size. Duration (B) and number (C) of cell cycle divisions vary dramatically in the mouse (pink) and the rhesus monkey (blue). (B) Cell cycle duration is significantly longer in the monkey at all stages of neurogenesis. (C) Black bars represent the length of gestation in the mouse (19 days) and the monkey (165 days). In the mouse (pink rectangle) neurogenesis lasts 6 days, from embryonic day (E) 11 to E17. In the monkey, neurogenesis lasts 60 days, from E40 to E100. The expanded duration and the increased number of cell cycles could be one mechanism involved in expansion of the primate neocortex. IZ, intermediate zone (white matter); M, marginal zone (layer I); SP, subplate zone. (Modified from Kornack and Rakic 1998; Kornack 2000.)

ventricular zone during neurogenesis (Chenn and Walsh 2002). These investigators demonstrate that transgenic mice that overexpress a truncated form of β-catenin have an exaggerated horizontal growth of the cortex (without a change in cortical thickness). Indeed, the increased size of the cortical sheet was so massive that the normally lissencephalic cortex of the mouse became gyrencephalic. This enlargement of the cortical sheet was the result of a twofold increase in the proportion of progenitor cells that re-entered the cell cycle and continued mitotic division. Thus, the disproportionate increase in cortical sheet size in some mammals could be regulated in part by β-catenin or other proteins, whose temporal and spatial patterns of expression vary slightly in different lineages.

The inordinate increase in the size of the cortical sheet in some cetaceans, such as odontoceti (dolphins and toothed whales) and their distant cousins, proboscidea (elephants) rivals that of humans. Because no other extant mammal exhibits such a disproportionate increase, and because the primate and cetacean lineages are very distantly related, the most parsimonious interpretation is that this cortical expansion has been independently achieved in these separate lineages. By comparing the temporal and spatial expression patterns of genes, or gene products such as Bcl-2 and β-catenin, in the developing ventricular zone of primates, dolphins, whales, and elephants, one could determine which genes specifically regulate the process of cortical sheet expansion. In doing so, we could determine whether this process occurs via homologous genetic mechanisms, or if there is more than one way in which the neocortex can change in size.

3.3 Genetic regulation of cortical domains and cortical fields

Accumulating developmental and comparative data indicate that both genes and neuronal activity regulate the organization and connectivity of the developing neocortex. However, the extent to which the emergence of cortical domains and individual cortical fields and their connectivity is genetically specified is not clear. There is ample evidence indicating that genes play a significant role in specifying the gross geometric anatomical relationships of the cortex.

Recent work in the field of molecular neurobiology has demonstrated that patterning or signaling centers exist in particular portions of the developing brain. These signaling centers are specific portions of neural tissue which express particular genes or gene products, and serve as morphogens. In turn, these signaling centers induce the fate or specification of nearby neural tissue, and contribute to the cellular architecture, type of neurotransmitter utilized, connectivity, and ultimate function of developing neurons. The role of signaling centers in allocating large portions of the central nervous system has been recognized for some time. For instance, major subdivisions of the brain, such as the telencephalon, diencephalon, midbrain, hindbrain, and spinal chord, are specified by either graded or abrupt patterns of gene expression during development. The homeobox genes *Emx1*, *Emx2*, *Otx1*, and *Otx2* are expressed in rostral portions of developing embryonic brains, and their expression domains are contained within each other (Simeone *et al.* 1992*a, b*). The boundaries of expression domains or particular overlap zones coincide with the boundaries of major brain structures, such as the telencephalon and diencephalon (see Boncinelli *et al.* 1995 for a review). At a finer level of detail, expression domains of genes such as *Otx1*, *Otx2*, and *Wnt3* within a particular structure such as the diencephalon, coincide with anatomical divisions within the diencephalon such as the dorsal and ventral thalamus and pretectum, and are involved in specifying these large subdivisions of the central nervous system, as well as smaller subdivisions therein (Marin and Rubenstein 2002). Because the neocortex is composed of multiple parts (cortical fields) with boundaries that are often abrupt, a situation analogous to the structural borders of subcortical structures and the smaller subdivisions described above, it is tempting to speculate that the same rules of specification apply to the developing neocortex. That is, genes or particular spatial and temporal combinations of gene expression strictly control cortical field emergence, organization, architecture, and connections.

There is evidence that particular genes and proteins serve as signaling centers and mark general axes of the cortex, such as rostro-caudal and dorso-ventral, and that particular spatial and temporal combinations of their expression patterns serve as a coordinate system for incoming thalamocortical axons. For instance, mounting evidence suggests that genes such as sonic hedgehog (*Shh*; Chiang et al. 1996) and some genes in the *Wnt* family (Grove et al. 1998) mark ventral telencephalic structures and the dorsal edge of the telencephalon, respectively, and proteins such as bone morphogenic protein (BMP; Furuta et al. 1997) may assign the dorsal telencephalon (see Levitt et al. 1997; Rubenstein et al. 1999; Marin and Rubenstein 2002 for reviews). Recent studies by Bishop et al. (2000) demonstrate that regulatory genes such as *Emx2* and *Pax6* are also involved in specifying the anterior–posterior axis of the cortex, since the deletion of such genes results in a caudal or rostral shift of thalamocortical afferents, respectively, and presumably the associated cortical fields (Fig. 3.4; Bishop et al. 2000). Fukuchi-Shimogori and Grove (2001) demonstrated that electroporation of the molecule FGF8, which is thought to serve as a signaling marker of rostral cortex, results in a posterior shift of anterior cortical fields and an antero-posterior elongation of cortical fields. FGF8 appears to function in part through repression of *Emx2* expression (Crossley et al. 2001). These data indicate that particular molecules (regulated by intrinsic gene patterning) may contribute to the emergence of cortical fields, although how these abrupt and graded patterns of gene expression would be altered to produce new cortical fields is not yet clear.

Recent studies indicate that these and other signaling centers can operate independently of peripheral activity. For instance, studies in mutant mice that fail to develop thalamocortical axons (Miyashita-Lin et al., 1999; Nakagawa et al. 1999; *Gbx2* −/−; *Mash1*−/−), and thereby have no access to patterned activity from peripheral sensory arrays, still have graded and abrupt patterns of gene expression (Fig. 3.5). These expression patterns are proposed to mark boundaries of cortical areas, but there is no direct evidence to support this contention.

These data are compelling in that they clearly demonstrate that an anterior-posterior/dorsoventral coordinate system is likely to be intrinsically mediated. Thus, the general location of primary fields and some aspects of the fields themselves may be specified by intrinsic genetic patterning which operates independently of thalamocortical input. Further, these studies demonstrate that cortical domains and primary cortical fields can be shifted when genes and molecules are manipulated via mutations and electroporation.

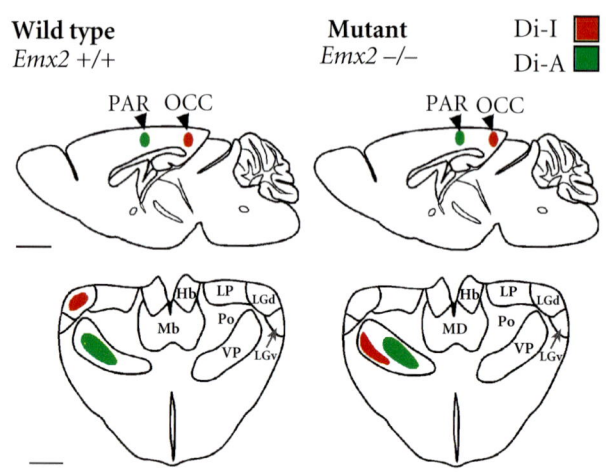

Fig. 3.4 Thalamocortical projections in *Emx2* wild-type (+/+) and mutant (–/–) mice as revealed by anatomical tracers placed into the cortex. *Emx2* is a regulatory gene which is expressed in a low rostral to high caudal gradient in mouse cortex during the late embryonic period. In both wild type (top left) and *Emx2*-deficient (mutant; top right) mice, the post-mortem tracer Di-A (green) implanted into the somatosensory cortex (PAR) retrogradely labeled cells in the ventroposterior nucleus of the somatosensory thalamus (VP, bottom). Di-I (red) implanted into visual cortex (OCC) of the mutant mice revealed retrogradely labeled cells in VP (red oval, bottom right), rather than in the normal location, in the dorsal lateral geniculate nucleus (LGd; red oval, bottom left). These differences in thalamocortical projections indicate that in the *Emx2*-deficient mice, there was a caudal shift in the thalamocortical projection patterns and presumably somatosensory cortical fields. The top row is an illustration of a lateral view of the brain, rostral is to the left. Green and red ovals in the top row represent Di-A and Di-I injection sites into the parietal (PAR) and occipital (OCC) regions of the neocortex, respectively. The bottom row depicts areas in which retrogradely labeled cells were observed in coronally sectioned thalamic tissue (data used to construct this figure is from Bishop *et al.* 2000). Scale bar = 1 mm. Top row rostral is left and medial is to the top. Bottom row, dorsal is to the top.

It should be noted that all current developmental studies examine arealization of primary sensory fields exclusively. Therefore, if indeed there are intrinsic signaling centers that specify a cortical field, this may only be true for primary fields. This notion is supported by recent comparative, embryonic, genetic, and immunohistochemical analyses indicating that all of the neocortex may not be of the same phylogenic origin. Specifically, medial portions (which contain primary sensory cortices) may have different phylogenetic precursors (Butler and Molnár 2002; Molnár and Butler 2002). It has been proposed that medial neocortex (and the sensory cortices therein) arise from the corticostriatal junction, and that these fields are homologous to the anterior portion of the dorso-ventricular ridge of sauropsids (birds and reptiles), while the more lateral portions, which contain non-primary fields, have different origins. This suggests that

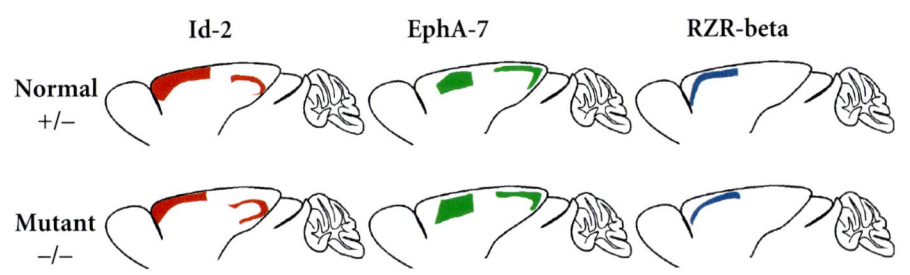

Fig. 3.5 Expression patterns of genetic markers in normal (+/–; top row) and *Gbx2* mutant (–/–; bottom row) mice at postnatal day 0. Illustrations depict a lateral view of the neocortex, rostral is to the left. In *Gbx2*-deficient (–/–) mice, thalamic axons do not form connections with the cortex. In a study by Miyashita-Lin *et al.* (1999), expression patterns of region-specific genes in the cortex were analyzed in *Gbx2* mutants. In the normal animal, *Id-2* (left column), *EphA-7* (middle column) and *RZR-beta* (right column) are expressed in discrete cortical regions and layers. Despite the lack of input from the thalamus in the mutant mice, expression patterns of these three gene markers were not different than in normal animals. These results demonstrate that thalamic input (and the patterned activity it relays to cortex) is not necessary for patterned expression of particular genes in the cortex (data used to construct this figure is from Miyashita-Lin *et al.* 1999). Rostral is left and medial is up.

the rules of arealization for primary fields and non-primary fields may be different, and each may be more or less influenced by activity versus genes.

3.4 Comparative studies of the neocortex: Peripheral morphology and activity-dependent regulation of cortical domains and cortical fields

While the evidence for genetic specification of cortical areas is strong, the concept of a strict genetic specification of cortical fields is at odds with an enormous amount of comparative data. Studies in a variety of mammals indicate that the assignment of cortical domains, the number of cortical fields within a domain, and the internal organization of a particular cortical field are dependent on peripheral morphology and the activity generated by particular sensory receptor arrays. This is best illustrated in mammals with an exaggerated or specialized morphological feature or sensory receptor array. There are three striking features of cortical organization in these animals. The first is the relationship between cortical domains and peripheral receptors; the second is a cortical magnification within a cortical field of the specialized receptor arrays; and the third is the generation of isomorphic substructures within a magnified representation which is directly related to peripheral receptor type, number, density of innervation, and use.

The first feature of organization that appears to be dictated by peripheral inputs and activity is the cortical domain territories assigned to a particular sensory system.

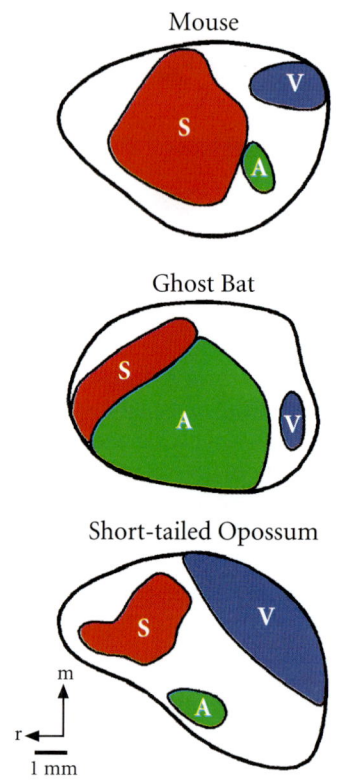

Fig. 3.6 Primary cortical areas in three species of mammals that have approximately the same size cortical sheet, but different amounts of cortex allotted to different sensory systems. These cortical differences are related to differences in peripheral sensory systems and use of particular sensory receptor arrays. For example, in the mouse, which relies heavily on tactile inputs from the whiskers for survival, the primary somatosensory cortex (red) and the rest of somatosensory cortex is enlarged, and the portion of cortex representing the whiskers is magnified, compared with that of the ghost bat and short-tailed opossum. Similarly, the primary auditory cortex and surrounding fields in the cortex of the echolocating ghost bat (green) is expanded, while the primary visual area (blue) and somatosensory area is relatively small. Finally, the cortex of the highly visual short-tailed opossum is dominated by V1 (blue) and other visual areas. Although the size, shape, and the details of internal organization of particular cortical fields vary, certain aspects of organization are conserved in these brains, such as the relative location of cortical domains and fields therein, and the general pattern of thalamocortical projections. Medial is up and rostral is to the left, scale bar = 1 mm.

Figure 3.6 illustrates three mammalian neocortices in which sensory domain assignment is remarkably different, despite the fact that the size of the neocortical sheet is approximately the same in each animal. For example, in the mouse, most of the cortical sheet is devoted to processing somatic inputs, particularly from the whiskers. In the echolocating ghost bat, most of the cortical sheet is devoted to processing auditory inputs, and in the highly visual short-tailed opossum, most of the cortex is devoted to processing inputs from the retina. In all of these mammals, there is an enlargement in the cortical territory occupied by the dominant sensory system, and this occurs at the expense of the remaining sensory domains.

At a more detailed level, peripheral innervation is reflected in the cortical magnification of specialized body parts and the organization within a cortical field. For instance, the duck-billed platypus has a large, highly innervated bill with interdigitating parallel rows of mechanosensory and electrosensory receptors. This striking morphological specialization, accompanied by the evolution of an electrosensory receptor, manifests in cortex as an enormous representation of the bill (Fig. 3.7A). This type of peripheral modification, coincident with the enlargement of sensory domains and cortical representations of the specialized body part, can be observed in all sensory and motor systems in a variety of mammals. The star-nosed mole, for example, has a large amount of cortical territory devoted to processing inputs from its specialized nose (Fig. 3.7B; Catania and Kaas 1995). In human and non-human primates, the somatosensory cortex is largely devoted to processing inputs from the remarkably specialized forepaw or hand (Kaas *et al.* 1979), the primary visual area contains an enlarged representation of the fovea (which has a higher density of retinal ganglion cells), and in humans the motor and premotor cortex contains an exaggerated motor representation of the lips, tongue, oral structures, larynx, and associated musculature (commonly referred to as Broca's area).

Finally, within a cortical field, anatomical and functional isomorphic representations of very specific peripheral morphologies can be identified, including barrel fields in some rodents, digit subdivisions in several primates, ray or follicle patterns in star-nosed moles, and electrosensory/mechanosensory stripes in the duck-billed platypus (Figs 3.7 and 3.8). The relationship between such detailed anatomical and functional subdivisions within a cortical field and its peripheral counterpart has been clearly demonstrated by Welker and Van der Loos (1986). In mice selectively bred to have an extra whisker or row of whiskers, extra barrels or rows developed within the barrel fields in the neocortex (Fig. 3.8A). The authors noted that the relationship between peripheral innervation density and cortical isomorph was not linear, and suggested that other factors, such as patterned activity, contribute to some aspects, such as size of the isomorphic representation.

More recent studies in the star-nosed mole by Catania and Kaas (1997a, b) support the findings of Welker and Van der Loos (1986). In star-nosed mole that naturally posses an additional nose appendage or ray, there is an extra isomorph of this appendage in the neocortex. These authors extend these initial observations by demonstrating a clear use-dependent construction of some aspects of cortical isomorphs by documenting a differential magnification of some of the nose rays compared to others (Fig. 3.7B). The eleventh, ventromedial ray is preferentially used in tactile exploration. Although it is the smallest ray, with the fewest number of sensory

(A)
Duck-billed platypus

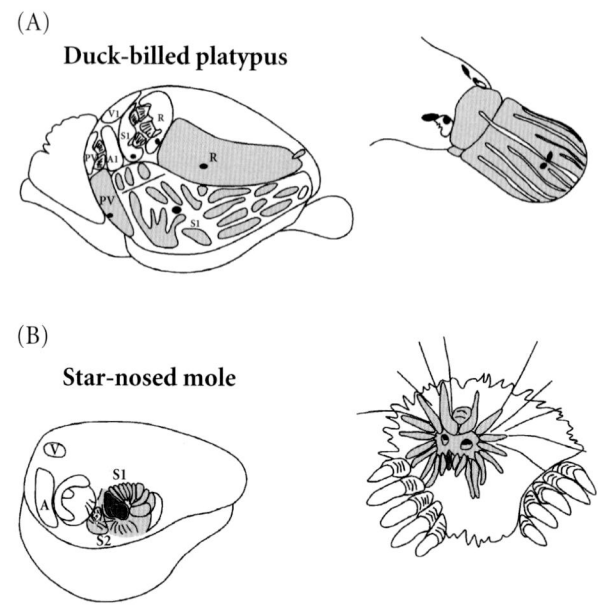

(B)
Star-nosed mole

Fig. 3.7 Cortical representation of the bill of the duck-billed platypus (A) and the nose of the star-nosed mole (B). Both mammals have evolved specializations in peripheral morphology and use of specialized body parts, which are accompanied by changes in cortical organization. The large bill of the platypus has an enormous representation in the cortex that spans several cortical fields (blue). Within S1, both electrosensory and mechanosensory inputs are arranged in bands that form isomorphic representations of the striped arrangement of receptors on the bill. The star-nosed mole has a structure that consists of an array of 22 appendages (rays), 11 on each side that are arranged around the nostrils. These rays are used to explore food items and the immediate surroundings, and have been likened to a fovea. In the cortex, these rays form isomorphic representations that appear band-like in both S1 and S2 (blue). One of the rays of the star-nosed mole (number 11—dark blue) is utilized preferentially compared to the other rays, and has an even larger representation in the cortex (dark blue) than its counterparts. The unusual morphological specializations in these mammals and the cortical magnification of the regions devoted to processing inputs from these appendages are striking demonstrations of the impact of peripheral morphology on organization of the neocortex. (A, based on Krubitzer 1998; B, modified from Catania and Kaas 1997a.) Medial is up and rostral is to the right.

end organs, it has the largest sensory representation in S1 of the neocortex, and the greatest area of cortical innervation relative to size of any of the other rays (Fig. 3.7B).

This clear relationship between peripheral morphology and use and cortical domain assignment, cortical field magnification, and generation of isomorphic representations as observed in comparative studies is difficult to reconcile with proposed intrinsic mechanisms of cortical arealization described earlier. Indeed, some of the results from both groups appear to be in direct conflict. For instance, studies in which FGF8 was

THE EVOLUTION OF HUMAN NEOCORTEX | 71

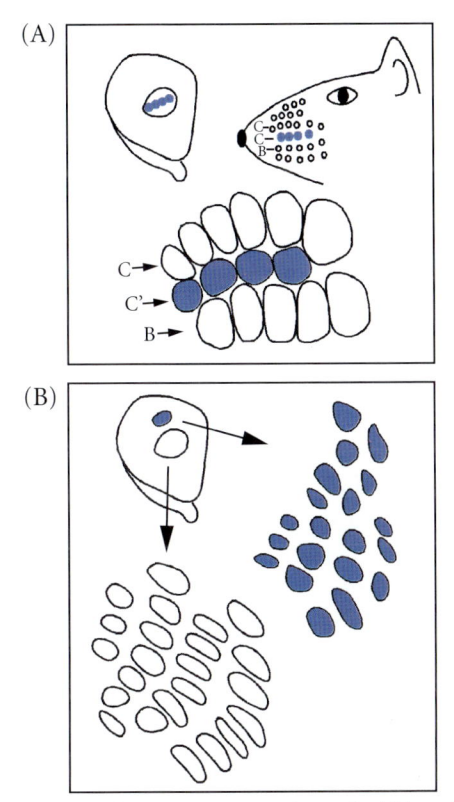

Fig. 3.8 Two potential ways in which extra representations of whiskers (barrels) may be generated in cortex. The first method (A) was demonstrated over a decade ago by Welker and Van der Loos. Mice that were selectively bred to have and extra row of whiskers (top right) had an extra row of whisker representations in the cortex (top left and bottom of A). A second method of inducing barrel formation in the neocortex is by artificially changing the pattern of intrinsic signaling centers (such as the molecule FGF8) early in development (B). In this study this FGF8 was electroporesed into a selected location caudal to its normal location of expression. An ectopic barrel field formed caudal to the barrel field in S1. (Data used to construct this figure is from Welker and Van der Loos 1986; and Fukuchi-Shimogori and Grove 2001.)

electroporesed into a caudal region of cortex clearly demonstrate the emergence of a new, ectopic barrel field (Fig. 3.8A; Fukuchi-Shimogori and Grove 2001), while other studies in mice that possess an extra row of whiskers (Welker and Van der Loos 1986), demonstrate additional rows of barrels in the cortex (Fig. 3.8B). The former study suggests a strict genetic specification of cortical fields, while the latter study indicates that peripheral innervation and use play a direct role in specification of cortical fields in development.

The issue, of course, is not how the cortex can be manipulated to produce alterations in cortical fields, but how specification of cortical fields naturally occurs in evolution, and how intrinsic and activity-dependent mechanisms operate together under normal conditions to produce a particular cortical phenotype. If one considers both the genetic manipulation studies and the comparative studies, a clearer picture of the genetic and activity-dependent contributions to the phenotype begin to emerge. For instance, both sets of data indicate that there are some features of cortical organization that are genetically mediated and highly constrained in evolution. The first is the gross geographic relationship of primary cortical areas to each other. Indeed, the relative position of fields is invariant across mammals. The second is thalamocortical connectivity, particularly the connections between major sensory nuclei such as the lateral geniculate nucleus (LGN), medial geniculate nucleus (MGN), and the ventro-posterior nucleus of the somatosensory thalamus (VP), and primary sensory areas such as V1, A1, and S1, respectively. Finally, some aspects of cortical architecture, such as the presence of a koniocellular layer and myelin density of primary sensory fields, are likely to be genetically regulated, and certainly appear to be constrained in evolution. On the other hand, comparative data indicate that several features of cortical organization are not genetically constrained and vary with changes in peripheral morphology and with the patterned activity associated with such morphology. These features include the total extent of a particular sensory domain (not its general location), the size and shape of a cortical field, the details of the internal organization of a cortical field, and some aspects of thalamocortical and cortical connectivity. These types of changes are driven by modifications to peripheral morphology, including changes in the size of an appendage or structure and the receptor type, number, and density within the structure. These peripheral modifications may be, but are not necessarily, genetically mediated.

3.5 Testing theories of cortical domain specification: Studies of bilaterally enucleated opossums

We can test the extent to which peripheral input and associated activity contribute to specifying cortical domains by making changes to the peripheral receptors, similar to the types of changes that occur naturally in evolution. As noted in Section 3.4, important features of cortical organization are associated with distinct peripheral morphologies and behaviors. The obvious conclusion from these comparative studies is that peripheral morphology and patterned use play a large role in cortical field specification in development. One way to test the total extent to which peripheral receptors can alter cortical domain territories is to increase or decrease the size of the sensory receptor array of a specific sensory system and examine the resulting neocortex using electrophysiological recording and anatomical techniques.

In a recent set of experiments in the South American short-tailed opossum (*Monodelphis domestica*), we eliminated visual input very early in development, prior to the formation of the retino-geniculo-cortico pathway (Dunn *et al.* 2001; Kahn and Krubitzer 2002). Electrophysiological recordings in these enucleated opossums after they reached adulthood revealed that 'visual cortex' was substantially reduced (Fig. 3.9).

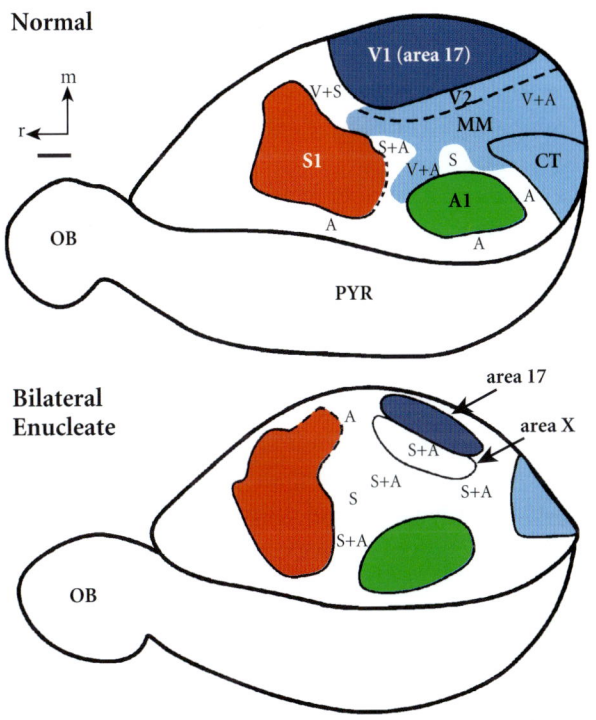

Fig. 3.9 Illustration of a dorsal view of the neocortex of a normal (top) and bilaterally enucleated (bottom) adult short-tailed opossum. Bilateral enucleations were performed early in development, prior to the formation of the retino-geniculo-cortical pathway. Despite the absence of input from retinal receptors in the enucleated animal, there was still an area in the caudomedial portion of the cortex that was anatomically similar to V1 (area 17) in the normal animals, although substantially smaller. The geographic relationship of S1 (red), A1 (green) and area 17 (blue; V1 in normal animals) was maintained despite the complete lack of activity from visual receptors. In the bilateral enucleate, area 17, which contains neurons that normally respond to visual stimulation, contained neurons responsive to somatosensory (S) and auditory (A) stimulation. Thus, when activity from visual receptors was experimentally eliminated, visual cortex was captured by the somatosensory and auditory systems. These changes in the configuration of cortical domains, and size of cortical fields indicate that some features of cortical organization are mediated by activity from peripheral receptors. Dark blue in the top figure (normal) represents the primary visual area, and in the bottom the primary visual area as defined architectonically. Light blue (top) represents other visually responsive cortical regions in the normal animal. Red = somatosensory, green = auditory, MM = multimodal area, CT = caudotemporal area, OB = olfactory bulb, PYR = pyriform cortex. Rostral is to the left and medial is up; scale bar = 1 mm. (Modified from Kahn and Krubitzer 2002.)

Further, cortical regions normally involved with visual processing, including area 17 or V1, now contained neurons that responded to a different sensory modality compared to normal animals. Thus, there were dramatic shifts in cortical domain territories, as large, or larger, than those produced by genetically modifying intrinsic signaling centers (discussed in Section 3.3).

However, there were also a number of features of the neocortex that remained unchanged, despite this massive loss of sensory input. For instance, examination of the bilateral enucleate brains using neuroanatomical tracing techniques indicated that cortico-cortical and thalamocortical connections of area 17 (V1 in normal animals) were largely preserved (Kahn and Krubitzer 2001). In addition, gross positional organization in terms of medio-lateral and rostro-caudal organization of the cortex was maintained. Finally, although V1 appeared to be substantially reduced in size, its cortical architecture was similar to that of normal animals. These results indicate that peripheral input plays a large role in assigning cortical domains, and that dramatic changes in the organization and the size of cortical fields can be determined by peripheral input. On the other hand, position, shape, architecture, and some aspects of connectivity of at least primary fields are likely to be mediated by intrinsic genetic signals (for review of related literature see Kahn and Krubitzer 2002). These observations are similar to those made in mammals that naturally have a reduced or absent visual system due to miniaturization or loss of the eye. For instance in the blind mole rat, the eyes are microophthalmic and covered with skin. In these animals, as in the bilateral enucleated animals, a geniculo-cortical pathway is still present (Bronchti *et al.* 1991; Cooper *et al.*, 1993), and neurons in the lateral geniculate nucleus (LGN) and 'visual' cortex respond to auditory stimulation (Bronchti *et al.* 1989).

These studies support observations from comparative studies in that the global geographic relationships of primary sensory fields are maintained, the architectonic features of primary cortical fields can be identified, and some aspects of connectivity are maintained even in the absence of use or loss of a sensory system. The preservation of global relationships of sensory cortical fields and of some aspects of connectivity in animals that have extreme specializations, like the platypus, or loss or reduction of a sensory system, like the blind mole rat and bilateral enucleate, fit well with data from developmental studies described earlier in this chapter. All of these studies are consistent with the view that intrinsic signaling centers (e.g. *Wnt*, *Shh*, BMP) provide positional information for incoming thalamocortical afferents, and relative location of cortical fields with respect to other cortical fields. These genes (and likely others) arose early in evolution, and certainly constrain the evolution of the mammalian neocortex.

Despite these consistencies across data sets, there are still several outstanding questions generated by comparative, molecular and developmental manipulation studies that need to be addressed. For example, how do intrinsic cortical mechanisms act in concert with activity-dependent mechanisms to allocate cortical domains and cortical fields that faithfully represent sensory receptor arrays? A second question is how are the dynamics of particular developmental mechanisms altered over larger time scales to produce variable phenotypes? The large-scale dynamics of evolution are rarely considered in the smaller context of individual developmental cascades. In particular, how are developmental regimes altered to produce a new cortical field? A third question is related to the co-evolution of the motor system with particular sensory system morphologies. As noted above, specialized peripheral morphologies are associated with specialized use; the receptor array is never stationary but is very specifically interfaced with the environment. Particular motor sequences, such as reaching and grasping, saccadic and smooth pursuit eye movements, whisking, and head orientation, have co-evolved with these receptor arrays. Thus, the motor system is an integral part of sensory reception. How are sensory and motor systems interfaced in development? How does the evolution of one affect the evolution of the other? Finally, at the cellular level, what are the changes in the pre-and postsynaptic elements that allow for the types of activity-dependent modifications observed in extant mammals? Are these cellular changes heritable? Are they only expressed in particular environmental contexts? Some of these questions can be addressed by considering specific cellular mechanisms that are influenced by activity. As discussed in the following section, accumulating evidence indicates that several features of synaptic architecture and function may indeed be context dependent, and thus highly variable across species.

3.6 Neurotrophins and activity-dependent changes to the nervous system

There have been several activity-dependent, molecular mechanisms proposed to account for the structural and functional changes that occur in the developing nervous system. One of the best candidates for such changes involves a class of proteins called neurotrophins. Neurotrophins are likely mediators for activity-dependent changes that occur during development, for several reasons. Activity regulates their levels and secretion and is in turn regulated by them, they are expressed in portions of the neuron that undergo changes (e.g. synapses), they regulate morphological changes in both the pre- and postsynaptic element (for reviews, see McAllister *et al.* 1995, 1999; McAllister 2001), and they trigger local protein synthesis at the dendrite (Aakalu *et al.* 2001; see Zhang and Poo 2001 for a review).

One way in which activity can ultimately affect the structural configuration and function of neurons via neurotrophin release, in particular by the release of brain-derived neurotrophic factor (BDNF), is through calcium channels. Neural activity increases intracellular calcium and, through a cascade of intracellular molecular events, induces activation of the cyclic AMP pathway that phosphorylates a transcription factor, cyclic AMP response element (CRE)-binding protein or CREB (for reviews, see Finkbeiner and Greenburg 1998; West *et al.* 2001). Phosphorylated CREB can bind to the regulatory region of a gene and induce the initiation, elongation, and translation of RNA transcripts. An RNA transcript is a complementary strand of DNA that is used as a template to translate the DNA code into a protein or peptide. In this way, activity can alter gene expression, by transcribing the code for neurotrophins such as BDNF.

Neurotrophins, such as BDNF, nerve growth factor (NGF), NT3, and NT4/5 play a critical role in the development of the nervous system and carry out a range of functions. At a very gross level, neurotrophins such as BDNF and NGF mediate both positive and negative rates of neuronal survival during development (for reviews, see Levi-Montalcini 1987; Miller and Kaplan 2001), and stimulate cell migration of neurons out of proliferative zones (Borghesani *et al.* 2002). Neurotrophins also influence the growth of axons and dendrites (Segal *et al.* 1995; Carter *et al.* 2002), exerting very specific effects on neuronal differentiation. Further, recent studies in hippocampal slices of adult animals indicate that BDNF stimulates protein synthesis in dendrites of hippocampal neurons (Aakalu *et al.* 2001). Thus, local production of particular proteins may be involved in determining dendritic and spine morphology as well as synaptic function.

The entire process by which activity promotes structural and functional changes in a neuron is intricate, and much of the evidence for exactly how activity alters structure is indirect, often correlational, and in some instances unknown. However, the important point for our discussion of phenotypic variability is that activity can induce cellular and systems level changes in the developing nervous system via calcium-induced alterations in gene expression. Such alterations in gene expression promote peptide and protein synthesis (of neurotrophins and many other proteins), which in turn generates structural and functional modifications throughout the cell. Thus, one can have changes in gene expression, alterations in connectivity, and ultimately large phenotypic changes that are not heritable. However, these modifications can masquerade as evolution as long as the physical and social environment that led to the generation of the particular patterned activity, which induced changes in gene expression and the resulting phenotype, is static. As discussed below, some phenotypic characteristics, including some features of cortical organization and connectivity, exist only within specific environmental contexts.

3.7 What are the genetic and activity-dependent mechanisms that give rise to features associated with complex brains?

We have discussed some of the features of complex brains that are likely to be under genetic control, and in some instances, the specific genes or proteins associated with a particular feature. First, the size of the cortical sheet is likely to be under genetic control, and simple regulation of cell-cycle kinetics in the ventricular zone can account for an exponential increase in the size of the cortical sheet. Proteins such as β-catenin appear to regulate some aspects of the cell cycle, particularly the fraction of cells that remain in the progenitor pool. Another feature of mammalian brains that appears to be genetically regulated is the anterior-posterior and dorso-ventral coordinate system of the neocortex. Intrinsic signaling genes and molecules such as *Wnt*, *Shh*, BMP, and fibroblast growth factor-8 (FGF8) may set up a combinatorial coordinate system that serves as a scaffold for incoming thalamocortical axons. Although not discussed in detail, changes in peripheral morphology that ultimately control the types of patterned activity that the CNS can access are likely to be under genetic control. Features such as the size and shape of an appendage, and the sensory receptor type, number, density, and location may also be genetically regulated. Finally, the intracellular machinery that allows for activity-dependent changes in the developing nervous system are likely to be genetically regulated and heritable, although the specific phenotype they generate is not.

The contribution of patterned activity to the construction of a complex phenotype is also critical. Although not discussed in this chapter, it is certainly worth mentioning that both passive environmental influences as well as active influences play a large role in nervous system construction. Passive influences can have resounding effects on the development of both the somatic and nervous system phenotype. Some types of passive influences include diet, toxins, pH, and temperature. As an extreme example, the phenotype of a nervous system that develops in the presence of alcohol is dramatically different from a normal phenotype, yet still viable. In these cases, gross morphological structure, organization and, we suspect, even connections are significantly modified. This change is not adaptive (but analogous modifications may well be) and is not heritable.

Active influences include changes in the relative activity patterns across sensory receptor arrays, and patterned activity associated with specialized morphology. Activity can alter indirectly the temporal and spatial patterns of gene expression, possibly via neurotrophins which, in turn, can alter the structure and function of neurons and their connections. These types of alterations can masquerade as evolution, because they are genetically mediated and the resulting phenotype can be dramatically altered. However, they are not heritable, but, rather, situation dependent. The examples we have provided are easily related to peripheral morphology and use, and include the bill of a platypus, the hand of a primate, and the lips, tongue, oral structures, and larynx of humans. However, one can also consider active influences that are not strictly tied to a particular sensory receptor array or associated behavior, such as language and skill acquisition, and social and cultural learning. These types of active influences can fundamentally alter the phenotype by changing patterns of synaptic efficacy, connectivity, and ultimately the organization and function of the neocortex. It is likely that much of the human neocortex that does not include the primary and secondary sensory and motor areas is largely shaped by such active influences, and the organization of these cortical fields is only expressed in a particular environmental context. This makes defining such fields across species difficult, since the stimuli that ultimately shape the field are complex, multifaceted, often multimodal, and are variable for different species.

It may seem that the extended discussion of genes, activity, and peripheral input somehow strays from the question that serves as a title for this paper: Is the human brain fundamentally different than that of other mammals?

The answer, of course, is no. The human brain is enslaved by the same genetic constraints and shaped by the same activity-dependent mechanisms as the brain of other mammals. Consequently, its future evolution will follow predictable paths. Although the precise specializations that may emerge cannot be known, one can speculate with a fair degree of accuracy the types of change possible, and the mechanisms by which changes will be achieved. If we consider the human brain and its evolution in this light, then our current ideas regarding derived or specialized areas, which we believe endow us with our uniqueness, need to be reconsidered. Morphology and use alone do not specify brain structure and therefore do not solely constitute the differences observed in the brains of humans and other primates. On the other hand, human brains are not chimpanzee brains with a few new parts added on (e.g. Broca's area, more prefrontal cortex, Wernicke's area, fusiform face area) via very specific genetic changes. Rather, these derivations are much like those observed in non-human mammals, in that several are tied to changes in peripheral morphology and use (such as the language areas), and the cortex in which they reside is likely to be an expanded and connectional specialized version of homologous cortical areas in other primates. While most current work in the human brain focuses on primary and secondary sensory and motor fields, these fields may be more genetically constrained, less variant, and their functional and anatomical attributes more predictable. However, that portion of cortex that we (as a species) are particularly interested in—the regions traditionally referred to as association cortex, including inferotemporal, posterior parietal, and prefrontal cortex—may be less genetically constrained than primary and secondary sensory fields, and the ultimate phenotype of cortical fields within these regions may only occur within a particular environmental context.

Acknowledgements

We wish to thank Deborah Hunt, Jeff Padberg, Marie Burns, Bruno Olshausen, and Elizabeth Disbrow for helpful comments on this manuscript.

References

Aakalu, G., Smith, W. B., Nguyen, N., Jiang, C., and Schuman, E. M. (2001). Dynamic visualization of local protein synthesis in hippocampal neurons. *Neuron* **30**, 489–502.

Allman, J. M. (1999). *Evolving brains*. W.H. Freeman and Co., New York NY.

Bishop, K. M., Goudreau, G., and O'Leary, D. D. M. (2000). Regulation of area identity in the mammalian neocortex by *Emx2* and *Pax6*. *Science* **288**, 344–9.

Boise, L. H., Gonzalez-Garcia, M., Postema, C. E., Ding, L., Lindsten, T., Turka, L. A., Mao, X., Nunez, G., and Thompson, C. B. (1993). Bcl-x, a bcl-2-related gene that functions as a dominant regulator of apoptotic cell death. *Cell* **74**(4), 597–608.

Boncinelli, E., Gulisano, M., Spada, F., and Broccoli, V. (1995). Emx and Otx gene expression in the developing mouse brain. In Bock, G. Cardew, G. (Eds.), *Ciba Foundation Symposium; Development of the cerebral cortex* pp. 100–16. CIBA Pharmaceutical Company, New Jersey NJ.

Borghesani, P. R., Peyrin, J. M., Klein, R., Rubin, J., Carter, A. R., Schwartz, P. M., Luster, A., Corfas, G., and Segal, R. A. (2002). BDNF stimulates migration of cerebellar granule cells. *Development* **129**(*Supp*), 1435–42.

Bronchti, G., Heil, P., Scheich, H., and Wollberg, Z. (1989). Auditory pathway and auditory activation of primary visual targets in the blind mole rat (*Spalax ehrenbergi*): I. 2-Deoxyglucose study of subcortical centers. *Journal of Comparative Neurology* **284**, 253–74.

Bronchti, G., Rado, R., Terkel, J., and Wollberg, Z. (1991). Retinal projections in the blind mole rat: WGA-HRP tracing study of a natural degeneration. *Developmental Brain Research* **58**, 159–70.

Butler, A. B. and Molnár, Z. (2002). Development and evolution of the collopallium in amniotes: A new hypothesis of field homology. *Brain Research Bulletin* **57**, 475–9.

Carter, A. R., Chen, C., Schwartz, P. M., and Segal, R. A. (2002). Brain-derived neurotrophic factor modulates cerebellar plasticity and synaptic ultrastructure. *Journal of Neuroscience* **22**, 1316–27.

Catania, K. C. and Kaas, J. H. (1995). Organization of the somatosensory cortex of the star-nosed mole. *Journal of Comparative Neurology* **351**, 549–67.

Catania, K. C. and Kaas, J. H. (1997*a*). The mole nose instructs the brain. *Somatosensory and Motor Research*, **14**, 56–8.

Catania, K. C. and Kaas, J. H. (1997*b*). Somatosensory fovea in the star-nosed mole: Behavioral use of the star in relation to innervation patterns and cortical representation. *Journal of Comparative Neurology* **387**, 215–33.

Chenn, A. and Walsh, C. A. (2002). Regulation of cerebral cortical size by control of cell cycle exit in neural precursors. *Science* **297**, 365–9.

Chiang, C., Litingtung, Y., Lee, E., Young, K. E., Corden, J. L., Westphal, H., and Beachy, P. A. (1996). Cyclopia and defective axial patterning in mice lacking Sonic hedgehog gene function. *Nature* **383**, 407–13.

Clark, D. A., Mitra, P. P., and Wang, S. S.-H. (2001). Scalable architecture in mammalian brains. *Nature* **411**, 189–93.

Cooper, H. M., Herbin, M., and Nevo, E. (1993). Visual system of a naturally microphthalmic mammal: The blind mole rat, *Spalax ehrenbergi*. *Journal of Comparative Neurology* **328**, 313–50.

Crossley, P. H., Martinex, S., Ohkubo, Y., and Rubenstein, J. L. R. (2001). Evidence that coordinate expression of Fgf8, Otx2, Bmp4, and Shh in the rostral prosencephalon defines patterning centers for telencephalic and optic vesicles. *Neuroscience* **108**, 183–206.

Dunn, C. A., Kahn, D. M., and Krubitzer, L. (2001). Development of retinal connections in the short-tailed opossum (*Monodelphis domestica*). *Society for Neuroscience Abstracts* **27**, 1523.

Finkbeiner, S. and Greenberg, M. E. (1998). Ca^{2+} channel-regulated neuronal gene expression. *Journal of Neurobiology* **37**, 171–89.

Finlay, B. L. and Darlington, R. B. (1995). Linked regularities in the development and evolution of mammalian brains. *Science* **268**, 1578–84.

Fukuchi-Shimogori, T., and Grove, E. A. (2001). Neocortex patterning by the secreted signaling molecule FGF8. *Science* **294**, 1071–4.

Furuta, Y., Piston, D. W., and Hogan, B. L. M. (1997). Bone morphogenetic proteins (BMPs) as regulators of dorsal forebrain development. *Development* **124**, 2203–12.

Grove, E. A., Tole, S., Limon, J., Yip, L.-W., and Ragsdale, C. W. (1998). The hem of the embryonic cerebral cortex is defined by the expression of multiple *Wnt* genes and is compromised in *Gli3*-deficient mice. *Development* **125**, 2315–25.

Kaas, J. H., Nelson, R. J., Sur, M., Lin, C. S., and Merzenich, M. M. (1979). Multiple representations of the body within the primary somatosensory cortex of primates. *Science* **204**, 521–3.

Kahn, D. M. and Krubitzer, L. (2001). Development of retinal connections in the short-tailed opossum (*Monodelphis domestica*). *Society for Neuroscience Abstracts* **27**, 1523.

Kahn, D. M. and Krubitzer, L. (2002). Massive cross-modal cortical plasticity and the emergence of a new cortical area in developmentally blind mammals. *Proceedings of the National Academy of Sciences of the United States of America* **99**, 11429–34.

Kornack, D. R. (2000). Neurogenesis and the evolution of cortical diversity: Mode, tempo, and partitioning during development and persistence in adulthood. *Brain Behavior and Evolution* **55**, 336–44.

Kornack, D. R. and Rakic, P. (1998). Changes in cell-cycle kinetics during the development and evolution of primate neocortex. *Proceedings of the National Academy of Sciences of the United States of America* **95**, 1242–6.

Krubitzer, L. (1998). What can monotremes tell us about brain evolution? *Philosophical Transactions of the Royal Society of London B Biological Sciences* **353**, 1127–46.

Krubitzer, L. (1995) The organization of neocortex in mammals: Are species differences really so different? *Trends in Neurosciences* **18**, 408–17.

Krubitzer, L. and Huffman, K. J. (2000) Arealization of the neocortex in mammals: Genetic and epigenetic contributions to the phenotype. *Brain Behavior and Evolution* **55**, 322–35.

Kuida, K., Haydar, T. F., Kuan, C-Y., Gu, Y., Taya, C., Karasuyama, H., Su, M.S.-S., Rakic, P., and Flavell, R. A. (1998). Reduced apoptosis and cytochrome c-mediated caspase activation in mice lacking caspase 9. *Cell* **94**, 325–37.

Levi-Montalcini, R. (1987). The nerve growth factor 35 years later. *Science* **237**, 1154–62.

Levitt, P., Barbe, M. F., and Eagleson, K. L. (1997). Patterning and specification of the cerebral cortex. In W. M. Cowan (Ed.), *Annual review of neuroscience*, pp. 1–24. Annual Reviews Inc., Palo Alto CA.

Marin, O. and Rubenstein, J. L. R. (2002). Patterning, regionalization, and cell differentiation in the forebrain. In J. Rossant, P. P. L. Tam (Eds.), *Mouse development; patterning, morphogenesis, and organogenesis*, pp. 75–106. Academic Press, San Diego CA.

McAllister, A. K. (2001). Neurotrophins and neuronal differentiation in the central nervous system. *CMLS Cellular and Molecular Life Sciences* **58**, 1054–60.

McAllister, A. K., Lo, D. C., and Katz, L. C. (1995). Neurotrophins regulate dendritic growth in developing visual cortex. *Neuron* **15**, 791–803.

McAllister, A. K., Katz, L. C., and Lo, D. C. (1999). Neurotrophins and synaptic plasticity. *Annual Review of Neuroscience*, **22**, 295–318.

Miller, F. D. and Kaplan, D. R. (2001). Neurotrophin signaling pathways regulating neuronal apoptosis. *CMLS Cellular and Molecular Life Sciences* **58**, 1045–53.

Miyashita-Lin, E. M., Hevner R., Wassarman, K. M., Martinez, S., and Rubenstein, J. L. R. (1999). Early neocortical regionalization in the absence of thalamic innervation. *Science* **285**, 906–9.

Molnár, Z. and Butler, A. B. (2002). The corticostriatal junction: A crucial region for forebrain development and evolution. *BioEssays* **24**, 530–41.

Motoyama, N., Wang, F., Roth, K. A., Sawa, H., Nakayama, K.-I., Nakayama, K., Negishi, I., Senju, S., Zhang, Q., Fujii, S., and Loh, D. Y. (1995). Massive cell death of immature hematopoietic cells and neurons in Bcl-x-deficient mice. *Science* **267**, 1506–10.

Nakagawa, Y., Johnson, J. E., and O'Leary, D. D. M. (1999). Graded and areal expression patterns of regulatory genes and cadherins in embryonic neocortex independent of thalamocortical input. *Journal of Neuroscience* **19**, 10877–85.

Peifer, M. and Polakis, P. (2000). Wnt signaling in oncogenesis and embryogenesis-a look outside the nucleus. *Science* **287**, 1606–9.

Recanzone, G. H. (2000). Cerebral cortical plasticity: Perception and skill acquisition. In: M. Gazzaniga (Ed.), *The new cognitive neurosciences* pp. 237–47. MIT Press, Cambridge MA.

Recanzone, G. H., Merzenich, M. M., Jenkins, W. M., Grajski, K. A., and Dinse, H. R. (1992). Topographic reorganization of the hand representation in cortical area 3b of owl monkeys trained in a frequency-discrimination task. *Journal of Neurophysiology* **67**, 1031–56.

Recanzone, G. H., Schreiner, C. E., and Merzenich, M. M. (1993). Plasticity in the frequency representation of primary auditory cortex following discrimination training in adult owl monkeys. *Journal of Neuroscience* **13**, 87–103.

Roth, K. A., Kuan, C.-Y., Haydar, T. F., D'Sa-Eipper, C., Shindler, K. S., Zheng, T. S., Kuida, K., Flavell, R. A., and Rakic, P. (2000). Epistatic and independent functions of Caspase-3 and Bcl-XL in developmental programmed cell death. *Proceedings of the National Academy of Sciences of the United States of America* **97**, 466–71.

Rubenstein, J. L. R., Anderson, S., Shi, L., Miyashita-Lin, E., Bulfone, A., and Hevner, R. (1999). Genetic control of cortical regionalization and connectivity. *Cerebral Cortex* **9**, 524–32.

Segal, R. A., Pomeroy, S. L., and Stiles, C. D. (1995). Axonal growth and fasciculation linked to differential expression of BDNF and NT3 receptors in developing cerebellar granule cells. *Journal of Neuroscience* **15**, 4970–81.

Simeone, A., Acampora, D., Gulisano, M., Stornaiuolo, A., and Boncinelli, E. (1992*a*). Nested expression domains of four homeobox genes in developing rostral brain. *Nature* **358**, 687–90.

Simeone, A., Gulisano, M., Acampora, D., Stornaiuolo, A., Rambaldi, M., and Boncinelli, E. (1992*b*). Two vertebrate homeobox genes related to the *Drosophila* empty spiracles gene are expressed in the embryonic cerebral cortex. *EMBO Journal* **11**, 2541–50.

Stephan, H., Frahm, H., and Baron, G. (1981). New and revised data on volumes of brain structures in insectivores and primates. *Folia Primatology* **35**, 1–29.

Tao, W. and Lai, E. (1992). Telencephalon-restricted expression of BF-1, a new member of the HNF-3 fork head gene family, in the developing rat brain. *Neuron* **8**, 957–66.

Welker, E. and Van der Loos, H. (1986). Is areal extent in sensory cerebral cortex determined by peripheral innervation density? *Experimental Brain Research* **63**, 650–4.

West, A. E., Chen, W. G., Dalva, M. B., Dolmetsch, R. E., Kornhauser, J. M., Shaywitz, A. J., Takasu, M. A., Tao, X., and Greenberg M. E. (2001). Calcium regulation of neuronal gene expression. *Proceedings of the National Academy of Sciences of the United States of America* **98**, 11024–31.

Zhang, L. I. and Poo, M-M. (2001). Electrical activity and development of neural circuits. *Nature Neuroscience* **4**(*Suppl*), 1207–14.

Chapter 4

Analysis of topographically organized patterns of response in fMRI data: Distributed representations of objects in the ventral temporal cortex

James V. Haxby

Abstract

The ventral object vision pathway in the human brain has a complex functional architecture, characterized by topographically organized patterns of differential responses to faces and different object categories that can be measured with functional magnetic resonance imaging (fMRI). We have developed a new method for analyzing these patterns of response, topographic pattern analysis. This method is a major departure from previous, standard methods for analyzing functional neuroimaging data. Whereas previous methods were designed to find clusters of voxels with similar response properties, topographic pattern analysis is designed to detect reliable patterns of differences among the responses of voxels. Previous methods increase signal-to-noise by averaging the responses across voxels, obscuring between-voxel differences. Topographic pattern analysis increases sensitivity by examining patterns of between-voxel differences across a large number of voxels. Topographic pattern analysis is illustrated with an experiment on face and object recognition (Haxby et al. 2001). The results show that this method can detect distinct patterns of response for a large number of different object categories, and that category-related information in these patterns is carried by both strong and weak responses. These results suggest that information about the visual appearance of objects is represented by a widely distributed population response, in which both strong and weak responses may play an integral role. More generally, this demonstration of topographic pattern analysis shows that patterns of between-voxel differences in response carry important information about the functional organization of the brain that was not detected by previous methods of analysis.

4.1 Introduction

Functional brain imaging research on the ventral object vision pathway has revealed a complex architecture. Studies have identified occipitotemporal cortical regions that respond maximally to faces, buildings and scenes, and human body parts (Kanwisher et al. 1997; McCarthy et al. 1997; Epstein et al. 1998; Aguirre et al. 1998; Haxby et al. 1999; Ishai et al. 1999; Downing et al. 2001). Other regions have been identified that respond preferentially to animals, tools, and even some categories of manmade objects (Chao et al. 1999a, b; Ishai et al. 1999; Haxby et al. 2001). The region that responds maximally to faces also has been implicated in expert recognition of non-face objects (Gauthier et al. 1999, 2000). The expanse of ventral occipitotemporal cortex that contains these regions also has been shown to have a coarse retinotopic organization, such that the regions associated more with face perception and expert visual recognition have a more foveal representation and the regions associated with recognition of scenes and buildings have a more peripheral representation (Hasson et al. 2002; Malach et al. 2002).

We have proposed that the functional architecture of the ventral object vision pathway may not be best characterized by regions with specialized functions (Ishai et al. 1999; Haxby et al. 2000, 2001). Instead, we have proposed that recognition of any object category may draw upon representations of information about object appearance from several or all of these subregions. The development of this proposal was motivated by our appraisal of the results of several fMRI experiments (Chao et al. 1999a, b; Haxby et al. 1999; Ishai et al. 1999, 2000), in which we showed that a large variety of different object categories each evoked a different pattern of response in the ventral temporal cortex. Moreover, we found that in all of the subregions that showed a consistent preference for one category, as manifested by a maximal response, all of the other categories did elicit significant, albeit weaker, responses.

We considered the possibility that these weaker responses might also carry information that contributes to the representation of object appearance (Ishai et al. 1999; Haxby et al. 2000), but standard methods for analyzing functional brain imaging data (Friston et al. 1995; Haxby et al. 1997) are not well suited to investigating this possibility. Consequently, we developed a new method for analyzing patterns of response in functional brain imaging data (Haxby et al. 1999, 2001). This method analyzes a type of information in functional imaging data sets that is not revealed by traditional methods. Rather than analyzing the pattern of response independently for each voxel and then looking for clusters of voxels that respond similarly, our new method analyzes the pattern of differences between voxel responses. The result is a method that detects topographically organized patterns of response in a defined cortical expanse. Because this method focuses on differences between voxels, rather than similarities of responses across voxels, it is more tuned to high spatial frequency features in patterns of neural response than are more traditional methods of brain image analysis.

In this chapter I will use the term 'topographic pattern analysis' to refer to both the specific method we developed for analyzing patterns of response and to other potential methods designed to look at this same type of information. The implementation of topographic pattern analysis presented here uses simple statistics, namely correlations, and was designed only to detect patterns in specified sets of voxels. It will certainly be overtaken by more sophisticated methods that will have increased sensitivity, will identify the voxels that contribute to response patterns and those that do not, and will provide more information about the structure of the topographic patterns.

We have applied topographic pattern analysis to data from a new experiment that was designed to investigate how widely distributed the representation of information about visually perceived objects is in ventral extrastriate cortex (Haxby *et al.* 2001). I will summarize the results of that study here to illustrate the use of this new method of analysis to demonstrate distributed representations. Then I will amplify those results to illustrate that topographic pattern analysis reveals a type of information that is not limited to tests of distributed representations and is not incompatible with more modular models of brain organization. It reveals patterns of response that may indicate how information is represented within a module as well as across a broader cortical expanse. These patterns may reflect the organization of a representation that is based on a population response. Such a representation has advantages for computational modeling of face and object recognition. A representation based on a topographically organized pattern of response also has the capacity to generate different representations for an unlimited number of object categories or individual objects or faces. Thus, this method of analysis has the potential to investigate directly *how* object and face appearance is represented, not just *where* it is represented.

4.2 Experimental methods

4.2.1 Subjects, stimuli, and task

Six subjects participated in a functional magnetic resonance imaging (fMRI) study of visual recognition of faces and objects. Whole-brain volumes, consisting of 40 3.5-mm-thick sagittal echoplanar images (TR = 2500 ms, TE = 30 ms, FA = 90, FOV = 24 cm) were obtained while subjects viewed grayscale pictures of faces, houses, cats, shoes, scissors, bottles, chairs, and phase-scrambled images, and performed a one-back repetition detection task. Stimuli were blocked by category. In each of 12 time series, blocks of all eight categories were presented. Each block consisted of 12 stimuli, each presented for 1 second with a 1 second interstimulus interval. Blocks were separated by 12-second periods of rest.

Details about the experimental methods can be found in Haxby *et al.* (2001).

4.2.2 Data analysis

The data were analyzed to test whether different stimulus categories evoke patterns of response that are distinct and reproducible. In each subject, two independent estimates of the patterns of response to each category were calculated, one based on the odd-numbered time series and the other based on the even-numbered time series. The similarity between independent measures of the patterns of response to each category was indexed by correlations. The significance of this measure of similarity was demonstrated by contrasting within-category correlations between response patterns to between-category correlations.

Multiple regression was used to identify voxels that responded differentially to object categories ($P < 10^{-6}$) and to calculate the amplitudes of the responses to each category in each voxel in each half of the data. The amplitudes of response were normalized by subtracting the mean response to all stimulus categories from the response to one category in one-half of the data in each voxel. This normalization step increased the extent to which correlations between patterns of response reflected shared category-specific variance and not a shared generic response to visual stimuli. Analysis was restricted to patterns of response in object-selective voxels in ventral temporal cortex, which was defined anatomically to include the lingual, parahippocampal, fusiform, and inferior temporal gyri.

Correlations were calculated to index the similarity of patterns of response in object-selective ventral temporal cortex derived from the even- and odd-numbered time series (Fig. 4.1). For example, correlations were calculated between the pattern of response to faces on even-numbered time series and the pattern of response to each stimulus category on odd-numbered time series. The resulting correlation matrix was then analyzed to determine whether each within-category correlation was larger than each of the between-category correlations that involved that category. For example, the correlation of the patterns of response to faces on even-numbered and odd-numbered time series was compared to the correlation between the response to faces on even-numbered time series and the response to chairs on odd-numbered time series. Such pairwise comparisons indicated whether the pattern of response to faces on odd-numbered time series could be distinguished correctly from the pattern of response to chairs on odd-numbered time series, based on the similarity of those responses to the response to faces on even-numbered time series. If the pairwise comparison was in the expected direction, it was considered a correct identification of the category being viewed. If the comparison was in the unexpected direction, it was considered an identification error.

Different sets of voxels were defined to test how distributed the category-related patterns were in the ventral temporal cortex. In addition to the analysis of all object-selective cortex, analysis for each pairwise comparison was restricted to cortex that did not respond maximally in either half of the data to either of the two categories being compared. This analysis was conducted to determine whether the category being viewed could still be identified based on the pattern of response to that category in cortex that responded maximally to other categories. A final analysis was conducted to determine whether the category-related patterns of response had a spatial frequency that was

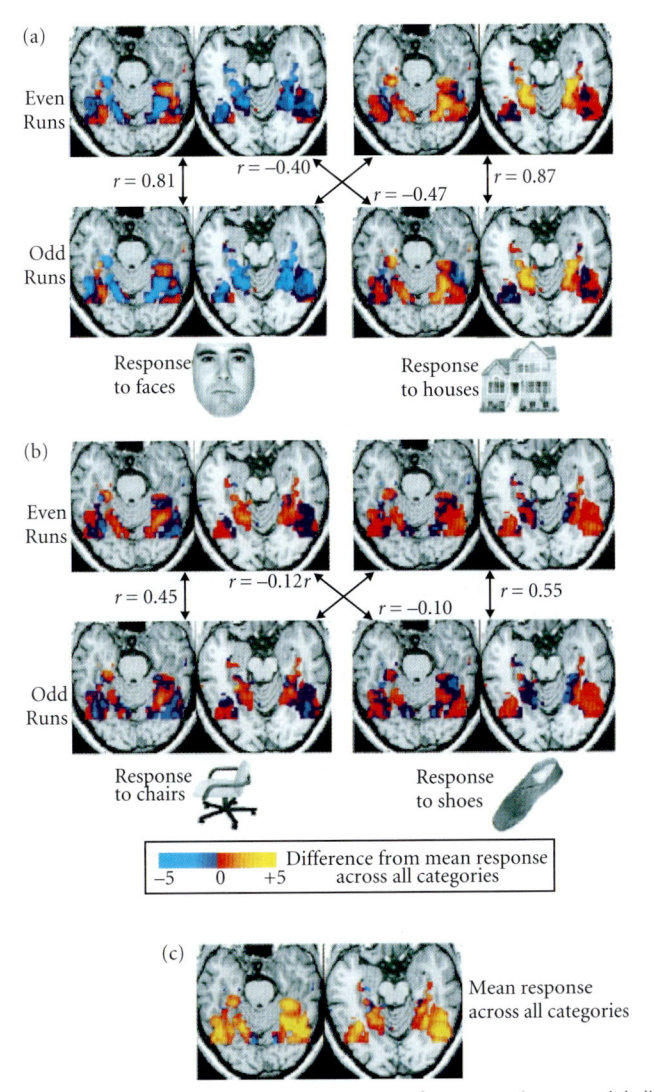

Fig. 4.1 Brain images showing the normalized patterns of response in two axial slices in a single subject. The left side of the brain is on the left side of each image. Responses in all object-selective voxels in the ventral temporal cortex are shown. For each pairwise comparison, the within-category correlation is compared to one between-category correlation. (a) Comparisons between the patterns of response to faces and houses in one subject. Note that the within-category correlations for faces ($r = 0.81$) and houses ($r = 0.87$) are both markedly larger than the between-category correlations, yielding correct identifications of the category being viewed. (b) Comparisons between the patterns of response to chairs and shoes in the same subject. Note again that the category being viewed was identified correctly for all comparisons. (c) Mean response across all categories relative to a resting baseline. (From Haxby et al. 2001.)

higher than that defined by regions that respond maximally to one category. This analysis restricted the calculation of correlations to voxels that responded maximally to a single category. Separate analyses were completed for voxels that responded maximally to faces in both halves of the data, voxels that responded maximally to houses, and voxels that responded maximally to cats.

4.3 Results

The category being viewed could be identified with high accuracy based on the pattern of response in all object-selective ventral temporal cortex, as well as in only the cortex that responded maximally to other categories (Fig. 4.2 and Table 4.1). The pattern of response in all object-selective ventral temporal cortex correctly identified the category being viewed in 96% of pairwise comparisons. Excluding the cortex which responded maximally to the categories being compared had only a minimal effect on the accuracy of identifying the category being viewed, reducing overall accuracy to 94%. These results indicate that information about the face or object being viewed is also carried by neural responses in cortical regions that respond more to other objects.

Even within regions that respond maximally to a single category, patterns of response could be demonstrated that distinguished between other categories. The category being viewed could be identified by patterns of response within the relatively large region that responded maximally to houses (93% overall accuracy). Even within the much smaller regions that responded maximally to faces and cats, patterns of response could correctly identify the category being viewed with accuracies that were better than chance (83% and 85%, respectively). These results indicate that the topographic pattern analysis used here was sensitive to higher spatial frequencies in the patterns of response than the spatial frequency defined by regions that respond maximally to a single stimulus category.

4.4 Discussion

4.4.1 Distributed representations of faces and objects

The results of this experiment show that the patterns of response to visually perceived faces and objects in the ventral temporal cortex are specific to a large number of object categories, including various categories of small manmade objects. By investigating the topographically organized arrangement of differences between large and small responses, rather than only the locations of subregions that respond maximally to a category, we have shown how the neural responses in these cortices can represent a virtually unlimited variety of object categories. Furthermore, our results show that these category-specific patterns exist at a spatial resolution that can be probed with current fMRI methods.

The analysis of fragments of this representation showed that information that is specific to a category is also carried by neural responses in cortices that do not respond maximally to that category. Moreover, category specificity was detected in patterns of response within regions that responded maximally to a single category, indicating that

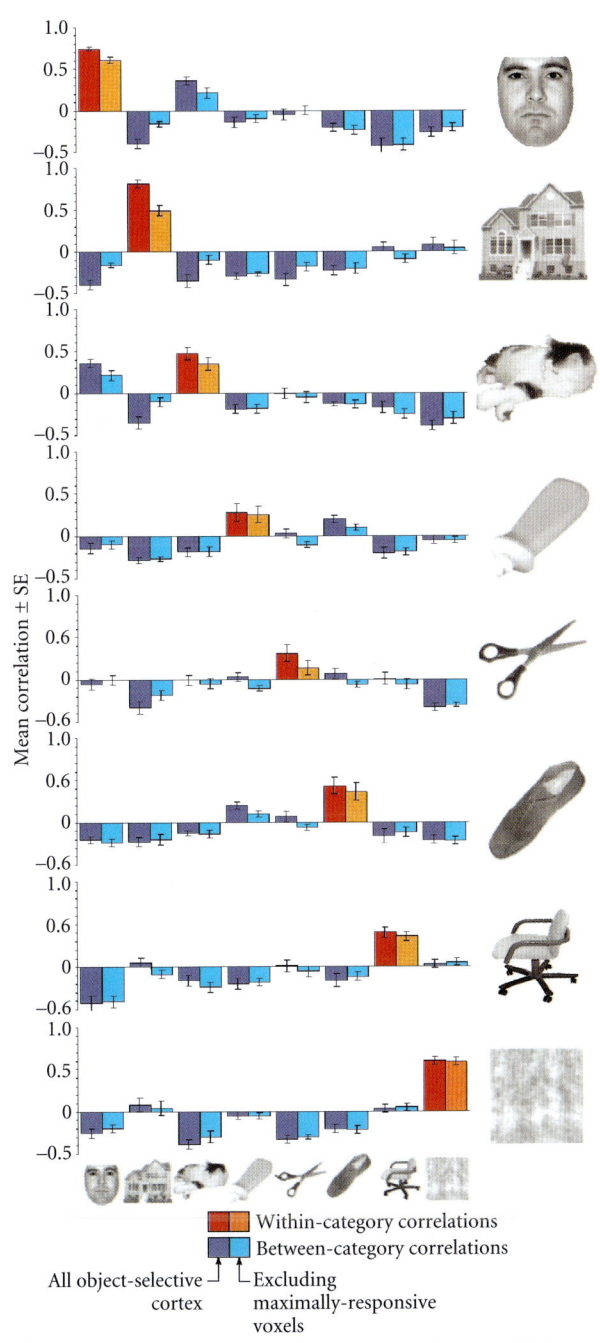

Fig. 4.2 Mean within-category and between-category correlations (± SE) between patterns of response across all subjects for all ventral temporal object-selective cortex (red and dark blue) and for ventral temporal cortex excluding the cortex that responded maximally to either of two categories being compared (orange and light blue). (From Haxby et al. 2001.)

Table 4.1 Accuracy of identification of the category being viewed based on the patterns of response evoked in ventral temporal cortex. Accuracies are the percentage of comparisons between two categories that correctly identified which category was being viewed

Region	Volume (cm³ ± SE)	Identification accuracy								
		Faces	Houses	Cats	Bottles	Scissors	Shoes	Chairs	Scrambled
All ventral temporal object-selective cortex	22.9 ± 2.8	100‡	100‡	98 ± 2‡	90 ± 6‡	92 ± 6‡	92 ± 7‡	96 ± 2‡	100‡
Minus regions that were maximally responsive to categories being compared	15.4 ± 1.8	100‡	100‡	95 ± 2‡	89 ± 6‡	85 ± 9†	90 ± 8†	98 ± 1‡	100‡
Regions maximally responsive to:									
Faces	3.1 ± 0.9	94 ± 7‡	99 ± 1‡	76 ± 13*	81 ± 14*	77 ± 9*	70 ± 16	77 ± 11*	92 ± 7‡
Houses	9.6 ± 1.8	100‡	100‡	88 ± 5‡	85 ± 10†	81 ± 6†	96 ± 2‡	94 ± 3‡	100‡
Cats	2.6 ± 0.4	96 ± 4‡	96 ± 2‡	82 ± 8†	65 ± 11	69 ± 5†	76 ± 9*	95 ± 4‡	100‡

Differs from chance (50%): * $P < 0.05$, † $P < 0.01$, ‡ $P < 0.001$

category-related patterns of response include spatial frequencies that are higher than that defined by such regions.

Our study has been replicated recently by Spiridon and Kanwisher (2002), who added some conditions that demonstrate the robustness of these distributed patterns of response. The study by Spiridon and Kanwisher used the same method of topographic pattern analysis and replicated the identification of the category being viewed based on all object-responsive voxels and based only on the voxels that responded submaximally to that category. They showed, further, that the category-specificity of these patterns was not due to the low-level features in the stimuli. The category being viewed could still be identified with high accuracy when different exemplars were shown in the two halves of the data and when the responses to grayscale photographs were compared to the responses to line drawings of the same objects. In our study (Haxby *et al.* 2001), the same exemplars were used for both halves of the data. We also reanalyzed an older data set (Ishai *et al.* 1999) and demonstrated the similarity between patterns of response to photographs and line drawings, but this demonstration was for a limited set of categories (faces, houses, and chairs). Spiridon and Kanwisher, however, were not able to replicate our finding that patterns of response within the region that responded maximally to faces discriminated between non-face objects, and that patterns of response within the region that responded maximally to houses discriminated between objects other than houses. Their results indicated that the patterns of response in these areas distinguished the category that elicited the maximal response (faces or houses) from other objects, but did not distinguish among other objects. The reasons for the discrepancy between their results and ours are not clear, but it highlights the need to develop better methods to probe the roles played by different parts of distributed representations in face and object representation. Overall, however, the report by Spiridon and Kanwisher replicates the most important points from our study, namely the demonstration that topographic pattern analysis can detect distinct patterns of response for an apparently unlimited number of object categories, and the demonstration that submaximal responses to an object carry information about its appearance.

4.4.2 Topographic pattern analysis

The method used to analyze the data from this experiment was developed to search for topographically organized patterns of response that discriminate between stimulus categories. Previous methods were not suited to this application. Previous standard methods analyzed the temporal pattern of response to changing cognitive states in each voxel independently and then looked for clusters of voxels that showed similar temporal responses (Friston *et al.* 1995; Haxby *et al.* 1997). These methods were designed to address a certain set of questions:

1. What brain areas are 'activated' by a cognitive operation relative to a baseline that does not involve that operation?

2. How is activity in a region modulated by attention or memory?
3. How is activity in a region modulated by activity in other regions (functional connectivity)?
4. What type of stimulus or process elicits the maximal response in a region?

Consequently, these methods are biased towards producing results that support a model of brain organization that is based on functionally homogeneous regions that are large enough to span several voxels.

By contrast, we wished to find patterns that are based on between-voxel response differences. The results suggest that topographic pattern analysis is actually very sensitive to stimulus-related neural activity, perhaps even more sensitive than previous standard methods. One reason for this sensitivity is that topographic pattern analysis pools data over large sets of voxels. Standard methods also pool data over multiple voxels, but this is usually done by averaging the responses over voxels in a functionally defined region, like the fusiform face area (FFA), or by spatial smoothing, which averages responses over neighboring voxels to increase signal-to-noise ratios. These methods of pooling data explicitly obscure differences between voxel responses. Topographic pattern analysis exploits these differences in the way it pools data. The second reason for the sensitivity of topographic pattern analysis, therefore, is that these patterns of differences apparently carry information, suggesting that topographic patterns are an important dimension in functional brain organization.

Topographically organized patterns of response are well established in sensory and motor cortices. A sophisticated, and now standard, method for mapping sensory topographies with functional brain imaging is phase-encoded mapping (Sereno *et al.* 1995; Engel *et al.* 1997). Phase-encoded mapping looks for a specific type of between-voxel difference in the temporal pattern of response to slowly changing stimuli, namely phase lags that indicate how a wave of activation moves across a cortical surface. Phase-encoded mapping explicitly reveals the nature of the information that underlies the topography—retinotopic location for visual cortex—whereas our method of topographic pattern analysis only detects whether a pattern exists and reveals nothing about how information is organized within that pattern. However, phase-encoded mapping may not be an appropriate method for analyzing more abstract representations such as object form topography. Whereas retinotopic mapping was based on detailed prior knowledge about retinotopic topographies from studies in animals, object form topography was previously unknown. Furthermore, it is unclear whether a slowly changing stimulus would elicit a gradual movement of activity through object form topography, as a slowly moving stimulus moves activity through a retinotopic topography. For example, morphing an object from one category to another may not translate into a gradual change of neural response because object recognition tends to be categorical. Even if a slow morph between categories produced a gradual movement of response, it is unclear how such a movement could reveal more than the endpoints associated with the two categories.

4.4.3 The information represented by topographically organized patterns

Whereas sensory topographies indicate that different parts of a sensory field have spatially discrete representations in the sensory cortex, the representation of an object in object form topography appears to be distributed across much, or even all, of the object-selective cortex. Such a distributed representation suggests that the integrated percept of object form is based on a population response in which both strong and weak responses contribute. Traditional methods of analysis, which use strength of response as an index of degree of participation, could not lead to such a conclusion. By contrast, a pattern of strong and weak responses is more detectable with topographic pattern analysis than is a pattern of uniformly strong responses. Topographic pattern analysis is therefore biased towards producing results that suggest a distributed representation based on a population response, just as traditional methods are biased towards producing results that suggest a network of connected processing modules.

The role of submaximal responses in a population response representation of the visual appearance of objects is poorly understood. While it may be possible that these responses do not play a central role in recognition, submaximal responses do play a critical role in the representation of simpler visual features, such as color and motion direction.

Color vision provides a clear example of a visual representation based on a population response (Hering 1964; Hurvich and Jameson 1974). The perception of hue is based on the combined responses of opponent process cell populations. The percept of a hue is dependent not only on the strongest responses in these populations, such as a maximal red response, but also on weaker responses in the other populations. For example, the percept of reddish-orange to reddish-violet hues depends on a weak response in the yellow–blue cell populations (Fig. 4.3).

Computational models and psychophysical results now provide some intriguing support for the idea that face recognition may also be based on the combined responses

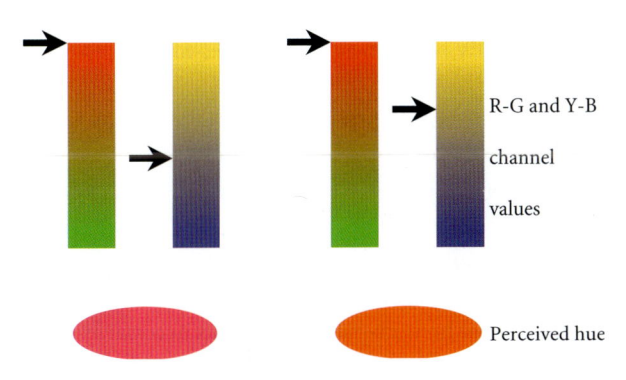

Fig. 4.3 Schematic diagram illustrating how population responses, involving cell populations representing two opponent processes, produce an integrated percept of hue. Note that the responses in the yellow–blue channel still play a critical role determining the perceived hue even though they are 'weak', meaning close to the central tendency.

of opponent cell populations (O'Toole *et al.* 1993; Blanz *et al.* 2000; Leopold *et al.* 2001). As compared to the representation of color, the opponent processes that may underlie face recognition are not as clearly defined and are certainly more numerous. However, principal components analyses of faces, have demonstrated how such opponent processes, referenced to a population average face, could represent constellations of correlated visual features, the integrated product of which can exactly specify the appearance of an individual face. As in the representation of color vision, weak responses on opponent process face components are as important as strong responses in representing the appearance of a face.

The means by which a representation based on a population response could represent the appearance of all objects is more speculative. The example of face representation suggests, however, that complex attributes of form could be represented by a limited set of opponent process cell populations. The more varied configurations of features that define other object categories, however, may not be accommodated by an analogous set of opponent processes. A population response based on cell populations with more unipolar response properties, expressing the similarity of an object to a given feature or configuration, is also not inconsistent with the results of our experiment.

4.4.4 Population responses and functionally defined processing areas

Our initial application of topographic pattern analysis, described here, was designed to investigate the extent to which the representations of faces and other object categories were distributed and overlapping. However, the method of analysis, and even the results, are not necessarily incompatible with accounts for the functional architecture that are based on cortical areas defined by functional specificity, such as a leading role in face perception or a sensitivity to holistic configurations that might mediate expert recognition. The finding of topographically organized representations may be orthogonal to the organization of the ventral temporal cortex into specialized regions, just as retinotopic organization coexists with the areal organization of early visual cortex. Object form topography may provide clues about *how* faces and objects are represented in these areas, rather than simply indicating *where* faces and objects are processed.

Our results do indicate, however, that the representation of faces and various object categories is more distributed across the ventral temporal cortex than previously believed. Topographic pattern analysis can identify the category of objects being viewed even when the analysis is restricted to cortex that responds maximally to other categories, indicating that information about object appearance is carried by the submaximal responses in these areas. However, the spatial resolution of fMRI does not allow us to conclude that these submaximal hemodynamic responses necessarily indicate submaximal neural responses. Each voxel in our fMRI studies probably contains roughly 50–100 cortical columns. It is possible that a subset of these columns is responding maximally to the viewed category, but a larger subset responds maximally to a different category. Definitive demonstration that weak neural responses carry information that is integral to face and object representation, as discussed in the previous section, will require higher resolution imaging.

Hemodynamic imaging, such as fMRI, also has limited temporal resolution. The temporal characteristics of responses to faces and other objects may reveal greater specialization. For example, the early and more automatic response to faces called the N170 or N200, which is known to originate in a region close to the FFA (as well as in the superior temporal sulcus) (Allison *et al.* 1999; Halgren *et al.* 2000), may be generated by a small and anatomically discrete cell population that plays a critical and specialized role in initiating face processing.

The discovery of a coarse retinotopic organization in the ventral temporal extrastriate cortex has led to the proposal that the arrangement of areas that respond maximally to certain categories is determined by the resolution needs for recognition of different types of objects (Levy *et al.* 2001; Hasson *et al.* 2002; Malach *et al.* 2002). This hypothesis is supported by the finding that the maximally face-responsive and letter-responsive regions lie in cortex that has a more central visual field representation, whereas the maximally house-responsive region lies in the more peripheral representation. However, a single, retinotopic dimension cannot account for the level of category specificity that our topographic pattern analysis has revealed. We find that category-related patterns exist within areas, such as the face-responsive region, that presumably have a very restricted retinotopic organization. However, the consistency across subjects of the topographic organization of the ventral temporal object-selective cortex has only been demonstrated for the relative locations of regions that respond maximally to faces, houses, letters, tools, and chairs. It is unclear whether the full patterns of strong and weak responses will be consistent across subjects. Retinotopic organization may determine a coarse organization of maximally responsive regions that is consistent across subjects, but the finer details of these patterns may develop more idiosyncratically.

In conclusion, we have developed a new method for analyzing topographically organized patterns of response in functional brain imaging data sets. Unlike previous methods for fMRI analysis, which look for sets of voxels that have similar responses to experimental manipulations of cognitive activity, topographic pattern analysis looks for patterns of between-voxel differences. Consequently, this method is sensitive to high spatial frequencies in the anatomical distribution of neural responses. Topographic pattern analysis tends to be biased towards finding patterns that suggest a distributed representation based on a population response in which both strong and weak responses contribute. The results of our experiment show that numerous object categories, including different categories of small manmade objects, evoke distinct patterns of response, and that activity that carries information about each category is widely distributed, overlapping the activity that carries information about other categories. These findings demonstrate how a limited cortical space can represent an unlimited variety of objects. However, these results are not necessarily inconsistent with models that posit the existence of specialized regions or cell populations within this broader representation that play a critical role in organizing processing of specific categories, such as faces, or in representing information that plays a more critical role in certain classes of perceptual processing, such as expert recognition.

References

Aguirre, G.K., Zarahn, E., and D'Esposito, M. (1998). An area within human ventral cortex sensitive to 'building' stimuli: evidence and implications. *Neuron*, **21**, 373–83.

Allison, T., Puce, A., Spencer, D.D., and McCarthy, G. (1999). Electrophysiological studies of human face perception. I: Potentials generated in occipitotemporal cortex by face and non-face stimuli. *Cerebral Cortex*, **9**, 415–30.

Blanz, V., O'Toole, A.J., Vetter, T., and Wild, H.A. (2000). On the other side of the mean: The perception of dissimilarity in human faces. *Perception*, **29**, 885–91.

Chao, L.L., Martin, A., and Haxby, J.V. (1999*a*). Are face-responsive regions selective only for faces? *Neuroreport*, **10**, 2945–50.

Chao, L.L., Haxby, J.V., and Martin, A. (1999*b*). Attribute-based neural substrates in posterior temporal cortex for perceiving and knowing about objects. *Nature Neuroscience*, **2**, 913–19.

Downing, P.E., Jiang, J., Shuman, M., and Kanwisher, N. (2001). Cortical area selective for visual processing of the human body. *Science*, **293**, 2470–3.

Engel, S.A., Glover, G.H., and Wandell, B.A. (1997). Retinotopic organization in human visual cortex and the spatial precision of functional MRI. *Cerebral Cortex*, **7**, 181–92.

Epstein, R., and Kanwisher, N. (1998). A *Nature* cortical representation of the local visual environment, **392**, 598–601.

Friston, K.J., Holmes, A.P., Poline, J.-B., Grasby, P.J., Williams, C.R., and Frackowiak, R.S.J. (1995). Analysis of fMRI time-series revisited. *Neuroimage*, **2**, 45–53.

Gauthier, I., Tarr, J.J., Anderson, A.W., Skudlarski, P., and Gore, J.C. (1999). Activation of the middle fusiform 'face area' increases with expertise in recognizing novel objects. *Nature Neuroscience*, **2**, 568–73.

Gauthier, I., Skudlarski, P., Gore, J.C., and Anderson, A.W. (2000). Expertise for cars and birds recruits brain areas involved in face recognition. *Nature Neuroscience*, **3**, 191–97.

Halgren, E., Raij, T., Marinkovic, K., Jousmäki, V., and Hari, R. (2000). Cortical response profile of the human fusiform face area as determined by MEG. *Cerebral Cortex*, **10**, 69–81.

Hasson, U., Levy, I., Behrmann, M., Hendler, T., and Malach, R. (2002). Eccentricity bias as an organizing principle for human high-order object areas. *Neuron*, **34**, 479–90.

Haxby, J.V., Maisog, L.M., and Courtney, S.M. (1997). Multiple regression analysis of effects of interest in fMRI time series (online). http://lbc.nimh.nih.gov/pubs.html

Haxby, J.V., Ungerleider, L.G., Clark, V.P., Schouten, J.L., Hoffman, E.A., and Martin, A. (1999). The effect of face inversion on activity in human neural systems for face and object perception. *Neuron*, **22**, 189–99.

Haxby, J.V., Ishai, A., Chao, L.L., Ungerleider, L.G., and Martin, A. (2000). Object form topology in the ventral temporal lobe. *Trends in Cognitive Sciences*, **4**, 3–4.

Haxby, J.V., Gobbini, M.I., Furey, M.L., Ishai, A., Schouten, J.L., and Pietrini, P. (2001). Distributed and overlapping representations of faces and objects in ventral temporal cortex. *Science*, **293**, 2425–30.

Hering, E. *Outlines of a theory of the light sense*, (1964). Harvard University Press, Cambridge MA.

Hurvich, L.M. and Jameson, D. (1974). Opponent processes as a model of neural organization. *American Psychologist*, **29**, 88–102.

Ishai, A., Ungerleider, L.G., Martin, A., Schouten, J.L., and Haxby, J.V. (1999). Distributed representation of objects in the human ventral visual pathway. *Proceedings of the National Academy of Sciences, USA*, **96**, 9379–84.

Ishai, A., Ungerleider, L.G., Martin, A., and Haxby, J.V. (2000). The representation of objects in the human occipital and temporal cortex. *Journal of Cognitive Neuroscience*, **12** (Supplement 2), 35–51.

Kanwisher, N., McDermott, J., and Chun, M.M. (1997). The fusiform face area: A module in human extrastriate cortex specialized for face perception. *Journal of Neuroscience*, **17**, 4302–11.

Leopold, D., O'Toole, A. J., Vetter, T., and Blanz, V. (2001). Prototype-referenced shape encoding revealed by high-level after effects. *Nature Neuroscience*, **4**, 89–94.

Levy, I., Hasson, U., Avidan, G., Hendler, T., and Malach, R. (2001). Center-periphery organization of human object areas. *Nature Neuroscience*, **4**, 533–9.

Malach, R., Levy, I., and Hasson, U. (2002). The topography of high-order human object areas. *Trends in Cognitive Sciences*, **6**, 176–84.

McCarthy, G., Puce, A., Gore, J.C., and Allison, T. (1997). Face-specific processing in the human fusiform gyrus. *Journal of Cognitive Neuroscience*, **9**, 605–10.

O'Toole, A.J., Abdi, H., Deffenbacher, K.A., and Valentin, D. (1993). Low dimensional representation of faces in high dimensions of the space. *Journal of the Optical Society of America A*, **10**, 405–10.

Sereno, M.I., Dale, A.M., Reppas, J.B., Kwong, K.K., Belliveau, J.W., Brady, T.J., Rosen, B.R., and Tootell, R.B. (1995). Borders of multiple visual areas in humans revealed by functional magnetic resonance imaging. *Science*, **268**, 889–93.

Spiridon, M., and Kanwisher, N. (2002). How distributed is visual category information in human occipito-temporal cortex? An fMRI study. *Neuron*, **35**, 1157–65.

Chapter 5

Time as coding space in the cerebral cortex

Wolf Singer

Abstract

Evidence is reviewed which suggests that the cerebral cortex applies, in parallel, complementary strategies to enhance the saliency of neuronal responses selected for further joint processing and to encode relations among the activities of distributed neurons. Saliency is enhanced by increasing discharge rate and/or by synchronizing the discharges of neurons. Relations between distributed neurons are encoded by integrating their responses in conjunction with specific neurons or by associating the neurons in functionally coherent cell assemblies that are distinguished by synchronous firing. It is proposed that selecting responses and encoding relations through synchronization is particularly well suited for the processing of rare or novel relations for which no conjunction-specific neurons have been implemented. As changes in synchrony need not be associated with changes in discharge rate, analysis techniques that capture variations in discharge rate may be insensitive to processes that rely primarily on temporal coding. Therefore, it is examined how well common non-invasive methods for the analysis of brain activity capture processes that rely on rate and temporal codes, respectively.

5.1 Introduction

Psychophysical and neurophysiological evidence indicates that the brain identifies perceptual objects in several consecutive steps. It decomposes the objects into components according to the feature selectivity of sensory neurons, analyzes the relations among the respective components and then represents, in a combined code, the components and their specific relations. This is an efficient strategy to cope with combinatorial complexity. First, it permits unambiguous descriptions of a virtually unlimited number of perceptual objects with a limited set of symbols for components and relations. Secondly, it can be scaled and applied also for the description of complex constellations, i.e. for the infinite variety of contextual configurations in which perceptual objects can occur. Linguistic descriptions follow the same principle. By recombining in ever-changing configurations a rather limited set of symbols for components, properties, and relations, a virtually inexhaustible universe of constellations can be encoded. However, there is an interesting trade-off between the complexity of symbolic descriptors and the syntactic rules required for the definition of relations among those descriptors.

In principle, a sentence describing a complex constellation of components can be substituted by a single symbol that captures the full meaning of the sentence. Implementation of such comprehensive symbols may be advantageous for the description of particularly important or frequently occurring constellations. It would make little sense, however, to substitute all possible sentences by specific symbols, as this would lead to a combinatorial explosion of the required vocabulary. Conversely, if symbols address solely elementary components, only few different symbols are required because the number of elementary components is orders of magnitude smaller than the number of possible configurations among components: a few dozens of different atoms suffice to generate a universe of molecules and four nucleotides suffice to spell out the alphabet of the genome that instructs the development of organisms. However, if symbols exist only for low-level components, descriptions of complex objects may become intolerably long and require highly sophisticated syntactic structures in order to cope with the large number of nested relations that need to be specified. Biological systems can be expected to have evolved optimally adapted compromises in this trade-off between the sophistication of descriptors and the complexity of relation-defining mechanisms, and this compromise is likely to differ for different coding tasks. Variables that are likely to be traded against each other are the costs of individual descriptors, the costs and hardware constraints of relation-defining mechanisms, and the reliability and speed of the respective coding strategies.

If nervous systems have evolved towards such a compromise between coding strategies, one expects to find both neurons that code for very elementary features and neurons that are sensitive to rather complex conjunctions of features. In addition, one expects the implementation of mechanisms that permit a flexible and context-dependent definition of relations among the responses of these neurons.

The numerous investigations of single-cell responses performed over the past decades have provided robust support for the existence of neurons in sensory areas of the cerebral cortex that signal the presence of certain features and some of their conjunctions. The former prevail at early, and the latter at later, stages of the processing streams. However, only few investigations have been devoted to the analysis of mechanisms permitting flexible and context-dependent definitions of relations among the responses of these neurons. Reasons were mainly conceptual. First, the discovery of neurons tuned to highly complex conjunctions of features suggested that relations are encoded mainly by conjunction-specific neurons, which acquire their specificity by recombination of input from neurons responsive to elementary features. This mechanism of serial recombination, if iterated sufficiently often, can, in principle, cope with the definition of all conceivable relations, but, as discussed above, it requires a large number of relation-encoding, conjunction-specific neurons. Another reason why a systematic search for alternative relation-coding mechanisms has not been attempted is the considerable explanatory power of models which assume that most of the information conveyed by a neuron is encoded in the amplitude of its response, i.e. in its discharge rate. Accordingly, it is commonly held that all relevant information is retrievable by recording the event-related responses of single cells. The numerous and excellent correlations between the response amplitudes of individual cells and particular perceptual and motor functions provided ample support for this notion. For these reasons there was little incentive to record from more than one unit at a time or to investigate the possibility that additional information is contained in the temporal relations among the discharges of simultaneously active neurons.

The systematic search for mechanisms that are complementary to conjunction-specific units and that permit the dynamic definition of relations among distributed responses began after the accidental finding that neurons in the cat visual cortex can engage in oscillatory firing patterns and synchronize their discharges in a context-dependent way, with a precision in the millisecond range (Gray and Singer 1987, 1989). This finding indicates that the cerebral cortex can coordinate the temporal fine structure of neuronal discharge sequences, and it suggested the hypothesis that this ability is exploited to define relations among the responses of distributed neurons in a more flexible way than is possible with hard-wired conjunction neurons. In the following, some of the theoretical considerations are reviewed which led to the hypothesis that temporal synchronization could serve as a strategy to define relations among distributed neurons that is complementary to the encoding of relations by conjunction-specific neurons. Subsequently, some of the experiments are discussed that were designed to test predictions derived from the synchronization hypothesis.

5.2 Smart neurons and assemblies

Figure 5.1 illustrates how relations among features can be encoded by conjunction-specific neurons which acquire their selectivity for specific constellations of features through selective recombination of input from lower-order feature detectors. Representing relations among components by conjunction units has two undisputed advantages: it permits rapid processing because it can be realized in feedforward architectures and it is unambiguous because the response of a particular cell always signals the same relation (labeled line coding). However, if not complemented by additional, more dynamic and context-sensitive mechanisms for the definition of relations, this strategy poses problems. First, excessively large numbers of conjunction units are required to cope with the manifold intra- and cross-modal relations between the features of real-world objects. Secondly, it is hard to see how the entirely new relations among the features of novel objects can be represented, as there cannot be an exhaustive

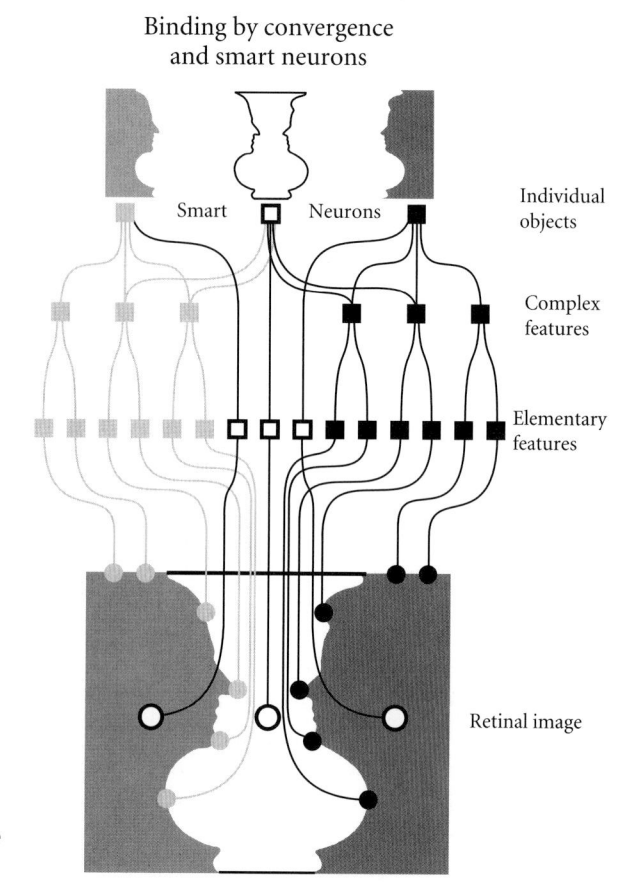

Fig. 5.1 Schematic wiring diagram of a hierarchically organized feedforward network that generates conjunction-specific neurons which respond selectively to different perceptual objects. Note that the neurons representing the faces and the vase, respectively, receive input from partially the same feature-specific neurons.

repertoire of a priori specified conjunction units for all possible feature constellations. Thirdly, unresolved problems arise with the specification of the nested relations that need to be defined to represent composite objects or scenes containing numerous objects. (For a more detailed review of the arguments see Gray 1999; Singer 1999; von der Malsburg 1999.)

These shortcomings can be overcome by assembly coding. As indicated in Fig. 5.2, elementary features and some of their frequently occurring conjunctions are still encoded by individual neurons but this process of serial recombination is not pursued beyond a certain level. Rather, the idiosyncratic constellations of features characterizing individual perceptual objects are represented by the simultaneous responses of cells tuned to frequently occurring feature constellations of intermediate complexity. As such feature constellations are, per definition, common to numerous objects, individual neurons can then contribute at different times to the representation of different objects by forming ensembles with varying partners (Fig. 5.2). This reduces dramatically the number of required conjunction units. It also solves the problem of representing the

Assembly coding

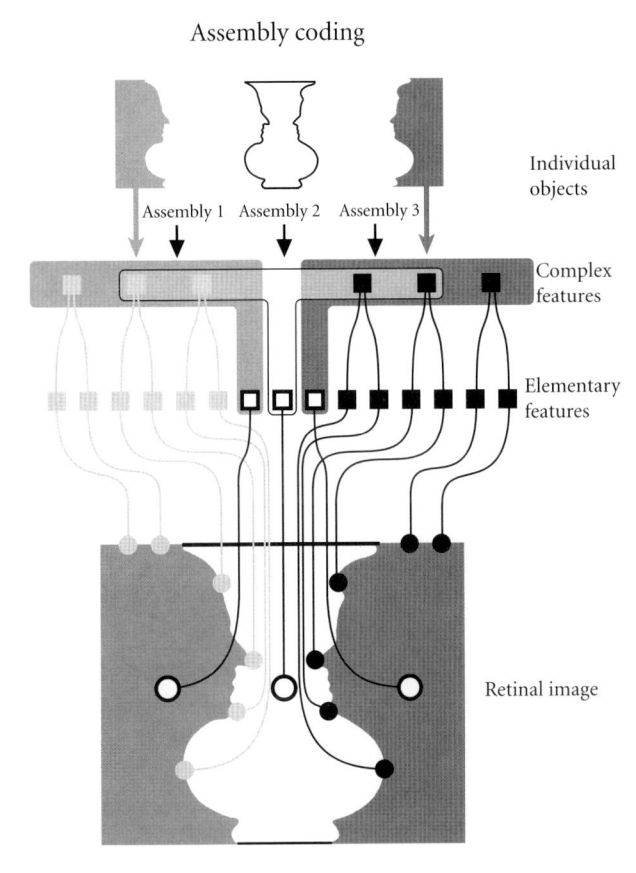

Fig. 5.2 Schematic wiring diagram of neuronal architectures serving the representation of perceptual objects by assemblies. Note that the assembly representing the vase shares neurons with the assemblies representing the faces. To ensure stability of the respective assemblies, additional reciprocal connections among neurons constituting an assembly are required (shaded regions) that bind responses of neurons belonging to the same assembly.

novel relations among the features of unfamiliar objects because cells representing features of intermediate complexity can be grouped dynamically in ever-changing constellations and then represent as an assembly the particular combination of features characteristic for the novel object. However, an essential prerequisite for this coding strategy is a dynamic binding mechanism that can group cells into assemblies and tag their responses as related so that they are recognizable as belonging together by other centers in the brain. Such dynamic binding of neurons into functionally coherent assemblies requires context-sensitive self-organization among large numbers of neurons that have to interact cooperatively with one another both within and across processing levels. Therefore, assembly coding cannot be realized in architectures containing only feedforward connections.

5.3 Synchrony as a code for relations

An unambiguous signature of relatedness is absolutely crucial in assembly coding because, unlike in labeled line codes, the meaning of responses changes with the context in which they are embedded. It needs to be assured that the responses of the neurons

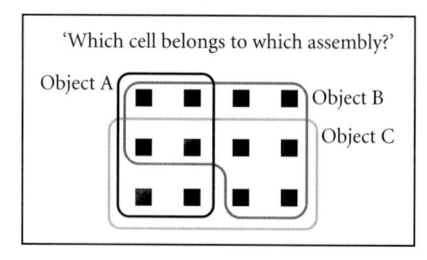

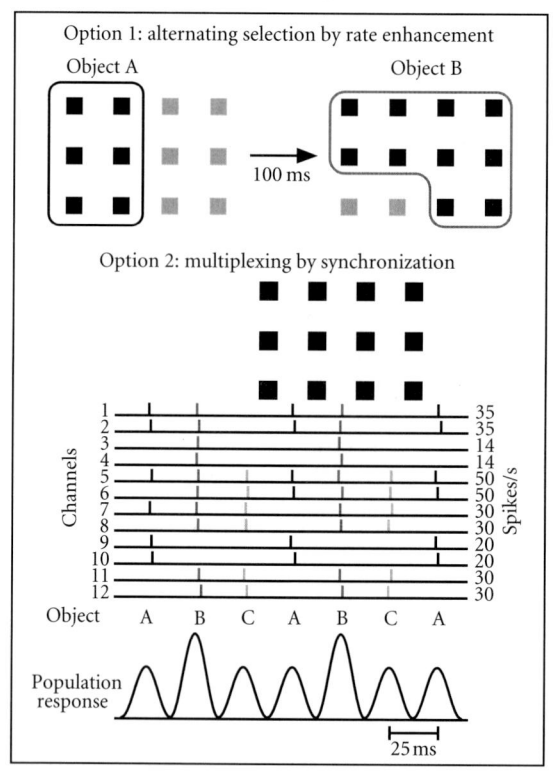

Fig. 5.3 Options for the solution of the superposition problem. Superposition problems arise if perceptual objects are present whose corresponding assemblies share partly the same neurons (upper box). In this case, different assemblies need to be segregated in time to avoid false conjunctions. One option is to raise successively the saliency of responses belonging to the respective assemblies by enhancing the discharge rate of the corresponding responses (lower box, option 1). An alternative solution is to enhance the saliency of responses belonging to a particular assembly by making the discharges of the respective neurons coincident in time (option 2). This permits rapid multiplexing of the different assemblies because coincidence can be evaluated within short time intervals, as it does not require temporal summation. Here it is assumed that the different assemblies alternate at intervals of approximately 25 ms.

that constitute an assembly are processed and evaluated together at subsequent processing stages and are not confounded with other, unrelated responses. The only way to select and group responses for further joint processing is to raise jointly and selectively the relative saliency of the respective responses, and there are three options. First, exclusion of unrelated responses from further processing by inhibition; secondly, increasing the discharge frequency of the selected responses; and, thirdly, synchronizing the discharges of the selected cells with a precision in the millisecond range. All three mechanisms enhance the relative impact of the selected responses and can therefore be used to tag them as related. However, inhibition and rate enhancement cannot be used as binding mechanisms when several assemblies need to be formed at the same time and need to recruit partly the same neurons—a condition that is likely to occur when a scene contains several objects that have subsets of features in common. This problem has been addressed as the superposition problem and can only be resolved by segregating assemblies in time (Fig. 5.3). The options to inhibit all assemblies but one in alternation, or to raise the discharge frequencies of the different assemblies successively, are not optimal because these strategies preclude rapid succession of assemblies. The reason is that evaluation of differences in discharge rates takes time because it requires integration over a sequence of discharges. In this respect labeling responses as related by synchronization is superior because it raises selectively and instantaneously the saliency of coincident discharges. Synchronization can, therefore, define relations with much higher temporal resolution than rate modulation. In principle, assemblies based on coincident firing can be configured on a spike-by-spike basis. If, as suggested by data and summarized in Fig. 5.3, synchronization of discharges occurs on the basis of population oscillations in the gamma frequency range (around 40 Hz), assemblies could alternate at intervals of about 25 ms without getting confounded. Thus, within 100 ms up to four different assemblies could be configured in a given array of neurons, even if these assemblies share a substantial number of neurons.

Another, potentially important advantage of using synchronization as a selection mechanism and synchrony as a tag of relatedness is that relations can be specified independently of firing rate. Discharge rates depend on numerous variables, such as, for example, the physical energy of stimuli or the match between stimulus and receptive field properties, or the context in which stimuli are embedded. In a scene containing multiple, partially overlapping objects whose contours have inhomogeneous contrast because of shadows, amplitudes of responses to contours of the same object may differ substantially. Similarity of response amplitudes is, thus, a poor indicator of relatedness.

Note that this temporal structure can be, but does not have to be, obvious in the discharge sequences of individual neurons (channels 1–12), but that the spike density function of the population response shows an oscillatory modulation in the 40 Hz range. Note also that the constellation of neurons contributing spikes to the oscillatory population response changes from cycle to cycle.

As synchrony can be modulated by temporal regrouping of discharges and, thus, can be varied independently of firing rates, synchronicity and rate can be used as orthogonal codes. Signals indicating the presence and the properties of visual features can thus be kept separate from signals indicating how these features are related.

Another advantage of using synchronization as a tag of relatedness is that synchronized input is transmitted with minimal latency jitter (Abeles 1982; Diesmann *et al.* 1999). Thus, signatures of relatedness can be relayed with great reliability across processing stages, which contributes to reducing the risk of false conjunctions. Finally, synchronization also enhances processing speed by accelerating synaptic transmission *per se* because synchronized excitatory postsynaptic potentials (EPSPs) trigger action potentials with minimal delay.

These features make response synchronization an attractive strategy for the encoding of relations. However, in order to be effective, several additional constraints need to be met:

1. Neuronal networks must be able to modify and to coordinate the temporal patterning of spike trains in a context-dependent manner.
2. The constellations of synchronously discharging neurons must be changeable within a few tens of milliseconds in order to be compatible with processing speed.
3. The temporal signatures of coordinated discharge sequences must be preserved with millisecond precision across processing stages.
4. Neuronal networks must be able to distinguish synchronous from temporally dispersed activity with a precision in the millisecond range, i.e. synchronous input must have a stronger impact on neurons than asynchronous input.
5. Rapidly changing constellations of synchronously firing cells must be recognized as different by subsequent processing stages, i.e. integration intervals for synaptic events must be shorter than the succession rate of assemblies.
6. If relations are encoded by synchrony, then Hebbian learning, i.e. the use-dependent modifications of synaptic gain, must also depend on the synchronicity of pre- and postsynaptic discharge patterns and not only on the mere covariation of rate changes.
7. Finally, the occurrence and structure of synchronous discharge patterns needs to be related in a meaningful way with perceptual or motor processes that require selection and dynamic grouping of responses.

5.4 Experimental evidence for temporal precision in neuronal processing

There is growing evidence that neuronal networks are capable of transmitting temporal patterns with high precision and able to distinguish between coincident (synchronous) and non-coincident (temporally dispersed) input signals. Psychophysical experiments indicate that the visual system is sensitive to stimulus onset asynchronies of less than 10 ms (Leonards *et al.* 1996; Alais *et al.* 1998; Usher and Donnelly 1998; Lee and Blake 1999), supporting the notion that temporally modulated responses can be transmitted over several processing stages with a precision in the millisecond range. Supportive evidence comes from electrophysiological investigations. Cross-correlation analysis between simultaneously recorded responses of retinal ganglion cells, relay neurons in the thalamus and cells of the visual cortex shows that the oscillatory patterning of synchronized retinal responses is reliably transmitted to the cortex (Castelo-Branco *et al.* 1998). Given the high frequency of the retinal oscillations (up to 100 Hz), this implies high temporal fidelity of synaptic transmission over several stages. The well-synchronized cortical responses to flicker stimuli point in the same direction (Rager and Singer 1998). As indicated by the precise temporal modulation of responses in area MT, a higher area of the monkey visual cortex, temporal fidelity of synaptic transmission also holds for cortico-cortical connections (Buracas *et al.* 1998). Simulation studies indicate that such precision is readily obtained with neurons that operate with conventional time constants. The only prerequisite is that transmission occurs in parallel channels that interact through diverging and converging axon collaterals (Diesmann *et al.* 1999). Once neurons at the same processing level have synchronized their discharges, the highly coherent pulse packets are conveyed with minimal dispersion across several synaptic stages, as postulated for synfire chains (Abeles 1991).

Evidence is also available that neuronal networks can shift very rapidly between different synchronization patterns and synchronous states. Simulations with spiking neurons revealed that networks of appropriately coupled integrate and fire units can undergo sudden transitions from uncorrelated to synchronized states and vice versa within a few tens of milliseconds (Gerstner 1996).

The postulate that synchronized activity should have a stronger impact in target structures than temporally dispersed firing is also supported by data. Simultaneous recordings from coupled neuron triplets along thalamo-cortical and intracortical pathways in the visual system have revealed that EPSPs synchronized within intervals below 2 or 3 ms are much more effective than EPSPs dispersed over longer intervals (Alonso *et al.* 1996; Alonso and Martinez 1998; Usrey and Reid 1999). Multielectrode recordings from several sites of the cat visual cortex and retinotopically corresponding loci in the superior colliculus indicated that the impact that a particular group of cortical neurons has on target cells in the colliculus increases substantially whenever the cortical cells synchronize their discharges with other cortical cell groups projecting to the same site in the tectum (Brecht *et al.* 1998). Enhanced saliency of synchronized responses can also be inferred from experiments in amblyopic cats which showed a close correlation between reduced synchrony in primary visual cortex and a loss of responses in higher visual areas (Roelfsema *et al.* 1994; Schröder *et al.* 2002). Similar conclusions are suggested by simulation studies (Niebur and Koch 1994; Bair and Koch 1996) and *in vitro* experiments (Stevens and Zador 1998; Schiller *et al.* 2000).

Further evidence that precise temporal relations among neuronal discharges matter in cortical processing comes from investigations of synaptic plasticity. Varying the temporal relations between presynaptic EPSPs and postsynaptic spike responses in simultaneously recorded coupled cortical cells revealed that long-term potentiation (LTP) results when the EPSP precedes the postsynaptic spike within intervals of 10 ms or less, while the polarity of the modification reverses to long-term depression (LTD) as soon as the EPSP follows the spike (Markram *et al.* 1997). Thus, shifts of a few milliseconds in the timing relations between pre- and postsynaptic discharges suffice to invert the polarity of use-dependent synaptic modifications. This raises the question whether timing relations remain critical when pre- and postsynaptic neurons discharge at high frequencies (up to 40 Hz) and EPSPs are both preceded and followed by postsynaptic spikes at short intervals. Recent results indicate that sensitivity to precise timing relations is preserved when neurons engage in oscillatory activity. Pyramidal cells of rat visual cortex slices were made to discharge tonically at 20 or 40 Hz by injecting sinusoidally modulated current through a patch pipette. Simultaneously, EPSPs were evoked at 20 Hz by electrical stimulation of excitatory afferents. Changing the phase relations between pre- and postsynaptic activity revealed that the stimulated input tended to undergo LTP when the EPSPs were coincident with the spikes, while afferents consistently underwent LTD when the EPSPs fell in the troughs of the membrane potential oscillations. Thus, phase shifts of about 12 ms between individual EPSPs and spikes reversed the polarity of the synaptic modifications (Wespatat *et al.* submitted).

In conclusion, there is converging evidence from different experimental approaches that cortical networks can handle temporal patterns with high precision, and that precise timing relations among the discharges of distributed neurons are computationally relevant. Thus, the major constraints for the use of synchrony as a selection mechanism and a relation-defining code are met. But does the brain exploit this option?

5.5 Properties of response synchronization

Several properties make response synchronization, as it has been observed in the visual cortex, a good candidate mechanism for response selection and the definition of relations:

1. Synchronization is sufficiently precise, in particular when it is associated with β- and γ-oscillations, to raise the saliency of the synchronized responses and to be computationally effective.

2. Joint firing is not simply the result of coherent, stimulus-locked variations in discharge rate, nor is it a trivial reflection of anatomical connectivity, such as shared input. Rather, synchronization is generated by dynamic interactions within the cortical network, and its spatio-temporal patterning depends on a large number of variables: among these are the stimulus configuration (Gray *et al.* 1989; Engel *et al.* 1991c),

the architecture of the intracortical synchronizing connections (Engel *et al.* 1991*a*; Löwel and Singer 1992), the activation state of the cortical network (Munk *et al.* 1996; Herculano-Houzel *et al.* 1999), and attention-dependent effects (Roelfsema *et al.* 1997; Steinmetz *et al.* 2000; Fries *et al.* 2001*b*). These multiple influences endow synchronization with the required context sensitivity.

3. Synchronization can be established across different cortical areas (Engel *et al.* 1991*b*; Roelfsema *et al.* 1997) and even hemispheres (Engel *et al.* 1991*a*). This is required for a relational code.

4. Synchrony has been shown to vary independently of rate changes (Riehle *et al.* 1997; Herculano-Houzel *et al.* 1999) which is advantageous for the encoding of relations (see above).

5. Synchronization is not an all or none phenomenon. When populations of cells engage in synchronous oscillatory activity, individual cells can skip cycles (Buzsaki and Chrobak 1995; Buzsaki 1996) and cells participating in population oscillations of different frequency can engage in partial correlations (Jensen and Lisman 1998). This could be exploited to express graded and nested relations.

6. When cells engage repeatedly in synchronous oscillatory activity, adaptive changes occur in cortical circuits which enhance the probability that the same cells synchronize in subsequent trials (Herculano-Houzel *et al.*, in prep.). This makes it possible, in principle, to store information about relations by changing synchronization probability.

In the following, some of these features of response synchronization will be discussed in more detail, whereby emphasis is laid on synchronization phenomena in the visual system and on relations between synchrony and cognitive processes.

5.6 Synchrony and perceptual grouping

Scene segmentation and perceptual grouping are typical examples of low-level visual functions requiring flexible definition of relations. If internal synchronization of discharges serves to tag responses as related and to assure their joint processing, synchronization probability among neurons in early visual areas should reflect some of the basic Gestalt criteria according to which the visual system groups related features during scene segmentation. Multielectrode studies designed to test this prediction revealed that neurons distributed across different columns within the same or different visual areas, and even across hemispheres, tend to synchronize their responses with close to zero phase lag when activated with a single contour, but fire more independently when stimulated simultaneously with two different contours (Gray *et al.* 1989; Engel *et al.* 1991*a*, *b*, *c*; Freiwald *et al.* 1995; Kreiter and Singer 1996). Further analysis of the dependence of synchrony on receptive field and stimulus configurations confirmed that the probability and strength of response synchronization reflected elementary

Gestalt criteria for perceptual grouping such as continuity, proximity, similarity in the orientation domain, colinearity and common fate (for a review see Singer 1993; Singer et al. 1997; Gray 1999). These early experiments were performed in anesthetized animals but more recent multielectrode recordings from awake cats and monkeys indicate that these synchronization phenomena are not artifacts of anesthesia but are even more pronounced when the animals are awake and attentive (Kreiter and Singer 1992, 1996; Frien et al. 1994; Fries et al. 1997; Gray and Viana Di Prisco 1997; Friedman-Hill et al. 2000; Maldonado et al. 2000). It is noteworthy that these systematic changes in synchronization probability can be, but need not be, associated with systematic stimulus-dependent changes of the neurons' discharge rate. In the light of the abundant evidence for rate enhancement of responses to stimuli that are the target of focused attention, this suggests the possibility that synchronization and response enhancement are complementary strategies for the selection and grouping of responses. This makes sense because synchronization of responses at one level of processing can, by virtue of enhancing the saliency of these responses, cause an increase in firing rate of the corresponding target cells at the subsequent processing stage.

A particularly close correlation between neuronal synchrony and perceptual grouping has recently been observed in experiments with plaid stimuli. These consist of two superimposed moving gratings that differ in orientation and direction of motion and give rise to ambiguous percepts. Such plaid stimuli may be perceived either as two surfaces, one being transparent and sliding on top of the other (component motion), or as a single checkerboard surface that moves in a direction intermediate to the component vectors (pattern motion) (Adelson and Movshon 1982; Stoner et al. 1990; Albright and Stoner 1995). Which percept dominates depends on the way the visual system groups the responses to the two gratings. If all responses are bound together, one perceives pattern motion; if only responses are bound that are evoked by the same grating, one perceives component motion. If this grouping is achieved by selective synchronization, i.e. if the relations between the stimulus components are encoded by the degree of response synchronization, a set of testable predictions can be formulated. One of these is that the respective groups of neurons that are tuned to the different gratings should synchronize their responses if the gratings are perceived as part of a single surface. By contrast, the same groups of cells should not synchronize with one another if the gratings are perceived as two independently moving surfaces. Cross-correlation analysis of responses from cell pairs recorded from various visual areas of lightly anesthetized cats confirmed this prediction, as well as a number of others. Cells synchronized their activity if they responded to contours that are perceived as belonging to the same surface, and these changes in synchronization patterns were not associated with systematic variations in discharge rate (Castelo-Branco et al. 2000). This is strong support for the hypothesis that synchronization serves to encode relations among simultaneous neuronal responses in a context-dependent way.

5.7 Synchrony and perceptual selection

If synchronization serves to raise the saliency of responses, one should expect that it is used as a mechanism for response selection that is complementary to response selection by rate increases.

An involvement of response synchronization in stimulus selection has been documented in experiments on binocular rivalry (Fries *et al.* 1997, 2002). When different stimuli that cannot be fused into a single percept are presented simultaneously to the two eyes perception always alternates between the two eyes. This can be exploited to investigate how neuronal responses to constant stimuli change if they pass from being selected and perceived to being suppressed and excluded from perception, and vice versa (Fig. 5.4). The outcome of these experiments was surprising because the responses in early visual areas (areas 17 and 18 in cat) were not enhanced in amplitude when they supported perception and were not attenuated when they were excluded from supporting perception. However, a close and highly significant correlation existed between changes in the strength of response synchronization and the outcome of rivalry. Cells mediating responses of the eye that won in interocular competition increased the synchronicity of their responses upon presentation of the rivalrous stimulus to the other, losing eye, while the reverse was true for cells driven by the eye that became suppressed. This agrees with rivalry experiments in awake, behaving monkeys which showed no systematic relation between the strength of visual responses and perception in early visual areas but a clear correlation between perceptual suppression and loss of neuronal responses in higher visual areas (Logothetis and Schall 1989; Leopold and Logothetis 1996; Sheinberg and Logothetis 1997). This is what one expects if the saliency of the responses from the two eyes is adjusted at early processing stages by modulating synchronization rather than discharge rates.

5.8 Dependency of synchronization on central states and attention

A characteristic feature of response synchronization is its marked state dependency. Precise spike synchronization and oscillatory patterning of responses in the γ-frequency range are particularly prominent when the cortex is in an activated state, i.e. when the EEG is desynchronized and exhibits high power in the β- and γ-frequency range (Munk *et al.* 1996; Herculano-Houzel *et al.* 1999). Especially, spike synchronization over long distances that requires oscillatory patterning of responses, breaks down completely when the EEG gets 'synchronized' and exhibits high power in the low-frequency range (<10 Hz). This close correlation between the occurrence of response synchronization, on the one hand, and EEG states characteristic for the aroused and performing brain, on the other hand, may be taken as support for a functional role of precise synchrony in cortical processing.

The magnitude and precision of synchronization in the β- and γ-frequency range also varies in the fully aroused brain and is then correlated with fluctuations of attention.

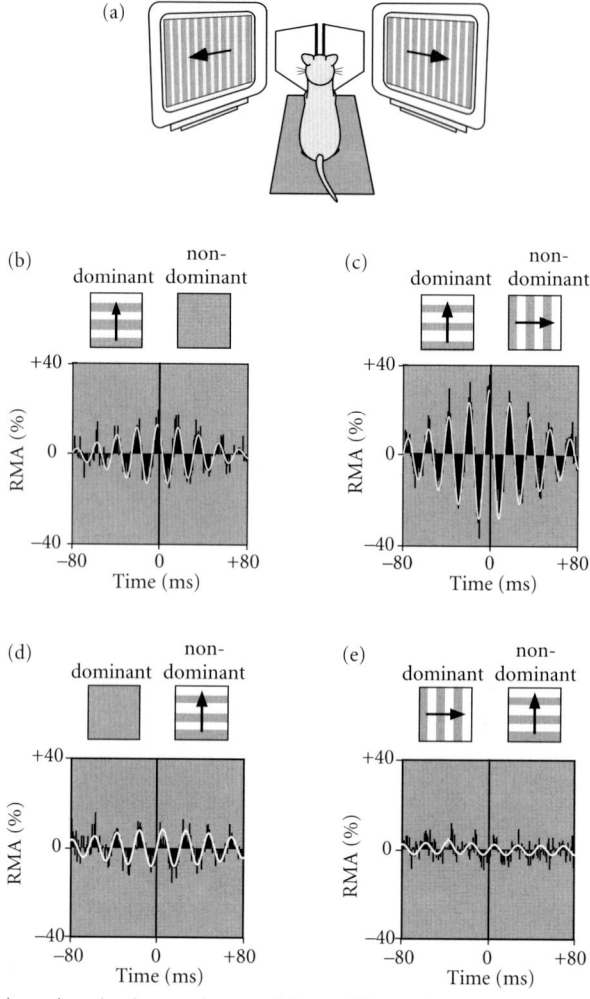

Fig. 5.4 Neuronal synchronization under conditions of binocular rivalry. (a) Using two mirrors, different patterns were presented to the two eyes of strabismic cats. Panels (b)—(e) show normalized cross-correlograms for two pairs of recording sites activated by the eye that won (b, c) and lost (d, e) in interocular competition, respectively. Insets above the correlograms indicate stimulation conditions. Cross-correlograms were computed from pairs of recording sites that responded to the same eye and had direction preferences corresponding to the stimulus shown to that eye. Under monocular stimulation (b), cells driven by the winning eye show a significant correlation which is enhanced after introduction of the rivalrous stimulus to the other eye (c). The reverse is the case for cells driven by the loosing eye (compare conditions d and e). The white continuous line superimposed on the correlograms represents a damped cosine function fitted to the data. RMA, relative modulation amplitude of the center peak in the correlogram, computed as the ratio of peak amplitude over offset of correlogram modulation. This measure reflects the strength of synchrony. (Modified from Fries *et al.* 1997.)

Such modulation of synchrony has been observed in cats trained to perform a visually triggered motor response. The visual, association, somatosensory, and motor areas involved in the execution of the task synchronized their oscillatory activity in the β-frequency range as soon as the animals focused their attention on the relevant stimulus. However, once the reward was available and the animals engaged in consumatory behavior, these coherent patterns collapsed and gave way to low-frequency oscillatory activity that did not exhibit any consistent relations with regard to phase and areal topology (Roelfsema et al. 1997). These results suggest that an attention-related process had imposed a coherent temporal patterning on the activity of cortical areas required for the execution of the task. Such anticipatory enhancement of coherence could facilitate rapid synchronization of responses, both within as well as across areas, once the stimulus appears, thereby accelerating selection and grouping of responses.

Anticipatory synchronization patterns would be particularly effective if they exhibited some topological specificity and reflected the architecture of intracortical connections. In that case, they could serve as a read-out mechanism that translates the grouping criteria defined by intracortical association connections into dynamic patterns of coherent activity against which incoming signals can be matched. Recent evidence supports such a scenario. Measurements of fluctuations in response latency revealed that self-generated oscillatory activity in the γ-frequency range exhibits a specific patterning that reflects intracortical connectivity. Response latencies of neurons in striate cortex fluctuate considerably for identical, repeatedly presented stimuli, but these fluctuations are not random. They are correlated across cortical columns sharing certain functional properties (Fries et al. 2001a) and these correlations are due to synchronous oscillatory activity that causes coherent shifts in response latency. The effect is that responses of cells located in coherently oscillating columns become synchronized right from the beginning due to latency adjustment. Physiological data suggest that these synchronous oscillations are mediated by cortico-cortical connections (Engel et al. 1991a; Löwel and Singer 1992; König et al. 1993), and anatomical evidence indicates that the cortico-cortical association connections link preferentially columns coding for related features that tend to be grouped perceptually (Ts'o and Gilbert 1988; Gilbert and Wiesel 1989; Malach et al. 1993; Schmidt et al. 1997). Thus, spontaneously occurring oscillatory activity could serve to continuously translate the functional architecture of the association connections into coordinated fluctuations of neuronal excitability which, in turn, would lead to fast, context-dependent synchronization of neuronal responses.

Indications on how ongoing oscillatory activity can lead to latency adjustments of responses to incoming EPSP trains are provided by *in vitro* experiments with slices of the visual cortex. In cells subject to an oscillatory modulation of their membrane potential, the latency of action potentials depends crucially on the phase relation between EPSP arrival and oscillation cycle. Because of an *N*-methyl-D-aspartate

(NMDA)-receptor-dependent mechanism, spikes are generated around the depolarizing peak of the oscillations irrespective of the time of EPSP arrival (Volgushev *et al.* 1998). Thus, discharges become phase locked to the membrane potential oscillations and, if these are synchronized among neurons, the discharges of these neurons also become synchronized.

In addition, the ongoing oscillatory patterns are modified by top-down influences from higher cortical areas and by immediately preceding changes of sensory input. Both effects would be equivalent to the functions commonly attributed to attentional mechanisms, the selection and binding of distributed neuronal responses, either as a consequence of bottom-up priming or of intentional top-down selection. The observed fluctuations could thus be equivalent with the system's updated expectancy that is determined by the fixed, locally installed grouping rules, by top-down influences, and preceding sensory input. Seen in this context, ongoing activity assumes the function of a predictor against which incoming activity is matched (Grossberg 1999). One of the effects of matching these predictions with incoming signals is a rapid temporal regrouping of output activity.

Direct evidence for an attention-dependent anticipatory modulation of synchronous γ-oscillations has recently been obtained in area V4 of the monkey visual cortex (Fries *et al.* 2001*b*). While the monkey expected a visual stimulus that needed to be discriminated, neuronal synchronization increased in the γ-frequency range. This increase was specific for the site where the relevant stimulus was expected to appear, and was not associated with a modulation of discharge rate. When the stimulus appeared, responses exhibited better coherence when the stimulus was attended than when attention was directed to another, distracting stimulus. Similar, attention-dependent enhancement of synchrony has been observed among neurons in the somatosensory cortex of monkeys engaged in a tactile discrimination task (Steinmetz *et al.* 2000).

These results suggest that cognitive processes consist not simply of the extraction and recombination of features, but of an active matching operation. Afferent sensory activity is matched against self-generated activity that exhibits an oscillatory time structure and spatially distinct patterns of synchronization (for a detailed review of relations between attention and synchronous oscillations, see Engel *et al.* 2001).

In conclusion, response synchronization appears to serve as a selection and grouping mechanism, both in preattentive bottom-up and attention-dependent top-down processing. In the former case, the dynamic self-organization process leading to the synchronization of responses is determined essentially by the stimulus constellation and the resonance properties of local networks, which are, in turn, specified by the functional architecture of intracortical association connections. In the latter case, these local processes are further influenced by top-down activity that shapes the local dynamics as a function of expectancy. In normal processing, both mechanisms are likely to be inseparably interlinked because of re-entry loops that assure continuous and mutual interactions between bottom-up and top-down effects.

5.9 Response synchronization and short-term memory

Another cognitive function requiring flexible binding of multiple contents is storage in short-term memory (STM). In this case, it is often necessary to remember conjunctions of several items after these have been selected and bound together for storage by an attentional or grouping mechanism. To remember simultaneously the yellow lemon in the right upper corner of the display and the red tomato in the left lower corner requires binding and retention of information about object identity, specific attributes (here color), and location. As these features can occur in a virtually unlimited number of different constellations that might actually never repeat, it would be uneconomical to devote sets of conjunction-specific neurons for the encoding of the specific relations. Accordingly, experiments have been performed in search of an involvement of response synchronization in STM. In a pioneering study, Tallon-Baudry and Bertrand (1999) obtained evidence that retention of visual stimuli in a delayed matching to sample task is associated with an increase of γ-power in the EEG recorded from the scalp. In a subsequent study, Tallon-Baudry et al. (2001) demonstrated, in patients with subdural electrodes, that retention of shape information in STM was associated with enhanced synchronization of local field potential (LFP) oscillations in the β-frequency range between spatially segregated regions in the temporal cortex. In a recent fMRI study, we investigated the effect of increasing the number of items and locations to be retained in STM and found a load-dependent increase of the BOLD signal in prefrontal cortical areas and a load-dependent biphasic modulation of the BOLD signal in parietal cortex (Linden et al., in prep.). In a parallel EEG study, we could demonstrate that these load-dependent increases are associated with enhanced γ-oscillations over parietal and prefrontal regions (Waltz et al. 2001). Synchronization in the γ-frequency range increased at the beginning of the retention period, attained a plateau that was maintained throughout the retention interval and then decreased to baseline levels when the test stimulus appeared. The amplitude of the γ-activity during the plateau phase increased markedly and significantly with increasing memory load. If EEG power increases in a narrow frequency band, this indicates that large assemblies of distributed neurons engage in oscillatory activity in the respective frequency and synchronize their discharges. Thus, the close correlation between changes in narrow-band EEG power and STM load suggests that increasing numbers of neurons are recruited in synchronously oscillating assemblies as memory load increases. These results suggest that precise synchronization of oscillatory discharges in the γ-frequency range plays a role in the selection and binding of distributed responses for storage in STM.

The question remains why, in the experiments reported by Tallon-Baudry et al. (2001), interareal synchronization occurred on the basis of an oscillatory patterning in the β- rather than the γ-frequency range. One possibility is that, in STM tasks, the neurons whose responses need to be bound together are widely distributed across parietal, temporal, and prefrontal regions when different items have to be chunked for storage in STM. In this case, it may not be possible to achieve synchrony in the γ-frequency range because coordination of high-frequency oscillations may be difficult over large distances because of long conduction delays. Support for this possibility comes from a study in which attention- and task-dependent synchronization was investigated in cats performing a visuomotor task (Roelfsema *et al.* 1997). Here, synchronization within visual areas occurred in the γ-frequency range, but synchronization across areas was prominent in the β-range. Another possibility is that oscillation frequencies decrease as the size of synchronously oscillating assemblies increases. Evidence also supports this interpretation. Stimulating the mesencephalic reticular formation (MRF) increases the probability that neurons become recruited into synchronously oscillating assemblies (Herculano-Houzel *et al.* 1999). Such recruitment, which is reflected by enhanced LFP oscillations, is often associated with a decrease in the frequency of γ-range oscillations.

5.10 The detectability of temporal codes with non-invasive methods

The evidence reviewed in this chapter suggests that important operations in cortical information processing may be based on the temporal coordination of firing patterns and, therefore, need not be associated with systematic changes in the firing rate of neurons. This raises the question whether the commonly applied non-invasive methods for the assessment of neuronal activity are sensitive to processes relying primarily on temporal coding. The answer is an unconditional 'yes' for methods that measure the electrical activity of the brain, such as the EEG and the MEG. Actually, these methods are particularly sensitive to changes in the temporal patterning of neuronal activity. As they average over the activity of very large neuron populations, only those activities that exhibit some temporal coherence contribute to the recorded signal. The amplitude of EEG and MEG signals depends on, and increases with, the precision of synchrony and the number of cells engaging in synchronous activity. By contrast, sustained increases in global spiking activity contribute only little to EEG or MEG signals. Such changes in global activation can only be detected in DC-recordings of the EEG, increases in cortical activation leading to a slowly rising, low-amplitude negativity. To which extent the BOLD signal used in fMRI imaging is sensitive to changes in discharge rate and/or temporal patterning is still unclear. Optical recordings of the intrinsic signal, which has a similar origin as the BOLD signal, revealed a close correlation between the discharge rate of neurons and blood oxygenation levels (Malonek and Grinvald 1996), and a direct correlation between BOLD signal amplitude and neuronal firing has recently been established in a seminal study by Logothetis *et al.* (2001). However, in this study, the best correlation between the BOLD signal and electrophysiological variables was found for the amplitude of the LFP fluctuations. The LFP, as the EEG, reflects the average transmembrane currents caused by synaptic events and, to a lesser degree, by action potentials in large populations of neurons. Accordingly, the amplitude of LFP fluctuations depends not only on the number of active elements but, to a crucial extent, also on the synchrony of the contributing electrical events (synaptic potentials and spikes). Thus, the BOLD signal could reflect not only the average discharge rate of neurons in a region, but also the degree of temporal coordination. One possibility is that synchronized activity is metabolically more demanding than temporally dispersed activity of similar amplitude. Synchrony favors synaptic cooperativity and, as a consequence, triggers large and partly active dendritic responses (Schiller *et al.* 2000) that lead to burst firing and Ca^{2+} entry. The latter, in turn, initiates a large number of molecular signaling cascades that upregulate not only housekeeping processes but also control metabolically expensive long-term modifications of synaptic efficacy. If the neuronal signal that triggers vasodilatation were coupled to Ca^{2+} entry rather than to the average number of action potentials, the BOLD signal could indeed capture changes in the temporal patterning of responses, such as synchronization and burst firing. This could account for the finding that cognitive processes that are probably not associated with major increases in firing rate are associated with an enhanced BOLD signal. Such processes are shifts in attention in the absence of sensory stimulation (Kastner *et al.* 1999; Hopfinger *et al.* 2000), conditions of rivalry (Polonsky *et al.* 2000; Tong and Engel 2001), imagery (Goebel *et al.* 1998; Formisano *et al.* 2002), and hallucinations (Dierks *et al.* 1999). In the first two cases, electrophysiological studies have revealed little or no modulation of discharge rates in early visual areas (Fries *et al.* 2001*b*, 2002), but fMRI studies show increases of the BOLD signal. In the latter cases, electrophysiological data are not available, but it is unlikely that spontaneous activity in primary sensory areas undergoes substantial modifications in the absence of sensory stimulation. However, it is likely that all of these cognitive processes are associated with a distinct temporal patterning of activity, because in the absence of major rate changes this is the only way to structure activity.

5.11 Conclusion

Taken together, the data and arguments given in this chapter support the notion that neuronal networks are capable of evaluating, with a precision in the millisecond range, the temporal relations among the discharges of neuronal populations, and that precise timing relations are computationally relevant, both in the context of signal processing and synaptic plasticity. Evidence suggests further that cortical networks exploit this ability, not only to transmit with high temporal fidelity the temporal features of stimuli across processing stages, but also to impose temporal signatures on neuronal activity that can be used in a variety of ways. Frequently observed signatures are an oscillatory patterning and precise synchronization of discharges. Evidence available so far suggests that the oscillatory patterning serves to synchronize responses at variable time scales and that synchronization enhances, with high temporal resolution, the saliency of the synchronized discharges. This, in turn, appears to be used for a variety of different functions, requiring selection of responses for further joint processing. Changes in synchrony correlate both with preattentive switches in stimulus selection—as indicated by the rivalry experiments—as well as with attention-dependent stimulus selection. In the first case, synchronization appears to result from a self-organizing process that is driven essentially by the conflicting sensory signals (bottom-up) and determined by interactions among neurons which encode the conflicting features. In the second case, the synchronization of cell assemblies appears to depend on top-down signals from higher centers, which bias the probability that certain cell groups engage in synchronous activity. Furthermore, synchronization appears to be used to label responses as related in the context of both signal processing and use-dependent synaptic modifications. This relation-defining function results from the fact that synchrony enhances jointly the relative saliency of synchronized responses. It has the characteristics of a binding mechanism that permits rapid and context-dependent definition of relations in ever-changing constellations. It is therefore ideally suited to serve as a selection mechanism in assembly coding that associates distributed responses with one another and assures their joint processing. Assembly coding, in turn, appears necessary in order to cope with the representation of the astronomical number of possible relations that can exist among the features of real-world objects and among the objects constituting visual scenes.

Thus, both theoretical considerations and experimental evidence converge in the conclusion that the cerebral cortex applies, with all likelihood, two complementary strategies in order to encode relations: First, an explicit representation of relations by the responses of conjunction-specific neurons, and secondly, an implicit representation of relations by dynamically associated assemblies that are characterized by the transient synchronization of discharges of the participating neurons. The first strategy seems to be applied for the representation of a limited set of conjunctions and is, with all likelihood, reserved for items that occur very frequently and/or are of particular behavioral importance. The second strategy seems to be reserved for the representation of novel conjunctions and all those relations that do not warrant explicit representation, either because these would require too many neurons or because the contents to be represented are too infrequent to justify the implementation of specialized, and henceforth committed, neurons. Thus, relations among the responses of neurons appear to be encoded in two non-exclusive and complementary ways: by the implementation of hard-wired conjunction-specific neurons and by the generation of dynamically configured cell assemblies that are characterized by their synchronized activity. A similar complementarity appears to exist for the mechanisms of response selection. The saliency of the selected responses is enhanced either by increasing the discharge rate of the neurons or by synchronizing their responses. The first mechanism can, in principle, act on a single cell and operates with low temporal resolution. The second mechanism can select only joint responses, but can do so with high temporal precision. These two selection mechanisms can, of course, be used in parallel as firing rates and the degree of synchrony can be modulated independently. Moreover, synchronization codes can be converted into rate codes because synchronous activity at one processing stage drives neurons at subsequent stages more effectively than asynchronous activity. Thus, if a complex perceptual object needs to be represented, whose components are familiar and encodable by conjunction units but occur in novel relations which are not (yet) covered by conjunction units, both representational strategies have to be exploited simultaneously. The presence of the familiar components is then signaled by the rate-coded responses of the conjunction-specific neurons and the relations between these components are expressed by synchronization.

Space did not allow me to review the growing literature on correlations between cognitive processes and synchronous β- and γ-oscillations in human subjects and in non-mammalian animal species. Nor was it possible to discuss the numerous *in vitro* studies that have contributed to our understanding of the mechanisms responsible for the oscillatory patterning and the synchronization of cortical responses. These aspects are dealt with in the recent reviews by Tallon-Baudry and Bertrand (1999) and by Whittington *et al.* (2000).

References

Abeles, M. (1982). Role of the cortical neuron: integrator or coincidence detector? *Israel Journal of Medical Sciences*, **18**, 83–92.

Abeles, M. (1991). *Corticonics. Neural Circuits of Cerebral Cortex*. Cambridge, Cambridge University Press.

Adelson, E.H. and Movshon, J.A. (1982). Phenomenal coherence of moving visual patterns. *Nature*, **300**, 523–5.

Alais, D., Blake, R., and Lee S.-H. (1998). Visual features that vary together over time group together over space. *Nature Neuroscience*, **1**, 160–4.

Albright, T.D. and Stoner, G.R. (1995). Visual motion perception. *Proceedings of the National Academy of Sciences USA*, **92**, 2433–40.

Alonso, J.-M. and Martinez, L.M. (1998). Functional connectivity between simple cells and complex cells in cat striate cortex. *Nature Neuroscience*, **1**, 395–403.

Alonso, J.-M., Usrey, W.M., and Reid, R.C. (1996). Precisely correlated firing in cells of the lateral geniculate nucleus. *Nature*, **383**, 815–19.

Bair, W. and Koch, C. (1996). Temporal precision of spike trains in extrastriate cortex of the behaving macaque monkey. *Neural Computation*, **8**, 1185–202.

Brecht, M., Singer, W., and Engel, A.K. (1998). Correlation analysis of corticotectal interactions in the cat visual system. *Journal of Neurophysiology*, **79**, 2394–407.

Buracas, G., Zador, A., Deweese, M., and Albright, T. (1998). Efficient discrimination of temporal patterns by motion-sensitive neurons in primate visual cortex. *Neuron*, **20**, 959–69.

Buzsaki, G. (1996). The hippocampo-neocortical dialogue. *Cerebral Cortex*, **6**, 81–92.

Buzsaki, G. and Chrobak, J.J. (1995). Temporal structure in spatially organized neuronal ensembles: a role for interneuronal networks. *Current Opinion in Neurobiology*, **5**, 504–10.

Castelo-Branco, M., Neuenschwander, S., and Singer, W. (1998). Synchronization of visual responses between the cortex, lateral geniculate nucleus, and retina in the anesthetized cat. *Journal of Neuroscience*, **18**, 6395–410.

Castelo-Branco, M., Goebel R., Neuenschwander, S., and Singer, W. (2000). Neural synchrony correlates with surface segregation rules. *Nature*, **405**, 685–9.

Dierks, T., Linden, D.E., Jandl, M., Formisano, E., Goebel, R., Lanfermann, H., and Singer, W. (1999). Activation of Heschl's gyrus during auditory hallucinations. *Neuron*, **22**, 615–21.

Diesmann, M., Gewaltig, M.-O., and Aertsen, A. (1999). Stable propagation of synchronous spiking in cortical neural networks. *Nature*, **402**, 529–33.

Engel, A.K., König, P., Kreiter, A.K., and Singer, W. (1991a). Interhemispheric synchronization of oscillatory neuronal responses in cat visual cortex. *Science*, **252**, 1177–9.

Engel, A.K., Kreiter, A.K., König, P., and Singer, W. (1991b). Synchronization of oscillatory neuronal responses between striate and extrastriate visual cortical areas of the cat. *Proceedings of the National Academy of Sciences USA*, **88**, 6048–52.

Engel, A.K., König, P., and Singer, W. (1991c). Direct physiological evidence for scene segmentation by temporal coding. *Proceedings of the National Academy of Sciences USA*, **88**, 9136–40.

Engel, A.K., Fries, P., and Singer, W. (2001). Dynamic predictions: oscillations and synchrony in top-down processing. *Nature Review Neuroscience*, **2**, 704–16.

Formisano, E., Linden, D.E., Di Salle, F., Trojano, L., Esposito, F., Sack, A.T., Grossi, D., Zanella, F.E., and Goebel, R. (2002). Tracking the mind's image in the brain I: time-resolved fMRI during visuospatial mental imagery. *Neuron*, **35**, 185–94.

Freiwald, W.A., Kreiter, A.K., and Singer, W. (1995). Stimulus dependent intercolumnar synchronization of single unit responses in cat area 17. *Neuroreport*, **6**, 2348–52.

Friedman-Hill, S., Maldonado, P.E., and Gray, C.M. (2000). Dynamics of striate cortical activity in the alert macaque: I. Incidence and stimulus-dependence of gamma-band neuronal oscillations. *Cerebral Cortex*, **10**, 1105–16.

Frien, A., Eckhorn, R., Bauer, R., Woelbern, T., and Kehr, H. (1994). Stimulus-specific fast oscillations at zero phase between visual areas V1 and V2 of awake monkey. *Neuroreport*, **5**, 2273–7.

Fries, P., Roelfsema, P.R., Engel, A.K., König, P., and Singer, W. (1997). Synchronization of oscillatory responses in visual cortex correlates with perception in interocular rivalry. *Proceedings of the National Academy of Sciences USA*, **94**, 12699–704.

Fries, P., Neuenschwander, S., Engel, A.K., Goebel, R., and Singer, W. (2001a). Rapid feature selective neuronal synchronization through correlated latency shifting. *Nature Neuroscience*, **4**, 194–200.

Fries, P., Reynolds, J.H., Rorie, A.E., and Desimone, R. Modulation of oscillatory neuronal synchronization by selective visual attention. (2001b). *Science*, **291**, 1560–3.

Fries, P., Schröder, J.-H., Singer, W., and Engel, A.K. (2002). Oscillatory neuronal synchronization in primary visual cortex as a correlate of perceptual stimulus selection. *Journal Neuroscience*, **22**, 3729–54.

Gerstner, W. (1996). Rapid phase locking in systems of pulse-coupled oscillators with delays. *Physiological Review Letters*, **76**, 1755–8.

Gilbert, C.D. and Wiesel, T.N. (1989). Columnar specificity of intrinsic horizontal and cortico-cortical connections in cat visual cortex. *Journal of Neuroscience*, **9**, 2432–42.

Goebel, R., Khorram-Sefat, D., Muckli, L., Hacker, H., and Singer, W. (1998). The constructive nature of vision: direct evidence from functional magnetic resonance imaging studies of apparent motion and motion imagery. *European Journal of Neuroscience*, **10**, 1563–73.

Gray, C.M. (1999). The temporal correlation hypothesis of visual feature integration: still alive and well. *Neuron*, **24**, 31–47.

Gray C.M. and Singer, W. (1987). Stimulus-dependent neuronal oscillations in the cat visual cortex: a cortical functional unit. *Society of Neuroscience Abstract*, **13**, 404.3.

Gray, C.M. and Singer, W. (1989). Stimulus-specific neuronal oscillations in orientation columns of cat visual cortex. *Proceedings of the National Academy of Sciences USA*, **86**, 1698–1702.

Gray, C.M., König, P., Engel, A.K., and Singer, W. (1989). Oscillatory responses in cat visual cortex exhibit inter-columnar synchronization which reflects global stimulus properties. *Nature*, **338**, 334–7.

Gray, C.M. and Viana Di Prisco, G. (1997). Stimulus-dependent neuronal oscillations and local synchronization in striate cortex of the alert cat. *Journal of Neuroscience*, **17**, 3239–53.

Grossberg, S. (1999). The link between brain learning, attention, and consciousness. *Consciousness and Cognition*, **8**, 1–44.

Herculano-Houzel, S., Munk, M.H.J., Neuenschwander, S., and Singer, W. (1999). Precisely synchronized oscillatory firing patterns require electroencephalographic activation. *Journal of Neuroscience*, **19**, 3992–4010.

Herculano-Houzel, S., Singer, W., and Munk, M.H.J. (in preparation). Use-dependent long-term modification of neuronal synchronization.

Hopfinger, J.B., Buonocore, M.H., and Mangun, G.R. (2000). The neural mechanisms of top-down attentional control. *Nature Neuroscience*, **3**, 284–91.

Jensen, O. and Lisman, J.E. (1998). An oscillatory short-term memory buffer model can account for data on the Sternberg task. *Journal of Neuroscience*, **18**, 10688–99.

Kastner, S., Pinsk, M.A., De Weerd, P., Desimone, R., and Ungerleider, L.G. (1999). Increased activity in human visual cortex during directed attention in the absence of visual stimulation. *Neuron*, **22**, 751–61.

König, P., Engel, A.K., Löwel, S., and Singer, W. (1993). Squint affects synchronization of oscillatory responses in cat visual cortex. *European Journal of Neuroscience*, **5**, 501–8.

Kreiter, A.K. and Singer, W. (1992). Oscillatory neuronal responses in the visual cortex of the awake macaque monkey. *European Journal of Neuroscience*, **4**, 369–75.

Kreiter, A.K. and Singer, W. (1996). Stimulus-dependent synchronization of neuronal responses in the visual cortex of the awake macaque moneky. *Journal of Neuroscience*, **16**, 2381–96.

Larkum, M.E., Zhu, J.J., and Sakmann, B. (1999). A new cellular mechanism for coupling inputs arriving at different cortical layers. *Nature*, **398**, 338–41.

Lee, S.-H. and Blake, R. (1999). Visual form created solely from temporal structure. *Science*, **284**, 1165–8.

Leonards, U., Singer, W., and Fahle, M. (1996). The influence of temporal phase differences on texture segmentation. *Vision Research*, **36**, 2689–97.

Leopold, D.A. and Logothetis, N.K. (1996). Activity changes in early visual cortex reflect monkeys' percepts during binocular rivalry. *Nature*, **379**, 549–53.

Linden, D.E.J., Bittner, R.A., Muckli, L., Waltz, J.A., Kriegeskorte, N., Goebel, R., Singer, W., and Munk, M.H.J. (2003; in press). Cortical capacity constraints for visual working memory: Dissociation of fMRI load effects in a fronto-parietal network. *Neuroscience*.

Logothetis, N.K. and Schall, J.D. (1989). Neuronal correlates of subjective visual perception. *Science*, **245**, 761–3.

Logothetis, N.K., Pauls, J., Augath, M., Trinath, T., and Oeltermann, A. (2001). Neurophysiological investigation of the basis of the fMRI signal. *Nature*, **412**, 150–7.

Löwel, S. and Singer, W. (1992). Selection of intrinsic horizontal connections in the visual cortex by correlated neuronal activity. *Science*, **255**, 209–12.

Malach, R., Amir, Y., Harel, M., and Grinvald, A. (1993). Relationship between intrinsic connections and functional architecture revealed by optical imaging and *in vivo* targeted biocytin injections in primate striate cortex. *Proceedings of the National Academy of Sciences USA*, **90**, 10469–73.

Maldonado, P.E., Friedman-Hill, S.R., and Gray, C.M. (2000). Dynamics of striate cortical activity in the alert macaque: II. Fast time scale synchronization. *Cerebral Cortex*, **10**, 1117–31.

Malonek, D. and Grinvald, A. (1996). Interactions between electrical activity and cortical microcirculation revealed by imaging spectroscopy: implications for functional brain mapping. *Science*, **272**, 551–4.

Markram, H., Lübke, J., Frotscher, M., and Sakmann, B. (1997). Regulation of synaptic efficacy by coincidence of postsynaptic APs and EPSPs. *Science*, **275**, 213–15.

Munk, M.H.J., Roelfsema, P.R., König, P., Engel, A.K., and Singer, W. (1996). Role of reticular activation in the modulation of intracortical synchronization. *Science*, **272**, 271–4.

Niebur, E. and Koch, C. (1994). A model for the neuronal implementation of selective visual attention based on temporal correlation among neurons. *Journal of Computational Neuroscience*, **1**, 141–58.

Polonsky, A., Blake, R., Braun, J., and Heeger, D.J. (2000). Neuronal activity in human primary visual cortex correlates with perception during binocular rivalry. *Nature Neuroscience*, **3**, 1153–9.

Rager, G. and Singer, W. (1998). The response of cat visual cortex to flicker stimuli of variable frequency. *European Journal of Neuroscience*, **10**, 1856–77.

Riehle, A., Grün, S., Diesmann, M., and Aertsen, A. (1997). Spike synchronization and rate modulation differentially involved in motor cortical function. *Science*, **278**, 1950–3.

Roelfsema, P.R., König, P., Engel, A.K., Sireteanu, R., and Singer, W. (1994). Reduced synchronization in the visual cortex of cats with strabismic amblyopia. *European Journal of Neuroscience*, **6**, 1645–55.

Roelfsema, P.R., Engel, A.K., König, P., and Singer, W. (1997). Visuomotor integration is associated with zero time-lag synchronization among cortical areas. *Nature*, **385**, 157–61.

Schiller, J., Major, G., Koester, H.J., and Schiller, Y. (2000). NMDA spikes in basal dendrites of cortical pyramidal neurons. *Nature*, **404**, 285–9.

Schmidt, K.E., Goebel, R., Löwel, S., and Singer, W. (1997). The perceptual grouping criterion of colinearity is reflected by anisotropies of connections in the primary visual cortex. *European Journal of Neuroscience*, **9**, 1083–9.

Schröder, J.-H., Fries, P., Roelfsema, P.R., Singer, W., and Engel, A.K. (2002). Ocular dominance in extrastriate cortex of strabismic amblyopic cats. *Vision Research*, **42**, 29–39.

Sheinberg, D.L. and Logothetis, N.K. (1997). The role of temporal cortical areas in perceptual organization. *Proceedings of the National Academy of Sciences USA*, **94**, 3408–13.

Singer, W. (1993). Synchronization of cortical activity and its putative role in information processing and learning. *Annual Review of Physiology*, **55**, 349–74.

Singer, W. (1999). Neuronal synchrony: a versatile code for the definition of relations? *Neuron*, **24**, 49–65.

Singer, W., Engel, A.K., Kreiter, A.K., Munk, M.H.J., Neuenschwander, S. and Roelfsema, P.R. (1997). Neuronal assemblies: necessity, signature and detectability. *Trends in Cognitive Sciences*, **1**, 252–61.

Steinmetz, P.N., Roy, A., Fitzgerald, P.J., Hsiao, S.S., Johnson, K.O., and Niebur, E. (2000). Attention modulates synchronized neuronal firing in primate somatosensory cortex. *Nature*, **404**, 187–90.

Stevens, C.F. and Zador, A.M. (1998). Input synchrony and the irregular firing of cortical neurons. *Nature Neuroscience*, **1**, 210–17.

Stoner, G.R., Albright, T.D., and Ramachandran, V.S. (1990). Transparency and coherence in human motion perception. *Nature*, **344**, 153–5.

Tallon-Baudry, C. and Bertrand, O. (1999). Oscillatory gamma activity in humans and its role in object representation. *Trends in Cognitive Sciences*, **3**, 151–62.

Tallon-Baudry, C., Bertrand, O., and Fischer, C. (2001). Oscillatory synchrony between human extrastriate areas during visual short-term memory maintenance. *Journal of Neuroscience* **21**, RC 177 (1–5).

Tong, F. and Engel, S.A. (2001). Interocular rivalry revealed in the human cortical blind-spot representation. *Nature*, **411**, 195–9.

Ts'o, D.Y. and Gilbert, C.D. (1988). The organization of chromatic and spatial interactions in the primate striate cortex. *Journal of Neuroscience*, **8**, 1712–27.

Usher, M. and Donnelly, N. (1998). Visual synchrony affects binding and segmentation in perception. *Nature*, **394**, 179–82.

Usrey, W.M. and Reid, R.C. (1999). Synchronous activity in the visual system. *Annual Review of Physiology*, **61**, 435–56.

Volgushev, M., Chistiakova, M., and Singer, W. (1998). Modification of discharge patterns of neocortical neurons by induced oscillations of the membrane potential. *Neuroscience*, **83**, 15–25.

von der Malsburg, C. (1999). The what and why of binding: the modeler's perspective. *Neuron*, **24**, 95–104.

Waltz, J.A., Linden, D.E.J., Prvulovic, D., Singer, W., and Munk, M.H.J. (2001). Joint time-frequency analysis of EEG activity in humans performing a delayed discrimination rask: The effect of short-term memory load. *Journal of Cognitive Neuroscience Supplement*, **13**, 88.

Wespatat, V., Tennigkeit, F., and Singer, W. (Submitted). Phase sensitivity of Hebbian modifications in oscillating cells of rat visual cortex.

Whittington, M.A., Traub, R.D., Kopell, N., Ermentrout, B., and Buhl, E.H. (2000). Inhibition-based rhythms: experimental and mathematical observations on network dynamics. *International Journal of Psychophysiology*, **38**, 315–36.

Chapter 6

Separate modifiability and the search for processing modules

Saul Sternberg

Abstract
One approach to understanding a complex process or system starts with an attempt to divide it into *modules*: parts that are independent in some sense, and functionally distinct. In this chapter I discuss a method for the modular decomposition of neural and mental processes that reflects recent thinking in psychology and cognitive neuroscience. This process-decomposition method, in which the criterion for modularity is *separate modifiability*, is contrasted with task comparison and its associated subtraction method. Four illustrative applications of process decomposition and one of task comparison are based on the event-related potential (ERP), transcranial magnetic stimulation (TMS), and functional magnetic resonance imaging (fMRI).

6.1 Modules and modularity

The first step in one approach to understanding a complex process or system is to attempt to divide it into *modules*: parts that are independent in some sense, and functionally distinct.[1,2] In the present context the complex entity may be a *mental process*, a *neural process*, the *brain* (an anatomical processor), or the *mind* (a functional processor). Four corresponding senses of 'module' are:

Module$_1$: A part of a *mental process*, functionally distinct from other parts, and investigated with behavioral measures, supporting a functional analysis.

Module$_2$: A part of a *neural process*, functionally distinct from other parts, and investigated with brain measures, supporting a neural-process analysis.

Module$_3$: A *neural processor*[3] (part of the brain), a population P_α of neurons that is functionally specialized to implement a particular neural process α. If it is also *localized* (the sole occupant of a delimited brain region) and the only population that implements α, one may find selective task impairment from localized damage of P_α, and selective activation of P_α by tasks that require α.[4]

Module₄: A *mental processor* or *faculty* (part of the mind), functionally specialized ('domain specific'), informationally isolated from (some) other processors ('encapsulated'), and a product of evolution (see Spelke, Chapter 2, this volume).

Most of the present discussion will be concerned with the first two senses. In what follows, I describe and illustrate an approach to the decomposition of mental and neural processes into *Modules₁* and *Modules₂* that reflects recent thinking in psychology and cognitive neuroscience.[5] This *process-decomposition* method and three illustrations based on the event-related potential (ERP) are described in Section 6.2. I contrast process decomposition with the more familiar *task-comparison* method in Section 6.3, describe an example of the latter based on the effects of repetitive transcranial magnetic stimulation (rTMS), and discuss the subtraction method as an embodiment of task comparison. Unlike task comparison, which is often used in a way that requires modularity to be assumed without test (Shallice 1988, Chapter 11; Sternberg 2001, Appendix A.1), the process-decomposition method incorporates such a test. I consider the use of fMRI activation maps in process decomposition for the discovery of *Modules₂* in Section 6.4, list some desiderata in such applications, and present an example. I briefly consider the question of task-general modules (Section 6.5) and the relation between *Modules₂* and *Modules₃* (Section 6.6), and close with a few questions (Section 6.7).

6.2 The process decomposition method

6.2.1 Separate modifiability

Much thinking by psychologists and brain scientists about the decomposition of complex processes appeals implicitly to *separate modifiability* as a criterion for modularity: two (sub)processes **A** and **B** of a complex process (mental or neural) are modules iff each can be changed independently of the other. To demonstrate separate modifiability of **A** and **B**, we must find experimental manipulations (factors) F and G that influence them selectively, i.e. such that **A** is influenced by F but is invariant with respect to G, whereas **B** is influenced by G but is invariant with respect to F.[6] Such double dissociation of subprocesses should be distinguished from the more familiar double dissociation of tasks (Sternberg 2003).

6.2.2 Processes and their measures, pure and composite

How do we demonstrate that a process is influenced by a factor, or invariant with respect to it? We know only about one or more *measures*, M_A, of process **A**, not about the process as such. Depending on the available measures, there are two ways to assess separate modifiability of **A** and **B**. Suppose we have *pure measures* M_A and M_B of the hypothesized modules: a pure measure of a process is one that reflects changes in that process only. Examples include the sensitivity and criterion parameters of signal-detection theory (reflecting sensory and decision processes), and the durations of two different neural processes. To show that *F* and *G* influence **A** and **B** selectively, we must demonstrate their selective influence on M_A and M_B. The influence and invariance requirements are both critical. Unfortunately, it is seldom appreciated that persuasive evidence for invariance cannot depend solely on failure of a significance test of an effect: such a failure could merely reflect variability and low statistical power.[7]

Instead of pure measures, suppose we have a *composite measure* M_{AB} of the hypothesized modules—a measure to which they both contribute. To demonstrate selective influence in this case we must also know or confirm a *combination rule*—a specification of how the contributions of the modules to the measure combine. Examples of composite measures are the ERP at a particular point on the scalp (which may reflect several ERP sources), and mean reaction time, $\overline{RT}$ (which may depend on the durations of several processes). Whereas factorial experiments are desirable for pure measures,[8] they are essential with a composite measure; unfortunately they are rare.

A given measure may be pure or composite, depending on the hypothesized modules of interest. However, rather than being determined a priori, this attribute of a measure is one of the components of a theory that is tested as part of the process-decomposition method.[9]

6.2.3 Three examples of decomposition of neural processes with ERPs

Here I provide brief summaries of three applications of these ideas, in which the brain measures are derived from ERPs.[10]

6.2.3.1 Parallel modules for selecting a response and deciding whether to execute it

Osman *et al.* (1992) investigated a task in which a speeded response was required to two of four equiprobable stimuli. The location of the stimulus (left versus right) indicated the correct response (left hand versus right hand); its category (letter versus digit) determined whether that response should be executed (Go versus NoGo trials). The two factors were the stimulus–response mapping (*SRM*, spatially compatible versus incompatible), and Go–NoGo (letter–digit) discriminability (*GND*, easy versus hard). The two hypothesized pure measures depend on the lateral asymmetry of the motor-cortex voltage versus time for Go and NoGo trials, $A_{MC}(t, Go)$ and $A_{MC}(t, NoGo)$.[11] One measure (M_α) is the time interval from stimulus onset to when the sum $A_{MC}(t, Go) + A_{MC}(t, NoGo)$ exceeds zero, which reflects the duration of α, the hypothesized response-selection module. The other measure (M_β) is the interval from stimulus onset to when the difference $A_{MC}(t, Go) - A_{MC}(t, NoGo)$ exceeds zero, which reflects the duration of β, the hypothesized execution-decision module. They found that M_α is influenced by *SRM* (effect = Δ = 121 ± 17 ms, *n* = 6) but negligibly by *GND* (Δ = 2.5 ± 5.0 ms, *n* = 6),[12] whereas M_β is influenced by *GND* (Δ = 43 ± 14 ms) but negligibly by *SRM* (Δ = 3.3 ± 8.8 ms), evidence for the separate modifiability of α and β.[13] Other aspects of the data indicate that α and β operate in parallel.

6.2.3.2 Serial modules for interpreting a stimulus and initiating the response

Smulders *et al.* (1995) investigated a task in which a digit stimulus indicated which hand had to execute a speeded response. The two factors were stimulus quality (*SQ*, two levels) and response complexity (*RC*, a single keystroke versus a string of three keystrokes). The two hypothesized pure measures were M_α, the duration of process α (from the stimulus to the onset of motor cortex asymmetry), and M_γ, the duration of process γ (from the onset of motor cortex asymmetry to the response). They found that M_α is influenced by *SQ* (Δ = 34 ± 6 ms, *n* = 14) but negligibly by *RC* (Δ = 4 ± 8 ms), whereas M_γ is influenced by *RC* (Δ = 21 ± 7 ms) but negligibly by *SQ* (Δ = 1 ± 8 ms), evidence for two neural processing modules arranged as stages.[14]

6.2.3.3 Two modules in word classification

Kounios (1999, 2002) required subjects to classify each of a sequence of spoken nouns by meaning. Most of the words required no response, while 5% were targets (names of body parts) that called for a manual response. The words consisted of *primes* and *probes*. The factors (two levels each) were the semantic relatedness (*REL*) of the probe to the preceding prime, and the semantic satiation (*SAT*) of that prime (number of immediate repetitions of the prime before the probe). The data were the ERPs elicited by the non-target probes at several locations on the scalp. For present purposes, a composite measure, $M_{\alpha\beta}$, is defined for each location as the mean ERP amplitude at that location during the epoch from 600 to 800 ms after probe onset. Consider the following theory, with three components:

(a) *Subprocesses*: The complex process of recognizing the probe as a non-target contains (at least) two subprocesses, α and β, carried out by different neural processors, P_α and P_β.

(b) *Selective influence*: α is influenced by *SAT* but not *REL*, whereas β is influenced by *REL* but not *SAT*.

(c) *Combination rule*: Each process is an ERP *source*; physics tells us that at any location the combination rule for sources is *summation*.

It can be shown that this theory implies that the effects of *SAT* and *REL* on $M_{\alpha\beta}$ will be additive at all scalp locations.[15] Kounios found such additivity (mean main effects of *REL* and *SAT* were 1.3 ± 0.2 µV and 2.1 ± 0.4 µV, respectively, while the mean interaction contrast was 0.01 ± 0.3 µV, *n* = 36), supporting the above theory and hence the modularity of α and β during the 600 to 800 ms epoch.[16] Also, the topographies of the two effects (their relative sizes across locations) differ markedly, indicating different locations in the brain for P_α and P_β.[17]

6.3 Process decomposition versus task comparison

The cases above exemplify a process-decomposition method, whose goal is to divide the complex process by which a particular task is accomplished into modular subprocesses, a method that has been used to find *Modules*₁ and *Modules*₂. The factor manipulations are not intended to produce 'qualitative' changes in the complex process (such as adding new operations, or replacing one operation by another), which may be associated with a change in the task, just 'quantitative' ones that leave it invariant.[18] The task-comparison method is a more popular approach to understanding the structure of complex processes. Here one determines the influence of factors on performance in different tasks, rather than on different parts of the complex process used to carry out one task. The data pattern of interest is the selective influence of factors on tasks, i.e. the single and double dissociation of tasks. (A classical factor used in brain studies is the amount, usually presence versus absence, of damage in a particular brain region.) Although it may achieve other goals, task comparison is inferior to process decomposition for discovering the modular subprocesses of a complex process: The interpretation of task comparison often requires assuming a theory of the complex process in each task (specification of at least the set of subprocesses) and a theory of their relationship (which subprocesses are identical across tasks); the method includes no test of such assumptions. In contrast, process decomposition requires a theory of only one task, and, as illustrated by the examples above, incorporates a test of that theory.

6.3.1 An example of task comparison: Effects of magnetic brain stimulation

An elegant example of task comparison is provided by Merabet *et al.* (2003) in their experiment on the effects of rTMS on subjective numerical scaling of two tactile perceptual dimensions, based on palpation of a set of tactile dot arrays by the fingers of one hand. The two dimensions were distance (between dots) and roughness. Where rTMS had an effect, it reduced the sensitivity of the obtained scale values to differences among dot arrays. One measure of relative sensitivity is the slope, b, of the linear regression of post-rTMS scale values on non-rTMS scale values. If there were no effect, we would have $b = 1.0$. The data (Fig. 6.1) indicate that performance in the roughness-judgment task is influenced by rTMS of the contralateral somatosensory cortex (rTMS$_s$, Panel A1) but negligibly by rTMS of the contralateral occipital cortex (rTMS$_o$, Panel A2), while performance in the distance–judgment task is influenced by rTMS$_o$ (Panel B2) but negligibly by rTMS$_s$ (Panel B1), a double dissociation of the two tasks.[19] Plausible modular theories of the two tasks might include modules for control of stimulus palpation in each task (α_d, α_r), for generation of a complex percept (β_d, β_r), for extraction of the desired dimension (γ_d, γ_r), and for conversion of its value into a numerical response (δ_d, δ_r). Any or all of these might differ between the tasks. The striking findings indicate that the members of one or more of these pairs of processes depend on different regions of the cortex. A weak pair of task theories might assert that α_d and α_r depend on occipital and somatosensory cortex, respectively,[20] but say nothing about the other processes. A stronger pair of task theories might include the assumptions that α_d and α_r are identical ($\alpha_d = \alpha_r = \alpha$), that β_d and β_r are identical ($\beta_d = \beta_r = \beta$), and that δ_d and δ_r are identical ($\delta_d = \delta_r = \delta$). Given the results we could then

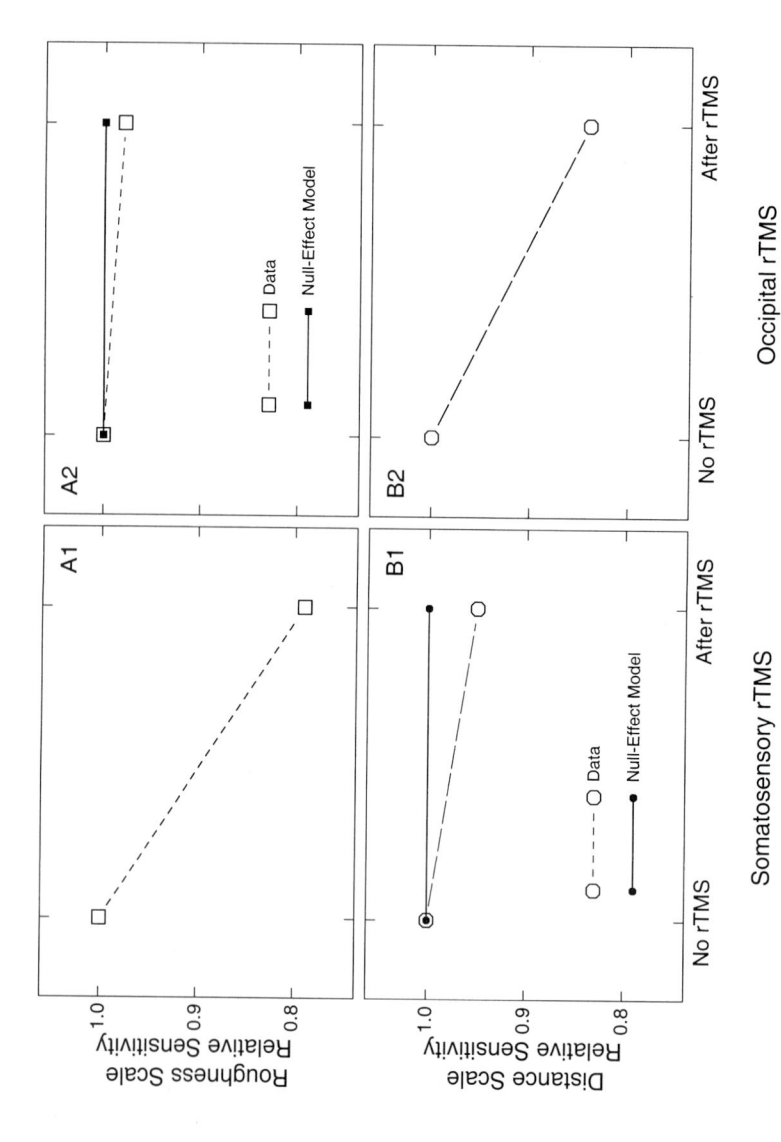

Fig. 6.1 Selective effects on two subjective scaling tasks of repetitive transcranial magnetic stimulation of two brain regions. Mean sensitivity of scale values from 11 subjects relative to their non-rTMS scales are shown for the scaling of roughness (Panels A1, A2) and distance (Panels B1, B2), and for rTMS of somatosensory (rTMS$_s$, Panels A1, B1) and occipital (rTMS$_o$, Panels A2, B2) cortex. Also shown are null-effect models in Panels A2 and B1. Effects on roughness scaling, measured by $1-\overline{b}$, are $1-\overline{b}_{rs} = 0.21 \pm 0.07$ (Panel A1, $p = 0.02$) and $1-\overline{b}_{ro} = 0.02 \pm 0.03$ (Panel A2). Effects on distance scaling are $1-\overline{b}_{do} = 0.16 \pm 0.07$ (Panel B2, $p=0.04$) and $1-\overline{b}_{ds} = 0.05 \pm 0.04$ (Panel B1).

conclude that it is γ_d and γ_r that must be implemented by processors in the different regions. And the results would then also suggest that none of processes α, β, or δ is sensitive to either rTMS_s or rTMS_o, perhaps indicating that they are implemented by processors in neither of the stimulated regions. But unfortunately the findings do not bear on the validity of such hypothesized task theories, weak or strong, or even on the question whether the operations in either task can be decomposed into modular subprocesses.

6.3.2 The subtraction method: Task comparison with a composite measure

One variety of task comparison, devised by Donders (1868) for the RT measure, has also often been used with brain activation measures (e.g. Petersen *et al.* 1988). Suppose we are interested in studying a subprocess β of a complex process. If β were implemented by a localized neural processor P_β in region R_β, then the level of activation of R_β might be a pure measure of the subprocess. However, suppose instead that P_β is not localized (Haxby, Chapter 4 this volume), and what we have is a composite measure that reflects contributions from more than one subprocess. Under these conditions the subtraction method is sometimes used. This method requires three hypotheses. In a simple case they are: *H1* (Task Theory 1): Task 1 is accomplished by process α; *H2* (Task Theory 2): Task 2 is accomplished by α and β; *H3* (Combination Rule): contributions u_α of α and u_β of β to measure $M_{\alpha\beta}$ combine by *summation*. (Possible justifications of this combination rule include, for a brain-activation measure, an assumption that α and β are implemented by different populations of neurons that contribute independently to the measure; and for an RT measure, an assumption that α and β are arranged as stages.) Let the $M_{\alpha\beta}$ measures in the two tasks be M_1 and M_2. The hypotheses imply that M_1 and $M_2 - M_1$ are estimates of u_α and u_β, respectively, and can thus play the roles of pure measures of α and β. But having these measures provides no test of the hypotheses.[21] If summation proves to be incorrect as the combination rule, then other strategies may be available. For example, suppose measured activation were shown to be a decelerating function of the amount of neural activity, in particular, a logarithmic function. Then we would have $M_{\alpha\beta} = \log(u_\alpha + u_\beta)$, and the subtraction method could be applied to the transformed activation measure $M'_{\alpha\beta} = \exp(M_{\alpha\beta}) = u_\alpha + u_\beta$.

6.4 Neural processing modules inferred from activation maps

Modular neural subprocesses can be discovered by applying process decomposition to the kinds of activation measures provided by PET and fMRI. Suppose localization of function, such that two such subprocesses, α and β, are implemented by different processors, P_α and P_β, in non-overlapping regions R_α and R_β. Then activation levels in R_α and R_β are pure measures of α and β, and, with sufficiently precise data and factors that influence the subprocesses selectively, separate modifiability is easy to test.[22] However, if α and β are implemented by different neural processors, P_α and P_β (or by the *same* processor $P_{\alpha\beta}$) in *one* region, $R_{\alpha\beta}$, then the activation level in $R_{\alpha\beta}$ is a composite measure that depends on both α and β, and to test separate modifiability we must know or show how their contributions to the activation measure are combined.[23]

6.4.1 Some desiderata for process decomposition using fMRI

1. The subject should be *performing a task* while measurements are taken. Even in sensory studies enough evidence has emerged favoring task effects at early levels of cortical processing so it is no longer appropriate merely to present stimuli to a passive observer.[24]

2. To increase the likelihood of discovering modules, the subject should be performing the *same* task as factor levels are varied. By 'same task' I mean that a persuasive argument can be made that for all combinations of factor levels, the same set of processing operations is involved, varying only 'quantitatively'.[25]

3. Because the *invariance* of one measure across levels of factor F is at least as important in our inferences as the *influence* of F on another measure, it is critical that we have some index of precision (such as a confidence interval) for the size of an effect when that effect is claimed to be null.[26]

4. Because the effects of factors on activation levels of selected voxels are the quantities of interest, the (mean) activation levels should be reported, rather than only quantities (such as the *t*-values of 'statistical maps') which amalgamate the means and variances of such activation levels.

5. While factorial experiments are not required with pure measures, they are desirable, to assess the generality—hence persuasiveness—of the pattern of effects, as in the example below.

6. Each factor should be studied at more than two levels. The resulting tests of generality protect against being misled by patterns fortuitously associated with particular levels, and the data provide evidence about qualitative task invariance.

7. Especially if process decomposition can be based on behavioral data, such data should be taken concurrently with the fMRI data, to permit comparisons and mutual validation of the two kinds of decomposition, and to investigate the relations among $Modules_1$, $Modules_2$, and $Modules_3$.

6.4.2 An example of process-decomposition using fMRI: Number comparison

Several of these desiderata are satisfied in a study by Pinel *et al.* (2001) that appears to involve pure measures. Subjects had to classify visually displayed numbers, k, as being greater or less than 65. One factor was notation (N), which could be arabic numerals (e.g. '68') or number names (e.g. 'SOIXANTE-HUIT'). The other was numerical proximity (P), $|k − 65|$, with three levels. A similar study (Dehaene 1996) had shown additive effects of N and P on $\overline{RT}$ (a composite measure); this was interpreted to indicate two modular subprocesses arranged as stages: encoding (**E**), influenced by N, which determines the meaning of the stimulus and is slower for number names than numbers, and comparison (**C**), influenced by P, which performs the comparison and is slower for greater (closer) proximities. In the new study, most of the brain regions reported whose activiation is influenced by N or by P are influenced significantly by only one of them, consistent with two separately modifiable neural processes ε and γ that are implemented by separately localized processors.[27] Averaging absolute effect sizes and SEs over the regions of each type, for the nine N-sensitive regions the N effect was $0.17 \pm 0.05\%$ (median p-value $= 0.01$), while the P effect was $0.06 \pm 0.08\%$; for the seven P-sensitive regions the P effect was $0.32 \pm 0.10\%$ (median p-value $= 0.01$), while the N effect was $0.04 \pm 0.04\%$.[28] fMRI data from three well-behaved regions are shown in Fig. 6.2, Panels B, C, and D. The concurrently collected RT data (Fig. 6.2, Panels A) replicated the earlier study, suggesting that we associate the neural modules ($Modules_2$) ε and γ with the functional modules ($Modules_1$) **E** and **C**, respectively; it is important that the functional and neural modules were selectively influenced by the same factors. However, while the direction of the P effect was the same in all the brain regions it influenced, the direction of the N effect was not: the change from numeric to verbal notation increased activation in some regions (e.g. Fig. 6.2, Panels C) and decreased it in others (e.g. Fig. 6.2, Panels D).[29] This is consistent with different neural populations implementing the ε process, depending on the level of N.[30]

6.5 Task-general processing modules

One plausible expectation is that different tasks are accomplished by different subsets of a small set of 'basic' modular processes. To test this expectation, we need a reasonable number of tasks for which persuasively successful decompositions have been achieved.[31] On the other hand, to get adequate data we require subjects to learn a task to a point of stable performance. With such intensive practice, it seems possible that the brain is sufficiently flexible that special-purpose routines would be developed that are specific to that task. Thus, an alternative plausible expectation is that at least some modular subprocesses are task-specific rather than task-general. In that sense, perhaps there is no 'fundamental architecture of the mind', but rather a flexible architect, who has some stylistic tendencies worth studying.

6.6 Processes and processors

What is the relation between $Module_3$ (a localized neural processor in region R that implements a particular process) and $Module_2$ (a modular subprocess of a complex neural process)? To answer this, consider one kind of evidence used to establish a $Module_3$: T_a and T_b are two classes of tasks, such that brain region R_α is activated during T_a, but not during T_b (or is activated *more* during one than the other), and such that we

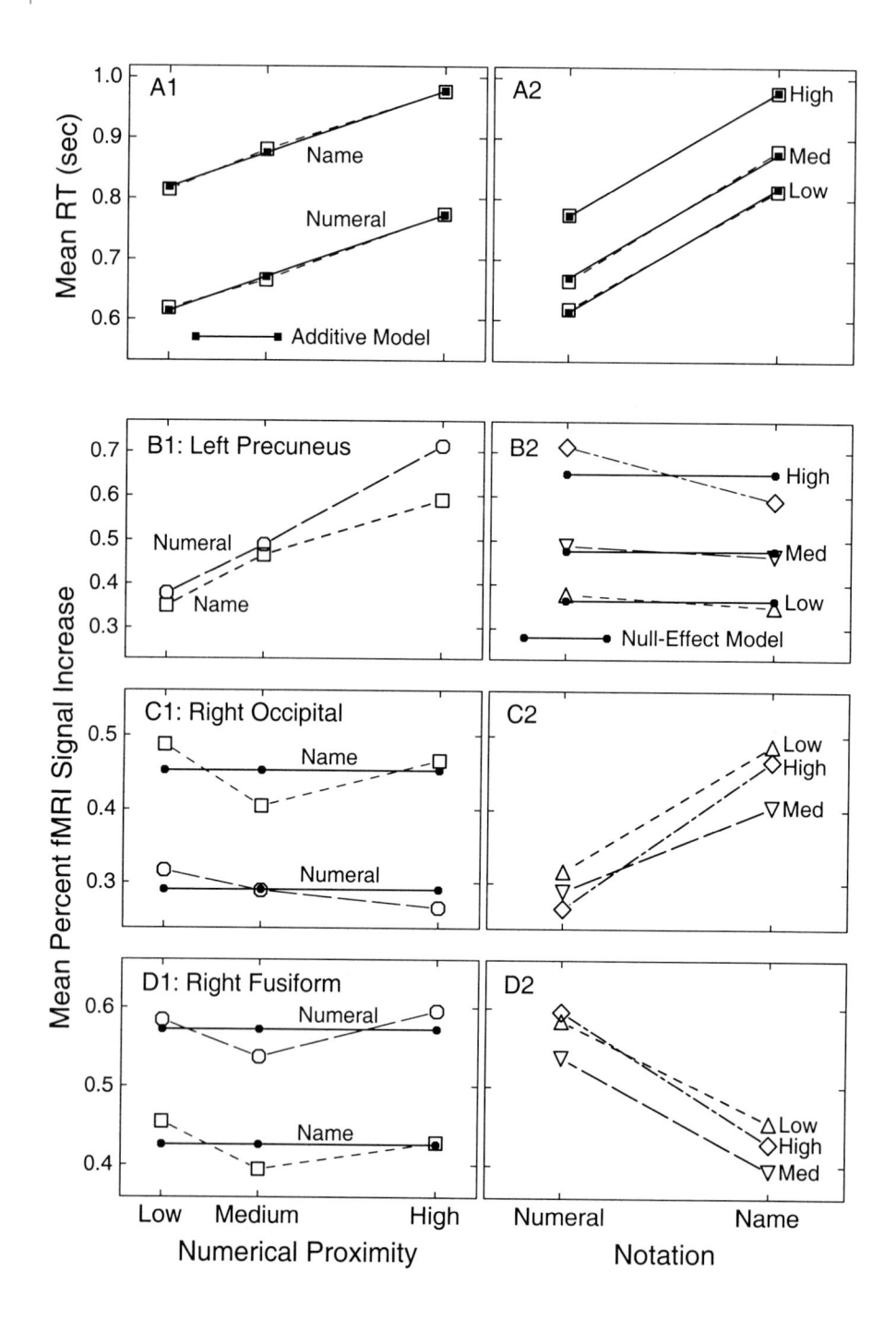

134 | FUNCTIONAL NEUROIMAGING OF VISUAL COGNITION

are willing to assume that all tasks T_a require a particular process, α, to be carried out, whereas none of tasks T_b do. While it may seem plausible, such task-specificity of R_α does not imply that the process α that it implements in a given task is a modular subprocess in the sense of being modifiable separately from other subprocesses in that task.[32]

6.7 Some questions

1. Is separate modifiability too strong or too weak to be a useful criterion for partitioning a process? What are the relative merits of alternative criteria for modularity, and alternative approaches to module identification? Is the weaker differential modifiability[33] more useful than separate modifiability?

2. Consider modular functional processes (*Modules*₁) in a task (supported by behavioral evidence) and modular neural processes (*Modules*₂) in that task (supported by brain measurements). Does either of these imply the other? On which psychophysical–physiological 'linking propositions' (Teller 1984) does the answer to this question depend? It would be helpful to have more studies (such as Pinel *et al.* 2001 and Smulders *et al.* 1995, summarized above) in which both brain and behavioral measures are taken, both directed at process decomposition.[34]

3. Is the encapsulation of Module 4 equivalent to separate modifiability? Given a mental faculty (*Module*₄), must there be a corresponding specialized neural processor (*Module*₃)?

Acknowledgements

I thank Janice Hamer, Nancy Kanwisher, Allen Osman, Seth Roberts, two anonymous reviewers, and especially Teresa Pantzer for comments on the manuscript, and Nancy Kanwisher, David E. Meyer, and Allen Osman for helpful discussions. For providing

---

Fig. 6.2 Reaction-time and selected brain-activation data from Pinel *et al.* (2001). The same data are plotted on the left as functions of *P* (proximity), with *N* (notation) the parameter, and on the right as functions of *N*, with *P* the parameter. Means over subjects of median *RT*s for correct responses are shown in Panels A, with a fitted additive model. The three levels of *P* have been scaled to linearize the main effect of *P* on $\overline{RT}$; this effect, from low to high *P*, is 159±24 ms, while the main effect of *N* is 204±34 ms. SEs are based on variability over the nine subjects. The difference across levels of *N* between the simple effects of *P* from low to high (a measure of interaction) is a negligible 4±20 ms. (The SE may be inflated by unanalyzed condition-order effects.) Activation measures from three sample brain regions are shown in Panels B, C, and D, accompanied by fitted null-effect models in Panels B2, C1, and D1. Shown in Panels B1, C1, and D1, the main effects of *P* (from low to high, using fitted linear functions) are 0.29±0.09% ($p \approx 0.01$), −0.03±0.03%, and 0.00±0.04%, respectively. Shown in Panels B2, C2, and D2, the main effects of *N* are −0.06±0.06%, 0.16±0.05% ($p \approx 0.01$), and −0.15±0.05% ($p \approx 0.02$), respectively.

unpublished details of their data, I thank Stanislaus Dehaene, John Kounios, Lotfi Merabet, Alvaro Pascual-Leone, Philippe Pinel, and Fren Smulders.

Notes

1. Heuristic arguments for the modular organization of complex biological computations have been advanced by Simon (1962) and, in his 'principle of modular design', by Marr (1976).
2. A module may itself be composed of modules.
3. *Processes* occur over time; their arrangement is described by a flow-chart. They are often confused with *processors* (parts of a machine), whose arrangement can be described by a circuit diagram.
4. By 'selective impairment' I mean impairment of tasks that require α and not of tasks that don't. By 'selective activation' I mean activation of P_α by tasks that require α and not by tasks that don't. In practice, *selective* activation is sometimes taken to mean the weaker *differential* activation (Kanwisher *et al.* 2001), akin to the tuning curves for simple features.
5. In Sternberg (2001) I discuss and defend the method, describe its antecedents, illustrate it with a dozen applications to mental and neural processes, and explicate its inferential logic.
6. Separate modifiability of **A** and **B** is also evidence for their functional distinctness (Sternberg 2001, p. 149); information about what a process does is provided by the sets of factors that do and do not influence it.
7. In evaluating the claim that an effect is null, it is important to have at least an index of precision (such as a confidence interval) for the size of the effect. An alternative is to apply an *equivalence test* (Berger and Hsu 1996; Rogers, Howard, and Vessey 1993) that reverses the asymmetry of the standard significance test. In either case we need to specify a critical effect size (depending on what we know and the particular circumstances) such that it is reasonable to treat the observed effect as null if, with high probability, it is less than that critical size.
8. See Sternberg (2001), Appendix A.9.
9. See Sternberg (2001), Sections 2 and 3 and Appendix A.2.3.
10. These examples are treated in detail in Sternberg (2001) in Sec. 6, Appendix A.6, and Sec. 14.
11. A_{MC} is the amplitude difference between the scalp ERPs associated with the parts of the motor cortex that control the left and right hands, taken in the direction that favors the response signaled by the stimulus location; an increase in A_{MC} from baseline is sometimes called the 'lateralized readiness potential'.
12. A cautionary note on the meaning of 'negligible', using this example. With SE = 5.0 ms, 5 df, and a mean of 2.5 ms, a 95% confidence interval based on the *t*-statistic indicates that the true *GND* effect may be as large as 15 ms. Another way of indicating the precision of the data is that a *GND* effect would have to be as large as ±13 ms to be detected, in the sense of differing significantly from zero at the $p = 0.05$ level. See Note 7.
13. SE estimates are based on between-subject variability. However, the SE values provided for the second and third examples are likely to be overestimates because balanced effects, such as those of condition order, were treated as error variance.
14. *SQ* and *RC* also had additive effects on concurrently measured $\overline{RT}$ (a composite measure), consistent with their selectively influencing two functional modules, **A** and **C**, that are arranged as stages (Main effects of *SQ* and *RC* were 34±3 ms and 25±7 ms, respectively; their interaction was a negligible 2±5 ms). Together with the similarity of effect sizes in the neural and behavioral analyses, this suggests that **A** and **C** are implemented by α and β, respectively.

15. In the present context, the effect of a factor on a measure is the change in the measure produced by changing the level of that factor. Letting (i, j) indicate the levels of REL_i and SAT_j, additivity (non-interaction) of the effects of SAT and REL means that $M_{\alpha\beta}(2, 2) - M_{\alpha\beta}(1, 1) = [M_{\alpha\beta}(2, 1) - M_{\alpha\beta}(1, 1)] + [M_{\alpha\beta}(1, 2) - M_{\alpha\beta}(1, 1)]$.

16. Support for the theory is support for all of its three components. However, because the combination rule is given by physics in this application, there is no need to test component (c).

17. In this application, modularity appears to change over time: During an earlier epoch (400 to 600 ms after probe onset) the two effects had similar topographies, but they interacted substantially.

18. Qualitative task changes should be avoided because they reduce the likelihood of discovering modules. Evidence is required to assert qualitative task invariance. One kind of evidence is the pattern of factor effects: for each factor, each change in level should influence the same operations and leave the same other operations invariant. The usefulness of such evidence is one of several reasons for using factors with more than two levels. Unfortunately, few studies (and none of the three examples above) have done so. See Sternberg (2001), Appendices A.2.1 and A.9.2.

19. Subscripts d and r refer to the two tasks; subscripts s and o refer to the two stimulated brain regions. SEs are based on between-subject variability. Also supporting the claim of double dissociation, the differences, $b_{ro} - b_{rs}$ and $b_{ds} - b_{do}$ are significant, with $P = 0.01$ and $P = 0.04$, respectively. However, because non-rTMS measurements were made only before rTMS, rather than being balanced over practice, straightforward interpretation of the slope values requires us to assume negligible effects of practice on those values.

20. Palpation for a distance judgement, but not a roughness judgement, might be associated with covert eye movements.

21. One way to test the set of hypotheses is to extend it by finding two additional tasks that satisfy H4 (Task 3 is accomplished by α and γ) and H5 (Task 4 is accomplished by α, β, and γ) and to extend H3 by including γ. The extended set of hypotheses can then be tested by confirming its prediction that $M_4 - M_3 = M_2 - M_1$.

22. Such tests require no assumptions about whether a change in factor level causes an increase or decrease in activation. This contrasts with the assumption, sometimes used to infer $Modules_3$ (Kanwisher et al. 2001), that stimuli more prototypical of those for which a processor is specialized will produce greater activation.

23. For example, if the combination rule is *summation* (often assumed without test) and if factors F and G influence α and β selectively, then the effects of F and G will be additive. Finding such additivity in a factorial experiment would support the combination rule as well as selective influence. If summation is assumed erroneously, selective influence would be obscured: the effect of each factor would appear to be modulated by the level of the other.

24. For example, with passive observing, different stimuli may attract attention differentially, which could influence activation measures.

25. See Note 18.

26. See Note 7 and Sternberg (2001), Sec. 1.5 and Appendix A.11.2. The same desideratum applies when an interaction (the modulation by one factor of the effect of another) is claimed to be null.

27. The way in which regions were selected my have contributed to this finding: regions found to be sensitive to P appear to have been selected only if the effect of P was not modulated by N. However, among the nine N-sensitive regions, the $N \times P$ interaction was significant in two, and the P effect significant in a third, in my analyses. How to interpret the coexistence of some regions showing selective influence with others showing joint influence (especially in combination with such persuasive RT data) is an important unresolved issue.

28. Effects are on the peak response over time in an event-related design (Pinel *et al.* 2001) for the 'best' voxel in each region, measured as a percentage of the intertrial activation level. (The mean peak response for the 16 voxels was 0.28%.). The 'best' voxel in an $N(P)$-sensitive region is the one whose $N(P)$ effect is most significant. SEs are based on between-subject variability over the nine subjects.
29. Two difficulties are created when an effect can have either sign, as in this case: (a) there may be voxels within which there is a mixture of effects in two directions, such that the effect appears to be null; and (b) even if such cancellation can be assumed not to occur, claims of null effects are harder to support statistically.
30. Without requiring it, this finding invites us to consider that there are two qualitatively different encoding processes, $\mathbf{E}_v$ and $\mathbf{E}_n$, one for each notation, rather than 'one' process whose settings depend on N. If so, we have a case where a change in the level of a factor (here, N) induces a task change (one operation replaced by another; see Section 6.3), but evidence for modularity emerges nonetheless. Whereas activation data (multidimensional) from such a simple (two-factor) experiment can support a claim of operations replacement, based on the idea that the processes implemented by different processors are probably different, RT data (unidimensional) that might support such a claim would require a more complicated experiment.
31. For speeded tasks in which processes are arranged as stages, Sanders (1998, Chapter 3) has amassed some suggestive evidence for a small set of functional modules.
32. Suppose, for example, that α provides a motivational or attentional resource that is required by one or more other processes γ that differ across tasks T_a. A change in α would then induce a change in γ, so they would not be separately modifiable.
33. If differential modifiability obtains, one can find factors F and G such that both factors influence both processes **A** and **B**, but for **A** (**B**) the effect of F (G) is the larger.
34. One starting point would be to take cases where behavioral data already exist that persuasively favor a modular decomposition, as was done by Pinel *et al.* (2001), and ask whether there is a corresponding decomposition based on brain data into modular neural processes that are influenced by the same factors, and invariant with respect to the same other factors.

References

Berger, R.L. and Hsu, J.C. (1996). Bioequivalence trials, intersection-union tests and equivalence confidence sets. *Statistical Science* **11**, 283–319.

Dehaene, S. (1996). The organization of brain activations in number comparison: Event-related potentials and the additive-factors method. *Journal of Cognitive Neuroscience* **8**, 47–68.

Donders, F.C. (1868). Over de snelheid van psychische processen [On the speed of mental processes]. *Onderzoekingen gedaan in het Physiologisch Laboratorium der Utrechtsche Hoogeschool, 1868–1869,* Tweede reeks, II, 92–120. Transl. by W.G.Koster (1969). In W.G. Koster (Ed.), *Attention and performance II. Acta Psychologica* **30**, 412–31.

Kanwisher, N., Downing, P., Epstein, R., and Kourtzi, Z. (2001). Functional neuroimaging of visual recognition. In R. Cabeza and A. Kingstone (Eds.), *Handbook of functional neuroimaging of cognition* pp. 109–51. MIT Press, Cambridge MA.

Kounios, J. (1999). The additive-amplitude method—Neural modules revealed by ERPs. *International Journal of Psychophysiology* **33**, 42–3.

Kounios, J. (2002). Functional neurocognitive modularity revealed by event-related brain potentials. Manuscript submitted for publication.

Marr, D. (1976). Early processing of visual information. *Philosophical Transactions of the Royal Society of London B,* **275**, 483–524.

Merabet, L., Thut, G., Murray, B., Andrews, J., Hsiao, S., and Pascual-Leone, A. (2003). Feeling by sight or seeing by touch? Manuscript submitted for publication.

Osman, A., Bashore, T.R., Coles, M.G.H., Donchin, E., and Meyer, D.E. (1992). On the transmission of partial information: Inferences from movement-related brain potentials. *Journal of Experimental Psychology: Human Perception and Performance* **18**, 217–32.

Petersen, S.E., Fox, P.T., Posner, M.I., Mintun, M., and Raichle, M.E. (1988). Positron emission tomographic studies of the cortical anatomy of single-word processing. *Nature* **331**, 585–9.

Pinel, P., Dehaene, S., Rivière, D., and LeBihan, D. (2001). Modulation of parietal activation by semantic distance in a number comparison task. *NeuroImage* **14**, 1013–26.

Rogers, J.L., Howard, K.I., and Vessey, J.T. (1993). Using significance tests to evaluate equivalence between two experimental groups. *Psychological Bulletin*, **113**, 553–65.

Sanders, A.F. (1998). *Elements of human performance: Reaction processes and attention in human skill.* Erlbaum, Mahwah NJ.

Shallice, T. (1988). *From neuropsychology to mental structure.* Cambridge University Press, Cambridge.

Simon, H.A. (1962). The architecture of complexity. *Proceedings of the American Philosophical Society* **106**, 467–82.

Smulders, F.T.Y., Kok, A., Kenemans, J.L., and Bashore, T.R. (1995). The temporal selectivity of additive factor effects on the reaction process revealed in ERP component latencies. *Acta Psychologica*, **90**, 97–109.

Sternberg, S. (2001). Separate modifiability, mental modules, and the use of pure and composite measures to reveal them. *Acta Psychologica* **106**, 147–246.

Sternberg, S. (2003). Process decomposition from double dissociation of subprocesses. *Cortex* **39**, 180–2.

Teller, D. (1984). Linking propositions. *Vision Research* **24**, 1233–46.

Part 2

Investigations of the ventral visual pathway: Extracting and representing visual objects

Chapter 7

Approaches to visual recognition

Shimon Ullman

Abstract

Visual recognition is a task that the human brain performs effortlessly and efficiently, in a manner that far surpasses the ability of any artificial system or computational model proposed to date. Recognition can be performed at different levels of specificity, ranging from general classification to the identification of individual objects. For humans, classification is generally easier and more natural than individual identification. In contrast, current artificial systems typically cope significantly better with identification compared with general classification. In this chapter I review and compare leading theoretical approaches to visual recognition, focusing on visual classification. The chapter also discusses some of the main unsolved problems, as well as promising future directions.

7.1 Introduction

Visual recognition is one of the most fascinating problems in the study of vision, and in the study of cognition in general. It is a central aspect of visual perception, since the fast recognition of scene components, such as faces, objects, familiar locations, written text, and the like, are crucial to many of our activities. Recognition is also a prime example of a natural task that the brain performs effortlessly and efficiently, in a manner that surpasses the ability of any artificial system or existing computational model. It is an intriguing observation that in tasks that appear to require significant intelligence and cognitive capacity, such as chess playing, computer systems can compete with the best human experts (albeit by employing different strategies), whereas in natural object recognition and classification, even a three-year-old outperforms current artificial systems.

The contrast is particularly striking in the area of general classification rather than identification. In classification, the task is to assign a viewed object to a general category, such as a face, house, dog, and the like. In identification, the task is to recognize a particular individual, such as an individual face, or a specific make of a car. Both are examples of visual recognition, but identification is performed at a more specific level than classification. For humans, classification is usually easier and more natural than individual identification: it is easier to recognize a car, for example, than a specific make of a car. For current artificial systems and computational models, the opposite is true. Some artificial systems can perform relatively well in the case of identifying individual objects under well-defined conditions. However, general classification remains an elusive task (Ullman 1996). These differences in performance between the human visual system and current modeling efforts suggest that a better understanding of visual recognition is likely to provide new insights on the mechanisms and computations used by the brain, and what separates them from current systems and models.

In this paper I will describe a number of leading approaches to visual recognition, with emphasis on classification, and discuss several basic issues in the study of classification. The first section discusses the main difficulties in performing visual classification. The next several sections describe and compare approaches to classification, including feature spaces, structural descriptions, neural network models, and fragment-based classification. I will then turn to discuss briefly the use of functional brain imaging in the study of classification, both past contributions and potential directions for further studies. Finally, I will discuss a number of basic open issues for future studies, both empirical and theoretical.

7.2 Variability and generalization

The difficulty in performing visual recognition comes primarily from the variability in the images that belong to the same object or object-class. Even a single object, such as a face, can have a large number of significantly different images, depending on such parameters as the viewing angle, illumination conditions, and facial expressions. In the case of classification, the variability is even larger, since it includes the variability in shape between individual objects within the same class.

To perform successful recognition and classification, the visual system must somehow handle this intra-class variability. The simplest conceivable approach to the problem is the use of the so-called 'direct approach' to object recognition. In the direct approach, a novel input image is recognized by a direct image comparison between the input image and a large set of previously stored images of different objects under different viewing conditions. The storage requirements for the direct approach are daunting, but from a biological standpoint methods belonging to this general approach may still appear appealing because they make extensive use of stored patterns combined with simple computations. Mathematical models of neural networks have shown that such networks can form efficient associative memories with a very large storing capacity that depends on the number of elements (neurons) and their interconnections (synapses). With some reasonable bounds on these numbers, the storage capacity becomes huge, sufficient, for example, to store an image every minute over 100 years.

Given this capacity, it may be argued that the direct approach can still offer a viable approach to biological object recognition. Some recognition models using either neural networks or optical devices (Abu-Mostafa and Psaltis 1987) have, in fact, used variations of this direct approach to perform object recognition. However, the main limitation of the direct approach does not arise merely from the reliance on extensive memory, but from its limited ability to generalize to novel conditions. This limited capacity of direct image comparison to generalize was studied quantitatively in the context of face images under changes in viewing angle and illumination direction (Adini *et al.*, 1997). The study used 25 faces, with five images per face, and showed that even under modest changes in illumination direction or viewing angle, recognition by direct images comparison leads to large recognition errors. Over 100 methods for image comparison were evaluated, in all of them recognition errors for novel images of the faces in the data set exceeded 20%, which is much poorer than human performance under the same conditions (less than 3% error). The same was true when the images were first processed by a number of standard image processing methods that are used to reduce the sensitivity of the images to the viewing conditions, such as the use of edge maps, which reduces image variations arising from varying illumination conditions.

The main shortcoming of the direct approach is therefore its ability to generalize to new conditions. Using this approach, we will successfully recognize a novel object only if we have already seen the same object under almost identical conditions in the past. Humans do not share this limitation: they can generalize to new viewing conditions and to new members of a familiar class even when the new image is substantially different from all images of the same object or class seen in the past. It is this ability, to extract the essential aspects and ignore the irrelevant ones, and generalize broadly to new conditions, that characterizes human perception and makes recognition a challenging problem.

In the next several sections I will review some of the main approaches that have been proposed for visual object recognition. The focus is on classification, although some of the methods are potentially applicable to recognition at different levels of specificity. The question of whether, in human vision, different levels of specificity are processed differently is an open one, and will be discussed briefly later.

A large number of different approaches have been proposed in the past to these tasks, and it will not be possible, and not necessarily insightful, to review a long list of different methods. Instead, I will focus on a number of methods that represent different conceptual frameworks, and that have been relatively typical or influential. For each method I will discuss briefly how it approaches the main problem of recognition, namely, generalization to novel conditions, and what are its main advantages and limitations.

7.2.1 Features used for classification

A recognition scheme typically first extracts from the image some primitive constructs, or measurements, that are called 'features', and then proceeds to construct a representation of the object and classify it based on its feature representation. A feature can be defined as a function from images to the real numbers: it can result from any measurement applied to the image, such as the angle between two lines, the length of a contour, the distance between two parts, the response to a gray-level templates, and the like. Sometimes the features have just a binary value, 1 if the feature is found in the image, and 0 if it is absent. As we will see below, a major difference between different recognition methods is in the type of the basic features they extract and use. The definition of a feature is, to some degree, a matter of convenience. In an extreme case one may reduce the entire recognition problem to the detection of a single feature—'1' if the object is present in the image and '0' otherwise. In this case, the complexity of the feature detection task is of course simply equivalent to the entire recognition task. A desirable goal is therefore to find a useful decomposition of the recognition task into a feature extraction stage, where the extraction of each feature is simple compared with the overall task, followed by a decision stage that is simplified by the use of the features, compared with the original recognition task.

The first general approach reviewed below uses simple features, and then focuses on the feature combination and decision stage. Some of the approaches reviewed next devote more effort to the extraction of useful features, in the hope that the final classification stage will be simplified. We will return to this issue of feature versus combination complexity later below.

7.3 Feature spaces and separating functions

If each feature extraction produces a measurement, which can be summarized as a single number, then a set of k different measurements produces a set of k values that can be thought of as a point in some k-dimensional feature space. Each object will be represented by a point in this space, and different objects will generally be mapped into different points. The next stage is to define a rule for distinguishing points that represent different object classes. This classification rule can be visualized as a surface in this high-dimensional space that separates points that belong to different classes. Almost any classification method can be described in these general terms, but the focus of feature space classifiers is on schemes that use a large number of simple features, and the main problem faced by the classifier is in finding an appropriate separating surface.

An extreme example of using multiple simple features is to use the raw intensity values of the input image (or the intensity values following a normalization of the intensity mean and standard deviation). In this case, each feature is the intensity value at a single point, and the number of features equals the number of pixels in the input image. This space is of a very high dimension, but, as described further below, some feature spaces use even higher dimensional spaces. Other examples of simple features used by feature space classifiers include edge, blob, and corner detectors (Mel 1977) and different types of local image measurements, such as wavelets or local Gabor filters (Field 1994). These can be thought of as local receptive fields, similar to simple and complex receptive field units found in the primate primary visual cortex. Classification is then obtained by using the output of multiple units of this type, tuned to different orientations and scales, and covering the entire visual field (e.g. Schneiderman and Kanade 1998).

At the other end of the spectrum are global features that cover full-object views. The most widely used method in this direction is to represent a class of images by their principal components. To apply this method, images are represented as one-dimensional vectors of gray-level values. It is typically crucial in such applications that the images of similar objects within a class will be aligned carefully. Using standard methods of linear algebra, a set of principal components is extracted from the original set of images. New images within the class are then represented by linear combinations of the basic images. The method was applied, for example, to face recognition using a set of basic images called 'eigenfaces' (Turk and Pentland 1990). The features in this approach are class-based rather than generic. Variability within the class is handled by combinations of the basic images, which are combined by gray-level superposition to generate novel class members.

Following the feature extraction stage, the object is described by a vector **v** of k measurements, and the next problem is to assign it to one of the known classes. For simplicity, we can consider a binary classification task where the problem is to decide whether or not the feature vector belongs to a given class, C. In feature space approaches, the main stage is to use training examples in order to derive a classification rule for separating the class from non-class examples. This is typically obtained by using the training examples to construct a function, f, such that the value $f(x)$ when applied to class members will be significantly higher than when applied to non-class images, for example, $f(x)$ will be always positive for x in the class, but negative otherwise.

The simplest case in this framework arises when the two populations (class and non-class) can be separated by a plane (or hyper-plane, in a high-dimensional space), in which case they are said to be 'linearly separable'. This form of classification has been studied extensively in the past, and a number of methods have been devised to find the separating plane between given populations, starting with the so-called Perceptron learning scheme (Rosenblatt 1958) and a number of subsequent modifications, such as the Winnow (Littlestone 1988) scheme. These schemes usually start with some initial hyper-plane, which is then incrementally modified whenever a classification mistake is made. They are guaranteed to eventually converge to a separating plane if such a plane, in fact, exists.

A more recent extension of these separation schemes is the method called Support Vector Machine (SVM, Vapnik 1995). Unlike previous methods, SVM seeks to discover not just any plane separating the classes, but an optimal separating plane. In previous methods, the separating plane between two classes is non-unique and may end up passing close to one class and far from the other. In the SVM method, the plane is placed optimally in the middle between the two classes, thereby maximizing the distance to the classes, and in this manner minimizing the probability of misclassifying future examples. The location of the optimal plane is determined by a subset of the training data, which are the points that are closest to the separating plane; the ones that are further away do not affect the location of the optimal plane. This set of closest points is called the 'support vectors', hence the name of this method. If x_i are the support vectors in the feature space, it turns out that the mathematical form of the SVM solution can be expressed as $f(x) = \Sigma w_i <x_i, x> - b$. The feature vector, x, we wish to classify is multiplied (scalar product) by the support vectors x_i with weights w_i and the sum is compared to a threshold b. If the result is positive, then the feature vector x belongs to the class, otherwise it is a non-class example.

The situation becomes more complex when the classes are not linearly separable in the original feature space. In this case, a natural extension is to consider some more complex and non-linear forms of the separating surface between the classes. An alternative approach is to embed the points of the feature space in a new feature space, of even larger dimensionality, but where the classes become linearly separable.

I will not elaborate on these methods except to note the main idea that is at the basis of a recent approach called 'kernel methods'. It can be noted that in the expression above for SVM classification, the support vectors x_i are used only to compute their scalar product with the input vector x. Suppose that we can obtain linear separability by embedding the feature points within a higher dimensional space and form the classification there. All this embedding and mapping to a new space can be quite complex, but if x_i' and x' are the coordinates in the new feature space, all we really need to know is the products $<x_i', x'>$; we do not really need to know the actual coordinate values in the new space. The kernel method can bypass the difficulties of the embedding in the new space by supplying a function ('kernel') $k(x,y)$, which gives us directly the scalar product of x',y' in the larger space. Not every embedding has such a simplifying kernel, but if it exists, it becomes possible to work efficiently with high-dimensional spaces and seek in them linear separation between classes of interest. Using such methods the dimensionality of the feature space can become even higher than the number of pixels in the raw image, but still remain manageable for subsequent classification (Schölkopf *et al.* 1998). As can be seen from the description above, the feature space methods typically use a large number of simple features, and develop sophisticated ways of determining optimal separating function in high-dimensional spaces based on training examples, which will capture the differences that separate class from non-class images. As we will see below, classification can become difficult if the appropriate features have not been used. Some of the alternative methods therefore place most of the burden on the extraction of the most useful features, rather than the complex combination of many simple ones.

7.4 Object-centered structural representations

One of the best-known and most influential approaches to object classification has been the use of three-dimensional, object-centered representation, such as Marr's theory using generalized cylinders (Marr 1982), and Biederman's approach of Recognition by Components (RBC; Biederman 1985). To deal with the variability in the object's appearance, this approach seeks to construct a representation of the object that is based on its main three-dimensional parts and their spatial relationships, and to anchor this representation in the object itself rather than the viewer's point of view. A table, for example, may be described as a flat surface supported from below by four cylinder-like structures. Such a description will apply to many different tables, of different shapes, orientations, and sizes.

More generally, the structural description approach assumes a fixed set of basic parts that are used to construct all possible objects, and a set of possible relations between these parts. In Marr's recognition theory the parts were generalized cylinders (cylinders with different cross-sections), and the RBC scheme uses as parts a set of so-called geons, that include simple three-dimensional volumes such as cylinders, boxes, cones, and the like. The relations between parts are qualitative spatial relations such as 'above', 'next-to', or 'touching'. These relations are supposed to be primarily 'object centered'. This means that they are determined by the object itself and are independent of the viewing direction. For example, the object's axes of elongation or symmetry can be used to define intrinsic coordinates that can, in turn, define relations such as 'above' in a view-independent manner. An object representation extracted from the image in this manner can be formally expressed as a graph structure, where the nodes represent the constituent parts, and the arcs between nodes describe their spatial relations. Similar graph structures representing different object classes are also stored in recognition memory. During visual recognition, the representation extracted from the image is matched against stored representation by some graph matching computation.

This approach has a number of appealing characteristics. First, it breaks the problem of describing complex objects into the sub-problems of extracting simpler parts and their relations. Secondly, the structural descriptions produced from the image are supposed to capture the essential properties of the structure, while ignoring irrelevant details such as the exact shape of the parts that may reflect intra-class variations. Thirdly, the descriptions are object-centered and, in principle, independent of the viewing conditions. Because of these advantages, the structural description approach enjoyed significant popularity as a theory of shape representation in human as well as computational vision. However, the approach suffers from a number of major drawbacks. First, the assumption that all objects are composed of a small set of simple generic parts is problematic. Secondly, the extraction of the object's three-dimensional parts and their object-centered relations is often impractical in natural images. For example, an animal such as a cat can be approximated to some degree by a collection of boxes and cylinders, but these are difficult to extract, for example, from the image of a sleeping cat. Thirdly, the theory attempts to distinguish between relevant and irrelevant aspects of the shape in a class-independent manner: certain aspects are always ignored and replaced by simplified descriptions of shape and qualitative relations.

But distinctions between natural classes, such as between horses, cows, and donkeys, require some precise details, which are irrelevant for other classes. From the point of view of human vision, many experiments were conducted in an attempt to test the use of structural descriptions for the purpose of recognition. For example, a number of experiments attempted to the study the question of view-invariance in human classification and identification. The results are still inconclusive, however, in my view, due to the limitations mentioned above, the structural description approach as outlined is unlikely to be a major part of the basic classification mechanism in human vision. From a computational vision standpoint, the robust extraction of structural descriptions is difficult, and therefore the method is therefore not used in practice by almost any computer vision recognition system.

7.5 Neural network models

One of the main goals of neural networks and connectionist modeling is to be able to capture some of the cognitive capacities inherent in biological systems. A fascinating possibility explored by such models is that the incorporation into the models of some basic characteristics of biological neuronal networks, such as rich interconnectivity and adaptive synaptic modification, will give rise to an emergent behavior, which can imitate biological perceptual systems. In particular, this will allow the systems to deal with natural inputs marred by inherent variability and noise, to separate essential from irrelevant information, and to perform useful classification into natural categories. Biological neural networks are, however, highly complex and only partially understood systems. All current models of such systems necessarily make substantial simplifications, and can incorporate only selected aspects of the real systems. It is therefore an open research problem whether different types of network models can successfully capture the essential computational properties of the real systems.

One of the earliest and best-known neural models of visual recognition is the use of associative memory models (Willshaw *et al.* 1969; Hopfield, 1982; Amit 1989). The main property of these networks is that they can store a large number of patterns (Pi), and then, given a corrupted version Pk' of one of the stored patterns, Pk, they quickly converge to the original uncorrupted pattern, Pk. The stored patterns become 'attractors' for nearby patterns, and therefore some versions of these networks are also called 'attractor neural networks'. Extensions of associative memory networks (Elliffe *et al.* 2000) grouped together multiple views of the same object within a single attractor, so that the different views have a shared representation within the network. Associative memory networks can store a large number of previously seen patterns, but their generalization capacity remains limited. For an associative memory that stored, for example, different car views, a noisy version of a stored image and an image of a novel car are both considered 'corrupted' inputs. They are treated in the same manner, and can be correctly recognized provided that the discrepancy between the novel input and one of the stored representations is not too large. A good recognition system would treat these cases differently, and would tolerate large deviations from the stored images as long as the changes conform to some rules that govern shape changes within the class (Adini *et al.* 1997).

The most widely used neural network model for visual recognition tasks has been the back-propagation model. Back-propagation networks have been used for different classification tasks, including faces, cars, ships, and medical imagery. As the network is trained on examples, the net's hidden layers learn to extract features for the task, and the features are combined in the output layer to produce the final classification. A large-scale effort to use back-propagation for classification is the handwritten character recognition model developed at AT&T (LeCun *et al.* 1989, 1995). A number of different classification networks were developed as a part of this effort, and they served as a test bed to experiment with different extensions to the basic back-propagation scheme, such as weight-sharing and tangent-propagation.

The network models based on back-propagation showed a number of useful properties for the purpose of classification. The intermediate layers of the net sometimes discovered useful and meaningful features for the classification task. The networks often showed substantial initial improvement in performance, based on a straightforward training procedure that does not require explicit instruction.

However, in my view, networks of this type have not provided a satisfactory model for natural object classification. Over the long run, the performance of such networks is typically limited, and increased training is insufficient to bring them to human level performance. In the handwritten character recognition example, extensive training and extensions to the basic network led to some improvements in performance, but the final performance was similar to that achieved by alternative schemes, and still substantially below human performance, even in this limited domain. In classifying natural classes, the gap between human performance and the demonstrated performance of artificial neural networks is wider. The hope from these network models was that human-like capacity for generalization and classification would emerge spontaneously from the collective behavior of the system, by incorporating a number of basic characteristics of the cortical networks; in particular, a high degree of connectivity, neuronal-type computation (modeled as a linear threshold function) and some version of Hebbian synaptic modification. It appears, however, that these general characteristics, as expressed in current neural network models, do not yet capture the spontaneous generalization capacity that characterizes biological systems. However, the search for better network models remains an important research direction, in an attempt to better model the mechanisms used by the human visual system.

7.6 Fragment-based classification

In this section I will describe an approach to visual classification based on representing shapes within a class by a combination of shared substructures called fragments. The fragments are taken from example views and used as building blocks to represent full object views within a given class of shapes.

The use of the combination of image fragments to deal with intra-class variability is based on the notion that images of different objects within a class have a particular structural similarity—they can be represented by combinations of common substructures. Roughly speaking, the idea is to approximate a new image of a face, say, by a combination of images of partial regions, such as eyes, hairline, and the like, of previously seen faces. Unlike approaches that use generic parts to represent different classes, such as RBC, the fragments are class-based features. For each class, the appropriate building blocks are discovered, and then used to distinguish the objects within the class from objects in different classes.

This discussion of this classification method will be somewhat more detailed than the previous methods because it will be used to raise several general issues related to feature selection and classification. In the next section I will discuss the problem of selecting a set of fragments that are best suited for representing a class of related objects, given a set of example images. I will then illustrate the use of these fragments to perform classification and deal with the variability in shape between different objects of the same class.

7.6.1 Informative class fragments

Assuming that we want to use common class-based fragments for classification, what object fragments would be the most useful? To distinguish class from non-class objects, the basic features are required to be distinctive. If we consider the class of face images, for example, a fragment, F, is a distinctive feature for the class if it is likely to be found within this class of images, but not in images of non-faces. This can be measured by the likelihood ratio $p(F|C)/p(F|NC)$. The ratio compares the likelihood of the feature F to appear in images from the class C, with its likelihood to appear in non-class images, NC. Fragments with a high likelihood ratio are highly distinctive in the sense that they provide strong evidence for the presence of a face. However, highly distinctive features are not necessarily useful fragments for representing the class of faces. The reason is that a fragment can be highly distinctive, but very rare. For example, a global template depicting a particular individual face is highly distinctive: its presence in the image means that a face is virtually certain to be present in the image. However, the probability of finding this particular fragment in a novel image, and using it for making classification, is low. On the other hand, a simple local feature, such as the contour of an eyebrow, will appear in many different face images, but it will appear in non-face images as well. The most useful features are therefore fragments that are both frequent and distinctive. These two requirements can be combined by measuring the amount of information supplied by the fragment about the class in question. The usefulness of a feature F for classifying a class C can therefore be expressed by the mathematical notion mutual information, defined as:

$$I(C,F) = H(C) - H(C/F) \quad (6.1)$$

In this expression, $I(C,F)$ is the information conveyed by F about the class C, and H denotes entropy, which is a measure of uncertainty, in this case, about the identity of the class in the image. This is a natural measure because it expresses how the uncertainty about the presence of the class C in the image is reduced by the possible presence of the fragment F in the image. The exact form of the measure is not crucial, but it is important that it will combine the effects of both frequency and specificity. The mutual information is defined by the frequency of detecting the fragment within and outside the class C, and these frequencies are measured using training examples of class and non-class images.

To detect a given fragment F in an image, the fragment is compared to the image at all possible locations by using a gray-level similarity measure. If the similarity at any location exceeds a certain threshold θ, then F has been detected in the image—an example of a useful similarity measure in a normalized cross-correlation of the gray-level values, which is insensitive to the overall illumination level of the image. This measure is somewhat sensitive to geometrical distortions, and therefore a number of alternative measures have been explored in different recognition models.

A technical but significant point is that the threshold θ used for detection is not arbitrary, but can be selected in an optimal manner, guided by the goal of extracting informative fragments. The setting of θ determines the frequency of detecting F both within and outside the class: as the value of the threshold increases, fewer instances of the fragment will be detected. Each value of θ will therefore result in a different value of the computed mutual information, and the optimal value for the threshold θ is the one that maximizes the extracted information. In this manner it is possible to select good features for classification, and at the same time determine the optimal parameters for their use.

7.6.2 Selecting a set of fragments

We examined above the information content of an individual feature: if we were to perform classification based on a single feature, then selecting a fragment with the highest information content would be a good choice. We next turn to the question of selecting a larger set of features from class and non-class examples. It is still desirable for the set of features to be informative, but this by itself is insufficient for classification. In terms of information content, the original image, considered as a set of intensity values, contains all the information we can hope to use, and any derived representation can only lose some of this information. However, the original images form a high-dimensional space where the separation of the classes of interest is complex, and cannot be obtained by simple classifiers such as separating planes. We therefore seek to represent the image using a set of features that will still remain highly informative, and at the same time will support an efficient classification procedure. The goal of the selection process is therefore to produce a representation that simplifies the classification task, and at the same time preserves as much as possible of the original information for the purpose of classification. This is obtained by a process that seeks to maximize the amount of information about the class contained by a limited number of informative fragments that are as independent as possible. The section below describes briefly a computational procedure for selecting a set of class-based fragments from training examples based on these considerations. The specific algorithm described is not intended to be a biological model of the selection process, but mainly to provide means for studying the selected features and their capacity to perform natural classification tasks (for a more detailed description of the process, see Ullman *et al.* 2001).

First, a large pool of candidate fragments is generated, and a smaller set of informative fragments is then selected from this pool. Generating the initial pool is straightforward, the process considers a large number of sub-images taken from the training class examples, at different positions, sizes, and resolutions. Fragments are then selected successively. At each stage, the fragment that contributes the largest amount of additional information is added to the set of selected fragments.

The first stage is therefore simple: we measure the information contributed by each fragment F_i individually, and select the fragment F_1 with maximal contribution. The next step is similar, except that now we seek to maximize the amount of additional information, $I(C; F_1, F_i)$, with the first fragment F_1 already given. The procedure continues iteratively: the $n + 1$ feature is selected so as to have a maximal increment in the amount of delivered information with respect to the previous n fragments. The computation that involves a large number of features simultaneously is difficult, and was approximated in practice by a simpler process that takes into account only limited statistical interactions between features (Ullman *et al.* 2001).

Examples of informative fragments selected for two classification tasks are shown in Fig. 7.1. One group of fragments was extracted by classifying images as 'cars' or 'non-cars'. The second group of fragments was extracted for classifying face images as belonging to either a 'Korean' or a 'Westerner' face class.

In general, the most informative features found for different classification tasks were mostly of intermediate complexity, including intermediate size at high resolution and larger size at intermediate resolution. It was also found that the proportion of high- and low-resolution optimal fragments depends on the class: for car classification, for instance, the relative proportion of low-resolution fragments was higher compared with that required for face classification.

7.6.3 Performing classification

The set of fragments extracted from the training images during the learning stage are then used to classify new images. In performing classification, the task is to assign the image to one of a known set of classes, or decide that the image does not depict any known class. For simplicity, we consider a single class, such as a face or a car, and the task is then to decide whether or not the input image belongs to this class. This binary decision can be extended to deal with multiple classes in a straightforward manner.

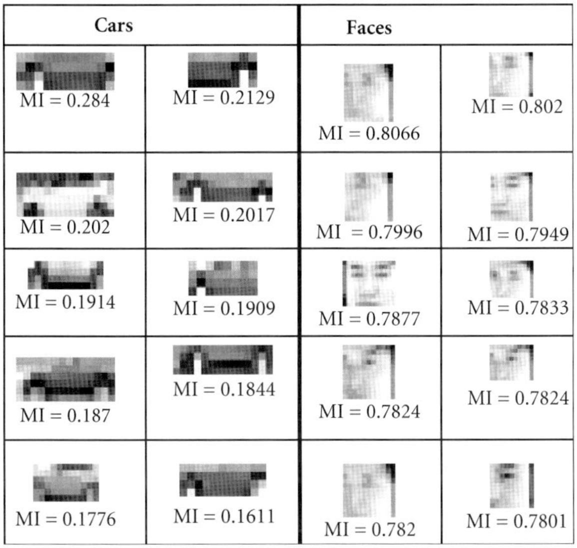

Cars		Faces	
MI = 0.284	MI = 0.2129	MI = 0.8066	MI = 0.802
MI = 0.202	MI = 0.2017	MI = 0.7996	MI = 0.7949
MI = 0.1914	MI = 0.1909	MI = 0.7877	MI = 0.7833
MI = 0.187	MI = 0.1844	MI = 0.7824	MI = 0.7824
MI = 0.1776	MI = 0.1611	MI = 0.782	MI = 0.7801

Fig. 7.1 Classification fragments selected for two classification tasks. On the left, fragments for detecting cars in low-resolution images. On the right, fragments for classifying face images as belonging to either a 'Korean' or 'Westerner' class. Under each fragment, the mutual information (MI) it supplies about the class, measured in bits, is indicated.

The image may contain multiple objects at unknown positions, and consequently the task requires a search over the image. We can view such a classification task also as a detection task, for instance, deciding whether the input image contains a face (one or more), and locating the position of the face if it is detected in the image.

The fragment-based classification requires two main stages: detecting fragments in the image, followed by a decision stage that combines the evidence from the detected fragments. In the first stage, fragments are detected by comparing the image at each location with stored fragment views. The comparison is based on a similarity measure between the fragment and the image, such as the normalized cross-correlation mentioned above. A biological model of the fragment detection stage might be based on populations of cells tuned to the different fragments at different retinal positions and scales, and selecting the maximal response (Riesenhuber and Poggio 1999), but a full network model for this process is yet to be developed.

Following the detection of the fragments in the image, the evidence from the detected fragments is combined to reach a final decision. There are different theoretical options for performing this combination stage. I will describe first a simple and biologically plausible combination scheme, and then comment on more complex alternatives.

Once a fragment F has been detected in the image, the strength of the evidence it supplies concerning the presence of an object of class C can be measured by the likelihood ratio, R, of finding F inside and outside C:

$$R(F) = \frac{P(F|C)}{P(F|\overline{C})} \tag{6.2}$$

This ratio is commonly used in signal detection, and it is an optimal detection criterion for the presence of an object from C given the fragment F. For computational reasons, it turns out to be more convenient to use $w=\log_2(R(F))$, rather that the ratio R itself, as the 'weight' of the fragments. It is worth noting that weight and the mutual information of a feature are two different criteria. A fragment can have, for example, a high weight but still have low mutual information. It will be efficient for the visual system to use features with high mutual information, because then a limited collection of features will be sufficient for good classification, but the features should then be used according to their weights. To decide whether or not the image represents the class in question, the fragments can then be combined by simply adding together the weights of the class fragments detected in the image:

$$\Sigma w_i F_i > \theta \tag{6.3}$$

If the sum exceeds a predetermined threshold θ, then the final decision is that the image contains a face. The value of the threshold determines the trade-off between the two types of mistakes that the classifier can make: missed detections and false alarms. As the value increases, the probability of false alarms decreases, but the probability of missed detections increases.

This simple combination scheme is justified under some conditions; in particular, when the features involved in the decision are conditionally independent. The features are certainly not entirely independent: if eyes are present in the image, the likelihood of a nose-fragment increases. However, within the face-class itself, both eyes and nose are highly likely (and the presence of one does not alter the likelihood of the other), and outside the class both are highly unlikely, and this is essentially what conditional independence means.

It is also interesting to note that the simple combination scheme used in the model above relies on the presence of the individual features, but does not use explicitly the spatial relations between features. This raises the potential problem of confusing a given shape with a shape constructed from the same fragments, but arranged in a different configuration. However, it turns out that the representation using multiple-scale, overlapping fragments, is useful for enforcing the correct overall arrangement of the features. When an image is scrambled and re-arranged, some of the local fragments will survive, but others will inevitably be destroyed. Simulations have shown that the use of overlapping fragments at multiple scales causes correct configurations to be consistently preferred over rearranged patterns (Ullman *et al.* 2002).

7.6.4 Feature complexity and combination complexity

Different models for classification have used more elaborate combination schemes that take into account pairwise or higher statistical interdependencies between the features (Amit *et al.* 1997; Mel 1997; Riesenhuber and Poggio 1999). Simulations using the fragment-based scheme show that the simplest combination scheme was almost as powerful as the more elaborate schemes (Ullman *et al.* 2001). One reason for this is that the basic features used for classification are selected to be highly informative, and consequently their presence already provides most of the relevant evidence for the decision. This can be contrasted with classification schemes that are based on simple local image features, such as short oriented lines, corners, wavelet basis functions, or local texture patches. Such features are generic in nature, that is, common to all visual classes. Consequently, the combination scheme must rely not only on the presence in the image of particular features, but also on their configurations, their spatial relations, and high-order statistical interdependencies between the features.

This points to an essential trade-off in classification between the use of simple visual features that require a complex combination scheme, and the use of more complex features, but with a simpler combination scheme. Most computational approaches to classification use simple features and employ sophisticated combination processes to reach a final decision. The alternative, which appears to me advantageous for biological modeling, is first to spend considerable learning time to extract more complex and highly informative features, specifically selected for each class. It then becomes possible to employ a simple combination scheme, because the features themselves already provide good evidence about the presence of the class in question.

7.6.5 Classification and feature selection

We can see from the above discussion that the selection and use of appropriate visual features is a crucial aspect of recognition and classification. Different schemes discussed above use different types of features, which vary from simple to complex and from generic to class-specific. The computational evidence suggests that the use of simple generic features is inefficient for natural classification. The task of visual classification becomes more accessible by using highly informative intermediate features of moderate complexity (Ullman *et al.* 2002). This also seems to be consistent with the strategy adopted by the visual cortex: classification does not proceed in a single step from V1 to object representation, but goes through a number of intermediate levels with increasing feature complexity. Single-unit studies in the monkey have revealed some properties of these intermediate representations in visual areas V4 (Gallant *et al.* 1993) and inferotemporal (IT) cortex (Fujita *et al.* 1992; Tanaka 1993; Logothetis *et al.* 1995), but the characterization of these features is still tentative. It will be interesting to try to devise ways of exploring questions related to the nature of the features used and the process of their selection with brain-imaging tools.

Computational results also suggest that features based on object fragments and combinations of such fragments provide a rich set of potential features, from which optimal ones can be effectively selected. This can be contrasted with two of the main classification schemes discussed above. Compared with the structural description approach, which uses a limited set of generic three-dimensional primitives, the fragment-based approach uses a much larger set of image-based primitives that are tailored to each class, and that are easier to extract directly from the image. Compared with back-propagation and other network models, such models typically start from randomly selected features and then seek to improve them locally by small changes, and consequently perform a local search in the very large space of all possible shapes. In contrast, the fragment-based scheme performs a more global search in a restricted space of features, composed of combinations of common object parts. The more global search results in more informative features (Ullman *et al.* 2002), probably because the unconstrained search converges to a local optimum that is lower in information than the features obtained by fragment selection. This suggests a general view of feature selection that is partly evolutionary and partly experience-based. As a result of evolutionary processes, the cortex is wired to produce families of potentially useful features, from which the most informative and useful ones are subsequently selected by specific individual experience.

7.7 Functional brain imaging and classification

Human brain imaging, in particular the use of fMRI and PET, has provided useful information in recent years regarding cortical regions involved in object recognition and classification. In this section I will review briefly the main issues addressed by functional imaging studies in relation to recognition and classification, and list some interesting future directions. For fuller recent reviews of studies and findings, see, for example, Kourtzi and Kanwisher (2000) and Malach *et al.* (2002).

In the monkey, especially the macaque, cortical regions involved in object recognition have been studied in detail, using single-unit and lesion studies. These studies revealed a number of cortical areas in the ventral processing stream involved in shape and object processing, including areas TEO and TE in the inferotemporal cortex. The analog areas in human vision involved in object processing were roughly estimated from lesion studies, but were not well delineated. Human brain-imaging studies revealed a number of cortical areas involved in different aspects of object recognition. These regions span a large cortical territory in the occipitotemporal cortex, which includes a number of areas, given different terms by different authors. A region that includes a number of subdivisions is located at the lateral and ventral parts of the occipital lobe, usually termed the lateral occipital cortex (LOC; Malach *et al.* 1995; Kanwisher *et al.* 1997*b*). Within this area, a region has been identified that shows particularly strong responses to face stimuli, termed the fusiform face area (FFA), located in the ventral part of the LOC, in the posterior fusiform gyrus (Kanwisher *et al.* 1997*a*). A neighboring region was shown to be particularly responsive to certain images of buildings and scenes, and was termed the parahippocampal place area (PPA), located somewhat medial to the LOC, in the collateral sulcus and the parahippocampal gyrus (Epstein and Kanwisher 1998).

A number of general properties revealed by fMRI studies implicate these cortical regions with different aspects of object recognition and classification. For example, the LOC region is activated effectively by object images, but not by images of non-object textures (Malach *et al.* 1995). It is activated by objects depicted by different visual cues, it shows substantial position and size invariance, and the activation shows correlation with recognition performance (Grill-Spector *et al.* 1999, 2000).

One of the central issues raised by these studies is the differential activation of object-related regions by different object categories. Notable examples mentioned above are the so-called face and place regions. Other object categories, such as tools, animals, and the human body (Martin *et al.* 1996; Downing *et al.* 2001), were also reported to show specific activation patterns within the object-related regions. The functional interpretation of the category-related differential activation is, however, still unclear, and is the subject of ongoing research. There are two related questions surrounding this issue. One is the degree of modularity in the organization of the classification system that can be inferred from the imaging study. The observed differences in activation patterns could reflect a modular representation of different categories in different cortical sites. Another possibility is that the cortical representations of different categories are broadly distributed and highly overlapping, but the peak activation is somewhat different for different categories (Ishai *et al.* 1999; Haxby *et al.* 2001). A second question has to do with the origin and functional aspects of the category-specific activation. The separation of categories to different cortical locations may be based on differences in the shapes represented within each category, or on functional difference, namely, that different categories are used in different visual tasks. Another possibility that has been raised is that different cortical locations within the object-processing areas are involved with recognition at different levels of specificity. According to this view, the face-related area, for example, may be associated more generally with recognition tasks that require expertise in objects identification based on fine details (Gauthier *et al.* 1999).

The main experimental approach used by most of the fMRI imaging studies of visual recognition relied on the comparison of spatial patterns of cortical activation in different tasks. For example, comparisons of the spatial patterns of elevated activity during face and non-face object recognition were used to identify cortical sites that are involved preferentially in face recognition. Another example along this line is the comparison of mental rotation tasks, with object recognition across different viewing directions (Gauthier *et al.* 2002). The study found significantly different activation patterns for the two tasks, providing evidence that mental rotation is not the main process used for achieving view-invariant recognition. As these examples illustrate, locating the cortical sites associated with different aspects of the recognition process provides useful information beyond the topographic charting of different cortical subregions. Functional localization can be used to break down the recognition process into functional components, and to study how the different components are used in different tasks.

Brain fMRI studies of recognition have not been limited to the comparison of sites of elevated activity. An example of a different experimental approach is the adaptation technique, based on the observation that fMRI signals in object-selective regions are reduced with repetitive presentation. By presenting similar objects, but of different sizes and at different locations, it was possible to study position- and size-invariant responses of subregions of LOC (Grill-Spector *et al.* 1999).

As can be seen from the brief discussion above, brain-imaging studies have already made substantial contributions to central questions in the study of visual recognition. There are many additional fundamental issues in the study of visual recognition and classification to which imaging studies could make fundamental contributions. The next section lists a number of basic issues open for future study.

7.8 Major issues open for study

The study of visual recognition and classification is still at an early stage from both theoretical and biological standpoints. From the theoretical standpoint, the creation of computational schemes for performing efficient and reliable recognition under natural conditions is still a challenging goal. From the biological perspective, we have only a limited understanding of the high-level cortical regions involved in object recognition and the stream of processing that leads to recognition and classification. In this section I will list briefly a number of basic open problems for future empirical and theoretical studies.

7.8.1 Classification and identification

One general issue has to do with the relationships between classification and specific object recognition. Are these two different processes, or just the same process, applied at different levels of specificity? The identification of an individual object can be viewed as a limiting case of classification, where the class consists of a single object. However, from a computational standpoint, methods that deal successfully with object identification tasks usually do not extend well to object classification (Ullman 1996). It is still not clear whether a single scheme, using the same or similar algorithms, can deal successfully with both aspects of the recognition problem. In the brain, classification and identification may be implemented as different processes that use different (though perhaps partially overlapping) machineries. Some evidence from brain lesions and psychophysical studies (e.g. Farah 1990) is consistent with this view. Alternatively, they could use the same sets of neurons, with response properties that sharpen over time, leading to temporal rather than spatial separation of the two processes.

The empirical study of this issue involves a combination of temporal and spatial aspects, and therefore would benefit in the future from the integration of experimental methods that give good temporal resolution (such as MEG and EEG) together with methods that currently give high spatial but poor temporal resolution.

7.8.2 Illumination, three-dimensional pose, and shape variations

A related question is whether the different sources of image variations in recognition are handled by distinct mechanisms. For example, whether changes in pose (three-dimensional viewing direction), illumination, and changes in shape (e.g. for articulated objects) are handled by the same or by a different mechanism. There is some evidence that restricted brain lesions can lead to specific difficulties with compensating for illumination changes (Weizkranz 1990) or dealing with unusual viewing directions (Warrington and James 1986). Brain-imaging studies could be used to provide empirical data related to this problem using the intact system, rather than lesion studies. However, the problem is not only to resolve the cortical sites of these mechanisms, but also to obtain more information about their function, namely, how does the visual system compensate for differences in location, size, pose, illumination, and shape?

7.8.3 View-dependent or independent representation

An ongoing controversy in the study of recognition is whether the representations used by the visual system for the purpose of recognition are view-dependent or -independent. The structural description models derive from the image an object-centered representation that does not depend on the particular viewing direction. Other models, such as view-interpolation methods (Poggio and Edelman 1990), assume that the brain first learns specific object views from different viewing directions, and then handles intermediate orientations by an internal interpolation process. A number of psychophysical studies have explored this issue, with conflicting results (Biederman and Gerhardstein 1993; Tarr *et al.* 1998). It appears that humans can generalize under some conditions to new viewing directions and fail under other conditions, but the interpretation in terms of the representations and processes used to compensate for viewing directions is still not entirely resolved.

7.8.4 Three-dimensional information versus two-dimensional appearance

A related problem is whether the representations used for recognition use three-dimensional object models, or are based on their two-dimensional appearance. The RBC model, for example, uses three-dimensional primitives, whereas other models, such as the Eigenface model, use two-dimensional image information only. The problem of two- versus three-dimensional information is often equated with the problem of view dependence, under the assumption that the use of three-dimensional information implies a view-independent recognition method. However, the two issues are distinct problems. A recognition system could use three-dimensional information about an object as seen from a particular viewing direction; for example, the system could use the relative depth to different object points as useful information for identification from a given viewing direction, but different viewing directions will be associated with different depth maps (Edelman and Bulthoff 1992).

7.8.5 Different object categories, are faces special?

Some of the problems raised above are related to the issue of modularity in visual recognition, namely, whether different aspects of the task are handled by the same or different processes. Another question related to modular organization is whether different object classes, such as animals, man-made objects, or places, are treated by the recognition system in a uniform manner, or by different (though probably related) processes. The best example of a distinct object class that may be handled by some special mechanisms is the class of faces. As mentioned above, there is empirical evidence from lesion and brain-imaging studies for a cortical site specializing in face recognition. Other examples include the representation of places, tools, animals, or the human body. As discussed above, the exact nature of this cortical specialization is still not entirely clear. The main open questions in this area are the degree of modularity in the organization of the classification system, and the origin and functional aspects of the category-specific representations, whether they are related to differences in shape, task, or degree of specificity.

7.8.6 Built-in versus learned invariances

Recognition can be viewed as a problem of dealing with invariances: recognizing an object, or an object class, as the same, despite changes in size, position, viewing direction, illumination, certain shape changes, and the like. Some of these changes, such as position and size, are simpler than others because they are object independent. If we see the image of an object at a particular location and size, we know how the image will look like at a different position and at a different scale. This does not hold for rotation in space, for instance, since the changes induced by three-dimensional rotation depend on the three-dimensional shape of the object and not on the image alone. The image-based transformations, including position, size, and image orientation, can, in principle, be compensated by some built-in object-independent mechanisms that may be wired into the visual system, independent of visual experience. Computational models of recognition typically contain position and size normalization stages that are intended to achieve position and size invariance. Similar normalization processes have also been suggested in biological models of human visual recognition, for example, the dynamical routing model (Olshausen *et al.* 1993). In these models, when an object is learned at one position and at a particular size, it will be recognized automatically at all other locations and scales. Although some studies (Biederman and Cooper 1991) have demonstrated generalization to new locations, others (Nazir and O'Regan 1990; Dill and Fahle 1997) have shown severe limitations in generalizing recognition learned at one position to new locations as close 4° of visual angle.

The alternative view is that the visual system does not contain general built-in mechanisms for position invariance, but acquires such invariance with time and practice with different objects (Foldiak 1991; Parga and Rolls 1998). However, it is clear that we do not have to learn to recognize each new object at every possible location. In the fragment approach discussed above, position invariance to the shape fragments is built into the system by past experience, and shift invariance of complex shapes is then obtained from the invariance of their constituent fragments (Ullman and Soloviev 1999).

On this view, position invariance is acquired rather than built-in, but it still generalizes to new objects based on common fragment representation.

7.8.7 The nature of the features used

The review above of different approaches to classification illustrates that a general and fundamental question in visual recognition has to do with the nature of the features used to represent objects in the object-related areas. Hopefully, ways will be found to use imaging studies to probe the nature of these features, whether, for example, they are generic three-dimensional shape primitives such as geons, or based on a hierarchy of object fragments as postulated in one of the proposals above.

7.8.8 Acquiring new features and object categories

Related to this is the problem of acquiring new features and learning new classes. Single-unit studies support the views that visual features in the monkey IT cortex are shaped by classification experience. For example, following training for visual classification using wireframe objects (Logothetis *et al.* 1995) or fractal patterns (Miyashita and Chang 1988), units were found in IT cortex with preferential responses for some of the trained stimuli. However, the strategy and time course of novel feature extraction remain an open problem. Imaging studies combined with tasks involving the learning of new categories may shed light on these important issues.

7.8.9 Recognition and segmentation

Recognition is complicated by the fact that objects are usually not seen in isolation, but embedded in complex scenes, surrounded, and sometimes partially occluded, by other objects. Traditional approaches to recognition assume that the recognition process is preceded by a figure-ground segmentation stage that separates the structures in the image that are likely to correspond to scene objects. Computational experiments have shown that such bottom-up segmentation is exceedingly difficult, and often fails to separate correctly figure from ground. Some psychophysical evidence (Peterson and Gibson 1994) has suggested that segmentation and classification are, in fact, tightly coupled and do not proceed in a simple sequential manner. A computational model of class-based classification (Borenstein and Ullman 2002) has shown how segmentation and classification can, in fact, be obtained simultaneously within the fragment-based approach. The basic notion is that the figure-ground assignment of the individual fragments is assumed to be known from experience, and the segmentation of complex objects is then inherited from the component fragments. The relation between segmentation and recognition in the human visual system is still an open problem, which could benefit from future imaging studies.

7.8.10 Cortical circuitry and the role of bi-directional computation

As mentioned in the discussion of neural networks above, current network models attempt to incorporate some of the basic characteristics of the cortical circuitry, such as high parallelism, neuron-like elements, rich interconnectivity, and synaptic connection modified by experience. However, these properties by themselves have proved insufficient to reproduce the capacity of the human system to generalize and classify natural stimuli. A fundamental open question therefore is what aspects of the cortical circuitry are still missing from current modeling. One property that deserves close attention is the use of bi-directional processing in the visual cortex. Anatomical studies have shown that cortical connections between visual areas include both forward and backward projections (Ungerleider and Mishkin 1982; Felleman and van Essen 1991). The forward connections terminate mainly in layer 4 of the target cortical area, while the back projections go mainly to superficial and also to deep layers, but always avoid layer 4. Overall, the back projections are as rich in quantity and connectivity as the forward projections, and the connections are typically (though not always) reciprocal, in the sense that if a forward connection exists from visual area A to another area B, then a descending projection typically exists from B back to A. This pattern of bi-directional connectivity stands in contrast to many neural network models, such as back-propagation, that rely on feed-forward processing (the back-propagation is used for learning only), and other network models, such as the Hopfield model, that use high interconnectivity but no bi-directional structure. Some general models for bi-directional computation have been proposed (Ullman 1995, 1996), but the full use of such processing in visual classification is still unclear. It seems likely to me that this aspect of the cortical structure will prove to have important computational implications, but the exact nature of these computations remain for future empirical and computational studies.

The problems listed above have been central questions in the study of recognition, and the challenge for functional brain imaging studies will be to come up with new ways and techniques to explore them. Clearly, any single approach on its own, including brain imaging, cannot supply full answers to the questions of recognition and classification. Meaningful answers will require the integration of different approaches, including human and non-human fMRI, techniques providing temporal information (MEG, EEG), single-cell and optical recordings, psychophysical studies, and computational modeling.

Acknowledgement

This research was supported by the Israel Ministry of Science grant number 2097199, and by funds from the Moross Laboratory at the Weizmann Institute of Science.

References

Abu-Mostafa, Y.S. and Psaltis, D. (1987). Optical neural computing. *Scientific American*, **256**, 66–73.
Adini, Y., Moses, Y., and Ullman, S. (1997). Face recognition: the problem of compensating for changes in illumination direction. *IEEE Transactions Pattern Analysis and Machine Intelligence*, **19**(7), 721–32.

Amit, D.J. (1989). *Modeling brain function.* Cambridge University Press, Cambridge.

Amit, Y., Geman, D., and Wilder, K. (1997). Joint induction of shape features and tree classifiers. *IEEE Transactions on Pattern Analysis and Machine Intelligence*, **19**(11), 1300–6.

Biederman, I. (1985). Human image understanding: Recent research and a theory. *Computer Vision, Graphics, and Image Processing*, **32**, 29–73.

Biederman, I. and Cooper, E.E. (1991). Evidence for complete translational and reflectional invariance in visual object priming. *Perception*, **20**, 585–93.

Biederman, I. and Gerhardstein, P.C. (1993). Recognizing depth-rotated objects: evidence and conditions for three-dimensional viewpoint invariance. *Journal Experimental Psychology, Human Perception and Performance*, **6**, 1162–82.

Borenstein, E. and Ullman, S. (2002). Class-specific top-down segmentation. *Proceedings of the 7th European Conference of Computer Vision ECCV, 2, Copenhagen, Denmark*, 109–22.

Dill, M. and Fahle, M. (1997). The role of visual field position in pattern-discrimination learning. *Proceedings Royal Society, B*, **264**, 1031–6.

Downing, P.E., Jiang Y., Shuman M., and Kanwisher N. (**2001**). A cortical area selective for visual processing of the human body. *Science*, **293**, 2470–3.

Edelman, S. and Bulthoff, H.H. (1992). Orientation dependence in the recognition of familiar and novel views of three-dimensional objects. *Vision Research*, **32**, 2385–400.

Elliffe, M.C.M., Rolls, E.T., Parga, N., and Renart, A. (2000). A recurrent model of transformation invariance by association. *Neural Networks*, **13**, 225–37.

Epstein, R. and Kanwisher, N. (1998). A cortical representation of the local visual environment. *Nature*, **392**, 598–601.

Farah, M.J. (1990). *Visual agnosia: Disorders of object recognition and what they tell us about normal vision.* MIT Press, Cambridge MA.

Felleman, D.J. and van Essen, D.C. (1991). Distributed hierarchical processing in primate visual cortex. *Cerebral Cortex*, **1**, 1–47.

Field, D.J. (1994). What is the goal of sensory coding? *Neural Computation*, **6**, 559–601.

Foldiak, P. (1991). Learning invariance from transformation sequences. *Neural Computation*, **3**, 194–200.

Fujita, I., Tanaka, K., Ito, M., and Cheng, K. (1992). Columns for visual features of objects in monkey inferotemporal cortex. *Nature*, **360**, 343–6.

Gallant, J.L., Braun, J., and van Essen, D.C. (1993). Selectivity for polar, hyperbolic, and cartesian gratings in macaque visual cortex. *Science*, **259**, 100–3.

Gauthier, I., Tarr, M.J., Anderson, A.W., Skudlarski, P., and Gore, J.C. (**1999**). Activation of the middle fusiform 'face area' increases with expertise in recognizing novel objects. *Nature Neuroscience*, **2**, 568–73.

Gauthier, I., Hayward, W.G., Tarr, M.J., Anderson, A.W., Skudlarski, P., and Gore, J.C. (2002). BOLD Activity during Mental Rotation and Viewpoint-Dependent Object Recognition. *Neuron*, **34**, 161–71.

Grill-Spector, K., Kushnir, T., Edelman, S., Avidan, G., Itzchak, Y., and Malach, R. (**1999**). Differential processing of objects under various viewing conditions in the human lateral occipital complex. *Neuron*, **24**, 187–203.

Grill-Spector, K., Kushnir, T., Hendler, T., and Malach, R. (2000). The dynamics of object-selective activation correlate with recognition performance in humans. *Nature Neuroscience*, **3**, 837–43.

Haxby, J.V., Gobbini, M.I., Furey, M.L., Ishai, A., Schouten, J.L., and Pietrini, P. (**2001**). Distributed and overlapping representations of faces and objects in ventral temporal cortex. *Science*, **293**, 2425–30.

Hopfield, J.J. (1982). Neural networks and physical systems with emergent collective computational abilities. *Proceedings National Academy of Science, USA*, **79**, 2554–8.

Ishai, A., Ungerleider, L.G., Martin, A., Schouten, J.L., and Haxby, J.V. (1999). Distributed representation of objects in the human ventral visual pathway. *Proceedings National Academy Science, USA*, **96**, 9379–84.

Kanwisher, N., McDermott, J., and Chun, M.M. (1997a). The fusiform face area: a module in human extrastriate cortex specialized for face perception. *Journal Neuroscience*, **17**, 4302–11.

Kanwisher, N., Woods, R.P., Iacoboni, M., and Mazziotta, J.C. (1997b). A locus in human extrastriate cortex for visual shape analysis. *Journal of Cognitive Neuroscience*, **9**, 133–42.

Kourtzi, Z. and Kanwisher, N. (2000). Cortical regions involved in perceiving object shape. *Journal Neuroscience*, **20**, 3310–8.

LeCun, Y., Boser, B., Denker, J.S., Henderson, D., Howard, R.E., Hubbard, W., and Jackel, L.D. (1989). Backpropagation applied to handwritten zipcode recognition. *Neural Computation* **1**(4), 541–51.

LeCun, Y., Jackel, L.D., Bottou, L., Brunot, A., Cortes, C., Denker, J.S., Drucker, H., Guyon, I., Muller, U.A., Sackinger, E., Simard, P., and Vapnik, V. (1995). Comparison of learning algorithms for handwritten digit recognition. In F. Fogelman and P. Gallinari, (Eds.) *International Conference on Artificial Neural Networks*, pp. 53–60. EC2&Cie, Paris.

Littlestone, N. (1988). Learning quickly when irrelevant attributes abound: A new linear-threshold algorithm. *Machine Learning*, **2**, 285–318.

Logothetis, N.K., Pauls, J., Bulthoff, H.H., and Poggio, T. (1995). Shape representation in the inferior temporal cortex of monkeys. *Current Biology*, **5**, 552–63.

Malach, R., Levy, I., and Hasson, U. (2002). The topography of high-order human object areas. *Trends in Cognitive Science*, **6**(4), 176–84.

Malach, R., Reppas, J.D., Benson, R., Kwong, K.K., Jiang, H., Kennedy, W.A., Ledden, P.J., Brady, T.J., Rosen, B.R., and Tootell, R.B.H. (1995). Object-related activity revealed by functional magnetic-resonance-imaging in human occipital cortex. *Proceedings of the National Academy of Science USA*, **92**, 8135–9.

Marr, D. (1982). *Vision*. Freeman, San Francisco CA.

Martin, A., Wiggs, C.L., Ungerleider, L.G., and Haxby, J.V. (1996). Neural correlates of category-specific knowledge. *Nature*, **379**, 649–52.

Mel, W.B. (1997). SEEMORE: Combining color, shape and texture histogramming in a neurally inspired approach to visual object recognition. *Neural Computation*, **9**, 777–804.

Miyashita, Y. and Chang, H.S. (1988). Neuronal correlate of pictorial short-term memory in the primate temporal cortex. *Nature*, **331**, 68–70.

Nazir, T. and O'Regan, J.K. (1990). Results on translation invariance in the human visual system. *Spatial Vision*, **5**, 81–100.

Olshausen, B.A., Anderson, C.H., and van Essen, D.C. (1993). A neurobiological model of visual attention and invariant pattern recognition based on dynamic routing of information. *Journal Neuroscience*, **13**(1), 4700–19.

Parga, N. and Rolls, E. (1998). Transform-invariant recognition by association in a recurrent network. *Neural Computation*, **10**(6), 1507–25.

Peterson, M.A. and Gibson, B.S. (1994). Must shape recognition follow figure-ground organization? An assumption in peril. *Psychological Science*, **5**, 253–9.

Poggio, T. and Edelman, S. (1990). A network that learns to recognize three-dimensional objects. *Nature*, **343**, 263–6.

Riesenhuber, M. and Poggio, T. (1999). Hierarchical models of object recognition in cortex. *Nature Neuroscience*, **2**(11), 1019–25.

Rosenblatt, F. (1958). The perceptron: A probabilistic model for information storage and organization in the brain. *Psychological Review*, **65**, 386–407.

Schneiderman, H. and Kanade, T. (1998). Probabilistic modeling of local appearance and spatial relationships for object recognition. In *Proceedings of IEEE Conference on Computer Vision and Pattern Recognition.* pp. 45–51. Santa Barbara CA.

Schölkopf, B., Smola, A., and Müller, K.R. (1998). Nonlinear component analysis as a kernel eigenvalue problem. *Neural Computation,* **10,** 1299–1319.

Tanaka, K. (1993). Neuronal mechanisms of object recognition. *Science,* **262,** 685–8.

Tarr, M.J., Williams, P., Hayward, W.G., and Gauthier, I. (1998). Three-dimensional object recognition is viewpoint dependent. *Nature Neuroscience,* **1**(4), 275–7.

Turk, M. and Pentland, A. (1990). Eigenfaces for recognition. *Journal of Cognitive Neuroscience,* **3,** 71–86.

Ullman, S. (1995). Sequence seeking and counter streams: a model for bi-directional information flow in the visual cortex. *Cerebral Cortex,* **5**(1), 1–11.

Ullman, S. (1996). *High-level vision: Object recognition and visual cognition.* MIT Press, Cambridge MA.

Ullman, S. and Soloviev, S. (1999). Computation of pattern invariance in brain-like structures. *Neural Networks,* **12,** 1021–36.

Ullman, S., Sali, E., and Vidal-Naquet, M. (2001). A fragment-based approach to object representation and classification. In C. Arcelli, L.P. Cordella, and G. Sanniti di Baja, (Eds.) *Visual form 2001: Lecture notes in computer science,* pp. 85–100. Springer, Berlin.

Ullman, S., Vidal-Naquet, M., and Sali, S. (2002). Visual features of intermediate complexity and their use in classification. *Nature Neuroscience,* **5**(7), 1–6.

Ungerleider, L.G. and Mishkin, M. (1982). Two cortical visual systems. In D.J. Ingle, M.A. Goodale, and R.J.W. Mansfield, (Eds.) *Analysis of visual behavior.* MIT Press, Cambridge MA.

Vapnik, V.N. (1995). *The nature of statistical learning theory.* Springer-Verlag, New York NY.

Warrington, E.K. and James, M. (1986). Visual object recognition in patients with right-hemisphere lesions: Axes or feature? *Perception,* **15,** 355–66.

Weiskrantz, L. (1990). Visual prototypes, memory, and the inferotemporal lobe. In E. Iwai and M. Mishkin, (Eds.) *Vision, memory and the temporal lobe,* pp. 13–28. Elsevier, New York NY.

Willshaw, D.J., Buneman, O.P., and Longuet-Higgins, H.C. (1969). Non-holographic associative memory. *Nature,* **222,** 960–2.

Chapter 8

The functional organization of the ventral visual pathway and its relationship to object recognition

Kalanit Grill-Spector

Abstract

Humans recognize objects at an astonishing speed and with remarkable ease. However, the functional organization of the system that enables this remarkable human ability is not well understood. Here we examine whether the human ventral stream is organized more around stimulus content or recognition task. We scanned subjects while they performed one of two tasks: object detection (objects versus textures) or subordinate identification (e.g. pigeons versus birds). In order to limit success at recognition, pictures were presented briefly and then masked. For each subject we searched for cortical regions where activity was correlated with correct answers, separately for each category and task. Analysis by task revealed that, for each category, regions correlated with correct detection and correct identification were similar. However, analysis by stimulus revealed that different patterns of activation across occipitotemporal areas were correlated with successful identification of different categories. Analysis of regions whose activity was correlated with face recognition revealed a higher signal for faces (compared to birds and guitars) only in trials in which faces were perceived, but not in trials in which faces were not detected. Overall, these results indicate that the functional organization of higher-order areas in the human ventral stream is organized more around stimulus content than recognition task. These results provide new insights into the representations underlying our ability visually to recognize objects.

8.1 Introduction

Humans recognize objects and faces instantly and effortlessly. What are the underlying neural mechanisms in our brains that allow us to detect and discriminate among objects so efficiently?

Multiple ventral occipitotemporal regions anterior to retinotopic cortex (Grill-Spector *et al.* 1998) respond preferentially to various objects compared to textures (Malach *et al.* 1995). Indeed, several studies provide evidence that the activation of occipitotemporal object areas is correlated with subjects' perception of objects in a variety of experimental paradigms and tasks (Tong *et al.* 1998; Grill-Spector *et al.* 2000; James *et al.* 2000; Hasson *et al.* 2001; Kleinschmidt *et al.* 2002). Further evidence suggests that these regions play a critical role in object recognition, since lesions to the fusiform gyrus and occipitotemporal junction produce various recognition deficits (Damasio 1990; Damasio *et al.* 1990; Farah 1992), and electrical stimulation of these regions interferes with recognition (Puce *et al.* 1999).

Functional imaging studies have revealed that some of these regions respond maximally to specific object categories, such as faces (Puce *et al.* 1995; Kanwisher *et al.* 1997), places (Aguirre *et al.* 1998; Epstein and Kanwisher 1998), body parts (Downing *et al.* 2001), letter strings (Puce *et al.* 1996), tools (Martin *et al.* 1996), and animals (Martin *et al.* 1996; Chao *et al.* 1999). These results suggest that areas that elicit a maximal response for a particular category are dedicated to the recognition of that category. However, there are many difficulties underlying this idea. First, comparing activation between a handful of object categories is problematic because it depends on the choice of categories. Secondly, while there is maximal activation to one category the activation to other categories is not negligible (Ishai *et al.* 1999; Haxby *et al.* 2001). Thirdly, comparing the amplitude of activation to object categories does not exclude the possibility that the underlying representation might not be of whole objects (Fujita *et al.* 1992; Grill-Spector *et al.* 1998; Lerner *et al.* 2001; Tsunoda *et al.* 2001). Finally, objects from different categories differ in many dimensions and it is possible that the source of higher activation for a category is not restricted to visual differences.

How is the functional organization of these object-selective regions related to our ability to recognize objects? One view, proposed by Kanwisher (2000), is that the ventral temporal cortex contains a limited number of modules specialized for the recognition of special categories, such as faces, places, and body parts, and the remaining cortex, which exhibits little selectivity for particular object categories, is a general-purpose mechanism for the perception of any shape of any kind of visually presented object. A second model, proposed by Haxby *et al.* (2001), is an 'object form topography' in which occipitotemporal cortex has a topographically organized representation of form attributes. The representation of an object is reflected by a distinct pattern of response across ventral cortex, and this distributed activation produces the visual percept. In contrast to the Kanwisher model, Haxby and colleagues propose that submaximal activations across the ventral stream may be as important as the maximal activations (see also Avidan *et al.* 2002). A third view, posited by Tarr and Gauthier (2000), is that the organization is based according to the perceptual processes carried out and not by the content of information processed. Here the fundamental idea is that different cognitive processes require different computations, that are instantiated in different parts of the visual cortex.

Here we examine whether the functional organization of object-selective regions in occipitotemporal cortex is based on the content of information processed or on computations dedicated to specific perceptual processes. To distinguish between these alternatives we scanned subjects while they performed different recognition tasks: object detection or subordinate-level identification (Rosch *et al.* 1976) on three object categories—faces, birds, and guitars. We asked two experimental questions: (1) within each task, are the same or different regions correlated with success at recognizing different categories; and (2) for each stimulus category, are the same or different regions correlated with success at different recognition tasks?

8.2 Parsing object recognition into component stages

What are the stages of processing involved in visual object recognition? Hierarchical models of object recognition suggest that recognition involves several processing steps, proceeding from low-level stages that extract local visual information about features (Livingstone and Hubel 1988; Gallant *et al.* 1993), contours (von der Heydt *et al.* 1984), and boundaries (Lamme 1995; Zhou *et al.* 2000), to high-level stages that perform recognition by matching the incoming visual stimulus to stored representations of objects. Many models of recognition posit an intermediate stage at which the object is segmented from the rest of the image. Underlying this idea is the intuition that an efficient object recognition system should not operate indiscriminately on just any region of an image, because most such regions will not correspond to distinct objects. Instead, researchers have argued that stored object representations should be accessed only for image regions selected as candidate objects by a prior image segmentation process (Rubin 1958; Nakayama *et al.* 1995; Driver and Baylis 1996). However, other evidence (Peterson 1994) suggests that object recognition may influence segmentation, and may perhaps precede image segmentation.

To tease apart processing stages involved in visual recognition we varied stimulus exposure duration and measured behavioral performance on three different recognition tasks, each designed to tap into a different candidate stage of object recognition: detection, categorization, and identification. The *detection task* was designed to be a minimalist test of object segmentation that does not require recognition. Here subjects were asked to decide whether a gray-scale image contained an object or not. They were told that they did not have to recognize the object to report its presence. Half of the trials contained objects from 10 categories and half of the trials consisted of texture patterns created by randomly scrambling object images to 225 squares to equate mean luminance and local low-level features (see Fig. 8.1a). The second task was *object categorization*, in which subjects were required to categorize the object in the picture from a set of 10 possible categories: face, bird, dog, fish, flower, house, car, boat, guitar,

172 | FUNCTIONAL NEUROIMAGING OF VISUAL COGNITION

(a) Detection task: 'object' versus 'not an object'

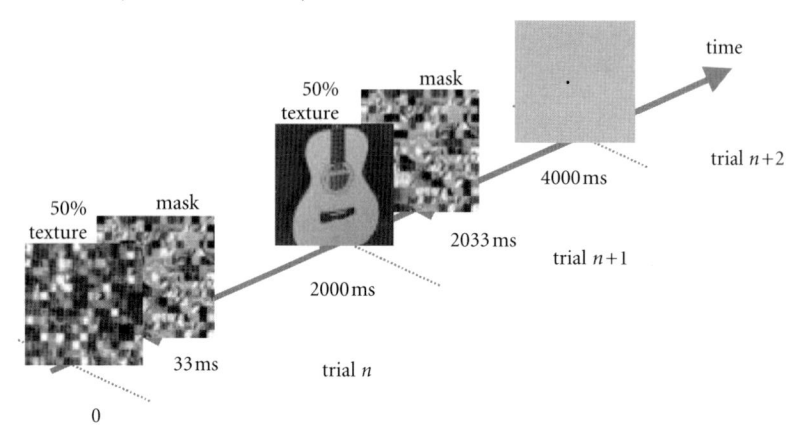

(b) Identification task: 'electric guitar' versus 'not an electric guitar'

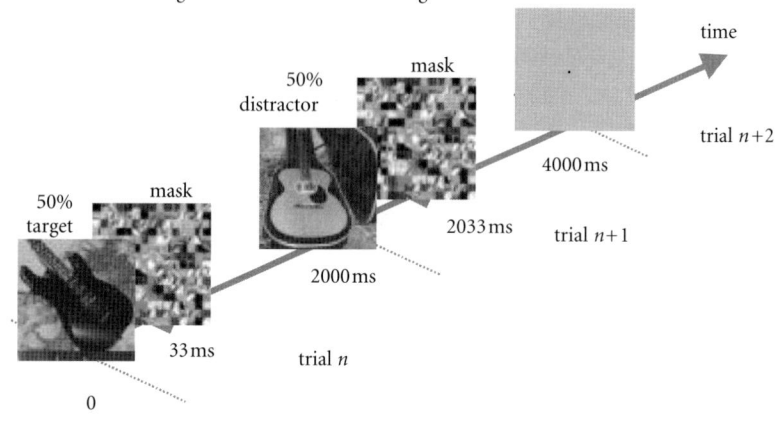

(c) Examples of stimuli

Fig. 8.1 Experimental design; (a) Detection task: in each trial (of duration 2 s) an image was presented briefly and then masked. Subjects had to respond during the duration of the mask, whether the image contained an object or not. In each scan, subjects were presented with objects from one category (in this case guitars) but the subjects were not told in advance the content of the pictures. Trials of objects, textures, and blanks (no visual stimulation) were counterbalanced. (b) Identification task. Here all pictures contained objects from a single basic

Table 8.1 Experiment 1: Behavioral data

	Detection	**Identification**
Accuracy corrected for guessing		
faces	67 ± 6	32 ± 6
birds	70 ± 6	40 ± 6
guitars	67 ± 6	32 ± 5
Reaction times (ms)		
faces	550 ± 24	652 ± 24
birds	594 ± 29	722 ± 40
guitars	586 ± 29	711 ± 48

Accuracy at identification was significantly lower than detection, as verified via an across-subject t-test (faces, $P < 10^{-8}$; birds, $P < 10^{-6}$; guitars, $P < 10^{-7}$) and reaction times were significantly longer (faces, $P < 10^{-6}$; birds, $P < 10^{-4}$; guitars, $P < 10^{-5}$). There was no statistical significance between identification or detection performance across categories (verified via a t-test).

or trumpet. In the *identification task* subjects were instructed to discriminate a particular subordinate member of a category from other members of that category. Possible answers were: Harrison Ford, pigeon, German shepherd, shark, rose, barn, VW beetle, sailboat, electric guitar versus 'other', i.e. other male faces, other birds, other cars, etc. In all three tasks, the frequency of each category was 10% and for each basic-level category half of the images were from a single subordinate class. Images were presented in five durations (between 17 and 167 ms) in a counterbalanced order.

The behavioral data revealed that longer stimulus exposures were required for subjects to reach the same accuracy levels in the identification task compared to the other two tasks (see Table 8.1). Lower accuracy at identification compared to categorization occurred for each of the object categories tested. This indicates that identification occurs after detection and categorization. Surprisingly, the curves relating performance to stimulus duration were nearly identical for the categorization and detection tasks despite the greater complexity of the ten-alternative forced choice categorization task compared to the two-alternative forced choice object detection task (Fig. 8.2). Thus, as soon as subjects detected an object, they already knew its category (Grill-Spector and Kanwisher 2003).

These data suggest that there are at least two main processing stages involved in object recognition: *detection* and *identification*, with detection preceding identification. The question that we will address in the following sections is whether the extra processing needed for identification compared to detection reflects additional processing

---

level category and subjects were required to identify a particular subordinate category. In this example, subjects had to respond whether the object was an electric guitar or not; distractors were other kinds of guitar. (c) Examples of the stimuli used in these experiments.

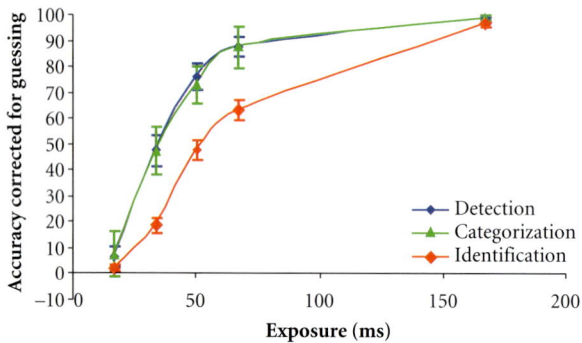

Fig. 8.2 Behavioral data. Accuracy in both the detection and categorization tasks was significantly higher than identification for stimulus exposures of 33–68 ms (*t*-test, *P* < 0.001); vertical axis denotes accuracy (corrected for guessing) on object pictures. Blue, detection; green, categorization; red, identification. Error bars indicate SEM (standard error of the mean) across 13 subjects.

within the same cortical area, or whether it requires additional processing at subsequent stages.

8.3 Regions correlated to detection or identification of three object categories

Our behavioral data indicates that subjects' performance in detection and categorization is similar. Therefore, in the fMRI experiments, we chose to use only two tasks: *detection* and *identification*. Here we asked subjects to perform either a detection task (see Fig. 8.1a) that required subjects to detect the presence of an object without having to recognize it, or an identification task (Fig. 8.1b) that required subjects to discriminate between objects belonging to the same basic level category (Rosch *et al.* 1976). We manipulated subjects' ability to recognize objects by presenting pictures very briefly, for 33 or 50 ms[1] and then masking the images with a texture pattern. Due to the brief visual presentation, in some trials subjects could identify or detect objects, and in others they could not. When subjects viewed this display, we measured both behavioral performance and brain activation in a rapid event-related design experiment, using a 3T fMRI scanner.[2] In contrast to conventional fMRI experiments, in which areas are defined based on their amplitude of activation to different types of stimuli, here we localized areas that were correlated with detection or identification object, using individual subjects' behavioral data.

In the *detection task*, subjects were asked to decide whether or not a gray-scale image contained an object. Half the trials contained objects from one basic level category and half the trials contained texture patterns. The *identification task* required subjects to discriminate a particular subordinate member of a category (e.g. electric guitar) from other members of that category (e.g. other guitars) that share a common structure. Here all images belonged to one category (in this example guitars). Half of the images were different pictures of the target subordinate category and half of the images were other objects from the same basic level category. In separate scans subjects were asked to perform one of the two tasks (detection or identification) on one of three object categories: faces, guitars, or birds.

Importantly, in these experiments, in each trial subjects viewed an image they had never seen before, so performance could not be affected by prior knowledge of particular images. Further, objects from each category and subordinate class were depicted in various viewing conditions and with different backgrounds (see Fig. 8.1c) to reduce the probability that subjects would use a small set of low-level features to perform these tasks.

The behavioral performance of subjects is given in Table 8.1. Similar to the ten-category experiment, for all object categories accuracy was significantly lower and reaction times significantly longer for the identification task compared to the detection task. Reaction times for detection were on the average 125 ms shorter than identification, suggesting that it occurs prior to identification. This occured for all three categories, including faces. The differences in performance in the detection and identification tasks indicate that subjects were indeed performing different perceptual tasks during these scans.

8.4 Detection experiment

For each subject we searched for regions that were correlated with successful detection separately for each category. Thus, we ran a statistical test searching for regions that showed a higher signal in trials in which objects were present and subjects successfully detected their presence (hits) compared to trials in which objects were present but subjects failed to detect their presence (misses). Texture stimuli were not included in the statistical analysis. This analysis was performed individually for each subject on a voxel by voxel basis. Importantly, in all trials a picture of an object from the same basic level category was shown for the same exposure duration. The only difference between trials was whether subjects succeeded or failed to detect the object.

For each of the categories, we found regions correlated with successful detection (examples of maps for faces and guitars are given in Fig. 8.3). These included the lateral occipital cortex (LO) and ventral occipitotemporal (VOT) areas including the occipitotemporal sulcus (OTS) and the fusiform gyrus. While we found for each object category regions that were correlated with successful detection, the pattern of activation across the human ventral stream was different for different categories. Hence, different subregions within higher-level areas were correlated with the detection of different categories.

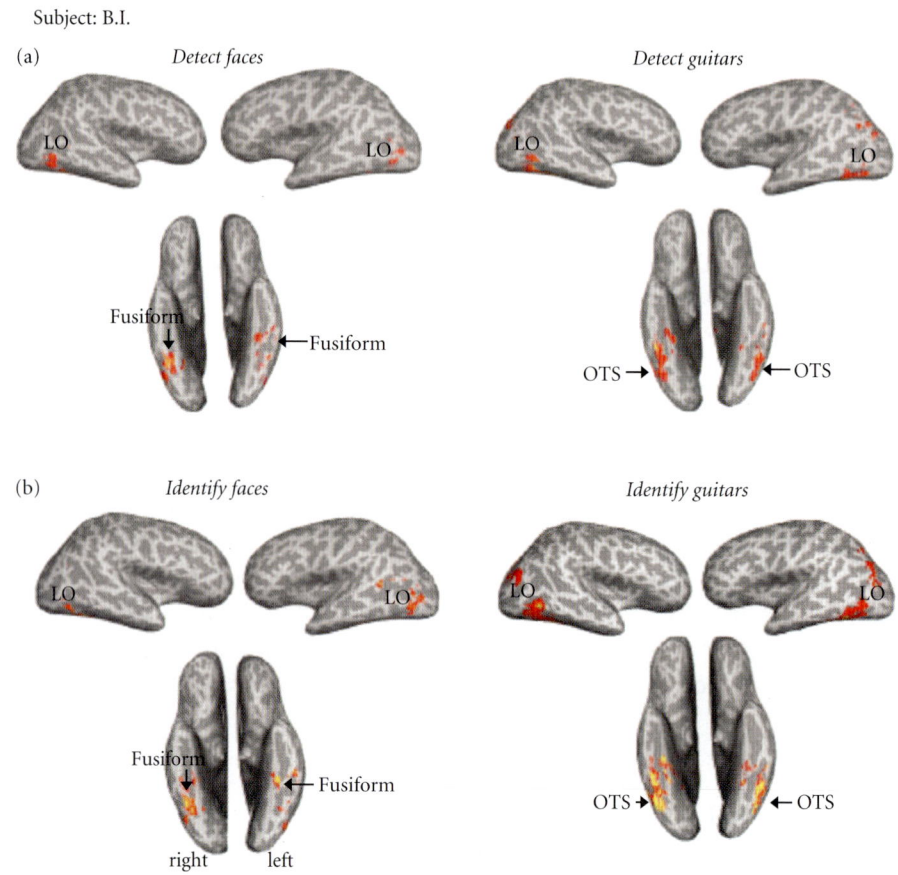

176 | FUNCTIONAL NEUROIMAGING OF VISUAL COGNITION

Subject: B.I.

Fig. 8.3 Areas correlated with correct detection and identification. (a) Areas correlated with hits > misses in the detection task shown on the inflated brain for one representative subject for two of the categories tested. Statistical analysis was performed using FS-fast software developed at MGH and brainalyzer software written by K.G.S. Brain reconstruction was performed using Freesurfer (Dale et al. 1999; Fischl et al. 1999, 2001). Color code indicates statistical significance (yellow, $P < 10^{-4}$; red: $P < 10^{-2}$). (b) Areas correlated with hits > misses in the identification task shown on the inflated brain of the same subject shown in (a). Color code indicates statistical significance (yellow, $P < 10^{-4}$; red, $P < 10^{-2}$).

8.5 Identification experiment

We performed a similar analysis for the identification experiment. Here we searched for regions that showed a higher signal in trials in which subjects were successful at identification. 'Hits' were defined as trials in which the target subordinate category was present and subjects answered correctly (e.g. electric guitar present and subjects responded 'electric guitar'). 'Misses' were trials in which the target subordinate category was present, but subjects answered incorrectly (e.g. electric guitar present but subject responded 'not an electric guitar'). Here catch trials contained other objects from the same basic level category (e.g. other guitars), but were not included in the statistical analysis.[3] This analysis was performed independently for each subject and category. Results are given in Fig. 8.3. Similar to the detection experiment, for each object category we found regions in the human ventral stream that were correlated with successful identification. Again, the pattern of activated areas across the human ventral stream that was correlated with successful identification was different for different categories.

8.6 Question 1: When we keep the category constant, are the same or different regions correlated with success at different tasks?

For each category we superimposed the maps of regions correlated with successful identification and successful detection and tested for conjunction effects. This superposition shows that when the category was kept constant there was a large degree of overlap between voxels that were correlated with detection and identification (yellow voxels in Fig. 8.4a). Across subjects, 60 ± 7% of the activated voxels were correlated with success at both object detection and object identification. Thus, most of the voxels were correlated with both identification and detection for each of the object categories. Thus, when the category was kept constant, similar regions across the human ventral stream were correlated with successful identification and detection.

8.7 Question 2: When we keep the task constant are the same or different regions correlated with recognition of different object categories?

Next we superimposed maps of areas correlated with identification of faces, guitars, and birds, to test whether these areas are overlapping or distinct (see Fig. 8.4b). When we superimposed maps of areas correlated with the identification of two object categories, most of the voxels were not overlapping. Across subjects, only 33 ± 10% of the activated voxels were correlated with the identification of two different categories. Taliarach coordinates for the center of activated regions in the identification task are given in Table 8.2. Thus, when the task was kept constant, largely different regions across the ventral stream were correlated with success at identifying different object categories.

The results of the superposition analysis reveal that there was approximately twice as much overlap between areas that were correlated with detection and identification of a particular category, than the amount of overlap between areas that were correlated with identification of different object categories. This suggests that the organization in the ventral stream is more around visual content rather than perceptual task.

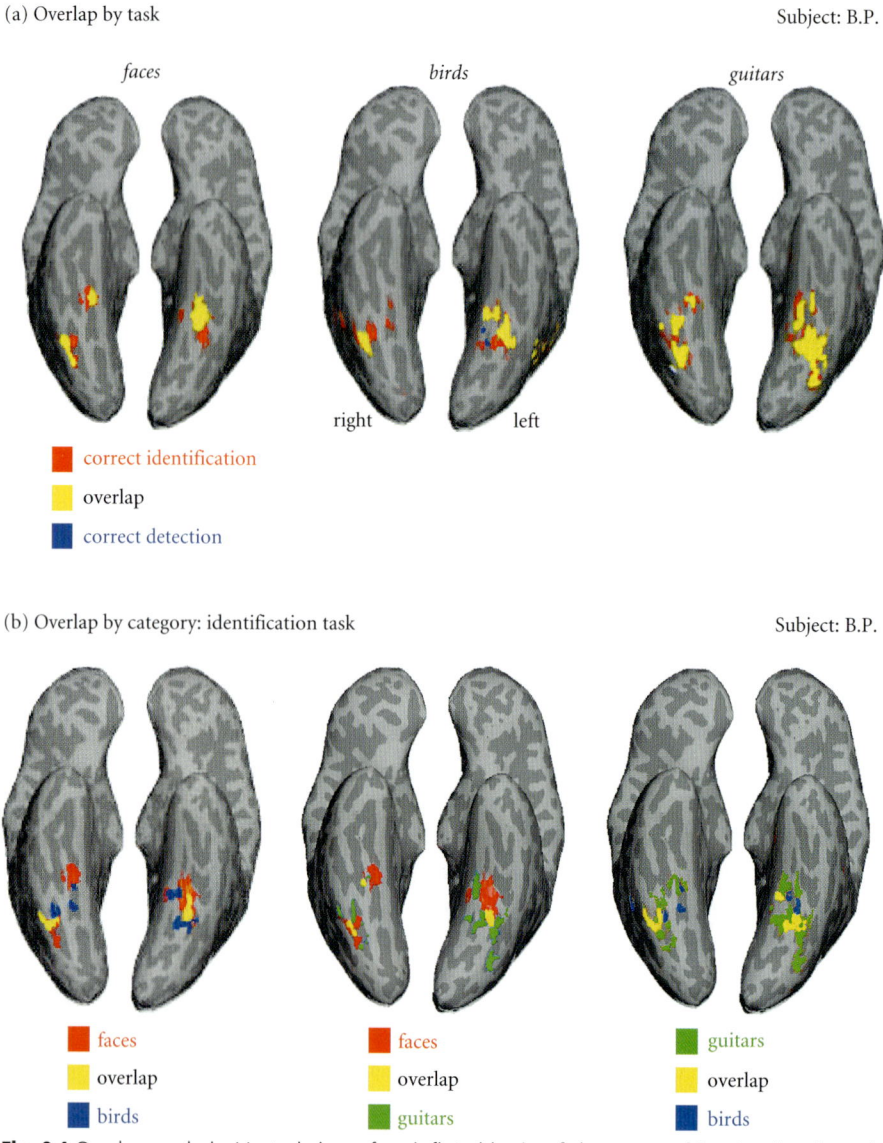

Fig. 8.4 Overlap analysis. Ventral view of an inflated brain of the same subject in all tasks and categories. (a) Overlap by task: yellow, areas that were correlated with both correct identification and correct detection of a given object category; red, areas that were correlated only with correct identification; blue, areas that were only correlated with correct detection.
(b) Overlap by category: identification task: yellow, areas that were correlated to successful identification of two categories; red, areas that were correlated only with successful face identification; blue, areas that were correlated only with bird identification; green, areas that were correlated only with guitar identification.

Table 8.2 Talairach coordinates

	Right			Left		
Ventral occipto-temporal (VOT)						
Faces	39 ± 3	−49 ± 37	−16 ± 5	−37 ± 4	−50 ± 7	−14 ± 5
Birds	41 ± 5	−52 ± 10	−20 ± 3	−38 ± 4	−55 ± 10	−17 ± 4
Guitars	46 ± 2	−56 ± 8	−15 ± 6	−41 ± 6	−59 ± 8	−15 ± 5
Lateral occipital (LO)						
Faces	45 ± 3	−77 ± 6	2 ± 8	−48 ± 3	−76 ± 6	6 ± 3
Birds	48 ± 5	−71 ± 10	−9 ± 3	−46 ± 2	−74 ± 8	6 ± 8
Guitars	45 ± 6	−74 ± 5	−4 ± 8	−41 ± 4	−73 ± 5	−5 ± 8

Nevertheless, there were regions that seemed to be correlated with only one recognition task. When we compared between tasks, most of the non-overlapping areas were regions that were correlated with successful identification but not detection (red voxels in Fig. 8.4a). This can be accounted for partially by the larger number of voxels that passed the statistical threshold in the identification task compared to the detection task. The critical question is whether these non-overlapping regions are dedicated to one recognition task?

Another concern in interpreting the overlap between detection and identification is the possibility that subjects were able to identify some of the objects in the detection experiment, even though they were not required to do so. We therefore conducted another experiment, but here we asked subjects to respond for each picture whether they could: (1) identify the object; (2) detect the object but not identify it; or (3) not detect it at all.

8.8 Comparing identification and detection directly within the same scan

In this set of experiments, five subjects saw in each scan different pictures from one object category. Half of the images were of the target subordinate category (e.g. different pictures of pigeons) and the rest of the pictures were other images from the same basic level (e.g. other birds). Subjects were asked to answer for each picture whether it was the target subordinate category (e.g. pigeon), or an object but not the target, or not an object. Subjects' behavioral performance is given in Table 8.3.

We first searched for regions that showed a higher signal for detected (but not identified objects) i.e. detection hits, versus not detected objects i.e. detection misses, independently for each object category. In contrast to the first detection experiment, here the detection hits consist of trials in which subjects could detect the presence of an object but could not identify it at the subordinate level. Consistent with the previous experiments, different patterns of activation across the human ventral stream were correlated with detection of different object categories. The time courses extracted from

Table 8.3 Experiment 2: Behavioral data, percent responses out of all target trials

	Identified	**Detected**	**Not detected**
Faces	31 ± 6	41 ± 5	28 ± 6
Birds	28 ± 6	38 ± 3	34 ± 6
Guitars	26 ± 6	39 ± 3	35 ± 6

these areas showed a higher signal amplitude for trials in which objects were detected compared to trials in which objects were not detected (see Fig. 8.5a, left). Importantly, the signal from these voxels was highest for trials in which subjects were successful at identification, even though these trials were not included in the statistical analysis. The higher signal for identification hits versus detection hits reached significance only for birds ($P<0.01$).

The design of this experiment also enabled us to search directly for regions that showed a higher signal for identification hits versus detection hits, or vice versa. Regions that showed a higher signal for identification hits versus detection hits tended to be located in more anterior ventral regions along the fusiform and the OTS, which we refer to here as VOT regions. However, the time-course analysis (Fig. 8.5a, right) revealed that while the signal was maximal for identification hits, the signal was statistically significantly higher for detection hits than detection misses (significance verified via a t-test: faces, $P<10^{-8}$; birds, $P<10^{-3}$; guitars, $P<10^{-3}$). We also searched for regions that showed a higher signal for 'identified' versus 'detected' for two or more categories. This revealed some activation in the fusiform gyrus. However, time-course analysis of these ROIs revealed that while the signal was highest for identification hits, it was also significantly higher for detection hits than detection misses ($P<10^{-3}$). Thus, regions that were correlated with successful identification, were also correlated with successful detection. We did not find any region that showed only a significant success effect for identification but not for detection. These results provide direct evidence that areas that were correlated with correct object identification were also correlated with correct object detection.

Previous studies (Grill Spector *et al.* 1999) indicate that there are functional differences between lateral object selective foci (LO) and anterior ventral occipitotemporal foci.[4] We therefore defined for each category regions of interest (ROIs) in LO and VOT[5] to test whether there are functional differences between these areas. Note that different foci in the vicinity of LO and VOT were correlated with successful recognition of these object categories. Both LO and VOT foci showed a main effect of success for both detection and identification performance. The difference between ventral and lateral foci was the magnitude of the effect across tasks. VOT foci exhibited a significant signal increase from detection misses to detection hits and also a significant signal increase from detection hits to identification misses. Both of these effects were statistically significant for all categories ($P < 0.01$). In LO the signal increased significantly from not detection misses to detection hits for all categories ($P < 0.001$), but the increase in the signal strength from detection hits to identification hits was statistically significant only for birds ($P < 0.01$). These results may suggest a hierarchy within the ventral stream, with LO regions contributing more to object detection/segmentation and VOT foci involved both in object detection and object identification.

Finally, we directly compared overlap by task and overlap by category in this experiment. Here, we performed the overlap analysis at two threshold levels[6] to ensure that our results do not depend on the choice of the threshold (see Fig. 8.5b). The results demonstrate that for both threshold levels the overlap by task was far greater than the overlap by category. At low thresholds the overlap by task was greater by twofold compared to overlap by category, and in higher thresholds the overlap by task was greater by threefold. These differences were statistical significant (LO, high threshold, $P < 10^{-5}$; low threshold, $P < 10^{-6}$; VOT, high threshold, $P < 10^{-6}$; low threshold, $P < 10^{-7}$). Importantly, at both threshold levels the majority of voxels that were correlated with correct recognition of one category were not the same voxels that were correlated with successful recognition of another category. Note that increasing the threshold level decreases the extent of activated areas. The finding that overlap by category is smaller at higher thresholds suggests that smaller ROI are more homogeneous and thus display a higher degree of category specificity. Thus, this analysis strengthens the conclusion that there is a higher degree of overlap by task compared to overlap by category and further show that this result does not depend on the choice of the threshold.

8.9 False identification

Our data so far suggest that activation in higher-order areas is predictive of success at object recognition. These regions were not activated to the same degree whenever an object stimulus was present; rather they were activated when the stimulus was there and the subject could report its presence consciously. The false-alarm data provide another demonstration that the activation was correlated with subjects' reports rather than with the presence of the stimulus.

In the first set of experiments we did not have a sufficient number of false-alarm trials to be able to measure a reliable signal. In the second set of experiments, 2 out of 5 subjects produced a sufficient number of false-alarm trials (17% and 22% of the non-target trials) to obtain a measurable signal. False alarms consisted of trials in which objects from the same basic level category as the target category (e.g. birds) were incorrectly identified as the target subordinate category (e.g. pigeons). Interestingly, the signal for falsely identified trials (green in Fig. 8.6) was almost as high as identification

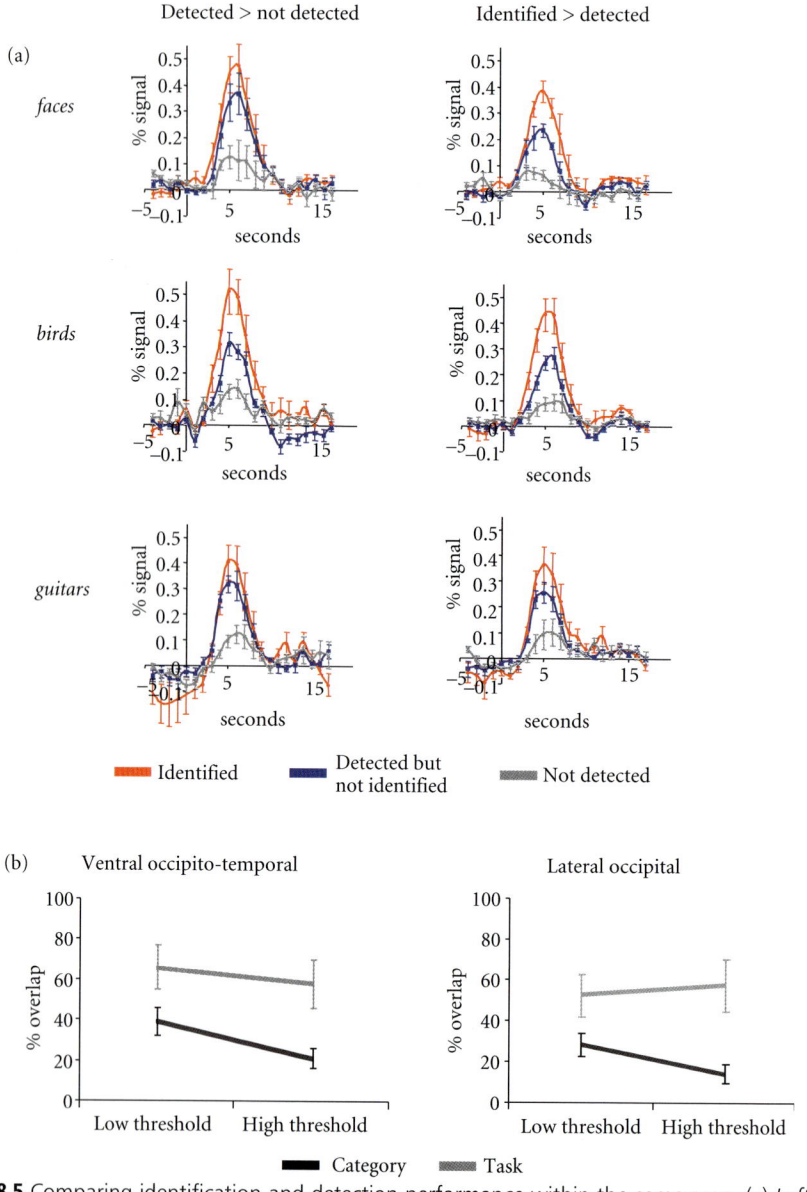

182 | FUNCTIONAL NEUROIMAGING OF VISUAL COGNITION

Fig. 8.5 Comparing identification and detection performance within the same scan. (a) *Left*: time courses from areas that passed the statistical threshold for detected > not detected averaged across five subjects. Areas were defined independently for each subject and category, with a threshold of $P < 0.01$. Trials were not included in the statistical analysis, but the signal for these trials is highest. *Right*: time courses from areas that passed the statistical threshold for identified > detected. Areas were defined independently for each subject and category with a threshold of $P < 0.01$. Trials, although 'not detected' trials were not included in the statistical analysis (faces, $P < 10^{-8}$; birds, $P < 10^{-3}$; guitars, $P < 10^{-3}$). (b) We analyzed

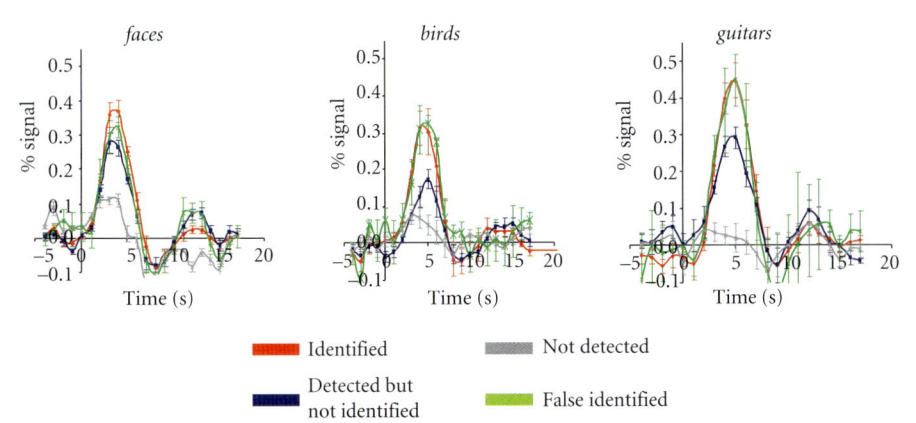

Fig. 8.6 False identification. Time courses averaged across two subjects for areas that were correlated with identification and detection of the target subordinate category. Falsely identified trials were not included in the statistical analysis. The curve for falsely identified was similar to the curve for identified targets and was higher than the curve for detected but not identified targets. Differences between false identification and identification hits did not reach significance.

hits (red in Fig. 8.6) and significantly stronger than detection hits (blue in Fig. 8.6). Thus, the signal in higher-order visual areas was higher when subjects reported that they identified an object (whether or not their answer was correct) compared to when they detected the presence of an object but could not identify it. This result further supports the idea that the activity in higher-order visual cortex correlates with what subjects reported they perceived rather than what was physically present.

8.10 Is the FFA a module for subordinate recognition?

One of the major debates regarding specialization based on content or process focuses on the specialization of the fusiform face area (FFA). Two main hypotheses for the role of the FFA in recognition have been proposed. Kanwisher and colleagues have suggested that this is a region specialized for face recognition (Kanwisher et al. 1997; Kanwisher 2000). Others, in particular Tarr and Gauthier (2000), have suggested that this is a region for subordinate recognition (of multiple categories) that is

---

separately the amount of overlap by task and overlap by category from lateral foci and ventral occipitotemporal foci in two threshold levels. Numbers are given in percent overlap and are averaged across tasks or categories. Threshold values were: low threshold, $P < 0.01$; high threshold, $P < 0.001$.

automated by expertise. These authors argue that faces (and expert categories) are automatically accessed at the subordinate level and therefore face recognition recruits the FFA.

These two hypotheses posit different outcomes for FFA activation in our experiments. The first hypothesis that FFA is a module for face recognition predicts that the activation of the FFA should be correlated with both face identification and face detection but not with identification of other object categories. In contrast, the second hypothesis, that the FFA is involved in subordinate identification, predicts that the activation of the FFA should be linked to successful identification of all categories, but should not be correlated with correct detection of faces (or other objects).

To directly test these hypotheses for each subject, we used an independent localizer scan to define the FFA based on a face selectivity test. In the localizer scan we used a block design experiment in which subjects passively viewed pictures of faces, cars, novel objects (abstract sculptures) in outdoor scenes and textures. The FFA was defined for each subject as the regions in the fusiform gyrus that showed a higher signal for faces compared to cars and novel objects, with a significance level of $P < 10^{-4}$ (see blue in Fig. 8.7).

First, we compared the locus of regions activated by correct versus incorrect identification (from the previous experiment) with face selective areas (faces versus objects). This analysis was performed separately for each object category (faces, birds, and guitars). For all five subjects, the regions that were correlated with successful identification of faces (yellow and red voxels in Fig. 8.7a) were similar to regions that were defined as face-selective in the independent localizer scan (blue contours in Fig. 8.7a). This indicates that face-selective regions were involved in face identification. In the bird experiments, there was some degree of overlap between areas correlated with bird identification and the FFA. In contrast, there was very little correspondence between face-selective regions and areas correlated with guitar identification (see Fig. 8.7c). We found partial overlap between the left FFA and areas correlated with guitar identification only in two subjects. This analysis revealed that face-selective regions were always involved in face identification, but not in guitar identification.

We then extracted the fMRI time course from the FFA in all experiments (see Fig. 8.8a). The signal from the FFA was correlated with both face identification (identified > detected; $P < 10^{-2}$) and face detection (detected > not detected; $P < 10^{-3}$). While the overall amplitude of the FFA signal was lower in the bird experiments, it showed correlation with both bird identification (identified > detected; $P < 10^{-2}$) and detection (detected > not detected; $P < 10^{-2}$). In contrast, the signal in the FFA was not correlated with success at guitar identification. The difference in the signal amplitude between trials in which guitars were identified compared to trials in which guitars were not identified did not reach statistical significance. Thus, the fMRI signal from the FFA does not correlate with success at guitar identification. Critically, the lack of a differential signal between trial types in the guitar experiment does not seem to stem from a

(a) Activation maps of areas correlated to face identification

(b) Activation maps of areas correlated to bird identification

(c) Activation maps of areas correlated to guitar identification

■ FFA - defined by localizer experiment:
faces > cars and novel objects $P < 10^{-4}$

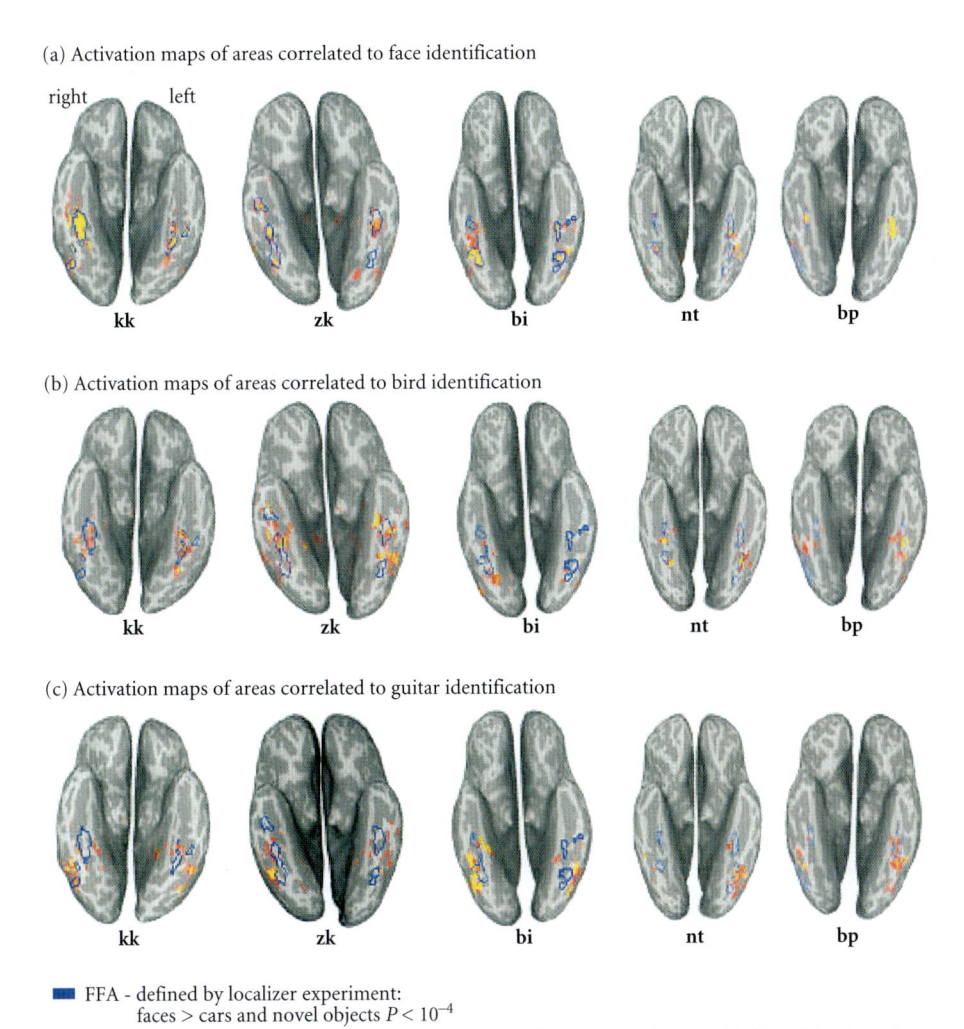

Fig. 8.7 Relation between the fusiform face area and areas correlated with identification of different object categories. The FFA was mapped independently for each subject in a separate localizer scan and was defined as areas in the fusiform gyrus that responded more strongly to faces compared to cars and novel objects, with a threshold of $P < 10^{-4}$. The blue contours indicate the boundaries of ventral face selective regions for each subject. We superimposed statistical maps of areas that were correlated with successful identification of: (a) faces, (b) birds, (c) guitars. Color of the statistical map indicates significance level (red, $P < 0.01$; yellow, $P < 0.0001$).

lack of a measurable signal, since the amplitude of the fMRI signal was not zero and the signal was not noisier than other experiments. The outcome of this analysis reveals that face selective regions were not correlated to success at identifying all subordinate-level object categories.

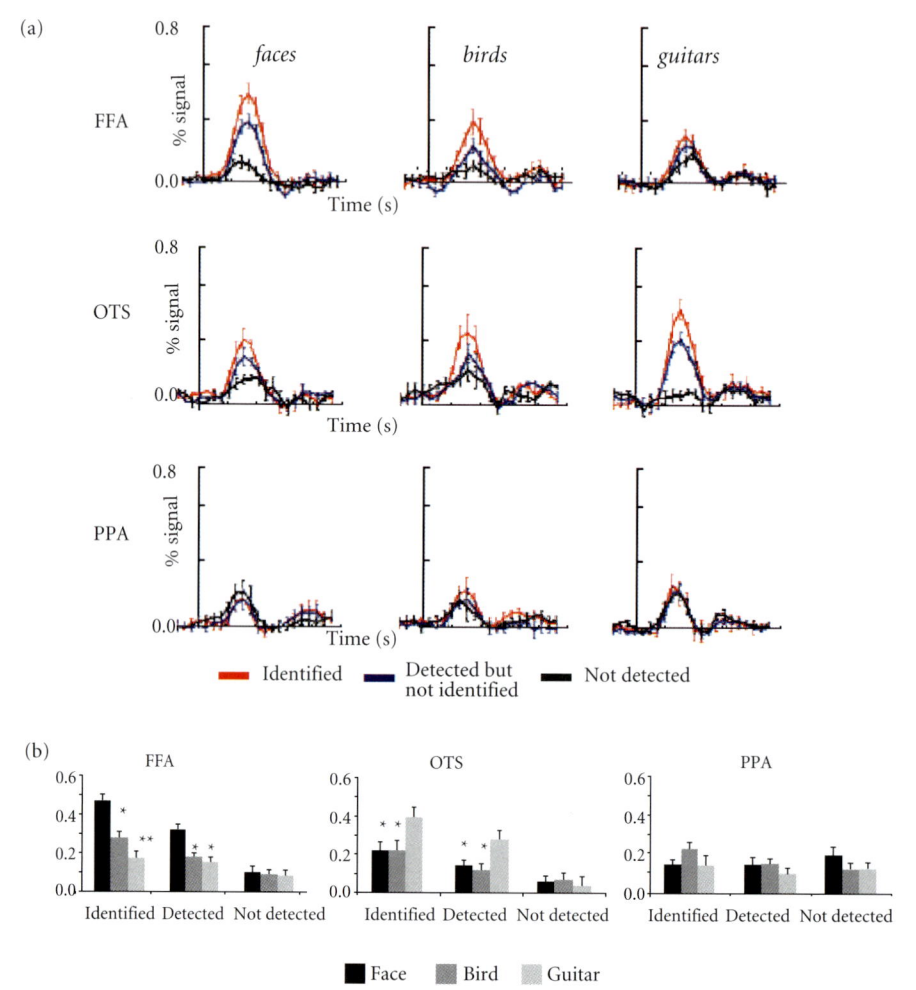

Fig. 8.8 Activation in the fusiform face area (FFA), occipitotemporal sulcus (OTS), and parahippocampal place area (PPA) across experiments. (a) Raw time courses extracted from the FFA, OTS, and PPA averaged across five subjects for three experiments: *left,* faces; *middle,* birds; *right,* guitars. The FFA was defined by the localizer scan (see blue contours in Fig. 8.7). OTS voxels were defined as voxels in the OTS that were correlated with guitar recognition and did not overlap with the FFA. PPA was defined in the localizer scan as regions in the parahippocampal gyrus that showed higher activation for outdoor scenes containing sculptures versus faces and cars. (b) Mean activation amplitudes for five subjects averaged across three time points around the peak of activation (4-6 s after trial onset). Error bars indicate SEM. In both FFA and OTS (but not the PPA) there was a main effect of category and success. FFA, asterisks indicate significantly lower activation for guitars and birds than faces. OTS, asterisks indicate significantly less activation for faces and bird than guitars. $^*P < 10^{-2}$; $^{**}P < 10^{-4}$. In the PPA the differences did not reach statistical significance. We verified the main effects of success, category, and interaction between them via a two-way ANOVA analysis.

The activation maps in Fig. 8.7 indicate that for all subjects there was a region in the OTS lateral to the FFA that was correlated with guitar identification. For each subject we defined ROIs in the OTS that were correlated with guitar identification but did not overlap with the FFA. Time courses extracted from the OTS revealed an opposite profile of activation relative to FFA activation (see Fig. 8.8a). While OTS voxels reveal a main effect of success, the higher signal for identification hits versus detection hits in the OTS reached significance ($P < 10^{-3}$) for guitars but did not reach significance for faces or birds. Surprisingly, OTS voxels demonstrated a main effect of category preference for guitars (see Fig. 8.8b). This category preference was revealed despite the fact that the selection of these voxels was not based on category selectivity.

The category preference in the FFA (for faces) and OTS (for guitars) was not automatic. On trials in which objects were present, but not detected, the fMR signal from both the FFA and OTS was lowest. Importantly, in this condition there was no statistically significant difference between the activations to different categories, in either the FFA or OTS (see Fig. 8.8b). However, on trials in which objects were detected or identified, the signal was significantly higher in the FFA for faces compared to both birds and guitars (and the converse was true for OTS activation). This indicates that the higher activation for the preferred category does not occur automatically but only when a percept occurs.

Is the FFA a module dedicated to subordinate identification? The data presented here indicate that the FFA is not a module for subordinate categorization of all categories. First, the signal from the FFA is predictive of success at both detection and identification of faces (and birds). Secondly, the signal from the FFA was correlated with successful identification of faces but not guitars. Thus, it is not correlated with identification of all object categories. Further, we found a different region in the OTS that was correlated with successful identification of guitars, showing that our failure to find this effect in the FFA was not due to some artifact. These results argue against the hypothesis suggested by Tarr and Gauthier (2000) that the role of the FFA is fine-grained subordinate discrimination between objects of any category.

8.11 Discussion

In sum, our data show that when the category was held constant but subjects performed different recognition tasks (fine-grained identification or object detection) similar regions in the human ventral stream were activated. However, when the task was kept constant and subjects were required to identify different object categories,

---

FFA, the interaction between faces and guitars was found for 'detected' compared with 'not detected' ($F > 6$; $P < 0.01$) and for 'identified' compared with 'detected' ($F > 4$; $P < 0.05$). OTS: 'detected' compared with 'not detected', $F > 6$, $P < 0.01$; 'identified' compared with 'detected', $F > 6.7$; $P < 0.001$.

different regions of the human ventral stream were activated. Even at the lowest thresholds there was twice as much overlap by task than content. This suggests that the human ventral stream is organized more around visual content than visual process (at least the processes treated here). Furthermore, we have shown that areas that were correlated with correct recognition of a category were also selective to that category, but only when the objects were detected or identified.

Our data reveal that similar regions within higher order areas were correlated with correct identification and detection when the object category was held constant. Surprisingly, even the most anterior regions along the ventral stream showed correlation with success at both detection and identification. This occurred despite the lack of necessity for explicit visual recognition in the detection task. In all areas that showed correlation with success at recognition, the signal was stronger for trials in which objects were successfully identified compared to trials in which objects were detected but not identified and lowest when objects were not detected at all. One possible explanation for the higher signal for identification compared to detection within the same regions is that identification requires longer processing times, and thus is a consequence of more neural processing. Since the bold signal sums up all the neural activations (Logothetis *et al.* 2001) this will be measured as a larger fMRI signal. Thus, one implication of these results is that the additional processing necessary for identification compared to detection occurs within the same cortical regions.

We found some differences between lateral-occipital and ventral occipitotemporal regions that were correlated with successful visual recognition. While the overlap by task analysis revealed that most of the voxels in LO and VOT were correlated with both identification and detection of an object category, some of the voxels in LO and VOT were correlated with detection but not with identification. Overall, LO regions contained a higher percentage of voxels that were correlated primarily with detection and not identification compared to VOT. Thus, if we consider a hierarchy of visual areas involved in object recognition, LO seems to be a candidate for a processing stage prior to VOT.

While there was substantially more segregation by category than by task, we did not find 100% overlap or segregation by either task or category. However, what clearly emerges is that the level of visual processing (at least those tested here) is not the major guideline for differentiating between regions in the ventral stream; object type seems to play a more critical role in differentiating among subdivisions of the human ventral stream. What still remains unknown is the precise nature of object representation in these regions, which could be whole objects, object fragments or even complex features and feature conjunctions.

Our data show that the recognition of a category is correlated with a distributed, distinct, and replicable pattern of activation across higher-order brain areas. This result is consistent with an 'object form topography' (Edelman *et al.* 1998; Haxby *et al.* 2001).

However, there are several important differences between the current results and those of Haxby *et al.* (2001). In their study Haxby *et al.* (2001) examined where there is information about object categories in ventral cortex. They showed that the response to a given category could be determined by the distributed pattern of activation across all ventral occipitotemporal cortex. Importantly, their analysis revealed that it was possible to predict the category of the object even when regions that showed maximal activation to a particular category were excluded. In contrast, the experiments described here examined which areas are used for visual recognition tasks. Our data suggest that the activation that is correlated with successful recognition of a category does not extend across all VOT. Rather, subregions within the VOT tend to be correlated with successful recognition of a category, and these were the same regions that showed higher activation for that category.

One of the predictions from Haxby *et al.* (2001) is that regions that respond maximally for houses contain information about non-preferred categories (such as faces). To test directly whether our subjects used house-selective regions in the parahippocampal place area (PPA) to recognize objects, we extracted time courses from PPA ROIs.[7] Unlike the activation in the OTS and FFA, the time courses extracted from the PPA did not show either a main effect of success (see Fig. 8.8a) or category (see Fig. 8.8b). Thus, the signal from the PPA was not correlated with successful recognition of these three object categories. Thus, while Haxby *et al.* (2001) suggest that activation in the PPA conveys information about all object categories, our data indicate that the PPA does not seem to be utilized for the recognition of several categories. Moreover, the fact that there exists a region within the ventral stream that is not correlated with recognition performance further strengthens the conclusion that not all regions within the VOT contribute equally to visual recognition of specific categories.

Another mystery that remains unresolved is why the higher activation for specific categories is localized and replicable across subjects? Malach *et al.* (2002) have suggested that category preference emerges from resolution needs, which are tightly linked to eccentricity bias (Levy *et al.* 2001; Hasson *et al.* 2002; Malach, *et al.* 2002). However this explanation does not fully account for the experiments described here. Here, all identification experiments required high visual acuity, yet different regions of the ventral stream were correlated with the identification of different categories. One possibility (that remains to be examined) is that regions along the OTS have a foveal bias. Another possibility is that eccentricity bias is not the only parameter that governs object-form topography in the human ventral stream.

However, this explanation does not fully account for the experiments described here. Here, all identification experiments required high visual acuity—yet different regions of the ventral stream were correlated with the identification of different categories. One possibility (that remains to be examined) is that regions along the OTS have a foveal bias. Another possibility is that eccentricity bias is not the only parameter that governs object-form topography in the human ventral stream.

Finally, we address the issue of modularity. While our data argue against the hypothesis that the FFA is a module for subordinate recognition, they also pose some constraints on the hypothesis that the FFA is a module dedicated solely for face recognition. While the signal from the FFA was highest for faces, it was also correlated with success at bird identification and detection. This correlation between the FFA signal and success at bird detection and identification occurred despite the fact that our subjects were not bird experts. One possibility is that neurons within the FFA are selective to both human and animal faces or face features. Birds (and other animals) have faces and face parts such as eyes, mouths, etc. Thus one possible interpretation of these results is that the axis of differentiation between subregions within ventral cortex is animate/inanimate rather than faces/objects, and the activation of FFA may be necessary for recognition of animate categories that contain faces. However, this remains to be verified.

Another difficulty in interpreting the function of the FFA as a module for face recognition is that the signal in an adjacent region in the OTS showed a success effect for faces, even though the signal was smaller for faces compared to guitars. One possible explanation is that this was caused by partial voluming artifacts. Another alternative is that the OTS contains some general shape processors that are used for both object and face recognition. Hopefully, in the near future we should be able to distinguish between these two alternatives by imaging the brain at higher resolutions.

To conclude, our data shows that the functional organization of object and face selective regions in the ventral visual pathway is organized on stimulus content rather than object recognition task (detection vs. identification). The experiments described here provide important insights to the functional organization of higher-order visual areas and their role in visual recognition.

Notes

1. For each subject we determined the exposure of images for the fMRI experiments by running a behavioral experiment prior to the scan, in which we showed masked images for various durations. The exposure duration was defined as the minimal exposure duration in which each subject could detect at least 50% of the images. This duration was fixed for a given subject and varied between 33 and 50 ms across subjects who participated in the fMRI scans.
2. Scanning parameters: MGH, Siemens allegra, head-only scanner; 10 oblique slices, $3.125 \times 3.125 \times 4$ mm; covering the occipital, posterior parietal and temporal lobes; FOV = 20 cm; TR = 1 s; TE = 43 ms; flip angle = 60°.
3. Correct rejects are not plotted here since the behavioral data on the non-target object images is ambiguous. One possibility is that these objects were rejected because subjects correctly identified them as a different subordinate category. However, it is possible that subjects could not identify them at all, and therefore these pictures were rejected.
4. Here we refer to these regions as VOT, in previous publications we referred to them as LOa.
5. VOT foci were defined as foci within the OTS and fusiform gyrus that were correlated with successful recognition. LO foci were located around the lateral occipital sulcus (LOS) and inferior occipital gyrus (IOG), lateral to a lower meridian representation. Both VOT and LO foci lie beyond retinotopic cortex (see Grill-Spector *et al.* 2000, for details).

6. The thresholds used here were chosen to be the lowest and highest thresholds possible in the conditions of these experiments. The low threshold used was $P < 0.01$, similar to that of Experiment 1. Decreasing the threshold to lower values resulted in detecting voxels outside the brain that were typically noise. The higher threshold was a value of $P < 0.001$.
7. For most subjects further increasing the value of threshold resulted in detecting few active voxels. Here we used a somewhat a non-standard test to define the PPA, and we used it as a *post-hoc* analysis method to define the relevant ROI. We searched for regions that showed a higher signal for outdoor scenes containing abstract sculptures compared to faces, cars, and textures in the localizer scan. This contrast activates regions in LO and in a lower thresholds region in the parahippocampal gyrus. For this analysis we took only the regions within the parahippocampal gyues that corresponded anatomically with the PPA (Epstein and Kanwisher 1998) that were statistically significant in this test.

Acknowledgements

Many thanks to Nancy Kanwisher for invigorating discussions and for her continuous support of this research. This research has been funded by HSFP LT0670 to K.G.S.

References

Aguirre, G. K., Zarahn, E., and D'Esposito, M. (1998). An area within human ventral cortex sensitive to 'building' stimuli: evidence and implications. *Neuron,* **21**(2), 373–83.

Allison, T., Puce, A., Spencer, D. D., and McCarthy, G. (1999). Electrophysiological studies of human face perception. I: Potentials generated in occipitotemporal cortex by face and non-face stimuli. *Cereb Cortex,* **9**(5), 415–30.

Avidan, G., Hasson, U., Hendler, T., Zohary, E., and Malach, R. (2002). Analysis of the neuronal selectivity underlying low fMRI signals. *Curr Biol,* **12**(12), 964–72.

Chao, L. L., Haxby, J. V., and Martin, A. (1999). Attribute-based neural substrates in temporal cortex for perceiving and knowing about objects. *Nat Neurosci,* **2**(10), 913–19.

Dale, A. M., Fischl, B., and Sereno, M. I. (1999). Cortical surface-based analysis I: Segmentation and surface reconstruction. *NeuroImage,* **9**, 179–94.

Damasio, A. R. (1990). Category-related recognition defects as a clue to the neural substrates of knowledge. *Trends Neurosci,* **13**(3), 95–8.

Damasio, A. R., Tranel, D., and Damasio, H. (1990). Face agnosia and the neural substrates of memory. *Annu Rev Neurosci,* **13**, 89–109.

Downing, P. E., Jiang, Y., Shuman, M., and Kanwisher, N. (2001). A cortical area selective for visual processing of the human body. *Science,* **293**(5539), 2470–3.

Driver, J. and Baylis, G. C. (1996). Edge-assignment and figure-ground segmentation in short-term visual matching. *Cognit Psychol,* **31**(3), 248–306.

Epstein, R. and Kanwisher, N. (1998). A cortical representation of the local visual environment. *Nature,* **392**(6676), 598–601.

Farah, M. J. (1992). Agnosia. *Curr Opin Neurobiol,* **2**(2), 162–4.

Fischl, B., Sereno, M. I., and Dale, A. M. (1999). Cortical surface-based analysis II: Inflation, Flattening, a surface-based coordinate system. *NeuroImage,* **9**, 195–207.

Fischl, B., Liu, A., and Dale, A. M. (2001). Automated manifold surgery: Constructing geometrically accurate and topologically correct models of the human cerebral cortex. *EEE Transactions on Medical Imaging,* **20**(1), 70–80.

Fujita, I., Tanaka, K., Ito, M., and Cheng, K. (1992). Columns for visual features of objects in monkey inferotemporal cortex. *Nature,* **360**(6402), 343–6.

Gallant, J. L., Braun, J., and Van Essen, D. C. (1993). Selectivity for polar, hyperbolic, and Cartesian gratings in macaque visual cortex. *Science,* **259**(5091), 100–3.

Gauthier, I., Tarr, M. J., Anderson, A. W., Skudlarski, P., and Gore, J. C. (1999). Activation of the middle fusiform 'face area' increases with expertise in recognizing novel objects. *Nat Neurosci,* **2**(6), 568–73.

Grill-Spector, K. and Kanwisher, N. (2003, submitted). Visual recognition: as soon as you see it you know what it is.

Grill-Spector, K., Kushnir, T., Hendler, T., Edelman, S., Itzchak, Y., and Malach, R. (1998). A sequence of object-processing stages revealed by fMRI in the human occipital lobe. *Hum Brain Mapp,* **6**(4), 316–28.

Grill-Spector, K., Kushnir, T., Hendler, T., and Malach, R. (2000). The dynamics of object-selective activation correlate with recognition performance in humans. *Nat Neurosci,* **3**(8), 837–43.

Hasson, U., Hendler, T., Ben Bashat, D., and Malach, R. (2001). Vase or face? A neural correlate of shape-selective grouping processes in the human brain. *J Cogn Neurosci,* **13**(6), 744–53.

Hasson, U., Levy, I., Behrmann, M., Hendler, T., and Malach, R. (2002). Eccentricity bias as an organizing principle for human high-order object areas. *Neuron,* **34**(3), 479–90.

Haxby, J. V., Gobbini, M. I., Furey, M. L., Ishai, A., Schouten, J. L., and Pietrini, P. (2001). Distributed and overlapping representations of faces and objects in ventral temporal cortex. *Science,* **293**(5539), 2425–30.

Ishai, A., Ungerleider, L. G., Martin, A., Schouten, J. L., and Haxby, J. V. (1999). Distributed representation of objects in the human ventral visual pathway. *Proc Natl Acad Sci USA,* **96**(16), 9379–84.

James, T. W., Humphrey, G. K., Gati, J. S., Menon, R. S., and Goodale, M. A. (2000). The effects of visual object priming on brain activation before and after recognition. *Curr Biol,* **10**(17), 1017–24.

Kanwisher, N. (2000). Domain specificity in face perception. *Nat Neurosci,* **3**(8), 759–63.

Kanwisher, N., McDermott, J., and Chun, M. M. (1997). The fusiform face area: a module in human extrastriate cortex specialized for face perception. *J Neurosci,* **17**(11), 4302–11.

Kleinschmidt, A., Buchel, C., Hutton, C., Friston, K. J., and Frackowiak, R. S. (2002). The neural structures expressing perceptual hysteresis in visual letter recognition. *Neuron,* **34**(4), 659–66.

Lamme, V. A. (1995). The neurophysiology of figure-ground segregation in primary visual cortex. *J Neurosci,* **15**(2), 1605–15.

Lerner, Y., Hendler, T., Ben-Bashat, D., Harel, M., and Malach, R. (2001). A hierarchical axis of object processing stages in the human visual cortex. *Cereb Cortex,* **11**(4), 287–97.

Levy, I., Hasson, U., Avidan, G., Hendler, T., and Malach, R. (2001). Center-periphery organization of human object areas. *Nat Neurosci,* **4**(5), 533–9.

Livingstone, M. and Hubel, D. (1988). Segregation of form, color, movement, and depth: anatomy, physiology, and perception. *Science,* **240**(4853), 740–9.

Logothetis, N. K., Pauls, J., Augath, M., Trinath, T., and Oeltermann, A. (2001). Neurophysiological investigation of the basis of the fMRI signal. *Nature,* **412**(6843), 150–7.

Malach, R., Reppas, J. B., Benson, R. R., Kwong, K. K., Jiang, H., Kennedy, W. A., Ledden, P. J., Brady, T. J., Rosen, B. R., and Tootell, R. B. (1995). Object-related activity revealed by functional magnetic resonance imaging in human occipital cortex. *Proc Natl Acad Sci USA,* **92**(18), 8135–9.

Malach, R., Levy, I., and Hasson, U. (2002). The topography of high-order human object areas. *Trends Cogn Sci,* **6**(4), 176–84.

Martin, A., Wiggs, C. L., Ungerleider, L. G., and Haxby, J. V. (1996). Neural correlates of category-specific knowledge. *Nature,* **379**(6566), 649–52.

Nakayama, K., He, Z. J., and Shimojo S, (1995). Visual surface representation: a critical link between lower-level and higher-level vision. In Kosslyn, S. M. and Osherson, D. N. (Eds.), *An invitation to cognitive science: visual cognition.* MIT press, Cambridge, MA.

Peterson, M. A. (1994). Shape recognition can and does occur before figure-ground organization. *Curr Direct Psychol Sci,* **3**, 105–11.

Puce, A., Allison, T., Gore, J. C., and McCarthy, G. (1995). Face-sensitive regions in human extrastriate cortex studied by functional MRI. *J Neurophysiol,* **74**(3), 1192–9.

Puce, A., Allison, T., Asgari, M., Gore, J. C., and McCarthy, G. (1996). Differential sensitivity of human visual cortex to faces, letterstrings, and textures: a functional magnetic resonance imaging study. *J Neurosci,* **16**(16), 5205–15.

Rosch, E. M., Mervis, C. B., Gray, M. D., Johnson, D. M., and Boyes-Braem, P. (1976). Basic objects in natural categories. *Cognit Psychol,* **8**, 382–439.

Rubin, E. (1958). Figure and ground. In W. M. Beardslee (Ed.), *Readings in perception.* New York: Van Nostrand.

Tanaka, J. (2001). The entry point of face recognition: evidence for face expertise. *J Exp Psychol Gen,* **130**(3), 534–543.

Tarr, M. J. and Gauthier, I. (2000). FFA: a flexible fusiform area for subordinate-level visual processing automatized by expertise. *Nat Neurosci,* **3**(8), 764–9.

Tong, F., Nakayama, K., Vaughan, J. T., and Kanwisher, N. (1998). Binocular rivalry and visual awareness in human extrastriate cortex. *Neuron,* **21**(4), 753–9.

Tsunoda, K., Yamane, Y., Nishizaki, M., and Tanifuji, M. (2001). Complex objects are represented in macaque inferotemporal cortex by the combination of feature columns. *Nat Neurosci,* **4**(8), 832–8.

von der Heydt, R., Peterhans, E., and Baumgartner, G. (1984). Illusory contours and cortical neuron responses. *Science,* **224**(4654), 1260–2.

Zhou, H., Friedman, H. S., and von der Heydt, R. (2000). Coding of border ownership in monkey visual cortex. *J Neurosci,* **20**(17), 6594–611.

Chapter 9

The cartography of human visual object areas

R. Malach, G. Avidan, Y. Lerner,
U. Hasson, and I. Levy

Abstract

A central theme underlying primate visual cortex research is the attempt to gain insight into cortical function through its parcellation, i.e. identifying the rules that subdivide the cortical surface into differentiable entities. With the advent of non-invasive functional imaging of the human brain, rapid progress has been obtained in such research. Here we will discuss our work on this issue, focusing on high-order, ventral stream, visual areas. We can identify functional division lines across all cortical scales. Most globally, on the dorso-ventral axis, ventral stream areas can be segregated from dorsal stream ones by modulating the subject's task, e.g. attending to the orientation versus the identity of an object will activate dorsal versus ventral stream regions, respectively. On the anterior–posterior axis, object areas can be distinguished through their hierarchically increasing correlation to object perception. At a finer level of resolution, within ventral occipitotemporal cortex, division lines are arranged according to visual field eccentricity associated with object category differentiations. Finally, within individual voxels, 'echoes' of shape-selective subgroups can be indirectly revealed. Overall, this complex organization points to a hierarchy of functional properties that map cortical regions at finer and finer topographical scales. Detailed understanding of these properties will point to the types of computations performed, at different scales, by human ventral stream visual areas.

9.1 Object-selective activation is extremely widespread in the human cortex

How many neurons are involved in the recognition of objects in the human brain? This straightforward question has by no means a straightforward answer—and, in fact, is extremely difficult to resolve experimentally. A much easier question is 'How many neurons are activated by the presentation of object images?' With the advent of fMRI it becomes possible to obtain at least a qualitative impression of that number, and Fig. 9.1, which shows the activity pattern produced by the presentation object images in the human visual cortex, gives a clear-cut answer: an extremely widespread territory shows enhanced activation to object images compared to textures. Note that this activity runs in an elongated strip, which extends dorsally to the intraparietal sulcus, and ventrally all the way to the collateral sulcus. Considering that a measurable fMRI signal likely depends on the activation of millions of neurons, such a widespread activation appears incompatible with models that predict a highly selective and sparse neuronal representation of objects in the brain. However, it should be noted that the BOLD activation integrates activity over fairly long temporal epochs—so that the widespread activation might also reflect intermediate processing stages, which ultimately lead to more restricted neuronal representations.

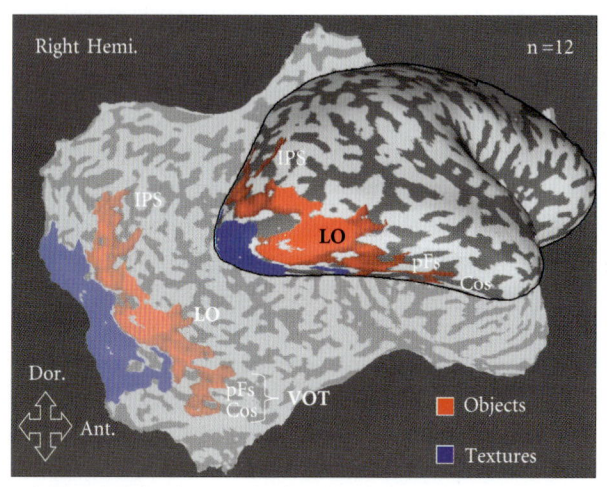

Fig. 9.1 Object and texture activations in the human visual cortex. An averaged map of 12 subjects depicting the preferential fMRI activation to images of objects compared to textures. Regions that showed higher activation to objects versus textures are indicated in red, while regions that showed preferential activation to textures versus objects are colored in blue. Data are shown from a postero-lateral view of an 'inflated' right hemisphere, and an unfolded version of the same hemisphere. Early retinotopic cortex is preferentially activated by textures, while high-order object areas show preferential activation to objects. Note the wide extent of preferential object activation, ranging from parietal cortex dorsally, to inferotemporal cortex ventrally. Major subdivisions of this band include the lateral occipital region (LO), the more ventrally located posterior fusiform gyrus (pFs), intra parietal sulcus (IPS), and the collateral sulcus (Cos). Note that the activation in LO to common objects is higher (brighter colors). It is further proposed that the pFs and collateral regions may be part of a larger structure, termed the ventral occipitotemporal cortex (VOT).

It can be also counter-argued, and rightly so, that the overall activity produced by object images is not the correct measure for recognition processes—we don't know how much of this brain activity is truly involved in recognizing objects. Viewing a picture of an object might initiate a large set of visual processes—such as identifying the texture of the object, identifying its location in space, whether it is moving or not, etc.

The theme we would like to propose here is that by searching for functional topographies within this widespread activity pattern, i.e. by searching for its 'functional architecture', we could obtain data that are relevant not only to the neuro-anatomy of human object-areas, but also for computational and physiological models of object representations in the human brain.

We will focus here on our results from ventral stream areas situated in occipitotemporal cortex, although interesting issues are still unaccounted for regarding the nature of object selectivity in parietal and frontal cortex.

9.2 Two-dimensional topographical organization in human object-related cortex

Examining occipitotemporal cortex, we can identify two orthogonal dividing lines of this expanse—both indicated in Fig. 9.1. First, along the posterior–anterior axis we see a hierarchy-related transition from a more local-feature representation posteriorly, to a more global, abstract representation anteriorly. Originally (Malach *et al.* 1995; Tootell *et al.* 1996) we identified a marked functional differentiation along this axis between early retinotopic cortex and the more lateral and anterior object-related cortex. In particular, object cortex manifested a relatively higher level of position and size invariance compared to early visual areas.

In Fig. 9.1 the transition between early retinotopic cortex and object-related areas is reflected in the change from preferential activation by visual textures—which are rich in local visual detail, but do not segment into any three-dimensional object (blue regions) into the more anterior, preferential object activation (red regions). By correlating these object-selectivity maps to more conventional retinotopy, we find that the posterior border of object-selective regions is close to the anterior border of retinotopic cortex (not shown). In subsequent studies we have demonstrated that the hierarchical transformation is actually gradual and that the object selectivity is built in small steps along the posterior–anterior axis (Grill-Spector *et al.* 1998; Lerner *et al.* 2001). Such functional hierarchy has been well documented in the primate visual cortex and has parallels both in single neuron recordings, and in extrinsic and intrinsic neuronal connectivity patterns (Rockland and Pandya 1979; Felleman and Van Essen 1991; Amir *et al.* 1993).

At the orthogonal axis we find a dorso-ventral differentiation that divides the occipitotemporal object-related cortex into a region we have termed LO (lateral occipital), located dorsally, overlapping the posterior–dorsal part of the occipitotemporal gyrus and the lateral occipital sulcus. More ventrally there is a region, which we originally termed the posterior fusiform gyrus (pFs), largely corresponding to the fusiform face area (Kanwisher *et al.* 1997; Grill-Spector *et al.* 1999; Grill-Spector and Malach 2001). More recent findings (see below) strongly suggest that this cortex may, in fact, be part of a larger organization which includes also the collateral sulcus. We have termed this larger organization the ventral occipitotemporal cortex (VOT) (Malach *et al.* 2002).

One functional distinction between LO and VOT is already evident in Fig. 9.1; that is, LO is more strongly activated by common objects compared to VOT. Additional differentiation, which applies at least to the difference between LO and pFs, and which has been described by our group and others, includes a higher sensitivity in LO for position and size changes (Grill-Spector *et al.* 1999), finer retinotopy (Levy *et al.* 2001) and motion sensitivity (Beauchamp *et al.* 2002; Kourtzi *et al.* 2002). Note that although LO is more sensitive to position and size when compared to VOT, it is still far more invariant to these dimensions when compared to early retinotopic cortex (Malach *et al.* 1995).

Although the exact significance of these functional differences is not fully established, they do suggest that the pFs contains a more abstract representation compared to the more motion-related and metric characteristics of LO. This hypothetical differentiation may gain support from another recent finding (Amedi *et al.* 2001) which shows that LO contains regions activated by visual and haptic stimuli, but not by auditory object-related inputs. Note that a common feature of somatosensory and visual representations, which distinguishes them from auditory stimuli, is their metric representation of objects. Thus, the exclusion of auditory activation in LO may suggest a more geometrical nature to its representation.

9.3 Topographic organization within ventral occipitotemporal cortex

Going into finer detail, we can examine the functional subdivisions of the VOT itself. First, several previous studies (Puce *et al.* 1995; Martin *et al.* 1996; Kanwisher *et al.* 1997; Epstein and Kanwisher 1998; Halgren *et al.* 1999) have demonstrated striking functional subdivisions of the VOT related to activation by object category, such as face-related and building/scene-related specializations. What could underlie this consistent segregation? While many dimensions have been discussed in the literature, e.g. shape factors, different tasks, expertise effects (see Malach *et al.* 2002), we have recently found an unexpected dimension—visual field eccentricity, i.e. distance from the fovea (Fig. 9.2). More specifically, we found that the VOT constitutes a map of eccentricity-biases, which are less retinotopic compared to early visual cortex, yet show a similar, consistent organization. Thus, the medially located collateral sulcus shows a peripheral visual field bias, while the more lateral, fusiform gyrus has a central-field bias. Considered from a wider perspective, this mapping is not surprising—since the entire

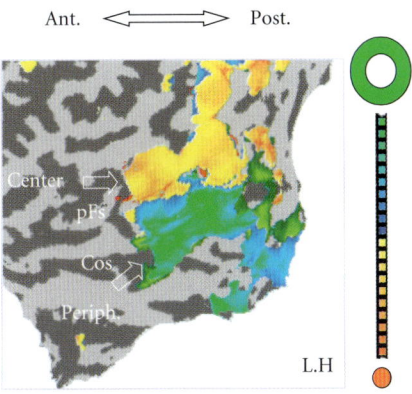

THE CARTOGRAPHY OF HUMAN VISUAL OBJECT AREAS | 199

Fig. 9.2 Eccentricity maps in ventral occipitotemporal cortex. An unfolded map of the visual cortex of a single subject. Relative activation to central visual field stimulation (1.8°, orange-yellow colors) and to peripheral stimulation (11.5–20°, green-blue colors). The segregation according to visual field eccentricity is evident. Note that the collateral sulcus shows a peripheral bias, while the more lateral fusiform gyrus shows a central-field bias.

Fig. 9.3 Relationship between object preference and eccentricity bias. Unfolded maps of the right and left hemispheres. Averaged activation of 11 subjects. Blue and green colors indicate central (3°) and peripheral (11.5–20°) visual field biases, respectively. Activations showing preferential activation to faces, letters, and building images (3° image size, presented centrally) are indicated by contours of orange, purple, and red, respectively. Note the clear association of faces and letters with central field and that of buildings with peripheral-field biases.

early visual cortex is organized around the eccentricity dimension. More interesting, perhaps, is the finding that the center/periphery segregation corresponds substantially (albeit perhaps not in the most anterior portions of the VOT) to the segregation of certain object categories (Fig. 9.3) (Levy *et al.* 2001; Hasson *et al.* 2002). In particular, faces show a preferential activation to the center-biased lateral activation in the posterior fusiform gyrus, mainly in the right hemisphere (orange contour in Fig. 9.3), while letters and words show preferential activation overlapping the center-biased representation in the left hemisphere (purple contour in Fig. 9.3). Finally, buildings occupy a more medial, periphery-biased location in the collateral sulcus of both hemispheres (red contour in Fig. 9.3).

9.4 Why should visual field eccentricity be tied at all to various object images?

A striking specialization that goes along with eccentricity is the retinal sampling density, or magnification factor. Starting from the retina, foveal regions are represented at a much higher density of neuronal elements compared to peripheral regions. Thus, the eccentricity specialization naturally relates to different recognition processes that might tap into high- or low-resolution sampling. Indeed, faces often involve extremely demanding visual tasks (e.g. detecting facial expressions). This raises the possibility that the various VOT subdivisions mediate different recognition processes, which depend differentially on sampling density. It should be emphasized that the notion proposed here is that the central/peripheral specialization is a by-product of the different demands on magnification factor. Thus, the association of buildings with peripheral bias does not imply that buildings are viewed exclusively with peripheral vision but that recognition of building images is based on integration of feature elements that are situated at large distances from each other, which are thus not processed by the small, central fovea.

Interestingly, this hypothesis gains support from an entirely different direction, that of computational models of object recognition. Recently, the group of Ullman and colleagues (Ullman *et al.* 2002) has developed a model of object recognition that is based on detecting optimal object fragments. These fragments are selected by optimizing two opposing constraints—their ability to differentiate the represented object category from others, and their probability of being present in the object images. For the current discussion, an important finding of the studies of Ullman's group is that when comparing the distribution of optimal car and face fragments the latter showed a bias towards high-resolution fragments. This finding is nicely compatible with the fact that car images activate more medial (and hence more peripherally biased) VOT locations compared to face images (for example, see Halgren *et al.* 1999).

It will be interesting to test this model further by examining the fragment distribution of house and word images—here the prediction is that words should contain higher-resolution fragments compared to house images.

Finally, we can move into a finer scale of analysis within, say, the fusiform face area. At this level the issue of topography becomes a controversy by itself. It is still not clear yet whether there are sharp lines of division between functionally homogeneous modules at this scale of cortical organization.

A fundamental question here concerns the issue of modularity. That is, to what extent a cortical region that shows preferential activation to a particular object category, such as faces, consists purely of face-selective neurons, and to what extent it might contain within it small subsets of neurons that are sensitive to other object categories, considered 'non-optimal' for this region. A straightforward way to address this issue could be through improvements in the spatial resolution of the imaging method, up to columnar resolution (Menon and Goodyear 1999; Cheng *et al.* 2001).

However, here we would like to suggest an indirect approach to this question which circumvents the fMRI spatial resolution limits. The approach is based on the adaptation paradigm that we and others have studied extensively (Grill-Spector and Malach 2001; Kourtzi and Kanwisher 2001; Huk and Heeger 2002). In this paradigm we use the fact that neuronal activation declines if the same stimulus is presented repeatedly. The advantage of the adaptation paradigm is that it does not depend on spatial isolation of neuronal populations, but uses repetition effects to 'tag' separate sets of neurons which are otherwise intermingled within the imaged voxel. The physiological process that underlies fMR adaptation is most likely the ubiquitous effect of neuronal fatigue, i.e. a temporal high-pass filtering operation that removes monotonous, repetitive information.

Recently we have provided indirect evidence that the presence of the adaptation process can be used to gauge the activity level of small sets of neurons embedded in the imaged voxel (Avidan *et al.* 2002). More specifically, we have found that a global reduction in neuronal activation through contrast reduction led to a concomitant reduction in fMR adaptation. In the experiment, we have presented line drawings of cars and faces in three contrast levels (high, medium, and low) and in two types of blocks— either a block of 12 repetitions of a single image (repeated condition) or a block of 12 different images (different condition). The result showed that as the image contrast was reduced there was a reduction in fMRI activation. Importantly, the fMRI adaptation level was also lowered, so that at low activation level, the repeated and different conditions were much closer in signal magnitude compared to the high activation level. For example, in LO, a drastic lowering of image contrast reduced activation from 1.3 to 0.4% and the adaptation level (measured as 1−repeated/different conditions) was concomitantly reduced from 0.6 to 0.2. This effect was highly significant, and was found across all object cortex.

A likely interpretation of this result is that the adaptation level is correlated to the activity level of the neurons prior to the initiation of the adaptation effect. Thus, the level of adaptation may be taken as an indirect measure of neuronal activity level. Note that this measure is not dependent on a global activation of the entire neuronal population within the imaged voxel, so even small sets of highly active neurons could be detected through their adaptation effect.

Given these results, one can translate the question of the homogeneity of the neuronal population into an adaptation question, i.e. will a non-optimal stimulus, such as building images in the fusiform face area, produce a robust adaptation effect? Figure 9.4

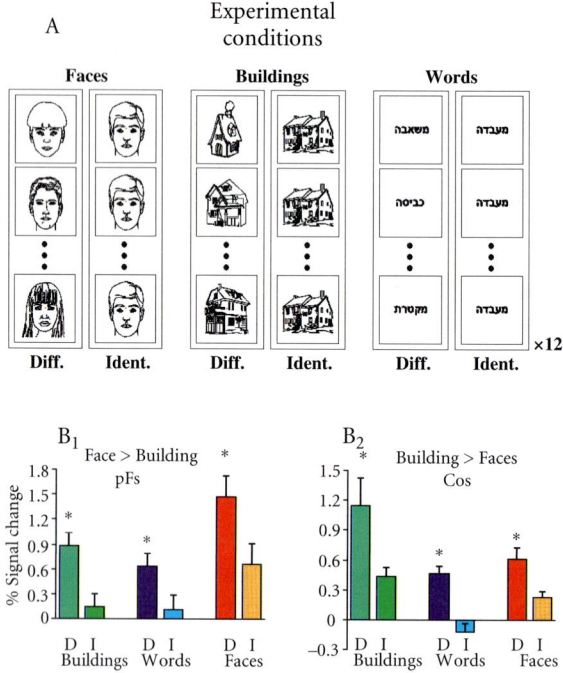

Fig. 9.4 Object-related adaptation effects. In (a) are shown the images used in the object adaptation experiment. Epochs in which different line drawings of either faces, buildings and different Hebrew words (Diff.) were contrasted with epochs in which images of an identical item from these categories was repeated throughout the epoch (Ident.). Epoch lasted 12 seconds, each image was presented for 800 ms followed by 200 ms blank. (b) The percentage signal change obtained in ventral occipitotemporal cortex preferentially active to faces (B_1, pFs) and to buildings (B_2, Cos). Note that the overall activation to different images indeed manifested a clear selectivity, i.e. a higher activation to faces in pFs and to buildings in Cos. However, the adaptation level, i.e. the reduced signal in the identical condition compared to the different condition, was as significant for the 'non-optimal' stimuli as it was for the optimal ones. We hypothesize that these adaptation effects are due to small subsets of neurons in each object-selective region which are actually selective to the non-optimal stimuli.

shows that the answer is clear-cut and positive. Note that despite the fact that the overall fMRI activation in a cortical region is markedly reduced by 'non-optimal' images, none the less it undergoes a robust adaptation by the same 'non-optimal' stimuli. We interpret these results to suggest that even within voxels that are highly selective for an object category, there are small sets of neurons that are tuned to other, 'non-optimal' objects.

9.5 Conclusion

The theme that we have tried to develop here concerns the importance of topography to our understanding of high-level visual areas. Given the immense level of neuronal neighborhood interactions in the cortex, the rules that govern the topographic layout of neurons are bound to be crucial for optimizing the specific computations performed in each cortical area. Thus, our finding of the eccentricity organization in occipitotemporal cortex can be viewed from this perspective as highlighting the importance of the magnification factor in recognition processes. Further understanding of the topographical rules in high-order areas will thus provide us with new insights into their function.

Acknowledgements

This study was funded by JSMF 99–28 CN-QUA.05, GIF I-0576–040.01/98 and Israel Academy 8009 grants. We thank M. Harel for the brain flattening, E. Okon for technical help, D. Ben Bashat and Y. Assaf, for their help in the fMRI set-up, and T. Hendler for help in conducting experiments.

References

Amedi, A., Malach, R., Hendler, T., Peled, S., and Zohary, E. (2001). Visuo-haptic object-related activation in the ventral visual pathway. *Nature Neuroscience,* **4**(3), 324–30.

Amir, Y., Harel, M., and Malach, R. (1993). Cortical hierarchy reflected in the organization of intrinsic connections in macaque monkey visual-cortex. *Journal of Comparative Neurology,* **334**(1), 19–46.

Avidan, G., Hasson, U., Hendler, T., Zohary, E., and Malach, R. (2002). Analysis of the neuronal selectivity underlying low fMRi signals. *Current Biology,* **12**, 1–20.

Beauchamp, M. S., Lee, K. E., Haxby, J. V., and Martin, A. (2002). Parallel visual motion processing streams for manipulable objects and human movements. *Neuron,* **34**(1), 149–59.

Cheng, K., Waggoner, R. A., and Tanaka, K. (2001). Human ocular dominance columns as revealed by high-field functional magnetic resonance imaging. *Neuron,* **32**(2), 359–74.

Epstein, R. and Kanwisher, N. (1998). A cortical representation of the local visual environment. *Nature,* **392**, 598–601.

Felleman, D. J. and Van Essen, D. C. (1991). Distributed hierarchical processing in the primate cerebral cortex. *Cerebral Cortex,* **1**(1), 1–47.

Grill-Spector, K. and Malach, R. (2001). fMR Adaptation: A tool for studying the functional properties of cortical neurons. *Acta Psychologica,* **107**, 293–321.

Grill-Spector, K., Kushnir, T., Hendler, T., Edelman, S., Itzchak, Y., and Malach, R. (1998). A sequence of object-processing stages revealed by fMRI in the human occipital lobe. *Human Brain Mapping,* **6**(4), 316–28.

Grill-Spector, K., Kushnir, T., Edelman, S., Avidan, G., Itzchak, Y., and Malach, R. (1999). Differential processing of objects under various viewing conditions in the human lateral occipital complex. *Neuron,* **24**(1), 187–203.

Halgren, E., Dale, A. M., Sereno, M. I., Tootell, R. B. H., Marinkovic, K., and Rosen, B. R. (1999). Location of human face-selective cortex with respect to retinotopic areas. *Human Brain Mapping,* **7**(1), 29–37.

Hasson, U., Levy, I., Behrmann, M., Hendler, T., and Malach, R. (2002). Eccentricity bias as an organizing principle for human high-order object areas. *Neuron,* **34**(3), 479–90.

Huk, A. C. and Heeger, D. J. (2002). Pattern-motion responses in human visual cortex. *Nature Neuroscience*, **5**(1), 72–5.

Kanwisher, N., McDermott, J., and Chun, M. M. (1997). The fusiform face area: A module in human extrastriate cortex specialized for face perception. *Journal of Neuroscience*, **17**(11), 4302–11.

Kourtzi, Z. and Kanwisher, N. (2001). Representation of perceived object shape by the human lateral occipital complex. *Science*, **293**(5534), 1506–9.

Kourtzi, Z., Bulthoff, H. H., Erb, M., and Grodd, W. (2002). Object-selective responses in the human motion area MT/MST. *Nature Neuroscience*, **5**(1), 17–18.

Lerner, Y., Hendler, T., Ben-Bashat, D., Harel, M., and Malach, R. (2001). A hierarchical axis of object processing stages in the human visual cortex. *Cerebral Cortex*, **11**(4), 287–97.

Levy, I., Hasson, U., Avidan, G., Hendler, T., and Malach, R. (2001). Center-periphery organization of human object areas. *Nature Neuroscience*, **4**(5), 533–9.

Malach, R., Reppas, J. B., Benson, R. R., Kwong, K. K., Jiang, H., Kennedy, W. A., Ledden, P. J., Brady, T. J., Rosen, B. R., and Tootell, R. B. H. (1995). Object-related activity revealed by functional magnetic-resonance-imaging in human occipital cortex. *Proceedings of the National Academy of Sciences of the United States of America*, **92**(18), 8135–9.

Malach, R., Levy, I., and Hasson, U. (2002). The topography of high-order human object areas. *Trends in Cognitive Sciences*, **6**(4), 176–84.

Martin, A., Wiggs, C. L., Ungerleider, L. G., and Haxby, J. V. (1996). Neural correlates of category-specific knowledge. *Nature*, **379**(6566), 649–52.

Menon, R. S. and Goodyear, B. G. (1999). Submillimeter functional localization in human striate cortex using BOLD contrast at 4 tesla: Implications for the vascular point-spread function. *Magnetic Resonance in Medicine*, **41**(2), 230–5.

Puce, A., Allison, T., Gore, J. C., and McCarthy, G. (1995). Face-sensitive regions in human extrastriate cortex studied by functional MRI. *Journal of Neurophysiology*, **74**(3), 1192–9.

Rockland, K. S. and Pandya, D. N. (1979). Laminar origins and terminations of cortical connections of the occipital lobe in the rhesus monkey. *Brain Research*, **179**, 3–20.

Tootell, R. B. H., Dale, A. M., Sereno, M. I., and Malach, R. (1996). New images from human visual cortex. *Trends in Neurosciences*, **19**(11), 481–9.

Ullman, S., Vidal-Naquet, M., and Sali, S. (2002). Visual features of intermediate complexity and their use in classification. *Nature Neuroscience*, **5**(7), 1–6.

Chapter 10

The neural bases of subliminal priming

Stanislas Dehaene

Abstract

Psychologists have long reported that words that are made invisible by forward and backward masking can nevertheless cause behavioral priming effects. Functional neuroimaging can now be used to explore the neural bases of masked priming. Subliminal priming causes reduced activation in multiple areas (fusiform gyrus, intraparietal sulcus, and motor cortex), in direct correspondence with behavioral manifestations of priming at the orthographic, semantic, and motor level. This implies that a whole stream of processors can operate unconsciously. The neural code in each area can be assessed by varying prime-target relations. A simple mathematical framework is proposed that tentatively relates priming at the voxel level with the shape of the tuning curves of single neurons in the underlying tissue. Priming thus provides a general method to study the fine microcode in each brain region (the 'priming method').

10.1 **Introduction**

Visual masking refers to the reduction of the visibility of a visual stimulus when it is preceded or followed, in close spatial and temporal proximity, by the presentation of another visual stimulus called the 'mask'. Many different sorts of masking exist (Enns and Di Lollo, 2000). In this chapter, I concentrate on pattern masking, a technique that has been used extensively in psychology to study the levels of representation of words. Under appropriate conditions, words that are followed by a masking pattern can be rendered invisible, even to trained observers. Yet, behavioral priming experiments have repeatedly indicated that the masked words are nevertheless processed unconsciously. Behavioral manifestations of masked priming have been observed at various levels including the orthographic, the phonological, and possibly the semantic level (see, for example, Marcel 1983; Bowers *et al.* 1984; Forster and Davis, 1984; Ferrand and Grainger, 1994; Greenwald *et al.* 1996; Neely and Kahan, 2001). Thus, masking appears as a useful paradigm to study two central questions in cognitive neuroscience: the various levels of representations of words, and the nature of the differences between conscious and unconscious processes.

In this chapter, I review a number of studies that have used masked priming to examine the neural activity evoked by subliminal words. I first examine which brain areas are affected by masked priming effects, and what this tells us about the level of coding of words in those areas. I then propose a mathematical formalism to explain how priming at the single-cell level can translate into macroscopic changes in fMRI activation. Finally, I discuss what changes in brain activation patterns may occur when a word crosses the threshold of consciousness and becomes visible.

10.2 General logic of subliminal priming studies

In a typical subliminal priming experiment (Fig. 10.1), each trial consists in the consecutive presentation, at the same screen location, of a random configuration of symbols or geometrical shapes (pre-mask), a first word (the prime), another random configuration of symbols or shapes (post-mask), and a second word (the target). The prime word is presented very briefly, typically for 10–50 ms. The target is presented for a much longer duration, typically 500 ms. Finally, the stimulus onset asynchrony between the prime and target is short (typically 60–120 ms). Under those conditions, subjects report seeing only the masks and the target word, but not the prime.

This subjective invisibility can be confirmed by asking subjects to perform an explicit task on the primes. In recent experiments, my colleagues and I have used a two-alternatives forced-choice identification task in which subjects have to select, amongst two alternative words, the one that matches the prime. As illustrated in Fig. 10.1, with a 29-ms prime presentation duration, performance typically does not differ from the chance level of 50% success (Dehaene *et al.* 2001). Other tasks such as prime

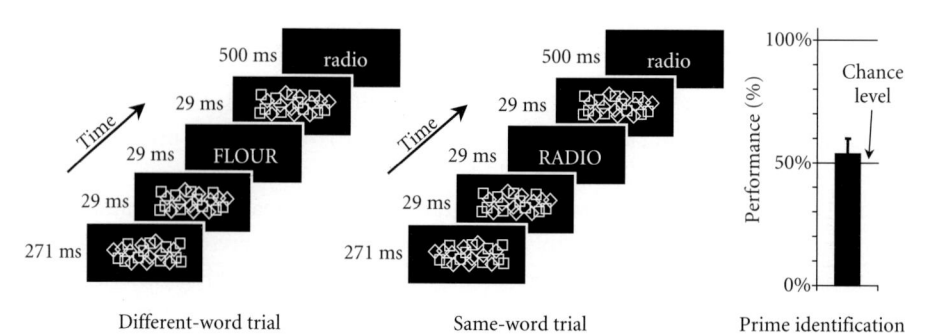

Fig. 10.1 Design of a typical repetition priming study (redrawn from Dehaene *et al.* 2001). The sequence of stimuli (left) typically includes a long pre-mask (here, two successive random geometrical patterns) as well as an optional brief post-mask surrounding the prime word. Chance-level performance is obtained on a two-alternative forced-choice test of prime identification (right).

presence–absence judgment (Dehaene *et al.* 1998*b*) or prime categorization (Naccache and Dehaene 2001*b*), yield similar results.

In spite of their subjective invisibility, the prime words impact on the processing of target words. This can be demonstrated by varying the relation between the prime and target. The simplest design compares trials in which the same word is presented twice, as both prime and target, with trials in which different words are presented as prime and target (Fig. 10.1). In this condition of repetition priming, response times to the target are consistently shorter on repeated than on non-repeated trials (e.g. Forster and Davis 1984; Forster 1998).

By systematically varying the physical, phonological, or even semantic proximity between the prime and target, it is then possible to probe which levels of word processing underlie this facilitation effect. As described below, in combination with brain imaging, this becomes a powerful method to probe the brain areas that are traversed by the wave of activation induced by subliminal primes, and the level of representation associated with each of them.

10.3 Subliminal priming in the visual word form system

Dehaene *et al.* (2001) first examined the cerebral bases of subliminal word repetition priming. fMRI data were collected in a fast event-related paradigm while subjects performed a bimanual semantic classification task on visual words. Unbeknownst to them, each target word was preceded by a subliminal prime (Fig. 10.1). I used a 2×2 design in which the prime and target could be the same word or different words, and could appear in the same or different case, thus defining four types of events. The amount of activation in each of those events was identified relative to a fifth 'null' event, in which only the masks were presented and no response was required.

The behavioral results showed that response times were faster on repeated trials, whether or not the words shared the same case (Fig. 10.2). In searching for the cerebral bases of this effect, two distinct types of brain regions were identified. The right extrastriate occipital cortex showed repetition suppression only for physically identical primes and targets, suggesting a role for right visual areas in coding the precise visual features of the letters (Marsolek *et al.* 1992). The left fusiform gyrus, however, showed repetition suppression whenever the same word was repeated, whether in the same case or not (Fig. 10.2). Thus, this region appears to encode the word string in a case-independent fashion.

This region may provide the cerebral substrate of the visual word form system (Warrington and Shallice, 1980), a structural representation of visual words as an ordered sequence of abstract letter identities or multi-letter graphemes, invariant for size, font, and case (Cohen *et al.* 2000, 2002; Dehaene *et al.* 2002). It is roughly symmetrical to the right-hemispheric fusiform face area, and may play for visual word recognition the same role that similar or neighboring regions of the fusiform and lingual

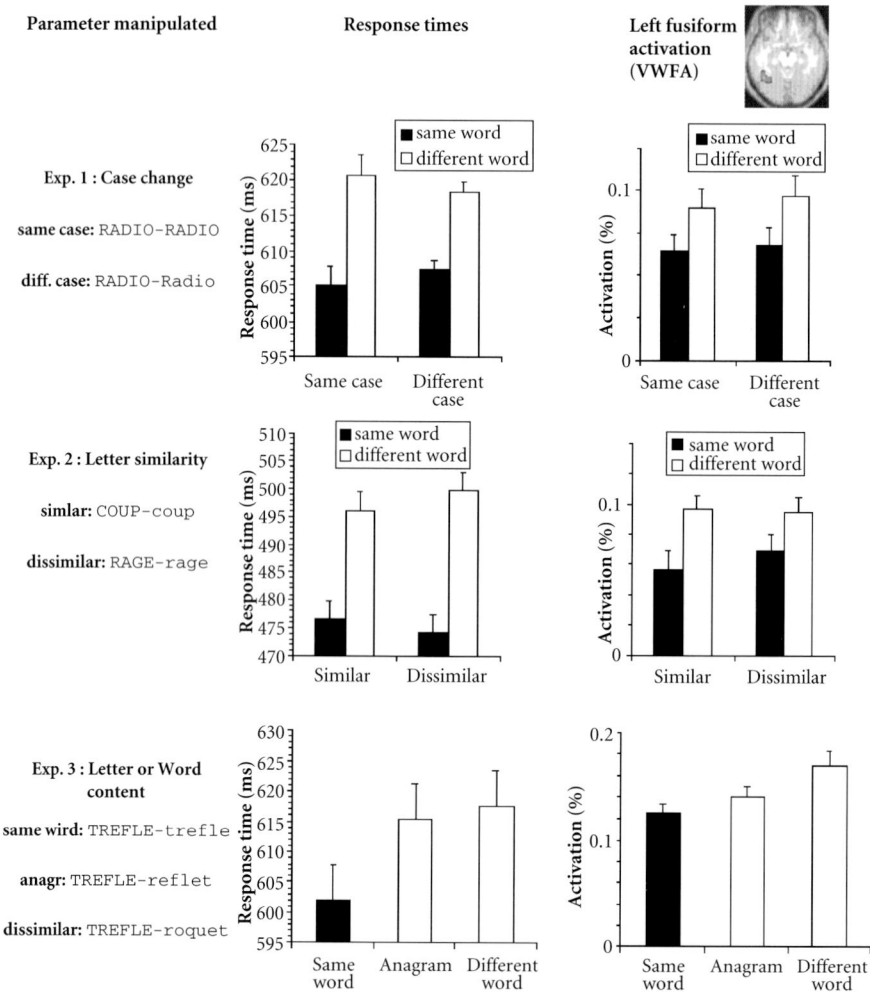

Fig. 10.2 Parallels and differences between behavioral and neuroimaging measures of word priming. In the visual word form area of the left fusiform gyrus (inset), repetition suppression resists case change (top) and is independent of letter similarity (middle), paralleling behavioral priming. However, the fusiform gyrus appears insensitive to the whole-word configuration of letters and thus shows priming even for anagrams, whereas behavioral priming shows whole-word selectivity (redrawn from data in Dehaene et al. 2001; Dehaene 2003). In this and subsequent figures, error bars represent standard errors calculated across subjects after removal of overall intersubject variability common to all conditions.

gyri play for other visual objects, such as faces, objects, or places (Haxby, Chapter 4 this volume). In adults, it has become partially attuned to a specific script, as shown by its greater response to real words than to consonant strings of similar arbitrary shape (Cohen et al. 2002). Indeed, in children this area activates in direct proportion to the child's reading skills, and its response is absent in dyslexic readers who have not developed expertise in word recognition (Shaywitz et al. 1998; 2002; Paulesu et al. 2001).

In the rest of this paper, I refer to this area of left fusiform activation during word reading as the visual word form area (VWFA). It should be clear that this label does not imply that this area responds only to words, or even maximally to words as opposed to other categories of visual stimuli. Indeed, specificity for words is unlikely, given that reading is a recent cultural invention that appears to 'recycle' neural tissue initially engaged in broader objet recognition functions. The issue of category specificity is not addressed by my research, which is rather aimed at understanding what type of code for visual words is present in this region.

In order to further specify the exact nature of the word representation attained by subliminal primes in the VWFA, I recently performed two more word repetition priming experiments, the results of which can be reviewed only briefly here (Dehaene et al. 2003). The first examined whether visual features alone could explain priming in this region. To this end, the visual similarity of upper- and lowercase letters was manipulated (Fig. 10.2). Half of the prime-target pairs comprised words made of letters that are highly similar in upper and lower case (e.g. Oo, Pp). The other half used only highly dissimilar letters (e.g. Aa, Gg) for which the uppercase–lowercase association is essentially arbitrary. The same strip of left fusiform cortex as in Experiment 1 showed replicable subliminal repetition suppression that was present even when the letters were visually dissimilar (similar to the response times observations; see Fig. 10.2). This confirmed that this region is not solely concerned with visual shapes, but encodes letter strings using a culturally acquired abstract letter-identity code.

A second experiment examined whether single letters or larger units, such as graphemes or whole words, are encoded in the VWFA. In order to repeat letters without repeating words, anagrams were used. For instance, by priming the French target word 'REFLET' with the prime 'trefle', almost all of the middle letters (r, e, f, l, e) could be repeated. By moving the prime one letter position relative to the target (e.g. 'trefle' followed by 'REFLET', it was even possible to repeat those letters at the same screen location without repeating the same word. By comparing this to a word-repeated trial, with or without a shift in letter position, the nature and position invariance of the neural codes underlying priming could be tested. Would priming depend on letter repetition, word repetition, or both? The results revealed an interesting dissociation between posterior and anterior areas. The posterior portion of the VWFA ($y = -68$ in the Talairach coordinate system) showed repetition suppression only when the same letters were repeated at the same location, regardless of their case. This region thus holds a case-invariant but position-selective letter code. More anteriorily ($y = -56$), location-independent priming was found for both repeated words and anagrams compared to a control, non-repeated condition (Fig. 10.2). Thus, this region encodes a case- and position-invariant representation of visual units that are smaller than the whole word. Finally, in a still more anterior fusiform region ($y = -48$), priming became greater for repeated words than for anagrams, thus revealing a case- and position-independent whole-word code, or at least a code sensitive to the larger graphemic units that distinguish a word from its anagram. Behavioral response times seem to be sensitive only to this whole-word code (Fig. 10.2).

Two conclusions may be drawn from those studies at the visual word form level. First, behavioral priming effects provide only a coarse indication of the levels of representations traversed by a subliminal prime. Functional imaging reveals a much richer variety of priming effects, ranging from feature-based to letter- or whole-word-based priming. Secondly, subliminal primes can be processed quite far along the ventral visual identification pathway. Although several studies have identified a tight correlation between ventral fusiform activity and the contents of visual consciousness (e.g. Grill-Spector *et al.* 2000; Bar *et al.* 2001), fusiform activity is not sufficient for conscious reportability. Exactly what more is needed will be addressed in the final discussion.

10.4 Subliminal priming at the motor level

How far beyond early vision can a subliminal word travel without entering into consciousness? To address this issue, Lionel Naccache and I studied another priming paradigm in which semantic and motor components of priming could be assessed (Fig. 10.3). Subjects viewed number words or digits that they had to compare with 5. 'Larger' or 'smaller' responses were made by depressing a left-hand or right-hand button, with response assignment changing in the middle of the experiment. Unbeknownst to the subjects, each target number was preceded by a numerical prime. Various prime–target relations could be tested with this paradigm (Fig. 10.3): the notation of the prime and target could be the same or different; they could represent the same or different quantity; and finally they could yield the same motor response (both larger or both smaller than 5) or a different motor response (one larger and the other smaller).

Dehaene *et al.* (1998*b*) showed that the latter factor had a measurable impact on both behavior and brain activation. Behaviorally, subjects were faster by 24 ms when the prime-induced and target-induced responses were congruent than when they differed. At the brain level, measures of motor activation obtained with both fMRI and ERPs revealed activation of motor cortices on the side that would have been appropriate for responding to the prime. This was revealed by computing a lateralized response index based on the difference in motor activity from the hemisphere involved in programming the correct target-induced response and from the other hemisphere (this measure is called the 'lateralized readiness potential' (LRP) in the ERP literature, and we termed it the 'lateralized BOLD response' (LBR) for fMRI). This index was larger on motorically congruent than on incongruent trials. With the higher temporal resolution afforded by ERPs, it was possible to identify two successive stages, an early one in which the LRP was only induced by the subliminal prime and a later one in which this small motor bias was overcome by the much larger motor response induced by the supraliminal target. In summary, the activation induced by a masked prime can travel all the

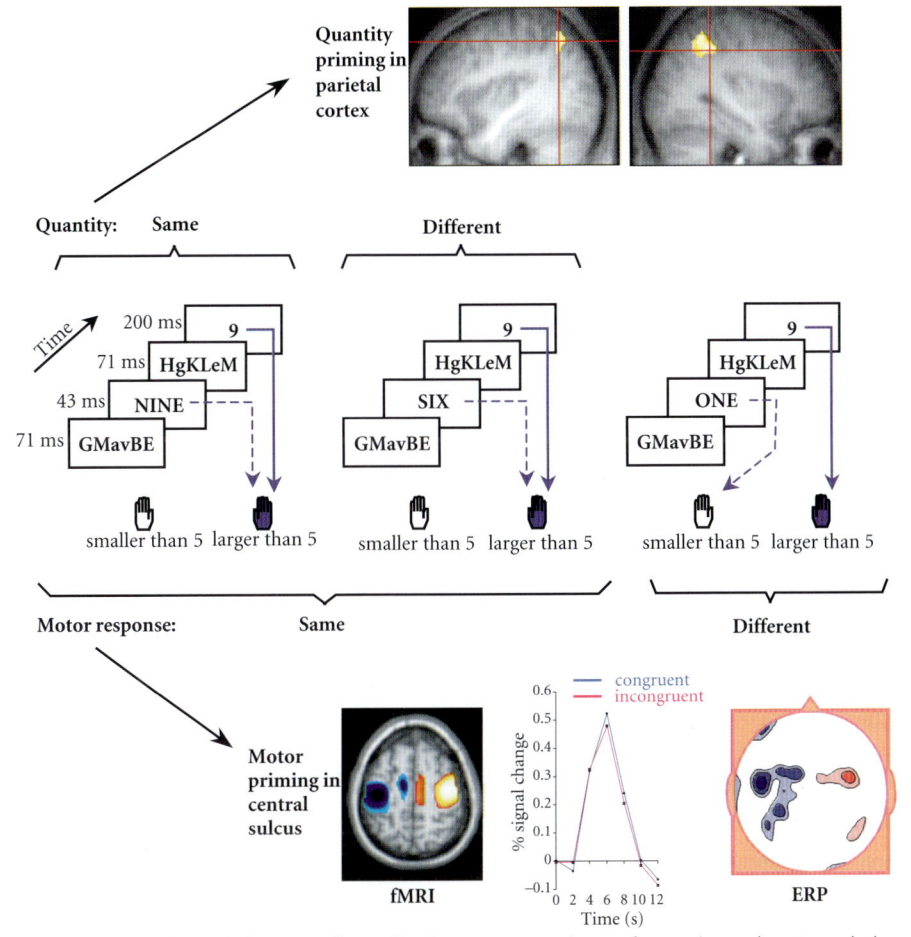

Fig. 10.3 A number priming paradigm affords separate analyses of quantity and motor priming (redrawn from Dehaene *et al.* 1998; Naccache and Dehaene 2001; see text for details).

way to the regions involved in programming a motor response. Similar results have been obtained using simpler visual stimuli such as arrow shapes (Neumann and Klotz 1994; Eimer and Schlaghecken 1998; Schmidt 2002; see also Jaskowski *et al.* 2002).

A crucial aspect of those finding is that motor priming is obtained although the assignment of responses to the left and right hands is arbitrary and is changed during the experiment. This implies that subliminal primes are processed along neural pathways that are temporarily established to comply with experimental instructions. Naccache *et al.* (2002) further confirmed the effect of top-down influences on subliminal priming by manipulating subjects' temporal attention in the number-priming paradigm. Across three experiments, the same prime–target pairs were presented in contexts in which their onset was either predictable or unpredictable. In Experiment 1, temporal predictability was manipulated by presenting trials at a fixed or variable lag with respect to trial onset. In Experiment 2, visual cues were presented or omitted prior to the prime–target pair. Finally, in Experiment 3, valid or invalid verbal cues specified when the prime–target pair was likely to appear. In all three situations, priming was found to depend critically on the deployment of temporal attention. Behavioral priming was reproducibly observed whenever subjects' attention could be focused on the prime–target pair. However, priming disappeared when the prime–target pair appeared at an unexpected time. Both motor and repetition priming were eliminated.

These results can be interpreted as showing that the focusing of temporal attention on the target plays a permissive role that benefits the processing of a temporally contiguous prime. In the absence of such attentional amplification, the processing of subliminal primes is considerably reduced or eliminated. Thus, subliminal priming is not independent of attention. This challenges the classical view of subliminal priming, based on automatic spreading activation theory (Neely 1991), according to which the unconscious processing of primes is passive and automatic (Posner and Snyder 1975; Schneider and Shiffrin 1977). If a processing pathway is prepared by top-down attention and intentions, it can be applied automatically to subliminal stimuli (Dagenbach *et al.* 1989; Hommel, 2000; Dehaene and Naccache 2001). However, if the organism is not prepared to process them, subliminal primes induce little or no spreading of activation.

10.5 Can subliminal priming occur at the semantic level?

Our finding of motor priming in the number comparison task was initially interpreted as a clear, though indirect, proof of *semantic* processing of subliminal number symbols. That subjects could activate the motor cortex of the hand that would have been appropriate for responding to the prime seemed to imply that subjects had unconsciously categorized the prime as larger or smaller than 5, thus implying semantic access.

However, an alternative interpretation was that the observed motor activation was due to direct motor specification (Neumann and Klotz 1994). Because a very small number of stimuli—the digits 1, 4, 6, 9, and the corresponding words—were repeatedly used as both primes and targets, subjects could have learned to associate each visual stimulus with the corresponding response, thus bypassing semantic access. Use of such a direct visuomotor route was recently demonstrated by Abrams and Greenwald (2000). In an affective categorization task, they showed that new primes that were made of fragments of previously seen targets yielded subliminal motor priming solely based on visual fragments, not on whole-word meaning. For instance, the prime word SMILE, created from the targets words SMUT and BILE, ended up paradoxically priming the negative rather than the positive response. Thus, although the task required semantic categorization, and although a priming effect was observed, the primes only received a shallow, non-semantic analysis of their component letters and the associated motor responses (see also Damian 2001).

Fortunately, new experiments and reanalyses have now demonstrated that the number priming results do not fall prey to a similar non-semantic interpretation. We have now replicated the original behavioral experiment with novel numbers that are only presented as primes, never as targets (Naccache and Dehaene 2001*b*). Because those numbers are never seen consciously and are never responded to, they cannot be associated with motor responses. Yet in two different experiments, those novel primes were found to cause significant motor priming, indicating that at least part of the motor priming effect arises from a genuinely semantic route. This positive effect of novel numerical primes has now been replicated and extended by others (Greenwald *et al.* 2003; Reynvoet *et al.* 2002), although it varies with the task contest and conscious instructions (Greenwald *et al.* 2003; Kunde *et al.* 2003).

Further analyses also demonstrated that the motor priming effect was present in the first block trial and that both motor priming and the classical semantic distance effect did not change with practice, which is inconsistent with the idea that the task is increasingly being performed using a non-semantic route.

Most crucially, another priming effect was identified that could only be interpreted at the semantic level. This effect was termed 'quantity priming' (Fig. 10.3). Subjects were faster on repeated trials (e.g. prime 9, target 9) than on congruent non-repeated trials (e.g. prime 6, target 9). Because in both cases the response induced by the prime is the same and is congruent with the target, motor priming cannot contribute to this effect. The effect is found with the same magnitude even when the notation is changed (e.g. prime NINE, target 9), suggesting that it occurs at an abstract level. Furthermore, behavioral experiments have demonstrated that the amount of quantity priming varies monotonically with the semantic distance between the prime and the target (e.g. 9–9 versus 8–9, 7–9, or 6–9), thus confirming that it originates at the level of quantity coding (Koechlin *et al.* 1999; Reynvoet and Brysbaert 1999).

Finally, at the neural level, fMRI demonstrated that quantity priming was associated with repetition suppression in the left and right intraparietal sulci (Naccache and Dehaene, 2001*a*). Those regions are consistently activated in a variety of number processing and calculation tasks, whenever subjects have to manipulate the quantity associated with numerical symbols (e.g. Dehaene *et al.* 1999). They are thought to represent numerical quantities on an internal continuum analogous to a mental 'number line'. The fact that their activation is reduced when the same quantity is repeated twice, possibly in different notations, confirms that they encode particular numerical quantities and can be activated unconsciously.

In summary, priming effects with subliminal numbers have been observed at both the semantic (intraparietal) and motor levels. This provides a clear indication that semantic-level processing of masked primes is possible. It should be noted that digits are some of the most frequent visual symbols and are semantically unambiguous.

The ease and speed of visual-to-semantic transduction may explain why it seems easier to obtain semantic priming with numbers than with other types of words (Abrams and Greenwald 2000; Damian 2001).

10.6 **The 'priming method': priming as a neuroimaging tool**

The above examples illustrate how priming can be used in combination with neuroimaging to separate out distinct stages of the cortical coding of words. In occipital cortex, priming depends on repeating the same marks at the same location on the retina, suggesting a position-specific feature-level code. In fusiform cortex, the code for words is increasingly more abstract, as priming can be obtained when repeating the same letters in a different case, and sometimes even at a different location. In intraparietal cortex, the code is even more elaborate, as priming can be elicited by repeating the same numerical quantity using different symbols, such as TWO and 2.

Inferences based on neuroimaging observations of priming are powerful because they indirectly reveal both the *precision* and the *abstraction* of a neural code. When repetition suppression is observed in the intraparietal area for, say, 9 followed by 9 as opposed to 6 followed by 9, this implies that the stimuli 9 and 6 are distinguished within this area: there are partially distinct populations of neurons encoding those two stimuli. I refer to such stimulus-selectivity as the 'precision' of the neural code. On the other hand, when the amount of repetition suppression is found not to differ for, say, NINE followed by 9 as opposed to 9 followed by 9, this implies that the neurons in this area recognize the similarity between the stimuli NINE and 9. Thus, there are populations of neurons that abstract away from the notation used to convey a number. I refer to this as the degree of 'abstraction' of the neural code.

Neither precision nor abstraction can be inferred on the basis of other neuroimaging designs, such as the subtraction, parametric, or conjunction methods. For instance, direct subtraction of the brain activations to the digits 6 and 9 would most likely result in an absence of detectable differences at the scale of resolution typical of most fMRI studies (2–4 mm). Furthermore, although a statistical conjunction or masking operation can be used to demonstrate that the word SIX and the digit 6 cause similar patterns of activation in the intraparietal cortex, this would not prove that they are being coded by the same neurons. At present, only the priming method affords such non-invasive inferences about the neural microcode.

Use of the priming method does not require that the primes be subliminal, or that primes and targets appear in close succession. Indeed, a very similar approach, often called the fMRI adaptation method, was initially applied to both blocked and event-related studies of mid- to long-term stimulus repetition, where it yielded detailed information on the coding of words and objects (e.g. Grill-Spector *et al.* 1999; Buckner *et al.* 2000; Kourtzi and Kanwisher 2000; Wagner *et al.* 2000; Vuilleumier *et al.* 2002). However, the use of subliminal primes is advantageous because it ensures that the results are not contaminated by attentional or strategic biases. For instance, novel stimuli may be more interesting or attention-grabbing than old ones. Conversely, reduced fMRI activation on consciously repeated trials may be due to lesser attention.

Conscious stimulus repetition may also result in large-scale task reorganization. For instance, subjects may merely repeat a previous response instead of computing it *de novo*, resulting in large-scale changes in brain activation on repeated trials (Raichle et al., 1994). Crucially, such attentional or strategic activation changes may occur in brain regions distant from those involved in the initial recognition that the stimulus was repeated, and hence may no longer be informative about the local neural microcode. Subliminal primes effectively prevent such strategies.

Finally, subliminal primes have only a very temporary effect, typically lasting less than 500 ms, and so effectively act as a *tracer* of existing representations. As will be argued further below, the priming method is effective under the assumption that neuronal tuning curves are not being changed by exposure to the primes. Consciously visible primes may lead to long-lasting learning-induced changes that no longer warrant the inference that repetition suppression reflects the pre-existing code in a given area.

A clear drawback of subliminal primes, however, is that they result in small and often hard-to-detect activation. The effects reported earlier were very small and required averaging across subjects. To study the code in higher-level areas, where prime-induced activation barely penetrates, conscious priming may be the only possibility. Still, it is desirable to prevent subject's awareness of repetitions, especially since it has been observed that attention to repetition may disrupt the repetition suppression phenomenon (Henson *et al.* 2002).

10.7 A mathematical formalism linking priming at the cellular and neuroimaging levels

I now elaborate a minimal mathematical formalism of the priming method, in order to formulate more precisely its predictions according to the nature of the experimental design. Consider the brain-imaging signal $I(s)$ measured, for instance with fMRI, in response to a single stimulus, s. In any voxel of a few cubic millimeters, this signal typically reflects the summed activation of several million neurons, and can thus be written as:

$$I(s) = h\left(\sum_i f_i(s)\right) \tag{10.1}$$

where $f_i(s)$ is the firing rate of neuron i in response to stimulus s, and h represents the hemodynamic response function which transforms firing rates into measurable changes in blood flow and oxygenation.[1] (For simplicity, I neglect here the time dimension of brain-imaging signals; a more complete treatment would treat h as a temporal convolution operator.)

The firing rate, $f_i(s)$, depends on a measure of the distance between the actual stimulus, s, and the neuron's preferred stimulus s_i^{pref}. This can be written as

$$f_i(s) = F\left(\left\|s - s_i^{\text{pref}}\right\|\right) \tag{10.2}$$

where $\|\ldots\|$ is a distance metric, and F is a monotonically decreasing function (possibly a Gaussian) specifying how firing rate varies with proximity in similarity space to the preferred stimulus. The metric $\|\ldots\|$ characterizes the neurons' tuning curves and how they weight different dimensions of stimulus variation. It is expected to vary considerably between different areas. Many neurons in visual area V1, for instance, prefer a certain stimulus location and orientation, but are blind to stimulus variation along other dimensions, such as object identity. A different metric would describe neurons in inferotemporal cortex, where a neuron might respond preferentially to a certain face while being insensitive to its exact illumination or location on the retina. Determination of the stimulus preference metric is crucial to understanding the function of a given brain area. In what follows, I examine to what extent brain-imaging measurements, $I(s)$, can reveal the preference metric of the underlying neurons.

10.7.1 The subtraction method

In this method, one simply examines the response of a given voxel to different stimuli, s_1 and s_2. Then two different cases can occur. In the simplest case, one stimulus may be a better stimulus than the other for a majority of the neurons: $\left\|s_1 - s_i^{\text{pref}}\right\| < \left\|s_2 - s_i^{\text{pref}}\right\|$ for most neurons, i. Then it follows that $I(s_1) > I(s_2)$: one can measure a stronger response of the whole voxel to the preferred stimulus that to the non-preferred stimulus. This situation occurs when one is probing the neural 'macrocode' of a given cortical region, for instance the retinotopy of the primary visual cortex. At some level, it is generally possible to identify a parameter that is preferred by most, if not all, neurons in the considered area (e.g. a specific retinal location for a V1 voxel, or faces versus objects for a given fusiform voxel, etc.).

When one is probing the neural 'microcode', however, it is generally not the case that all the neurons have the same global preference. Rather, populations of neurons with different preferences are intermixed at a sub-millimetric spatial scale. In any given voxel, there might be just as many neurons that prefer s_1 over s_2, than there are neurons that prefer s_2 over s_1. It follows that $I(s_1) \cong I(s_2)$: the signals recorded to stimuli s_1 and s_2 cannot be distinguished. It is important to note that this can occur even though the stimuli s_1 and s_2 vary along a dimension which is relevant to the neurons in the considered area, and even though they excite different populations of neurons. The spatial intermingling of these populations, together with the coarse size of the voxels relative to the neural scale, renders them impossible to detect with current imaging methods.

10.7.2 The priming method

The priming method measures the response to the same target stimulus, s, but varies the context in which it is presented by preceding it with variable primes (Fig. 10.4).

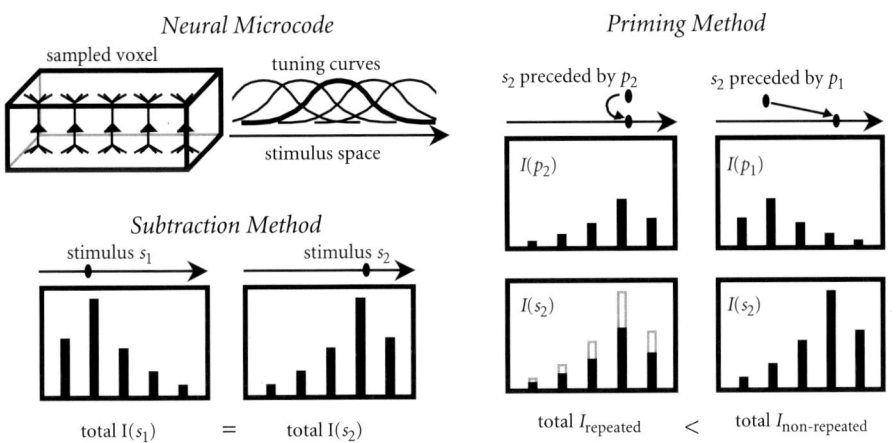

Fig. 10.4 Advantages of the priming method over the classical subtraction method in neuroimaging. An fMRI voxel typically contains over a million neurons, each tuned to a particular preferred stimulus in perceptual or abstract space (top left). Although two stimuli may be encoded by different populations of neurons within this voxel, activation averaged over the entire population of neurons is likely to fail to reveal any measurable difference between them (bottom left). However, thanks to the reduced level of activation associated with repetition suppression, differences may be observed by comparing the activation evoked by the same stimulus s_2 when preceded either by the same prime stimulus p_2, or by a different stimulus p_1 (right). This is true only if the neural code in this voxel distinguishes between p_1 and p_2.

For simplicity, we consider here only the simplest case, in which a single presentation of a prime stimulus, p, precedes the presentation of the target stimulus, s. Assuming a linear fMRI response, the measured signal is a simple combination of the signals induced by p and s:

$$I(p,s) = \alpha I(p) + \beta I(s) \qquad (10.3)$$

In this equation, the attenuation factors α and β characterize how the measured signal differs from the simple sum of the activations $I(p)$ and $I(s)$ that would be found if p and s were presented in isolation. The α term reflects the fact that the prime-induced activation can be attenuated, for instance because the prime p is presented for a shorter duration, or much in advance of the target s, etc. At one extreme, $\alpha = 0$ if the time interval between p and s is sufficiently long that one can measure the activation induced by s without any contamination by the activation induced by p (e.g. in long-term priming experiments).

However, the β term is more important. This term is introduced to take into account the repetition suppression effect of p on s. I tentatively propose that, at the single neuron level, the neural response of neuron i to stimulus s preceded by prime p can be written as:

$$f_i(p,s) = \left(1 - A\left(\|p - s\|\right)\right) F\left(\|s - s_i^{\text{pref}}\|\right) \qquad (10.4)$$

where A is a function that decreases monotonically from A_0 to zero. A_0 measures the amount of habituation to a repetition of the same stimulus, expressed as a percentage of the normal response to that stimulus when it is not repeated. The function A, which specifies the shape and amount of repetition effect, is expected to vary with the details of the priming paradigm. For instance, it would be expected to decrease exponentially as a function of the prime-target asynchrony in a subliminal priming situation.

Equation 10.4 makes several assumptions that remain to be verified. First it assumes that repetition suppression has a multiplicative effect on neuronal firing: the curve indexing the tuning of the cell to the target is unchanged by priming, but the intensity of its firing is modulated as a function of the distance between the prime and target. Secondly, it assumes that repetition suppression is a continuous phenomenon: there can be graded levels of adaptation as a function of the degree of similarity between prime and target, rather than a mere categorical distinction between repeated and non-repeated situations. Thirdly, and crucially, equation 10.4 assumes that the *same metric* underlies stimulus preference and repetition suppression. The hypothesis here is that what counts as a repetition must be measured along the same dimensions that characterize tuning curves in the area of interest. If, say, an inferotemporal neuron prefers a given face while being 'blind' to variations in illumination, equation 10.4 predicts that repetition suppression should be sensitive to exactly the same variables, such that the neuron would show exactly the same degree of habituation if the preferred face was repeated with a different illumination or with the same illumination. This putative property of repetition suppression remains to be confirmed at the single-cell level.

Although speculative, equation 10.4 has the advantage of making explicit quantitative predictions that may guide the much-needed further electrophysiological research into the characteristics of the repetition suppression phenomenon. At present, the biological mechanisms of repetition suppression are unknown. One possibility is that the amount of repetition suppression is determined locally, for instance as a direct reflection of how much the neuron has fired in the recent past. However, such a simple habituation mechanism is unlikely to be sufficient. It would predict that habituation at the single-neuron level is just a function of the similarity between the prime and the neuron's preferred stimulus, rather than between the prime and the target, as proposed in equation 10.4. This is a significant distinction that could be separated experimentally. The present proposal that repetition suppression depends on a global measure of prime–target similarity suggests that one should rather look for a mechanism at the neuronal population level, for instance in the speed with which a set of neurons converges to a stable population code.

From equation 10.4, one can derive a version of equation 10.3, which makes explicit the repetition suppression effect at the voxel, rather than at the single-neuron, level:

$$I(p,s) = \alpha I(p) + \left(1 - A\left(\|s - p\|\right)\right)I(s) \qquad (10.5)$$

This equation shows clearly that the metric underlying the tuning curves of individual neurons in the measured voxel is directly accessible in the brain-imaging measure, I.

10.7.3 Does repetition priming always result in reduced activation?

As seen from equation 10.5, $I(p,s)$ reflects both stimulus preference and priming. Can we systematically predict the direction of the resulting effect? Suppose that an experiment is set up to compare the activation induced by repeating the same stimulus twice, $I_{repeated} = I(s,s)$, with the activation induced on non-repeated trials $I_{non\text{-}repeated} = I(p,s)$ ($p \neq s$). Furthermore suppose that non-repeated trials cause no repetition suppression ($A(\|s-p\|)=0$). The predicted difference then reduces to:

$$I_{repeated} - I_{non\text{-}repeated} = \alpha(I(s)-I(p)) - A_0 I(s) \quad (10.6)$$

Equation 10.6 shows that, in general, repetition priming need not always result in a reduced activation. Rather, the measured effect of priming on the fMRI signal is the sum of two terms. The first term reflects the stimulus preference of the observed region and can be positive or negative, while the second term reflects repetition suppression and is always negative. In general, there is no guarantee that the second term will always win. If the activation induced by the non-repeated prime, $I(p)$, is smaller than the one induced by the repeated target $I(s)$, then it is possible for activation to be greater on repeated trials than on non-repeated trials.[2]

However, there is a remarkable complementarity between the priming method and the classical subtraction method. As noted previously, the subtraction method does not work when the contrasted stimuli p and s are such that $I(p) = I(s) = I$. Yet, it is precisely in this situation that the priming method works best. In that case, equation 10.6 reduces to:

$$I_{repeated} - I_{non\text{-}repeated} = -A_0 I \quad (10.7)$$

which is necessarily negative. More generally, the observed activation on a given trial with prime p and target s, according to equation 10.5, reduces to:

$$I(p,s) = \left(\alpha + 1 - A(\|s-p\|)\right) I \quad (10.8)$$

Thus, the observable $I(p,s)$ is found to depend *only* on the measure of similarity between prime p and target s for the voxel under consideration.

10.7.4 The 2 × 2 priming design

When stimuli p and s do not cause identical levels of activation ($I(p) \neq I(s)$), or when it is unknown whether they do, it is still possible to obtain a brain-activation measure which is proportional to the similarity between s and p, and which is guaranteed to decrease on repeated trials. This is achieved by using a two-by-two stimulus design in which the same two stimuli, s_1 and s_2, serve both as primes and as targets.

Four prime–target pairs are possible, two repeated ones s_1–s_1 and s_2–s_2, and two non-repeated ones s_1–s_2 and s_2–s_1. The following activation levels are predicted:

	Prime s_1	Prime s_2
Target s_1	$I(s_1, s_1) = \alpha I(s_1) + (1 - A_0) I(s_1)$	$I(s_2, s_1) = \alpha I(s_1) + \left(1 - A(\|s_1 - s_2\|)\right) I$
Target s_2	$I(s_1, s_2) = \alpha I(s_2) + \left(1 - A(\|s_1 - s_2\|)\right) I$	$I(s_2, s_2) = \alpha I(s_2) + (1 - A_0) I(s_2)$

The interaction term of this matrix then provides a necessarily negative measure proportional to s_1–s_2 similarity:

$$\text{Interaction} = I_{\text{repeated}} - I_{\text{non-repeated}} = -\left(A_0 - A(\|s_1 - s_2\|)\right) \frac{I(s_1) + I(s_2)}{2} \quad (10.9)$$

Furthermore, the influence of $I(s_1)$ and $I(s_2)$ can be removed by normalizing the interaction term with respect to the amount of activation obtained on repeated trials:

$$Q(s_1, s_2) = \frac{I_{\text{repeated}} - I_{\text{non-repeated}}}{I_{\text{repeated}}} = -\frac{A_0 - A(\|s_1 - s_2\|)}{\alpha + 1 - A_0} \quad (10.10)$$

The resulting quantity, Q, is always negative. Note that Q does not depend on $I(s_1)$ and $I(s_2)$, and hence is devoid of all stimulus preference effects. Q provides a direct estimate of how similar the stimuli s_1 and s_2 appear to neurons in the studied voxel. By varying the nature of the relation between s_1 and s_2, and measuring Q using a 2 × 2 priming design, it becomes possible to directly evaluate the metric or 'neural code' by which neurons in the considered area classify various stimuli as similar or dissimilar. All of the experiments reviewed earlier used such a design, with the same stimuli appearing with the same frequency as primes and targets while only the prime–target relation varied.

10.8 Conclusion: subliminal and supraliminal processing

Neuroimaging studies supplement behavioral studies of priming by identifying the neural structures successively activated by the primes, and their associated neural codes. The results indicate that subliminal primes activate abstract visual, semantic, and even motor levels of representation. Given such an unexpected depth of processing, why do the subliminal primes fail to reach consciousness? And what additional neural events occurs when the threshold of consciousness is reached? In this conclusion, I have space only for a few theoretical and empirical considerations on these issues.

The neuronal workspace model (Dehaene et al. 1998a; Dehaene and Naccache 2001) proposes that access to consciousness is associated with the sudden, coordinated firing of many neurons distributed throughout the brain, though particularly concentrated in parietal, prefrontal, and cingulate cortices, and linked by a dense network of long-distance axons. Dense recurrent connections allow information that reaches this level to become quickly available to many different processes, including categorization, evaluation, memorization, and intentional action. This contrasts with the encapsulation of subliminal information to narrow and highly specialized processors. Because workspace neurons are densely interconnected by multiple recurrent connections, the hypothesis predicts the existence of a dynamical threshold for masked perception. Above a critical level, neural activation becomes self-amplifying and suddenly jumps to a much higher and longer-lasting level ('ignition'). Below this threshold, activation is weak and transient, though it can still propagate in a feedforward manner.

According to this theory, masking acts by shutting down the transient activation of the prime before the interplay of bottom-up and top-down connections has time to amplify it and render it accessible at a whole-brain scale. This view is compatible with neurophysiological recordings in inferotemporal cortex or frontal eye field while awake monkeys were presented with brief masked or unmasked stimuli (Rolls and Tovee 1994; Kovacs et al. 1995; Thompson and Schall 1999). Masking was found to leave the initial phasic response of the neurons unchanged, while preventing almost entirely the later, more sustained part of the firing train (see also Super et al. 2001). I obtained analogous evidence from fMRI and ERPs using masked or unmasked words (Dehaene et al. 2001). Compared to word-absent trials, masked words caused only a small transient activation which was increasingly smaller as one moved from extrastriate cortex to fusiform gyrus and precentral cortex. Unmasking the words greatly enhanced activation in the same areas and created a large-scale activation which included distant parietal, inferior prefrontal, and midline precentral/cingulate cortices. Unmasking also enhanced the long-distance correlation between those sites and, in ERP recordings, was associated with a enhanced late positive complex (P300) that was absent or greatly reduced in the masked situation. Further work should examine whether this predicted correlation between large-scale amplification and conscious access continues to hold on a single-trial basis during masking as well as other paradigms, such as change blindness or the attentional blink (see Vogel et al. 1998; Beck et al. 2001).

Notes

1. The possibility that the fMRI signal may reflect synaptic potentials rather than spikes should not fundamentally alter the present mathematical formalism because during fast bottom-up stimulus processing, local firing rate and local synaptic activity are expected to be highly correlated (for recent discussions, see Logothetis, 2003; Smith et al. 2002).
2. Other factors may also contribute to the observation of repetition enhancement rather than repetition suppression. In particular, my equations suppose that the neuronal tuning curves remain unchanged by priming, and that priming acts only as a temporary and reversible multiplicative factor on neural firing rates. If, however, the presentation of the primes leads to learning-induced changes in tuning curves, this may compensate or even overturn repetition-suppression effects (see, for example, Henson et al. 2000).

References

Abrams, R. L. and Greenwald, A. G. (2000). Parts outweigh the whole (word) in unconscious analysis of meaning. *Psychol Sci*, **11**(2), 118–24.

Bar, M., Tootell, R. B. H., Schacter, D. L., Greve, D. N., Fischl, B., Mendola, J. D., Rosen, B. R., and Dale, A. M. (2001). Cortical mechanisms specific to explicit visual object recognition. *Neuron*, **29**, 529–35.

Beck, D. M., Rees, G., Frith, C. D., and Lavie, N. (2001). Neural correlates of change detection and change blindness. *Nature Neurosci*, **4**, 645–50.

Bowers, J. S., Vigliocco, G., and Haan, R. (1998). Orthographic, phonological, and articulatory contributions to masked letter and word priming. *J Exp Psychol Hum Percept Perform*, **24**(6), 1705–19.

Buckner, R. L., Koutstaal, W., Schacter, D. L., and Rosen, B. R. (2000). Functional MRI evidence for a role of frontal and inferior temporal cortex in amodal components of priming. *Brain*, **123**, 620–40.

Cheesman, J. and Merikle, P. M. (1984). Priming with and without awareness. *Percept Psychophys*, **36**, 387–95.

Cohen, L., Dehaene, S., Naccache, L., Lehéricy, S., Dehaene-Lambertz, G., Hénaff, M. A., and Michel, F. (2000). The visual word form area: Spatial and temporal characterization of an initial stage of reading in normal subjects and posterior split-brain patients. *Brain*, **123**, 291–307.

Cohen, L., Lehericy, S., Chochon, F., Lemer, C., Rivaud, S., and Dehaene, S. (2002). Language-specific tuning of visual cortex? Functional properties of the Visual Word Form Area. *Brain*, **125**(Pt 5), 1054–69.

Dagenbach, D., Carr, T. H., and Wilhelmsen, A. (1989). Task-induced strategies and near-threshold priming: Conscious effects on unconscious perception. *J Mem Lang*, **28**, 412–43.

Damian, M. F. (2001). Congruity effects evoked by subliminally presented primes: automaticity rather than semantic processing. *J Exp Psychol Hum Percept Perform*, **27**(1), 154–65.

Dehaene, S., Jobert, A., Naccache, L., Ciuciu, P., Poline, J. B., Le Bihan, D. *et al.* (2003). Letter binding and invariant recognition of masked words: Behavioural and neuroimaging evidence. *Psychological Science*, in press.

Dehaene, S. and Naccache, L. (2001). Towards a cognitive neuroscience of consciousness: Basic evidence and a workspace framework. *Cognition*, **79**, 1–37.

Dehaene, S., Kerszberg, M., and Changeux, J. P. (1998a). A neuronal model of a global workspace in effortful cognitive tasks. *Proc Natl Acad Sci USA*, **95**, 14529–34.

Dehaene, S., Naccache, L., Le Clec'H, G., Koechlin, E., Mueller, M., Dehaene-Lambertz, G., van de Moortele, P. F., and Le Bihan, D. (1998b). Imaging unconscious semantic priming. *Nature*, **395**, 597–600.

Dehaene, S., Spelke, E., Stanescu, R., Pinel, P., and Tsivkin, S. (1999). Sources of mathematical thinking: Behavioral and brain-imaging evidence. *Science*, **284**, 970–4.

Dehaene, S., Naccache, L., Cohen, L., Le Bihan, D., Mangin, J. F., Poline, J. B., and Rivière, D. (2001). Cerebral mechanisms of word masking and unconscious repetition priming. *Nature Neurosci*, **4**, 752–8.

Dehaene, S., LeClec'H, G., Poline, J. B., LeBihan, D., and Cohen, L. (2002). The visual word form area: A prelexical representation of visual words in the fusiform gyrus. *NeuroReport*, **13**(3), 1–5.

Eimer, M. and Schlaghecken, F. (1998). Effects of masked stimuli on motor activation: behavioral and electrophysiological evidence. *J Exp Psychol Hum Percept Perform*, **24**(6), 1737–47.

Enns, J. T. and Di Lollo, V. (2000). What's new in visual masking. *Trends Cog Sci*, **4**, 345–52.

Ferrand, L. and Grainger, J. (1994). Effects of orthography are independent of phonology in masked form priming. *Q J Exp Psychol [A]*, **47**(2), 365–82.

Forster, K. I. (1998). The pros and cons of masked priming. *J Psycholinguist Res,* **27**(2), 203–33.

Forster, K. I. and Davis, C. (1984). Repetition priming and frequency attenuation in lexical access. *J Exp Psychol Learn Mem Cogn,* **10**, 680–98.

Greenwald, A. G., Draine, S. C., and Abrams, R. L. (1996). Three cognitive markers of unconscious semantic activation. *Science,* **273**(5282), 1699–702.

Greenwald, A. G., Abrams, R. L., Naccache, L., and Dehaene, S. (2003). Long-term semantic memory versus contextual memory in unconscious number processing. *J Exp Psychol Learn Mem Cogn,* **29**, 235–47.

Grill-Spector, K., Kushnir, T., Edelman, S., Avidan, G., Itzchak, Y., and Malach, R. (1999). Differential processing of objects under various viewing conditions in the human lateral occipital complex. *Neuron,* **24**(1), 187–203.

Grill-Spector, K., Kushnir, T., Hendler, T., and Malach, R. (2000). The dynamics of object-selective activation correlate with recognition performance in humans. *Nature Neurosci,* **3**(8), 837–43.

Henson, R., Shallice, T., and Dolan, R. (2000). Neuroimaging evidence for dissociable forms of repetition priming. *Science,* **287**(5456), 1269–72.

Henson, R. N. A., Shallice, T., Gorno-Tempini, M.-L., and Dolan, R. J. (2002). Face repetition effects in implicit and explicit memory tests as measured by fMRI. *Cereb Cortex,* **12**, 178–86.

Hommel, B. (2000). The prepared reflex: Automaticity and control in stimulus-response translation. In S. Monsell and J. Driver (Eds.), *Control of cognitive processes: Attention and Performance, Vol. XVIII* pp. 247–73. MIT Press, Cambridge MA.

Jaskowski, P., van der Lubbe, R. H., Schlotterbeck, E., and Verleger, R. (2002). Traces left on visual selective attention by stimuli that are not consciously identified. *Psychol Sci,* **13**(1), 48–54.

Koechlin, E., Naccache, L., Block, E., and Dehaene, S. (1999). Primed numbers: Exploring the modularity of numerical representations with masked and unmasked semantic priming. *J Exp Psychol Hum Percept Perf,* **25**, 1882–1905.

Kourtzi, Z. and Kanwisher, N. (2000). Cortical regions involved in perceiving object shape. *J Neurosci,* **20**(9), 3310–18.

Kovacs, G., Vogels, R., and Orban, G. A. (1995). Cortical correlate of pattern backward masking. *Proc Natl Acad Sci USA,* **92**, 5587–91.

Kunde, W., Kiesel, A., and Hoffmann, J. (2003). Conscious control over the content of unconscious cognition. *Cognition,* **88**(2), 223–42.

Logothetis, N. K. (2003). The underpinnings of the BOLD functional magnetic resonance imaging signal. *J Neurosci,* **23**(10), 3963–71.

Marcel, A. J. (1983). Conscious and unconscious perception: Experiments on visual masking and word recognition. *Cogn Psychol,* **15**, 197–237.

Marsolek, C. J., Kosslyn, S. M., and Squire, L. R. (1992). Form-specific visual priming in the right cerebral hemisphere. *J Exp Psychol Learn Mem Cogn,* **18**, 492–508.

Naccache, L. and Dehaene, S. (2001*a*). The priming method: imaging unconscious repetition priming reveals an abstract representation of number in the parietal lobes. *Cereb Cortex,* **11**(10), 966–74.

Naccache, L. and Dehaene, S. (2001*b*). Unconscious semantic priming extends to novel unseen stimuli. *Cognition,* **80**, 215–29.

Naccache, L., Blandin, E., and Dehaene, S. (2002). Unconscious masked priming depends on temporal attention. *Psychol Sci,* **13**, 416–24.

Neely, J. H. (1991). Semantic priming effects in visual word recognition: A selective review of current findings and theories. In D. Besner and G. W. Humphreys (Eds.), *Basic processes in reading. Visual word recognition.* pp. 264–336. Erlbaum, Hillsdale NJ.

Neely, J. H. and Kahan, T. A. (2001). Is semantic activation automatic? A critical re-evaluation. In H. L. Roediger, J. S. Nairne, I. Neath, and A. M. Surprenant (Eds.), *The nature of remembering: Essays in honor of Robert G. Crowder.* pp. 69–93. American Psychological Association, Washington DC.

Neumann, O. and Klotz, W. (1994). Motor responses to non-reportable, masked stimuli: Where is the limit of direct motor specification. In C. Umiltà and M. Moscovitch (Eds.), *Attention and Performance XV: Conscious and non-conscious information processing.* pp. 123–50. MIT Press, Cambridge MA.

Paulesu, E., Demonet, J. F., Fazio, F., McCrory, E., Chanoine, V., Brunswick, N., Cappa, S. F., Cossu, G., Habib, M., Frith, C. D., and Frith, U. (2001). Dyslexia: cultural diversity and biological unity. *Science,* **291**(5511), 2165–7.

Posner, N. I., and Snyder, C. R. R. (1975). Attention and cognitive control. In S. R. L. (Ed.), *Information processing and cognition: The Loyola symposium.* Erlbaum, Hillsdale NJ.

Raichle, M. E., Fiez, J. A., Videen, T. O., MacLeod, A. K., Pardo, J. V., Fox, P. T., and Petersen, S. E. (1994). Practice-related changes in human brain functional anatomy during non-motor learning. *Cereb Cortex,* **4**, 8–26.

Reynvoet, B. and Brysbaert, M. (1999). Single-digit and two-digit Arabic numerals address the same semantic number line. *Cognition,* **72**(2), 191–201.

Reynvoet, B., Caessens, B., and Brysbaert, M. (2002). Automatic stimulus-response associations may be semantically mediated. *Psychon Bull Rev,* **9**, 107–12.

Rolls, E. T. and Tovee, M. J. (1994). Processing speed in the cerebral cortex and the neurophysiology of visual masking. *Proc R Soc Lond B Biol Sci,* **257**(1348), 9–15.

Schmidt, T. (2002). The finger in flight: real-time motor control by visually masked color stimuli. *Psychol Sci,* **13**(2), 112–8.

Schneider, W., and Shiffrin, R. M. (1977). Controlled and Automatic Human Information Processing. In *I*(Vol. 84, pp. 1–66).

Shaywitz, S. E., Shaywitz, B. A., Pugh, K. R., Fulbright, R. K., Constable, R. T., Mencl, W. E., Shankweiler, D. P., Liberman, A. M., Skudlarski, P., Fletcher, J. M., Katz, L., Marchione, K. E., Lacadie, C., Gatenby, C., and Gore, J. C. (1998). Functional disruption in the organization of the brain for reading in dyslexia. *Proc Natl Acad Sci USA,* **95**(5), 2636–41.

Shaywitz, B. A., Shaywitz, S. E., Pugh, K. R., Mencl, W. E., Fulbright, R. K., Skudlarski, P., Constable, R. T., Marchione, K. E., Fletcher, J. M., Lyon, G. R., and Gore, J. C. (2002). Disruption of posterior brain systems for reading in children with developmental dyslexia. *Biol Psychol,* **52**, 101–10.

Smith, A. J., Blumenfeld, H., Behar, K. L., Rothman, D. L., Shulman, R. G., and Hyder, F. (2002). Cerebral energetics and spiking frequency: the neurophysiological basis of fMRI. *Proc Natl Acad Sci USA,* **99**(16), 10765–70.

Super, H., Spekreijse, H., and Lamme, V. A. F. (2001). Two distinct modes of sensory processing observed in monkey primary visual cortex (V1). *Nature Neurosci,* **4**, 304–10.

Thompson, K. G. and Schall, J. D. (1999). The detection of visual signals by macaque frontal eye field during masking. *Nature Neurosci,* **2**, 283–8.

Vogel, E. K., Luck, S. J., and Shapiro, K. L. (1998). Electrophysiological evidence for a postperceptual locus of suppression during the attentional blink. *J Exp Psychol Hum Percept Perform,* **24**(6), 1656–74.

Vuilleumier, P., Henson, R. N., Driver, J., and Dolan, R. J. (2002). Multiple levels of visual object constancy revealed by event-related fMRI of repetition priming. *Nat Neurosci,* **5**(5), 491–9.

Wagner, A. D., Koutstaal, W., Maril, A., Schacter, D. L., and Buckner, R. L. (2000). Task-specific repetition priming in left inferior prefrontal cortex. *Cereb Cortex,* **10**(12), 1176–84.

Warrington, E. K. and Shallice, T. (1980). Word-form dyslexia. *Brain,* **103**, 99–112.

Chapter 11

Audio-visual associative learning enhances responses to auditory stimuli in visual cortex

Désirée Gonzalo and Christian Büchel

Abstract

This study aimed to test whether audio-visual learning is mediated by modulation in sensory cortices (i.e. visual or auditory cortex). During fMRI scanning, subjects were exposed to repeated presentations of three tones and three visual stimuli. In 50% of presentations, one tone was paired with a face, and another one with a moving noise pattern. The third sound was not paired with any visual stimulus and was the target in an incidental detection task. Over time, activation in extrastriate visual cortex (V4) and the fusiform face area (FFA) increased in response to the sound that predicted the face, but not in response to the other two sounds. In bilateral area MT/V5, the tone predictive of the face and the unpaired tone led to marked decrease of activation over time. Compared to these responses, the tone predictive of the motion stimulus showed a relative increase over time. These results were obtained for the unpaired sounds, i.e. when they were not followed by a visual stimulus. In right prefrontal cortex, activation evoked by all tones was initially high, but then decreased over time in accord with previous reports linking this area to novelty detection. These results indicate that sensory cortex plays a crucial role in cross-modal learning and suggest a high degree of functional specificity in relation to the type of material being learnt.

11.1 Introduction

Much evidence has accumulated to support the belief that visual cortex is tuned to selectively respond to visual input (Rakic 1988; Levitt *et al.* 1997). Moreover, within visual cortex further functional specialization has been observed. An example is an area in the fusiform gyrus that responds selectively to human faces (Damasio *et al.* 1982;

Allison *et al.* 1994; Haxby *et al.* 1994; Puce *et al.* 1995; Puce *et al.* 1996; Kanwisher *et al.* 1997), or areas like MT/V5 which are sensitive to visual motion (Zeki *et al.* 1991; Tootell and Taylor 1995). However, some cells in visual cortex of the normally developed cat and rabbit brain can respond in addition to non-visual stimuli (Thompson *et al.* 1963; Voronin and Skrebitsky 1965; Bental *et al.* 1968; Spinelli *et al.* 1968). In humans, cross-modal effects between the tactile and visual modalities have been shown in blind (Sadato *et al.* 1996; Büchel 1998) and sighted subjects (Zangaladze *et al.* 1999; Macaluso *et al.* 2000; Amedi *et al.* 2001). With respect to the visual and auditory modalities, it has been shown that regional cerebral blood flow in occipital cortex in normal brains can be increased by auditory stimulation alone after having learnt that a sound predicts a visual event (McIntosh *et al.* 1998).

More specifically, inferences from studies of cross-modal integration suggest a possible mechanism underlying audio-visual learning, involving increased activity in sensory cortices over the duration of learning. Support for this hypothesis comes from a study showing that, for instance, visual cortex becomes increasingly responsive to an auditory stimulus in the course of audio-visual learning (McIntosh *et al.* 1998). The present experiment used event-related fMRI to examine this possibility in the context of on-line learning of audio-visual associations, whereby stimulus presentation in one sensory modality becomes predictive of consecutive presentation of a stimulus in a different modality. A positive contingency should develop if both stimuli are presented in a paired fashion (i.e. temporally correlated) often enough to establish an associative pattern.

We hypothesized that the visual cortex would be activated to single presentations of auditory stimuli after learning that certain tones are associated with specific images, and thus become predictive of visual presentations. More specifically, we tested the hypothesis that visual areas with a high degree of functional specialization, such as face-responsive areas in the fusiform gyrus and the motion-sensitive area V5/hMT (human MT) (Zeki *et al.* 1991; Tootell and Taylor 1995), would become increasingly responsive to a sound which became predictive of a face or a moving visual stimulus, respectively. Furthermore, we tested for the specificity of the association, i.e. whether an auditory stimulus is predictive of a specific visual stimulus, cross-modally activating the respective visual area.

11.2 Methods

11.2.1 Subjects

Twelve healthy volunteers (three females) with a mean age of 28.5 years (range 18–34 years) participated in the study, which was approved by the local ethics committee. All volunteers were free of neurological or psychiatric illness, had normal hearing and either normal or corrected to normal vision. All subjects provided informed consent and were free to withdraw from the study at any time. In a behavioral experiment, five additional subjects were tested outside the scanner for the purpose of demonstrating behavioral associative learning effects with our paradigm.

11.2.2 Materials and procedure

Subjects were presented with pictures of neutral faces, a moving visual noise stimulus, and sounds. All visual stimuli were back-projected on to a screen on top of the head coil. This screen was viewed through a 45° mirror. Sounds were presented through MR-compatible piezoelectric headphones. Visual stimuli consisted of grayscale photographs of neutral faces. Auditory stimuli were pure sine tones. The visual motion stimulus consisted of a random, black and white noise pattern moving leftwards at 2.8° per second, visible through a circular aperture (7.6° diameter) in the middle of a gray background.

In total, two different faces and one moving pattern were used. One face was presented alone all of the time (target); the other face was presented paired with a high-frequency tone (5 kHz) half of the time (50% partial reinforcement), and the moving pattern was paired to a low-frequency tone (200 Hz) half of the time (50% partial reinforcement) (Büchel *et al.* 1998). An additional middle frequency tone (1 kHz) was always presented alone (auditory control). The high and low tones were also presented alone half of the time. Six event types were defined: middle-frequency tone alone (A), face 1 alone = target (B), low-frequency tone alone (C), low-frequency tone plus moving pattern (D), high-frequency tone alone (E), and high-frequency tone plus face 2 (F) (Fig. 11.1). These events were presented as single trials and block-randomized, i.e. A–C–E–F–D–B, D–E–A–F–B–C, etc.

All auditory stimuli were presented for 2 s. In the case of paired stimuli, the tone always preceded the visual stimulus, and presentation overlapped during the last second. The intertrial interval was randomly jittered and varied between 8 and 12 s. There were 26 repetitions of each event type. Subjects were scanned while performing an incidental task, namely pressing a button on a keypad every time they saw the target

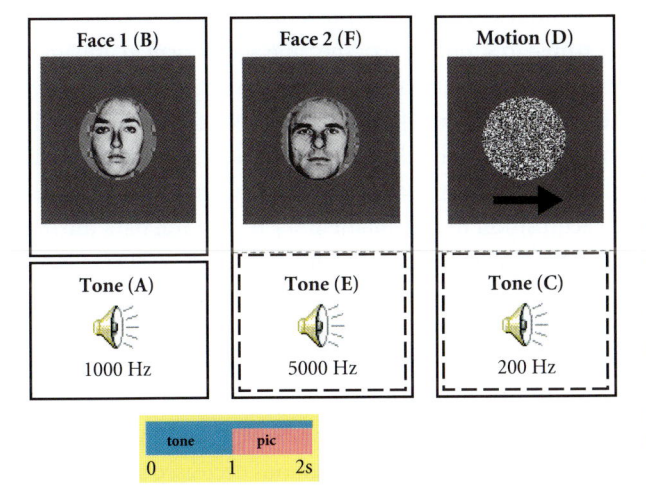

Fig. 11.1 Stimuli used in the experiment. Two experimental stimuli were a face paired to a tone (F and E) and a visual motion pattern (D) paired with a specific tone (C). We also used two control stimuli, a face not paired to a tone (B) and a tone not paired with a visual stimulus (A).

stimulus (unpaired face 1; A). The duration of the entire experiment was approximately 30 min.

Functional MRI data were acquired on a 1.5 Tesla MR scanner (Magnetom Siemens VISION whole-body MRI system, Erlangen, Germany) equipped with a standard transmit/receive head coil. Contiguous axial multislice T_2^*-weighted echoplanar images were acquired with an echo time of 40 ms, acquisition time of 2.5 s., flip angle 90°, voxel size 3 × 3 mm, matrix 64 × 64, 32 × 3 mm thick slices with 1 mm gap, positioned to cover the whole of the cerebrum. A total of 720 measurements were acquired from each subject in a single session. Subsequently, a high-resolution (1 × 1 × 1 mm voxel size) T_1-weighted structural MRI was acquired for each subject using a 3D-FLASH sequence.

Data analysis was performed using the general linear model as implemented in SPM99 (Friston *et al.* 1995) (http://www.fil.ion.ucl.ac.uk/spm). Scans were realigned, slice-timing corrected, normalized into stereotactic anatomical space as defined by the templates provided by SPM and smoothed with an 8 mm full width at half maximum isotropic Gaussian kernel. Each event type was modeled as delta functions convolved with a canonical hemodynamic response function as implemented in SPM99. In addition, the effect of learning was modeled as a time by condition interaction. This approach allowed us to assess whether signal intensity changed over the course of the experiment. Subsequently, the data were statistically analyzed using a random effects model in order to make population-based inferences. BOLD signal changes were tested with *t*-statistics and displayed as a statistical parametric map. The significance level was set at $P < 0.05$ corrected for the whole brain volume. With regard to activation in the fusiform face area, for which we had an *a priori* hypothesis, correction was performed for a region of interest only (sphere with 6 mm radius).

Additionally, for the purpose of acquiring behavioral data, another set of five subjects was tested outside the scanner with a slightly different version of the paradigm. They were presented with the same conditions but asked to press one of three buttons every time they saw face 1 (which was paired to a tone 50% of the time and presented unpaired the rest of the time), face 2 (which was always unpaired), or the moving stimulus (which was paired to a tone 50% of the time and presented unpaired the rest of the time). In this fashion, it was possible to calculate the difference in reaction times in response to presentation of the different visual stimuli during the course of the experiment. This procedure was adopted because the acquisition of complementary behavioral data during scanning would have confounded the data with sensorimotor activation.

11.3 Results

11.3.1 Behavioral data

In the behavioral experiment performed outside the scanner, we found that reaction times in response to the visual stimuli which were paired to sounds decreased significantly over time, whereas reaction times to the control unpaired visual stimulus remained almost unchanged (Fig. 11.2). The observed decrease in reaction times in response to the visual stimuli preceded by auditory stimuli is interpreted as a facilitatory effect produced by the tones which allow faster recognition of the paired visual stimuli. This implies that, as the audio-visual associations were learnt, the presentation of the sounds functioned as a cue to the associated visual stimuli. This effect is likely to result from associative learning, which is the process of interest in the fMRI exploration. This observation was statistically tested by means of an ANOVA. The interaction of condition by time was significant both for the visual motion condition (50% paired to the tone) relative to the control condition (unpaired face) $[F(4,16) = 3.8; P < 0.05]$, as well as for the face condition (50% paired to tone) relative to the control condition $[F(4,16) = 8.3; P < 0.05]$.

With respect to the fMRI experiment, subjects were explicitly tested after the scanning phase, both for awareness of the audio-visual associations in the paired conditions and for vigilance from consistency of target detection. All participating volunteers reported awareness of the pairings. Reaction times to the target visual stimulus became faster over time and indicated that subjects remained alert throughout the experiment.

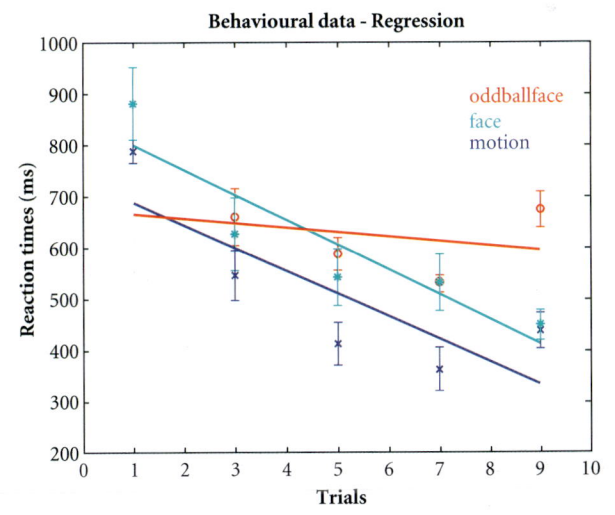

Fig. 11.2 Plot illustrating the behavioral data acquired outside the scanner as regression lines corresponding to each condition. The data points correspond to the group mean reaction times averaged over time every two trials. Subjects were exposed to 10 repetitions of the material. Note the interaction between conditions and time, whereby a significant difference in slope can be observed for the paired conditions ('face' and 'motion') relative to the unpaired condition ('oddballface'). This change is due to learning predictive associations between tones and the subsequently presented visual stimuli.

11.3.2 Neuroimaging data

The aim of this study was to test whether specialized areas in ventral visual cortex could specifically respond to auditory stimuli which became predictive of either faces or motion patterns. In this respect, it was hypothesized that ventral visual areas in the fusiform gyrus and MT/V5 would show increasing activation with time as an expression of learning the tone–face and tone–motion associations, respectively.

We specifically asked: (1) whether activation in FFA shows a significantly greater increase in activation over time for the tone predictive of a face; and (2) whether activation in V5/hMT elicited by the associated tone increases over time as compared to the non-predictive tone. For this purpose a contrast was performed in which the regressors assessing the time by condition interactions were compared and the result masked with the main effect of face-related responses, in the case of the tone–face association. This resulted in a differential time by condition interaction, addressing the question of where there is a stronger increase for the tone predicting the face or motion pattern as opposed to the non-predictive tone.

Masking with the main effect of faces (i.e. face versus rest) served to restrict the search volume to areas that respond to faces *per se*. The activation profile in posterior fusiform area ($x = 21$, $y = -75$, $z = -3$; $t(11) = 3.41$; $P < 0.05$ corrected; Fig. 11.3a) was very specific to the tone predictive of the face, because no increase was seen for the unpaired tone and for the tone paired with the motion stimulus. This can be seen by the plots showing how evoked activity changes as a function of time (Fig. 11.3). The specificity of the effect is indicated by increases to the consistently paired sound (Fig. 11.3b) whereas activity evoked by the two other tones remained constant over time (Fig. 11.3c and d). In comparison to previous studies of ventral occipital areas, the location of this activation is most likely to be within area V4 (Kastner et al. 1998).

More interestingly, we found a similar activation pattern in an area with coordinates comparable to published coordinates of the fusiform face area ($x = 45$, $y = -69$, $z = -3$; $t(11) = 3.46$; $P < 0.05$ corrected; Fig. 11.4a) (Kanwisher et al., 1997). Again, the effect was very specific to the tone predictive of the face, because no increase in this area was seen for the unpaired tone or for the tone paired with the motion stimulus (Fig. 11.4b–d).

A significant differential time by condition interaction was also obtained for left V5/hMT ($x = -48$, $y = -78$, $z = 0$; $t(11) = 3.34$; $P < 0.05$ corrected) and a trend for right V5/MT ($x = 48$, $y = -75$, $z = -6$; $t(11) = 1.83$; NS). However, in contrast to V4 or FFA, activation did not increase as a function of time for the tone paired with the motion stimulus, but tended to decrease over time with the unpaired tone and the tone paired with the face.

Another finding was a very significant exponentially decreasing response over time observed in the right inferior frontal gyrus ($x = 57$, $y = 21$, $z = 21$; $t(11) = 7.3$; $P < 0.05$ corrected; Fig. 11.5). This result arose from analyzing time-dependent signal changes common to both associated tones when presented alone (50% of presentations). This effect,

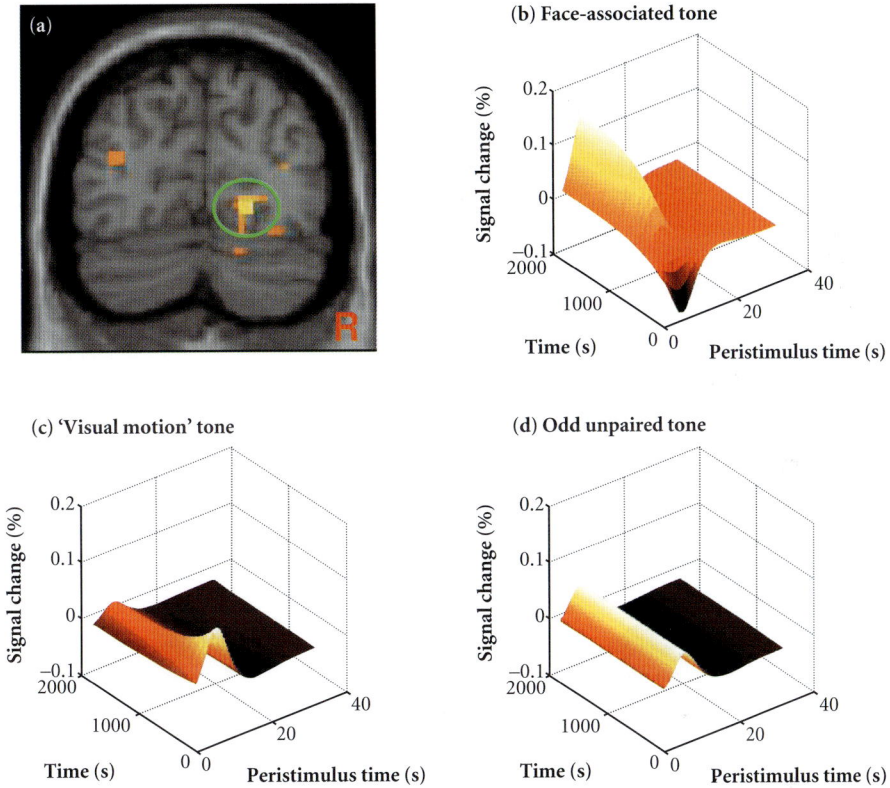

Fig. 11.3 Time by condition interaction in area V4. (a) Coronal section highlighting activation in the right posterior fusiform gyrus ($y = -75$), thresholded at $P < 0.05$ corrected. (b) Graphic plot showing evoked hemodynamic responses to the tone that was paired with the face in 50% of presentations in relation to the unpaired stimulus ('peristimulus time (s)' versus 'signal change (%)'). The additional third dimension ('scans') illustrates how this evoked response changes throughout the experiment. For the tone paired to a face in 50% of presentations, the ventral visual cortex shows a significant increase of evoked responses, indicated by smaller or negative amplitudes at the beginning and greater amplitudes towards the end of the experiment. This, however, was not the case for the tone that was paired to visual motion in 50% of presentations (c) or for the unpaired tone (d), which shows that this effect is specific.

however, was not specific since a similar decrease in activation over time was observed for the unassociated control tone (Fig. 11.5).

11.4 Discussion

Brain-adaptive mechanisms can function to a remarkable degree of specialization. We have shown that pairing stimuli in different sensory modalities in temporal proximity can render otherwise unimodal (visual) cortex to respond to a different modality. A very selective increase in activity was seen to a tone predictive of a face in ventral

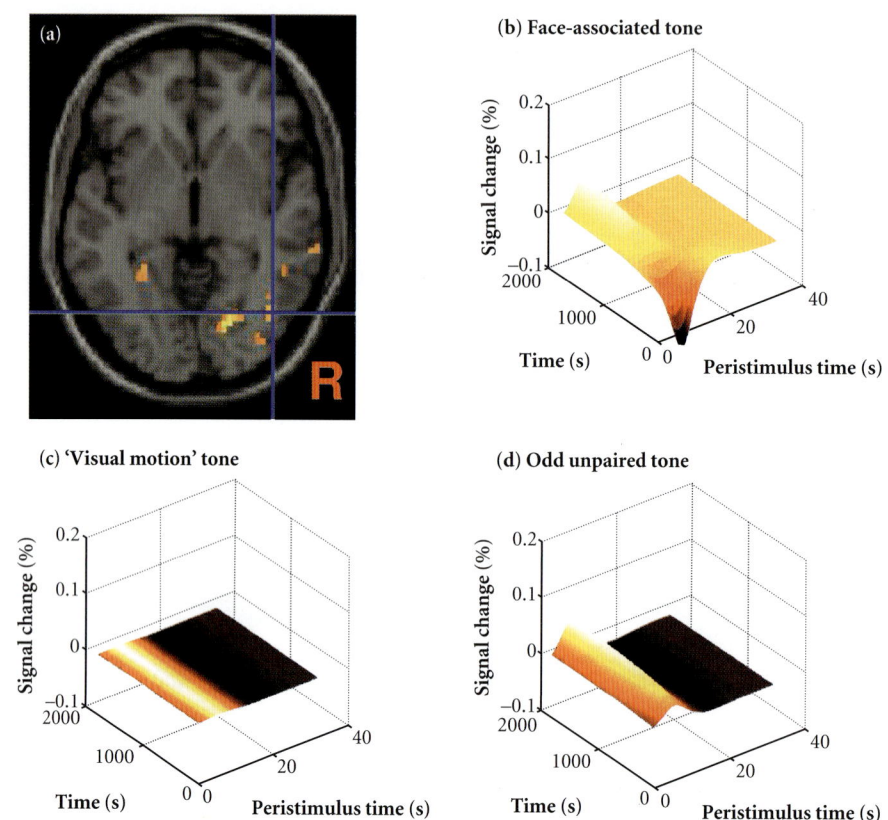

Fig. 11.4 Time by condition interaction in the fusiform face area. (a) Axial section highlighting activation in right posterior fusiform gyrus ($y = -69$), thresholded at $P < 0.05$ corrected. (b) Graphic plot showing evoked hemodynamic responses to the tone which was paired with the face in 50% of presentations in relation to the unpaired stimulus ('peristimulus time (s)' versus 'signal change (%)'). The additional third dimension ('scans') illustrates how this evoked response changes throughout the experiment. For the tone paired to a face for 50% of presentations, the ventral visual cortex shows a significant increase of evoked responses, indicated by smaller or negative amplitudes at the beginning and greater amplitudes towards the end of the experiment. This, however, was not the case for the tone that was paired to visual motion in 50% of presentations (c) or for the unpaired tone (d), which shows that this effect is specific.

occipital cortex (FFA). Only activity associated with the tone predictive of the face increased as a function of time. The other two tones (paired to a visual motion stimulus and unpaired) showed no such increase. Additional behavioral data acquired outside the scanner showed a significant decrease in reaction times to the sounds that were paired half of the time as compared to the unpaired sound. This selective decrease in RT is indicative of cross-modal learning of specific associations between sounds and images.

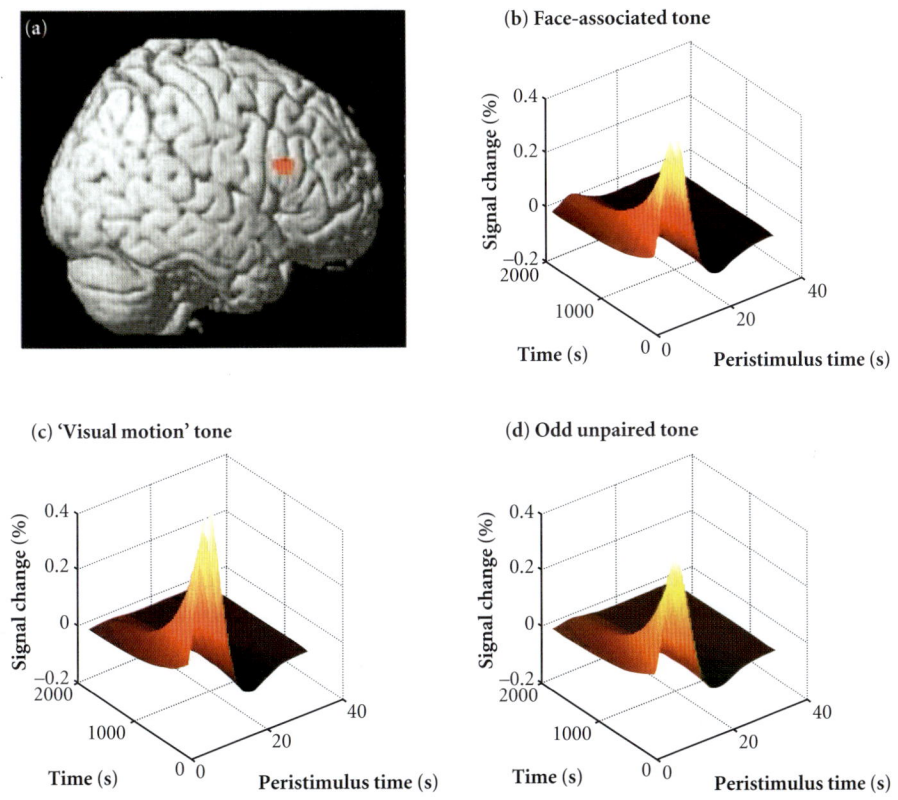

Fig. 11.5 (a) Rendered view of a brain, highlighting activation in right inferior frontal gyrus in response to the paired tones. Graphic plots showing evoked hemodynamic responses to the tone that was paired with the face in 50% of presentations (b), to the tone paired with the motion pattern in 50% of presentations (c), and to the tone which was never paired with a visual stimulus (d), in relation to the unpaired stimulus ('peristimulus time (s)' versus 'signal change (%)'). The additional third dimension ('scans') illustrates how this evoked response changes throughout the experiment. For all tones, this area shows a significant decrease of evoked responses, indicated by greater or positive amplitudes at the beginning and smaller amplitudes towards the end of the experiment.

Previous work in this area, using a similar paradigm, i.e. associative learning whereby a tone signals a visual stimulus, and subsequent functional connectivity analysis, have shown interregional covariation between occipital cortex and other brain regions, including superior temporal cortex (McIntosh *et al.* 1998).

Our data show that through associative learning, extrastriate visual cortex has the potential to respond to stimuli in a sensory modality other than vision, in this case audition. Furthermore, this was observed in a functionally specialized manner, such that brain areas which normally respond to certain visual stimuli, e.g. faces, become responsive to non-face stimuli, which have been paired to faces through associative learning. More specifically, our experiment highlights the possibility that certain adaptive mechanisms allow these areas to respond to single presentations of sounds, which have previously been learnt to predict the appearance of a face.

Similar findings have been reported in the ventral stream of monkeys. For instance, one study demonstrated that auditory stimuli could evoke delayed activity in inferotemporal cortex (IT, an area known to respond to complex visual patterns, such as faces and hands (Kobatake and Tanaka 1994) if the animal expected a visual stimulus to follow (Colombo and Gross 1994). In another experiment a cross-modal delayed matching-to-sample (DMS) task led to specific long-term memory associations between visual and auditory stimuli (Gibson and Maunsell 1997). The results, in this case, showed neurons in IT contributing to abstract memory representations that can be activated by input from other sensory modalities. Additionally, two further studies have suggested that responsiveness of IT neurons to tones plays a role in attentional modulation, where an auditory stimulus acts as a warning signal, i.e. in anticipation, concretely of a visual stimulus (Iwai *et al.* 1987; Ringo and O'Neill 1993).

In humans it has been shown that in the context of attention, area V4 and the FFA show a strong attentional modulation possibly coming from parietal areas (Kastner *et al.* 1998; Wilkinson *et al.* 2000). More interestingly, in V4, attention can lead to activation even in the absence of a stimulus (Kastner *et al.* 1999). There is thus a striking analogy to the data presented here. The cue in the study of Kastner *et al.* (1999) can be compared to the tone associated to a certain visual stimulus. After the presentation of this cue, activation in V4 increases in expectation of the stimulus, similar to our experiment where activation in V4 and FFA increases after the specific tone is presented.

One possible mechanism mediating this type of plasticity might consist of facilitation in activation of already existing neural connections between the two sensory cortices involved in processing the associated material, in this case visual and auditory cortices (Fig. 11.6). Monkey neuroanatomical data have demonstrated neuronal connections between inferotemporal cortex and the superior temporal sulcus by means of anterograde tracing (Saleem *et al.* 2000). A possible mechanism could be reciprocal reinforcement of selective neuronal projections from auditory areas through Hebbian learning. In the visual area we assume two populations of neurons, one tuned to respond to faces more than to other visual stimuli (filled circles in Fig. 11.6). Connections from peri-auditory areas terminate at both types of neurons and show initially the same connection strength. Hearing tone A together with the visual presentation of the associated face leads to a strengthening of this projection, but not to a strengthening of other connections. Finally this connection reaches an efficacy that is strong enough to lead to a reciprocal activation of the visual area by the tone alone. It is thus possible that in our study, already existing anatomical connectivity between visual and auditory association cortices (as described above) might mediate the increase in activation over time observed in the fusiform gyrus.

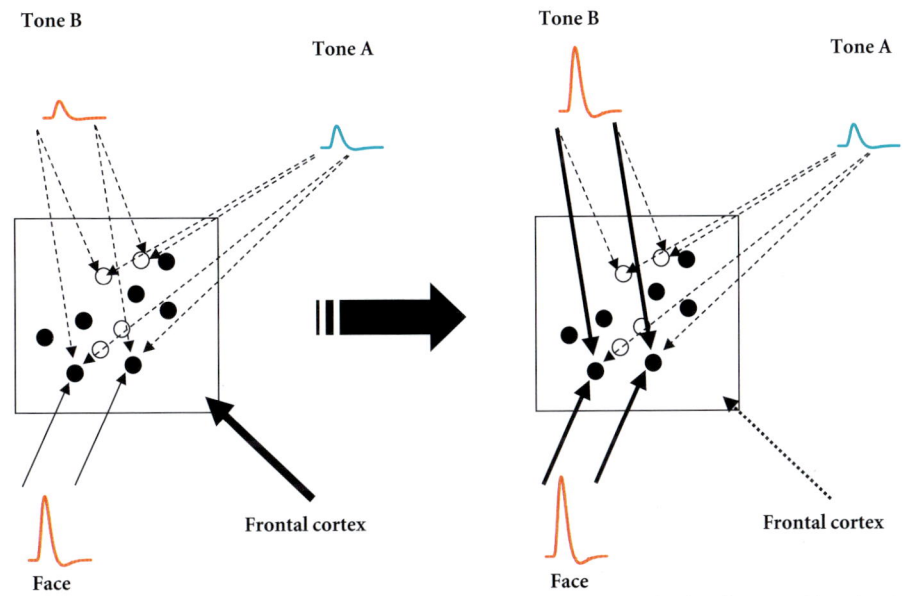

Fig. 11.6 A possible mechanism for cross-modal auditory responses in visual cortex. The visual area (box) contains neurons tuned to specific features. The filled black circles denote neurons that respond strongly to faces. Projections from auditory cortex selective for tone A and B terminate at both types of neurons. Initially, input through these projections (tone B and A) leads to no, or only weak, activation of the visual area (small hemodynamic response; cartoon in the left upper part). After strengthening of the projection between neurons in auditory cortex representing tone B and face-responsive neurons in the visual area, the tone alone can evoke activity in the visual area (small hemodynamic response; cartoon in the right upper part for tone B). No such strengthening occurs for projections from neurons representing tone A, because it has not been consistently paired with neuronal responses in this area. Frontal cortex activation is stronger at the beginning when novelty is maximal and diminishes over time as the associations are learnt. This might suggest a modulatory role of frontal cortex on associative learning in FFA, as hypothesized in the figure.

A dichotomy has been observed between a neuronal population in anteroventral TE (an area important for object vision, i.e. discrimination and recognition of visual images of objects) projecting to the lower rostral bank of superior temporal sulcus (STS), which is exclusively visual (Seltzer and Pandya 1978; Desimone and Gross 1979; Bruce et al. 1981; Baylis et al. 1987), and a population in anterodorsal TE projecting preferentially to the upper bank of STS, which is polysensory (Seltzer and Pandya 1978; Baizer et al. 1991; Barnes and Pandya 1992). Further monkey data, obtained by methods of silver impregnation and autoradiographic techniques, have shown that auditory association cortex in the superior temporal gyrus projects to a strip in the upper bank of STS (Seltzer and Pandya 1978). These findings suggest intricate connectivity between auditory association cortex and the ventral occipital cortex in monkeys, probably mediated by STS. Similarly, in humans, anatomical data show that ventral visual cortex projects to the adjacent multimodal cortex, which extends from the parieto-occipital junction and the intraparietal sulcus, through the angular gyrus in the caudal superior temporal gyrus, into STS (Nieuwenhuys et al. 1988). The auditory association cortex in the superior temporal gyrus also projects to the multimodal cortex in STS.

Cross-modal studies involving non-visual stimuli (Amedi et al. 2001) or studies in which a visual stimulus is expected (Kastner et al. 1999) always face the confound of visual imagery. This is also possible in our design, given that through the consistent pairing of a specific tone with a face, it is possible that the subjects become aware of this contingency and, after the tone is presented, actively imagine a face, which consequently activates the fusiform face area, as shown previously (O'Craven and Kanwisher 2000). Although we cannot refute the possibility that mental imagery accounts at least partially for the result obtained, one of the findings is difficult to explain by mental imagery. Active mental imagery can, by definition, only occur after subjects have become aware of the pairing. Since awareness for the stimulus contingency is a binary (either/or) process, one would expect an abrupt stepwise increase in FFA and V4 activity, and not a smooth increase, as we have observed.

In order to exclude the confound of mental imagery, a suggested experiment would consist of repeating this cross-modal associative learning paradigm, but avoiding explicit knowledge of the pairing, i.e. an implicit cross-modal learning paradigm.

We also observed a significant differential time by condition interaction for the tone consistently paired to the visual motion stimulus. However, the time courses revealed that activation for the unpaired tone and the tone paired to the face in area V5/hMT decreased as a function of time. In contrast to this decrease for the control stimuli, a constant activation for the tone paired with the visual motion stimulus resembles a relative increase in activation over time. However, more data are needed to confirm this finding, since it is possible that this activation pattern might be related to the weak visual motion stimulus used in our experiment. Furthermore, in this study we could not assess the main effect of motion, because the motion stimulus was always presented in association with a tone.

Furthermore, we found very significant activation in right inferior frontal gyrus, especially at the beginning of the experiment when the associations were newly learnt. This activation decayed over time as the associations became more familiar and less novel. Recently, it was reported that activation in right dorso-lateral prefrontal cortex (DLPFC) in the context of new learning was greater the more unpredictable associations were, became attenuated as a function of learning or increasing familiarity over time, and was evoked again by surprise violations of the learnt association (Fletcher et al. 2001). As stimuli become more familiar, activation in this region decreases (Raichle et al. 1994; Fletcher et al. 2000, 2001). Our focus of activation lies at the border of ventrolateral and dorsolateral prefrontal cortex and extends dorsally into the middle frontal gyrus. More detailed exploration of the data revealed that the decrease in activation over time observed in this region generalized to all auditory conditions, and was thus not specific to cross-modal associative learning. Although this decrease was observed for all tones, i.e. was not specific to the learnt association, it is still possible that this activation reflects a facilitatory influence which, in combination with consistent pairing of stimuli, might lead to cross-modal associative learning (i.e. an interaction of novelty detection and consistent stimulus pairings). This suggests that this activation might resemble a mechanism of novelty detection, possibly modulating the increase in connectivity between auditory and visual areas (Fig. 11.6).

11.5 Conclusion

Our data show that extrastriate visual areas can respond to auditory stimuli which have become predictive of visual stimuli through associative learning. We found activation within this region, specifically in the fusiform face area, in response to a paired tone, which increased over time as a function of learning. This effect was neither observed for a tone associated to a visual control stimulus nor for an unpaired sound, highlighting the specificity of the result in V4 and the fusiform face area. Furthermore, activation of a region in right lateral prefrontal cortex present in the early stages of associative learning decreases over time, in accord with a putative modulatory function for this area in cross-modal associative learning.

Acknowledgements

This work was supported by the Volkswagen Stiftung. We thank Michael Rose, Thomas Wolbers, Ulrike Bingel, and René Knab for their technical assistance.

References

Allison, T., McCarthy, G., Nobre, A., Puce, A., and Belger, A. (1994) Human extrastriate visual cortex and the perception of faces, words, numbers, and colors. *Cereb Cortex;* **4**: 544–54.

Amedi, A., Malach, R., Hendler, T., Peled, S., and Zohary, E. (2001) Visuo-haptic object-related activation in the ventral visual pathway. *Nat Neurosci;* **4**: 324–30.

Baizer, J. S., Ungerleider, L. G., and Desimone, R. (1991) Organization of visual inputs to the inferior temporal and posterior parietal cortex in macaques. *J Neurosci;* **11**: 168–90.

Barnes, C. L. and Pandya, D. N. (1992) Efferent cortical connections of multimodal cortex of the superior temporal sulcus in the rhesus monkey. *J Comp Neurol;* **318**: 222–44.

Baylis, G. C., Rolls, E. T., and Leonard, C. M. (1987) Functional subdivisions of the temporal lobe neocortex. *J Neurosci;* **7**: 330–42.

Bental, E., Dafny, N., and Feldman, S. (1968) Convergence of auditory and visual stimuli on single cells in the primary visual cortex of unanesthetised unrestrained cats. *Exp Neurol;* **20**: 341–51.

Bruce, C., Desimone, R., and Gross, C. G. (1981) Visual properties of neurons in a polysensory area in superior temporal sulcus of the macaque. *J Neurophysiol;* **46**: 369–84.

Büchel, C. (1998) Functional neuroimaging studies of Braille reading: cross-modal reorganization and its implications. *Brain;* **121**: 1193–4.

Büchel, C., Morris, J., Dolan, R. J., and Friston, K. J. (1998) Brain systems mediating aversive conditioning: an event-related fMRI study. *Neuron;* **20**: 947–57.

Colombo, M. and Gross, C. G. (1994) Responses of inferior temporal cortex and hippocampal neurons during delayed matching to sample in monkeys (*Macaca fascicularis*). *Behav Neurosci;* **108**: 443–55.

Damasio, A. R., Damasio, H., and Van Hoesen, G. W. (1982) Prosopagnosia: anatomic basis and behavioral mechanisms. *Neurology;* **32**: 331–41.

Desimone, R. and Gross, C. G. (1979) Visual areas in the temporal cortex of the macaque. *Brain Res;* **178**: 363–80.

Fletcher, P. C., Shallice, T., and Dolan, R. J. (2000) 'Sculpting the response space'– an account of left prefrontal activation at encoding. *Neuroimage;* **12**: 404–17.

Fletcher, P. C., Anderson, J. M., Shanks, D. R., Honey, R., Carpenter, T. A. Donovan, T., *et al.* (2001) Responses of human frontal cortex to surprising events are predicted by formal associative learning theory. *Nat Neurosci;* **4**: 1043–8.

Friston, K. J., Holmes, A. P., Poline, J-B., Grasby, P. J., Williams, S. C. R., Frackowiak, R. S. J., et al. (1995) Analysis of fMRI time-series revisited. *NeuroImage;* **2**: 45–53.

Gibson, J. R. and Maunsell, J. H. (1997) Sensory modality specificity of neural activity related to memory in visual cortex. *J Neurophysiol;* **78**: 1263–75.

Haxby, J. V., Horwitz, B., Ungerleider, L. G., Maisog, J. M., Pietrini, P., and Grady, C. L. (1994) The functional organization of human extrastriate cortex: a PET-rCBF study of selective attention to faces and locations. *J Neurosci;* **14**: 6336–53.

Iwai, E., Aihara, T., and Hikosaka, K. (1987) Inferotemporal neurons of the monkey responsive to auditory signal. *Brain Res;* **410**: 121–4.

Kanwisher, N., McDermott, J., and Chun, M. M. (1997) The fusiform face area: A module in human extrastriate cortex specialized for face perception. *J Neurosci;* **17**: 4302–11.

Kastner S, DeWeerd P., Desimone R., and Ungerleider LC. (1998) Mechanisms of directed attention in the human extrastriate cortex as revealed by functional MRI. *Science;* **282**: 108–11.

Kastner, S., Pinsk, M. A., De Weerd, P., Desimone, R., and Ungerleider, L. G. (1999) Increased activity in human visual cortex during directed attention in the absence of visual stimulation. *Neuron;* **22**: 751–61.

Kobatake, E. and Tanaka, K. (1994) Neuronal selectivities to complex object features in the ventral visual pathway of the macaque cerebral cortex. *J Neurophysiol;* **71**: 856–67.

Levitt, P., Barbe, M. F., and Eagleson, K. L. (1997) Patterning and specification of the cerebral cortex. *Annu Rev Neurosci;* **20**: 1–24.

Macaluso, E., Frith, C. D., and Driver, J. (2000) Modulation of human visual cortex by crossmodal spatial attention. *Science;* **289**: 1206–8.

McIntosh, A. R., Cabeza, R. E., and Lobaugh, N. J. (1998) Analysis of neural interactions explains the activation of occipital cortex by an auditory stimulus. *J Neurophysiol;* **80**: 2790–6.

Nieuwenhuys, R., Voogd, J., and Huijzen, V. (1988) *The human central nervous system: synopsis and atlas*. Berlin: Springer.

O'Craven, K. M. and Kanwisher, N. (2000) Mental imagery of faces and places activates corresponding stiimulus-specific brain regions. *J Cogn Neurosci;* **12**: 1013–23.

Puce, A., Allison, T., Gore, J. C., and McCarthy, G. (1995) Face-sensitive regions in human extrastriate cortex studied by functional MRI. *J Neurophysiol;* **74**: 1192–9.

Puce, A., Allison, T., Asgari, M., Gore, J. C., and McCarthy, G. (1996) Differential sensitivity of human visual cortex to faces, letterstrings, and textures: A functional magnetic resonance imaging study. *J. Neurosci.;* **16**: 5205–15.

Raichle, M. E., Fiez, J. A., Videen, T. O., MacLeod, A. M., Pardo, J. V., Fox, P. T., *et al.* (1994) Practice-related changes in human brain functional anatomy during nonmotor learning. *Cereb. Cortex;* **4**: 8–26.

Rakic, P. (1988) Specification of cerebral cortical areas. *Science;* **241**: 170–6.

Ringo, J. L. and O'Neill, S. G. (1993) Indirect inputs to ventral temporal cortex of monkey: the influence of unit activity of alerting auditory input, interhemispheric subcortical visual input, reward, and the behavioral response. *J Neurophysiol;* **70**: 2215–25.

Sadato, N., Pascual-Leone, A., Grafman, J., Ibanez, V., Deiber, M-P., Dold, G., *et al.* (1996) Activation of the primary visual cortex by Braille reading in blind subjects. *Nature;* **380**: 526–8.

Saleem, K. S., Suzuki, W., Tanaka, K., and Hashikawa, T. (2000) Connections between anterior inferotemporal cortex and superior temporal sulcus regions in the macaque monkey. *J Neurosci;* **20**: 5083–101.

Seltzer, B. and Pandya, D. N. (1978) Afferent cortical connections and architectonics of the superior temporal sulcus and surrounding cortex in the rhesus monkey. *Brain Res;* **149**: 1–24.

Spinelli, D. N., Starr, A., and Barrett, T. W. (1968) Auditory specificity in unit recordings from cat's visual cortex. *Exp Neurol;* **22**: 75–84.

Thompson, R. F., Smith, H. E., and Bliss, D. (1963) Auditory, somatic sensory and visual response interactions and interrelation in association and primary cortical fields of the cat. *J Neurophysiol;* **26**: 365–78.

Tootell, R. B. and Taylor, J. B. (1995) Anatomical evidence for MT and additional cortical visual areas in humans. *Cereb Cortex;* **5**: 39–55.

Voronin, L. L. and Skrebitsky, V. G. (1965) Spontaneous and induced potential of the cortex neurons in non-anaesthetised rabbits. *Abstr. 6th Intern. Congr. Electroencephalog. Clin. Neurophysiol.* 79.

Wilkinson, F., James, T. W., Wilson, H. R., Gati, J. S., Menon, R. S., and Goodale, M. A. (2000) An fMRI study of the selective activation of human extrastriate form vision areas by radial and concentric gratings. *Curr Biol;* **10**(22):1455–8.

Zangaladze, A., Epstein, C. M., Grafton, S. T., and Sathian K. (1999) Involvement of visual cortex in tactile discrimination of orientation. *Nature;* **401**: 587–90.

Zeki, S., Watson, J. D., Lueck, C. J., Friston, K. J., Kennard, C., and Frackowiak, R. S. (1991) A direct demonstration of functional specialization in human visual cortex. *J Neurosci;* **11**: 641–9.

Chapter 12

Neural representation of object images in the macaque inferotemporal cortex

Manabu Tanifuji, Kazushige Tsunoda, and Yukako Yamane

12.1 Introduction

Information about visually presented objects is transmitted from the primary visual cortex (V1) to the inferotemporal cortex (IT) through multiple prestriate areas in macaque monkeys. To understand neural mechanisms of perception and recognition of objects through their visual images, response properties of neurons in the anterior part of the IT cortex, defined architectonically as area TE, have been investigated extensively. Because of the position- and size-invariant response properties of those neurons, this area is believed to correspond to lateral occipital cortex (LOC) in the human ventral visual pathway (Malach *et al.* 1995).

Physiological recording experiments have shown neurons responding equally well to object images and to visual features that are geometrically less complex than the object images (Desimone *et al.* 1984; Tanaka *et al.* 1991; Kobatake and Tanaka 1994). In particular, Tanaka and colleagues explored the simplest visual feature that maximally activates individual neurons in area TE ('critical feature'), using anesthetized monkeys. They found that critical features of many neurons in area TE are moderately complex. Figure 12.1 shows a representative critical feature extracted according to their method. These results indicate that an object is represented as combinations of visual features extracted by these neurons. However, these response properties of neurons in area TE have been investigated by extracellular recording of neuronal firing, and only a small number of neurons can be examined simultaneously. Thus, our understanding of object representation by these neurons in area TE is still at an early stage.

On the other hand, some other studies reported on neurons responding specifically to visual images that are familiar to the animals, such as faces or body parts (Gross *et al.* 1972; Perrett *et al.* 1982; Desimone *et al.* 1984; see also Fig. 12.11). Furthermore, several studies have investigated visual responses of neurons in monkeys trained with a particular set of visual stimuli, and revealed that some neurons specifically responded to

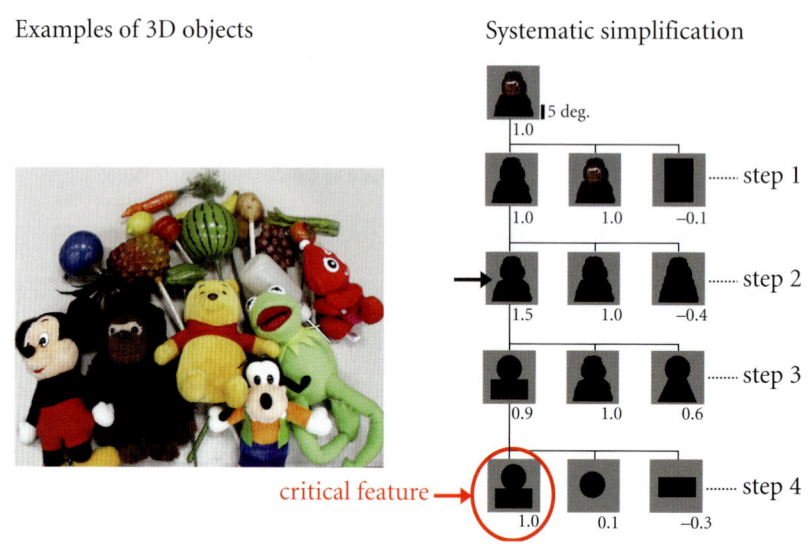

Fig. 12.1 The 'critical feature', the visual feature that maximally activates each cell, is determined by systematic stimulus simplification of the best object stimulus. First, we tested the cell with various three-dimensional objects, including faces, hands, stuffed animals, plastic fruits and vegetables, and paper mounts (left panel for some examples). After determining the best stimulus, we simplified it step by step to find the simplest stimulus that maximally activates the cell (right panel). For example, at step 1, we compared the best coloured object with its silhouette, and found that the silhouette activated the cell equally well. The rightmost rectangle was taken as a control stimulus. The numbers below each picture indicate the response amplitudes normalized to the response to the reference stimulus, the best object. The stimulus that evoked at least more than 70% of the response elicited by the best stimulus in the previous step, was again examined in the next step as the reference stimulus. At step 2, we examined the effect of the 'sharpness' of the corner at the junction of upper and lower parts (arrow), and found that the silhouette with the sharp corners was the most effective stimulus. From left to right, the stimuli were the silhouette with sharp corners, the silhouette that evoked the best response at the previous step, the silhouette without corners. Further simplification was carried out at step 3. Finally, we determined the critical feature as a combination of a circle and a rectangle because neither the upper nor lower part alone activated the cell.

visual stimuli that became familiar through training (Logothetis *et al.* 1995; Kobatake *et al.* 1998). The relationship between these neurons and those responding to visual features is also an issue that needs to be investigated.

Recently, intrinsic signal imaging has enabled us to reveal the spatial distribution of neurons activated by object images and to investigate response properties of neurons in characteristic sites revealed by intrinsic signal imaging. By this new approach, we have begun to understand more about object representation in area TE.

12.2 Intrinsic signal imaging

Neurons with similar response properties are clustered into a column in area TE (Gochin *et al.* 1991; Fujita *et al.* 1992). Thus, intrinsic signal imaging of columnar activation can be used to investigate spatial patterns of activation (Wang *et al.* 1996, 1998; Tsunoda *et al.* 2001). Intrinsic signal imaging measures the decrease in the degree of light reflection elicited by neural activation from the exposed cortical surface using a CCD camera (Fig. 12.2) (Grinvald *et al.* 1999). These reflection changes are due to metabolic changes elicited by neural activation, including deoxygenation of hemoglobin in capillaries (Grinvald *et al.* 1999).

Intrinsic signal imaging in area TE revealed multiple spots elicited by visual stimulation. The mean size of the 'active spots' (Fig. 12.2d) was 0.50 ± 0.13 mm along the longer axis and 0.35 ± 0.09 mm along the shorter axis ($n = 94$). These dimensions agreed well with the size previously reported for a column of cells with similar responsiveness in this area (Gochin *et al.* 1991; Fujita *et al.* 1992). Although these reflection changes are not a direct measure of neural activation, intrinsic signals coincide well with the activity of neurons examined by conventional extracellular recordings (Fig. 12.3) (Tsunoda *et al.* 2001).

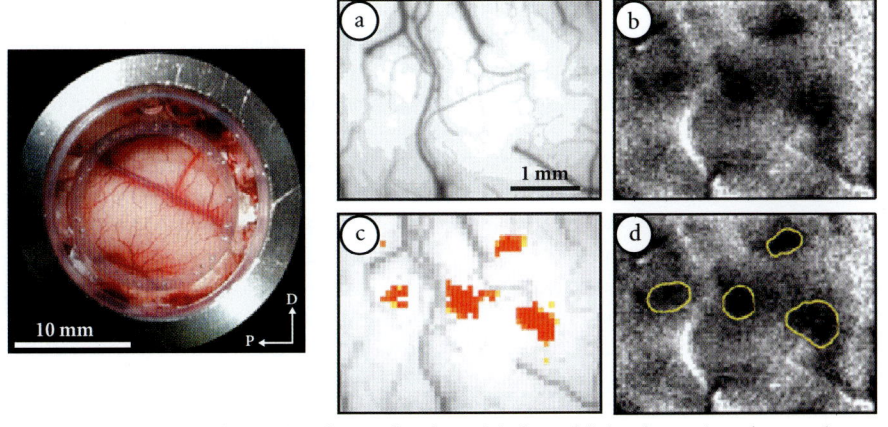

Fig. 12.2 Intrinsic signal imaging detects local modulation of light absorption changes in area TE. Surface view of the exposed portion of dorsal area TE (left panel). The thickest vessel running obliquely from the upper left to the lower right is along the superior temporal sulcus. The dorsal part of area TE is ventral to this vessel. (a) Portion of area TE where intrinsic signals were recorded. (b) A differential image showing a local increase in absorption. (c) Active regions, where the degree of reflection change evoked by the stimulus was significantly greater than that without the stimulus presentation. The region with the highest significance level is in red, that with the lowest significant level in yellow ($P < 0.05$). (d) Extracted active spots outlined by connecting pixels with 1/2 of the peak absorption value. (Modified from Tsunoda *et al.* 2001.)

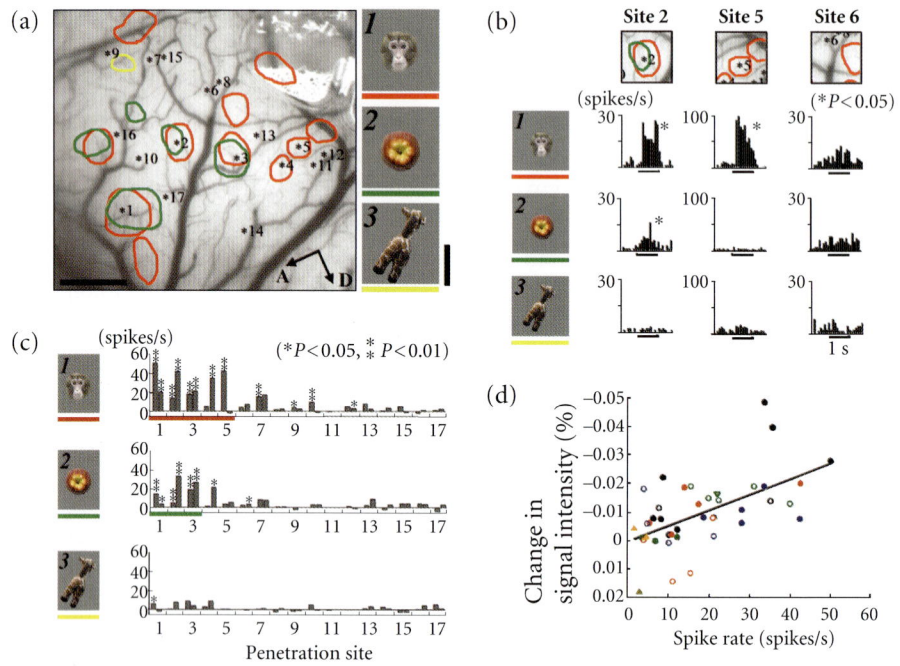

Fig. 12.3 Relationship between intrinsic signals and spike activity in area TE. (a) Active spots elicited by three different stimuli (*1*, *2* and *3*, on the right), and numbered electrode penetration sites. The color of individual contours indicates the active spots elicited by the stimulus underlined with the same color (same in all the figures showing spot distribution). A, anterior; D, dorsal. (b) Representative peristimulus–time histograms (PSTHs) showing the extracellular activity elicited by the three different stimuli, recorded at the sites indicated at the top of each column. Each row gives the PSTHs obtained following stimulation shown on the left side. Horizontal bars in the histograms indicate the one-second period for visual stimulation. (c) Mean firing rates evoked by the three different visual stimuli, for all 34 cells from 17 different sites. Two different cells separated by at least 200 μm were recorded at each penetration site. The penetration sites inside the active spots for a given stimulus were indicated by the colored bars in the top (red) and the second graphs (green). For stimulus 3, all the cells were recorded outside the active spot. (d) Correlation between the intensity of local changes of optical signal and evoked spike rates measured at the same cortical locations. Thirty-four cells were tested with 10 stimuli, and cellular activity with significant visual responses ($P < 0.05$) was plotted against the optical signal intensity obtained at the corresponding penetration sites for the individual stimuli. The activities of neurons from the same site were averaged. Different symbols indicate the responses at different sites. The regression line is given by $Y = -5.05e - 6 + 5.46e - 6X$ with a correlation coefficient of 0.57 ($n = 41$). Calibration bar, 10 degrees for the stimulus size in (a). Significance of difference determined by the Kolmogorov–Smirnov test for individual cell response is indicated by single ($P < 0.05$) or double ($P < 0.01$) asterisks in (b) and (c). (The same as in Fig. 12.5.) (Adapted from Tsunoda *et al.* 2001.)

To relate the optical response specificity to the neural activities, we recorded extracellular activity from 34 neurons in 17 sites located inside and outside the active spots (Fig. 12.3a). The representative peristimulus–time histograms (PSTHs) showed that the neuron in site 2, where both stimulus *1* and *2* elicited significant intrinsic signals, was activated by these two stimuli, but not by stimulus *3* (Fig. 12.3b, left-hand column). Similarly, only stimulus *1* significantly activated the cell in site 5 (Fig. 12.3b, middle column) and none of the stimuli activated cells in site 6 (Fig. 12.3b, right-hand column). In summary, among 34 neurons, 28 cells (82.4%) showed the same responsiveness to stimulus *1* as indicated by the optical responses: eight active cells inside the spot and twenty inactive cells outside (top row) (Fig. 12.3c). Similar results were obtained for 32 out of 34 cells (94.1%) for stimulus *2* (second row), and for 33 out of 34 cells (97.1%) for stimulus *3* (bottom row). A large variation of neuronal responses within the active spots (ex. Fig. 12.3c) did not alter the agreement between the intrinsic signals and extracellular responses. First, the optical response intensity and the firing rates of individual neurons showed a statistically significant positive correlation ($P < 0.001$) (Fig. 12.3d). Secondly, the average extracellular responses inside the spots (24.9 ± 17.2 and 16.8 ± 11.6 spikes/s for stimuli *1* and *2*, respectively) were significantly larger than those outside the spots (4.30 ± 4.95 and 3.32 ± 4.72 spikes/s for stimuli *1* and *2*, respectively) (Fig. 12.3c). These results indicate that the intrinsic signal coincides well with the firing activity of neurons examined by conventional extracellular recordings.

It should be mentioned that, in all the extracellular recordings, we recorded extracellular responses from superficial layers of cortex that are less than 1 mm in depth. Thus, strictly speaking, an 'active spot' revealed by intrinsic signal imaging may not necessarily correspond to a 'column', which usually means a cluster of neurons with similar response properties extending vertically from surface to the white matter. In the following sections, we use 'spots' for our results obtained by intrinsic signal imaging and 'columns' when we interpret our result in relation to previous studies.

12.3 Object representations revealed by intrinsic signal imaging

Intrinsic signal imaging revealed that different complex objects activated spots with different distribution patterns, together with some common active spots (Fig. 12.3a). Assuming that each spot represented a particular visual feature, the spots activated by one object may represent visual features specific to that object, and the spots activated by multiple objects may represent features common among these objects. We examined this idea by comparing distribution patterns of active spots with those produced by systematically simplified stimuli (Fig. 12.4) (Tsunoda *et al.* 2001). For example, in one case, a 'black cat' (a, *1*) was simplified to its 'head' (a, *2*), and then to the 'silhouette of its head' (a, *3*) (Fig. 12.4a). The original image (a, *1*) elicited 14 spots, but presenting the 'head' (a, *2*) elicited only eight of the original 14 spots. The silhouette (a, *3*) only activated three (yellow) of the eight spots elicited by the head (a, *2*). Similarly, Fig. 12.4b shows that spots A and B disappeared but spot D remained when the 'handle' and

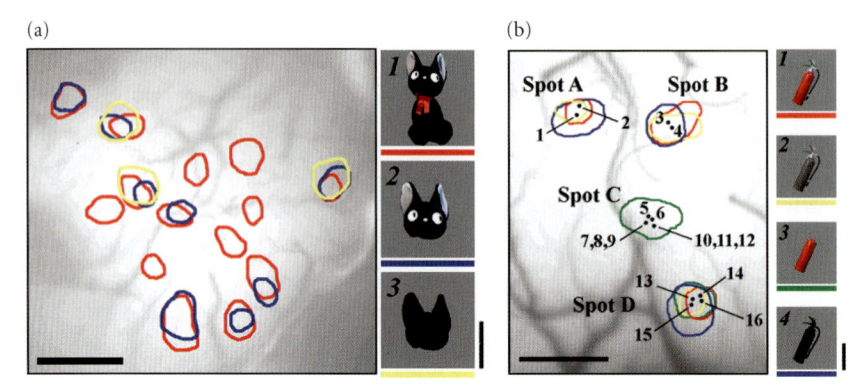

Fig. 12.4 Representation of complex object images and their simplification in area TE. (a) A case where simplified stimuli elicited only a subset of spots evoked by more complex stimuli. (b) Cases in which new spots appeared when the original stimulus was simplified. The numbers (1–16) indicate electrode penetration sites. (Modified from Tsunoda et al. 2001.)

'hose' were removed from the original stimulus, 'fire extinguisher'. In addition to the disappearance of spots, we found that new spots emerged by apparent simplification of an object: spot C appeared when the handle and hose were removed from the fire extinguisher. We have examined 12 pairs of activation patterns obtained before and after the simplification of the objects, and we observed changes in the distribution patterns consistent with either Fig. 12.4a or 12.4b for all of the pairs.

12.4 Visual features represented by individual spots

To directly address the neural mechanisms of appearance and disappearance of spots by object simplification, we recorded extracellular responses from 25 cells in the four spots shown in Fig. 12.4b, and analyzed the response properties of the cells in each spot. Figure 12.5 shows response properties of the representative cells in these spots. The difference in optical response patterns to stimuli *1* and *3* in Fig. 12.4b suggests that spots A and B represented visual features related to the handle and hose of the fire extinguisher. In fact, cells in spots A and B were significantly activated by the handle and hose in isolation (Fig. 12.5, a,*2* and b,*2*) as well as by the silhouette of the original fire extinguisher (Fig. 12.5, a,*1* and b,*1*). The cells in spot A were activated by the handle (Fig. 12.5, a,*3*) having protrusions, but not by the hose (Fig. 12.5, a,*4*). Furthermore, other stimuli with sharp protrusions, such as a 'hand' (Fig. 12.5, a,*5*) and cat's head (Fig. 12.5, a,*6*), also activated the cells. These cells seemed to require 'sharp protrusions' for activation. In contrast, cells in spot B were activated by the hose (Fig. 12.5, b,*4*), but neither by the handle (Fig. 12.5, b,*3*) nor a 'line segment' (Fig. 12.5, b,*5*). Thus, we determined the critical feature as an 'asymmetric arc' (Fig. 12.5, b,*4*). The neural responses of cells in spots C and D were consistent with the imaging results in Fig. 12.4b: cells in spot C were activated by the 'cylinder' but not by the original fire

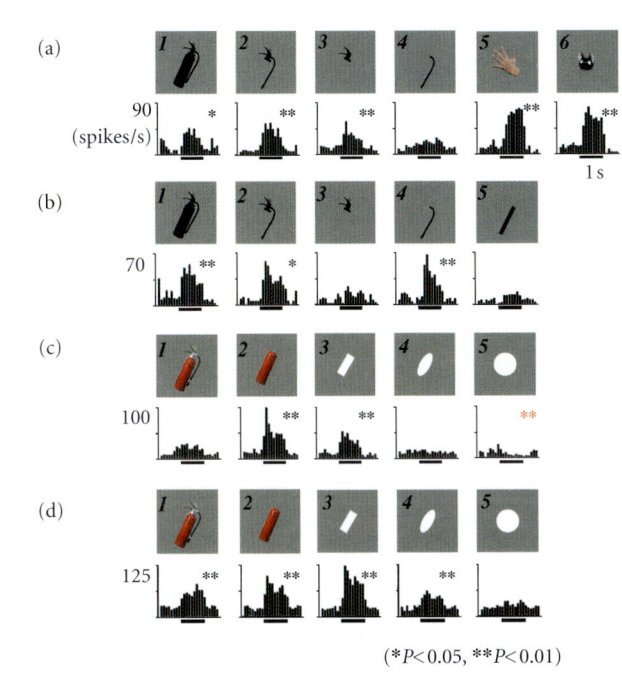

Fig. 12.5 Visual responsiveness of representative cells in spots A–D in Fig. 12.4b. (a), (b), (c), and (d) indicate responses in spots A (track 2, depth 620 μm), B (track 3, depth 540 μm), C (track 8, depth 280 μm), and D (track 16, depth 280 μm), respectively. Red asterisks indicate significant inhibition ($P < 0.01$). (Adapted from Tsunoda et al. 2001.)

($*P<0.05$, $**P<0.01$)

extinguisher (Fig. 12.5, c,*1* and c,*2*), and cells in spot D were significantly activated by both stimuli (Fig. 12.5, d,*1* and d,*2*). The critical feature for cells in spot D was a 'rectangular shape' (Fig. 12.5, d,*3*), but cells also responded significantly to an 'ellipse' (Fig. 12.5, d,*4*). Since there was no response to a 'circle' (Fig. 12.5, d,*5*), we determined the critical feature of the spots as an 'elongated structure'.

Similarly, the simplest visual feature that could activate the cells in spot C was a rectangular shape (Fig. 12.5, c,*3*). In contrast to the cells in spot D, however, there was no activation by an ellipse (Fig. 12.5, c,*4*). In addition, the cells were inhibited by a circle (Fig. 12.5, c,*5*). Thus, these results suggest that the response properties of the cells in spot C (Fig. 12.4b) are determined by the balance between excitatory and inhibitory inputs: the excitatory inputs were given by a feature related to a rectangular shape and the inhibitory inputs are given by a feature related to a circle. This explanation would account for the lack of activation by the fire extinguisher, where the hose (circular shape) attached to the rectangular cylinder makes the entire shape elliptical. These results suggest that some of the columns representing a particular feature are inactive when other features are presented together with that feature. This could explain the optical imaging results in which active spots appeared following simplification of the stimulus.

The extracellular recording shows that inhibitory mechanisms are involved in the responsiveness of the neurons in spot C (Fig. 12.5, c,*5*). Fujita and Fujita (1996) found intrinsic excitatory connections that extend parallel to the cortical surface for long distances. The underlying anatomical substrate of the inhibitory mechanisms could be combination of the excitatory connections and local inhibitory circuits within a target column. Alternatively, inhibitory neurons may extend their axons directly to distant target columns, as in the case of basket cells in the cat primary visual cortex (Kisvarday et al. 1994). One other possibility is that inhibitory interactions within area TEO is reflected in area TE through columnar projections from area TEO to area TE (Saleem et al. 1993).

12.5 Object representation with a combination of active and inactive columns

The combination of intrinsic signal imaging and extracellular recording can suggest the spatial layout of neural activity evoked by complex objects. Intrinsic signal imaging showed distributed representation of object images in area TE: object images are represented as combinations of multiple spots (Fig. 12.3a). In general, there are two concepts of distributed representation, depending on the fraction of neurons in the population: sparsely distributed and densely distributed representation (Foldiak and Young 1995). Although Fig. 12.3a shows many active spots elicited by some of the stimuli, the region activated by a single object image was, on average, only 3.3 ± 2.5% of the entire recording area (number of examined object images = 37). This low density suggests a sparsely distributed representation of objects. Another finding was that object simplification resulted in systematic changes in the distribution patterns of spots. We found that 32 out of 106 activity spots (30%) disappeared when part of the visual features were removed from the stimuli by object simplification (Fig. 12.4a). Extracellular recording showed that the optimal stimuli for neurons in these spots were visual features less complex than the original objects. These results suggest that there are spots specific for the representation of a particular visual feature within an object image, and that an object image is represented by a combination of spots specified for these visual features (Fig. 12.6a, b). However, among 106 activity spots, 18 spots appeared only after stimulus simplification (Fig. 12.4b, spot C). These results cannot be explained by this scheme, which implicitly assumes that all the spots related to a single feature in an object image are activated (Fig. 12.6b). Thus, we propose an extended scheme of distributed representation, where objects are represented not by the simple sum of feature columns but by combinations of active and inactive spots for individual features (Fig. 12.6c). According to this scheme, we think that the fire extinguisher in Fig. 12.4b was represented not only by the active spots representing the handle (spot A), the hose (spot B), and the cylinder (spot D), but by the absence of spot C representing the entire elliptical structure. Combinations of inactive as well as active columns increase the number of available activation patterns, and thus, in general, could increase the number of objects to be specifically represented (Fig. 12.6c). From a

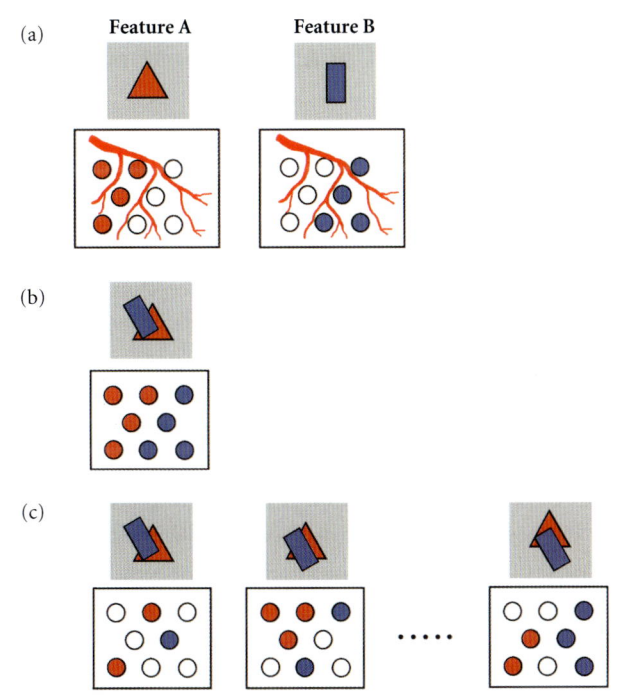

Fig. 12.6 Conventional and extended models of distributed representation of objects. Assuming that features A and B are represented by spots shown in red and blue, respectively, on the cortical surface (a), the visual stimulus consisting of these two features will activate all of these red and blue spots in models based on the conventional distributed representation (b). In our extended model, only a part of them are activated by the same stimulus (c). By this mechanism, stimuli consisting of the same features but arranged in different configurations can be represented by different activation patterns on the cortex. In each figure, the upper and lower panels indicate the visual stimulus and cortical area including spots responding to the stimulus.

physiological point of view, we consider the relationship between increased activation patterns and object representation in the following way:

In our scheme, we assume that a single visual feature maximally activates a set of columns (Fig. 12.6a). However, their tuning properties are not exactly the same, so that each column can represent additional information about visual images. For example, both spots C and D in Fig. 12.4b represent 'elongated structure'. But in addition, appearance of these spots depends on whether the given stimulus is elliptical or not. Then, objects differentiated by the additional information can be represented differently by activating a subset of these columns. One case would be representations of objects that consist of the same set of visual features but with a different arrangement of them. Another case would be representation of the same objects that look different under different vantage points, with partial occlusion, and with shading.

12.6 Representation of spatial arrangement of parts in object images

Examination of visual features represented by neurons in area TE suggested that at least some of the neurons in this area, represent 'local features' in object images, as neurons in spots A and B (Figs 12.4b and 12.5) represent 'protrusions' and 'asymmetric curvature', respectively. Since information about the spatial arrangement of 'local features' is necessary for the specific representation of object images, some of the other neurons may represent visual features related to the spatial arrangement of local features ('configurational information'). Here, we refer to 'local features' as visual features that occupy part of an object image and are distinguishable from other parts of an object image by their particular shapes, colors, or textures. 'Configurational information' is information about the spatial arrangement of 'local features' themselves or about spatial arrangement of parts including local features. For example, intrinsic signal imaging with extracellular recording suggests that the presence or absence of activity in spot C constrains the spatial arrangement of the hose and the cylinder ('local features') of the fire extinguisher (Fig. 12.4b and Fig. 12.5c). This spot was activated when the hose is attached to the side of the cylinder and makes the entire shape elliptical, but may not be if the hose is secured above the handle where the rectangular shape of cylinder is exposed. Thus, we consider that activity in spot C has information about configuration to some extent, although this may be one way of representing a particular 'configurational information'.

To further examine the representation of 'configurational information', we investigated spots activated by an object (original, Fig. 12.7, 1) and the same object with a gap introduced between parts of the object (Fig. 12.7, 4), but not by a part alone (Fig. 12.7, 2 and 3) (Yamane et al. 2001). We suggest that these spots do not simply represent local features in objects, because either part is not essential for activation. Moreover, activation by the stimulus with an introduced gap indicates that local features appearing at

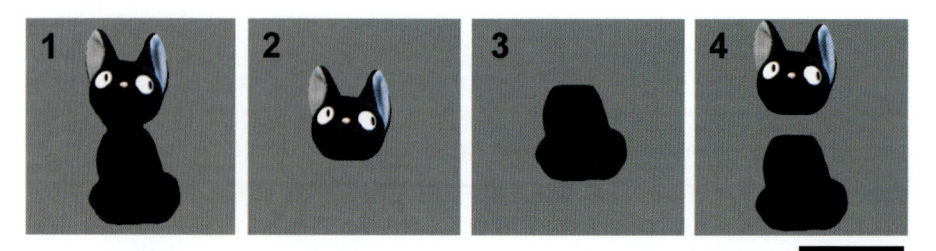

Fig. 12.7 A representative set of visual stimuli used in intrinsic signal imaging for examination of the representation of the spatial arrangement of parts. The response properties shown in Figs. 12.8 and 12.9 were obtained from a spot activated by stimuli 1 and 4, but not by stimuli 2 and 3.

the junction of two parts, such as sharp connecting corners in Fig. 12.7, stimulus 1, are also not essential. We recorded single-cell responses from a spot having stimulus selectivity described above, and found three characteristic response properties ($n = 14$):

(1) These cells had critical features which were combinations of two vertically aligned parts (Fig. 12.8a, see also Fig. 12.1). There was no activation by either part (Fig. 12.1, step 4).

(2) These cells were less sensitive to color, texture, and local shapes of either part. Three findings suggest this conclusion. First, there were no changes in the responses after removing color and texture during the stimulus simplification procedure (Fig. 12.1 and 12.8a). Secondly, changes in the shapes of the parts did not alter responses of these neurons very much. For example, a neuron, whose critical feature was determined as a combination of a circle and a rectangle, was also significantly activated by a combination of a circle and an ellipse (data not shown). Finally, these cells responded equally well to object images even having different color, texture, and local shapes, as long as they had the global shape similar to the critical features (Figs 12.8b and 12.9).

(3) These cells were highly selective to a particular position of the upper part relative to the lower part (Fig. 12.10).

These characteristic response properties enable these neurons to respond to visual stimuli regardless of the local features embedded in either part, but only when these local features are aligned in particular orientation, as in Fig. 12.10. These results, as well

Fig. 12.8 Effective stimuli for neurons in a spot identified by the stimuli in Fig. 12.7. (a) Representative critical features determined by stimulus simplification. Please note that, when color and texture are not essential, the stimulus was filled black (see Fig. 12.1). (b) The best object stimuli for these neurons, among 100 object stimuli examined before stimulus simplification. Scale bar, 5°.

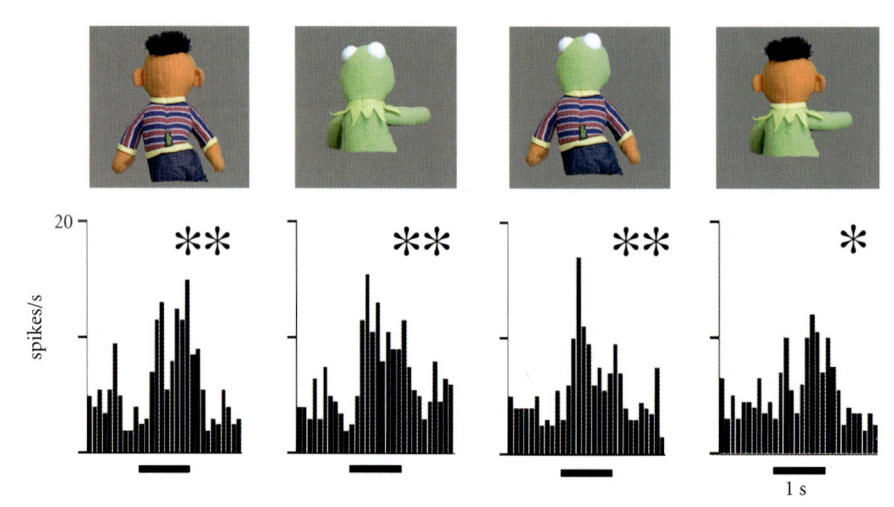

Fig. 12.9 Neural responses of one representative neuron to object stimuli. Upper panel shows visual stimuli, and lower panel indicates peristimulus–time histograms (PSTHs) showing responses of the neuron to the stimuli given above. The stimulus was presented for a period of 1 s, indicated by the horizontal line segment in each PSTH. These stimuli activated the cell equally well. *$P < 0.05$, **$P < 0.01$.

as neurons in spot C in Fig. 12.4b, suggest that neurons in area TE do not necessarily represent local features but also configurational information of the object image. Object images could be specifically represented by a combination of spots representing 'local features' and those representing 'configurational information'.

12.7 Face-specific neurons in area TE

'Face neurons' are representative neurons in area TE responding to the visual images of familiar objects. We consider that 'face neurons' also represent 'configurational information'. 'Face neurons' respond to 'faces', but these responses cannot be explained by specific responses to a part of the 'face' (Fig. 12.11). In this particular case, for example, a face without eyes did not activate the cell, but there was no activation by 'eyes' alone (Fig. 12.11b, c). Furthermore, previous studies have shown that a 'face' with scrambled facial parts does not activate these neurons (Desimone *et al.* 1984). There are two characteristic properties of 'face neurons'. First, many of them are broadly tuned to images of faces from a particular vantage point (Fig. 12.11a; Desimone *et al.* 1984; Perrett *et al.* 1991). Secondly, these cells have sensitivity to individual faces, but the tunings are broad (Perrett *et al.* 1984; Baylis *et al.* 1985; Yamane *et al.* 1988; Young and Yamane 1992). These response properties suggest that face neurons represent the 'configuration' specific to faces.

Intrinsic signal imaging showed that there are spots specifically activated by faces (Wang *et al.* 1996, 1998) (Fig. 12.12). Thus, face neurons, as well as neurons specifically

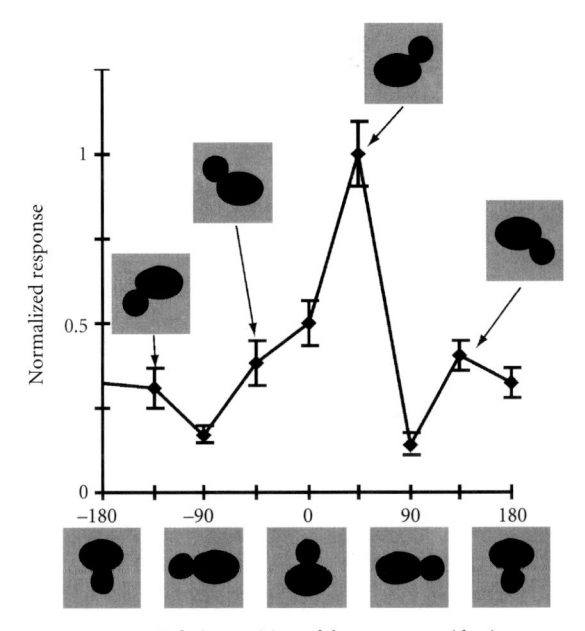

Fig. 12.10 Response specificity of a representative cell to the spatial arrangement of parts. The upper part of the critical feature of the cell was rotated relative to the lower part. The horizontal axis indicates the angle between the line connecting the center of upper and lower parts of each stimulus and that of the critical feature. The vertical axis indicates normalized value of stimulus evoked responses. In this particular case, the best response was elicited by the stimulus with 45°, but this is not only the case. Some other neurons respond maximally at 0°. It may be the case that neurons with different angles are located in close vicinity, as is the case for face columns (Fig. 12.12).

responding to visual features, are clustered together. Furthermore, activation patterns produced by images of faces from different vantage points revealed that the peaks of activity spots shift along the cortical surface as the face rotates from the left profile, to the right profile, through the front face. This representation of faces from different vantage points in close vicinity may be important for view-independent recognition of faces.

12.8 Summary and conclusions

In summary, our results from both optical and extracellular recordings provide evidence that a complex object is represented by combinations of columns in area TE, each of which represents visual features of the object image. The combinations are not simply based on summing up of the columns, but may instead rely on a combination of active and inactive columns to represent objects. These columns do not necessarily

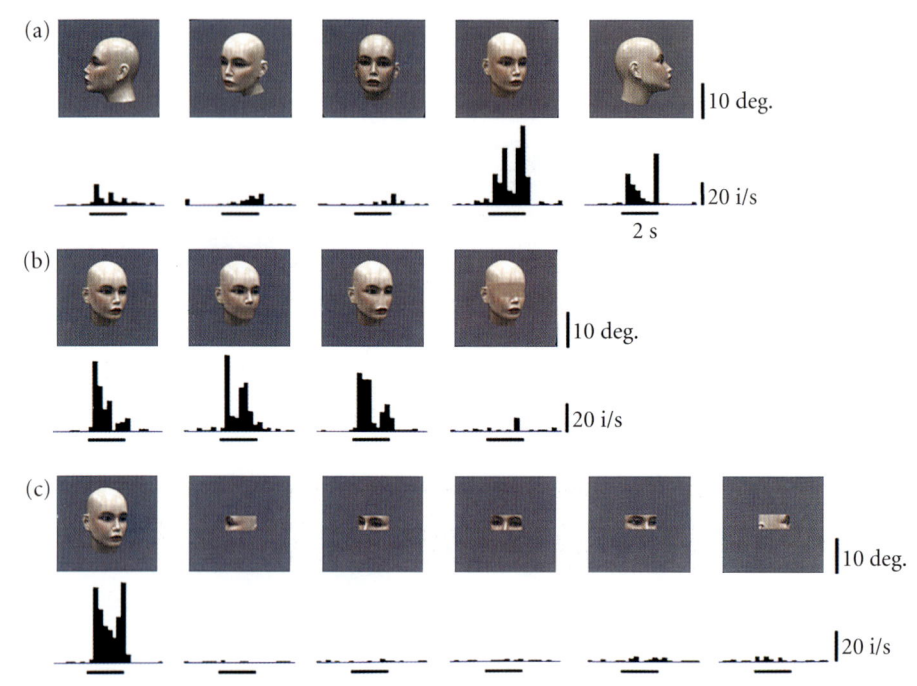

Fig. 12.11 Response properties of a typical face neuron shown by the PSTHs of neural responses evoked by stimuli given above. The recording consisted of three sessions. In the first session, (a), we examined selectivity of the neuron to a face at different vantage points. In the second session, (b), we examined the same neuron with face stimuli lacking different parts of the face. In the final session, (c), we examined responses to insolated eyes from different vantage points.

represent 'local features' but some of them may represent visual features related to 'configurational information'. We consider that neurons specific for images of familiar objects, such as 'faces', also represent visual features specific for these objects. For example, face neurons may represent 'configuration' specific for 'faces'. An image of a particular face could be represented by a combination of columns. Some of these columns represent 'local feature' specific to an individual, and others represent configurational information about the face at a particular vantage point.

Visual features represented in area TE seem to range from simple features such as 'rectangular shape' to highly complex visual features such as 'configuration of faces'. Although there is no direct evidence, this wide range of complexity in visual features may be produced through experience-dependent association of visual features. For example, neurons responding to vertically aligned parts (Fig. 12.8) may be generated because monkeys frequently experience objects having such a part arrangement. In support, in monkeys trained with a particular set of visual stimuli, previous studies revealed that some neurons specifically responded to visual stimuli that became familiar

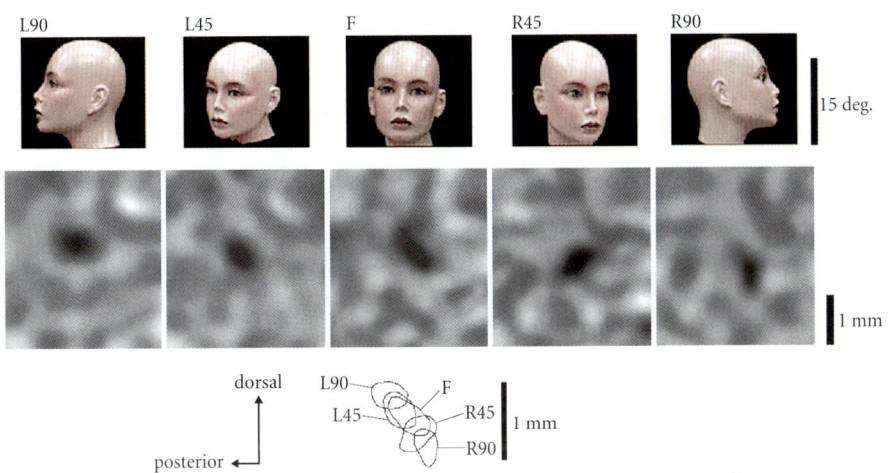

Fig. 12.12 Systematic shift of activation spots with rotation of the face. Images of the same cortical area (middle panels) obtained for five different views of the same mannequin face (top panels). The contours of the activity spots are superimposed at the bottom. (Adapted from Wang *et al.* 1996 Permission sought.)

through training (Logothetis *et al.* 1995; Kobatake *et al.* 1998). Those visual stimuli could represent 'local features' or 'configurational information' depending on the stimulus set and the task design.

Acknowledgements

We thank R. Uma Maheswari, Kathleen Rockland, and Charles Rockland for helpful comments on the manuscript. This work was partly supported by Research Fellowships of the Japan Society for the Promotion of Young Scientists to Y.Y.

References

Baylis, G.C., Rolls, E.T., and Leonard, C.M. (1985) Selectivity between faces in the responses of a population of neurons in the cortex in the superior temporal sulcus of the monkey. *Brain Res.* **342**, 91–102.

Desimone, R., Albright, T.D., Gross, C.G., and Bruce, C. (1984) Stimulus-selective properties of inferior temporal neurons in the macaque. *J. Neurosci.* **4**, 2051–62.

Fodiak, P. and Young, M.P. (1995) Sparse coding in the primate cortex. In M.A. Arbib (ed.) *The handbook of brain theory and neural networks* (pp. 895–8). MIT Press, Cambridge MA.

Fujita, I. and Fujita, T. (1996) Intrinsic connections in the macaque inferior temporal cortex. *J. Comp. Neurol.* **368**, 467–86.

Fujita, I., Tanaka, K., Ito, M., and Cheng, K. (1992) Columns for visual features of objects in monkey inferotemporal cortex. *Nature* **360**, 343–6.

Gochin, P.M., Miller, E.K., Gross, C.G., and Gerstein, G.L. (1991) Functional interactions among neurons in inferior temporal cortex of the awake macaque. *Exp. Brain Res.* **84**, 505–16.

Grinvald, A., Shoham, D., Shmuel, A., Glaser, D., Vanzetta, I., Shtoyerman, E., Slovin, H., Wijnbergen, C., Hildesheim, R., and Arieli, A. (1999) In-vivo optical imaging of cortical architecture and dynamics. In U. Windhorst and H. Johansson (Eds.), *Modern techniques in neuroscience research* (pp. 893–970). Springer, Berlin.

Gross, C.G., Rocha-Miranda, C.E., and Bender, D.B. (1972) Visual properties of neurons in inferotemporal cortex of the Macaque. *J. Neurophysiol.* **35**, 96–111.

Kisvarday, Z.F., Kim, D-S., Eysel, U.T., and Bonhoeffer, T. (1994) Relationship between lateral inhibitory connections and the topography of the orientation map in cat visual cortex. *Eur. J. Neurosci.* **6**, 1619–32.

Kobatake, E. and Tanaka, K. (1994) Neuronal selectivities to complex object features in the ventral visual pathway of the macaque cerebral cortex. *J. Neurophysiol.* **71**, 856–67.

Kobatake, E., Wang, G., and Tanaka, K. (1998) Effects of shape-discrimination training on the selectivity of inferotemporal cells in adult monkeys. *J. Neurophysiol.* **80**, 324–30.

Logothetis, N.K., Pauls, J., and Poggio, T. (1995) Shape representation in the inferior temporal cortex of monkeys. *Curr. Biol.* **5**, 552–63.

Malach, R., Reppas, J.B., Benson, R.R., Kwong, K.K., Jiang, H., Kennedy, W.A., Ledden, P.J., Brady, T.J., Rosen, B.R., and Tootell, R.B. (1995) Object-related activity revealed by functional magnetic resonance imaging in human occipital cortex. *Proc. Natl Acad. Sci. USA* **92**, 8135–39.

Perrett, D.I., Rolls, E.T., and Caan, W. (1982) Visual neurones responsive to faces in the monkey temporal cortex. *Exp. Brain Res.* **47**, 329–42.

Perrett, D.I., Smith, P.A.J., Potter, D.D., Mistlin, A.J., Head, A.S., Milner, A.D., and Jeeves, M.A. (1984) Neurons responsive to faces in the temporal cortex: studies of functional organization, sensitivity to identity and relation to perception. *Hum. Neurobiol.* **3**, 197–208.

Perrett, D.I., Oram, M.W., Harries, M.H., Bevan, R., Hietanen, J.K., Benson, P.J., and Thomas, S. (1991) Viewer-centered and object-centered coding of heads in the macaque temporal cortex. *Exp. Brain. Res.* **86**, 159–73.

Saleem, K.S., Tanaka, K., and Rockland, K.S. (1993) Specific and columnar projection from area TEO to TE in the macaque inferotemporal cortex. *Cerebr. Cortex* **3**, 454–64.

Tanaka, K., Saito, H., Fukada, Y., and Moriya, M. (1991) Coding visual images of objects in the inferotemporal cortex of the macaque monkey. *J. Neurophysiol.* **66**, 170–89.

Tsunoda, K., Yamane, Y., Nishizaki, M., and Tanifuji, M. (2001) Complex objects are represented in macaque inferotemporal cortex by the combination of feature columns. *Nat. Neurosci.* **4**, 832–8.

Wang, G., Tanaka, K., and Tanifuji, M. (1996) Optical imaging of functional organization in the monkey inferotemporal cortex. *Science* **272**, 1665–8.

Wang, G., Tanifuji, M., and Tanaka, K. (1998) Functional architecture in monkey inferotemporal cortex revealed by in vivo optical imaging. *Neurosci. Res.* **32**, 33–46.

Yamane, S., Kaji, S., and Kawano, K. (1988) What facial features activate face neurons in the inferotemporal cortex of the monkey? *Exp. Brain. Res.* **73**, 209–14.

Yamane, Y., Tsunoda, K., Matsumoto, M., Phillips, A., and Tanifuji, M. (2001) Decomposition of object images by feature columns in macaque inferotemporal cortex. *Neurosci. abstr* 399.6.

Young, M.P. and Yamane, S. (1992) Sparse population coding of faces in the inferotemporal cortex. *Science* **256**, 1327–31.

Chapter 13

Plasticity and functional brain development: The case of face processing

Mark H. Johnson

Abstract

This commentary addresses the issue of how specialization for visual—cognitive functions arise in the human cortex. Three viewpoints on this issue are reviewed: a maturational view, a skill-learning view, and an 'interactive specialization' approach. In considering the example of specialization for face processing, it appears that the interactive specialization view can potentially explain the widest range of phenomena.

13.1 Introduction

Many of the chapters in this book concern the extent and nature of specialization of perceptual and attention functions within the human adult cerebral cortex. This contribution, and some of those preceding it (see Chapter 2), focus on the more basic question of how such specialization of function within the cortex arises in the first place. In other words, what are the factors, both intrinsic and extrinsic to the cortex, that ensure that: (1) we develop particular types of specialized cognitive functions relevant for our survival, such as face and language processing; and (2) these specialized functions are usually located in approximately the same parts of cortex? The most obvious answer to these questions is that specific genes are expressed in particular parts of cortex and 'code for' patterns of wiring particular to certain computational functions. While this type of explanation appears to be valid for specialized computations within subcortical structures, a variety of genetic, neurobiological, and cognitive neuroscience evidence indicates that it is, at best, only part of the story for many human cognitive functions dependent on cerebral cortex. For example, in human adults, experience or practice in certain domains can change the extent of cortical tissue activated during performance of a task. In this commentary I consider these issues from the point of view of three different perspectives that have been taken on the postnatal development of human brain function.

13.2 Perspectives on human functional brain development

Much of the research attempting to relate brain to behavioral development in humans has been from a maturational viewpoint, in which the goal is to relate the 'maturation' of particular regions of the brain, usually regions of cerebral cortex, to newly emerging sensory, motor, and cognitive functions. Evidence concerning the differential neuroanatomical development of cortical regions can be used to determine an age when a particular region is likely to become functional. Success in a new behavioral task at this age is then attributed to the maturation of a new brain region. This approach is, in essence, a reverse of adult neuropsychology; instead of regions and their associated functions being lost, they are added-in to the brain one by one.

In one example of this approach, maturation within the frontal lobes has been related to advances in the ability to reach for desirable objects toward the end of the first year. Infants younger than 9 months often fail to accurately retrieve a hidden object after a short delay period if the object's location is changed from one where it was previously successfully retrieved. Instead, the infants perseverate by reaching to the location where the object was found on the immediately preceding trial (Piaget 1954). This error is similar to those made by human adults with frontal lesions and monkeys with lesions to the dorso-lateral prefrontal cortex (Diamond and Goldman-Rakic 1986, 1989), leading to the proposal that the maturation of this region in human infants allows them to retain information over space and time, and to inhibit prepotent response (Diamond 1991). In turn, these developments allow successful performance in object retrieval paradigms.

Despite successes of the maturational approach, and its intuitive simplicity, there are reasons to believe that it may not successfully explain all aspects of human functional brain development. For example, a view of human functional brain development in which regions mature sequentially cannot easily account for the dynamic changes in patterns of cortical activation observed during postnatal development, or for activity in frontal cortical regions during the first months. Further, evidence reviewed in the next paragraphs indicates that consideration of the emerging interactions between regions of the brain is at least as important as the development of connectivity within a region.

The second view on human functional brain development originates from recent neuroimaging evidence from adults that has highlighted changes in the neural basis of behavior resulting from acquiring perceptual or motor expertise. One hypothesis is that the regions active in infants during the onset of new perceptual or behavioral abilities are the same as those activated at the onset of skill acquisition in adults (Gauthier and Nelson 2001). This hypothesis predicts that some of the changes in the neural basis of behavior during infancy will mirror those observed during more complex skill acquisition in adults.

In contrast to more precocial mammals, one of the most striking features of human infants is their initial inability to perform simple motor tasks, such as reaching for an object. Work on complex motor skill learning tasks in adult primates shows that prefrontal cortex is often activated during the early stages of acquisition, but this activation recedes to more posterior regions as expertise is acquired (Rushworth *et al.* 1997; Shadmehr and Holcomb 1997; Miller 2000). In addition to the examples of prefrontal involvement described earlier, activity in this region, or at least within the frontal lobe, has been reported in a number of infancy studies where action is elicited, and damage to these structures has more severe long-term effects than damage to other cortical regions. For example, Csibra and colleagues (Csibra *et al.* 2001) examined the cortical activity associated with the planning of eye movements in 6-month-old infants. They observed eye-movement-related potentials over frontal sites, but not over the more posterior (parietal) sites, where they are normally observed in adults. Converging results are obtained when eye movement tasks are studied in infants with perinatal focal damage to the cortex. While infants with damage to the frontal quadrants of the brain show long-term deficits in visual orienting tasks, infants with the more posterior damage, which causes deficits in adults, are unaffected (Craft and Schatz 1994; Johnson *et al.* 1998).

With regard to perceptual expertise, Gauthier, Behrmann and colleagues have shown that extensive training of adults with artificial objects, 'greebles', eventually results in activation of a cortical region previously associated with face processing, the 'fusiform face area'. This suggests that the region is normally activated by faces in adults, not because it is prespecified for faces, but due to our extensive expertise with that class of stimulus, and encourages parallels with the development of face-processing skills in infants (Gauthier and Nelson 2001). The extent of the parallels between adult perceptual expertise and infant perceptual development remains unclear. However, in both cases event-related potential (ERP) studies have revealed effects of stimulus inversion only after substantial expertise has been acquired with faces or greebles (Rossion *et al.* 2000; de Haan *et al.* 2002). Future experiments need to trace in more detail changes in the patterns of cortical activation during training in adults and development in infants.

According to a third perspective, 'interactive specialization', it is assumed that postnatal functional brain development, at least within the cerebral cortex, involves a process of organizing inter-regional interactions (Johnson 2000). While this view stands in contrast to the maturational approach, it is not mutually exclusive with the skill learning view. By the interactive specialization approach, we should observe changes in the response properties of cortical regions during ontogeny as regions interact and compete with each other to acquire their role in new computational abilities. The onset of new behavioral competencies during infancy will be associated with changes in activity over several regions, and not just by the onset of activity in one or more additional region(s). In further contrast to the maturational approach, this view predicts that during infancy patterns of cortical activation during behavioral tasks may be more extensive than those observed in adults, and involve different patterns of activation. Within broad constraints, even apparently the same behavior in infants and adults could involve different patterns of cortical activation.

Recent evidence indicates that the same behavior in infants and adults can sometimes be mediated by different structures and pathways, and that there are dynamic changes in the cortical processing of stimuli during infancy. Experiments with scalp-recorded electrical potentials have suggested that, at least for word learning and face processing (Neville *et al.* 1992; de Haan *et al.* 2002), there is increasing spatial localization of selective processing with age or experience of a stimulus class. For example, for word recognition, differences between known words and control stimuli are initially found over large areas, but this difference narrows to the leads over the left temporal lobe only when vocabulary reaches around 200 words, irrespective of maturational age. In parallel with changes in the patterns of regional activation are changes in the 'tuning' of individual regions. This evidence for dynamic changes in cortical processing during infancy is consistent with a process in which inter-regional interactions help to shape intra-regional connectivity, such that several regions together come to support particular perceptual and cognitive functions.

13.3 The case of face processing

Currently, one of the best-studied cases of functional specialization in the human cortex is face processing (see, for example, Chapters 3 and 7, this volume). How does this specialization arise, and why do face-sensitive regions tend to be located in particular regions of cortex? I will suggest that the three viewpoints outlined above provide different answers to this question. According to the maturational view, specific genes are expressed within particular cortical regions (such as the fusiform 'face area') and pre-wire those areas for face processing. One of several problems with this argument is that differential gene expression within the mammalian cerebral cortex tends to be on a much larger scale than the functional regions identified in imaging studies (see Chapter 3, this volume). While the skill-learning approach can explain a portion of the currently available data, I propose that the interactive specialization view can better account for the majority of behavioral and imaging data currently available.

Returning to the issue of how specialization for visual-cognitive functions arises in the cortex, we can now examine in more detail the factors, both intrinsic and extrinsic to the cortex, that ensure that we (1) develop specialized processing of faces, and (2) that this specialized function usually becomes located in the same parts of cortex in adults. With regard to developing a cortex specialized for face processing, a number of studies have shown that newborns (in some studies within the first hour of life) preferentially look toward simple face-like patterns (e.g. Johnson *et al.* 1991; Valenza *et al.* 1996). While the exact visual cues that elicit this preference are still debated, it may be as simple as three high-contrast blobs in the location of eyes and mouth. Uni-dimensional psychophysical properties of the stimuli, such as their spatial frequency spectrum, have been ruled out (Morton and Johnson 1991). A number of lines of evidence suggest that this newborn preference is not mediated by the same cortical structures involved in face processing in adults, and may be due to subcortical structures such as the pulvinar. One purpose of this early tendency to fixate on faces may be to elicit bonding from adult caregivers. However, I suggest that an equally important purpose is to bias the visual input to plastic cortical circuits. This biased sampling of the visual environment over the first days and weeks of life may ensure the appropriate specialization of later developing cortical circuitry (Morton and Johnson 1991). In addition to these findings, recent work has shown that newborns prefer to look at faces that engage them in direct (mutual) eye gaze (Farroni *et al.* 2002). Maintaining mutual gaze with another's face ensures foveation of that stimulus, a fact that may be relevant to the eventual location of 'face areas' within the cortex (see Chapter 8). Thus, I suggest, this newborn bias effectively 'tutors' the developing cortex, and early foveation of faces may, in part, determine the location of face-sensitive areas within the cortex.

While the notion of one brain system 'bootstrapping' another is consistent with the interactive specialization approach, is there any evidence for dynamic changes in the extent of specialization or localization of processing in the cortex during the first months or years of life? The results from a series of ERP studies show that while adults have a scalp-recorded ERP response to human upright faces that is different to that seen with closely related stimuli such as inverted faces and monkey faces, in infants this response is much less specific (less finely tuned). For example, in 3- and 6-month old babies identical ERPs are generated to inverted faces as to upright faces, indicating common processing of the two stimuli. Evidence that the response becomes more localized comes from a recent functional magnetic resonance imaging (fMRI) study in which more extensive cortical activation resulted when children performed a simple face-matching task compared to adults (Passarotti *et al.* 2003).

13.4 Conclusions

In this brief discussion of plasticity and development, I reviewed three different perspectives on human functional brain development. These perspectives lead to different expectations about functional imaging of visual cognition during development. By a maturational view, effort is directed to establishing which areas of cortex become functionally mature at different ages. A general expectation is that the number of active areas will increase with development. By the 'skill learning' view, it is predicted that there will be similarities between patterns of brain activation in adults as they acquire a new skill, and those observed in infants and children as the tackle simpler versions of the same type of task. Finally, by an interactive specialization view, we expect dynamic changes in the patterns of cortical activation seen, even when behavioral performance is matched across different ages. In some cases, increasing specialization of function could lead to decreases in the number of active cortical areas during development.

The advent of new technologies for the non-invasive functional imaging of infants and children potentially opens new vistas for addressing basic questions about the plasticity and development of visual cognition.

Acknowledgements

I acknowledge financial support from the MRC (grant G9715587) and Birkbeck College.

References

Craft, S. and Schatz, J. (1994). The effects of bifrontal stroke during childhood on visual attention: Evidence from children with sickle cell anemia. *Developmental Neuropsychology,* **10**(3), 285–97.

Csibra, G., Tucker, L.A., and Johnson, M.H. (2001). Differential frontal cortex activation before anticipatory and reactive saccades in infants. *Infancy,* **2**(2), 159–74.

de Haan, M., Pascalis, O., and Johnson, M.H. (2002). Specialization of neural mechanisms underlying face recognition in human infants. *Journal of Cognitive Neuroscience,* **14**, 199–209.

Diamond, A. and Goldman-Rakic, P.S. (1986). Comparative development of human infants and infant rhesus monkeys of cognitive functions that depend on prefrontal cortex. *Neuroscience Abstracts,* **12**, 274.

Diamond, A. and Goldman-Rakic, P.S. (1989). Comparison of human infants and infant rhesus monkeys on Piaget's AB task: Evidence for dependence on dorsolateral prefrontal cortex. *Experimental Brain Research,* **74**, 24–40.

Diamond, A. Neuropsychological insights into the meaning of object concept development. In S. Carey and R. Gelman (Eds.), *The epigenesis of mind: Essays on biology and cognition,* pp. 67–110. Lawrence Erlbaum Associates, Hillsdale NJ.

Farroni, T., Csibra, G., Simion, F., and Johnson, M.H. (2002). Eye contact detection in humans from birth. *Proceedings of the National Academy of Sciences USA,* **99**, 9602–5.

Gauthier, I. and Nelson, C.A. (2001). The development of face expertise. *Current Opinion in Neurobiology,* **11**, 219–24.

Johnson, M.H. *et al.* (1991). Newborns' preferential tracking of face-like stimuli and its subsequent decline. *Cognition,* **40**, 1–19.

Johnson, M.H. *et al.* (1998). Visual attention in infants with perinatal brain damage: Evidence of the importance of left anterior lesions. *Developmental Science,* **1**, 53–8.

Johnson, M.H. (2000). Functional brain development in infants: Elements of an interactive specialization framework. *Child Development,* **71**, 75–81.

Miller, E.K. (2000). The prefrontal cortex and cognitive control. *Nature Reviews Neuroscience,* **1**(1), 59–65.

Morton, J. and Johnson, M.H. (1991). CONSPEC and CONLERN: A two-process theory of infant face recognition. *Psychological Review,* **98**(2), 164–81.

Neville, H.J., Mills, D. and Lawson, D. (1992). Fractionating language: Different neural sub-systems with different sensitive periods. *Cerebral Cortex,* **2**, 244–58.

Passarotti *et al.* (2003). The development of face and location processing: A fMRI study. *Developmental Science,* **6**(1), 100–17.

Piaget, J. (1954). *The construction of reality in the child.* Basic Books, New York NY.

Rossion, B. *et al.* (2000). The N170 occipito-temporal component is delayed and enhanced to inverted faces but not to inverted objects: an electrophysiological account of face-specific processes in the human brain. *NeuroReport,* **11**, 69–74.

Rushworth, M.F. *et al.* (1997). Ventral prefrontal cortex is not essential for working memory. *Journal Neuroscience,* **17**, 4829–38.

Shadmehr, R. and Holcomb, H. (1997). Neural correlates of motor memory consolidation. *Science,* **277**, 821–4.

Valenza, E. *et al.* (1996). Face preference at birth. *Journal of Experimental Psychology: Human Perception and Performance,* **22**, 892–903.

Part 3

Mechanisms of visual attention

Chapter 14

Neurobiology of human spatial attention: Modulation, generation and integration

Jon Driver, Martin Eimer, Emiliano Macaluso, and José van Velzen

Abstract

Studies of human spatial attention using neurobiological measures (i.e. functional imaging, event-related potentials, or magnetoencephalography) have transformed the field, in conjunction with single-cell studies of behaving primates. We review how such empirical work has altered the theoretical framework originally developed in pioneering information processing models. Modulation of sensory processing by spatial attention has now been shown at cortical stages even earlier than envisaged by 'early' selection, although attenuated information may still feed into higher stages. Activity in sensory brain areas may also be modulated by spatial attention prior to stimulus presentation. Such preparatory activity might be involved in subsequent modulations of the neural response to incoming stimuli. Finally, recent work has increasingly focused on the possible control structures that might impose (or 'generate') such modulations. Frontal/parietal/superior-temporal circuits have commonly been implicated, but direct causal demonstrations of their role in modulating sensory responses remain lacking. Our own work on cross-modal interactions in spatial attention further illustrates these themes. Attending to a particular location to judge one modality produces spatially corresponding modulation of sensory activity for other modalities also, illustrating the integrated response of anatomically distant brain areas that appear to be subject to common control processes.

14.1 Introduction

Selective attention was a central topic in early cognitive psychology (e.g. Broadbent 1958) and has been equally central to the emergence of cognitive neuroscience. Selective attention allows some stimuli to be processed more thoroughly than others. This is to some extent under endogenous voluntary control, but is also influenced by stimulus salience and history. Here we focus on spatial aspects of attention, and their impact on sensory processing, primarily for endogenous cases.

Although functional imaging with positron electron tomography (PET) or functional magnetic resonance imaging (fMRI) has only been available since the 1980s or 1990s, these new methods have already transformed the study of selective attention. Other neurobiological measures, including single-cell recording in animals, plus scalp recording of event-related potentials (ERPs) and magnetoencephalography (MEG) in humans, have also made seminal contributions. These advances can be illustrated by contrasting the current state of the field with classic information-processing models of selective attention that date back to the 1950s (see also Driver 2001).

14.2 Classic information-processing models of attention

Figure 14.1 provides a schematic summary of four influential information-processing models. Figure 14.1a illustrates Broadbent's (1958) filter theory. Initial sensory processing of elementary 'physical' properties (e.g. stimulus location) was held to proceed in parallel for all incoming stimuli, regardless of attentional state. A bottleneck was reached at a subsequent limited-capacity stage extracting 'deeper' properties (e.g. stimulus semantics). This limited capacity was protected by a filter that could be set to pass only stimuli with certain physical properties (e.g. from one specific location), for deeper processing. Figures 14.1b–d illustrate the family resemblance of subsequent information-processing models in this tradition. Figure 14.1b shows the traditional rival to Broadbent's account; a 'late-selection' view (e.g. Deutsch and Deutsch 1963; Duncan 1980) that has a similar parallel-then-serial feedforward architecture, but posits that all perceptual processing is unaffected by attention, with the capacity limit proposed to arise only at a stage of selection for action (and/or for awareness). The notion of attention as a gateway to awareness and action remains influential, but the claim that all perceptual processing is unaffected by attention has now been discredited. Figure 14.1c shows Anne Treisman's (1960, 1969) more subtle reformulation of Broadbent's account, on which unattended stimuli provide attenuated inputs to later stages, rather than being filtered out completely. For a stimulus that was expected or primed, this attenuated input might sometimes be sufficient to trigger substantial further processing. The attenuation model may survive recent neurobiological findings rather better than the models in Figs 14.1a and b, but with the caveat that attenuation for unattended stimuli (or enhancement for attended) may arise across a cascade of processing stages, rather than at just one stage. Finally, Fig. 14.1d illustrates Treisman's subsequent feature-integration

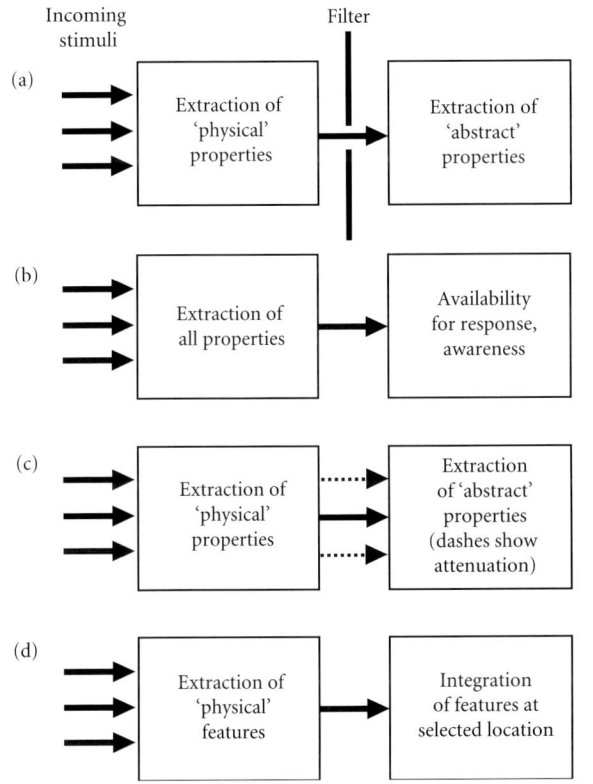

Fig. 14.1 Schematic depictions of four influential models of selective attention, in the information-processing tradition. (a) Broadbent's filter theory; (b) late-selection theory; (c) Treisman's attenuation theory; (d) Treisman's subsequent feature-integration theory. See main text for details. While the differences between these rival theories have traditionally been emphasized, our main point here is the many similarities between them in terms of the overall architecture, with a strong 'family resemblance' to Broadbent's original theory.

theory (e.g. Treisman 1988), which can be viewed as a reformulation of Broadbent's original theory, but proposing a more specific function for spatially selective attention (serial combination of features as a solution to the so-called binding problem).

For present purposes, we wish to emphasize the many similarities between Figs 14.1a–d, rather than the differences. All have a parallel-then-serial feedforward architecture, with selectivity arising at one particular point. All agree that substantial sensory processing (e.g. for 'physical' properties such as stimulus location, color, orientation, etc., in the case of vision) should proceed regardless of attention. None of these models has much to say about top-down or feedback influences. They are more concerned with the impact of setting a filter to pass only certain stimuli, or with the properties for which this might be set, than with the means by which such a filter might be voluntarily re-set (although Broadbent 1958, p.299, did tentatively sketch a possible feedback pathway to the proposed filter). Finally, it is striking that all of these flow diagrams now appear extremely abstract, and naively simple, in comparison with the many complex pathways in the brain. Some authors have seen this abstraction from neurobiological reality as a virtue of the box-model approach, but it is increasingly seen as a weakness.

14.3 Neurobiological measures reveal attentional modulation of sensory processing

Many of the assumptions that are common to the four models in Fig. 14.1 have been overturned or modified by recent research, with neurobiological measures (including functional imaging) playing a major role. One traditional issue has concerned the level(s) of stimulus processing at which selective attention can exert an influence. Pioneering ERP studies (e.g. Eason *et al.* 1969; Van Voorhis and Hillyard 1977) provided the first neurobiological evidence that this influence might begin (though not necessarily be completed) even earlier than Broadbent had envisaged. Neural responses to visual stimuli were studied by averaging voltage-fluctuations time-locked to each stimulus, at posterior electrodes on the human scalp. Eason (1981) found that these could show amplitude differences within 80–150 ms of stimulus onset, when the same stimuli were ignored versus attended (for more recent examples, see Hopfinger *et al.* 2000*a*; Luck *et al.* 2000; Chapter 19 this volume; and later sections of this chapter). Combining ERP methods with functional imaging and MEG has now revealed detailed information about the likely neural generators of such attentional effects on the amplitude of early sensory-specific ERP components (e.g. Heinze *et al.* 1994; Mangun 1995; Hopfinger *et al.* 2000*a*).

Pioneering single-cell studies in awake behaving primates (e.g. Bushnell *et al.* 1981; Moran and Desimone 1985) also reported that neural responses to visual stimuli can depend on attentional factors. The initial findings concerned activity in parietal or extrastriate cortex for macaque monkeys, but were subsequently extended to many other brain areas (see Maunsell 1995; Kastner and Ungerleider 2001). Early functional imaging studies of human selective attention also tested whether activity in sensory areas of the human brain can depend on which aspect of a given display is attended to (e.g. Roland 1982; Corbetta *et al.* 1990). In one example concerning spatial attention (Heinze *et al.* 1994), streams of visual characters were presented concurrently in left and right hemifields, with subjects discriminating those on just one or other side, while maintaining central fixation. Activity in ventral extrastriate visual cortex was higher when attending the contralateral side. Subsequent fMRI studies have shown that even primary visual cortex can be modulated by spatial attention in humans, under certain conditions (e.g. see Brefczynski and DeYoe 1999; Gandhi *et al.* 1999; Martinez *et al.* 1999; Somers *et al.* 1999; Kanwisher and Wojciulik 2000; Kastner and Ungerleider 2001). Modulation of primary and/or secondary sensory cortices has also been reported from some imaging studies of spatial attention for audition or touch (e.g. see Macaluso *et al.* 2000*a*, 2002*a*).

Combining fMRI with ERP and/or MEG, within the same or closely related visual studies, led to suggestions that modulation of primary visual cortex on fMRI measures may not reflect the initial volley of firing in this area, but rather later activity at the same site, which may depend on feedback influences (e.g. Martinez *et al.* 1999;

Hopfinger *et al.* 2000*a*; Noesselt *et al.* 2002). Several imaging studies have shown that attentional modulation may typically increase in successive visual areas (e.g. Kastner *et al.* 1999), especially along the ventral stream, analogous to some single-cell results (see Kastner and Ungerleider 2000, 2001). Thus, instead of arising at just one particular stage (as fondly imagined in Fig. 14.1), attentional modulation may arise in a widespread manner, affecting many levels of sensory processing to different degrees (e.g. see Chapters 15, 17, and 19, this volume).

Early imaging studies used the blocked designs required of PET. Their results were typically interpreted at the time as showing modulation of sensory responses to each incoming stimulus, analogous to findings from measures that are closely time-locked to stimulus onset (as in single-cell data, or ERP and MEG findings). However, because of their blocked designs, some of the initial imaging results might reflect tonic 'baseline-shifts' in activity, that endure throughout each block (reflecting the task performed or the associated preparatory state), rather than changes in phasic neural responses that are time-locked to each stimulus (see Rees *et al.* 1997). Subsequent work (e.g. Chawla *et al.* 1999; Kastner *et al.* 1999) has confirmed that attentional modulations can in fact arise in both forms, at least in some cases. Some modulations arise in a preparatory manner, prior to presentation of any stimulus; while others reflect modulation of the neural response that is time-locked to each stimulus onset. These influences can be distinguished in event-related designs (e.g. Hopfinger *et al.* 2000*b*). Importantly, both types of effect can influence brain areas traditionally regarded as sensory, such as primary and secondary cortices. For instance, Kastner *et al.* (1999) found spatially specific activation of visual cortex as a function of the retinal quadrant in which subjects anticipated attending to visual events that had yet to be presented. Such modulation of sensory cortex prior to stimulus presentation appears to show a pure 'top-down' attentional influence. This raises intriguing questions about how such preparatory activity in sensory cortex differs from activity that can produce sensations when a stimulus is actually presented; and also about how attentional preparation may relate to imagery (see Driver and Frith 2000).

To summarize thus far, the attentional state of the observer can modulate preparatory activity in sensory regions, as well as subsequent stimulus-locked responses when a stimulus is presented. It is possible that the former preparatory influences play some role in modulating sensory responses to the incoming stimulus when this arrives (e.g. see Chawla *et al.* 1999; Driver and Frith 2000), although it should be noted that preparatory activations in sensory areas are not always found (e.g. see Chapter 17, this volume). The findings of attentional modulation affecting sensory regions appear broadly consistent with recent psychophysical results showing that attention can affect many 'low-level' perceptual phenomena (e.g. Shulman 1990; Freeman *et al.* 2001). However, the results overturn many of the fundamental assumptions behind traditional information-processing models of selective attention. All of the models shown in Fig. 14.1 had assumed that initial perceptual processing (extracting basic attributes such as location, orientation, color, or motion, for the case of vision) is unaffected by attention. This 'preattentive' stage was thought to be followed, in a strictly feedforward manner, by a later 'attentive' stage. The neurobiological findings challenge all this, showing that selective attention affects multiple levels of sensory processing, which can even include primary cortex (though perhaps not its initial response). Moreover, the work on preparatory effects (e.g. Kastner *et al.* 1999) clearly demonstrates top-down (not feedforward) influences. Indeed, such attentional influences can actually precede what was traditionally considered to be 'pre-attentive' sensory processing!

14.4 An emerging issue: Identifying control processes, or the sources of attentional modulation

As described above, many studies using neurobiological measures have now found sensory modulations as a function of attentional state. There is now increasing interest in determining the *sources* generating such top-down attentional influences, as distinguished from the various *sites* where they modulate perceptual processing (the site/source terminology was originally introduced by Posner and Driver 1992). This issue is reminiscent of the many attempts to distinguish control processes from the processes they operate upon, throughout many years of research in the information-processing tradition that characterized previous Attention and Performance symposia. A similar issue now arises in neurobiological form.

Neuropsychology has, for many years, implicated regions of prefrontal cortex in executive control and attention, based on the effects of brain damage (for reviews, see Monsell and Driver 2000); plus areas of the parietal lobe (and temporo-parietal junction) for specifically spatial aspects of attention (e.g. see Driver and Vuilleumier 2001). However, the lesions can be diffuse in many patients with florid attentional deficits. The challenge of identifying the critical areas, and in particular how these may modulate processing in anatomically remote sensory areas, still remains (but see Barcelo *et al.* 2000; Hilgetag *et al.* 2001). Nevertheless, it is striking that numerous functional imaging studies have now also implicated diffuse fronto-parietal networks (sometimes including superior temporal cortex, and/or subcortical structures) as potential sources of attentional modulation, using a variety of different criteria (for precedents concerning such a network, see Mesulam 1981; Corbetta *et al.* 1993; and for recent reviews, see Kanwisher and Wojciulik 2000; Kastner and Ungerleider 2000, 2001; Corbetta and Shulman 2002).

Devising comparisons that might isolate activity in putative 'control' structures during functional imaging is more challenging than examining sensory responses. A bewildering variety of different criteria have been suggested for identifying putative control structures with imaging. For instance, they have been identified by comparing attention-shifting conditions to sustained attention at one location (e.g. Corbetta *et al.* 1993) or to rest baselines (Nobre *et al.* 1997). Anticipatory activity in areas that do not subsequently respond to an incoming stimulus has been used as another criterion (Kastner *et al.* 1999). Further criteria include: activity that is time-locked to an attention-directing cue, rather than to a subsequent target stimulus (e.g. Corbetta *et al.* 2000; Hopfinger *et al.* 2000*b*); activity that is sustained across blocks of an attention task, regardless of the number of stimuli presented (e.g. Rees *et al.* 1997); activity that varies with the rate of spatial attention shifts (e.g. Macaluso *et al.* 2001); or activity that is found on invalidly cued trials (that require re-orienting of spatial attention) but not on validly cued trials (e.g. Corbetta *et al.* 2000). It should be clear that each of these various comparisons makes somewhat different assumptions when seeking to isolate a putative 'attentional control' process; but, encouragingly, that the comparisons are becoming increasingly subtle. Direct comparisons of activations from the many different criteria have rarely been performed within a common study (though see Wojciulik and Kanwisher 1999). Nevertheless, it remains striking that many of the criteria have been found to activate diffuse frontal/parietal/superior-temporal networks, often found bilaterally but with stronger activity in the right hemisphere.

14.5 An emerging framework: 'Competition' as the neural instantiation of limited capacity?

Duncan, Desimone and their colleagues (e.g. Desimone and Duncan 1995; Duncan 1996) have put forward a general framework for thinking about selective attention, from a more neural perspective than the original information-processing models (c.f. Fig 14.1). Their 'biased competition' framework (sometimes referred to as 'integrated competition' instead, e.g. Duncan 1996) has three simple premises. First, multiple concurrent stimulus-inputs compete to drive neurons where these inputs meet (e.g. when falling in common or interconnected receptive fields). Such competition may be a general property of neural coding. Secondly, these competitions can be biased not only by intrinsic stimulus properties (e.g. intensity or salience lending an advantage), but also by top-down factors. The latter may even bias various competitions in advance of stimulus presentation, by preparatory states that prime representations of the target stimulus. The recent functional imaging data on 'baseline-shifts' and related phenomena, in the absence of stimulation (e.g. Kastner *et al.* 1999), may provide examples of such pure top-down influences. Finally, Duncan has proposed (e.g. 1996) that while competition between different stimuli (or between rival 'interpretations' of the incoming data) may initially be played out within separate modules (e.g. within color-selective or motion-selective areas of the visual system), the 'winner' of a competition in one module may then tend to become dominant in other modules also, via interconnections. As a result, activity in a diffuse network of different areas may come to converge on a common 'winner' (thus providing one possible explanation for various 'object-based' attentional effects; see Duncan 1996). As we shall discuss later, this third premise about competition becoming integrated is perhaps the least tested aspect of the emerging framework.

This is a potentially powerful framework, although in nascent form it may provide more of a general overview than a source of specific falsifiable predictions. While the notion of neural 'competition' seems capable of putting neuroscientific flesh on the old skeleton of 'limited capacity', it may beg some of the same questions (e.g. at what levels does competition arise, and why?). Moreover, at its most basic, the notion of 'biased competition' seems reminiscent of the traditional insight that capacity limits can arise (now due to 'competition'), and that top-down influences (now recast as 'biases') can determine which stimuli will have access to the limited capacities. Nevertheless, there has already been considerable progress in relating the framework to existing (non-attentional) ideas about neural circuitry for competitive interactions (e.g. Reynolds *et al.* 2000; see also Chapters 15 and 18, this volume). Moreover, Duncan's (1996) premise about the outcome of local competitions being integrated, so that a diffuse network comes to work on a common winner, has several novel implications that can be put to empirical test, as we discuss later in relation to our own work on attentional interactions between different sensory modalities.

The recent imaging data on sensory modulations, on preparatory baseline-shifts, and on putative control structures, when taken together with ideas from the biased-competition framework, have led to an emerging consensus about a likely overall neural architecture for selective attention. Control processes thought to be instantiated by fronto-parietal circuits (and perhaps the temporo-parietal junction also; see Macaluso and Driver 2001; Corbetta and Shulman 2002) generate preparatory states, that specify the current target(s) for selective attention. These preparatory states may induce baseline-shifts of activity in posterior sensory regions, presumed to be imposed via the fronto-parietal control circuits (although note that baseline-shifts in sensory regions are not always observed; see Chapter 17, this volume). Such preparatory states bias the system in advance, to favor some particular stimuli over others. The result is that phasic responses to incoming stimuli are modulated by the preparatory state. Note that this general scheme thus envisages a specific causal chain.

Many questions still remain about this general scheme, however. While fronto/parietal/superior-temporal circuits have indeed been activated by many comparisons thought to involve attentional control processes, we still know remarkably little about the exact functions of the activated control structures, nor whether distinct areas within frontal and parietal cortex play different roles (but see Corbetta and Shulman 2002, for some recent suggestions). Secondly, while the prevailing hypothesis appears to be that activity in fronto-parietal networks directly *causes* posterior modulations in sensory brain areas, there have been few, if any, demonstrations of such causality to date. Although single-cell work has shown that some neurons in prefrontal cortex appear ideally suited for the control of visual attention (e.g. Miller 2000), to our knowledge there have not as yet been any decisive studies showing directly that if such prefrontal populations are disrupted (e.g. by lesion, cooling, or muscimol), attentional modulations in posterior sensory neurons are correspondingly disrupted (but see Tomita *et al.* 1999, for a revealing disconnection study on frontal control of visual memory). Likewise, there have been relatively few studies to date of how lesions to putative control structures in humans (or transient disruptions such as transcranial magnetic stimulation (TMS)) may affect neural responses to incoming stimuli in (structurally intact) remote areas of posterior sensory cortex (but see Barcelo *et al.* 2000; Hilgetag *et al.* 2001; Moore and Fallah 2001). Moreover, no strictly causal relationship has as yet been established between anticipatory changes in posterior sensory regions (e.g. baseline-shifts), when found, and modulation of subsequent stimulus-locked responses there. Thirdly, attentional modulation of sensory responses can be observed even for a single stimulus presented in isolation (as in many ERP studies), when there is no direct competition with another stimulus.

Finally, Duncan's (1996) provocative notion of attentional competition becoming 'integrated' across anatomically distant subsystems has rarely been tested with neural measures (though see O'Craven *et al.* 1999). Recall that this integrative proposal suggests that the winner of one local competition will tend to become dominant in other subsystems also (due to interconnections), with the system as a whole thus tending to focus on one particular object or source of information across many subsystems. Here we propose that the study of *cross-modal* links in spatial attention, between different sensory modalities, may be particularly revealing on this issue, while also illustrating many of the other points raised above.

Below we describe our own recent work on cross-modal interactions in human spatial attention between vision, hearing, and touch. Because initial processing for each of these modalities is carried out within specialized areas that are anatomically remote from each other (i.e. in visual, auditory, or somatosensory cortex), studying cross-modal attention can test whether attentional effects may propagate between different subsystems, to produce 'integrated' outcomes. We shall demonstrate that this does apply for cross-modal links in spatial attention; and will propose that such effects reflect attentional control processes operating in a supramodal manner.

Some of the other chapters in this volume address the issue of identifying separable modules in the human mind/brain (see Part 1). We think it is also important to examine how the different components may interact; and how remote brain areas with different specializations may influence each other. Studies of cross-modal interactions in spatial attention may also offer a tractable handle on this very general issue.

14.6 Cross-modal links in attention: Behavioral evidence

Many studies of selective attention consider only a single modality at a time, treating each of the senses as if they were distinct modules, or sets of modules. However, there is now substantial evidence for cross-modal interactions in spatial attention, from both behavioral and neurobiological studies (for reviews, see Driver and Spence 1998; Eimer 2001; Eimer and Driver 2001; Macaluso and Driver 2001; Spence and Driver, in press).

Behaviorally, Driver and Spence (1994) found that in dual-task situations, auditory and visual streams from a common location could be judged more efficiently than when the two sources of task-relevant information were at different locations. This suggested that it may be difficult to attend separate locations in different modalities. Spence *et al.* (2000*b*) found, in a selective- rather than divided-attention setting, that visual distractors were harder to ignore if coming from the same location as auditory targets (for analogous tactile–visual effects, see Pavani *et al.* 2001). This suggests that it is difficult *not* to attend the same location in different modalities.

Spence and Driver (1996) found that directing attention to one location, in anticipation of *auditory* targets expected there, led to better *visual* as well as auditory performance at that location (even when visual targets were twice as likely elsewhere). Analogously, directing attention to one location in anticipation of a visual target affected auditory as well as visual performance, in a spatially corresponding manner. When subjects attempted to direct auditory attention to one side but visual attention to the other side concurrently, effects of spatial attention were reduced. Spence *et al.* (2000*a*) found analogous interactions in endogenous spatial attention between vision and touch.

Many cross-modal spatial interactions have also now been documented behaviorally for exogenous (or stimulus-driven) cases, as when a spatially non-predictive 'cue' event in one modality leads to better judgments for a target in another modality if appearing at around the same time in the same or similar location, as compared with a target appearing elsewhere (e.g. Spence *et al.* 1998). Exogenous cross-modal spatial-cueing effects of this sort have now been reported for all possible pairings of cue and subsequent target involving vision, hearing, and touch (see Spence 2001). Moreover, such effects can influence perceptual sensitivity (i.e. d-prime) in the target modality, for at least some cases (e.g. McDonald *et al.* 2000).

In many of these exogenous studies, a spatially non-predictive cue in one modality enhances discrimination performance for a subsequent target nearby in another modality, even though the cue itself carries no information whatsoever about the property that must be judged for the target modality. For instance, a sound can enhance judgments of the presence versus absence of a visual event nearby (McDonald *et al.* 2000); and a visual event can enhance judgments of whether a tactile vibration nearby has low versus high frequency (Gray and Tan 2002), or contains one versus two pulses (Spence *et al.* 1998). It thus appears that when a sudden event in one modality makes a particular location salient, other modalities (extracting their own stimulus properties) may also start to focus on that same location, in accordance with the general idea of 'integrated' competition discussed earlier.

The idea that attended or attention-capturing information in one modality might lead attention to be directed to particular information in other modalities, when this seems linked to it by location (and/or timing; see Vroomen and de Gelder 2000), has some parallels with previous observations about one class of so-called 'object-based' attention within purely visual studies. In such studies, attending to one visual property of a particular visual object or group (e.g. its color) may lead to other properties from the same object or group also being attended (e.g. its motion); see Duncan (1996) and O'Craven *et al.* (1999). Analogously, attending to one particular location (and/or point of time) in one sensory modality seems to have implications for other sensory modalities also.

14.7 Sensory modulation by cross-modal spatial attention; ERP evidence

Several ERP studies have now examined cross-modal interactions in spatial attention. ERP effects might reveal the level at which cross-modal attentional influences arise (e.g. whether early, sensory-specific components are affected, as in studies of unimodal attention; or only later components, possibly reflecting decision-related processes instead). Such studies have typically involved cross-modal extensions of ERP designs that were first established to study unimodal endogenous spatial attention. In a prototypical study (e.g. Hillyard *et al.* 1984; Eimer and Schröger 1998; Teder-Sälejärvi *et al.* 1999; Eimer and Driver 2000), a series of sensory events is presented one at a time, with each event on the left or right unpredictably. The task is to monitor just one side, in just one modality, to detect occasional deviant stimuli (e.g. longer than standard events) with an overt response, while maintaining central fixation. ERPs are examined for the standard stimuli that receive no overt response, to avoid contamination of stimulus-locked waveforms by motor activity. The well-established result from many unimodal studies with such designs (e.g. for the visual modality, see Mangun and Hillyard 1991; Eimer 1994; Hopfinger *et al.* 2000a) is that the amplitude of relatively early ERP components (e.g. visual P1 and N1, peaking at around 90–120 and 160–200 ms post-stimulus, respectively) is typically larger for a given stimulus when covert spatial attention is directed to it, than when the same stimulus is ignored while attention is directed to the other side (see Fig. 14.2a, c for examples of visual ERPs at occipital electrodes).

Cross-modal extensions of such a design can assess whether attending to one side to detect infrequent deviant stimuli in a particular modality might affect ERPs for stimuli in other modalities also. This can be addressed simply by including task-irrelevant stimuli in other modalities (e.g. presenting not only visual but also (separately) auditory events on the left or right, while requiring subjects to monitor only one side in just one modality, for occasional deviants). Several studies with such designs have now revealed cross-modal effects of endogenous spatial attention on ERPs (for reviews, see Eimer 2001; Eimer and Driver 2001). Figure 14.2b provides one example, from Eimer and Schröger (1998), where ERPs in response to task-irrelevant *visual* events were modulated as a function of which side was attended for an *auditory* monitoring task (see also Hillyard *et al.* 1984; Teder-Sälejärvi *et al.* 1999). It is noteworthy that this particular cross-modal effect influenced the amplitude of the visual N1 component, which

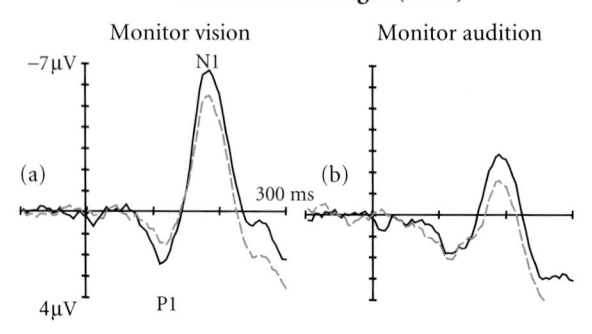

Eimer and Schröger (1998)

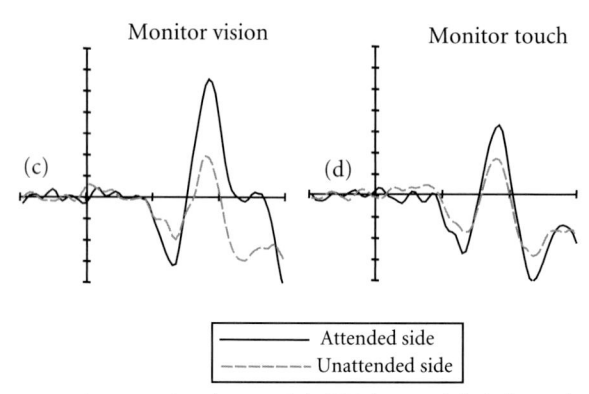

Fig. 14.2 Grand-averaged event-related potentials (ERPs) at occipital electrodes (pooled over OL and OR) contralateral to visual stimuli at attended locations (solid lines) and unattended locations (dashed lines). (a) and (b): data from Eimer and Schröger (1998), investigating audio-visual links. (c) and (d): Data from Eimer and Driver (2000), on tactile–visual links. Waveforms shown separately for conditions where attention was directed to one side to monitor visual events (a and c), or to monitor auditory (b), or tactile (d) events. There are three types of effect on ERP amplitudes. First, visual ERPs are larger when vision is task-relevant (a and c) versus irrelevant, reflecting intermodal attention. Secondly, P1 and N1 components are larger for visual stimuli on the side currently attended for visual monitoring (solid versus dashed lines within a and c), reflecting unimodal spatial attention. Thirdly, the visual N1 (and also P1 for d) is also larger for visual stimuli on the side currently attended for monitoring in a non-visual modality (solid versus dashed lines within b and d), reflecting cross-modal spatial attention.

is traditionally considered to be a sensory-specific component, probably reflecting activity from several regions of visually responsive cortex; the cross-modal attentional modulation arose at around 150 ms post-stimulus (Fig. 14.2b).

Figure 14.2d provides another example of visual ERPs being influenced cross-modally, but now by which side was attended for a *tactile* task. In this case (Eimer and Driver 2000), both the amplitude of the visual N1 component and also the earlier P1 component were affected cross-modally, by which side was monitored covertly for tactile deviants. The implication is that components arising shortly after stimulus presentation (e.g. around 100 ms post-stimulus for P1 in Fig. 14.2d), traditionally regarded as reflecting modality-specific sensory processing (probably within extrastriate visual cortex for P1; see Mangun 1995), can be modulated by where attention is directed for a *different* modality (here for touch). Such ERP results are consistent with the behavioral evidence for cross-modal links in endogenous spatial attention discussed earlier (e.g. Spence and Driver 1996; Spence *et al.* 2000*a*). But the ERP data go further in suggesting that relatively early stages of sensory processing can be affected by cross-modal spatial links, probably within brain areas that would traditionally be regarded as 'unimodal' (e.g. visual cortex).

Conceptually analogous effects have also been obtained for *auditory* rather than visual ERPs, with the auditory N1 showing amplitude-modulation not only due to where spatial attention is directed for unimodal auditory tasks, but also due to which side is attended for visual or tactile monitoring tasks (e.g. Eimer and Schröger 1998; Teder-Sälejärvi *et al.* 1999; Eimer *et al.* 2002). Somatosensory ERPs can also be affected by the direction of attention for tasks in other modalities (e.g. vision), but apparently only if touch remains potentially task-relevant (see Eimer and Driver 2000).

These ERP studies show that directing attention to one side for a task in one modality can affect sensory-specific ERP components for stimuli in other modalities also. However, a skeptic might argue that such cross-modal effects could conceivably reflect an optional choice by the subject to attend towards the same side in several modalities, even when performing a task for only one modality. Because the secondary modality was task-irrelevant in the cross-modal ERP studies described above (and moreover each stimulus was presented in isolation, in a relatively slow sequence), there might have been no penalty for the subject in choosing to attend within the secondary modality to the same side as for the primary modality.

Such an optional strategy seems unconvincing as an explanation for some of the *behavioral* cross-modal results we described earlier (e.g. the difficulty of dual tasks when streams of stimuli in different modalities come from different places (Driver and Spence 1994); the difficulty of ignoring distractors in an irrelevant modality when presented at the same location as target information in another modality (Spence *et al.* 2000); and the reduced effects of spatial attention on discrimination performance when trying to attend opposite sides rather than a common location in two modalities, (Spence and Driver 1996)). Nevertheless, the 'optional-choice' objection does require consideration for the ERP effects.

Eimer (1999, 2001) addressed this by requiring subjects to monitor one or other side in *two* modalities, to detect occasional deviants at the specified location(s) for each modality. In some blocks, the same location had to be monitored for both vision and audition (Eimer 1999), or for both vision and touch (Eimer 2001). In other blocks, a particular modality had to be monitored on one particular side, while the other modality had to be monitored on the *opposite* side. The ERP results (Fig. 14.3) showed that when the tasks required the same side to be monitored for both modalities, the usual amplitude-modulations by spatial attention were observed for early, sensory-specific components (e.g. the left panels in Fig. 14.3 show this for: (a) the visual P1 and N1; (b) the auditory N1 and its descending flank; and (c) the somatosensory N140). The critical result was found when opposite sides had to be monitored for two modalities. Modulations of early sensory components by spatial attention (i.e. when the stimulus appeared on the monitored versus unmonitored side for that modality) were now completely eliminated (see the three right panels in Fig. 14.3a–c). Spatial attention now affected only much later components (from around 200–300 ms post-stimulus). Thus attentional modulation of sensory components was eliminated when subjects were required to try and 'split' their attention to opposite sides for two different modalities. Only later components (probably decision-related) were now affected, as if subjects perceptually processed all stimuli on both sides to a similar extent in the 'split' situation, selecting a stimulus as potentially relevant for their decision only at a later stage. These results suggest that the cross-modal constraints on spatial attention that affect early sensory ERP components do not merely reflect optional, strategic decisions by the subject to attend to a common side in different modalities. Instead, they seem to reflect some relatively 'hard-wired' tendency for attention to be directed to a common location across different senses.

To summarize, cross-modal ERP studies of endogenous spatial attention show that relatively early stimulus-locked components, typically regarded as sensory-specific (or 'unimodal'), can be affected by cross-modal spatial constraints on attention (see Fig. 14.2 and left-hand side of Fig. 14.3). By contrast, later decision-related components are not restricted in the same way (see right-hand side of Fig. 14.3). The implication is that *cross-modal* links in endogenous spatial attention can affect perceptual processing that may arise in '*unimodal*' sensory cortex. This suggests that remote brain areas, each specializing in different modalities, come to focus on a common location together, in a manner consistent with the idea of 'integrated' attentional function. A later section will report converging evidence from functional imaging, showing that cross-modal effects can indeed influence brain areas traditionally regarded as unimodal.

Some analogous ERP results have now also been observed for *exogenous* cross-modal spatial-cueing situations, rather than the endogenous cases considered so far. Again, the cross-modal effects can modulate relatively early sensory components that would usually be considered modality-specific. Kennett *et al.* (2001) observed that a spatially non-predictive, task-irrelevant tactile cue to one hand could modulate ERPs to a visual event presented shortly afterwards, with enhanced visual N1 for stimuli near the same location as the preceding tactile cue (see also McDonald and Ward 2000, for a related audio-visual study). With the hands crossed over, this spatial-cueing effect on the visual N1 crossed correspondingly. The common location of events from different

(a) Visual ERPs

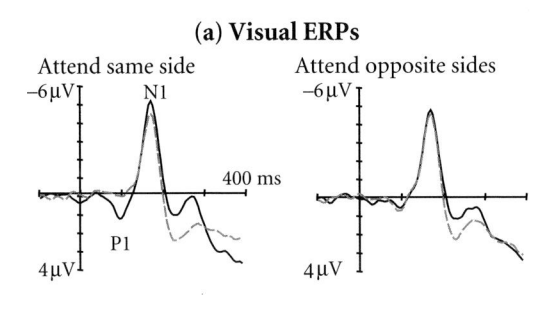

(b) Auditory ERPs

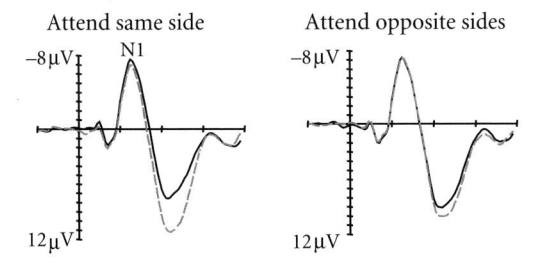

(c) Somatosensory ERPs

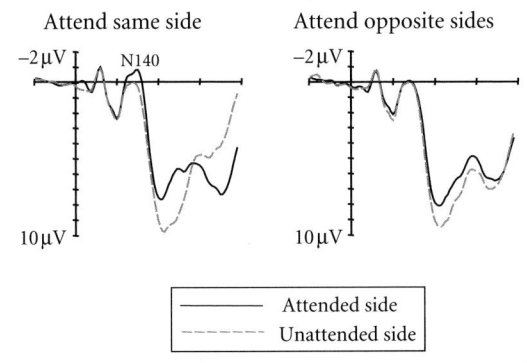

— Attended side
----- Unattended side

Fig. 14.3 ERPs in response to visual, auditory, or tactile stimuli at locations that were task-relevant (solid lines) or irrelevant (dashed lines) for that modality, either when attention was directed to the same side for two modalities (left-hand plots), or when attention had to be directed to opposite sides in the two modalities (right-hand plots); see main text. (a) ERPs elicited at parietal electrodes (PL/PR) contralateral to visual stimuli in an audio-visual study (Eimer 1999); (b) ERPs elicited at Cz for auditory stimuli in the same audio-visual experiment; (c) ERPs elicited at Cz for tactile stimuli in a visuo-tactile experiment summarized by Eimer (2001). Note that sensory-specific components (P1/N1 for vision; N1 and its descending flank for audition; N140 for touch) are modulated by spatial attention (solid versus dashed lines) in all three left-hand plots, but none of the right-hand plots. Attentional modulation arises only later in the three right-hand plots.

modalities in external space thus seems more important than their initial anatomical projections to one or other hemisphere. This suggests that tactile events may influence visual processing via communicating brain areas that also receive postural (e.g. proprioceptive) information.

14.8 Modulation of sensory areas by cross-modal spatial attention: Functional imaging data

Recent PET and fMRI studies provide further evidence that attending to a particular location or side for a task in one sensory modality has implications for other modalities also. These cross-modal spatial influences can affect sensory brain areas, including some that would traditionally be regarded as unimodal (e.g. areas of visual cortex may be affected by the location attended for a purely tactile task).

In one imaging study, Macaluso *et al.* (2000*a*) presented bilateral streams of stimuli in either vision or touch (i.e. unimodal stimulation only, at any one time). Unseen tactile stimulation was on the hands, which were uncrossed, and any visual events were near the hands. Subjects covertly attended one side, to discriminate stimuli (single- versus double-pulse) there in the currently stimulated modality, which was blocked. We compared activations for attending left minus attending right, and vice versa, separately for visual and tactile blocks. Critically, we also tested for any effects of attending to a particular side that were common across touch and vision ('multimodal' spatial attention effects) versus those found only for one of the two modalities ('unimodal' effects).

As expected, given previous studies of only one modality, unimodal effects of spatial attention were found in contralateral posterior occipital cortex for vision, and in contralateral somatosensory cortex for touch. More importantly, some multimodal effects of spatial attention (i.e. common to both vision and touch) were also detected (Macaluso *et al.* 2000*a*). These included a contralateral region of intraparietal sulcus. This seems likely to be a multimodal area, given findings of multimodal neurons responding to both visual and tactile stimuli, in and around the intraparietal sulcus of the monkey brain (e.g. Duhamel *et al.* 1998). Moreover, we confirmed with fMRI (Macaluso and Driver 2001) that this region activates in humans for passive stimulation on the contralateral side, in either vision or touch.

Macaluso *et al.* (2000*a*) also found *multimodal* effects of spatial attention in brain areas that would traditionally be regarded as *unimodal* visual areas. Specifically, lateral and superior occipital regions were not only more active when attending to the contralateral side for a visual task, but also when attending contralaterally for a tactile task. Thus, the direction of attention for a tactile task modulated activity in areas of visual cortex, analogously to the ERP findings reviewed above, where selecting one side for a tactile task could modulate visual ERPs.

A skeptic might again raise the 'optional-strategy' objection; perhaps subjects simply chose to direct visual attention towards the location selected for the tactile task, since this might not disrupt performance in the situation studied by Macaluso *et al.* (2000*a*). However, in a follow-up study, Macaluso *et al.* (2002*a*) implemented a similar design, but now stimulating both modalities at the same time, on both sides. The task was again to discriminate stimuli (double- versus single-pulses) on just one side (left or right) in just one modality (vision or touch). Distractors in the irrelevant modality could now be incongruent with the concurrent target (e.g. a double-pulse when the target was single). Hence the task now motivated subjects to try and ignore the currently irrelevant modality, rather than strategically shifting attention within it to the same side as for the relevant modality. Nevertheless, the imaging results largely replicated those of Macaluso *et al.* (2000*a*). Once again, multimodal effects of spatial attention (i.e. observed both when attending touch and when attending vision) were found in the anterior intraparietal sulcus, contralateral to the attended side (this particular contralateral effect was more pronounced in the left hemisphere for this study). More importantly, multimodal effects of spatial attention were again found in areas traditionally considered as unimodal visual cortex; middle and superior occipital gyri (see Fig. 14.4). These 'visual' areas also showed a main effect of attended modality, being most active during the visual task. The additive combination of an influence from attended side, and also from attended modality, meant that these regions were most active when attending to the contralateral side for a visual task. But they nevertheless still showed increased activation when attending contralaterally for the tactile task (see plots in Fig. 14.4), when the visual stimuli could serve only as distractors.

These two studies suggest that the direction of spatial attention for a tactile task can modulate activity not only in somatosensory and multimodal (intraparietal) cortex, but also in visual cortex. Anatomically remote brain areas specializing in processing different senses can apparently come to focus on a common spatial location. This is consistent with 'integration' of attentional competition between remote subsystems. While these studies addressed endogenous spatial attention, further work with event-related fMRI shows that exogenous cross-modal effects also reveal spatially specific tactile influences on visual cortex.

Macaluso *et al.* (2000*b*) presented single visual targets randomly in the left or right hemifield, unpredictably with or without concurrent touch on one hand (right hand near the right-visual-field target location for one group of subjects; see schematic at the top of Fig. 14.5a; left hand near the left-visual-field target for the other group of subjects, see Fig. 14.5b). This study examined whether multimodal stimulation at the same external location led to an enhanced response from contralateral visual cortex, as compared with purely visual stimulation, or adding touch at a different location. Activation in lingual and fusiform gyri showed this pattern (bottom of Fig. 14.5). Response to a contralateral visual stimulus was boosted here by adding a concurrent tactile event at the same external location, even though this region of visual cortex did not respond to the presence of tactile stimulation *per se* (i.e. did not show an overall main effect of adding tactile stimulation, only an interaction depending on the added

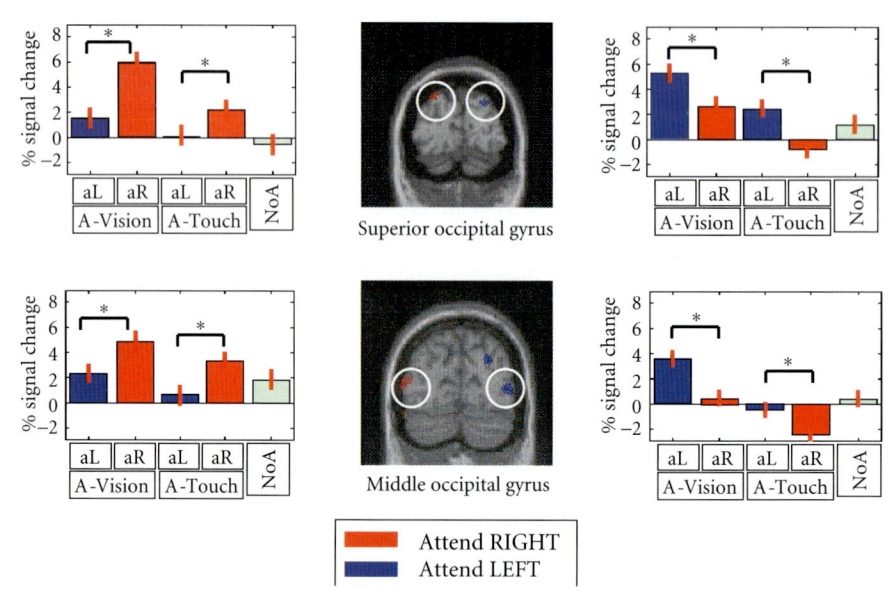

Fig. 14.4 Multimodal attentional modulation in visual cortex when attending covertly to one or other hemifield, during bimodal and bilateral visuo-tactile stimulation (adapted from Macaluso et al. 2002a). Plots refer to the voxel at the maxima of the activated cluster, always found contralateral to the attended side. Note that these occipital areas displayed differential activity depending on the attended side, not only for attend-vision blocks (bars 1 and 2, in each plot), but also during blocks of tactile attention (see bars 3 and 4, in each plot). Activities are expressed in percent signal change compared to a baseline condition without any peripheral stimulation. The NoA-condition (fifth bar in each graph) refers to brain activity in the presence of bilateral, bimodal stimulation, but with attention directed to a central task performed on the fixation point. (See Macaluso et al. 2002a, for details). Coronal sections are taken at $y = -90$ (top slice) and $y = -80$ (bottom slice). aL/aR indicate the attended side. *$P < 0.05$ corrected.

touch falling at the same location as the visual target to which this area responded, in the contralateral hemifield); see plots at bottom of Fig. 14.5.

In this study of Macaluso et al. (2000b), tactile and visual stimuli from the same external location projected initially to the same hemisphere, so a simple account in terms of 'hemispheric competition' (Kinsbourne 1970) might be invoked to explain the cross-modal enhancement. But in a follow-up by Macaluso et al. (2002b), direction of gaze was varied (as happens continuously in daily life). The right hand on which any tactile stimulation was now delivered could fall either in the right or left visual field (see schematic in Fig. 14.6a). Touch on the right hand boosted response for a concurrent right visual target (in left lingual/fusiform gyri) when it lay near that target; but boosted response for a concurrent *left* visual target instead (in right lingual/fusiform gyri) when it lay in the left visual field (see Fig. 14.6b). This reversal of the hemisphere

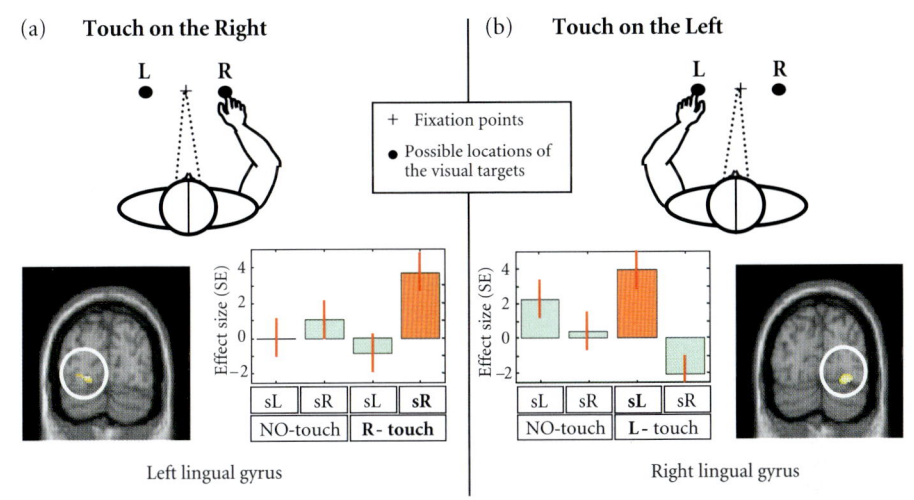

Fig. 14.5 Effect of multimodal spatial correspondence on visual responses in occipital cortex (see Macaluso et al. 2000b). For each group (a: group receiving tactile stimulations to the right hand; b: group with touch to the left hand) we depict the spatial relation between possible visual and tactile input (top), and brain activity in the four experimental conditions (vision only: left or right; bimodal stimulation: in same or opposite hemifields). The sections ($y = -82$ and $y = -80$, for left and right hemisphere activation, respectively) show the anatomical location of the cross-modal interaction observed in the hemisphere contralateral to the location where spatially congruent multimodal stimulation could be presented. The signal plots show the amplification of visual responses when vision and touch were stimulated at the same-minus-contralateral location (see red bars in both graphs). Effect sizes are expressed in standard error (SE) units. sL/sR refers to the side of the visual target (left or right). For display purposes, the anatomical sections shown were thresholded at P uncorrected $= 0.05$.

in which right-hand touch boosted visual activity arose even when the hand was occluded. This indicates that eye-position signals can modulate spatial interactions between touch and vision, and can even reverse which hemisphere is affected.

Once again, the implication from these fMRI studies of exogenous spatial-cueing is that spatial information can be shared between remote brain areas, with the location of a tactile event boosting responses in visual cortex to a concurrent visual target at the same external location. Analysis of 'effective connectivity' between brain areas in the fMRI data, testing for regions showing stronger covariation during multimodal stimulation at the same rather than at different external locations, found that lingual gyrus contralateral to the visual target showed stronger coupling with somatosensory cortex contralateral to the stimulated hand, and with inferior-parietal/superior-temporal cortex, in the situation with multimodal spatial congruence.

Most of the above cross-modal imaging studies documented cross-modal effects involving touch upon visual cortex, rather than visual effects on tactile cortex. As noted by Eimer and Driver (2000) in their ERP study, while effects of visual attention on

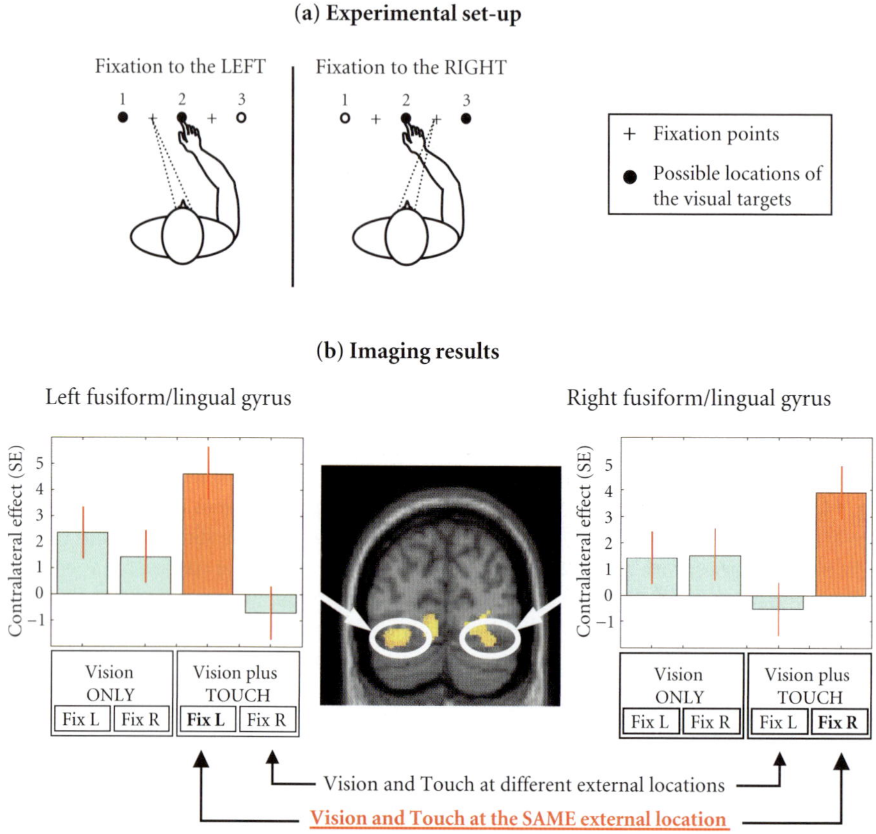

(a) Experimental set-up

Fixation to the LEFT   Fixation to the RIGHT

+ Fixation points
● Possible locations of the visual targets

(b) Imaging results

Left fusiform/lingual gyrus   Right fusiform/lingual gyrus

Vision and Touch at different external locations
Vision and Touch at the SAME external location

Fig. 14.6 Effect of gaze direction on cross-modal spatial interactions in visual cortex (Macaluso et al. 2002c). (a) Depiction of the experimental set-up (possible direction of gaze, possible location of visual targets, and position of the unseen right hand). In different blocks, participants fixated either leftward or rightward of the centrally placed right hand. During leftward fixation visual stimuli could be delivered in either position 1 or position 2. When a visual stimulus in position 2 was coupled with tactile stimulation of the right hand, the multimodal stimulation was spatially congruent. However, both stimuli projected to the left hemisphere, so any amplification of visual responses during leftward fixation might be due to intra-hemispheric effects, rather than the spatial relation of the stimuli in external space. The rightward fixation conditions allowed us to disambiguate between these two alternatives. During rightward fixation, multimodal spatial congruency occurred for simultaneous stimulation of the *left* visual field plus *right*-hand touch (both stimuli in position 2), with the two stimuli now projecting to different hemispheres. (b) Activity in ventral visual cortex for the different stimulation conditions and directions of gaze. Signal plots show the activity in the two hemispheres expressed as the difference between visual stimulation of the contralateral-minus-ipsilateral visual hemifield (i.e. contralateral effect, in standard error units (SE)). Maximal effects of contralateral minus ipsilateral visual stimulation were detected when the visual target was coupled with touch at the same external location (see red bars), in the hemisphere contralateral to the *external* location of the multimodal event (i.e. left hemisphere for leftward fixation, left panel; and right hemisphere for rightward fixation, right panel). Coronal section taken at $y = -78$; section threshold set to P uncorrected $= 0.01$.

somatosensory responses can, in fact, be observed, this may only arise when touch is potentially task-relevant. Further research is needed to determine the principles determining symmetry or asymmetry in cross-modal interactions.

14.9 Studying preparatory control processes cross-modally: ERP data

The preceding ERP and functional-imaging sections focused on sensory modulations. We next consider possible preparatory control processes that might be involved in generating such effects, beginning with ERP evidence. While most ERP studies on spatial attention focus on sensory modulations, a few pioneering unimodal (purely visual) studies examined preparatory components that might be associated with anticipatory shifts of spatial attention, prior to delivery of the imperative stimulus. Harter *et al.* (1989) studied ERP components time-locked to onset of a symbolic cue (central arrow) indicating which side should be attended for an upcoming target. They observed ERP components in the cue–target interval, that depended on whether attention was covertly shifted to the left or right in anticipation of the subsequent target. An early negative deflection at posterior electrodes (possibly a visual response to the central arrow), contralateral to the direction of the cued shift, was observed (see also Nobre *et al.* 2000). Subsequent to this, an enhanced contralateral positivity at posterior electrodes ('late-directing attention positivity', or LDAP) was found. In addition, Nobre *et al.* (2000) and Hopfinger *et al.* (2000*a*) observed enhanced negativities at frontal electrodes contralateral to the direction of cued attention-shifts, which arose earlier than the LDAP, at around 300–500 ms following onset of the central cue (anterior-directing attention negativity, or ADAN).

We recently tested whether such preparatory ERP components in the cue–target interval, time-locked to a symbolic cue indicating which side to attend for an upcoming target, may be specific to anticipating *visual* targets on the cued side (Eimer *et al.* 2002). Alternatively, they might be found regardless of whether the subject is preparing for a visual, auditory, or tactile target on the symbolically cued side. On each trial, a central arrow indicated whether subjects should attend covertly to the left or right for a subsequent target, presented 700 ms after cue onset. In different experimental halves, subjects judged only tactile or only auditory targets, with the central arrow thus indicating which side to attend *for only one relevant modality* (i.e. touch or audition). Trials with any horizontal eye movements detected in the horizontal electro-oculogram (HEOG), were excluded. (There was no significant residual HEOG deviation for attending one side versus the other; averaged across participants it was below 1 µV throughout the cue–target interval, corresponding to residual gaze deviations of less than 0.2°.)

We found a negativity contralateral to the attended side at anterior electrodes, in the period 350–500 ms after onset of the central cue (ADAN). In the subsequent final 200 ms of the cue–target interval, we observed a positivity contralateral to the attended side at posterior electrodes (the LDAP). Figure 14.7 illustrates these ERP lateralizations specific to the direction of attentional shifts. It shows difference-waveforms obtained from lateral frontal (top panels) and lateral posterior sites (bottom). These difference-waves were generated by first subtracting ERPs during attentional shifts to the right from those during leftward attention; and then subtracting the resulting difference-waveforms at right electrodes from the difference-waveforms at corresponding electrodes over the left hemisphere. After these double subtractions, a net negativity (ADAN) contralateral to the direction of an attentional shift is reflected by positive values (downward-going deflections in Fig. 14.7), while a net positivity (LDAP) at contralateral sites is reflected by negative values (upward deflections).

The critical new result in Fig. 14.7 is that both of these preparatory components (ADAN and LDAP) were found regardless of whether subjects where shifting attention to one side in anticipation of a tactile target, or an auditory target. Moreover, they closely resembled the ADAN and LDAP results found previously in purely visual studies. Our subsequent studies confirm that we obtain equivalent ADAN and LDAP results

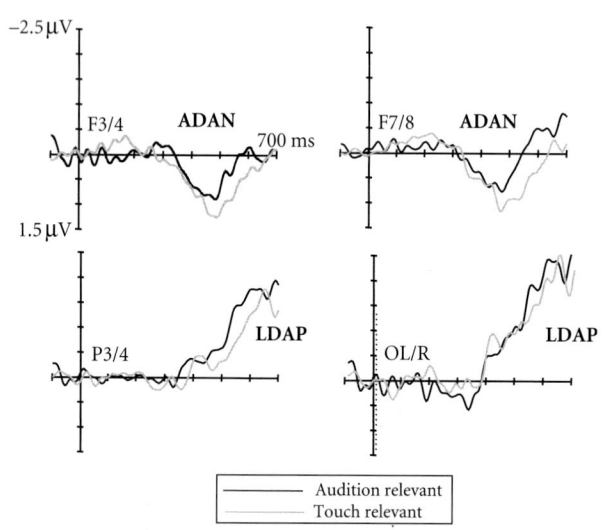

Fig. 14.7 Difference waveforms obtained at lateral anterior (F3/4, F7/8) electrode-pairs (top plots) and posterior (P3/4; OL/R) electrode pairs (bottom plots), in the 700 ms interval between onset of a central symbolic cue and the subsequent peripheral stimulus, reflecting lateralized ERP modulations sensitive to the direction of cued attentional shifts. These difference waves were generated by first subtracting ERPs recorded during rightward attentional shifts from ERPs elicited during leftward attentional shifts, and then subtracting the resulting difference waveforms at right electrodes from difference waveforms at corresponding electrodes over the left hemisphere. An anterior negativity contralateral to the direction of attentional shifts (ADAN) is reflected by positive amplitude values in the plots. The posterior contralateral positivity (LDAP) is reflected by negative values. Note the very similar results regardless of whether the relevant side was cued for an auditory task (thick black lines) or for a tactile task (thin gray lines). (Data from Eimer et al. 2002.)

from subjects preparing for a visual, auditory, or tactile target, even when they judge only one of these modalities throughout the experiment. Furthermore, scalp distributions for these preparatory effects are very similar for the three different modalities (e.g. see Eimer *et al.* 2002, their Figure 7), in addition to timing and polarity. Moreover, the ADAN and LDAP components do not depend closely on the type of central cue that is used either, being found not only when central visual arrows indicate which side to attend, but also when symbolic tones or musical sounds are used for this (see Eimer and Van Velzen 2002).

One possible interpretation of the similarity for the ERP findings in the cue–target interval, regardless of whether the subject prepares for a visual, auditory, or tactile discrimination on the symbolically cued side, is that they may reflect *supramodal* attentional control processes. These would be responsible for directing attention to a particular location or side, regardless of which modality is involved. One alternative is that the preparatory ERP components might always reflect *visual* shifts of attention, with these arising even for auditory or tactile tasks. In many studies of auditory or tactile tasks, the possible sources of stimulation are visible, so that visual information might provide a useful high-acuity means of directing covert attention to the appropriate location. We recently tested whether ADAN and LDAP components depend on locations being visible during a tactile task (Eimer *et al.* 2003). The experiment was conducted either in a lit environment, or in total darkness, with a central auditory cue symbolically indicating which side to attend for a tactile judgment. Equivalent ADAN and LDAP components were observed in both cases, indicating that these components do not require stimulus locations to be visible at the time when attention is covertly directed to the symbolically cued side.

Our working hypothesis is that ADAN and LDAP may reflect supramodal preparatory control processes. But as we discussed earlier, merely observing putative control activations, or preparatory states, falls short of demonstrating that they play a *causal* role in determining subsequent sensory modulations. Establishing causality usually requires direct manipulation (e.g. disrupting the putative control process, and then documenting the consequences of this, as might in principle be done with lesions or TMS; or by cooling or neuropharmacological manipulations (e.g. muscimol in animal studies)). For the case of anticipatory effects arising *substantially in advance* of an imperative stimulus, it may be possible to take a step closer to causality, by seeking correlations that have a specific direction in time. One can seek to correlate preparatory states with the sensory modulations that subsequently follow, even in non-disrupted brains.

We recently tested whether subject-by-subject variation in amplitude of the ADAN preparatory component during the cue–target interval, at anterior electrodes, might correlate with the size of attentional modulations for the visual P1 component, at occipital electrodes, when a visual target subsequently appears. Data were taken from Eimer *et al.* (2002) and Eimer and Van Velzen (2002). Figure 14.8a and c present new analyses showing that such a relationship can indeed be found, for two separate experiments

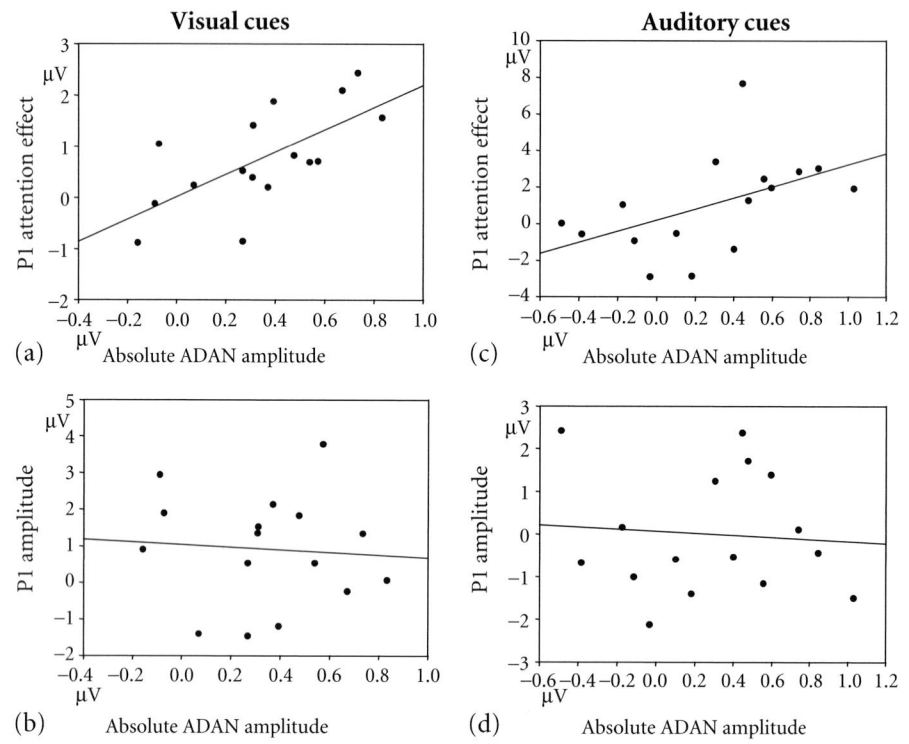

Fig. 14.8 Scatterplots showing the subject-by-subject relationship between the size of the ADAN component elicited in the cue–target interval, and parameters of the P1 component triggered in response to subsequent peripheral visual stimuli, from experiments employing either visual arrows as central cues (Eimer *et al.* 2002; left two plots here), or symbolic sounds (Eimer and Van Velzen 2002; right two plots here). *n* = 16 for both studies. Panels (a) and (c) show that absolute ADAN amplitude (i.e. difference in amplitudes during leftward and rightward attentional shifts at contralateral versus ipsilateral frontal electrodes, averaged across F3/4, F7/8, and FC5/6) was correlated with the size of the attentional P1 modulations at occipital electrodes (OL/OR) for subsequent peripheral visual stimuli in both studies. Panels (b) and (d) show that no such correlation was present between ADAN amplitudes and absolute P1 amplitudes (when disregarding attention).

that differed in whether a central visual arrow was used to cue the attended side for each trial (Eimer *et al.* 2002) (Fig. 14.8a), or a central symbolic sound (Eimer and Van Velzen 2002) (Fig. 14.8c). For both experiments, subjects showing a larger ADAN component tended also to be those showing larger modulation of the P1 by spatial attention ($r(14) = 0.67$ and 0.50; $P < 0.004$ and 0.05, respectively).

It might be objected that subject-by-subject correlations could merely indicate that some subjects may provide higher signal-to-noise for ERP recordings than others, even across a variety of electrode sites (frontal versus occipital here). However, analysis of a 'control' correlation (amplitude of the same ADAN component at anterior electrodes, against the overall amplitude (regardless of attention) for the visual P1 at the same occipital electrodes) showed no relationship (see Figs 14.8b and d; $r(15) = -0.072$ and -0.074, NS). Thus, subject-by-subject variation in the amplitude of the ADAN at anterior sites, in the cue–target interval, is specifically associated with the size of *attentional modulation* for the P1 component at occipital electrodes in response to subsequently presented visual targets. A challenge for the future is to uncover any possible relationships of this type (i.e. between preparatory components and subsequent stimulus-locked components) at the trial-by-trial level, rather than subject-by-subject; and to do so for other sensory modalities in addition to vision. As yet, we have found no subject-by-subject correlation between LDAP amplitude and subsequent sensory modulation, this has been observed only for the ADAN component.

14.10 Studying preparatory and control processes cross-modally: fMRI data

In a recent fMRI study (Macaluso *et al.* 2003) we sought to examine any preparatory brain activity when attending left versus right (or vice versa) that might be found in common regardless of whether the subject prepares for vision or for touch on the symbolically cued side. The issues addressed were thus similar to cross-modal ERP studies of the ADAN/LDAP components, but the paradigm differed in a few respects to optimize it for fMRI. Each trial commenced with a central symbolic cue (auditory pure tone over headphones), whose frequency indicated which side to attend for an upcoming visual or tactile task. On most trials (80%), this cue was followed (1200 ms later) by bilateral, bimodal stimulation (concurrent vision and touch on both sides). A discrimination (single- versus double-pulse) had to be made just for the cued side (varied trial-by-trial), in just the currently relevant modality (blocked). In 20% of trials, no stimulation followed the symbolic central cue, allowing a test for any preparatory activity that depended on the side to which attention was directed, but applied regardless of whether or not subsequent stimulation followed.

As Fig. 14.9a shows, such preparatory activity was found in anterior intraparietal sulcus, contralateral to the covertly attended side, both when touch and when vision was the modality to be judged. This region of intraparietal sulcus would traditionally be regarded as a 'multimodal' brain area. Moreover, as noted earlier, we have found that passive contralateral stimulation activates this region for both vision and touch (Macaluso and Driver 2001). Here we show that the same region can be modulated by endogenous spatial attention even in the absence of stimulation (c.f. Kastner *et al.* 1999).

Middle occipital gyrus also showed preparatory activation contralateral to attended side, for both visual and tactile tasks (Fig. 14.9b). This once again shows that regions traditionally regarded as unimodal visual areas can be affected by the direction of attention in tactile as well as visual tasks. Either of the two activations shown in Fig. 14.9 might conceivably relate to the ADAN or LDAP components found in cue–target

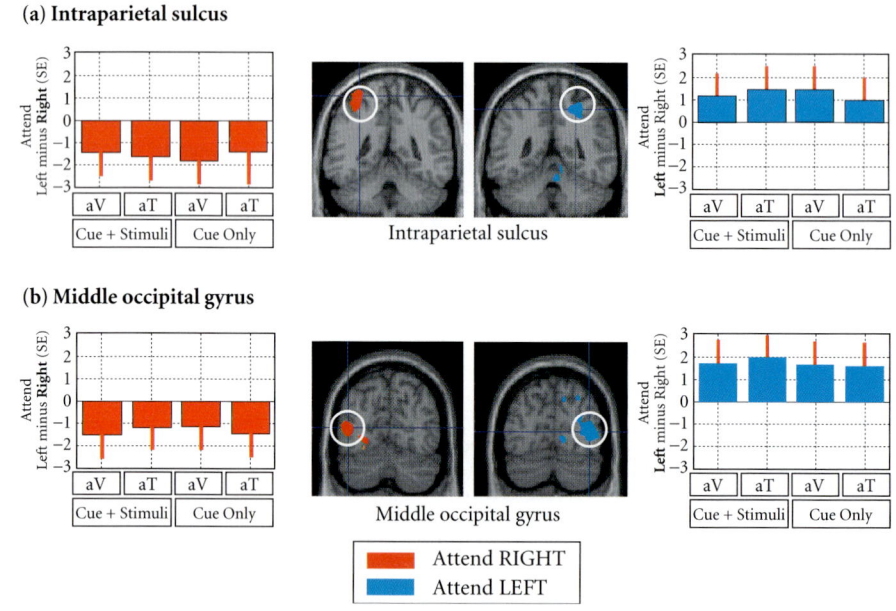

Fig. 14.9 Multimodal preparatory effects for spatial attention to one or the other hemifield. Activities are expressed as differences between attend-left minus attend-right (i.e. positive values indicate greater activity during leftward attention, while negative values indicate greater activity during rightward attention). As expected, spatial attention effects were contralateral to the attended side. The plots show that the spatial attention effects (i.e. differential brain activity between contralateral minus ipsilateral attention) were also present for cue-only trials (two rightmost bars in all signal plots), indicating preparatory activity in both intraparietal areas and middle occipital gyrus. Critically both these areas showed preparatory effects regardless of which modality was task-relevant, again indicating that multimodal spatial attentional effects can influence occipital cortex (here the middle occipital gyrus). Coronal sections are taken at $y = -44$ and $y = -48$ (for the left and right intraparietal activation, top); and $y = -70$ and $y = -76$ (for the left and right occipital activations, bottom). aV/aT indicate the task relevant (attended) modality. Display threshold set to P uncorrected $= 0.01$.

intervals with ERPs. However, resolving with this will require more closely matched ERP and fMRI studies to be conducted, with dense electrode-arrays and detailed source-localization for the ERPs (c.f. Chapter 19, this volume).

In the final fMRI study to be described here, the validity of a symbolic central cue with respect to the subsequent side of a peripheral target was manipulated (Macaluso *et al.* 2002c). The design was similar to a purely visual study by Corbetta *et al.* (2000), except that we used an unpredictable sequence of visual *or tactile* targets, to assess the multimodal issue. In 80% of trials with targets, the central symbolic cue (a digitized voice saying 'left' or 'right' over headphones) correctly indicated the side of the upcoming discrimination target, so spatial cueing was 'valid'. In 20% of trials with targets, the target unexpectedly appeared on the other side, to produce an 'invalid' trial. Such invalid targets may trigger an additional shift of attention to the actual target location (e.g. Posner 1980; Corbetta *et al.* 2000), but also involve a breach of expectancy (Nobre *et al.* 1998). Note that valid and invalid trials were not distinguished in our study until a target was presented in one particular modality. Hence modality-specific effects of invalid trials might in principle be found.

The results largely replicated the previous, purely visual findings of Corbetta *et al.* (2000) (see also Corbetta and Shulman 2002), but critically they extended these to show the multimodal nature of the activations. Invalid minus valid trials activated inferior-parietal/superior-temporal cortex and inferior premotor/prefrontal cortex, bilaterally but stronger in the right hemisphere (see red regions in Fig. 14.10). Common activation for valid and invalid trials highlighted a more superior frontal-parietal

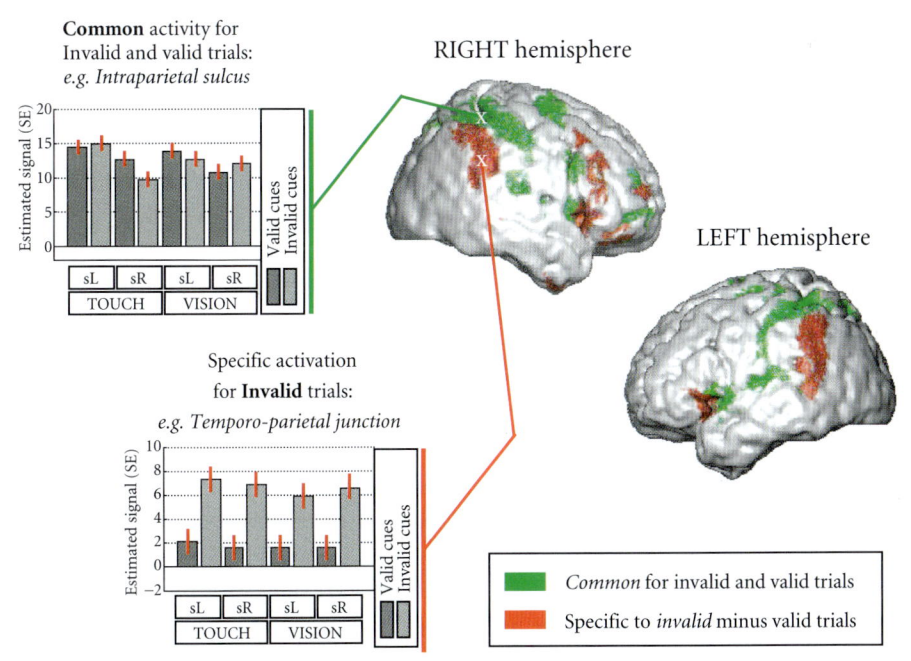

Fig. 14.10 Multimodal effects of spatial cueing for vision and touch, as reported in Macaluso *et al.* (2002b). The figure shows a rendered view of the overall effect of the spatial cueing task (in green), plus regions activated more for invalid than valid trials (when an additional shift of attention between cued hemifield and actual target location is thought to occur; activations shown in red). The plots show the pattern of activity in all experimental conditions for representative maxima in the intraparietal sulcus (*x, y, z* = 44, –42, 62; activated for both valid and invalid trials), or in the temporo-parietal junction (*x, y, z* = 64, –40, 32; more active for invalid than valid trials). In both these regions responses were independent of the modality (and side) of the target, indicating their possible multimodal involvement in the spatial cueing task. sL/sR: target occurs in the left or right hemifield. Display thresholds: *P* uncorrected = 0.001 for the direct comparison of invalid minus valid trials, *P* corrected = 0.001 for the overall effect of the attentional task.

network (c.f. Corbetta *et al.* 1993; Gitelman *et al.* 1999; Wojciulik and Kanwisher 1999), with diffuse activation all along the intraparietal sulcus (see green regions in Fig. 14.10). These findings confirm the differential involvement of an inferior network in invalidly cued trials, and the common involvement of the superior network in both trial types, as recently emphasized by Corbetta and Shulman (2002). The more important point for present purposes concerns the *multimodal* nature of both these networks, which we found to activate equivalently for visual and tactile targets (Macaluso *et al.* 2002c). The existence of multimodal control processes for spatial attention could explain the various cross-modal effects documented throughout this chapter, whereby the disposition of spatial attention for one modality has implications for other modalities also, impacting even on remote sensory-specific cortices and sensory-specific components. Nevertheless, the challenge of identifying the exact functions of the many potential 'control structures' identified in Fig. 14.10, and their direct causal role (if any) in modulating sensory processing elsewhere, still remains. For now, we simply assert that these areas apparently operate in a multimodal manner (i.e. equivalently for visual and tactile stimuli; Macaluso *et al.* 2002c).

14.11 Concluding remarks

Neurobiological measures have revealed:

(a) attentional modulation of sensory neural responses, at numerous levels;

(b) some anticipatory modulations of sensory areas prior to stimulus presentation; and

(c) putative control structures.

Direct causal links between these three types of effect (from c to b to a) have been proposed, but rarely demonstrated to date. A general framework of 'biased competition' or 'integrated competition' has emerged, but its integrative premise has rarely been tested with neural measures. In a series of ERP and functional imaging experiments, we find that directing spatial attention to a particular location in one modality can influence other modalities also, with sensory-specific ERP components, and activation in 'unimodal' brain areas, being influenced for one modality (e.g. vision) by the currently attended or stimulated location for another modality (e.g. touch). Such cross-modal interactions provided clear examples of the 'integration' of attentional selectivity between remote brain areas, possibly via supramodal control processes.

Acknowledgements

Our research is supported by the Medical Research Council (UK), the Wellcome Trust, and the McDonnell–Pew Foundation. J.D. holds a Royal Society–Wolfson Research Merit Award. We thank our many colleagues, including Chris Frith and Charles Spence for imaging and behavioral collaborations, respectively. Thanks also to John 'Broadbent' Duncan for being such an integrative thinker.

References

Barcelo, F., Suwazono, S., and Knight, R.T. (2000). Prefrontal modulation of visual processing in humans. *Nature Neuroscience*, **3**, 399–405.

Brefczynski, J.A. and DeYoe, E.A. (1999). A physiological correlate of the 'spotlight' of visual attention. *Nature Neuroscience*, **4**, 370–4.

Broadbent, D. (1958). *Perception and communication*. Pergamon Press, London.

Buchel, S., Josephs, O., Rees, G., Turner, R., Frith, C.D., and Friston, K.J. (1998). The functional anatomy of attention to visual motion. *Brain*, **121**, 1281–94.

Bushnell, M.C., Goldberg, M., and Robinson, D.L. (1981). Behavioural enhancement of visual responses in monkey cerebral cortex: 1. Modulation in posterior parietal cortex related to selective visual attention. *Journal of Neurophysiology*, **46**, 755–72.

Chawla, D., Rees, G., and Friston, K.J. (1999). The physiological basis of attentional modulation in extrastriate visual areas. *Nature Neuroscience*, **2**, 671–6.

Corbetta, M. and Shulman, G.L. (2002). Control of goal-directed and stimulus-driven attention in the brain. *Nature Reviews Neuroscience*, **3**, 201–15.

Corbetta, M., Miezin, F.M., Dobmeyer, S., and Shulman, G.L., et al. (1990). Attentional modulation of neural processing of shape, color, and velocity in humans. *Science*, **248**, 1556–9.

Corbetta, M., Miezin, F.M., Shulman, G.L., and Petersen, S.E. (1993). A PET study of visuospatial attention. *Journal of Neuroscience*, **13**, 1202–26.

Corbetta, M., Kincade, J.M., Ollinger, J.M., McAvoy, M.P., and Shulman, G.L. (2000). Voluntary orienting is dissociated from target detection in human posterior parietal cortex. *Nature Neuroscience*, **3**, 292–7.

Desimone, R. and Duncan, J. (1995). Neural mechanisms of selective visual attention. *Annual Review of Neuroscience*, **18**, 193–222.

Deutsch, J.A. and Deutsch, D. (1963). Attention: Some theoretical considerations. *Psychological Review*, **87**, 272–300.

Driver, J. (2001). A selective review of selective attention research from the past century. *British Journal of Psychology*, **92**, 53–78.

Driver, J. and Frith, C. (2000). Shifting baselines in attention research. *Nature Reviews Neuroscience*, **1**, 147–8.

Driver, J. and Spence, C. (1994). Spatial synergies between auditory and visual attention. In C. Umilta and M. Moscovitch, (Eds.) *Attention & Performance: Vol. XV*. Erlbaum, Hillsdale NJ.

Driver, J. and Spence, C. (1998a). Attention and the crossmodal construction of space. *Trends in Cognitive Sciences*, **2**, 254–62.

Driver, J. and Vuilleumier, P. (2001). Perceptual awareness and its loss in unilateral neglect and extinction. *Cognition*, **79**, 1–2, 39–88.

Duhamel, J.R., Colby, C.L., and Goldberg, M.E. (1998). Ventral intraparietal area of the macaque: congruent visual and somatic response properties. *Journal of Neurophysiology*, **79**, 126–36.

Duncan, J. (1980). The locus of interference in the perception of simultaneous stimuli. *Psychological Review*, **8**, 272–300.

Duncan, J. (1996). Cooperating brain systems in selective perception and action. In T. Inui, J.L. McClelland, *Attention and Performance: Vol XVI*. MIT Press, Cambridge MA.

Eason, R.G. (1981). Visual evoked potential correlates of early neural filtering during selective attention. *Bulletin of the Psychonomic Society*, **18**, 203–6.

Eason, R., Harter, M., and White, C. (1969). Effects of attention and arousal on visually evoked cortical potentials and reaction time in man. *Physiology and Behavior*, **4**, 283–9.

Eimer, M. (1994). 'Sensory gating' as a mechanism for visual-spatial orienting: Electrophysiological evidence from trial-by-trial cueing experiments. *Perception and Psychophysics*, **55**, 667–75.

Eimer, M. (1999). Can attention be directed to opposite locations in different modalities? An ERP study. *Clinical Neurophysiology*, **110**, 1252–9.

Eimer, M. (2001). Crossmodal links in spatial attention between vision, audition, and touch: Evidence from event-related brain potentials. *Neuropsychologia*, **39**, 1292–303.

Eimer, M. and Driver, J. (2000). An event-related brain potential study of cross-modal links in spatial attention between vision and touch. *Psychophysiology*, **37**, 697–705.

Eimer, M. and Driver, J. (2001). Crossmodal links in endogenous and exogenous spatial attention: Evidence from event-related brain potential studies. *Neuroscience and Biobehavioral Reviews*, **25**, 497–511.

Eimer, M. and Schröger, E. (1998). ERP effects of intermodal attention and cross-modal links in spatial attention. *Psychophysiology*, **35**, 313–27.

Eimer, M. and Van Velzen, J. (2002). Crossmodal links in spatial attention are mediated by supramodal control processes: Evidence from event-related brain potentials. *Psychophysiology*, **39**, 437–49.

Eimer, M., Van Velzen, J., and Driver, J. (2002). Crossmodal interactions between audition, touch, and vision in endogenous spatial attention: ERP evidence on preparatory states and sensory modulations. *Journal of Cognitive Neuroscience*, **14**, 254–71.

Eimer, M., Van Velzen, J., and Driver, M. (2003). Shifts of attention in light and in darkness: An ERP study of supramodal attentional control and crossmodal links in spatial attention. *Cognitive Brain Research*, **15**, 308–23.

Freeman, E., Sagi, D., and Driver, J. (2001). Lateral interactions between targets and flankers in low-level vision depend on attention to the flankers. *Nature Neuroscience*, **4**, 1032–6.

Gandhi, S.P., Heeger, D.J., and Boynton, G.M. (1999). Spatial attention affects brain activity in human primary visual cortex. *Proceedings of the National Academy of Sciences, USA*, **96**, 3314–9.

Gitelman, D.R., Nobre, A.C., Parrish, T.B., LaBar, K.S., Kim, Y.H., Meyer, J.R., and Mesulam, M. (1999). A large-scale distributed network for covert spatial attention: further anatomical delineation based on stringent behavioural and cognitive controls. *Brain*, **122**, 1093–106.

Gray, R. and Tan, H.Z. (2002). Dynamic and predictive links between touch and vision. *Experimental Brain Research*, **145**, 50–55.

Harter, M.R., Miller, S.L., Price, N.J., LaLonde, M.E., and Keyes, A.L. (1989). Neural processes involved in directing attention. *Journal of Cognitive Neuroscience*, **1**, 223–37.

Heinze, H.J., Mangun, G.R., Burchert, W., Hinrichs, H., Scholz, M., Munte, T.F., Gos, A., Scherg, M., Johannes, S., and Hundeshagen, H. (1994). Combined spatial and temporal imaging of brain activity during visual selective attention in humans. *Nature*, **372**, 543–6.

Hilgetag, C.C., Theoret, H., and Pascual-Leone, A. (2001). Enhanced visual spatial attention ipsilateral to rTMS-induced 'virtual lesions' of human parietal cortex. *Nature Neuroscience*, **4**, 953–7.

Hillyard, S.A., Simpson, G.V., Woods, D.L., Van Voorhis, S., and Münte, T.F. (1984). Event-related brain potentials and selective attention to different modalities. In F. Reinoso-Suarez, C. Ajmone-Marsan, (Eds.). *Cortical Integration* pp. 395–414. Raven Press, New York NY.

Hopfinger, J.B., Jha, A.P., Hopf, J.M., Girelli, M., and Mangun, G.R. (2000*a*). Electrophysiological and neuroimaging studies of voluntary and reflexive attention. In S. Monsell, J. Driver, (Eds.). *Attention and Performance XVII* pp. 125–53. MIT Press, Cambridge MA.

Hopfinger, J.B., Buonocore, M.H., and Mangun, G.R. (2000*b*). The neural mechanisms of top-down attentional control. *Nature Neuroscience*, **3**, 284–91.

Kanwisher, N. and Wojciulik, E. (2000). Visual attention: Insights from Neuroimaging. *Nature Reviews Neuroscience*, **1**, 91–100.

Kastner, S. and Ungerleider, L.G. (2001). The neural basis of biased competition in human visual cortex. *Neuropsychologia*, **39**, 1263–76.

Kastner, S., Pinsk, M., De Weerd, P., Desimone, R., and Ungerleider, L. (1999). Increased activity in human visual cortex during directed attention in the absence of visual stimulation. *Neuron*, **22**, 751–61.

Kastner, S. and Ungerleider, L.G. (2000). Mechanisms of visual attention in the human cortex. *Annual Review of Neuroscience*, **23**, 315–41.

Kennett, S., Eimer, M., Spence, C., and Driver, J. (2001). Tactile-visual links in exogenous spatial attention under different postures: Convergent evidence from psychophysics and ERPs. *Journal of Cognitive Neuroscience*, **13**, 462–78.

Kinsbourne, M. (1970). A model for the mechanism of unilateral neglect of space. *Transactions of the American Neurological Association*, **95**, 143–6.

Luck, S.J., Woodman, G.F., and Vogel, E.K. (2000). Event-related potential studies of attention. *Trends in Cognitive Sciences*, **4**, 432–40.

Macaluso, E. and Driver, J. (2001). Spatial attention and crossmodal interactions between vision and touch. *Neuropsychologia*, **39**, 1304–16.

Macaluso, E., Frith, C., and Driver, J. (2000a). Selective spatial attention in vision and touch: unimodal and multimodal mechanisms revealed by PET. *Journal of Neurophysiology*, **83**, 3062–75.

Macaluso, E., Frith, C.D., and Driver, J. (2000b). Modulation of human visual cortex by crossmodal spatial attention. *Science*, **289**, 1206–8.

Macaluso, E., Frith, C.D., and Driver, J. (2001). Mutlimodal mechanisms of attention related to rates of spatial shifting in vision and touch. *Experimental Brain Research*, **137**, 445–54.

Macaluso, E., Frith, C.D., and Driver, J. (2002a). Directing attention to locations and to sensory modalities: Multiple levels of selective processing revealed with PET. *Cerebral Cortex*, **12**, 357–68.

Macaluso, E., Frith, C.D., and Driver, J. (2002b). Crossmodal spatial influences of touch on extrastriate visual areas take current gaze direction into account. *Neuron*, **34**, 647–658.

Macaluso, E., Frith, C.D., and Driver, J. (2002c). Supramodal effects of covert spatial orienting triggered by visual or tactile events. *Journal of Cognitive Neuroscience*, **14**, 389–401.

Macaluso, E., Eimer, M., Frith, C.D., and Driver, J. (2003). Preparatory states in crossmodal spatial attention: Spatial specificity and possible control mechanisms. *Exp Brain Research*, **149**, 62–74.

Mangun, G.R. (1995). The neural mechanisms of visual selective attention. *Psychophysiology*, **32**, 4–18.

Mangun, G.R. and Hillyard, S.A. (1991). Modulations of sensory-evoked brain potentials indicate changes in perceptual processing during visual-spatial priming. *Journal of Experimental Psychology: Human Perception and Performance*, **17**, 1057–74.

Martinez, A., Anllo-Vento, L., Sereno, M.I., Frank, L.R., Buxton, R.B., Dubowitz, D.J., Wong, E.C., Hinrichs, H., Heinze, H.J., and Hillyard, S.A. (1999). Involvement of striate and extrastriate visual cortical areas in spatial attention. *Nature Neuroscience*, **2**, 364–9.

Maunsell, J.H.R. (1995). The Brain's Visual World: Representation of Visual Targets in Cerebral Cortex. *Science*, **270**, 764–9.

McDonald, J.J. and Ward, L.M. (2000). Involuntary listening aids seeing: Evidence from human electrophysiology. *Psychological Science*, **11**, 167–71.

McDonald, J.J., Teder-Salejarvi, W.A., and Hillyard, S.A. (2000). Involuntary orienting to sound improves visual perception. *Nature*, **407**, 906–8.

Mesulam, M.M. (1981). A cortical network for directed attention and unilateral neglect. *Annals of Neurology*, **4**, 309–25.

Miller, E. (2000). The neural basis of top-down control of visual attention in prefrontal cortex. In: S. Monsell, J. Driver, (Eds.). Control of Cognitive Processes: *Attention and Performance XVIII* pp. 511–34. MIT Press, Cambridge MA.

Monsell, S. and Driver, J. (2000). Control of Cognitive Processes: Attention and Performance XVIII. MIT Press, Cambridge MA.

Moore, T. and Fallah, M. (2001). Control of eye movements and spatial attention. *Proceedings of the National Academy of Sciences, USA*, **98**, 1273–6.

Moran, J. and Desimone, R. (1985). Selective attention gates visual processing in the extrastriate cortex. *Science*, **229**, 782–4.

Nobre, A.C., Sebestyen, G.N., Gitelman, D.R., Mesulam, M.M., Frackowiak, R.S., and Frith, C.D. (1997). Functional localization of the system for visuospatial attention using positron emission tomography. *Brain*, **120**, 515–33.

Nobre, A.C., Coull, J.T., Frith, C.D., and Mesulam, M.M. (1998). Orbitofrontal cortex is activated during breaches of expectation in tasks of visual attention. *Nature Neuroscience*, **2**, 11–12.

Nobre, A.C., Sebestyen, G.N., and Miniussi, C. (2000). The dynamics of shifting visuospatial attention revealed by event-related brain potentials. *Neuropsychologia*, **38**, 964–74.

Noesselt, T., Hillyard, S.A., Woldorff, M.G., Schoenfeld, A., Hagner, T., Jancke, L., Tempelmann, C., Hinrichs, H., and Heinze, H.J. (2002). Delayed striate cortical activation during spatial attention. *Neuron*, **35**, 575–87.

O'Craven, K., Downing, P., and Kanwisher, N. (1999). fMRI evidence for objects as the units of attentional selection. *Nature*, **401**, 584–7.

Pavani, F., Spence, C., and Driver, J. (2001). Visual capture of touch; Out-of-the-body experiences with rubber gloves. *Psychological Science*, **11**, 353–9.

Posner, M.I. (1980). Orienting of attention. *Quarterly Journal of Experimental Psychology*, **32**, 3–25.

Posner, M.I. and Driver, J. (1992). The neurobiology of selective attention. *Current Opinion in Neurobiology*, **2**, 165–9.

Rees, G., Frackowiak, R.S.J., and Frith, C. (1997). Two modulatory effects of attention that mediate object categorisation in human cortex. *Science*, **275**, 835–8.

Reynolds, J.H., Pasternak, T., and Desimone, R. (2000). Attention increases sensitivity of V4 neurons. *Neuron*, **26**, 703–14.

Roland, P.E. (1982). Cortical regulation of selective attention in man: A regional cerebral blood flow study. *Journal of Neurophysiology*, **48**, 1059–78.

Shulman, G.L. (1990). Relating attention to visual mechanisms. *Perception and Psychophysics*, **47**, 199–203.

Somers, D.C., Dale, A.M., Seiffert, A.E., and Tootell, R.B.H. (1999). Functional MRI reveals spatially specific attentional modulation in human primary visual cortex. *Proceedings of the National Academy of Sciences, USA*, **96**, 1663–8.

Spence, C. (2001). Crossmodal attentional capture: A controversy resolved? In: C.L. Folk, and B.S. Gibson, (Eds.). *Attention, Distraction and Action* pp. 231–62. Elsevier, Amsterdam.

Spence, C. and Driver, J. (1996). Audiovisual links in endogenous covert spatial attention. *J Exp Psychol Hum Percept Perform*, **22**, 1005–30.

Spence, C.J. and Driver, J. (1996). Audiovisual links in endogenous covert orienting. *Journal of Experimental Psychology: Human Perception and Performance*, **22**, 1005–30.

Spence, C. and Driver, J. (in press). *Crossmodal space and crossmodal attention*. Oxford University Press, Oxford.

Spence, C., Nicholls, M.E., Gillespie, N., and Driver, J. (1998). Cross-modal links in exogenous covert spatial orienting between touch, audition, and vision. *Perception and Psychophysics*, **60**, 544–57.

Spence, C., Pavani, F., and Driver, J. (2000a). Crossmodal links between vision and touch in covert endogenous spatial attention. Journal of Expermental Psychology: *Human Perception and Performance*, **26**, 1298–319.

Spence, C.J., Ranson, J., and Driver, J. (2000b). Crossmodal selective attention: On the difficulty of ignoring sounds at the locus of visual attention. *Perception and Psychophysics*, **62**, 410–24.

Teder-Sälejärvi, W.A., Münte, T.F., Sperlich, F.-J., and Hillyard, S.A. (1999). Intra-modal and cross-modal spatial attention to auditory and visual stimuli: An event-related brain potential (ERP) study. *Cognitive Brain Research*, **8**, 327–43.

Tomita, H., Ohbayashi, M., Nakahara, K., Hasegawa, I., and Miyashita, Y. (1999). Top-down signal from prefrontal cortex in executive control of memory retrieval. *Nature*, **401**, 699–703.

Treisman, A. (1960). Contextual cues in selective listening. *Quarterly Journal of Experimental Psychology*, **12**, 242–8.

Treisman, A. (1969). Strategies and models of selective attention. *Psychological Review*, **76**, 282–99.

Treisman, A. (1988). Features and objects: The fourteenth Bartlett memorial lecture. *Quarterly Journal of Experimental Psychology*, **40A**, 201–37.

Van Voorhis, S.T. and Hillyard, S.A. (1977). Visual evoked potentials and selective attention to points in space. *Perception and Psychophysics*, **22**, 54–62.

Vroomen, J. and de Gelder, B. (2000). Sound enhances visual perception: Cross-modal effects of auditory organization on visual perception. *Journal of Experimental Psychology: Human Perception and Performance*, **26**, 1583–90.

Wojciulik, E. and Kanwisher, N. (1999). The generality of parietal involvement in visual attention. *Neuron*, **23**, 747–64.

Chapter 15

Towards a neural basis of human visual attention: Evidence from functional brain imaging

Sabine Kastner

Abstract

Attentional mechanisms are needed to select relevant, and to filter out irrelevant, information from cluttered visual scenes. Evidence from single-cell physiology studies in monkeys and functional brain mapping studies in humans indicates that selective attention modulates neural activity in the visual system. Attentional modulation is not confined to cortical processing, but occurs already in the lateral geniculate nucleus, which may serve as an early gatekeeper in attentional gain control. At cortical processing stages, selective attention appears to operate by biasing interactions among multiple stimuli that compete for neural representation, thereby effectively filtering out irrelevant information from nearby distracters. Evidence in support of an attentional-load-dependent push–pull mechanism is reported that operates by facilitating responses evoked by attended stimuli and by suppressing responses evoked by unattended stimuli over large portions of the visual field. The overall view that emerges is that neural mechanisms of selective attention operate at multiple stages in the visual system and are determined by the visual processing capabilities of each stage. These attention signals are not generated in the visual system, but in a distributed network of higher-order areas in frontal and parietal cortex that exerts top-down control via feedback projections.

15.1 Introduction

Imagine yourself on a beach in Sicily on a sunny afternoon. You are looking for Nancy and John in the large crowd of people at the beach. Fortunately, Nancy is the only person wearing a red sun hat, which eases your search. However, it would take you all afternoon to find John if you could not rely on mechanisms of selective attention that help to identify him among the many other people. This example illustrates the capacity limit of the visual system to process information from multiple objects in parallel. Therefore, multiple objects present at the same time will compete for neural processing resources of the visual system. This constant competition can be influenced by stimulus driven ('bottom-up') mechanisms, such as stimulus salience (e.g. Nancy's hat), or by 'top-down' cognitive mechanisms, such as selective attention. Directing attention to a particular location in the scene will facilitate information processing at the attended location, and will help to filter out the irrelevant information from other parts of the scene. As a result, John will eventually be identified in the crowd of people.

Selective attention is a broad term that refers to a variety of different behavioral phenomena. Directing attention to a spatial location has been shown to improve the accuracy and speed of subjects' responses to target stimuli that occur in that location (e.g. Posner 1980). Attention also increases the perceptual sensitivity for the discrimination of target stimuli (Lu and Dosher 1998), increases contrast sensitivity (Cameron *et al.* 2002), reduces the interference caused by distracters (Shiu and Pashler 1995), and improves acuity (Yeshurun and Carrasco 1998). In this chapter, I will discuss functional magnetic resonance imaging (fMRI) studies from my laboratory on mechanisms of selective visual attention in the human brain and how they relate to single-cell physiology studies in monkeys. The results of these studies may account for many of the known behavioral effects of attention, thereby providing links between systems neuroscience and cognitive psychology.

One classical debate in the cognitive psychology of attention is that of 'early' versus 'late' selection. The early selection account (e.g. Broadbent 1958) proposed that the neural representations of attended and unattended stimuli are affected differently at early sensory processing stages. As a result, the processing of attended stimuli will be facilitated and that of unattended stimuli will be suppressed. The late selection account (e.g. Duncan 1980) suggested that both types of stimuli are more or less fully processed at sensory levels and that the rejection of unattended stimuli occurs at a stage beyond sensory coding (e.g. in memory systems). Two important questions emerge from this debate which can be addressed at the neural level. First, at which stage of the visual pathway does selective attention first affect neural processing? And secondly, how is attentionally selected information represented, and what is the neural fate of the unattended information? In the following sections, I will attempt to provide some answers to these questions. First, I will describe evidence that selective attention modulates visual processing at the earliest possible stage, that is, the lateral geniculate nucleus (LGN). Secondly, I will review mechanisms of attentional selection and spatial filtering of unwanted information at intermediate processing levels of visual cortex. And thirdly, I will describe the evidence for a distributed network of areas that exerts top-down control via feedback projections to visual cortex. The overall view that emerges from these studies is that neural mechanisms of selective attention operate at multiple stages in the visual system and are determined by the visual processing capabilities of each stage. Selective attention appears to increase neural response gain at the thalamic level and filters out irrelevant information by biasing competitive interactions at cortical processing stages.

15.2 The lateral geniculate nucleus: An early 'gatekeeper'

Converging evidence from single-cell recording studies in monkeys and functional brain mapping studies in humans shows that selective attention modulates neural activity in the visual system (Desimone and Duncan 1995; Kastner and Ungerleider 2000). Attentional response modulation was originally demonstrated in extrastriate cortex (e.g. Moran and Desimone 1985). Recent evidence has shown that neural activity in striate cortex can also be affected depending on certain task-related factors, such as the attentional demands or the need to integrate contextual information from areas beyond the classical receptive field (RF) (e.g. Motter 1993; Ito and Gilbert 1999; Martinez *et al.* 1999). Little is known, however, about the role of earlier, subcortical structures in attentional processing.

The lateral geniculate nucleus (LGN) is the thalamic station in the retinocortical projection and has traditionally been viewed as the gateway to the visual cortex (Jones 1985; Sherman and Guillery 2001). In addition to retinal afferents, the LGN receives input from multiple sources, including striate cortex, the thalamic reticular nucleus (TRN), and the brainstem. The LGN therefore represents the first stage in the visual pathway at which cortical top-down feedback signals could affect information processing. It has proven difficult to study attentional response modulation in the LGN in single-cell physiology due to the small RF sizes of LGN neurons and the possible confound of small eye movements. Several studies have failed to demonstrate attentional modulation in the LGN (e.g. Mehta *et al.* 2000). These negative results appear to support the notion that selective attention affects neural processing only at the cortical level. We recently challenged this view by investigating attentional modulation in the human LGN (O'Connor *et al.* 2002).

At the cortical level, selective attention has been shown to affect visual processing in (at least) three different ways. First, neural responses to attended visual stimuli are enhanced relative to the same stimuli when unattended (attentional enhancement; e.g. Moran and Desimone 1985; Corbetta *et al.* 1991). Secondly, neural responses to unattended stimuli are attenuated depending on the load of attentional resources engaged elsewhere (attentional suppression; Rees *et al.* 1997). And thirdly, directing attention to a location in the absence of visual stimulation and in anticipation of the stimulus onset increases neural baseline activity (attention-related baseline increases; Luck *et al.* 1997; Kastner *et al.* 1999*a*). We investigated these effects of selective attention in a series of three experiments, which were designed to activate optimally the human LGN. Flickering checkerboard stimuli of high or low contrast were used in all experiments, which activated the LGN (Chen *et al.* 1999) and areas in visual cortex, including V1, V2, V3/VP, V4, TEO, V3A and MT/MST, as determined on the basis of retinotopic mapping (Sereno *et al.* 1995; Kastner *et al.* 2001).

To investigate attentional response enhancement in the LGN, checkerboard stimuli were presented to the left or right hemifield while subjects directed attention to the stimulus (attended condition) or away from the stimulus (unattended condition). In the unattended condition, attention was directed away from the stimulus by having subjects count letters at fixation. The letter-counting task ensured proper fixation and prevented subjects from covertly attending to the checkerboard stimuli (Kastner *et al.* 1998). In the attended condition, subjects were instructed to covertly direct attention to the checkerboard stimulus and to detect luminance changes that occurred randomly in time at 10° eccentricity. In our statistical model, stimulation of the left visual hemifield was contrasted with stimulation of the right visual hemifield. Thereby, the analysis was restricted to voxels activated by the peripheral checkerboard stimuli and excluded foveal stimulus representations. Relative to the unattended condition, the neural activity evoked by both the high-contrast stimulus and the low-contrast stimulus increased significantly in the attended condition (Fig. 15.1A). The attentional response enhancement was shown to be spatially specific (see Exp. 4 in O'Connor *et al.* 2002). These results suggest that attention facilitates visual processing in the LGN by enhancing neural responses to an attended stimulus relative to those evoked by the same stimulus when ignored.

To investigate attentional-load dependent suppression in the LGN, high- and low-contrast checkerboard stimuli were presented to the left or right hemifield while subjects performed either an easy attention task or a hard attention task at fixation and ignored the peripheral checkerboard stimuli. During the easy attention task, subjects counted infrequent, brief color changes of the fixation cross. During the hard attention task, subjects counted letters at fixation. Behavioral performance was 99% correct on average in the easy attention task and 54% in the hard attention task, thus indicating the differences in attentional demands. Relative to the easy task condition, neural activity evoked by the high- and the low-contrast stimuli decreased significantly in the hard task condition (Fig. 15.1B). This finding suggests that neural activity evoked by ignored stimuli was attenuated in the LGN depending on the load of attentional resources engaged elsewhere.

To investigate attention-related baseline increases in the LGN, subjects were cued to covertly direct attention to the periphery of the left or right visual hemifield and to expect the onset of the stimulus. The expectation period was followed by attended presentations of a high contrast checkerboard stimulus during which subjects counted the occurrence of luminance changes. During the expectation period, fMRI signals increased significantly relative to the preceding blank period in which subjects were fixating but not directing attention to the periphery. Because the visual input, a gray blank screen, was identical in both conditions, the increase in baseline activity

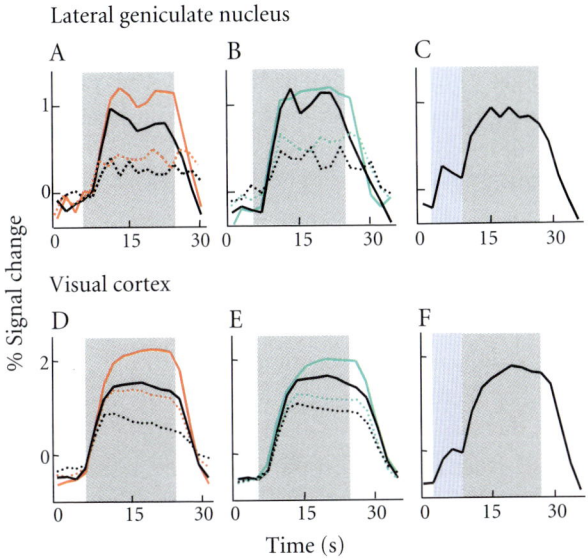

Fig. 15.1 Attentional response modulation in the LGN. Time series of fMRI signals in the LGN (A–C) and in visual cortex, averaged across all activated visual areas (D–F). Data from the LGN and visual cortex were combined across left and right hemispheres. A, D: Attentional enhancement. Responses to both a high-contrast stimulus (solid curves) and low-contrast stimulus (dashed curves) were enhanced during directed attention (red curves) relative to an unattended condition (black curves). B, E: Attentional suppression. During an attentionally demanding fixation task (black curves), responses evoked by both a high-contrast stimulus (solid curves) and low-contrast stimulus (dashed curves) were attenuated relative to a less-demanding attention task at fixation (green curves). C, F: Baseline increases. Baseline activity was elevated during directed attention to the periphery of the visual hemifield in expectation of the stimulus onset (blue shade), followed by a further response increase evoked by the presentation of the checkerboard stimulus. Gray shades indicate periods of checkerboard presentations. (Adapted from O'Connor et al. 2002.)

appeared to be related to directed attention and may be interpreted as a bias in favor of the attended location (see also Section 15.4). The baseline increase was followed by a further response increase evoked by the visual stimuli (Fig. 15.1C). It is important to note that, because of our statistical model, the increase in baseline activity was not related to the cue, which was presented at fixation. This finding suggests that neural activity in the LGN can be affected by attention-related top-down signals even in the absence of any visual stimulation whatsoever.

At the cortical level, qualitatively similar effects of attention were found, as shown in the time series of fMRI signals averaged across all activated areas in visual cortex (Fig. 15.1D–F). The attention effects found at the thalamic and at the cortical level were compared by normalizing the mean fMRI signals evoked in the LGN and in each activated cortical area. This analysis revealed two important results. First, and in accordance with previous findings (Kastner *et al.* 1998; Martinez *et al.* 1999; Mehta *et al.* 2000; Cook and Maunsell 2002), all attention effects increased from early to more advanced processing levels along both the ventral and dorsal pathways of visual cortex. Secondly, all attention effects tended to be stronger in the LGN than in striate cortex. This finding suggests that attentional response modulation in the LGN is unlikely to be due solely to corticothalamic feedback from striate cortex, but may be further influenced by additional sources of input. Alternatively, the differences in magnitude of the modulation between the LGN and V1 may be due to regional disparities underlying the blood oxygenation level dependent signal or non-linearities in thalamo-cortical signal transmission.

In summary, these studies indicate that selective attention modulates neural activity in the LGN by enhancing neural responses to attended stimuli, by attenuating those to ignored stimuli, and by increasing baseline activity in the absence of visual stimulation. These findings challenge the classical notion that attention effects are confined to cortical processing. Further, they suggest the need to revise the traditional view of the LGN as a mere gateway to the visual cortex. In fact, due to its afferent input, the LGN may be in an ideal strategic position to serve as an early 'gatekeeper' in attentional gain control. In addition to corticothalamic feedback projections from V1, the LGN receives inputs from the superior colliculus, which is part of a distributed network of areas controlling eye movements, and the TRN. For several reasons, the TRN has long been implicated in theoretical accounts of selective attention (Crick 1984). First, all feedforward projections from the thalamus to the cortex, as well as their reverse projections, pass through the TRN. Secondly, the TRN receives not only inputs from the LGN and V1, but also from several extrastriate areas and the pulvinar. Thereby, it may serve as a node where several cortical areas and thalamic nuclei of the visual system can interact to modulate thalamocortical transmission through inhibitory connections to LGN neurons (Guillery *et al.* 1998). And, thirdly, the TRN contains topographically organized representations of the visual field and can thereby modulate thalamocortical or corticothalamic transmission in spatially specific ways. Even though much remains to be learnt about the complex thalamic circuitry that may subserve attentional gain control in the LGN, the LGN appears to be the first stage in the processing of visual information that is modulated by attentional top-down signals.

15.3 Visual cortex: Spatial filtering of distracter information in areas V4 and TEO

At the thalamic level, attention may serve to amplify gain, thereby increasing neural signals relative to background noise. At the cortical level, one important function of attention is to filter out another type of noise that is induced by the vast majority of visual information: the unwanted information from distracters (Reynolds and Desimone 1999). Evidence from single-cell physiology and lesion studies in monkeys and fMRI studies in humans suggests that areas V4 and TEO are important sites where relevant information is selected and irrelevant information is filtered out (Kastner *et al.* 1998; DeWeerd *et al.* 1999; Reynolds *et al.* 1999). In this section, I will first describe the evidence that multiple stimuli in cluttered visual scenes compete for neural representation in visual cortex. Secondly, I will describe intra- and extra-RF mechanisms for the spatial filtering of unwanted information. And thirdly, I will propose a push–pull model of selective attention that may account for some of these empirical findings.

15.3.1 A neural basis of competition among multiple stimuli

Due to the limited processing capacity of the visual system, multiple objects present at the same time in visual scenes appear to compete for neural representation. What is the neural basis for competitive interactions among multiple objects in the visual field? In single-cell physiology studies, neural responses to a single visual stimulus presented alone in a neuron's RF were compared to the responses evoked by the same stimulus when a second one was presented simultaneously within the same RF. The responses to the paired stimuli were found to be smaller than the sum of the responses evoked by each stimulus individually and turned out to be a weighted average of the individual responses (Reynolds *et al.* 1999). This result suggests that multiple stimuli present at the same time within a neuron's RF are not processed independently, but interact with each other in a mutually suppressive way, indicating competition for neural representation. Competitive interactions among multiple stimuli present at the same time in the visual field have been found in several areas, including V2, V4, MT, MST, and IT (Miller *et al.* 1993; Recanzone *et al.* 1997; Reynolds *et al.* 1999).

Based on hypotheses derived from these monkey physiology studies, we investigated competitive interactions among multiple stimuli in the human cortex (Kastner *et al.* 1998, 2001). Colorful visual stimuli, which optimally activate ventral visual cortex, were presented in four nearby locations to the periphery of the visual field, while subjects performed a letter-counting task at fixation. The stimuli were presented under two different presentation conditions, sequential and simultaneous. In the sequential presentation condition, a single stimulus appeared in one of the four locations, then another appeared in a different location, and so on, until each of the four stimuli had been presented in the different locations. In the simultaneous presentation condition, the same four stimuli appeared in the same four locations, but they were presented together. Thus, integrated over time, the physical stimulation parameters were identical in each of the four locations in the two presentation conditions. However, suppressive (competitive) interactions among stimuli within RFs could take place only in the simultaneous, not in the sequential presentation condition. Based on the results from monkey physiology, we predicted that the fMRI signals would be smaller during the simultaneous than during the sequential presentation condition due to the presumed mutual suppression induced by the competitively interacting stimuli.

In ventral visual cortex, areas V1, V2, V4, and TEO were consistently activated during visual stimulation as compared to blank periods. As predicted by our hypothesis, simultaneous presentations evoked weaker responses than sequential presentations in all activated visual areas. The response differences were smallest in V1 and increased in magnitude towards ventral extrastriate areas V4 (Fig. 15.2A) and TEO. This increase in magnitude of the suppression effects across visual areas suggested that the competitive interactions were scaled to the increase in RF size of neurons within these areas. That is, the small RFs of neurons in V1 and V2 would encompass only a small portion of the visual display, whereas the larger RFs of neurons in V4 and TEO would encompass all four stimuli. Therefore, suppressive interactions among the stimuli within RFs could take place most effectively in these more anterior extrastriate visual areas. In V1 and V2, it is likely that surround inhibition from regions outside the classical RF contributed to the small sensory suppression effects (e.g. Kastner *et al.* 1999*b*). To rule out the possibility that the differential responses evoked by the two presentation conditions reflected differences in the rate of transient stimulus onsets, suppressive interactions were also demonstrated in a control experiment, in which the presentation rate was kept constant (see Kastner *et al.* 1998, 2001).

The idea that suppressive interactions are scaled to RF size was tested directly in a second study, in which the spatial separation among the stimuli was increased (Kastner *et al.* 2001). According to the RF hypothesis, the magnitude of the suppressive interactions should be inversely related to the degree of spatial separation among the stimuli.

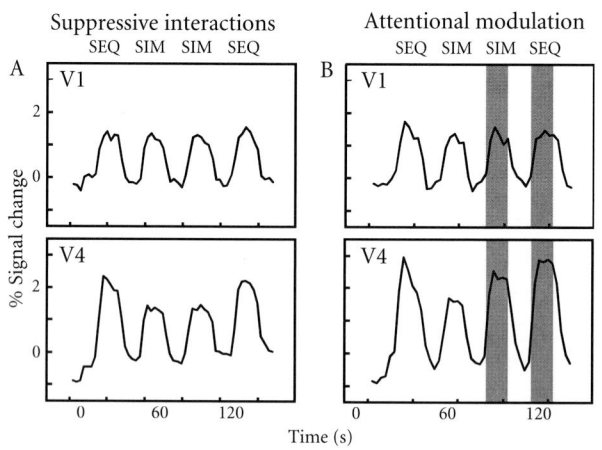

Fig. 15.2 Competitive interactions and attentional modulation in visual cortex. (A) Suppressive interactions in V1 and V4. Simultaneously presented stimuli (SIM) evoked less activity than sequentially presented stimuli (SEQ) in V4, but not in V1, suggesting that suppressive interactions were scaled to the RF size of neurons in visual cortex. (B) Attentional modulation of suppressive interactions. The suppression effect in V4 was replicated in the unattended condition of this experiment, when the subjects' attention was directed away from the stimulus display (unshaded time series). Spatially directed attention (shaded time series) increased responses to simultaneously presented stimuli to a larger degree than to sequentially presented ones in V4. (Adapted from Kastner *et al.* 1998.)

In agreement with this idea, separating the stimuli by 4° abolished suppressive interactions in V2, reduced them in V4, but did not affect them in TEO. Separating the stimuli by 6° led to a further reduction of suppression effects in V4, but again had no effect in TEO. These results confirmed our hypothesis that competitive interactions occur mainly at the level of the RF. Further, by systematically varying the spatial separation among the stimuli and measuring the magnitude of suppressive interactions, we estimated the average RF sizes at an eccentricity of about 5° to be less than 2° in V1, in the range of 2–4° in V2, about 6° in V4, and larger than 6°, but still confined to a quadrant, in TEO. These numbers may underestimate RF sizes in the human visual cortex due to additional suppressive influences from beyond the RF, which cannot be distinguished from interactions within RFs in our experimental paradigm. It was striking, however, that these estimates of RF sizes in human visual cortex as determined on the basis of hemodynamic responses, are similar to those measured in the homologous visual areas of monkeys as defined at the level of single cells (e.g. Desimone and Ungerleider 1989).

In summary, these fMRI studies have begun to establish a neural basis for competition among multiple stimuli present at the same time in the visual field. Importantly, the degree to which this competition occurs appears to depend critically on the RF sizes of neurons across visual cortical areas. This has important implications for the operations of spatial selective attention in visual cortex, as will be described in the next section.

15.3.2 Spatial filtering of distracter information: Intra-receptive field mechanisms

In single-cell recording studies, it has been demonstrated that spatially directed attention can influence the competition among multiple stimuli in favor of one of the stimuli by modulating competitive interactions. When a monkey directed attention to one of two competing stimuli within a RF, the responses in extrastriate areas V2, V4, and MT were as large as those to that stimulus presented alone, thereby eliminating the suppressive influence of the competing stimulus (Reynolds *et al.* 1999; Recanzone and Wurtz 2000). The attentional effects were less pronounced when the second stimulus was presented outside the RF, suggesting that competition for processing resources within visual cortical areas takes place most strongly at the level of the RF. These findings imply that attention may resolve the competition among multiple stimuli by counteracting the suppressive influences of nearby stimuli, thereby enhancing information processing at the attended location. This may be an important mechanism by which attention filters out information from nearby distracters (Desimone and Duncan 1995; Duncan 1996).

A similar mechanism appears to operate in the human visual cortex (Kastner *et al.* 1998). We studied the effects of spatially directed attention on multiple competing visual stimuli in a variation of the paradigm, described in the previous section. In addition to the two different presentation conditions, sequential and simultaneous, two different attentional conditions were tested, attended and unattended. During the unattended condition, attention was directed away from the peripheral visual display by having subjects count letters at fixation. In the attended condition, subjects were instructed to attend covertly to the peripheral stimulus location closest to fixation in the display and to count the occurrences of one of the four stimuli. Based on the results from monkey physiology, we predicted that attention should reduce suppressive interactions among stimuli. Thus, responses evoked by the competing, simultaneously presented stimuli should be enhanced more strongly than responses evoked by the non-competing, sequentially presented stimuli.

Directing attention to the location closest to fixation in the display enhanced activity to sequentially and to simultaneously presented stimuli in extrastriate areas V2/VP, V4 and TEO with increasing effects from early to later stages of visual processing; attentional effects were absent in V1 (Fig. 15.2B). In accordance with our prediction, directed attention led to greater increases of fMRI signals to simultaneously presented stimuli than to sequentially presented stimuli in areas V4 (Fig. 15.2B) and TEO. The magnitude of the attentional effect scaled with the magnitude of the suppressive interactions among stimuli, with the strongest reduction of suppression occurring in ventral extrastriate areas V4 (Fig. 15.2B) and TEO, suggesting that the effects scaled with RF size. These findings support the idea that directed attention enhances information processing of stimuli at the attended location by counteracting suppression induced by nearby stimuli, which compete for limited processing resources, thereby filtering out unwanted information from nearby distracters.

This filter mechanism is compatible with the idea that directed attention to a stimulus may cause the RF to shrink around the attended stimulus, thereby leaving the unattended stimuli in nearby locations outside the RF (Moran and Desimone 1985; Reynolds and Desimone 1999). Given that the magnitude of suppressive interactions scaled with RF size in our fMRI studies (Kastner *et al.* 2001), we estimated the RF sizes in V4 and TEO during directed attention to the display. The reduced suppressive interactions in V4 and TEO during directed attention were similar in magnitude to the suppressive interactions obtained in area V2 in the unattended condition. Hence, it is possible that directed attention caused a constriction of RFs in V4 and TEO from 4–8° to about 2°, thereby presumably enhancing spatial resolution. This interpretation is compatible with behavioral studies showing that spatial attention improves acuity (Yeshurun and Carrasco 1998).

In summary, areas at intermediate levels of visual processing, such as V4 and TEO, appear to be important sites for the filtering of unwanted information by counteracting competitive interactions among stimuli at the level of the RF. This notion has also been supported by studies in a patient with an isolated V4 lesion and in monkeys with lesions of areas V4 and TEO (De Weerd *et al.* 1999; Gallant *et al.* 2000). In these studies, subjects performed an orientation discrimination of a grating stimulus in the absence and in the presence of surrounding distracter stimuli. Significant performance deficits were observed in the distracter-present, but not in the distracter-absent condition, suggesting a deficit in the efficacy of the filtering of distracter information.

15.3.3 Spatial filtering of distracter information: Extra-receptive field mechanisms

The studies described thus far have provided intriguing evidence that selective attention operates by modulating competitive interactions among multiple stimuli for neural representation at the level of the RF. Given the RF sizes of neurons in areas V4 and TEO, this attention mechanism may serve to filter out distracters from nearby locations at a local scale of 4–8°. But how is information filtered out across larger portions of the visual field? One possibility is that large-scale spatial filtering operates by modulating competitive interactions in areas anterior to V4 and TEO with larger RFs, such as the lateral occipital cortex (LOC). However, the retinotopic maps of these areas are typically biased towards the central visual field and do not contain a concise map of the peripheral visual field (Malach *et al.* 2002). Another possibility is that attention operates by using extra-RF mechanisms for large-scale spatial filtering of unwanted information in areas with a detailed representation of the peripheral visual field, such as V4 and TEO. In single-cell recording studies, it has been shown that RFs of neurons are typically surrounded by extensive areas that are not responsive to a stimulus when presented alone, thereby excluding the surrounds from regions of the 'classical' RF. However, stimuli presented to these extra-RF surrounds often interact with stimuli presented to the RF. Such influences of surrounding stimuli on the stimulus in the RF can be suppressive or facilitatory, often depending on contextual factors, such as alignment (Kapadia *et al.* 1995). Contextual modulation is thought to be mediated by long-range horizontal connections and transcallosal connections, thereby affecting neural responses over extensive areas of the visual field (Desimone *et al.* 1993; Lamme 1995). Because attention has been shown to modulate contextual effects in V1 (Ito and Gilbert 1999), we hypothesized that large-scale filtering of unwanted information by attention may be mediated by extra-RF mechanisms.

We tested this idea in an experiment in which attended (target) stimuli were presented to the periphery of the upper right quadrant while ignored (distracter) stimuli were presented to a corresponding location of the contralateral hemifield (Pinsk and Kastner 2001). Thereby, neural activity evoked by target and distracter stimuli could be dissociated. Target stimuli evoked activity in left visual cortex, whereas distracter stimuli evoked activity in right visual cortex. Pairs of colorful patterned stimuli were faded into each other to provide a wide range of ambiguous stimuli that were perceptually difficult to discriminate. Subjects covertly directed attention to a series of peripherally presented stimuli and searched for targets either in a detection task of a non-ambiguous stimulus (Det) or in a 1-back matching task (Mat), while distracter stimuli appeared in the contralateral hemifield. In both tasks, stimuli were identical but appeared in a different sequence. Behavioral results showed that the matching task was more difficult than the detection task, as indicated by increased reaction times and error rates. Due to the differences in perceptual discrimination difficulty between the two tasks, the attentional load was higher in the matching than in the detection task. For the target-search related activity, we predicted that neural activity should increase with increasing attentional load. For the distracter-related activity, the prediction was less clear, and several different possibilities needed to be considered. First, neural responses evoked by distracter stimuli may be suppressed, depending on the attentional load of the task, as previously shown in dorsal extrastriate cortex (Rees *et al.* 1997). Secondly, neural responses evoked by distracter stimuli in the contralateral hemifield may not be affected by attention, as suggested by studies of patients suffering from visuo-spatial hemineglect. In these patients, neural responses evoked by object stimuli presented to the neglected hemifield were similar compared to those evoked by the same stimuli presented to the intact hemifield (Rees *et al.* 2000; Vuilleumier *et al.* 2001). And thirdly, both mechanisms may operate at different processing levels.

Areas V1, V2, V4, and TEO were consistently activated during visual presentations across all conditions relative to interleaved blank periods. In left V1 and V2, target-search related activity was enhanced to a similar degree during the detection and matching tasks relative to a control condition during which subjects ignored target stimuli and counted letters at fixation (Unatt; Fig. 15.3A). In left V4 and TEO, target-search related activity was also enhanced relative to the control condition, but the attentional response enhancement increased with increasing attentional load, supporting our hypothesis (Fig. 15.3A). In these areas, fMRI signals evoked by target stimuli reflected the subjects' behavioral performance. Because of the different patterns of attention effects found in early and intermediate visual areas, it is not likely that the effects simply reflected an addition of attention-related baseline increases to visually evoked activity. Rather, the enhancement may be at least partially due to the facilitation of responses evoked by attentionally selected stimuli. Interestingly, attention effects on responses evoked by single stimuli, which are typically absent or very small in single-cell physiology studies, were found across visual cortex. This raises the possibility that fMRI, which measures neural activity at a population level, may be better suited to uncover small modulatory effects that cannot be found reliably in single-cell physiology, but are revealed when summed across large populations of neurons. In right V1 and V2, distracter-related activity was not affected, suggesting that visual processing was not controlled by attention and was mediated by bottom-up mechanisms (Fig. 15.3B), in accordance with the results from neglect patients. In right V4 and TEO, however, distracter-related activity was suppressed, depending upon the attentional load (Fig. 15.3B), indicating that attentional-load-dependent suppression does not only operate in dorsal, but also in ventral extrastriate cortex. Thus, in areas at intermediate processing stages, the attentional-load-dependent suppression of distracter-related activity mirrored the attentional-load-dependent enhancement of target-search related activity. This finding presents direct evidence for a push–pull mechanism of

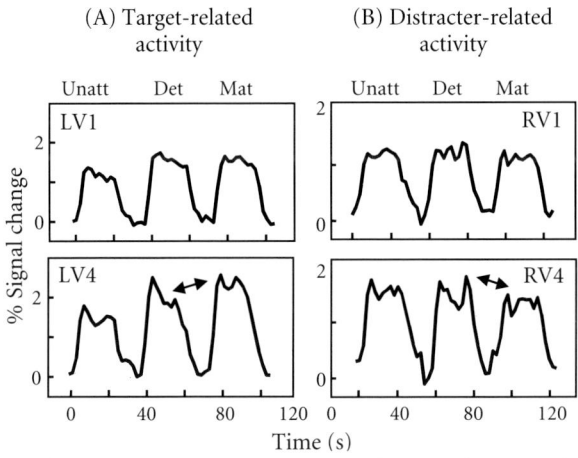

Fig. 15.3 Attentional facilitation and suppression in visual cortex. fMRI signals evoked by attentionally selected (target) stimuli were enhanced during both a 0-back detection task (Det) and a 1-back matching task (Mat) in left V1 relative to an unattended condition during which subjects counted letters at fixation (Unatt). The attention effects depended on the differences in attentional load between the two tasks in left V4. At the same time, fMRI signals evoked by ignored (distracter) stimuli in the contralateral hemifield were suppressed depending on attentional load in right V4, but not in right V1, thus mirroring the facilitatory effects. L = left, R = right.

selective attention that enhances neural activity evoked by selected stimuli and suppresses activity evoked by unattended stimuli. In summary, our findings suggest that attention filters out unwanted information, at least partially, over large spatial scales at intermediate processing levels of visual cortex.

15.3.4 Conclusion: A 'push–pull' model of spatial attention

The studies described above suggest that selective attention operates in areas V4 and TEO by using intra- and extra-RF mechanisms to filter out the vast majority of information from visual scenes: the unattended stimuli. At the level of the RF, attention appears to operate by counteracting competitive (suppressive) influences from nearby distracter stimuli; this intra-RF mechanism may be mediated by lateral (horizontal) connections. Beyond the RF, suppressive mechanisms of attention appear to range over large portions, if not the entire visual scene, possibly mediated by transcallosal connections in addition to lateral connections (Desimone *et al.* 1993). These intra- and extra-RF mechanisms for the spatial filtering of unwanted information can be combined into a unified framework, a 'push–pull' mechanism of spatially selective attention. I suggest that this push–pull mechanism operates by facilitating responses to attentionally selected stimuli and by suppressing responses to ignored stimuli at the same time. In this simple model, the facilitatory and suppressive attentional mechanisms are interdependent; that is, activation of the facilitatory mechanism will also lead to an activation of the suppressive mechanism at the level of the RF and beyond. The suppressive mechanisms occur most strongly at the level of the RF and are more moderate beyond the RF, decreasing gradually with increasing distance from the locus of attentional selection. The lateral extent of suppressive mechanisms that filter out irrelevant information depends on the strength with which the facilitatory mechanism is activated. The overall activation of this push–pull mechanism will depend on the attentional demands of the task at hand. Even though this model is certainly an oversimplification of the underlying mechanisms, it is able to explain a number of empirical findings. First, attention effects are typically smaller at early than at later stages of processing. This result is often taken as evidence for a top-down attentional feedback mechanism that uses cortico-cortical feedback projections and basically reverses the processing hierarchy of visual information. Alternatively, in accord with the model, this result can be explained by the smaller RFs and less extensive extra-RF surrounds of the early processing stages. Secondly, the effects of attentional response enhancement and suppression were mirroring each other at intermediate processing levels, which can be explained by the co-activation of the facilatory and suppressive mechanisms in the model. And thirdly, the magnitude of the attentional response enhancement and suppression depended on the attentional load, which is reflected in the model by the lateral extent to which the suppressive mechanisms can be activated and by the variable amplitude of the facilitatory mechanism. In summary, the facilitation of selected information and the filtering of unwanted information may be mediated by an attention mechanism that operates in a push–pull fashion at intermediate processing levels of visual cortex. Such a mechanism is likely controlled by a distributed network of higher-order areas, as described in the next section.

15.4 **Frontal and parietal cortex: Sorting out sources and sites**

There is evidence from studies in patients suffering from attentional deficits due to brain damage, and from functional brain imaging studies in healthy subjects performing attention tasks, that attention-related modulatory signals are not generated within the visual system, but rather derive from higher-order areas in parietal and frontal cortex and are transmitted via feedback projections to the visual system (Kastner and Ungerleider 2000; Kanwisher and Wojciulik 2000; Nobre 2001; Corbetta and Shulman 2002). For example, in our fMRI studies, activations of parietal and frontal cortex were investigated in addition to activations within visual cortex. Results for a single subject are shown in Fig. 15.4A. In this subject, the frontal eye fields (FEF) were activated bilaterally, together with the supplementary eye field (SEF) and the superior parietal lobule (SPL) during directed attention to the periphery. A network consisting of areas in the SPL, FEF, and the SEF was consistently activated across subjects and has been found

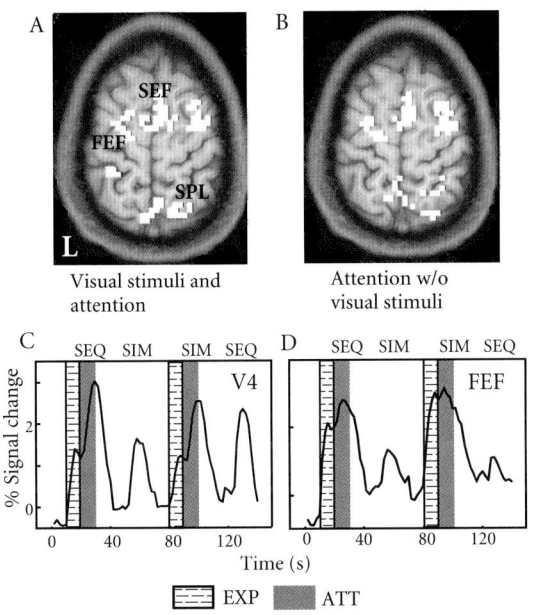

Fig. 15.4 A fronto-parietal network for spatial attention. Axial slice through frontal and parietal cortex. (A) When the subject directed attention to a peripheral target location and performed a discrimination task, a distributed fronto-parietal network was activated including the supplementary eye field (SEF), the frontal eye fields (FEF) and the superior parietal lobule (SPL). (B) The same network of frontal and parietal areas was activated when the subject directed attention to the peripheral target location in expectation of the stimulus onset. L indicates left hemisphere. (C) Time series of fMRI signals in V4. Directing attention to a peripheral target location in the absence of visual stimulation led to an increase of baseline activity (textured blocks), which was followed by a further increase after the onset of the stimuli (gray-shaded blocks). Baseline increases were found in both striate and extrastriate visual cortex. (D) Time series of fMRI signals in FEF. Directing attention to the peripheral target location in the absence of visual stimulation led to a stronger increase in baseline activity than in visual cortex; the further increase of activity after the onset of the stimuli was not significant.

to be activated in a variety of visuo-spatial tasks (for a meta-analysis, see Kastner and Ungerleider 2000). Thus, there appears to be a general attention network that operates independently of the specific requirements of the visuo-spatial task. However, from these studies it is not clear whether the activity found in areas of frontal and parietal cortex reflects complex processing of visual information or whether this activity reflects the attentional operations themselves. To sort out 'sources' that generate attentional feedback signals from 'sites' that receive top-down modulatory signals, we performed an experiment that probed the effects of spatially directed attention in the presence and in the absence of visual stimulation (Kastner *et al.* 1999*a*).

A third experimental condition was added to the design that was used to investigate competitive interactions and their modulation by spatial attention, as described above.

In addition to the two visual presentation conditions, sequential (SEQ) and simultaneous (SIM) and the two attentional conditions, unattended and attended, an expectation period preceding the attended presentations was introduced, during which subjects were required to direct attention covertly to the target location and instructed to expect the occurrences of the stimulus presentations. In this way, the effects of attention in the presence (ATT in Fig. 15.4C, D) and absence (EXP in Fig. 15.4C, D) of visual stimulation could be studied. The time courses of fMRI signals in activated regions of visual cortex, frontal and parietal cortex were analyzed to dissociate sources of attentional top-down feedback and sites of attentional modulation.

In visual cortex, as illustrated for area V4 in Fig. 15.4C, the fMRI signals increased during the expectation period (textured epochs), before any stimuli were present on the screen. This increase of baseline activity was followed by a further increase of activity evoked by the onset of the stimulus presentations (gray-shaded epochs). The baseline increase was found in all visual areas with a representation of the attended location, indicating that it was topographically specific. It was strongest in V4, but was also seen in early visual areas. It is noteworthy that baseline increases were found in V1, even though no significant attentional modulation of visually evoked activity was seen in this area.

These results from visual cortex are in accordance with single-cell recording studies showing that spontaneous (baseline) firing rates were 30–40% higher for neurons in areas V2 and V4 when the animal was cued to attend covertly to a location within the neuron's RF before the stimulus was presented there (Luck *et al.* 1997). The baseline increases found in human visual cortex may be subserved by increases in spontaneous firing rate similar to those found in these physiology studies, but summed over large populations of neurons. Baseline increases have been shown to depend on the anticipated task difficulty (Ress *et al.* 2000). The increases evoked by directing attention to a target location in anticipation of a behaviorally relevant stimulus are likely to reflect a top-down feedback bias in favor of the attended location in human visual cortex.

In parietal and frontal cortex, the same distributed network for spatial attention was activated during directed attention in the absence of visual stimulation as during directed attention in the presence of visual stimulation, consisting of the FEF, the SEF, and the SPL (Fig. 15.4A, B). A time-course analysis of the fMRI signals revealed that, as in visual cortical areas, there was an increase in activity in these frontal and parietal areas due to directed attention in the absence of visual input. However, first, this increase in activity was stronger in SPL, FEF, and SEF than the increase in activity seen in visual cortex (as exemplified for FEF in Fig. 15.4B), and secondly, there was no further increase in activity evoked by the attended stimulus presentations in these parietal and frontal areas. Rather, there was sustained activity throughout the expectation period and the attended presentations (Fig. 15.4B).

These results from parietal and frontal areas suggest that the activity reflected the attentional operations of the task and not visual processing. These findings therefore provide first evidence that these parietal and frontal areas may be the sources of feedback that generated the top-down biasing signals seen in visual cortex. The anatomical connections of SPL, FEF, and SEF put them in a position to serve as sources of top-down feedback signals that modulate neural processing in the visual system. In the monkey, FEF and SEF are reciprocally connected with ventral stream areas (Ungerleider *et al.* 1989; Webster *et al.* 1994) and posterior parietal cortex (Cavada and Goldman-Rakic 1989). The posterior parietal cortex is connected with ventral stream areas via the lateral intraparietal area (area LIP) (Webster *et al.* 1994).

15.5 Conclusions

Evidence from functional brain imaging reveals that attention operates at various processing levels within the visual system and beyond. First, the lateral geniculate nucleus appears to be the first stage in the processing of visual information that is modulated by attention, consistent with the idea that it may play an important role as an early gatekeeper in controlling neural gain. Secondly, areas at intermediate cortical processing levels, such as V4 and TEO, appear to be important sites at which attention filters out unwanted information by means of intra- and extra-receptive field mechanisms. These attention mechanisms operate in a push–pull fashion to facilitate the processing of attended, and to suppress the processing of unattended, stimuli. Thirdly, the attention mechanisms that operate in the visual system appear to be controlled by a distributed network of higher-order areas, which generate top-down signals that are transmitted via feedback connections to the visual system. Fourthly, attentional top-down signals may bias visual processing in favor of the attended location, as evidenced by increased baseline activity during directed attention in the absence of visual stimulation. The role of both early and higher areas of visual cortex in attention is less well understood. Future studies focusing on the role of attention in functions associated with these areas, such as grouping and scene segmentation or object-based selections, will further lead us towards a neural basis of visual attention in the human brain.

Acknowledgements

This paper is dedicated to the memory of Otto D. Creutzfeldt. I would like to thank Leslie G. Ungerleider, Robert Desimone, Peter De Weerd, Mark A. Pinsk, Daniel H. O'Connor, Glen M. Doniger, and Miki M. Fukui for their invaluable contributions to this work, and John Duncan, Anne Treisman and two anonymous reviewers for valuable comments on the manuscript. Supported by the National Institute of Mental Health and the Whitehall Foundation.

References

Broadbent, D.E. (1958). *Perception and communication.* Pergamon Press, London.
Cameron, E.L., Tai, J.C., and Carrasco, M. (2002). Covert attention affects the psychometric function of contrast sensitivity. *Vision Research,* **42**, 949–67.

Cavada C. and Goldman-Rakic, P.S. (1989). Posterior parietal cortex in rhesus monkey: II. Evidence for segregated cortico-cortical networks linking sensory and limbic areas with the frontal lobe. *Journal of Comparative Neurology*, **287**, 422–45.

Chen, W., Zhu, X.H., Thulborn, K.R., and Ugurbil, K. (1999). Retinotopic mapping of lateral geniculate nucleus in humans using functional magnetic resonance imaging. *Proceedings of the National Academy of Sciences USA*, **96**, 2430–4.

Cook, E.P. and Maunsell, J.H. (2002). Attentional modulation of behavioral performance and neuronal responses in middle temporal and ventral intraparietal areas of macaque monkey. *Journal of Neuroscience*, **22**, 1994–2004.

Corbetta, M. and Shulman, G.L. (2002). Control of goal-directed and stimulus-driven attention in the brain. *Nature Reviews: Neuroscience*, **3**, 201–15.

Corbetta, M., Miezin, F.M., Dobmeyer, S., Shulman, G.L., and Petersen, S.E. (1991). Attentional modulation of neural processing of shape, color, and velocity in humans. *Science*, **248**, 1556–9.

Crick, F. (1984). Function of the thalamic reticular complex: the searchlight hypothesis. *Proceedings of the National Academy of Sciences USA*, **81**, 4586–90.

Desimone, R. and Duncan, J. (1995). Neural mechanisms of selective visual attention. *Annual Review of Neuroscience*, **18**, 193–222.

Desimone, R. and Ungerleider, L.G. (1989). Neural mechanisms of visual processing in monkeys. In Boller, F., Grafman, J., (Eds) *Handbook of neuropsychology*, Volume 2. Elsevier, Amsterdam.

Desimone, R., Moran, J., Schein, S.J., and Mishkin, M. (1993). A role for the corpus callosum in visual area V4 of the macaque. *Visual Neuroscience*, **10**, 159–71.

De Weerd, P., Peralta, M.R. 3rd, Desimone, R., and Ungerleider, L.G. (1999). Loss of attentional stimulus selection after extrastriate cortical lesions in macaques. *Nature Neuroscience*, **2**, 753–8.

Duncan, J. (1980). The locus of interference in the perception of simultaneous stimuli. *Psychological Review*, **87**, 272–300.

Duncan, J. (1996). Cooperating brain systems in selective perception and action. In Inui, T., McClelland, J.L. (Eds.) *Attention and performance XVI*. pp. 549–578. MIT Press, Cambridge MA.

Gallant, J.L., Shoup, R.E., and Mazer, J.A. (2000). A human extrastriate area functionally homologous to macaque V4. *Neuron*, **27**, 227–35.

Guillery, R.W., Feig, S.L., and Lozsadi, D.A. (1998). Paying attention to the thalamic reticular nucleus. *Trends in Neuroscience*, **21**, 28–32.

Ito, M. and Gilbert, C.D. (1999). Attention modulates contextual influences in the primary visual cortex of alert monkeys. *Neuron*, **22**, 593–604.

Jones, E.G. (1985). *The thalamus*. Plenum Press, New York.

Kanwisher, N. and Wojciulik E. (2000). Visual attention: insights from brain imaging. *Nature reviews: Neuroscience*, **1**, 91–100.

Kapadia, M.K., Ito, M., Gilbert, C.D., and Westheimer, G. (1995). Improvement in visual sensitivity by changes in local context: parallel studies in human observers and in V1 of alert monkeys. *Neuron*, **15**, 843–56.

Kastner, S. and Ungerleider, L.G. (2000). Mechanisms of visual attention in the human cortex. *Annual Review of Neuroscience*, **23**, 315–41.

Kastner, S., De Weerd, P., Desimone, R., and Ungerleider, L.G. (1998). Mechanisms of directed attention in the human extrastriate cortex as revealed by functional MRI. *Science*, **282**, 108–11.

Kastner, S. Pinsk, M.A., De Weerd, P., Desimone, R., and Ungerleider, L.G. (1999*a*). Increased activity in human visual cortex during directed attention in the absence of visual stimulation. *Neuron*, **22**, 751–61.

Kastner, S., Nothdurft, H.C., and Pigarev, I.N. (1999b). Neuronal responses to orientation and motion contrast in cat striate cortex. *Visual Neuroscience*, **16**, 587–600.

Kastner, S., De Weerd, P., Pinsk, M.A., Elizondo, M.I., Desimone, R., and Ungerleider, L.G. (2001). Modulation of sensory suppression: Implications for receptive field sizes in the human visual cortex. *Journal of Neurophysiology*, **86**, 1398–411.

Lamme, V.A. (1995). The neurophysiology of figure-ground segregation in primary visual cortex. *Journal of Neuroscience*, **15**, 1605–15.

Lu, Z.L. and Dosher, B.A. (1998). External noise distinguishes attention mechanisms. *Vision Research*, **38**, 1183–98.

Luck, S.J., Chelazzi, L., Hillyard, S.A., and Desimone, R. (1997). Neural mechanisms of spatial selective attention in areas V1, V2, and V4 of macaque visual cortex. *Journal of Neurophysiology*, **77**, 24–42.

Malach, R., Levy, I., and Hasson, U. (2002). The topography of high-order human object areas. *Trends in Cognitive Neuroscience*, **6**, 176–84.

Martinez, A., Vento, L.A., Sereno, M.I., Frank, L.R., Buxton, R.B., Dubowitz, D.J., Wong, E.C., Hinrichs, H., Heinze, H.J., and Hillyard, S.A. (1999). Involvement of striate and extrastriate visual cortical areas in spatial attention. *Nature Neuroscience*, **2**, 364–9.

Mehta, A.D., Ulbert, I., and Schroeder, C.E. (2000). Intermodal selective attention in monkeys. I: distribution and timing of effects across visual areas. *Cerebral Cortex*, **10**, 343–58.

Miller, E.K., Gochin, P.M., and Gross, C.G. (1993). Suppression of visual responses of neurons in inferior temporal cortex of the awake macaque by addition of a second stimulus. *Brain Research*, **616**, 25–9.

Moran, J. and Desimone, R. (1985). Selective attention gates visual processing in the extrastriate cortex. *Science*, **229**, 782–4.

Motter, B.C. (1993). Focal attention produces spatially selective processing in visual cortical areas V1, V2, and V4 in the presence of competing stimuli. *Journal of Neurophysiology*, **70**, 909–19.

Nobre, A.C. (2001). The attentive homunculus: now you see it, now you don't. *Neuroscience and Biobehavioral Reviews*, **25**, 477–96.

O'Connor, D.H., Fukui, M.M., Pinsk, H.A., and Kastner, S. (2002). Attention modulates responses in the human lateral geniculate nucleus. *Nature Neuroscience*, **5**, 1203–9.

Pinsk, M.A. and Kastner, S. (2001). Selective attention and task difficulty: effects on target and distracter stimuli. *Society for Neuroscience Abstracts*, **574**.4.

Posner, M.I. (1980). Orienting of attention. *Quarterly Journal of Experimental Psychology*, **32**, 3–25.

Recanzone, G.H. and Wurtz, R.H. (2000). Effects of attention on MT and MST neuronal activity during pursuit initiation. *Journal of Neurophysiology*, **83**, 777–90.

Recanzone, G.H., Wurtz, R.H., and Schwarz, U. (1997). Responses of MT and MST neurons to one and two moving objects in the receptive field. *Journal of Neurophysiology*, **78**, 2904–15.

Rees, G., Frith, C.D., and Lavie, N. (1997). Modulating irrelevant motion perception by varying attentional load in an unrelated task. *Science*, **278**, 1616–9.

Rees, G., Wojciulik, E., Clarke, K., Husain, M., Frith, C., and Driver, J. (2000). Unconscious activation of visual cortex in the damaged right hemisphere of a parietal patient with extinction. *Brain*, **123**, 1624–33.

Ress, D., Backus, B.T., and Heeger, D.J. (2000). Activity in primary visual cortex predicts performance in a visual detection task. *Nature Neuroscience*, **3**, 940–5.

Reynolds, J.H. and Desimone, R. (1999). The role of neural mechanisms of attention in solving the binding problem. *Neuron*, **24**, 111–25.

Reynolds, J.H., Chelazzi, L., and Desimone, R. (1999). Competitive mechanisms subserve attention in macaque areas V2 and V4. *Journal of Neuroscience*, **19**, 1736–53.

Sereno, M.I., Dale, A.M., Reppas, J.B., Kwong, K.K., Belliveau, J.W., Brady, T.J., Rosen, B.R., and Tootell, R.B. (1995). Borders of multiple visual areas in humans revealed by functional magnetic resonance imaging. *Science*, **268**, 889–93.

Sherman, S.M. and Guillery, R.W. (2001). *Exploring the thalamus*. Academic Press, San Diego CA.

Shiu, L.P. and Pashler, H. (1995). Spatial attention and vernier acuity. *Vision Research*, **35**, 337–43.

Ungerleider, L.G., Gaffan, D., and Pelak, V.S. (1989). Projections from inferior temporal cortex to prefrontal cortex via the uncinate fascicle in rhesus monkeys. *Experimental Brain Research*, **76**, 473–84.

Vuilleumier, P., Sagiv, N., Hazeltine, E., Poldrack, R.A., Swick, D., Rafal, R. and Gabrieli, J.D.E. (2001). Neural fate of unseen faces in vsuospatial neglect: A combined event-related functional MRI and event-related potential study. *Proceedings of the National Academy of Sciences USA*, **98**, 3495–500.

Webster, M.J., Bachevalier, J., and Ungerleider, L.G. (1994). Connections of inferior temporal areas TEO and TE with parietal and frontal cortex in macaque monkeys. *Cerebral Cortex*, **4**, 470–83.

Yeshurun, Y. and Carrasco, M. (1998). Attention improves or impairs visual performance by enhancing spatial resolution. *Nature*, **396**, 72–5.

Chapter 16

The influence of scene organization on attention: Psychophysics and electrophysiology

Mitchell Valdes-Sosa, Maria A. Bobes, Valia Rodríguez, Yanelis Acosta, Alejandro Pérez, Jorge Iglesias, and Mayelin Borrego

Abstract

Many studies have examined the neural mechanisms of visual attention by recording event-related potentials (ERPs). In most of these studies, objects are abruptly presented to the observer, often in a rapid succession of events. Similar stimuli are used in the psychophysical technique known as rapid serial visual presentation (RSVP). We argue that in RSVP (and related ERP paradigms) it is difficult to evince object-based attentional effects and to study the influence of scene organization. Rapid serial object transformation (RSOT) is offered as an alternative. Therein, several objects are presented over seconds and are monitored to detect brief sequential mutations of their attributes (e.g. changes in motion direction affecting illusory transparent surfaces, or local form transformations of objects defined by shape and color). We show that when two events engage the same object in rapid succession, both can be attended to effectively. In contrast, events that affect different objects compete and produce an attentional blink (AB). The AB is associated with smaller ERPs that are possibly generated in early visual extrastriate cortex. This suggests that the competition between these visual events is played out at the level of features codes, but is flexibly guided by higher-order representations similar to an 'object-file'.

16.1 Competition for attention

Many, if not all, of the visual scenes encountered daily are full of numerous objects, and each object, in turn, can be comprised of multiple parts. Moreover, collections of objects can integrate perceptual groups. For each observed scene, these diverse entities compete for visual attention. Considerable progress has been made in understanding how exogenous cues (like stimulus salience), and endogenous factors (like a person's beliefs and goals), interact in the resolution of this competition (Desimone and Duncan 1995; Duncan 1996). For example, highly salient and more relevant objects are preferred over less salient or less relevant ones, allowing the winners preferential access to action and memory systems in the brain and into consciousness.

The biasing of competition in favor of some stimuli over others is particularly clear in experiments showing endogenous selection of the inputs coming from one location, with rejection of those originating in other sites of the visual field. This type of selection leads to faster and more accurate detection—or recognition—of stimuli presented at the privileged location in a number of different paradigms (e.g. Eriksen and Hoffman 1973; Jonides 1981; LaBerge 1983; Posner and Cohen 1984). Results of this nature, reported in a host of studies, have given support to the classical 'spotlight metaphor' of visuo-spatial attention.

More recently it has been understood that competition for attention is also influenced by the organization of the visual entities contained in a scene. The structure of the scene can constrain the allocation (and re-allocation) of attention. It has been shown that it is easier to divide attention between two features on one object, than to do so for two features of different objects, even if the two objects are in close spatial proximity (Duncan 1984; Vecera and Farah 1994; Lavie and Driver 1996; Behrmann *et al.* 1998; Valdes-Sosa *et al.* 1998*b*). In fact, attention to one attribute of an object—or surface—may entail some degree of obligatory attention to other attributes of the same object (Duncan 1984, 1996; Egly *et al.* 1994; He and Nakayama 1995). Moreover, competition between objects can turn into cooperation (Driver and Baylis 1989) if they are perceptually grouped (i.e. move in the same direction). However, within the same object a cost for dividing attention between clearly differentiable parts has been reported (Vecera *et al.* 2000).

16.2 Neural mechanisms of attention competition

Understanding of the neural mechanisms of attentional competition is rapidly increasing in several directions,[1] especially for the thoroughly studied situation in which spatial locations are selected. Neuroimaging has demonstrated that both pre-stimulus and stimulus-elicited activity in visual extrastriate areas are enhanced within the retinotopic sectors corresponding to the selected region of visual space (Mangun *et al.* 2000; Chapter 15, this volume). These effects may also be present in striate cortex (Brefczynski and DeYoe 1999; Gandhi *et al.* 1999).

Research in several labs (see reviews by Hillyard and Anllo-Vento 1998; Hillyard *et al.* 1998; and Chapter 19, this volume) has established that event-related potentials (ERPs), elicited by stimuli flashed at the attended locations, are enhanced in amplitude relative to ERPs elicited by stimuli presented elsewhere. The earliest components affected are P1 and N1. The P1 begins as early as 80 ms after the stimulus onset, and can be modeled by sources in early extrastriate cortex (Heinze *et al.* 1994; Mangun *et al.* 1997; Martinez *et al.* 1999; Di Russo *et al.* 2001). The N1 begins at about 150 ms after the stimulus, and is a more complex mixture of subcomponents, including several sources in extrastriate cortex and in more frontal regions (Di Russo *et al.* 2001).

Note that these early ERP effects are only obtained under conditions of large perceptual load (Lavie 1995; Luck and Hillyard 2000), in which the time to process potentially competing stimuli is limited. In other words, the early sensory modulation indexed by the P1 and N1 attentional effects are only present when it is difficult to switch attention in time between different input sources.

Despite convincing psychophysical evidence for object-based attention (cited above), much less work has been performed on this topic with neuroimaging techniques (e.g. Arrington *et al.* 2000). The paucity of ERP studies on the topic is even more acute, hence little is known about the timing of the processes involved. This, in part, is a consequence of the dominant form of stimulus presentation used in ERP research. In contrast with other neuroimaging methods, ERPs are best elicited by abruptly presented, brief, and discrete stimulus variations (Hillyard and Picton 1987). Furthermore, the successive stimuli must be presented under strict time limits to enhance perceptual load (if selective attention is of interest). It is not surprising that the stimuli used to study the sharing of attention over time and those used to elicit attentional ERP effects are so similar, an issue we examine next.

16.3 The attentional blink, RSVP, and RSOT

The temporal constraints in sharing attention between objects (see review by Egeth and Yantis 1997) have been studied using a technique known as rapid serial visual presentation (RSVP). In RSVP, objects are presented briefly, and generally one at a time. The objects can be words, letters, or pictures. In one variant (see Fig. 16.1A), a series of stimuli (say letters) are presented in rapid succession, usually at the same site. The stream is divided into targets and distracters. Recognition of a first target (T1) within the stream hampers processing of a subsequently appearing target (T2) until several hundred milliseconds have transpired, or until several distracters have intervened between the targets (Raymond *et al.* 1992; Shapiro *et al.* 1994; Chun, 1997). This phenomenon has been dubbed the attentional blink (AB). The attentional nature of the AB is attested by asking the observers to ignore T1 (thereby focusing attention on T2), in which case performance on T2 discrimination improves (Fig. 16.1B). This rules out sensory masking as an explanation of the interference (Egeth and Yantis 1997).

A more austere variant of RSVP uses only two stimuli (see Fig. 16.1A), that are masked after a short presentation. A long-lasting impediment in the identification of the second stimulus is also found in this paradigm, (Duncan *et al.* 1994; Ward *et al.* 1996). Again, the interference is present only when attention is paid to the first stimulus.

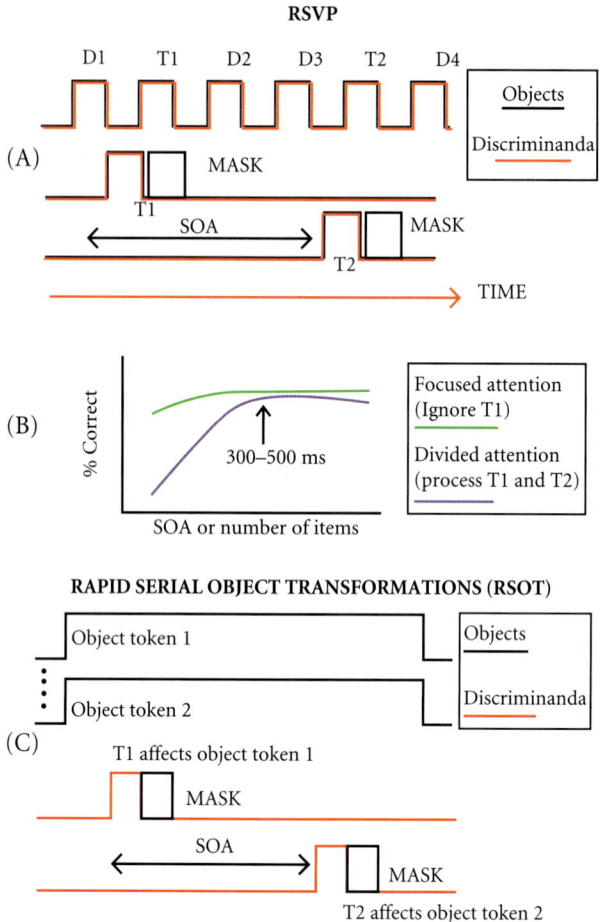

Fig. 16.1 Traditional variants of rapid serial visual presentation (RSVP). (A) A series of many objects presented in rapid succession, each substituting (and, if at the same site, also masking) its predecessor; in the other variant (below), only two objects are presented, each with a post-mask that interrupts processing and limits the temporal availability of stimulus information. (B) Idealized graph of basic RSVP findings. Accuracy is reported as the percent correct of T2 identification given correct T1 identification. Attending to T1 hampers processing of T2 up to stimulus onset asynchrony (SOA) of about 300–500 ms. This effect is eliminated if T1 is ignored. (C) Rapid serial object transformations (RSOT). Object tokens (e.g. small shapes or surfaces) are presented for longer lifetimes than in RSVP. Targets consist of brief events that transform the objects without destroying their spatio-temporal continuity, as shown in the timechart. Here masking of the event is accomplished by reverting the transformation and returning the object to its baseline condition.

The duration of this interference can be as long as half a second (but see Moore *et al.* 1996). The data from both paradigms indicate that resolving the attentional competition between objects can take a considerable time. An attractive aspect of RSVP is the controlled timing of visual input. This allows relatively direct measurements of the ability to re-allocate attention over time (Duncan *et al.* 1994), in contrast with the awkward assumptions needed when estimating this ability with visual search tasks (see Wolfe, 1998 for a discussion).

However, note that in all variants of RSVP the objects have short lifetimes (Fig. 16.1A). In the visual world, many (perhaps most) objects are relatively long lived within a scene. Also, an input of only one object per instant of time is perhaps infrequent in real-life scenarios. Furthermore, the creation of new objects may capture attention (Yantis 1998), and the disappearance of stimuli may disengage attention (Mackeben and Nakayama 1993) automatically. Consequently, the onset and offset of stimuli in RSVP may impose a peculiar temporal dynamics to the allocation of attention, not always present in natural scenes. These usually have a complex structure, where it may be necessary to shift attention between aspects of the same part of an object, different parts of the same object, as well as between different (and simultaneously present) objects or groups of objects.

Therefore, it is difficult to study the influence of scene organization on the attention with RSVP. RSVP shares important features with the experimental designs used in many ERP experiments on visual attention (e.g. the use of briefly flashed objects and fast stimulus presentation rates), which might explain why electrophysiological signatures of object-based attention have been so elusive. To observe these effects, a longer permanence of the objects giving structure to the scene is probably required. A different approach is needed, one that would allow us to combine multiple objects within a visual scene (so that perceptual organization can be examined) while conserving the controlled timing of visual input from RSVP (so that the temporal dynamics of attention can be studied).

One alternative, which we have named rapid serial object transformation (RSOT), consists of presenting several objects simultaneously in a display for relatively prolonged durations. Observers monitor the scene, and the accuracy in identifying events that transform these objects is measured. The critical aspect is how accuracy is affected by the duration of the interval between the events (see Fig. 16.1C). The objects may change shape, color, position, or any other attribute. Despite this mutability, each object token survives as an entity. The individuality of each object is not at issue. These stimuli instantiate the concept of an 'object-file' (Kahneman and Treisman 1984; Kahneman *et al.* 1992), discussed later in this chapter. RSOT captures certain aspects of real-life scenes, and many examples are readily available. Imagine talking with several colleagues (springing a new idea) while monitoring their faces for changes in expression. Or imagine tracking the position of several vehicles when driving in heavy traffic.

With RSOT (just as with RSVP) the timing of attentional shifts can be measured directly from the duration of the interference between the recognition of two events.

But in contrast to RSVP, perceptual organization can be manipulated by varying the relationships between groups of objects, objects, parts of objects, or even features of one object. Also, the events transforming the objects can be used as triggers to elicit ERPs. Note that the capture of attention by changes in pre-existing objects may not be mandatory (Hillstrom and Yantis, 1994). Object transformations have been used before in some studies of attention (e.g. Yantis and Jonides 1984; Sears and Pylyshyn 2000), and in studies of eye movements during reading or scene perception (e.g. Henderson and Ferreira 1990; Morris *et al.* 1990).

16.4 **RSOT under conditions of extreme competition: Transparent motion**

We have used RSOT of transparent motion, a situation characterized by strong perceptual competition between two illusory entities or 'objects-tokens'. When two sets of dots move in different directions within the same region of visual space, an illusion is generated. Two 'surfaces' that slide one across the other (separated in depth, with an ambiguous order, but seen as very close together) are seen. These surfaces are rivals for attention. We have shown that it is easier to divide attention between two features of the same transparent surface than between identical attributes for different surfaces. This holds for simultaneous judgments about direction and speed of motion (Valdes-Sosa *et al.* 1998*b*), and about direction of motion and the shape of the moving elements (Rodriguez *et al.* 2002). These results extend the early demonstration of object-based attention by Duncan (1984).

Building on these results, a RSOT paradigm was created. Two sets of dots (colored green and red respectively) were rotated in opposite directions around a colored fixation point (see Valdes-Sosa *et al.* 2000 for more details). After a baseline period of rotation, the dots from one surface briefly changed their direction of motion undergoing a translation (T1). The illusion is that the flow of rotating dots suddenly heads in another direction, and then recovers the original motion. After a variable T1-T2 stimulus onset asynchrony (SOA), either the same or the other surface was affected by an additional change in direction of motion (T2). The task was to report the direction of translation for both targets. The surface (red or green) affected by T1 was cued in advance by the color of the fixation point, whereas the surface affected by T2 was unpredictable. Note that the two illusory surfaces were continuously present (albeit with mutations in their direction of motion) and that the set of possible directions of translations was the same for both surfaces.

The first target event was discriminated accurately under all conditions (Fig. 16.2). If the second target event (T2) affected the same surface as T1, it was also judged accurately. In sharp contrast, a large impairment in performance was observed if T2 affected the other surface (Figure 16.2). This two-surface cost persisted until about 500 ms after T1. The difficulty was not only in discrimination, but also affected detection

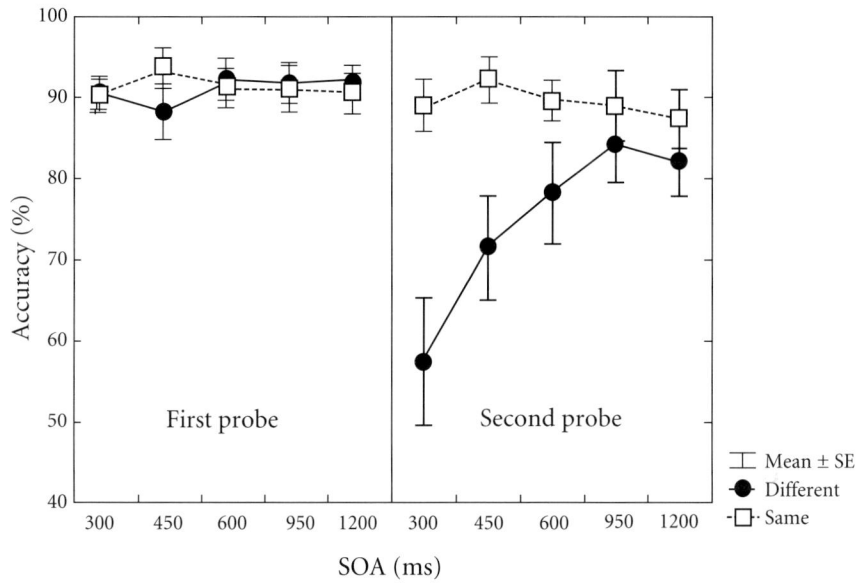

Fig. 16.2 Performance in the transparent motion RSOT paradigm. The mean accuracy for T1 discrimination for 10 observers is shown in the panel on the left, and the corresponding values for T2 are shown in the panel to the right. Error bars represent the standard error (reproduced with permission from Valdes-Sosa *et al.* 2000).

of the T2 probe motion (Pinilla *et al.* 2001). In other words, an AB was obtained when two events concerned different object tokens (surfaces), but not when they concerned the same object token (Fig. 16.2). Foreknowledge of the direction of the attentional shift did not ameliorate the two-surface cost (Cobo *et al.* 1999). Recently this pattern of results has been replicated independently (see Chapter 18, this volume).

Blaser *et al.* (2000) recently performed a related RSOT experiment with a pair of superimposed (transparent) Gabor patches that smoothly (but unpredictably) changed orientation, spatial frequency, and color (note the similarity to conditions for monocular rivalry). The attributes held by one Gabor could (over time) become assigned to the other Gabor. Unpredictably, small jumps in the pattern of change would occur, which were to be reported. In line with our results, carrying out pairs of judgments for the same Gabor patch was easier than performing the identical judgments but on two different patches.

16.5 Neural correlates of attentional competition in transparent motion

The neural basis of the AB in the transparent motion RSOT paradigm was studied by means of ERP recordings (Pinilla *et al.*, 2001). The responses elicited by the two target events (Fig. 16.3) both contained a negative wave known as N200, which has been previously described in relation to motion onsets or direction changes (Kuba and Kubová 1992; Bach and Ullrich 1994; Torriente *et al.* 1999). The N200 elicited by T2 (but not the preceding positivity) was significantly reduced in amplitude when the targets engaged different surfaces, relative to when they affected the same surface (Fig. 16.3). This suggests attenuated sensory processing of the T2 during the AB in this paradigm, since the N200 is considered an index of activity within motion-specific visual cortices (Kuba and Kubová 1992; Bach and Ullrich 1994). Interestingly, when observers sustain attention on the same illusory surface for a long period of time, and the perceptual load is increased, all of the ERP components are essentially suppressed (Valdes-Sosa *et al.* 1998*a*). This suggests that perhaps larger suppressive effects could be found at shorter T1–T2 SOAs (where the AB is larger).

A further characterization of the ERPs in the transparent motion RSOT paradigm has been recently carried out (Rodríguez *et al.* submitted). Niedeggen *et al.* (2002) have pointed out that if the AB is related to sensory suppression, then ERPs indexing activity in early visual areas should be smaller for trials on which identification fails. Therefore in this new study, the trials in which T2 was correctly and incorrectly identified were

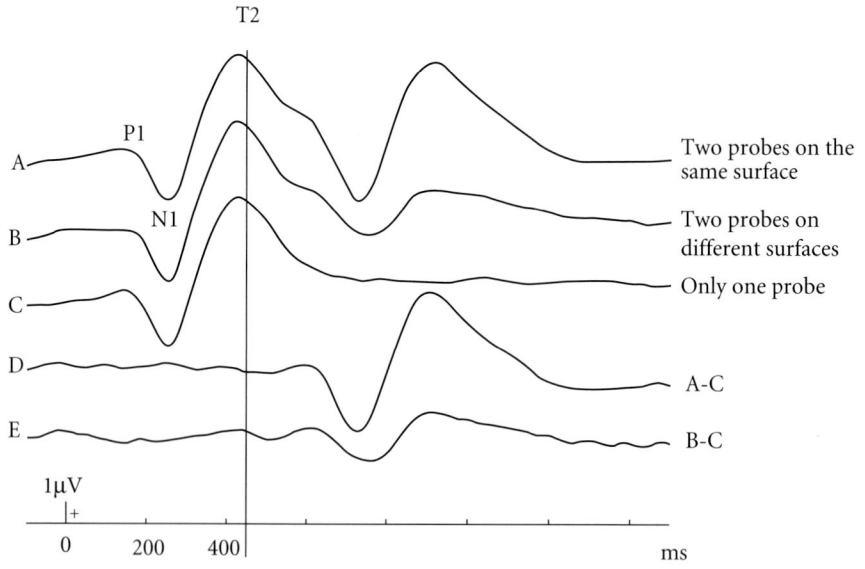

Fig. 16.3 Grand average ERPs in the transparent motion RSOT paradigm, for 10 observers, and recorded at the right posterior temporal electrode (T6). The timing of T1 corresponds to the origin of the time axis, whereas that of T2 is indicated by the vertical line. First row: ERPs from same-surface trials. Second row: ERPs from different-surface trials. Third row: ERPs from trials in which only T1 was presented. The fourth and fifth rows show the difference waveforms obtained by subtracting the T1-only response from the responses associated with same- and different-surface trials. In this and subsequent graphs positive deflections are plotted up. (Reproduced with permission from Pinilla *et al.* 2001).

averaged separately (correct T1 report was required in all cases). Also, high-density ERPs (120 channels) were recorded to order to ascertain the scalp distribution of the N200 (and of its attentional modulation) more exactly.

Larger N200 amplitudes (associated to T2) were again obtained for same- relative to different-surface trials. Additionally, we found that the N200 was much larger for trials with correct responses than for trials with incorrect responses (Rodríguez *et al.* 2003), in both the same- and different-surface conditions. This indicates that the N200 was smaller on trials in which the identification of T2 failed in both these conditions, and that the proportion of trials with such failures is simply higher when attention must switch between surfaces. The N200 attentional effect is therefore best estimated by subtracting the correct/same-surface response from the incorrect/different-surface response (instead of the subtraction used in Fig. 16.3). The scalp distribution of peak of the N200 modulation reflected in this difference waveform presented a maximum amplitude over right posterior temporal sites (see Fig. 16.4A).

The intracranial generators of this negativity were modeled with a data-driven method employing multiple distributed current sources, known as VARETA (for more complete descriptions see Muller *et al.* 1998; Picton *et al.* 1999; Bosch-Bayard *et al.* 2001). The scalp distribution is consistent with bilateral generators located in lateral

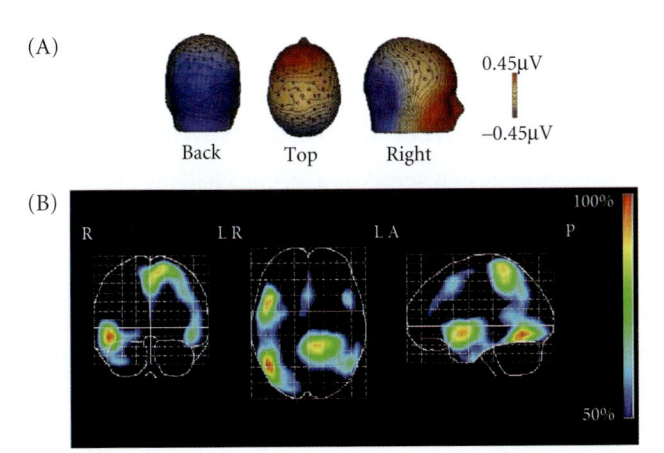

Fig. 16.4 (A) Scalp distribution of the attentional effect on the N200 in the transparent motion RSOT experiment (grand average of data from 10 observers). The data were obtained from a high-density electrode array (120 channels) and correspond to the voltage at the latency of the peak of the N200 component. The attentional effect was estimated by subtracting the ERPs related to incorrectly identified T2 stimuli in the different-surface condition from the ERPs related to correctly identified T2 stimuli in the same-surface condition. (B) Current sources for the scalp distribution shown in Fig. 16.4A, as estimated by the VARETA method and represented within the Talairach space in a 'glass brain' (for details, see Rodríguez *et al.* 2003). Note that the right side of the brain is on the left. The estimated current density at each unitary source is represented as a percentage of the magnitude of the voxel with the largest response (values lower than 50% are not shown).

ventral extrastriate cortex, and at the junction of the temporal and occipital lobes (Fig. 16.4B). There seems to be more activation on the right side with sources in lateral occipito-temporal extrastriate, parietal, and frontal cortices. The Talairach coordinates of the largest current source are near a site identified in PET and fMRI studies as the motion-processing area MT+ (see review by Culham *et al.* 2001). The somewhat larger involvement of the right hemisphere is consistent with previous ERP and fMRI studies of motion processing (for more details see Rodríguez *et al.* 2003).

These modeling efforts can only suggest that the AB in our paradigm is related to a reduction of activity in MT+ and other extrastriate areas, but this conclusion would be congruent with the object-based attentional suppression of activity (evinced with transparent motion) that has been demonstrated in similar cortical areas by several neuroimaging studies (O'Craven *et al.* 1997, 1999; Watanabe *et al.* 1998). Moreover, recordings of neurons in cortical area V4 have been obtained recently in monkeys trained to observe transparent surfaces similar to those described here. These data support the idea the suppressive interactions within extrastriate cortex are at play during the AB in our paradigm (see Chapter 18, this volume).

To summarize, there is suggestive evidence that the AB in the transparent motion RSOT task is associated with reduced neural activity in MT/MST and perhaps in V4, which are relatively early visual extrastriate areas. Hence, under some conditions 'early sensory filtering' (i.e. attenuation of sensory motion processing) may contribute to the AB.

16.6 RSOT based on object shape transformations

The transparent motion design serves as a strong challenge the classical 'spotlight' metaphor of attention. But, although objects do overlap and transparency is present in many natural scenes, the extreme form of perceptual competition described up to now is not the rule. It is therefore necessary to examine the influence of scene organization on attention when the competing stimuli are not transparently overlapped (and thus more separated in space), which represents the mainstream situation in visual attention experiments. Moreover, we would like to demonstrate that the object-based modulation of the AB found with transparent motion is not a quirk of the dorsal stream of visual extrastriate areas (Desimone and Ungerleider 1989), limited to situations where the object tokens are defined by relative motion and the target events consist of changes in motion direction.

Therefore we now turn to studies using stationary object tokens specified by shape and color, that do not overlap completely, and to transformations consisting of changes in local shape. Color and shape are thought to be preferentially processed in the ventral visual stream (Desimone and Ungerleider 1989). For this, we (Pinilla and Valdes-Sosa submitted) modified the stimuli designed by Behrmann and coworkers (Behrmann *et al.* 1998). Two overlapping bar shapes were presented (see Fig. 16.5A). The tips of the bars were shaped so that erasing one set of pixels would reveal a form

Fig. 16.5 (A) RSOT paradigm based on object shape transformations. Two overlapped bars (of different colors) were presented, and the tips could mutate into two or three bumps. Observers were cued about which object would be affected first by the color of the fixation point. The top row depicts a same-location trial (T1 and T2 affect the same tip). The middle row represents a same-object trial (T1 and T2 affect different tips from the same object). The bottom row represents a different-object trial (T1 and T2 affect tips from different objects). (B) Accuracy in identifying targets in the shape transformation RSOT paradigm, which was high for T1 in for all conditions and SOA values. This was also true for T2 identification when it affected the same tip as T1. There was an AB when T2 affected a different tip than the tip affected by T1; however, if the two tips belong to the same object, the AB was shorter than when the two tips belonged to different objects. (C) Control conditions. In the column to the left, the same shape transformations already described were used, and the response required was the same, but all the tips are isolated from each other (the colored pixels from the middle of the two bars were reallocated to disconnect them). In the column to the right, the transformations consisted in flashing a white rectangle upon the background bars. The observer had to determine which dimension (vertical or horizontal) was larger.

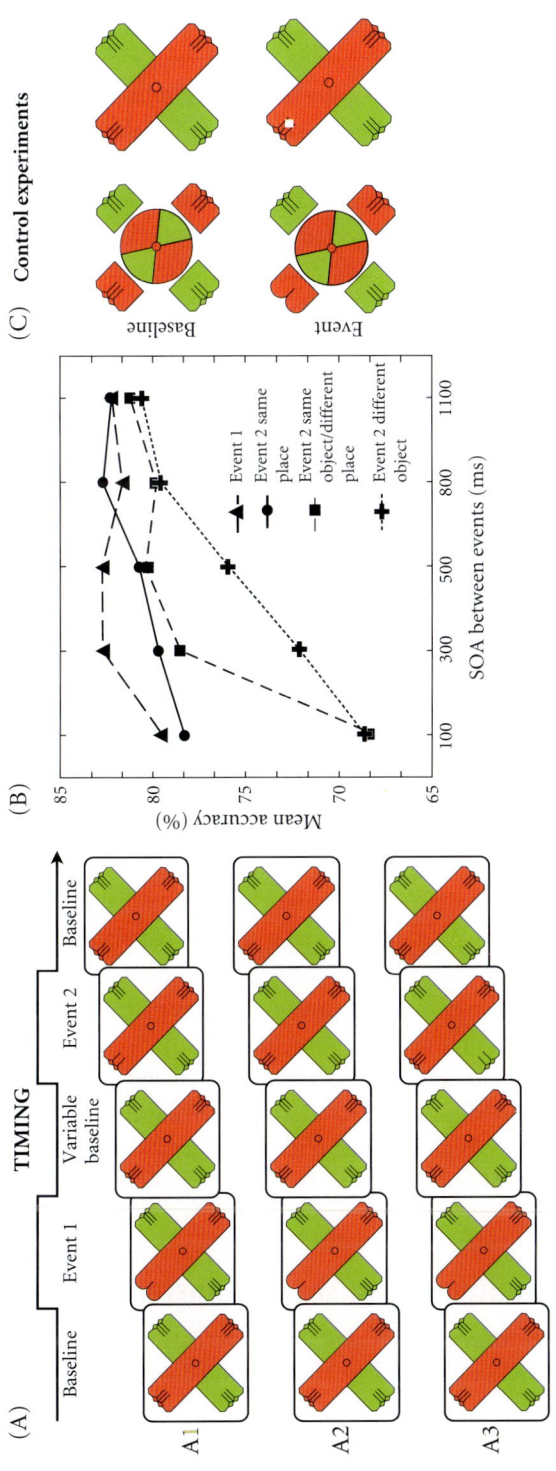

consisting of two bumps, whereas erasing another set of pixels would reveal three bumps. Thus two potential tip forms (with either two or three bumps) could emerge from the baseline stimulus. After the presentation of the baseline stimulus on each trial (see Fig. 16.5A), a brief change in shape (i.e. the unmasking of one of the two bump shapes) occurred at a randomly selected tip. This comprised the T1 event. The bar to be affected by T1 was pre-cued by the color of the fixation point. After a variable SOA, a second shape change took place at a randomly selected (and unpredictable) tip.

In all cases T1 was accurately identified. Also, T2 was accurately discriminated for all the SOAs explored when it engaged the same tip as T1 (Fig. 16.5B). In other words, the AB was absent for successive events taking place at the same object-part. In contrast, interference between T1 and T2 was found when they involved different tips of the same object. However, this AB only lasted until about 300 ms. Note that the elongated form of the bars possibly facilitated the segregation of the tips as distinct parts of the overall object (see Vecera *et al.* 2000 for congruent findings). Importantly, interference between T1 and T2 lasted longer when these events concerned the tips of distinct objects, with an AB lasting up to 500 ms (Fig. 16.5B). Note that the distance between tips on different object is shorter than between tips of the same object, hence the larger AB for between- versus within-object attentional shifts can not be explained by purely spatial mechanisms. The pattern of results was the same when trials with eye movements (monitored with infrared spectacles) were discarded from the analysis.

When the continuity of the bars was broken (see Fig. 16.5C), the speed for attentional shifts between all tip-pairs was equivalent. If the stimulus was changed to white rectangles briefly flashed on the bar tips (Fig. 16.5C), whose shape had to be discriminated, the object-based advantage was absent. A tendency for a shorter AB between events on tips of different objects was found, perhaps related to the shorter distance between these pairs of tips. The last finding is particularly relevant when designing ERP studies of object-based attention. Although perhaps optimal for eliciting large responses with adequate signal to noise ratio, sudden-onset stimuli (that may be perceived as new objects unrelated to the previous structure of the scene) are not useful in uncovering organizational constraints on the distribution of attention.

A related result has been described for a hybrid experiment, where in one condition the successive frames in a RSVP were different objects, whereas in another condition successive frames were perceived as a single rotating object (in effect RSOT). The AB was diminished for the single object- compared to the different-object condition (Raymond 2003).

16.7 **ERPs in an RSOT paradigm using shape transformations**

The findings of the previous section were confirmed in an alternative RSOT design, which also served to measure ERPs (Valdes-Sosa *et al.* 2003). In this design (schematized in Fig. 16.6A), five small diamonds are presented as 'holes' in two non-overlapping objects. The objects were of different colors and shapes. One diamond was in the center

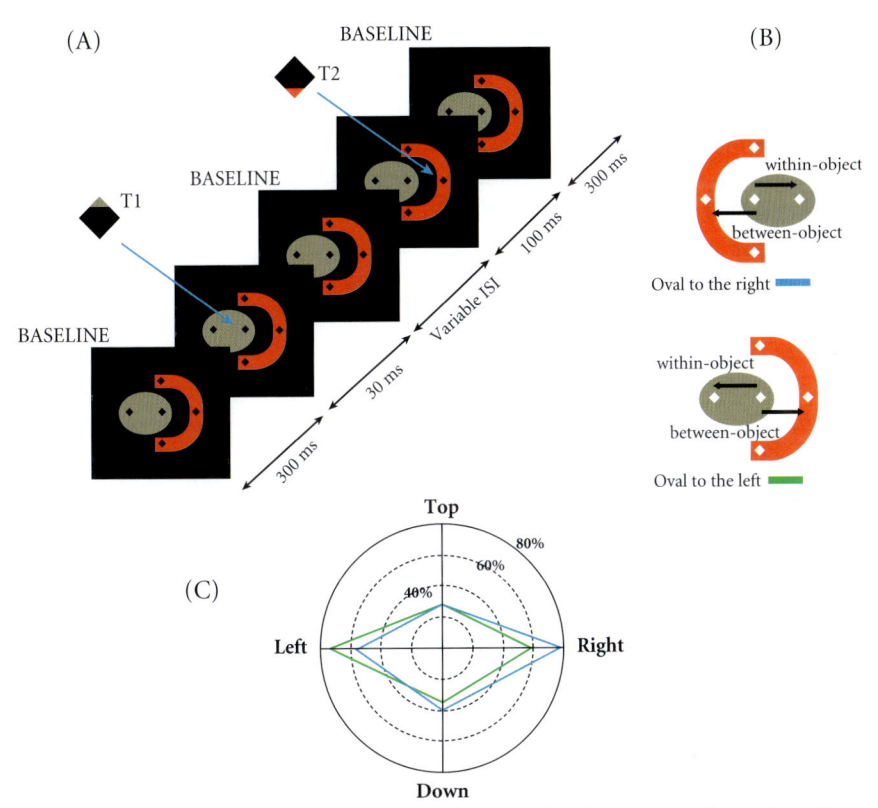

Fig. 16.6 (A) Timechart for an RSOT paradigm based on local shape transformations. Two global objects were depicted on a monitor screen for the duration of the complete trial. Target events were brief changes of one of five diamond-shaped 'holes' within the two objects: the loss of one corner (see inset). T1 was always a 30 ms modification of the central diamond, which also served as fixation point. T2, with a 100 ms duration, could affect any of the four eccentric diamonds (left, right, top and down), with equal probability. T2 event duration was selected to produce about 80% correct responses in most subjects when presented alone. In the example, T1 affects the central, and T2 affects the rightmost, diamond. (B) Two mirror layouts used. In one case, depicted above, the shift of attention to the left crosses an object boundary, whereas the shift to the right stays within the same object. In the layout depicted below, a leftward shift of attention remains within the object, whereas a rightward shift moves between objects. (C) Polar plots of accuracy in identifying T2 (given accurate T1 recognition). The mean percent correct for 10 observers is plotted for each of the eccentric diamonds at the corresponding angle. Results for the two stimulus configurations of Fig. 16.6B are overlaid. A 380 ms SOA between T1 and T2 was used.

of the display and served as the fixation point, as well as the medium for creating T1 on all trials. It always belonged to an oval that also included one of the diamonds on the horizontal meridian. The other three diamonds were included within a different curved object.

The centers of the four peripheral diamonds were placed at the same retinal eccentricity (about 1.6°), in a 'cross-like' layout. The serial transformations consisted in the brief disappearance of one of the four corners of a diamond (see inset in Fig. 16.6A), an alteration which observers had to describe by pressing the appropriate arrow keys of the computer keyboard. Each trial included an event at the center diamond (T1), and a subsequent event at one of the peripheral diamonds (T2). T1 was uninformative about the upcoming location of T2.

Two stimulus configurations were used (see Fig. 16.6B): a layout that included the central diamond in the same object (oval) as the rightmost diamond, and another layout which included the central diamond in the same object as the leftmost diamond. Accordingly, the local events were always the same, but the perceptual linkage between these events varied for the two configurations. The first question asked was if T2 was discriminated with equal accuracy when it concerned the same object, or a different object, as T1.

In an experiment with a fixed SOA between T1 and T2 (380 ms), accuracy of T1 recognition was uniformly high (over 96% in all conditions and observers). Note that overall accuracy is lower in this design than in the one from the previous section, perhaps due to the smaller size of the shape mutations. T2 discrimination was about 20% more accurate when the affected peripheral diamond was part of the same object containing the central diamond (the site of T1), than when it was not connected to the central diamond. In other words, the AB was smaller when the two target events concerned the same object (see polar plot in Fig. 16.6C), confirming the conclusions drawn in Section 16.6). Equivalent results were obtained when the trials with eye movements were discarded.

The temporal dynamics of the AB were measured in a second task with the same stimuli but with a variable T1–T2 SOA. The results are illustrated in Fig. 16.7, where data from both stimulus configurations are collapsed. One block was performed with focused attention (ignoring T1) and another block was performed with divided attention (attending to both T1 and T2). During focused attention, all T1 were the same in order to make the event easier to ignore. Blocks were presented in a counterbalanced manner across subjects.

As expected, T1 identification was accurate. For within-object shifts of attention, the mean accuracy of T2 recognition was slightly above 60% correct identification of shape changes (Fig. 16.7, left panel). This mean accuracy did not vary as a function of SOA, and was only slightly affected by division of attention. No cost was found for this condition at 100 ms, a finding seemingly at variance with the results from the previous section (see Fig. 16.5B). We believe that intra-object grouping cues (and thus intra-object facilitation) were stronger in the stimuli of this new experiment. Note that division into parts is difficult for the oval used here, but not for the bars used before (compare Figs 16.5A and 16.6A). However, this idea has to be tested systematically by varying the concavity of the region connecting the two diamonds in the oval.

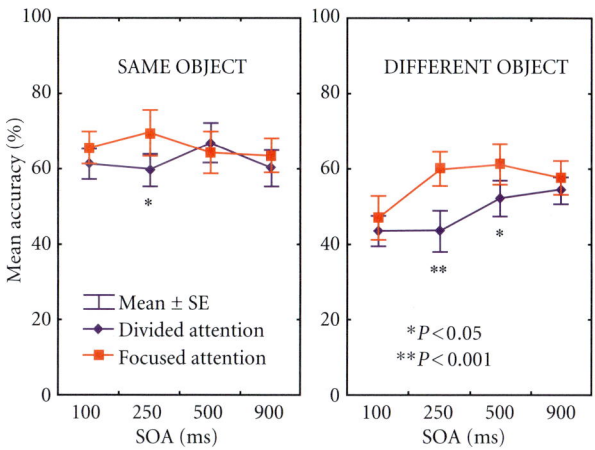

Fig. 16.7 Accuracy of T2 identification as a function of the T1–T2 SOA, and of the type of attentional shift (within- and between-object shifts as defined in Fig. 16.6B). Data for T2 collapsed over the right and left diamonds are shown, plotted separately for the blocks on which T1 was ignored or identified. Results for within-object attentional shifts are depicted in the left panel, and results for between-object shifts are shown in the right panel. Only trials with correct T1 identification were included.

On the other hand, under divided attention there was a substantial AB at short T1–T2 SOAs for between-object attentional shifts (Fig. 16.7, right panel). This impairment was ameliorated as the SOA increased. When T1 was ignored (focused attention), there was a facilitation of T2 identification, which was most pronounced at the SOA of 250 ms. This shows that sensory masking cannot explain the performance impairment for T2 discrimination during the AB. Interestingly, trying to ignore T1 produced little benefit for the SOA of 100 ms. Given the short time between T1 and T2, this outcome may be due to exogenous cueing, which automatically calls attention to visual entities and can not be voluntarily suppressed by the observer (Jonides 1981; Remington *et al.* 1992). Exogenous processes have been shown to operate on object-based attention (Macquistan 1997). We also demonstrated that when the targets in the RSOT consisted of shapes flashed briefly on top of the figures in the scene (new objects), then the object-based modulation of the AB was largely attenuated, in line with the results of Pinilla and Valdes-Sosa (2003).

Is the AB described in this section reflected in the ERP associated with T2? An experiment with eight observers was performed using high-density ERP recordings and a constant T1–T2 SOA of 380 ms, an interval at which the AB was expected to be large for the between-object attentional shifts as confirmed by the data (Valdes-Sosa *et al.* 2003). The ERPs related to T2 presented at the rightmost diamond, and from one electrode (left posterior temporal site) are shown in Fig. 16.8A. The N230 component elicited by T2 is significantly larger for the trials in which attention remained within the same object than for trials when attention shifted between objects (<0.01 in the latency range from 170 to 265). An equivalent result was obtained for the responses related to T2 at the leftmost diamond. Also, responses from trials in which T2 was correctly identified were significantly larger than responses from trials in which T2 was missed. This establishes a tight link between the AB and N230 amplitude.

The scalp distribution of the modulation of N230 was studied by obtaining the difference waveform resulting from the subtraction of the ERPs related to incorrectly

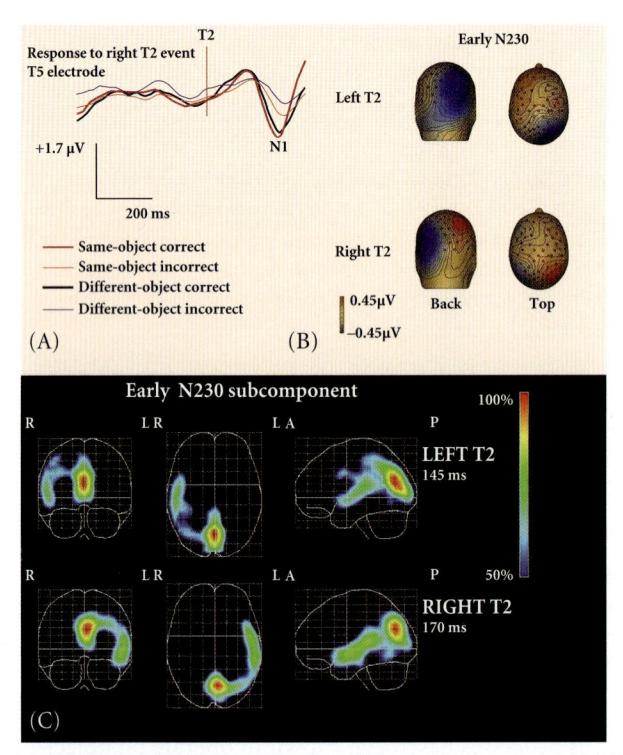

Fig. 16.8 Attentional effects on the ERPs related to T2 in the shape-change RSOT task (modified from Valdes-Sosa et al. 2003). (A) ERPs from the left posterior temporal electrode (T5 of the 10/20 system), plotted separately for trials with within- and between-object attentional shifts, and for correct and incorrect T2 identifications. Difference waveforms were obtained by subtracting the ERPs related to incorrectly identified between-object T2s from ERPs related to correctly identified within-object T2s (for details, see Valdes-Sosa et al. 2003). These waveforms therefore represent the modulation related to the AB. Recordings are the grand averages from eight observers and were obtained from 120 electrodes. (B) Scalp distributions of the voltage modulation affecting the early N230 subcomponent, presented separately for trials when T2 was presented at the leftmost and rightmost diamonds. (C) Current sources for the scalp distributions of the attentional effect on the early N230 subcomponent shown in Fig. 16.8B, as estimated by the VARETA method and represented with the same conventions as in Fig. 16.4B.

identified/different-object trial from ERPs related to correctly identified/same-object trials. The N230 contained at least three subcomponents. The earliest contribution to N230 was a large contralateral negativity with maximum amplitudes at posterior sites (see Fig. 16.8B). This subcomponent was present for the time region corresponding to the descending limb of N230, at about 145 ms for stimuli on the left and at about 170 ms for stimuli on the right. Additional, and somewhat later, contributions were found at ipsilateral and more frontal sites. This resembles the structure of the N1 elicited by the onset of the pattern stimuli used in previous studies of visuospatial attention (Di Russo *et al.* 2001). All of these subcomponents were affected by the AB.

The sources of the earliest N230 subcomponent, modeled with the VARETA method, are shown in Fig. 16.8C. The generators estimated as most active were located in several visual extrastriate areas contralateral to the side on which T2 was presented. A later time region (on the ascending limb of N230 at about 245 ms) was also examined. For this region, in which the ipsilateral and frontal subcomponents were more evident at the scalp, the modeled sources are displaced frontally and laterally (for more details see Valdes-Sosa *et al.* 2003).

If we accept that N230 amplitude reflects the strength with which sensory information is represented in visual extrastriate cortex, then these results indicate that the AB in the shape-change RSOT is related to early suppression of this information, in a manner similar to that demonstrated in the transparent motion design. The association of an AB with early sensory suppression is therefore not limited to objects and events defined by motion but seems a more general phenomenon.

16.8 Discussion and conclusions

By using RSOT designs, we have measured directly the timing of attentional interference within multiobject visual scenes. This timing depends on the perceptual linkage between successive target events. It is of very short duration and small magnitude when attention shifts within an object. It is larger, producing an AB, when attention shifts between objects. We can therefore generalize the principle that it is easier to divide attention to attributes within a single object than between different objects (Duncan 1984), to situations where the attributes are not simultaneously present.

The amplitude of the early N200 and N230 components was reduced during the object-based AB, a suppressive effect originally described as a signature for spatial attention (Hillyard and Anllo-Vento 1998). Therefore, the neural processes reflected by these ERP components are determined not only by local stimulation, but also by more global properties of the scene, including constraints on how attention can shift between different objects. These ERP effects are correlated with the subject's perceptual reports. Since the modulated components probably originate in early visual extrastriate cortex, this result suggests early suppression (or attenuation) of sensory information during the AB.

Previous ERP studies of the AB have obtained different results. One study using a traditional RSVP design (Luck *et al.* 1996; Vogel *et al.* 1998) found no modulation of the P1 or N1. In another study, using an RSOT with moving dots, during the AB a smaller N200 was found, but the amplitude of this component was not related to the observer's perceptual accuracy (Niedeggen *et al.* 2002).

These discrepancies could be due to several reasons. First, Luck *et al.* used a salient (abrupt-onset) probe superimposed on T2 to characterize neural reactivity during the AB. This type of stimulus may capture attention automatically (see Yantis 1998) and therefore always elicit a large P1 and N1. Secondly, the possibility remains that the AB found in more traditional RSVP paradigms is of a different nature from the AB found in RSOT paradigms, due to different attentional dynamics, or the use alphanumeric symbols as stimuli. On the other hand, the perceptual load in the study by Niedeggen and coworkers is lower than in our transparent motion task, which may lead to later attentional competition. Further research is needed to clarify the causes for the discrepancies.

We have presented evidence consistent with selective filtering of sensory information in early extrastriate areas during the AB (see also Chapter 18, this volume). But exactly what is filtered? The results in the shape-change paradigm might be explained by a modified spatial filter, as posited by the grouped array hypothesis (Vecera and Farah 1994; Lavie and Driver 1996; Arrington *et al.* 2000). By this view, the 'spotlight of attention' is warped to accommodate the shape of the attended object within a spatiotopic representation of the visual field. Therefore, at this moment a parsimonious account of these data can be reached by simply modifying location-based theories to accommodate an influence of perceptual organization. A critical test of this account will be to study the effects of displacing the objects within the scene.

However, we have argued in more detail elsewhere (Valdes-Sosa *et al.* 2000) that several candidates for attentional filtering (including representations of locations, elementary features, and two- or three-dimensional grouped arrays) are inconsistent with the results of the transparent motion paradigm (the same considerations are valid for the findings of Blaser *et al.* 2000). In brief, the spatial superposition of the competing transparent surfaces precludes selection of one surface based on locations, or a grouped array. A direct measurement of the difficulty in shifting attention between stationary sets of dots separated in depth, or of the additional difficulty produced by separating the illusory surfaces in transparent motion, reveals very weak effects (see Valdes-Sosa *et al.* 2000).

More importantly, these results also rule out fixed filters selecting specific values along some sensory dimension (such as direction of motion or color). For example, we presented evidence that signals in MT are attenuated during the AB in the transparent motion RSOT paradigm. This is achievable by inhibiting cortical columns selective for the direction of motion of T2 on that trial. Nevertheless, since possible directions for the T2 are identical for the same- and different-surface conditions, a flexible mechanism for setting the filters is needed, changing from trial to trial. This is even more necessary in the experiment reported by Blaser and coworkers, given that the features of their two Gabors could interchange, and thus any attribute (or combination of attributes) potentially belonged to either of the two objects present in the scene. In other words, it is necessary to re-specify the settings of lower-order filters for elementary attributes as an object changes its aspect.

Therefore, although the influence of scene organization on attention in our experiments is probably played out in early extrastriate cortex (where the different N1 components are generated), information from a higher-order representation is needed to explain the large flexibility in filter settings. A theoretical approach in consonance with these considerations postulates mental representations named 'object-files' (Kahneman and Treisman 1984; Kahneman *et al.* 1992). Object-files are codes for episodic information that bridge the variations in location and attributes of the same object entity. Therefore the identity and continuity of an object token are preserved. The object-file hypothesis stipulates that a mechanism is needed to update the attributes bound to a particular token.

A higher-order representation could control the activity of extrastriate areas in a 'top-down' manner, possibly mediated by the massive feedback projections these areas receive from other cortical sites (Lamme and Roelfsema 2000). This possibility is ignored by most theoretical accounts of the AB, that use strictly feedforward cognitive architectures (which are considered critically in the context of attention research in Chapter 14, this volume). This has been aptly elucidated by Chun and Wolfe (2003) with a 'conveyor-belt metaphor'. Imagine that perception delivers information, somewhat like dropping objects on to a conveyor-belt, at a rate faster than a subsequent stage can use (e.g. entry into visual short term memory), similar to a slow unloading of the belt. The difference in speed of the two operations would create the AB. But just as conveyor-belts move in only one direction, this type of bottleneck cannot explain the sensory suppression described here.

Our results can be accommodated (admittedly loosely at this point) within the framework of the integrated competition theory (Duncan 1996). When several objects are present in a scene we can suppose that an invariant neuronal representation of each token is set up. The appropriate local features are some how bound to each token code. When a change occurs in an object, new features appear and others disappear from the scene. Some neurons are activated and others are deactivated in different visual cortices. The new cells must somehow undergo binding to the representation of their object token, thus exchanging facilitation with other units already linked to the object and entering into inhibitory interactions with units representing other objects. This last idea is central to the integrated competition theory.

In our experiments, attention is first drawn to the object affording T1. This entails a momentary activation of cells representing new features, their binding to the object representation, and the ensuing activation of other units within this representation. Due to suppressive interactions, the neurons representing the other object are momentarily inhibited. Both effects would endure for a period after T1 offset. If a second event, T2, arrives before these changes dissipate, the ease with which the new features will be processed will be affected. If the new features are bound to the activated object, their coding units will benefit from its facilitation. If the features are bound to the suppressed objects, then the neurons representing them will inherit the corresponding inhibition.

Of course, this proposal begs the question as to how exactly 'object-files' are coded in the brain. The problem of how the brain represents objects is a complex issue (see Kanwisher and Treisman 1998, and Chapter 4, this volume), and there is no firm solution at hand. However, further studies with RSOT (in addition to capturing some interesting traits of real-life scenes) could perhaps further an understanding of the nature of object representations. This type of study could help to identify the neural activity that is invariant for the same object token, distinguishing it from the more volatile codes representing mutable aspects of the object, and thus contribute to unravel the puzzle of how objects are represented in the brain.

Acknowledgements

The authors thank Greysi Horta, Belkis Alonso, and Carlos Suarez-Murias for technical assistance, Lourdes Diaz Comas for software development, and Mitchell Valdes-Bobes for help with typing. This work was supported by a grant from the Human Frontier Science Program.

Notes

1 The recording of neurons in awake monkeys is contributing to the increased knowledge of the mechanisms of visual attention but is beyond the scope of this chapter (see Chapter 18, this volume).

References

Ahlfors, S.P., Simpson, G.V., Dale, A.M., Belliveau, J.W., Liu, A.K., Korvenoja, A. *et al.* (1999). Spatiotemporal activity of a cortical network for processing visual motion revealed by MEG and fMRI. *Journal of Neurophysiology,* **82,** 2545–55.

Arrington, C.M., Carr, T.H., Mayer, A.R., and Rao, S.M. (2000). Neural mechanisms of visual attention: Object-based selection of a region in space. *Journal of Cognitive Neuroscience,* **12,** 106–17.

Bach, M. and Ullrich, D. (1994). Motion adaption governs the shape of motion-evoked cortical potentials. *Vision Research,* **34,** 1541–7.

Behrmann, M., Zemel, R.S., and Mozer, M.C. (1998). Object-based attention and occlusion: Evidence from normal participants and a computational model. *Journal of Experimental Psychology: Human Perception and Performance,* **24,** 1011–36.

Blaser, E., Pylyshyn, Z., and Holcombe, A. (2000). Tracking an object through feature space. *Nature,* **408,** 196–9.

Bosch-Bayard, J., Valdes-Sosa, P., Virues-Alba, T., Aubert-Vázquez, E., John, E.R., Harmony, T., Riera-Díaz, J., and Trujillo-Barreto, N. (2001). 3D statistical parametric mapping of EEG source spectra by means of variable resolution electromagnetic tomography (VARETA). *Clinical Electroencephalography,* **32,** 47–61.

Brefczynski, J.A. and DeYoe, E.A. (1999). A physiological correlates of 'spotlight' of visual attention. *Nature Neuroscience*, **4**, 370–4.

Bundo, M., Kaneoke, Y., Inao, S., Yoshida, J., Nakamura, A., and Kakigi, R. (2000). Human visual motion areas determined individually by magnetoencephalography and 3D magnetic resonance imaging. *Human Brain Mapping*, **11**, 33–45.

Chun, M. (1997). Types and tokens in visual processing: a double dissociation between the attentional blink and repetition blindness. *Journal of Experimental Psychology: Human Perception and Performance*, **23**, 738–55.

Chun, M.M. and Potter, M.C. (1995). A two stage-stage model for multiple target detection in rapid serial visual presentation. *Journal of Experimental Psychology*, **21**, 109–127.

Chun, M.M. and Wolfe, J.M. (2001). Visual Attention. In E. B. Goldstein (Ed.), *Blackwell handbook of perception*, pp. 272–310. Blackwell, Oxford.

Clark, V.P. and Hillyard, S.A. (1996). Spatial selective attention affects early extrastriate but not striate components of the visual evoked potential. *Journal of Cognitive Neuroscience*, **8**, 387–402.

Cobo, A., Pinilla, T. and Valdes-Sosa, M. (1999). Attention to surfaces defined by transparent motion: measuring dwell time. *Brain and Cognition*, **40**, 85–90.

Culham, J., He, S., Dukelow, S., and Verstarten, F.A.J. (2001). Visual motion and the human brain: what has neuroimaging told us? *Acta Psychologica*, **107**, 69–94.

Desimone, R. and Duncan, J. (1995). Neural mechanism of selective visual attention. *Annual Review of Neuroscience*, **6**, 377–87.

Desimone, R. and Ungerleider, L.G. (1989). Neural mechanisms of visual processing in monkeys. In F. Boller and J. Grafman (Eds.), *Handbook of neuropsychology* pp. 267–99. Elsevier, Amsterdam.

Di Russo, F., Martinez, A., Sereno, M., Pitzalis, S., and Hillyard, S.A. (2001). Cortical sources of the early components of the visual evoked potential. *Human Brain Mapping*, **15**, 95–111.

Driver, J. and Baylis, G.C. (1989). Movement and visual attention: The spotlight metaphor breaks down. *Journal of Experimental Psychology: Human Perception and Performance*, **15**, 448–56.

Duncan, J. (1984). Selective attention and the organization of visual information. *Journal of Experimental Psychology: General*, **113**, 501–17.

Duncan, J. (1996). Co-ordinated brain systems in selective perception and action. In T. Unui and J.L McClelland (Eds.), *Attention and performance XVI*, pp. 549–78. MIT Press, Cambridge MA.

Duncan, J., Ward, R., and Shapiro, K. (1994). Direct measurements of attentional dwell time in human vision. *Nature*, **369**, 313–15.

Egeth, H.E. and Yantis, S. (1997). Visual attention: Control, representation and time course. *Annual Review of Psychology*, **48**, 269–97.

Egly, R., Driver, R., and Rafal, R.D. (1994). Shifting visual attention between objects and locations: Evidence from normal and parietal lesion subjects. *Journal of Experimental Psychology: General*, **123**, 161–77.

Eriksen, C.W. and Hoffman, J. (1973). The extent of processing of noise elements during selective encoding from visual displays. *Perception and Psychophysics*, **14**, 155–60.

Gandhi, S.P., Heeger, D.J., and Boynton, G.M. (1999). Spatial attention affects brain activity in human primary visual cortex. *Proceedings of National Academy of Sciences USA*, **96**, 3314–9.

Gomez-Gonzalez, C.M., Clark ,V.P., Fan, S., Luck, S.J., and Hillyard, S.A. (1994). Sources of attention-sensitive visual event-related potentials. *Brain Topography*, **7**, 41–51.

Grill-Spector, K. and Malach, R. (2001). fMRI-adaptation: a tool for studying the functional properties of human cortical neurons. *Acta Psychologica*, **107**, 293–321.

He, Z.J. and Nakayama, K. (1995). Visual attention to surfaces in three-dimensional space. *Proceedings of the National Academy of Sciences,* **92**, 11155–9.

Heinze, H.J., Mangun G.R., Burchert, W., Hinrichs, H., Scholz, M., Münte, T.F., Gös, A., Johannes, S., Scherg, M., Hundeshagen, H., Gazzaniga, M.S., and Hillyard SA. (1994). Combined spatial and temporal imaging of brain activity during visual selective attention in humans. *Nature,* **392**, 543–6.

Henderson, J.M. and Ferreira, F. (1990). Effects of foveal processing difficulty on the perceptual span in reading: Implications for attention and eye movement control. *Journal of Experimental Psychology: Learning, Memory and Cognition,* **16**, 417–29.

Hillstrom, A.P. and Yantis, S. (1994). Visual motion and attentional capture. *Perception and Psychophysics,* **55**, 339–411.

Hillyard, S.A. and Anllo-Vento, L. (1998). Event-related potentials in the study of visual selective attention. *Proceedings of National Academy of Science USA,* **95**, 781–7.

Hillyard, S.A. and Picton, T.W. (1987). Electrophysiology of cognition. In: F. Plum (Ed.), *Handbook of physiology: section 1. The nervous system,* vol 5, part 2 pp. 519–84. American Physiological Society, Baltimore MD.

Hillyard, S.A., Vogel, E.K., and Luck, S.J. (1998). Sensory gain control (amplification) as a mechanism of selective attention: Electrophysiological and neuroimaging evidence. *Philosophical Transactions of the Royal Society of London,* **353B**, 1257–70.

Jonides, J. (1981). Voluntary versus automatic control over the mind's eye. In J. Long and A. Baddeley (Eds.), *Attention and performance IX,* pp. 187–203. Lawrence Erlbaum Associates, Hillsdale NJ.

Kahneman, D. and Treisman, A. (1984). Changing views of attention and automaticity. In R. Parasuraman and D.R. Davies (Eds.), *Varieties of attention* pp. 29–61. Academic Press, Orlando FA.

Kahneman, D., Treisman, A., and Gibbs, B.J. (1992). The reviewing of object files: Object-specific integration of information. *Cognitive Psychology,* **24**, 175–219.

Kanwisher, N. G. and Treisman, A. (1998). Perceiving visually presented objects: recognition, awareness, and modularity. *Current Opinion in Neurobiology,* **8**, 218–26.

Kuba, M. and Kubora, Z. (1992). Visual evoked potentials specific for motion onset. *Documenta Opthalmologica,* **80**, 83–90.

La Berge, D. (1983). Spatial extents of attention to letters and words. *Journal of Experimental Psychology: Human Perception and Performance,* **9**, 371–9.

Lamme, V.A.F. and Roelfsema, P.R. (2000). The distinct modes of vision offered by feedforward and recurrent processing. *Trends in Neuroscience,* **23**, 571–9.

Lavie, N. (1995). Perceptual load as a necessary condition for selective attention. *Journal of Experimental Psychology: Human Perception and Performance,* **21**, 451–68.

Lavie, N. and Driver, J. (1996). On the spatial extent of attention in object-based visual attention. *Perception and Psychophysics,* **58**, 1238–51.

Luck, S.J. and Hillyard, S.A. (2000). The operation of selective attention at multiple stages of processing: Evidence from human and monkey electrophysiology. In M. Gazzaniga (Eds.), *The new cognitive neurosciences* pp. 687–700. MIT Press, Cambridge MA.

Luck, S.J., Vogel, E.K., and Shapiro, K.L. (1996). Words meanings can be accessed but not reported during the attentional blink. *Nature,* **383**, 616–18.

Mackeben, M. and Nakayama, K. (1993). Express attentional shifts. *Vision Research,* **33**, 85–90.

Macquistan, A.D. (1997). Object-based allocation of visual attention in response to exogenous but not endogenous spatial precues. *Psychonomical Bulletin and Review,* **4**, 512–15.

Mangun, G.R., Hopfinger, J.B., Kussmaul, C.L., Fletcher, E.M., and Heinze, H.J. (1997). Covariations in ERP and PET measures of spatial selective attention. *Human Brain Mapping*, **5**, 273–9.

Mangun, G.R., Jha, A.P., Hopfinger, J.B., and Handy, T.C. (2000). The temporal dynamics and functional architecture of attentional processes in human extrastriate cortex. In M. Gazzaniga (Ed.), *The new cognitive neurosciences* pp. 687–700. MIT Press, Cambridge MA.

Martínez, A., Anllo-Vento, L., Sereno, M.I., Frank, L.R., Buxton, R.B., Dubowitz, D.J., Wong, E.C., Hinrichs, H., Heinze, H.J., and Hillyard, S.A. (1999). Involvement of striate and extrastriate visual cortical, (1999). Involvement of striate and extrastriate visual areas in spatial attention. *Nature Neuroscience*, **2**, 364–9.

Moore, C.M., Egeth, H., Berglan, L.R., and Luck, S.J. (1996). Are attentional dwells times inconsistent with serial visual search? *Psychonomic Bulletin and Review*, **3**, 360–5.

Morris, R.K., Rayner, K. and Pollatsek, A. (1990). Eye movement guidance in reading: The role of parafoveal letter and space information. *Journal of Experimental Psychology: Human Perception and Performance*, **16**, 268–81.

Muller, M.M., Picton, T.W., Valdes-Sosa, P., Riera, J., Teder-Salejarvi, W.A., and Hillyard, S.A. (1998). Effects of spatial selective attention on the steady-state visual evoked potential in the 20–28 Hz range. *Cognitive Brain Research*, **6**, 249–61.

Niedeggen, M., Sahraie, A., Hesselmann, G., Milders, M., and Blakemore, C. (2002). Is experimental motion blindness due to sensory suppression? An ERP approach. *Cognitive Brain Research*, **13**, 241–7.

O'Craven, K.M., Rosen, B.R., Kwong, K.K., Treisman, A., and Savoy, R.L. (1997). Voluntary attention modulates fMRI activity in human MT–MST. *Neuron*, **18**, 591–8.

O'Craven, K.M., Downing, P.E., and Kanwisher, N. (1999). FMRI evidence for objects as the units of attentional selection. *Nature*, **401**, 584–7.

Picton, T.W., Alain, C., Woods, D.L., John, M.S., Scherg, M., Valdes-Sosa, P., Bosch-Bayard, J., and Trujillo, N.J. (1999). Intracerebral sources of human auditory-evoked potentials. *Audiology and Neurootology*, **4**, 64–79.

Pinilla, T. and Valdes-Sosa, M. (2003: submitted). Attentional shift time and scene organization: Not all blinks are equal. *Psychological Science*.

Pinilla, T., Cobo, A., Torres, K., and Valdes-Sosa, M. (2001). Attentional shifts between surfaces: Effects on detection and early brain potentials. *Vision Research*, **41**, 1619–30.

Posner, M.I. and Cohen, Y. (1984). Components of visual orienting. In: H. Bouma and D.G. Bouwhuis (Eds.), *Attention and performance X* pp. 55–66. Erlbaum, Hillside NJ.

Raymond, J. (2003). New objects, not new features, trigger the Attentional Blink. *Psychological Science*, **14**, 54–9.

Raymond, J., Shapiro, K.L., and Arnell, K.M. (1992). Temporary suppression of visual processing in an RSVP task: An attentional blink? *Journal of Experimental Psychology: Human Perception and Performance*, **18**, 849–60.

Remington, R.W., Johnson, J. C., and Yantis, S. (1992). Involuntary attentional capture by abrupt onsets. *Perception and Psychophysics*, **51**, 279–90.

Torriente, I., Valdes-Sosa, M., Ramirez, D., and Bobes, M.A. (1999). Visual evoked potentials related to motion-onset are modulated by attention. *Vision Research*, **39**, 4122–39.

Rodríguez, V., Valdes-Sosa, M., and Freiwald, W. (2002). Dividing attention between form and motion in transparent motion perception. *Cognitive Brain Research*, **13**, 187–93.

Rodríguez, V., Bobes, M., and Valdes-Sosa, M. (2003: submitted). Sensory suppression during shifts of attention between surfaces in transparent motion.

Sahraie, A., Milders, M., and Niedeggen, M. (2001). Attention induced motion blindness. *Vision Research*, **41**, 1613–7.

Sears, C.R. and Pylyshyn, Z.W. (2000). Multiple object tracking and attentional processing. *Canadian Journal of Experimental Psychology,* **54**, 1–14.

Shapiro, K.L., Raymond, J.E., and Arnell, K.M. (1994). Attention to visual pattern information produces the attention blink in RSVP. *Journal of Experimental Psychology: Human Perception and Performance,* **20**, 357–71.

Treisman, A. and Kanwisher, N.G. (1998). Perceiving visually presented objects: recognition, awareness, and modularity. *Current Opinion in Neurobiology,* **8**, 218–26.

Valdes-Sosa, M., Bobes, M.A., Rodríguez, V., and Pinilla, T. (1998a) Switching attention without shifting the spotlight: Object-based attentional modulation of brain potentials. *Journal of Cognitive Neuroscience,* **10**, 137–51.

Valdes-Sosa, M., Cobo, A., and Pinilla, T. (1998b). Transparent motion and object-based attention. *Cognition,* **66**, B13–B23.

Valdes-Sosa, M., Cobo, A., and Pinilla, T. (2000). Attention to object files defined by transparent motion. *Journal of Experimental Psychology: Human Perception and Performance,* **26**, 488–505.

Valdes-Sosa, M., Rodríguez, V., Iglesias, J., Acosta, Y., and Bobes, M.A. (2003: submitted). Perceptual grouping influences attentional competition within visual extrastriate cortex.

Vecera, S.P. and Farah, M.F. (1994). Does visual attention selects objects or locations? *Journal of Experimental Psychology: General,* **123**, 146–60.

Vecera, S.P., Behrmann, M., and McGoldrick, J. (2000). Selective attention to the parts of an object. *Psychonomic Bulletin and Review,* **2**, 301–8.

Vogel, E.K., Luck, S.J., and Shapiro, K. (1998). Electrophysiological evidence for a postperceptual locus of suppression during the attentional blink. *Journal of Experimental Psychology: Human Perception and Performance,* **24**, 1656–74.

Ward, R., Duncan, J., and Shapiro, K. (1996). The slow time-course of visual attention. *Cognitive Psychology,* **30**, 79–109.

Watanabe, T., Harner, A.M., Miyauchi, S., Sasaki, Y., Nielsen, M., Palomo, D. et al. (1998). Task-dependent influences of attention on the activation of human primary visual cortex. *Proceedings of the National Academy of Science USA,* **95**, 11489–92.

Watson, D.G. and Humphreys, G.W. (1997). Visual marking: prioritizing selection for new objects by top-down attentional inhibition of old objects. *Psychological Review,* **104**, 90–122.

Wolfe, J. (1998). Visual Search. In H. Pashler (Ed.), *Attention* pp. 223–56. Psychology Press, Hove, East Sussex.

Yantis, S. (1998). Control of visual attention. In H. Pashler (Ed.), *Attention* pp. 223–56. Psychology Press, Hove, East Sussex.

Yantis, S. and Jonides, J. (1984). Abrupt visual onset and selective attention: Evidence from visual search. *Journal of Experimental Psychology: Human Perception and Performance,* **10**, 601–21.

Chapter 17

Endogenous and stimulus-driven mechanisms of task control

Gordon L. Shulman and Maurizio Corbetta

Abstract

While cognitive processes can be prepared to carry out pre-specified tasks, they must be sufficiently flexible to respond to new contingencies. We propose that this flexibility is achieved through the interaction of two sets of regions in the human brain. One set is concerned with the endogenous control of tasks and primarily involves dorso-lateral prefrontal cortex, intraparietal sulcus (IPs), and frontal eye field (FEF). These regions are active during task preparation, generating signals for selecting task-relevant stimuli and linking them to responses, and maintain these signals during task performance. The other set is recruited by behaviorally important stimuli and involves the temporal–parietal junction (TPJ) and prefrontal cortex. These regions act as a stimulus-driven circuit breaker, interrupting ongoing task processes. In conjunction with dorsal frontal–parietal regions involved in task preparation, the TPJ allows new stimulus contingencies to be responded to appropriately.

17.1 Introduction

Any task involves an interaction between endogenous signals, reflecting goals, expectations, or memory, and stimulus-driven signals. If a person looks for an apple in the refrigerator, several processes may be prepared before the refrigerator door is opened. The shape and color of apples may be retrieved from memory to match against sensory information. Prior knowledge about where apples are kept may be retrieved to restrict the search to a part of the refrigerator. Motor structures for grasping the apple may be primed.

When the refrigerator door is opened, these prepared signals interact with stimulus-driven signals. Objects in the refrigerator may compete for attention, based on their distinctiveness or their match to the prepared apple template (Wolfe *et al.* 1989).

Detecting the apple terminates the search. But search might also be ended by an unexpected stimulus. The phone might ring. While cognitive processes can be prepared to carry out pre-specified tasks, they must be sufficiently flexible to respond to new contingencies.

We propose that this flexibility is achieved through the interaction of two sets of regions in the human brain (Corbetta and Shulman 2002). One set is concerned with the endogenous control of tasks and involves dorsal parietal and frontal cortex. Results presented in the first section of the chapter show that these regions are active during task preparation, generating signals for selecting the task-relevant stimulus, and maintaining these signals during task performance.

The other set of regions is recruited by behaviorally important stimuli and involves the temporal–parietal junction (TPJ) and prefrontal cortex, although this chapter will focus on the TPJ. These regions act as a circuit breaker that interrupts ongoing processes, allowing new contingencies to be responded to (Corbetta and Shulman 2002). The second part of the chapter presents several lines of evidence for this idea. The TPJ responds strongly to target stimuli that capture attention, but poorly to non-target stimuli. It is modulated by stimuli that violate an expectation and therefore require a change in a prepared task. Moreover, TPJ responses generalize over stimulus modality and the perceptual features or dimensions that define the expectation, indicating that TPJ globally signals when a new contingency has occurred.

The last section of this chapter examines how these two sets of regions interact when an unexpected stimulus requires a change in the current task. We suggest that while TPJ interrupts ongoing processes, dorsal parietal regions involved in task preparation are recruited by the new contingency, resulting in coactivation of TPJ and parietal cortex.

17.2 Task preparation

An important aspect of task preparation is indicating which input should control the response. In the example above, the grasping response should be controlled by apples, not by milk cartons. We provide evidence below that this process partly involves general processes, in the sense that the same cortical region encodes a variety of inputs. However, preparation for a specific stimulus or response may also involve pathways that are more specialized. For example, if motion is expected to the left, preparatory processes can aid detection of that motion (Ball and Sekuler 1980, 1981), perhaps by activating directionally selective channels. Similarly, if a subject knows in advance the effector to move, preparatory activity can facilitate the subsequent movement (Rosenbaum 1980). This preparatory activity may involve different cell populations, depending on which effector has been specified (Snyder *et al.* 1997).

We have observed both general and specialized signals for task preparation in a match-to-sample task (Shulman *et al.* 2003) (Fig. 17.1A). Subjects saw a cue word ('left', 'right', 'red', or 'green') for 480 ms. Following a blank interval of 3.84 s, a test display of moving colored dots was presented for 480 ms. Subjects indicated if the test

(a) **Match-to sample task**

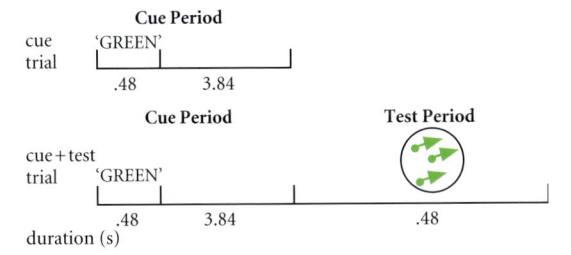

(b) **Search task**

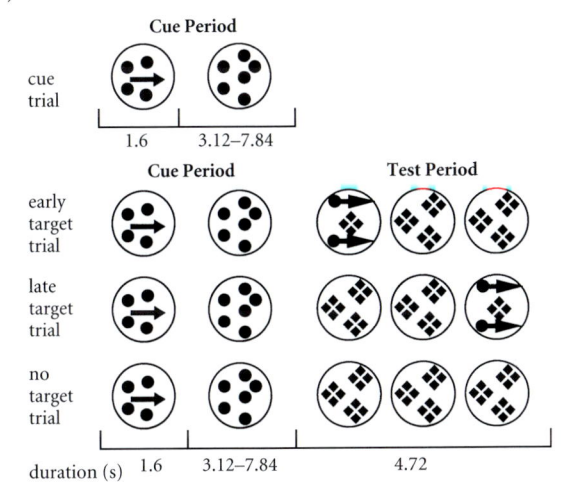

(c) **Posner task**

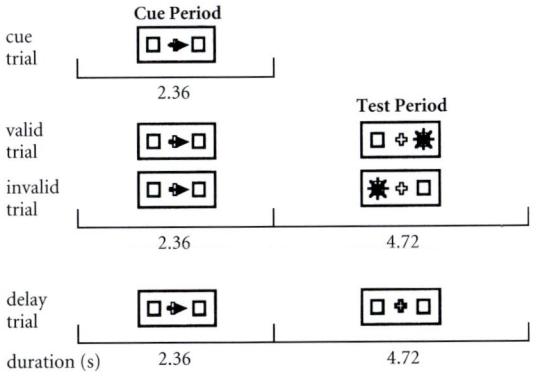

Fig. 17.1 Procedure for the match-to-sample, search, and Posner tasks. (Figures adapted from: Shulman et al. (1999) with permission, copyright 1999 by the Society for Neuroscience; Corbetta et al. (2000) with permission, copyright 2000 by Nature Publishing Group; and Shulman et al. (2001) with permission, copyright 2001 National Academy of Sciences, USA.)

display matched the cued feature, with right- and left-hand keypresses, respectively, for matching and mismatching displays. In order to make the task difficult, subjects were trained in a behavioral pre-session to associate each cue word with a particular standard feature. 'Red' for example, specified a particular shade of red. Non-matching displays following a 'red' cue contained dots of a slightly different red hue. Similarly, 'left' specified a particular oblique leftward motion (20° from horizontal). Non-matching displays following a 'left' cue involved a slightly more inclined leftward direction of motion. On any trial, the feature of the task-irrelevant dimension (e.g. color if the cue word was 'right') could be either standard or non-standard.

Since the cue specified a particular feature within a dimension, preparatory processes could modulate a specific input pathway. For example, prior physiological and anatomical studies indicate that parietal regions are involved in motion processing (Maunsell and Van Essen 1983; Ungerleider and Desimone 1986; Colby et al. 1993; Eskandar and Assad 1999) and might be preferentially activated by motion cues. There are strong projections from motion-selective regions such as MT and MST to parietal regions VIP and LIP (ventral and lateral intraparietal areas, respectively) (Maunsell and Van Essen 1983; Ungerleider and Desimone 1986). Correspondingly, directionally selective cells are common in VIP (Colby et al. 1993), while LIP cells maintain information concerning the direction of motion of a target during periods in which the target is occluded (Eskandar and Assad 1999). This capacity to represent the motion direction of an absent stimulus may be particularly relevant to direction-selective preparation.

We also manipulated whether the task-relevant dimension (color or motion) was blocked within a scan, with different scans involving color cues or motion cues, or mixed within a scan, so that any of the four cues might be presented. On blocked scans, information about the relevant dimension could be tonically maintained, but on mixed scans, this information needed to be encoded on each trial. Comparisons of blocked and mixed scans should reveal event-related signals related to specifying the task-relevant dimension.

In order to separate the blood-oxygenation level dependent (BOLD) signal during the cue and test periods, cue trials, in which the trial ended following the cue period, were randomly interspersed on 25% of the trials. The end of a trial was indicated by dimming the fixation point. We introduced this technique in order to separate task preparation from task execution, without assuming a hemodynamic response function (Shulman et al. 1999). Anatomical and functional images were collected on 19 subjects. Sixteen functional scans were collected, eight in the mixed condition, and four each in the blocked motion and color conditions. Each scan consisted of 128 magnetic resonance (MR) frames (TR = 2.36 s), where each frame contained an image of the entire brain (16, 8.0 mm slices; 3.75 × 3.75 mm in-plane resolution). Linear models estimated the time course of BOLD signals during the cue and test periods without assuming a hemodynamic response function (Ollinger et al. 2001). ANOVAs were conducted on the estimated time courses after they had been transformed to atlas space. ANOVAs treated subjects as a random effect and involved a whole-brain correction for multiple comparisons, based on a joint consideration of cluster size and z-score.

Behavioral data from the scanner session indicated that both color and motion tasks were difficult, with 77.2 and 78.2% correct, respectively, and reaction times of 1113 and 1192 ms. Performance on blocked and mixed scans did not differ, nor were there sequential dependency effects in which performance on a trial depended on the cue for the previous trial. The absence of a task-switching cost (Allport *et al.* 1994; Rogers and Monsell 1995) probably reflected three factors. The long cue period allowed for adequate preparation (Meiran *et al.* 2000) and was coupled with a long intertrial interval, allowing effects from the previous trial to dissipate (Meiran *et al.* 2000). Moreover, the limiting factor on performance was discriminability rather than the speed of stimulus–response translation. A significant congruency effect was observed in errors (79.4% versus 76.1%, respectively for displays in which the task-relevant and irrelevant features specified the same or a different response; $F(1, 18) = 16.2$, $P < 0.001$), but this effect did not differ significantly between blocked and mixed scans ($F(1, 18) = 1.02$).

Figure 17.2A shows a map of significant voxels in the left hemisphere during the cue and test periods. Functional data have been projected onto the gray matter of a partially inflated brain. Yellow (cue period) and blue (test period) cross-hatchings indicate voxels in which activation was greater on mixed than blocked scans, reflecting effects of specifying the task-relevant dimension. Red (cue period) and black (test period) cross-hatchings indicate voxels in which activation was greater following motion than color cues, reflecting motion-selective activity. No regions were more active on blocked or color scans.

The graphs in the left panel show the time course of the BOLD signal during the cue period in prefrontal and parietal (IPs, SPL/IPs) areas that were modulated by the experimental variables. Left prefrontal cortex showed larger signals on mixed than blocked scans. This effect was observed for both color and motion cues, indicating that this region coded information for both tasks.

A larger signal on mixed than blocked scans was also observed in left IPs, and this effect again generalized over the cue dimension. However, these signals were accompanied by motion-selective signals. On both blocked and mixed scans, left IPs showed a larger signal following motion cues than color cues.

Therefore, left parietal cortex carried two preparatory signals. One signal was involved in specifying which dimension was task-relevant and was observed for both color and motion tasks. Other studies (Kimberg *et al.* 2000; Sohn *et al.* 2000) have reported that when subjects switched between task sets involving letters and digits, activation related to the switch also occurred in left parietal cortex. Therefore, left parietal cortex codes information about the task-relevant stimulus for a variety of inputs. These preparatory signals were not observed on blocked scans, even though the cued feature changed over trials. Preparatory frontal activations were not observed on blocked

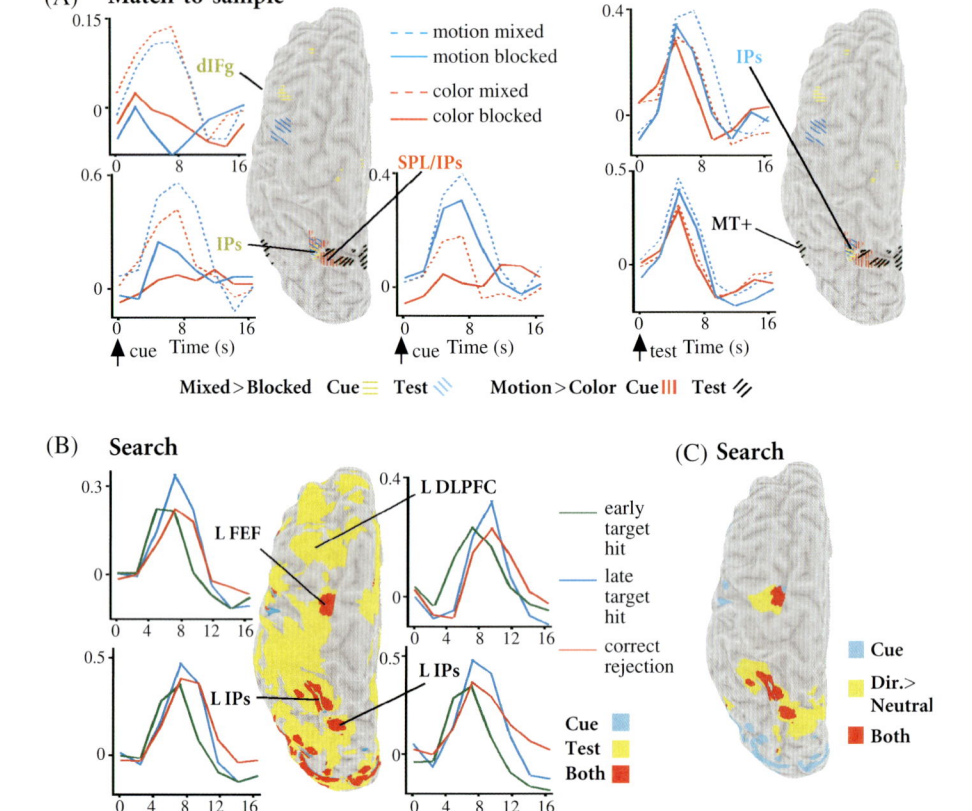

Fig. 17.2 (A) Match-to-sample task. Voxels showing a greater BOLD signal during motion than color trials and during mixed than blocked scans. Effects of these variables are shown separately for the cue and test periods, using cross-hatching of different colors. A dorsal view of the left hemisphere is presented. The three graphs on the left show the time course of the BOLD signal during the cue period. The two graphs on the right show time courses during the test period. On all graphs, the y-axis is percent signal change, the x-axis is time (s). (B) Search task. Voxels showing a significant BOLD signal during the cue period (blue), test period (yellow), or both periods (red). Graphs show the time course for trials in which motion was detected early or late, or no motion was presented and a response was correctly withheld. (C) Search task. Active voxels during the cue period of the event-related version of the search experiment (blue), voxels showing a greater BOLD signal on blocks of trials involving a directional (Dir.) cue than a neutral cue (yellow; from the blocked-design version of the search experiment), or voxels showing both effects (red). (Figures adapted from: Shulman et al. (1999) with permission, copyright 1999 by the Society for Neuroscience; Corbetta et al. (2000) with permission, copyright 2000 by Nature Publishing Group; and Shulman et al. (2001) with permission, copyright 2001 National Academy of Sciences, USA.)

scans, while preparatory parietal activity was not observed on blocked scans involving the color task. Since the test stimulus contained only one motion direction and hue, knowledge of the task-relevant dimension was sufficient to specify which feature should be matched to the cue. If two superimposed dot patterns had been presented (e.g. red dots moving left, green dots moving right), then knowledge of the task-relevant feature (e.g. left), not just the dimension, would have been necessary to match the proper attribute. Under these conditions, blocked and mixed scans might have yielded similar signals.

A second preparatory signal in left parietal cortex was motion-selective. Similar parietal regions were activated during the preparation period of search tasks that involved motion detection (see below). We suggest that the preparatory motion-specific component carried information about direction. This is consistent with the fact that motion-specific activity was observed in the blocked motion condition, in which only direction was manipulated. Feature-level information may augment preparatory modulation of input pathways. For example, if the cue had specified motion as the task-relevant dimension, but not a particular direction of motion, motion-selective signals might have been less evident.

The two experimental variables, blocked versus mixed presentation and color versus motion cues, did not interact in the ANOVA. Although additivity can be interpreted to mean that each variable affected an independent process, this interpretation involves assumptions about the appropriate scale of BOLD activity (e.g. linear or log). More conservatively, these data indicate that parietal cortex carried specialized (i.e. motion-selective) and general (i.e. active for both color and motion cues) preparatory signals, but whether or how they interacted was less clear. One possibility is that general task representations modulated more specialized pathways. During task preparation, prefrontal and perhaps parietal regions may have encoded the task-relevant dimension and sent signals to motion-selective regions. On trials involving motion cues, these signals facilitated the selection of motion information, while on trials involving color cues, these signals prevented the selection of motion information.

Both preparatory signals were maintained in parietal cortex during the test period (although voxels significantly sensitive to both signals no longer overlapped). Figure 17.2A shows that partly overlapping regions in left parietal cortex were more active for mixed than blocked scans during the cue and test periods. Similarly, partly overlapping regions were more active for motion than color cues during both periods. Therefore, left parietal cortex carried preparatory signals involved in specifying the task-relevant dimension and preparatory signals that were motion-selective, and maintained those signals as the task was performed. In contrast, many motion-selective regions were only modulated during the test period (e.g. MT$^+$; motion-selective activations were more evident in the right hemisphere, which is not shown). The instruction signals for these modulations may have originated in the parietal motion-selective region that was active during the cue period.

Parietal signals specific to the color task were not observed. Parietal activations have been reported for match-to-sample tasks involving surface orientation but not color (Shikata et al. 2001). Previous studies comparing hue and speed of motion have reported color-selective modulations in ventral cortex (Chawla 1999), but tasks involving motion direction as opposed to speed may also engage ventral regions. In the present work, ventral occipital cortex showed strong activity during both tasks. Ferrera and colleagues reported that V4 cells showed an equivalent dependence on the motion direction and color of the sample in match-to-sample tasks (Ferrera *et al.* 1994), while both the lingual and fusiform gyri showed significant activity during a match-to-sample task involving motion direction (Cornette *et al.* 1998).

The effects of the color-motion and mixed-blocked variables on preparatory activity in parietal cortex were only significant in the left hemisphere. Several studies of task-switching, which have not involved overt verbal cues, have also reported a left-hemisphere bias for parietal activations related to switching (Kimberg *et al.* 2000; Sohn *et al.* 2000). This bias, which may reflect symbolic coding of task information, indicates that these signals were not related to eye movements, which produce bilateral activations (Petit *et al.* 1997; Luna *et al.* 1998).

In summary, the match-to-sample experiment provided evidence for the role of dorsal frontal and parietal regions in the preparation and maintenance of task-related signals. Preparatory signals involved in specifying the task-relevant dimension were observed in left prefrontal and parietal cortex for both color and motion tasks. Regions in left parietal cortex, however, also showed more activity for motion than color cues, reflecting signals that were specific for a particular task. During the test period, both types of signals were maintained in left parietal cortex.

17.3 Stimulus-driven control of tasks

The results from the match-to-sample experiment emphasize the involvement of dorsal frontal and parietal regions in the preparation and maintenance of signals for carrying out a task. But as a task is performed, stimuli may occur which profoundly affect ongoing processes. The next sections consider regions that may detect these stimuli and interrupt ongoing processes.

17.3.1 Distinguishing salient from non-salient stimuli

Salient stimuli can potentially capture our attention. Those that do may be entered into working memory and prevent other stimuli from being detected or responded to (Duncan 1980). Stimulus salience is determined by both bottom-up and top-down factors. While highly distinctive stimuli, or stimuli of enduring biological significance, may grab our attention, stimuli also attract attention when they share target or display features specified by an active task (Folk *et al.* 1992). Regions involved in stimulus-driven control should respond well to salient stimuli but poorly to non-salient stimuli. In contrast, endogenous mechanisms may be active during task performance even for non-salient stimuli. Using visual search we have shown a strong dissociation between regions that are activated during the processing of non-target stimuli (i.e. non-salient stimuli) and regions that are not activated until a target is detected. We argue that several of these latter regions play a critical role in stimulus-driven control.

Thirteen subjects saw an arrow cue indicating a direction of motion (Fig. 17.1B) (Shulman *et al.* 2001). The arrow was presented for 1.6 s and then removed for the remainder of the cue period, which was 4.72, 7.08, or 9.44 s. During the subsequent test period, dynamic noise was presented for 4.72 s. On early-target trials, 300 ms of coherent motion was randomly presented between 0.4 and 1.8 s of the test period. On late-target trials, coherent motion was randomly presented between 2.8 and 4.2 s, while on no-target trials only dynamic noise was presented. The three trial types occurred with equal frequency and were presented in the context of an event-related design. Subjects pressed a key with their right hand if they detected motion and withheld a response if no motion was detected. The motion coherence percentage was adjusted for each subject in a behavioral pre-session, so that roughly 80% of targets were detected. During this pre-session, we verified that performance on the detection task was significantly better following an arrow cue that specified a direction of coherent motion than following a neutral plus-sign cue. Subjects used the direction cue, consistent with earlier work (Ball and Sekuler 1980, 1981).

We first show that preparing for visual search involved dorsal parietal regions. Event-related techniques were used to separate the activations during the cue and test periods. Voxels activated during the cue period are shown in blue (activated only during the cue period) and red (activated during both the cue and test periods) in Fig. 17.2B. Regions included IPs and the proposed human homologue of the primate frontal eye fields (FEF), which is typically activated by cues for location and motion (Shulman *et al.* 1999; Corbetta *et al.* 2000). FEF was also activated during the cue period of the match-to-sample experiment, but this activation was not significantly different for color and motion (there was a trend in left FEF for more activity for motion cues). The posterior portion of the IPs activation matched the motion-selective region observed in the match-to-sample experiment (Fig. 17.2A, red cross-hatching). The involvement of this region in preparing a directionally specific set for motion is further supported by results from a blocked-design version of the search experiment. In this experiment, blocks involving trials with directional cues were compared to blocks involving trials with neutral plus-sign cues (Shulman *et al.* 1999). Since this study involved a blocked design, the activations reflected processes during both cue and test periods. Moreover, because of the small sample size ($n = 7$), statistical analyses involved a fixed-effects model rather than a random-effects model, and therefore should be treated cautiously. However, Fig. 17.2C shows that the voxels more strongly activated by directional than neutral cues in the blocked-design study (yellow/red voxels in Fig. 17.2C), matched the voxels activated during the cue period of the event-related study (blue/red voxels in Fig. 17.2C). These results indicate that IPs is involved in preparing a search for a particular direction of motion.

We next consider the activations during the test period, in which subjects searched for and detected coherent motion targets in dynamic noise. The voxels activated by search and detection during the test period are shown in yellow and red in Fig. 17.2B. While IPs and FEF continued to be activated, signals were also observed in many other regions. The contrast between the restricted set of regions involved in task preparation and the large number activated by search and detection is striking. In order to determine which of these test-related regions might be involved in stimulus-driven control, we distinguished regions that were activated by the non-target stimulation (dynamic noise) during search, by the salient target stimulus (i.e. the coherent motion) during detection, or by both.

Since the target occurred at different times during the search interval, activity related to search and target detection could be separated. Regions involved in search of non-target stimuli should be active from the start of the test period, irrespective of whether targets were presented early, late, or not at all. The graphs in Fig. 17.2B show that signals in parietal cortex and FEF rose above baseline by 4.7 s for all three types of correct trials (early-target hit trials, in which a target was presented early and was detected, late-target hit trials, and correct-rejection trials). These 'early-onset' regions were engaged from the beginning of the test period. The frame at which the signal rose above baseline for each condition was determined by separate *t*-tests comparing each frame to a baseline frame. Regions had to simultaneously satisfy three *t*-tests in order to be considered 'early-onset' (for details, see Shulman *et al.* 2001).

IPs and FEF were also modulated by presence–absence detection (as well as many other regions). The signal on early target hit trials showed an increment at 4.7 s, corresponding to target detection. A similar increment was observed when the signals on early-target hit and miss trials were compared (not shown; see Shulman *et al.* 2001), indicating that this effect did not only reflect sensory stimulation from the low coherence target motion. Following detection on early-target hit trials, the signal fell off, reflecting the early termination of search. This fall-off supports the hypothesis that the activity in these regions was search related. If the BOLD signals at the beginning of the test period had been solely caused by sensory stimulation, which continued throughout the trial, then the fall-off in activity would not have occurred. Therefore, IPs and FEF maintained their activity during search among non-targets. This search-related activity is consistent with the conclusion from the match-to-sample experiment that dorsal frontal–parietal regions are involved in the endogenous control of tasks.

In contrast, regions involved in stimulus-driven control should respond poorly to non-target stimuli but strongly to target stimuli. The onset of activity should vary with the time of the presence–absence judgment. BOLD signals should occur early on early-target hit trials, but late on late-target hit or correct rejection trials. Many regions in prefrontal and temporal–parietal cortex showed this latter, 'late-onset' pattern (regions had to simultaneously satisfy five *t*-tests to be categorized as 'late-onset' (for details, see Shulman *et al.* 2001). Figure 17.2B shows the time course in a region of dorso-lateral prefrontal cortex (DLPFC) with a late-onset profile. Activity on early-target hit trials rose prior to activity on late-target hit or correct rejection trials.

Therefore, late-onset regions were not active during search among non-target stimuli. Instead, the BOLD signal was related to presence–absence detection. We argue below that one of these detection-related regions, the right TPJ, plays a particularly important role in stimulus-driven task control.

For the low-coherence motion in the search task, salience was primarily defined by top-down factors that specified a target. However, TPJ is also modulated when salience is defined by sensory distinctiveness in the absence of explicit task demands. Downar and colleagues (Downar *et al.* 2000) presented simultaneous visual, auditory, and tactile stimulus streams in a passive viewing paradigm. Within each stream, two types of stimuli were presented; for example two types of sounds. Transitions from one stimulus type to another strongly activated the TPJ, particularly in the right hemisphere, and this signal generalized over the modality of the stimulus change. These results suggest that the TPJ may respond to salient stimuli whether salience is defined by top-down factors that specify a target, or bottom-up factors based on sensory distinctiveness.

In summary, the search experiment isolated two sets of regions that showed different dependencies on stimulus salience. One set, including dorsal frontal–parietal regions, was active during search through non-targets (non-salient stimuli). Partially overlapping regions were also active during the cue period, in which the search was prepared, consistent with a role of these regions in the preparation and maintenance of endogenous signals for task performance. A second set of 'late-onset' regions was only active during presence–absence detection. Some of these regions, mainly in the TPJ and lateral prefrontal cortex, responded to salient stimuli whether salience was defined by top-down or bottom-up factors.

17.3.2 Detecting unexpected stimuli that are behaviorally relevant

Regions involved in stimulus-driven control should be modulated by expectancy as well as salience. Unexpected stimuli often require a prepared task to be interrupted or changed. When a stimulus is expected in one location but appears in another, for example, attention may be shifted to the new location. If the unexpected stimulus is behaviorally relevant but not part of the current task, as in the case of the telephone ringing, the task may be changed entirely.

We have studied the BOLD signals to targets that occurred at unexpected locations using the Posner task (Corbetta *et al.* 2000). Cueing paradigms are particularly useful for studying expectancy, since expectancy is not confounded with stimulus or response frequency. An arrow cue instructed subjects to shift attention to a location in space, and was followed by a test period in which subjects pressed a key to indicate detection of a brief stimulus (Fig. 17.1C). On valid trials (73% of trials involving a target), the target stimulus (an asterisk, presented for 100 ms) occurred in the cued location. On invalid trials (27% of trials in which a target was presented) the target occurred in the opposite uncued location. Targets occurred at random times between 1500 and

3000 ms from the start of the test period, which was indicated by the offset of the arrow cue. On delay trials, no targets were presented. The end of a trial was indicated by a change in the color of a small fixation point from green to red. At the start of the next trial, the color changed back from red to green.

The statistical map in Fig. 17.3A shows that right TPJ was more activated by invalid than valid targets. The color scale reflects the equivalent z-score of the activation, with

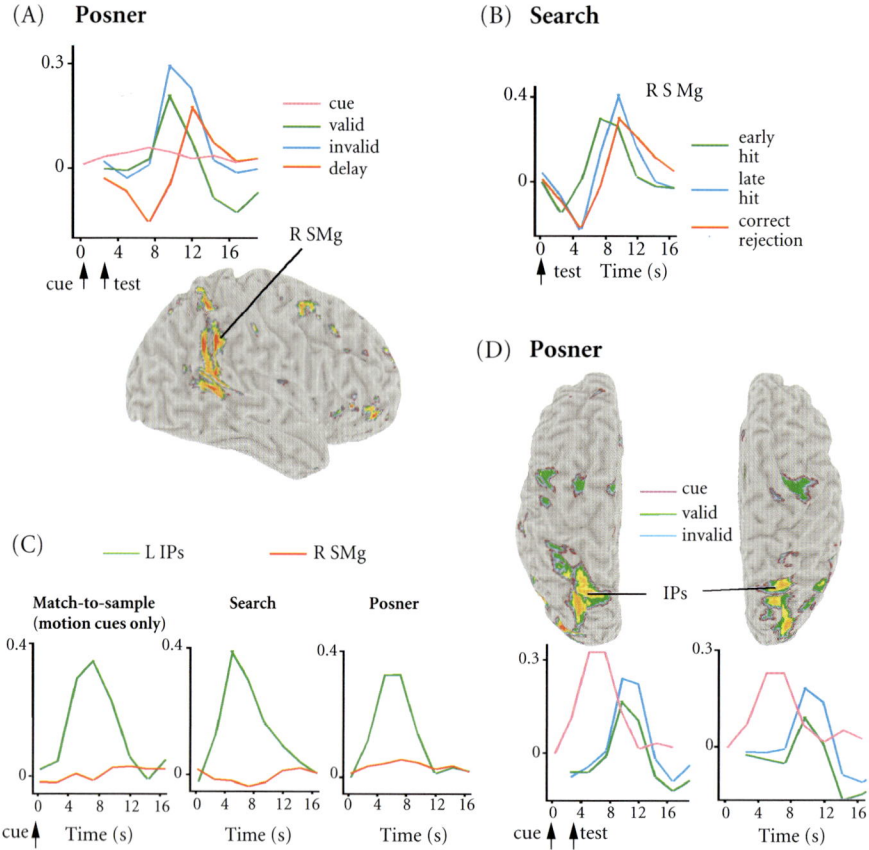

Fig. 17.3 (A) Posner task. Voxels showing a larger BOLD signal for invalid than valid targets. Colors indicate equivalent z-scores, with warmer colors reflecting greater z-scores. The graph shows the time courses during the cue period, and during the test period, of valid, invalid, and delay trials. (B) Search task. Time courses in right supramarginal gyrus (R SMg) for early-target, late-target and correct rejection trials. (C) Activity in left intraparietal sulcus (L IPs) and R SMg during the cue period of the three experiments discussed in the text. (D) Posner task. The statistical map shows the voxels active during the cue period. Warmer colors indicate larger z-scores. Dorsal views of the left and right hemispheres are shown. Graphs show the time course during the cue period (pink line), and during the test period of valid (green line) and invalid (blue line) trials. (Figures adapted from Corbetta et al. (2000) with permission. Copyright 2000 by Nature Publishing Group.)

'warmer' colors reflecting greater significance levels. The time course of the response to valid (green line) and invalid (blue line) targets, starting from the onset of the test period, is shown in the graph. Although right TPJ responded to both valid and invalid targets, the response to invalid targets was larger. The signal increment for invalid targets in right TPJ has been confirmed in several laboratories and occurs for tactile as well as visual stimuli (Arrington et al. 2000; Macaluso et al. 2002).

Validity effects in the behavioral literature have been modeled by parameters that weight information from different parts of the visual field (Shaw and Shaw 1977). The increased BOLD signal in the TPJ might reflect a detection-related process that occurs for both valid and invalid targets but has a longer duration for invalid targets because of the low weighting of the invalid location. However, the 45 ms increase in reaction time on invalid trials may not be sufficient to account for the increased BOLD signal.

Alternately, invalid targets may have triggered a separate process not engaged on valid trials. This is more consistent with the results of both transcranial magnetic stimulation (TMS) and lesion-behavior studies. TMS in the right angular gyrus increases reaction time to invalid targets, but not valid targets, while TPJ lesions increase RTs to invalid targets in the neglected field (Friedrich et al. 1998; Rushworth et al. 2001).

The additional process could reflect a stimulus-driven re-orienting to the spatial location of the target. However, other imaging studies indicate that TPJ activity is increased even when an unexpected stimulus does not initiate a shift of attention. These studies have typically involved 'oddball' paradigms in which a low-frequency target is presented in a series of 'standard' stimuli. In one study (Marois et al. 2000), subjects pressed one key to a standard stimulus of a particular shape at a particular location and pressed a second key to occasional stimuli of different shape and/or location. The TPJ signal was increased to a low-frequency shape, even if it occurred at the standard location. In this case, the unexpected stimulus did not generate a shift of spatial attention, yet still activated the TPJ. Similarly, low-probability foveal visual shapes or low-probability auditory pitches increase TPJ activity, although a right hemisphere dominance is not always reported (Linden et al. 1999; Braver et al. 2001; Kiehl et al. 2001). Similar results have been reported for evoked potentials. Electrical evoked potentials associated with the detection of infrequent targets are localized to and disrupted by lesions to the TPJ (Knight and Scabini, 1998).

TPJ activity is sometimes tied to spatial attention because of its association with spatial neglect (Friedrich et al. 1998), but these neuroimaging studies suggest a broader role. Neglect patients with TPJ lesions might respond poorly to low-frequency targets defined by non-spatial features, even at expected locations. If the increased TPJ signal on invalid trials reflected an additional process, that process was more general than spatial reorienting. We have suggested that the TPJ acts as a circuit-breaker that interrupts ongoing processes and leads to a change in set or attention (Corbetta and Shulman 2002).

An unexpected stimulus often requires a different response than an expected stimulus, but TPJ activity is probably not related to response conflict *per se*. Neuroimaging studies have emphasized the role of the anterior cingulate (Botvinick *et al.* 2001) in monitoring the presence of conflict. Braver and colleagues (Braver *et al.* 2001) measured the BOLD signal to high- and low-frequency targets in both go–no-go and choice response paradigms. In the go–no-go tasks, subjects pressed a key for the low-frequency stimulus but withheld a response to the high-frequency stimulus (or the reverse); while in the choice tasks, subjects pressed one key for the low-frequency stimulus and another key for the high-frequency stimulus. The anterior cingulate, as well as right TPJ, was more strongly activated by low- than high-frequency stimuli for all response mappings. However, while invalid targets in the Posner task differentially activated the TPJ, they did not differentially activate the anterior cingulate (Corbetta *et al.* 2000). Since both valid and invalid targets in the Posner task were mapped to the same response key, we suggest that the TPJ is activated by unexpected targets, irrespective of whether inappropriate responses are evoked. In contrast, the anterior cingulate is primarily engaged when an evoked response is inappropriate.

In summary, both cueing and oddball paradigms show that the TPJ responds particularly well to target stimuli when they are unexpected. These enhanced responses generalize over stimulus modality and occur for both spatial and non-spatial expectancies. In conjunction with evidence that TPJ is modulated by stimulus salience, these results support the view that TPJ acts as a stimulus-driven circuit breaker that can interrupt ongoing processes (Corbetta and Shulman 2002), leading to a change in set or attention.

17.3.3 The TPJ is deactivated during search

While we normally want to orient to a broad range of stimulus contingencies, we do not want irrelevant interruptions from a 'circuit-breaker' while we are focused on a particular task. Perhaps reflecting this, we have observed that the TPJ is deactivated during search through non-targets. Figure 17.3B shows the time course of right supramarginal gyrus (R SMg) during the test period of the search experiment, in which subjects monitored dynamic noise for a coherent motion target. The BOLD signal decreased during search through the noise, followed by a detection-related increment. More recent work has confirmed that the deactivation is search-related (Shulman *et al.* 2002*a*). Although the interpretation of deactivations is controversial, they may reflect a reduction of afferent input to an area. Studies indicate that the decrease in cerebellar blood flow that occurs following strokes in contralateral frontal cortex, reflects a loss of frontal inputs to the cerebellum (Gold and Lauritzen 2002).

What caused the deactivation during search? One possibility is that non-salient stimulation during search (e.g. dynamic noise) decreased the TPJ signal. However, a similar deactivation was evident in the TPJ during the delay condition of the Posner task, in which subjects monitored an unchanging display for a target. In the valid target condition (Fig. 17.3A, green function), the interval prior to target presentation was short and the signal was dominated by detection-related activity. In the delay condition (Fig. 17.3A, red function), this interval was lengthened and a deactivation was evident prior to a signal at the end of the trial (for a discussion of this latter signal, see Shulman et al. 2002b).

Therefore, TPJ was deactivated by a set for a target, rather than by non-target stimulation. We speculate that the deactivation reflected the effects of a top-down filter that restricted the range of stimuli activating the TPJ during search, reducing interruptions from task-irrelevant stimuli. When no task set is present, the filter is broadly tuned and the TPJ gives strong responses to distinctive stimuli (Downar et al. 2000), and perhaps to stimuli of biological significance, learned (e.g. our name) or otherwise. When a task set for a particular target is active, the filter is more sharply tuned and these stimuli are less likely to produce a TPJ signal (Downar et al. 2001).

17.4 Relationship between endogenous and stimulus-driven regions for task control

How do TPJ signals result in a response to a new stimulus contingency, such as an invalid target in the Posner task? We suggest that TPJ signals recruit parietal regions, including regions involved in task preparation. The response to an unexpected stimulus is unlikely to be mediated solely by the TPJ, since it is not active during the cue period when a linkage between a stimulus and a response is initially prepared. Figure 17.3C compares the responses in left IPs and the right SMg during the cue periods of the three experiments discussed in this paper: the match-to-sample experiment (motion cue conditions only), the search experiment (averaged over the three different cue durations), and the Posner task. In all cases, robust signals were observed in IPs, but not SMg. Note that SMg responses were not observed even though an abrupt onset that was behaviorally relevant occurred at the beginning of the cue period (i.e. the cue stimulus).

While IPs and FEF were active during task preparation and TPJ was inactive, both sets of regions showed an increased response to targets. Figure 17.2B shows the target-related response in left IPs (L IPs) and FEF, while Fig. 17.3B shows the response in right SMg (R SMg). Unexpected targets, which increased TPJ activity (Fig. 17.3A), also increased activity in IPs, although not as robustly. The statistical map in Fig. 17.3D shows the regions that were significantly activated during the cue period of the Posner task. The time course of activity in those regions, which are likely involved in directing attention to a location, is shown in the graphs in Fig. 17.3D. The pink function in the graphs confirms that IPs was active during the cue period. The green and blue functions show the time courses for IPs from the start of the test period on valid (green) and invalid (blue) trials. IPs was activated by both valid and invalid targets, but the response to invalid targets was larger. This increment was not significant in the voxel-level analyses, which involved conservative multiple-comparison corrections, but it was significant in uncorrected regional ANOVAs (L IPs, $F(7, 84) = 2.51$, $P < 0.05$; R IPs, $F(7, 84) = 3.03$, $P < 0.05$). Stimulus-driven spatial reorienting may involve dorsal parietal regions that are active during endogenous shifts of attention, in conjunction with TPJ activity (Posner *et al.* 1984).

These results indicate an asymmetrical relationship between the TPJ and dorsal frontal–parietal regions. During task preparation, IPs activity is increased and TPJ activity remains at baseline. When a display is monitored for a specific target, IPs maintains its activity and TPJ is deactivated, perhaps reducing interruptions from task-irrelevant stimuli. Detecting an unexpected target stimulus activates the TPJ and interrupts the set for the expected target. But responding to the unexpected target also involves some of the regions that prepared the original target set, resulting in coactivation of TPJ and IPs.

17.5 Conclusions

Task control involves an interaction between regions that prepare and maintain task signals and regions that respond when new contingencies arise. We have suggested that for simple, well-practiced visual tasks, task preparation primarily engages dorsal frontal–parietal regions. These regions specify which inputs are task-relevant and modulate feature-specific pathways. Stimuli that require a change in ongoing task processes, such as targets that violate an expectation, activate the TPJ, which acts as a stimulus-driven circuit-breaker. These TPJ signals generalize over stimulus modality and the perceptual features or dimensions that define the expectation. In conjunction with activity in dorsal frontal–parietal regions, TPJ activity allows new stimulus contingencies to be responded to appropriately.

References

Allport, A., Styles, E. A., and Hsieh, S. (1994). Shifting intentional set: Exploring the dynamic control of tasks. In C. Umilta and M. Moscovitch (Eds), *Attention and Performance XV* pp. 421–52. MIT Press, Cambridge MA.

Arrington, C. M., Carr, T. H., Mayer, A. R., and Rao, S. M. (2000). Neural mechanisms of visual attention: object-based selection of a region in space. *Journal of Cognitive Neuroscience*, **12**, 106–17.

Ball, K. and Sekuler, R. (1980). Models of stimulus uncertainty in motion perception. *Psychological Review*, **87**, 435–69.

Ball, K. and Sekuler, R. (1981). Cues reduce direction uncertainty and enhance motion detection. *Perception and Psychophysics*, **30**, 119–28.

Botvinick, M. M., Braver, T. S., Barch, D. M., Carter, C. S., and Cohen, J. D. (2001). Conflict monitoring and cognitive control. *Psychological Review*, **108**, 624–52.

Braver, T. S., Barch, D., Gray, J. R., Molfese, D. L., and Snyder, A. Z. (2001). Anterior cingulate cortex and response conflict: effects of frequency, inhibition, and errors. *Cerebral Cortex*, **11**, 825–36.

Chawla, D., Rees, G., and Friston, K. J. (1999). The physiological basis of attentional modulations in extrastriate visual areas. *Nature Neuroscience*, **2**, 671–6.

Colby, C. L., Duhamel, J.-R., and Goldberg, M. E. (1993). Ventral intraparietal area of the macaque: anatomic location and visual response properties. *Journal of Neurophysiology*, **69**, 902–14.

Corbetta, M. and Shulman, G. L. (2002). Control of goal-directed and stimulus-driven attention in the brain. *Nature Reviews Neuroscience*, **3**, 201–15.

Corbetta, M., Miezin, F. M., Dobmeyer, S., Shulman, G. L., and Petersen, S. E. (1991). Selective and divided attention during visual discriminations of shape, color, and speed: Functional anatomy by positron emission tomography. *Journal of Neuroscience*, **11**(8), 2383–402.

Corbetta, M., Kincade, J. M., Ollinger, J. M., McAvoy, M. P., and Shulman, G. L. (2000). Voluntary orienting is dissociated from target detection in human posterior parietal cortex. *Nature Neuroscience*, **3**, 292–7.

Cornette, L., Dupont, P., Rosier, A., Sunaert, S., Hecke, P., Hichiels, J., Mortelmans, L., and Orban, G. A. (1998). Human brain regions involved in direction discrimination. *Journal of Neurophysiology*, **79**, 2749–65.

Downar, J., Crawley, A. P., Mikulis, D. J., and Davis, K. D. (2000). A multimodal cortical netowork for the detection of changes in the sensory environment. *Nature Neuroscience*, **3**, 277–83.

Downar, J., Crawley, A. P., Mikulis, D. J., and Davis, K. D. (2001). The effect of task relevance on the cortical response to changes in visual and auditory stimuli: an event-related fMRI study. *Neuroimage*, **14**, 1256–67.

Duncan, J. (1980). The locus of interference in the perception of simultaneous stimuli. *Psychological Review*, **87**, 272–300.

Eskandar, E. N. and Assad, J. A. (1999). Dissociation of visual, motor and predictive signals in parietal cortex during visual guidance. *Nature Neuroscience*, **2**, 88–93.

Ferrera, V. P., Rudolph, K. K., and Maunsell, J. H. R. (1994). Responses of neurons in the parietal and temporal visual pathways during a motion task. *Journal of Neuroscience*, **14**(10), 6171–86.

Folk, C. L., Remington, R. W., and Johnston, J. C. (1992). Involuntary covert orienting is contingent on attentional control settings. *Journal of Experimental Psychology: Human Perception and Performance*, **18**(4), 1030–44.

Friedrich, F. J., Egly, R., Rafal, R. D., and Beck, D. (1998). Spatial attention deficits in humans: A comparison of superior parietal and temporal–parietal junction lesions. *Neuropsychology*, **12**(2), 193–207.

Gold, L. and Lauritzen, M. (2002). Neuronal deactivation explains decreased cerebellar blood flow in response to focal cerebral ischemia or suppressed neocortical function. *Proceedings of the National Academy of Sciences, USA*, **99**, 7699–704.

Kiehl, K. A., Laurens, K. R., Duty, T. L., Forster, B. B., and Liddle, P. F. (2001). Neural sources involved in auditory target detection and novelty processing: an event-related fMRI study. *Psychophysiology*, **38**, 133–42.

Kimberg, D. Y., Aguirre, G. K., and D'Esposito, M. (2000). Modulation of task-related neural activity in task-switching: an fMRI study. *Cognitive Brain Research*, **10**, 189–96.

Knight, R. T. and Scabini, D. (1998). Anatomic bases of event-related potentials and their relationship to novelty detection in humans. *Journal of Clinical Neurophysiology*, **15**, 3–13.

Linden, D., Prvulovic, D., Formisano, E., Vollinger, M., Zanella, F., Goebel, R., and Dierks, T. (1999). The functional neuoranatomy of target detection: An fMRI study of visual and auditory oddball tasks. *Cerebral Cortex*, **9**, 815–23.

Luna, B., Thulborn, K. R., Strojwas, M. H., McCurtain, B. J., Berman, R. A., Genovese, C. R., and Sweeney, J. A. (1998). Dorsal cortical regions subserving visually-guided saccades in humans: An fMRI study. *Cerebral Cortex*, **8**, 40–7.

Macaluso, E., Frith, C. D., and Driver, J. (2002). Supramodal effects of covert spatial orienting triggered by visual or tactile events. *Journal of Cognitive Neuroscience*, **14**, 389–401.

Marois, R., Leung, H. C., and Gore, J. C. (2000). A stimulus-driven approach to object identity and location processing in the human brain. *Neuron*, **25**, 717–28.

Maunsell, J. H. R. and Van Essen, D. C. (1983). The connections of the middle temporal visual ara (MT) and their relationship to a cortical hierarchy in the macaque monkey. *Journal of Neuroscience*, **3**, 2563–86.

Meiran, N., Chorev, Z., and Sapir, A. (2000). Component processes in task switching. *Cognitive Psychology*, **41**, 211–53.

Ollinger, J. M., Shulman, G. L., and Corbetta, M. (2001). Separating processes within a trial in event-related functional MRI I. The Method. *Neuroimage*, **13**(1), 210–17.

Petit, L., Clark, V. P., Ingeholm, J., and Haxby, J. V. (1997). Dissociation of saccade-related and pursuit-related activation in human frontal eye fields as revealed by fMRI. *Journal of Neurophysiology*, **77**, 3386–90.

Posner, M. I., Walker, J. A., Friedrich, F. J., and Rafal, R. D. (1984). Effects of parietal injury on covert orienting of attention. *Journal of Neuroscience*, **4**(7), 1863–74.

Rogers, R. D. and Monsell, S. (1995). The cost of a predictable switch between simple cognitive tasks. *Journal of Experimental Psychology: General*, **124**, 207–31.

Rosenbaum, D. A. (1980). Human movement initiation: Specification of arm, direction and extent. *Journal of Experimental Psychology: General*, **109**(4), 444–74.

Rushworth, M. F. S., Ellison, A., and Walsh, V. (2001). Complementary localization and lateralization of orienting and motor attention. *Nature Neuroscience*, **4**, 656–61.

Shaw, M. and Shaw, P. (1977). Optimal allocation of cognitive resources to spatial locations. *Journal of Experimental Psychology: Human Perception and Performance*, **3**, 201–11.

Shikata, E., Hamzel, F., Glauche, V., Knab, R., Dettmers, C., Weiller, C., and Buchel, C. (2001). Surface orientation discrimination activates caudal and anterior intraparietal sulcus in humans: an event-related study. *Journal of Neurophysiology*, **85**, 1309–14.

Shulman, G. L., Ollinger, J. M., Akbudak, E., Conturo, T. E., Snyder, A. Z., Petersen, S. E., and Corbetta, M. (1999). Areas involved in encoding and applying directional expectations to moving objects. *Journal of Neuroscience*, **19**, 9480–96.

Shulman, G. L., Ollinger, J. M., Linenweber, M., Petersen, S. E., and Corbetta, M. (2001). Multiple neural correlates of detection in the human brain. *Proceedings of the National Academy of Sciences, USA*, **98**, 313–18.

Shulman, G. L., McAvoy, M., Cowan, M. J., Astafiev, S. V., Tansy, A. P., and d'Avossa, G. (2002a). Modeling the bold signal during visual search. In *Society for Neuroscience*. Orlando FL.

Shulman, G. L., Tansy, A. P., Kincade, M., Petersen, S. E., McAvoy, M., and Corbetta, M. (2002b). Reactivation of networks involved in preparatory states. *Cerebral Cortex*, **12**, 590–600.

Shulman, G. L., d'Avossa, G., Tansy, A. P., and Corbetta, M. (2003). Two attentional processes in the parietal lobe. *Cerebral Cortex*, **12**, 1124–31.

Snyder, L. H., Batista, A. P., and Andersen, R. A. (1997). Coding of intention in the posterior parietal cortex. *Nature*, **386**, 167–70.

Sohn, M. H., Ursu, S., Anderson, J. R., Stenger, V. A., and Carter, C. S. (2000). Inaugural article: the role of prefrontal cortex and posterior parietal cortex in task switching. *Proceedings of National Academy of Sciences, USA*, **97**(24), 13448–53.

Ungerleider, L. G. and Desimone, R. (1986). Cortical connections of visual area MT in the macaque. *Journal of Comparative Neurology*, **248**, 190–222.

Wolfe, J. M., Cave, K. R., and Franzel, S. L. (1989). Guided search: an alternative to the feature integration model for visual search. *Journal of Experimental Psychology*, **15**, 419–33.

Chapter 18

The role of competitive circuits in extrastriate cortex in selecting spatially superimposed stimuli

John H. Reynolds and Mazyar Fallah

Abstract

Extracellular recording studies of attention in monkeys trained to perform visual tasks find that changes in neuronal responses differ qualitatively depending on whether one or more stimuli appear within the neuron's receptive field (RF). Attention to a single excitatory stimulus typically results in an increase in firing rate, regardless of whether the stimulus is a preferred or poor stimulus for the neuron. In contrast, attention to one of multiple stimuli within the RF can result either in increases or decreases in firing rate, depending on the neuron's relative preference for the stimuli in the RF. These findings can be interpreted within the context of a psychological theory that relates them to psychophysical and functional imaging studies of attention in humans. This interpretation leads to specific predictions about changes in neuronal responses in an object-based attention paradigm.

18.1 Introduction

Despite considerable progress in our understanding of the neural mechanisms of spatial attention, relatively little is known about the neural mechanisms that are involved in attending to and selecting out coherent objects for visual processing. Here we summarize the results of recent single-unit recording studies of attention in monkeys, which accord with the predictions of a psychological theory of attention that has been used to account for a range of attentional phenomena, including aspects of object-based attention. While this biased-competition theory has successfully predicted changes in neuronal responses with attention, the single-unit recording studies that have been carried out to test its predictions have all used stimuli that occupied separate spatial locations, thereby confounding spatial and object-based attention. Several psychophysical studies of object-based attention in humans have isolated object-based attention from spatial attention by using superimposed stimuli that occupied the same spatial location and could not, therefore, be selected by a purely spatial attention mechanism. Adapting these paradigms for use with monkeys holds the promise of elucidating mechanisms of object-based attention at the level of the neuron and the cortical circuit. One such paradigm, introduced by Valdes-Sosa and colleagues, shows that when one of two superimposed surfaces is selected by attention, this suppresses processing of the other surface for several hundred milliseconds. This impairment can be interpreted within the context of the biased-competition theory, and this leads to testable predictions at the behavioral and neuronal levels. We describe the results of psychophysical tests that have proven consistent with this interpretation, and outline predictions for changes in neuronal responses that should hold, if the biased-competition model is to be considered an adequate model of object-based attention.

18.2 The biased-competition model

Our inability to simultaneously process multiple visual stimuli is thought to reflect the limited capacity of some stage (or stages) of sensory processing, decision-making, or behavioral control. As a result of these computational bottlenecks, it is necessary to have neural mechanisms in place to ensure the selection of stimuli that are immediately relevant to behavior. According to the biased-competition account (Desimone and Duncan 1995; Reynolds *et al.* 1999; see also Bundesen 1990), the selective processing of behaviorally relevant stimuli is accomplished by two interacting mechanisms: competition among potentially relevant stimuli, and biases that determine the outcome of this competition. These biases include both stimulus-driven factors, such as differences in the relative salience of competing stimuli, and goal-directed factors that depend on the task at hand. For example, when an observer is asked to detect the appearance of a target at a particular location, this is thought to activate spatially selective feedback signals in cortical areas, such as parts of parietal cortex, that provide a task-appropriate spatial reference frame. These signals feed into extrastriate cortex, where they bias competition in favor of stimuli appearing at the cued location. Similarly, when an observer searches for an object in a cluttered scene, feature-selective feedback is thought to bias competition in favor of stimuli that share features in common with the searched-for object.

When competition is resolved, the responses of many neurons are determined primarily by the winning stimulus, enabling signals that provide information about the winning stimulus to propagate more strongly through the visual system. Competition is assumed to take place between objects that have already been integrated into wholes prior to entering into competition (Duncan 1998). Therefore, the model predicts that when attention is directed to one feature of an object, this will put the whole object at a competitive advantage. As a result, this object will tend to win the competition, and all of its features, including those features that are not immediately relevant to behavior, will be selected. For example, when attention is directed to the color of a stimulus, its orientation will also be selected 'for free'. The orientation of the object can therefore be discriminated without requiring an additional selection. In contrast, unselected objects are temporarily suppressed, explaining observers' difficulty in making simultaneous discriminations of two different objects.

Extracellular recording studies of attention in monkeys performing attention-demanding visual tasks have identified cortical circuits in the dorsal and ventral visual processing streams that have precisely the properties predicted by the biased-competition model. Moran and Desimone (1985) found that when two stimuli appear together within the RFs of neurons in the later stages of the ventral visual processing stream, the neuronal response to the pair depends on which of the two stimuli is attended. For each cell, they identified two stimuli: one that elicited a large response when presented alone (the cell's *preferred* stimulus) and another that elicited a small response (a *poor* stimulus for the cell). The response to the pair was larger when the monkey attended to the preferred stimulus than when the poor stimulus was attended. Consistent with the idea that this response difference results from the resolution of competition, Moran and Desimone and others have found much smaller changes in response when an attended stimulus appears alone in the RF, and there is no competition to be resolved (Motter 1993; Treue and Maunsell 1996; Luck *et al.* 1997; see, however, Seidemann and Newsome 1999).

Reynolds *et al.* (1999) tested whether these changes in response result from the resolution of competition, using a behavioral paradigm that isolated automatic competitive mechanisms from attentional modulation. As in the study of Moran and Desimone, this study recorded neuronal responses in the ventral stream when stimuli appeared alone or in pairs within the RF, and examined the effect of directing attention to one of the stimuli. Consistent with the earlier results, responses to the stimulus pair were larger when the preferred stimulus was attended than when the poor stimulus was attended. We reasoned that if, as posited by the biased-competition model, these changes in firing rate depend on automatic competitive mechanisms that are biased by goal-directed feedback signals, these competitive interactions should be evident even when attention was directed away from the RF of the neuron.

Consistent with this prediction, with attention directed away from a neuron's RF, the response to the preferred stimulus was typically suppressed by the addition of the poor stimulus. The magnitude of this suppression was determined by the neuron's selectivity for the two stimuli, such that a very poor stimulus was typically more suppressive than a stimulus that elicited an intermediate response (Reynolds *et al.* 1999; see also Recanzone *et al.* 1997). Further, changes in firing rate with attention depended on these competitive interactions. Directing attention to the poor stimulus magnified its suppressive effect, causing the neuron to respond at a rate that was comparable to the response elicited by the poor stimulus alone (Fig. 18.1A). Directing attention to the

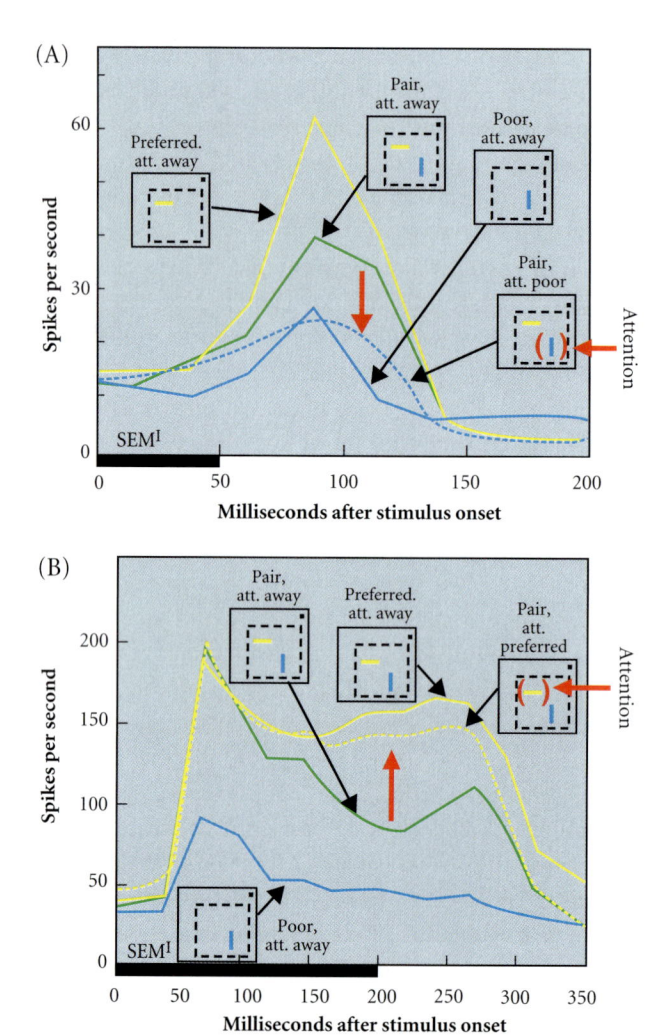

Fig. 18.1 Attention to one stimulus of a pair filters out the effect of the ignored stimulus. (A) The x-axis shows time (in milliseconds) from stimulus onset, and the thick horizontal bar indicates stimulus duration. Small iconic figures illustrate sensory conditions. Within each icon, the dotted line indicates the RF, and the small dot represents the fixation point. The location of attention inside the RF is indicated in red. Attention was directed away from the RF in all but one condition. The horizontal yellow bar indicates the preferred stimulus and the vertical blue bar indicates the poor stimulus. In fact, the identity of both stimuli varied from cell to cell. The yellow line shows the response to the preferred stimulus. The solid blue line shows the response to the poor stimulus. The green line shows the response to the pair with attention away from the RF. The dotted blue line shows the response to the pair, when attention was directed to the poor stimulus. The addition of the poor stimulus suppressed the response to the preferred stimulus. Attention to the poor stimulus (red arrow) magnified its suppressive effect, and drove the response down to a level that is similar to the response elicited by the poor stimulus alone. (B) Response of a second V2 neuron. The format is the same as in panel (A). As in the neuron in (A), the response to the preferred stimulus was

preferred stimulus strongly reduced the suppressive effect of the ignored poor stimulus (see Fig. 18.1B). Recanzone and Wurtz (2000) have found a very similar pattern of results in areas MT and MST in the dorsal stream.

Thus, consistent with the biased-competition model, with two stimuli in the RF, attention to one of them can cause either an *increase* or a *decrease* in response, and this depends on the neuron's selectivity for the two stimuli in the RF. These effects are qualitatively different from the effect of attention to a single stimulus appearing alone in the RF, which typically leads to an increase in firing rate (Bushnell *et al.* 1981; Mountcastle *et al.* 1987; Spitzer *et al.* 1988; Gottlieb *et al.* 1998; McAdams and Maunsell 1999). Such increases are observed regardless of whether the stimulus is a poor or a preferred stimulus for the cell (McAdams and Maunsell 1999; Reynolds *et al.* 2000). In addition, whereas attention effects with multiple stimuli in the RF appear to depend on suppressive sensory interactions, attending to a single RF stimulus typically causes an increase in response, even if there are no stimuli near enough to the RF to interact directly with the RF stimulus. That is, the increase in response when attention is directed to a stimulus appearing alone in the RF does not appear to depend on underlying sensory interactions, and may instead reflect an increase in the strength of the signal elicited by the attended stimulus.

Evidence for a goal-directed bias in favor of behaviorally relevant stimuli has been found in several extrastriate areas. In addition to the above-mentioned studies that have found an increase in firing rate when a single stimulus is attended, Luck *et al.* (1997) found that when attention is directed to a location within a neuron's RF prior to the appearance of a stimulus, this causes a spatially selective increase in the neuron's baseline firing rate. Feature-selective increases in baseline activity have also been reported in neurons that are selective for a stimulus that must be stored in memory in order to perform a search task (Chelazzi *et al.* 1993, 1998) or a match to sample task (Fuster *et al.* 1981; Miyashita and Chang 1988). Consistent with the idea that the elevated baseline activity biases competition between stimuli, competition tends to be resolved in favor of the neurons that showed elevated baseline activity during the delay interval.

Reynolds *et al.* (2000) found evidence that attention also increases the effective strength of a stimulus, which could help put it at a competitive advantage against unattended stimuli. This study measured contrast response functions of V4 neurons when monkeys either attended to a location inside the RF or away to a location far from the RF. As illustrated in Fig. 18.2, attention caused neurons to respond to stimuli that were too faint to elicit a response when unattended. The contrast response functions of individual

---

suppressed by the addition of the poor stimulus. Attention directed to the preferred stimulus filtered out this suppression, returning the neuron to a response similar to the response that was elicited when the preferred stimulus appeared alone inside the RF. (Adapted from Reynolds, J.H., Chelazzi, L., and Desimone, R. (1999) Competitive mechanisms subserve attention in macaque areas V2 and V4. *J Neurosci*, **19**(5), 1736–53. Copyright 1999 by the Society for Neuroscience).

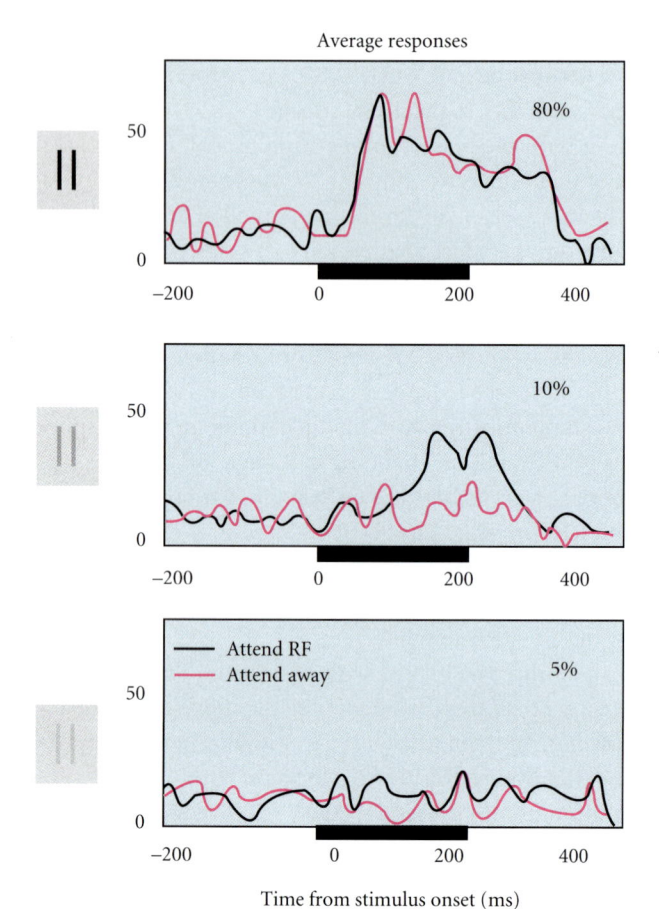

Fig. 18.2 Single neuron example: improved sensitivity with attention. Each of the three panels shows the responses that were elicited by the stimulus when it was presented at different levels of contrast, indicated by the gratings on the left of each panel. Contrast increases from 5% contrast in the bottom panel to 10% in the middle panel, up to 80% contrast in the top panel, illustrated by the grating stimuli on the left. The two lines in each panel show the mean response of a V4 neuron when attention was either directed away from the receptive field (red line) or when attention toward the receptive field (black line). The stimulus elicited a robust response when it was presented at 80% contrast, whether or not it was attended. The faintest stimulus (5% contrast) was too faint to be detected by the neuron, even when the monkey was attending to its location. However, attention enabled the cell to detect a 10% contrast stimulus, which it did not detect when attention was directed elsewhere. That is, attention shifted the neuron's response threshold, making it more sensitive to stimuli appearing at the attended location. (Adapted from *Neuron*, **26**(3), Reynolds, J.H., Pasternak, T., and Desimone, R., Attention increases sensitivity of V4 neurons, pp. 703–14, 2000. Reprinted with permission from Elsevier Science.)

neurons shifted to the left with attention. That is, when attention was directed to a stimulus, V4 neurons respond much as though the physical contrast of the stimulus had increased. This leftward shift in the contrast response function is illustrated in Fig. 18.3, which shows the average firing rates of V4 neurons to attended (solid black curve) and unattended (solid gray curve) stimuli, as a function of contrast. In a pilot

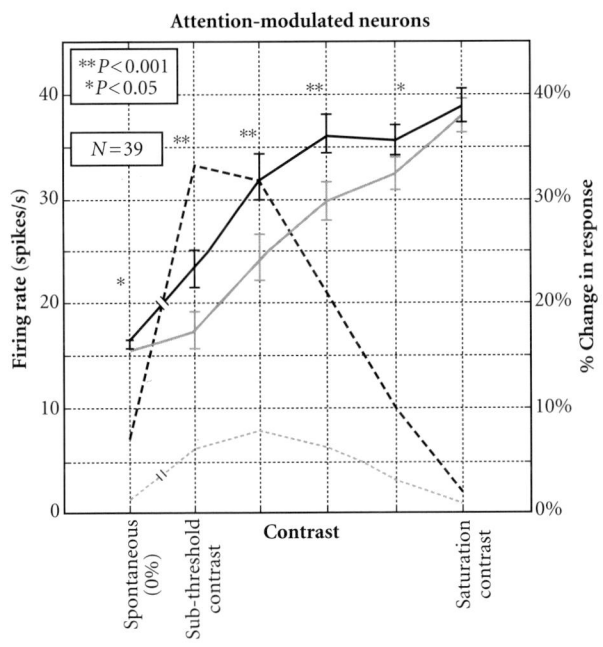

Fig. 18.3 Attention to a single stimulus increases its effective contrast. Average neuronal responses to attended and ignored stimuli for neurons that were significantly ($P < 0.01$) modulated by attention, according to a two-way ANOVA of firing rate, with the five contrast levels and attentional state (attend away, attend RF) as factors. Thirty-nine out of 84 (46.4%) neurons showed either a significant effect of attention or an interaction between attention and contrast. Contrasts increase from 0% (spontaneous firing rate, computed during the 250 ms prior to stimulus onset), on the left, to saturation contrast, on the right. Solid gray and black lines show mean firing rates elicited by ignored and attended stimuli, respectively, during the first 400 ms after stimulus onset. Firing rates are indicated on the left axis. The dashed line shows the percentage change in absolute response with attention (i.e. without subtracting away baseline response), with values indicated on the right axis. The dotted line shows the arithmetic difference in firing rate. Statistically significant differences are indicated by asterisks ($P < 0.05$) and double asterisks ($P < 0.001$). Error bars indicate ± two times the standard error of the mean difference in response to attended and ignored stimuli. Attention caused larger and more significant increases in response for intermediate contrast stimuli. There was a small increase in response with attention at the highest contrast tested, but this was not statistically significant. (Adapted from *Neuron*, **26**(3), Reynolds, J.H., Pasternak, T., and Desimone, R., Attention increases sensitivity of V4 neurons, pp. 703–14, 2000. Reprinted with permission from Elsevier Science.)

experiment, Nicholas *et al.* (1996) found a similar increase in sensitivity to texture-defined stimuli. Texture-defined shapes are not salient if they are composed of line elements that are similar in orientation to background elements. Attention substantially increased responses to such stimuli, but not to more salient texture-defined stimuli. These stimuli are an example of parts (texture elements) that cohere into a common whole. Consistent with the idea that the neurons processed these stimuli as wholes, neuronal responses typically were sensitive to the global shape, not the orientation of the local line elements.

Behavioral studies in monkeys (Schiller 1993; De Weerd *et al.* 1999) have revealed that when area V4 is lesioned, this strongly impairs the animal's ability to discriminate the orientation of a stimulus, but only if it is presented with highly salient distractors. In a related study, Braun (1994) found that when attention is partially withdrawn from a discriminandum, this selectively impairs judgments of the discriminandum if it is presented with highly salient distractors. Together, these results suggest that area V4 plays an important role in selecting low-salience targets from among high-salience distractors. Consistent with this interpretation, Reynolds and Desimone (2003), recording from individual neurons in area V4, found results that are consistent with the idea that the increase in the effective strength of attended stimuli biases competition. When attention is directed away from the RF, competitive interactions are biased in favor of the stimulus with the greatest luminance contrast. They also found that attention to the lower-contrast stimulus enabled it to overcome this competitive disadvantage and control the neuronal response. This is consistent with the idea that stimulus-driven biases (e.g. differences in relative contrast) and goal-directed biases (e.g. spatially selective feedback) bias the same competitive circuit. Thus, attention to a relevant stimulus can cause it to gain control over neuronal responses, even if it appears among more salient distractors.

Taken together, these studies provide compelling evidence that attention increases the effective strength of the attended stimulus, and that attentional selection then occurs through the resolution of competition in favor of the attended stimulus. However, they have not brought into focus what type of representation engages in competition. In these studies, competing stimuli appeared at separate locations, and this confounds selection of locations with selection of objects. The observed competitive interactions could therefore reflect either competitive interactions among locations or competition between coherent objects. However, psychophysical paradigms developed to distinguish between object-based and spatial attention have set the stage for neurophysiologists to identify the site of such integrated object representations.

18.3 **Psychophysical, neuropsychological, and brain-imaging studies of object-based attention**

Experiments carried out in a number of laboratories using different attentional paradigms provide converging support for *object-based* models of selective attention.

Early psychophysical studies showed that observers can attend to one stimulus while suppressing another, even when the two stimuli are superimposed and cannot therefore be selected by spatial attention alone (Neisser and Becklen 1975; Rock and Gutman 1981). Subsequent studies have found that when an observer makes a judgment about one feature of an object (e.g. its color), simultaneous judgments about other features of the same object (e.g. its orientation and motion) do not interfere with the first judgment. This suggests that when attention is directed to one feature of an object, all of the features that make up the object are automatically selected together (Duncan 1984; Valdes-Sosa *et al.* 1998, 2000; Blaser *et al.* 2000). Consistent with this interpretation, O'Craven *et al.* (1999) have found that discriminating one feature of an object results in increased cerebral blood flow in cortical areas that respond to the irrelevant features of the attended object, but not in areas that respond to features of an unattended overlapping object.

Whereas multiple features of a single stimulus are selected automatically, it is difficult to simultaneously monitor features of different stimuli. For example, Blaser *et al.* (2000) found that observers could easily report the color and orientation of one stimulus. They were much less accurate in judging the color of one stimulus and the orientation of another, overlapping stimulus. On trials in which subjects made correct judgments about one stimulus they were severely impaired in their judgments about the other stimulus. Similar results have been reported using overlapping line drawings (Duncan 1984) and overlapping surfaces (Valdes-Sosa *et al.* 1998, 2000). Vecera and Farah (1994) have found that this inability to simultaneously make accurate judgments about two different objects occurs regardless of whether the stimuli are overlapping or at separate locations. This suggests that attention can select spatially invariant object representations. Consistent with the notion that objects can be the target of attention, studies of inhibition of return (IOR), the impaired processing of a recently attended stimulus, have found IOR for objects even when they move from one location to another (Tipper *et al.* 1991).

Related results show that attention selects out objects, even when their parts are distributed in space. Baylis and Driver (1993) presented ambiguous displays that could be perceived either as containing one or two objects, and found that judging the relative location of two contours was more difficult when the contours were perceived as belonging to two objects rather than one. Egly *et al.* (1994) found that attention to one part of an object improves an observer's ability to detect a change in another part of the same object, relative to when the change appeared in an equidistant part of an unattended object. He and Nakayama (1995) have also provided evidence that attention to one part of an object improves processing of other parts of the same object. They found that attention could be directed efficiently to elements that were arranged to form a single surface, relative to when the elements did not cohere into a common surface.

In addition to these studies that show that attention can be allocated to whole objects or surfaces, a number of studies have shown that during visual search, attention can be directed to pre-attentively defined objects appearing among distractors. For example,

Wolfe (1994) found that whereas search for targets defined by conjunctions of two colors is typically inefficient (see Wolfe *et al.* 1990), attention can easily select a target defined by the same two colors if the target is pre-attentively grouped as an object composed of two differently colored parts. In a related study, Rensink and Enns (1998) provided evidence that attention selects pre-attentively grouped objects by showing that efficient search for an irregularly shaped object among squares can be made inefficient if the irregular shape of the target is perceived as resulting from occlusion.

Studies of attentional deficits among patients with right parietal lesions provide further evidence that objects can receive substantial processing before being selected by attention. In extinction, which is most often observed after damage to the right parietal lobe, patients can detect stimuli on either side of the visual field, but are often unaware of a contralesional stimulus if it is presented concurrently with a stimulus presented in the intact hemifield. Mattingley *et al.* (1997) found that extinction is reduced when right and left visual field inputs are grouped into a single surface. This finding provides evidence that extinction arises after visual inputs are integrated into objects or surfaces.

18.4 Competitive interactions among surfaces

The above-mentioned experiments show that attention can select objects for processing, even when two objects appear at the same location in space, and that selection of one object impairs processing of the other object. Valdes-Sosa and colleagues (Valdes-Sosa *et al.* 2000) have identified one of the clearest examples of such selection, using an ingenious paradigm designed to examine surface-based attention, isolated from the influence of feature-based or spatial attention. Their paradigm is illustrated in Fig. 18.4. On each trial, observers viewed two random dot patterns (one red, one green) that rotated rigidly around a common center in opposite directions, giving rise to the percept of two overlapping transparent surfaces. The fixation point color (red or green) acted as an endogenous cue that directed subjects to attend to the surface of the corresponding color. After a brief delay, the cued surface translated briefly in one of eight directions while the uncued surface continued to rotate. After this translation, both surfaces rotated until one of the two surfaces underwent a second brief translation. On each trial, observers reported the directions of the two shifts. The endogenous cue indicated which surface would shift first so observers could ignore one surface in order to reliably report the first shift.

Figure 18.5, adapted from a study that replicated the results of Valdes-Sosa and colleagues, shows average performance in this task. Observers were able to report the first translation accurately, and could also report the second translation of the *cued* surface accurately even when two successive translations occurred with an inter-stimulus interval (ISI) as short as 150 ms. If the *uncued* surface translated second, however, judgments were severely impaired, and this impairment lasted approximately 600 ms.

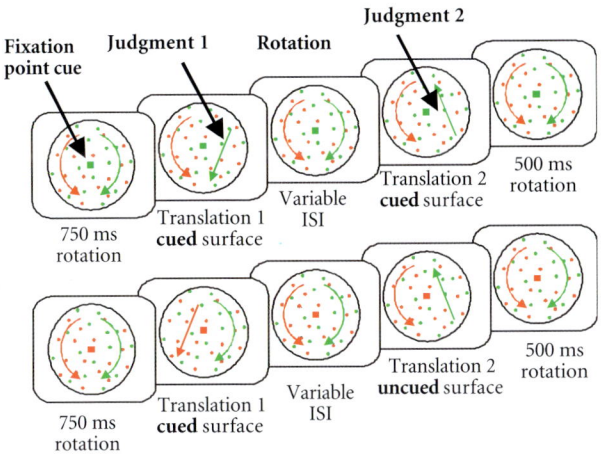

Fig. 18.4 Object-based attention task introduced by Valdes-Sosa and colleagues. Panels are arranged from left to right according to the sequence of events in each trial. On half of all cued trials, the fixation point was green (upper panels), indicating that the green surface would be the first to translate. On the remaining cued trials, the fixation point was red (lower panels), indicating that the red surface would translate first. The observer began each trial with a key press, resulting in a period of 750 ms in which the two surfaces rotated around the fixation point, in opposite directions. The cued surface then translated for 150 ms in one of the eight cardinal directions, while the other surface continued to rotate. Following this first translation, the two surfaces continued to rotate for a variable delay of 150–1050 ms, at which point one of the two surfaces, with equal probability, shifted for 150 ms. After this second shift, both surfaces continued to rotate for an additional 500 ms. Observers had to maintain fixation throughout the trial, and report the direction of each shift. (Adapted from *Vision Research*, **43**, Reynolds, J.H., Alborzian, S., and Stoner, G.R., Exogenously cued attention triggers competitive selection of surfaces, pp. 59–66, 2003. Reprinted with permission from Elsevier Science.)

These results can be interpreted as arising from the limited capacity of the visual system to process information about multiple objects, coupled with the slow dwell time of attention (Valdes-Sosa *et al.* 2000). According to this interpretation, the endogenous cue, which was 100% valid, enabled observers to attend to one of the two surfaces and their performance was therefore high. However, either surface could undergo the second translation with equal probability, and therefore, observers had to divide attention between the two surfaces. The extra cost of attending to two objects caused their performance on the second judgment to be poorer, on average, than their performance on the first judgment. The observation that this reduction in performance occurred primarily when judgments were of the surface that was not endogenously cued can be explained as resulting from the initial allocation of attention to the endogenously cued surface. Previous studies show that attention remains locked on a cued stimulus for a similar period of several hundred milliseconds, during which time judgments of other stimuli are impaired (Duncan *et al.* 1994).

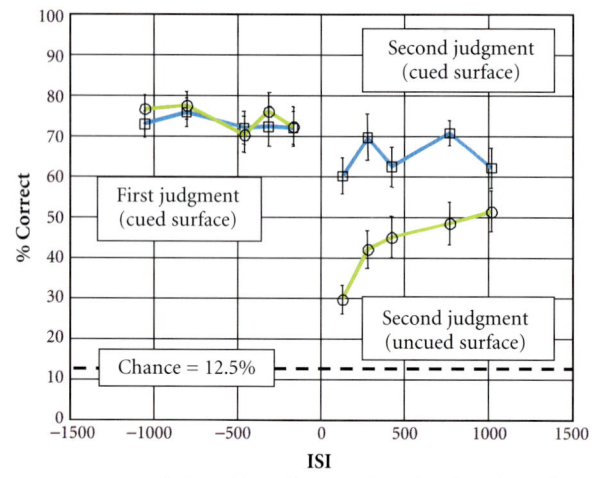

Fig. 18.5 Mean accuracy across eight subjects in reporting the direction of two successive translations, averaged across trials in which the fixation point color indicated which surface would translate first. Chance performance, indicated by the dashed horizontal line, was 12.5%. ISI (inter-stimulus interval) indicates the duration of the interval between the offset of the first translation and the onset of the second translation. By convention, negative ISIs correspond to judgments of the first translation, and positive ISIs correspond to judgments of the second translation. Line color indicates whether the first and second translations occurred on the same surface (blue) or different surfaces (green). Error bars indicate standard errors of mean (SEM) performance across subjects. Observers correctly reported the direction of the first translation (which was always on the surface cued by the fixation point), regardless of whether the second translation also occurred on the cued surface (blue line) or occurred on the other surface, and regardless of how soon after the first translation the second translation occurred. Subjects reported the second translation accurately if it occurred on the cued surface. However, subjects were severely impaired in making judgments about the second translation when it occurred on the uncued surface. This impairment was greatest at the shortest ISIs tested (150 ms) and gradually diminished over time. (Adapted from *Vision Research*, **43**, Reynolds, J.H., Alborzian, S., and Stoner, G.R., Exogenously cued attention triggers competitive selection of surfaces, pp. 59–66, 2003. Reprinted with permission from Elsevier Science.)

Reynolds *et al.* (2003) reasoned that if the behavioral impairment resulted from an inability to simultaneously attend to the two surfaces, removal of the endogenous cue, requiring attention to be allocated to both surfaces during the first translation, should cause a severe impairment of the first translation judgment. They tested this by repeating the original experiment while intermixing trials in which the endogenous cue (colored fixation point) was replaced with a non-informative gray fixation point. On these trials, the two translations were equated in that observers had no way to predict which surface would undergo either translation. As illustrated in Fig. 18.6, the removal of the endogenous cue had a very limited effect on observers' ability to judge the direction of the first translation, and the surface-specific impairment persisted in judging the

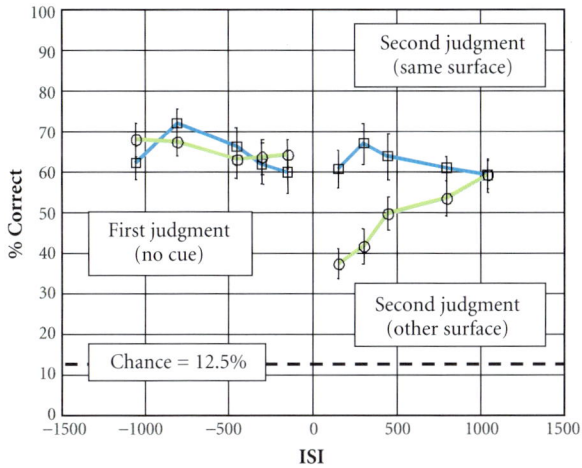

Fig. 18.6 Mean accuracy across eight subjects in reporting the direction of two successive translations, for trials in which the fixation point was gray. Conventions are identical to those used in Fig. 18.5. Despite the absence of the endogenous cue, observers were able to report the direction of the first translation on 65.1% of trials, and their performance did not depend on whether the second translation also occurred on the cued surface (blue line) or occurred on the other surface (green line), and did not depend on ISI. Subjects reported the second translation accurately if the same surface translated twice. However, if one surface translated, this severely impaired the observers' ability to report a shift of the other surface. As was the case in the cued condition, this impairment was greatest at the shortest ISIs tested (150 ms) and gradually diminished over time. (Adapted from *Vision Research*, **43**, Reynolds, J.H., Alborzian, S., and Stoner, G.R., Exogenously cued attention triggers competitive selection of surfaces, pp. 59–66, 2003. Reprinted with permission from Elsevier Science.)

second translation. Notably, when two translations occurred on the same surface, this did not lead to an *increase* in performance judging a second translation of the same surface, as might be expected if the first translation acted as a simple exogenous cue. Rather, the first translation led to a substantial impairment in judging translations of the other surface. Given their accurate performance in judging the first translation of either surface, why did observers not simply treat the second translation as a novel event, and judge it accurately, as well?

One plausible explanation is that the impairment is the result of automatic competitive selection circuits in extrastriate cortex. As noted earlier, when two stimuli appear together within a neuron's receptive field, they activate populations of neurons that automatically compete with one another. If the preferred stimulus for a neuron is presented with a second stimulus in the RF that does not activate the neuron, the added stimulus suppresses the neuronal response automatically, even when attention is directed to a distant location by an attention-demanding task. These competitive interactions can be biased by internally generated feedback signals that favor an endogenously cued stimulus.

They can also be biased by stimulus-driven biases such as differences in the relative salience of competing stimuli. For example, as noted by Reynolds and Desimone (2003), an increase in the luminance contrast of one stimulus can cause it to suppress responses elicited by lower contrast stimuli nearby, even when attention has been endogenously cued to a location far from the RF.

The first translation causes a momentary increase in the salience of the translating surface. Thus, a plausible explanation for the impairment when first one, then the other surface is translated is that this increase in salience may have caused competition to be resolved in favor of the surface that underwent the first translation. If so, the resulting suppression of neurons that encode the non-translating surface would be expected to have impaired processing of information about the uncued surface.

Although this is a plausible model, it is well known that discrimination of one stimulus momentarily impairs discrimination of subsequently presented stimuli, a phenomenon that is known as the attentional blink (Shapiro et al. 1994). This impairment might therefore depend on the subject making a judgment about the first translation. Reynolds and colleagues tested this possibility by removing the first translation, and instead delaying the onset of one of the two surfaces. A suddenly appearing stimulus is often processed preferentially (Yantis and Jonides 1984, 1990), and would be expected to cause a strong onset transient that would be expected to bias competitive interactions in favor of neurons selective for the newly appearing stimulus. Consistent with this, observers were markedly better at discriminating brief translations on the new surface, as compared to the old.

While these psychophysical results are compatible with the predictions of the biased-competition model, the duration of the required suppression is remarkably long, on the order 600–1000 ms. The most direct test of the model is to record the responses of neurons activated by the two surfaces, to determine whether neurons that respond to one of two superimposed stimuli are suppressed for such long durations, under conditions that lead to impaired behavioral performance. In a pilot experiment, Fallah et al. (2002) have recorded the responses of V4 neurons using the same stimuli employed by Reynolds et al. (2003). One set of dots was of the neuron's preferred color and the other was of an isoluminant non-preferred color. The response elicited by the preferred-color surface was consistently suppressed by the addition of the non-preferred surface. Moreover, the suppression caused by the non-preferred surface was greater when it appeared after the preferred surface. Remarkably, the duration of this differential suppression lasted for hundreds of milliseconds, with suppression being maximal immediately following the appearance of the poor stimulus, and falling off over a time course that was similar to the recovery time of the behavioral impairment observed by Valdes-Sosa et al. (2000) and Reynolds et al. (2003). These results show that competitive circuits in area V4 are not limited to mediating competition between spatial locations, and may play a role in object-based attention. The similarity of the time courses of the neuronal and behavioral impairments lends support to this proposal.

18.5 Conclusions

Single-unit recording studies of visual attention have consistently found that when multiple stimuli appear together, they appear to activate mutually suppressive interactions between neurons with contrasting response properties. In each area where these interactions have been observed, it has proven possible to bias neuronal responses in favor of a desired stimulus, providing a mechanism by which the nervous system can select a desired stimulus out of a cluttered visual world. Psychophysical experiments have provided evidence that competitive interactions occur among whole-object representations. However, all single unit studies of competitive circuits in the extrastriate visual cortex have used stimuli that appeared at separate spatial locations, thereby confounding spatial and object-based attention. It is therefore not yet known whether these circuits are the neural correlates of object-based attention, measured at the behavioral level. However, preliminary single-unit recording studies in area V4, where competitive circuits have been found to play a role in selecting spatially separated stimuli, find that competition also occurs among superimposed stimuli. This proves that these circuits are not limited to mediating competition among spatial locations. Moreover, when the onset of one of the surfaces is delayed, this causes competition to be resolved in favor of the newly appeared surface. The resulting suppression is similar in duration to the time course over which human observers, viewing the same stimuli, are impaired making perceptual judgments about the suppressed surface. While additional work is necessary to establish a link between neuronal responses and behavioral performance in the monkey, these results appear to suggest that competitive circuits in extrastriate cortex may mediate competition among whole objects.

Acknowledgements

Funding provided by grants from The McKnight Endowment Fund for Neuroscience, NEI grant 1R01EY13802–01 and PHS/33201A/T32MH2002.

References

Baylis, G.C. and Driver, J. (1993). Visual attention and objects: evidence for hierarchical coding of location. *J Exp Psychol Hum Percept Perform* **19**(3), 451–70.

Blaser, E., Pylyshyn, Z.W., and Holcombe, A.O. (2000). Tracking an object through feature space. *Nature* **408**(6809), 196–9.

Braun, J. (1994). Visual search among items of different salience: removal of visual attention mimics a lesion in extrastriate area V4. *J Neurosci* **14**(2), 554–67.

Bundesen, C. (1990). A theory of visual attention. *Psychol Rev* **97**(4), 523–47.

Bushnell, M.C., Goldberg, M.E., and Robinson, D.L. (1981). Behavioral enhancement of visual responses in monkey cerebral cortex. I. Modulation in posterior parietal cortex related to selective visual attention. *J Neurophysiol* **46**(4), 755–72.

Chelazzi, L., Miller, E.K., Duncan, J., and Desimone, R. (1993). A neural basis for visual search in inferior temporal cortex. *Nature* **363**(6427), 345–7.

Chelazzi, L., Duncan, J., Miller, E.K., and Desimone, R. (1998). Responses of neurons in inferior temporal cortex during memory-guided visual search. *J Neurophysiol* **80**(6), 2918–40.

Desimone, R. and Duncan, J. (1995). Neural mechanisms of selective visual attention. *Annu Rev Neurosci* **18**, 193–222.

De Weerd, P., Peralta, M.R. 3rd, Desimone, R., and Ungerleider, L.G. (1999). Loss of attentional stimulus selection after extrastriate cortical lesions in macaques. *Nature Neurosci* **2**(8), 753–8.

Duncan, J. (1984). Selective attention and the organization of visual information. *J Exp Psychol Gen* **113**(4), 501–17.

Duncan, J. (1998). Converging levels of analysis in the cognitive neuroscience of visual attention. *Phil Trans R Soc Lond B. Biol Sci* **353**(1373), 1307–17.

Duncan, J., Ward, R., and Shapiro, K. (1994). Direct measurement of attentional dwell time in human vision. *Nature* **369**(6478), 313–15.

Egly, R., Driver, J., and Rafal, R.D. (1994). Shifting visual attention between objects and locations: evidence from normal and parietal lesion subjects. *J Exp Psychol Gen* **123**(2), 161–77.

Fallah, M., Stoner, G.R., and Reynolds, J.H. (2002). Competitive selection of superimposed stimuli in V4. *Society for Neuroscience Abstracts* **28**, 418.5.

Fuster, J.M. and Jervey, J.P. (1981). Inferotemporal neurons distinguish and retain behaviorally relevant features of visual stimuli. *Science* **212**(4497), 952–5.

Gottlieb, J., Kusunoki, M., and Goldberg, M.E. (1998). The representation of visual salience in monkey parietal cortex. *Nature* **391**(6666), 481–4.

He, Z.J. and Nakayama, K. (1995). Visual attention to surfaces in three-dimensional space. *Proc Natl Acad Sci USA* **92**(24), 11155–9.

Luck, S.J., Chelazzi, L., Hillyard, S.A., and Desimone, R. (1997). Neural mechanisms of spatial selective attention in areas V1, V2, and V4 of macaque visual cortex. *J Neurophysiol* **77**(1), 24–42.

Mattingley, J.B., Davis, G., and Driver, J. (1997). Preattentive filling-in of visual surfaces in parietal extinction. *Science* **275**(5300), 671–4.

McAdams, C.J. and Maunsell, J.H.R. (1999). Effects of attention on orientation-tuning functions of single neurons in macaque cortical area V4. *J Neurosci* **19**(1), 431–41.

Moran, J. and Desimone, R. (1985). Selective attention gates visual processing in the extrastriate cortex. *Science* **229**(4715), 782–4.

Motter, B.C. (1993). Focal attention produces spatially selective processing in visual cortical areas V1, V2, and V4 in the presence of competing stimuli. *J Neurophysiol* **70**(3), 909–19.

Mountcastle, V.B., Motter, B.C., Steinmetz, M.A., and Sestokas, A.K. (1987). Common and differential effects of attentive fixation on the excitability of parietal and prestriate (V4) cortical visual neurons in the macaque monkey. *J Neurosci* **7**(7), 2239–55.

Miyashita, Y. and Chang, H.S. (1988). Neuronal correlate of pictorial short-term memory in the primate temporal cortex. *Nature* **331**(6151), 68–70.

Neisser, U. and Becklen, R. (1975). Selective looking: Attending to visually significant events. *Cognitive Psychology* **7**, 480–94.

Nicholas, et al. (1996). *Society for Neuroscience Abstracts* **22**, 634.2.

O'Craven, K.M., Downing, P.E., and Kanwisher, N. (1999). fMRI evidence for objects as the units of attentional selection. *Nature* **401**(6753), 584–7.

Recanzone, G.H. and Wurtz, R.H. (2000). Effects of attention on MT and MST neuronal activity during pursuit initiation. *J Neurophysiol* **83**(2), 777–90.

Recanzone, G.H., Wurtz, R.H., and Schwarz, U. (1997). Responses of MT and MST neurons to one and two moving objects in the receptive field. *J Neurophysiol* **78**(6), 2904–15.

Rensink, R.A. and Enns, J.T. (1998). Early completion of occluded objects. *Vision Res* **38**(15–16), 2489–505.

Reynolds, J.H. and Desimone, R. (2003). Interacting roles of attention and visual salience in V4. *Neuron* **37**(5), 853–63.

Reynolds, J.H., Chelazzi, L., and Desimone, R. (1999). Competitive mechanisms subserve attention in macaque areas V2 and V4. *J Neurosci* **19**(5), 1736–53.

Reynolds, J.H., Pasternak, T., and Desimone, R. (2000). Attention increases sensitivity of V4 neurons. *Neuron* **26**(3), 703–14.

Reynolds, J.H., Alborzian, S., and Stoner, G.R. (2003). Exogenously cued attention triggers competitive selection of surfaces. *Vision Research*, **43**(1), 59–66.

Rock, I. and Gutman, D. (1981). The effect of inattention on form perception. *J Exp Psychol Hum Percept Perform* **7**(2), 275–85.

Schiller, P.H. (1993). The effects of V4 and middle temporal (MT) area lesions on visual performance in the rhesus monkey. *Vis Neurosci* **10**(4), 717–46.

Seidemann, E. and Newsome, W.T. (1999). Effect of spatial attention on the responses of area MT neurons. *J Neurophysiol* **81**(4), 1783–94.

Shapiro, K.L., Raymond, J.E., and Arnell, K.M. (1994). Attention to visual pattern information produces the attentional blink in RSVP. *J Exp Psychol: Hum Percept Perform* **20**, 357–71.

Spitzer, H., Desimone, R., and Moran, J. (1988). Increased attention enhances both behavioral and neuronal performance. *Science* **240**(4850), 338–40.

Tipper, S.P., Driver, J., and Weaver, B. (1991). Object-centered inhibition of return of visual attention. *Q J Exp Psychol A* **43**(2), 289–98.

Treue, S. and Maunsell, J.H. (1996). Attentional modulation of visual motion processing in cortical areas MT and MST. *Nature* **1382**(6591), 539–41.

Valdes-Sosa, M., Cobo, A., and Pinilla, T. (1998). Transparent motion and object-based attention. *Cognition* **66**(2), B13–B23.

Valdes-Sosa, M., Cobo, A., and Pinilla, T. (2000). Attention to object files defined by transparent motion. *J Exp Psychol Hum Percept Perform* **26**(2), 488–505.

Vecera, S.P. and Farah, M.J. (1994). Does visual attention select objects or locations? *J Exp Psychol: General* **123**, 146–60.

Wolfe, J.M. (1994). Guided Search 2.0: A revised model of visual search. *Psychon Bull Rev* **1**(2), 202–38.

Wolfe, J.M., Yu, K.P., Stewart, M.I., Shorter, A.D., Friedman-Hill, S.R., and Cave, K.R. (1990). Limitations on the parallel guidance of visual search: color × color and orientation × orientation conjunctions. *J Exp Psychol Hum Percept Perform* **16**(4), 879–92.

Yantis, S. and Jonides, J. (1984). Abrupt visual onsets and selective attention: evidence from visual search. *J Exp Psychol Hum Percept Perform* **10**(5), 601–21.

Yantis, S. and Jonides, J. (1990). Abrupt visual onsets and selective attention: voluntary versus automatic allocation. *J Exp Psychol Hum Percept Perform* **16**(1), 121–34.

Chapter 19

The imaging of visual attention

Steven A. Hillyard, Francesco Di Russo, and Antigona Martinez

Abstract

Imaging studies reported in this book have made substantial advances in delineating the cortical networks responsible for allocating attention to locations and objects in the visual fields. Progress has also been made in specifying the multiple levels of the visual pathways at which incoming sensory information is selected or rejected. To reveal the dynamic mechanisms of attentional control and selection, it is necessary to employ imaging techniques having high temporal resolution (such as event-related potentials and magnetoencephalography) as well as techniques that offer anatomical precision (such as functional magnetic resonance imaging).

19.1 Introduction

Several contributors to this book have pointed out the fundamental distinction between the neural systems responsible for the control of selective attention and the sites within the sensory pathways where incoming information is actually modulated (Chapters 14, 15, and 17, this volume). Evidence from both neuroimaging and neuropsychological studies indicates that the control system for visual attention consists of a network of interconnected cortical regions that includes dorso-lateral frontal and posterior parietal areas. It is hypothesized that anatomical projections run from this control system to appropriate levels of the visual pathway and exert facilitatory and/or inhibitory influences on the neurons that encode incoming information. One version of this general scheme is illustrated in Fig. 19.1, where the red arrows represent attentional control signals and the blue arrows the flow of sensory information.

19.2 Levels of selection in spatial attention

A key research goal within this framework has been to identify the level(s) of the visual pathways where sensory inputs are modulated by the attentional control

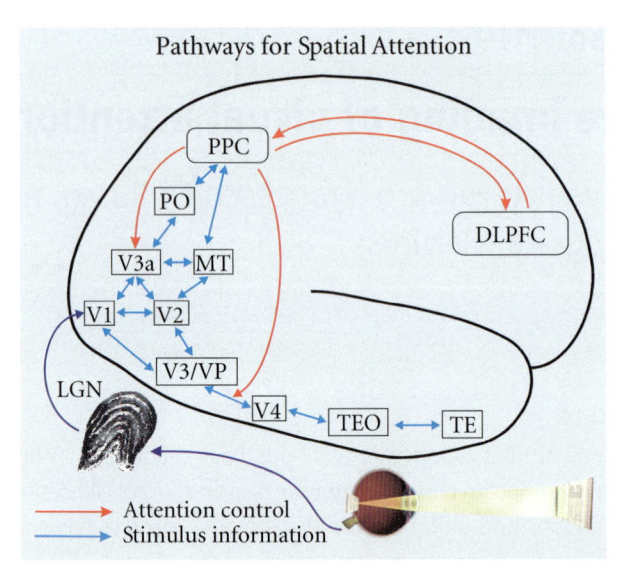

Fig. 19.1 Proposed pathways mediating spatial attention based on event-related potential (ERP), magnetoencephalography (MEG), and functional magnetic resonance imaging (fMRI) evidence. An attentional control network, consisting of interconnected dorso-lateral prefrontal (DLPFC) and posterior parietal (PPC) cortical areas, is proposed to modulate incoming visual information in both dorsal and ventral extrastriate areas. Evidence summarized by Martinez *et al.* (2001) suggests that attended inputs are enhanced dorsally in the region of areas V3/V3A/middle occipital gyrus and ventrally in the vicinity of area V4/fusiform gyrus in the time range 80–130 ms after stimulus onset. These enhanced signals are then fed forward to higher visual areas and back to lower areas, including V1.

network (Chapter 14, this volume). Neurophysiological studies in monkeys have documented strong attentional influences over visual processing in multiple extrastriate cortical areas (reviewed in Maunsell and McAdams 2000; Chapter 18, this volume), and recent experiments have found that neural activity in primary visual cortex (area V1) may also be modulated by attention under certain conditions (reviewed in Martinez *et al.* 2001). Neuroimaging studies have demonstrated that V1 activity may also be affected by spatial attention in humans; directing attention to a stimulus was found to increase neural activity at the retinotopic locus in area V1 corresponding to its location in the visual field (e.g. Tootell *et al.* 1998; Martinez *et al.* 2001). Kastner (Chapter 15, this volume) has presented new fMRI evidence that spatial attention can even influence neural activity at subcortical levels of the visual pathway. An increased BOLD signal was observed in the lateral geniculate nucleus (LGN) when subjects directed attention to a peripheral stimulus in comparison with counting letters at fixation. Kastner suggested that the LGN may serve as an 'early gatekeeper' that exerts a gain control over thalamo-cortical transmission of visual information.

In light of this evidence that spatial attention can influence neural activity in both the LGN and V1, one may well ask why Fig. 19.1 shows attentional control being exerted directly on higher, extrastriate levels of the visual–cortical pathway. The rationale for this comes from studies of event-related potentials (ERPs) and magnetoencephalography (MEG), which suggest that incoming visual information is first modulated at extrastriate levels rather than in V1 or the LGN. This apparent discrepancy between the results of ERP/MEG studies, on the one hand, and fMRI studies, on the other, may be resolved by considering that ERP/MEG recordings can determine the time course of attention-related neural activity with a millisecond-level of resolution, whereas blocked fMRI designs reveal neural activity patterns with high anatomical precision but give little information about when this activity occurs with respect to stimulus events.

19.3 Evidence from ERP/MEG experiments

Several ERP and MEG studies have investigated the effects of spatial attention on different components of the stimulus-evoked waveforms that have been ascribed to striate and extrastriate neural generators (reviewed in Martinez *et al.* 2001; Noesselt *et al.* 2002). In a recent study of this type (Di Russo *et al.* 2003), small, circular stimuli were flashed to right and left visual field positions in random order, either in the upper or lower quadrants (Fig. 19.2). The subject's task was to pay attention to the flashes on one side (ignoring the contralateral flashes) and to detect occasional target flashes of smaller diameter. The earliest ERP component (C1, onsetting at 50–60 ms after stimulus onset) was maximal over the midline parietal-occipital scalp (Fig. 19.2A) and was localized through dipole modeling to midline occipital cortex, which includes area V1 within the calcarine fissure (Fig. 19.2B, C). Significantly, the negative-going C1 was not affected by spatial attention. The earliest attentional modulation was an enhanced positivity (the P1 component onsetting at 70–80 ms) to attended stimuli that was largest over the lateral occipital scalp and was localized to generators in both dorsal and ventral extrastriate cortex (Martinez *et al.* 2001; Noesselt *et al.* 2002; Di Russo *et al.* 2003).

Several lines of evidence indicate that the attention-invariant C1 component does, in fact, represent early evoked neural activity in area V1. These include its being the earliest component in the visual ERP, its co-localization with fMRI activation in area V1 and, most importantly, its polarity inversion when comparing the ERPs to upper versus lower visual field quadrants. (Di Russo *et al.* 2003; Martinez *et al.* 2001). This inversion may be ascribed to the anatomical projections of the upper and lower visual fields to opposing cortical surfaces in area V1 within the calcarine fissure.

It remains to be explained why the C1 component that appears to represent early evoked activity in area V1 shows no modulation with attention, while the co-localized hemodynamic response seen with fMRI does indicate an attention-related increase in

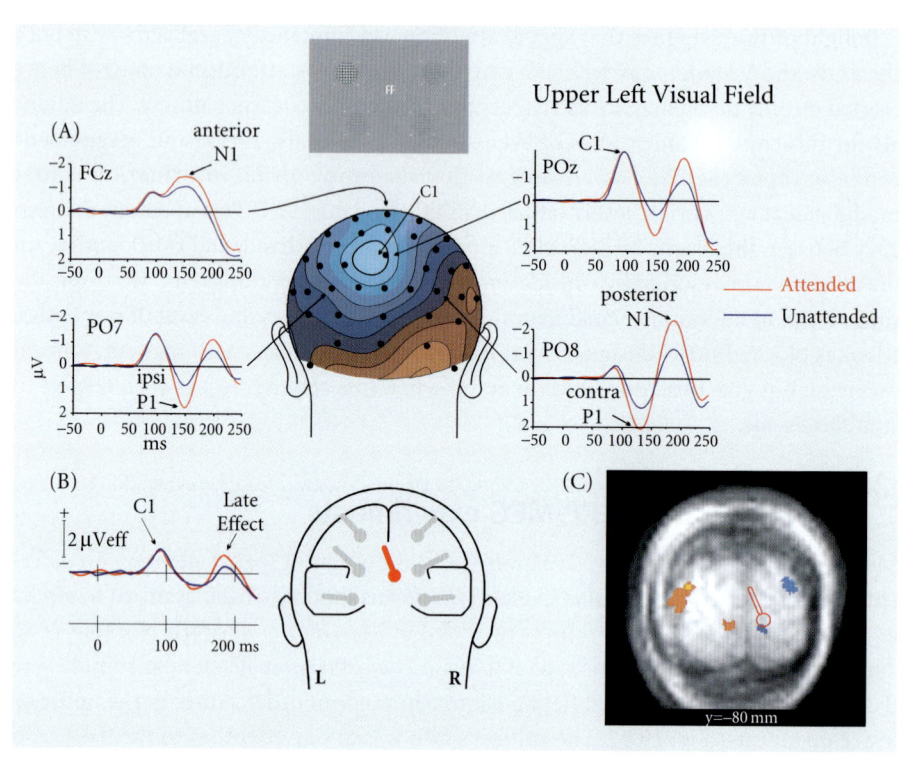

Fig. 19.2 (A) Averaged ERPs from four scalp sites in response to stimuli in the upper left quadrant in a spatial attention experiment by Di Russo et al. (2003). Superimposed ERP waveforms compare conditions when those stimuli were attended versus when stimuli in the opposite field were attended. Head map shows scalp distribution of the early negative C1 component, which is unaffected by attention (bluish shades are negative, reddish are positive). Note that P1 (80–130 ms) and N1 (130–200 ms) components are enhanced by attention. Averaged data from 23 subjects. (B) Dipole model of neural source of the C1 component in midline occipital cortex. Waveforms at the left show the time course of dipole source activity. Note enhanced source activity to attended stimulus at 160–220 ms (late effect). (C) C1 dipole is superimposed on a structural MRI averaged over seven subjects who performed the same task while undergoing fMRI. Blue/red pixels are those showing greater BOLD signals during attention to the left/right visual quadrants. Left hemisphere is on the left. Note spatial correspondence between the C1 dipole position and that of the attention-related BOLD signal.

neural activity in V1 (Martinez et al. 2001; Noesselt et al. 2002) (see Fig. 19.2C). Several possible explanations may be entertained to account for this dissociation:

(1) the attention-related increase in V1 activity is the result of delayed feedback from higher extrastriate areas;

(2) this activity represents an increase in the baseline firing rate of V1 neurons rather than an increase in stimulus-evoked activity (see Chapter 15, this volume);

(3) the V1 activity occurs in oscillatory bursts that are not time locked to the attended stimuli and hence do not appear in averaged ERP waveforms; or

(4) the V1 activity takes place in stellate neurons that do not produce far-field electrical or magnetic signals.

Recent studies in monkeys provide evidence for the first (delayed feedback) alternative. Single-unit recordings from V1 have shown that spatial attention typically enhances neural activity at fairly long latencies (80–100 ms or greater), well beyond the peak of the initial sensory evoked response (Vidyasagar 1999; Lamme et al. 2000). In addition, concurrent ERP recordings from implanted electrodes in multiple visual areas showed that evoked activity in higher extrastriate areas such as V4 was enhanced by attention at shorter latencies than was activity in area V1 (Schroeder et al. 2001).

ERP and MEG recordings in humans are also supportive of a delayed feedback to area V1 following attended stimuli. Dipole modeling of ERP and MEG attention effects have found that a source localized to area V1 shows a late (onset at 130–160 ms) but not an early modulation with attention (Martinez et al. 2001; Noesselt et al. 2002). This effect can be seen in Fig. 19.2B, where the source waveform of the calcarine dipole shows a delayed modulation with attention. Taken together, these animal and human studies suggest that spatial attention first enhances sensory inputs from attended locations in higher extrastriate areas, and these augmented signals are then conveyed back to area V1 by re-entrant feedback projections (see Fig. 19.1). Such feedback signals are proposed to improve figure/ground segregation and the salience of stimuli within the spotlight of spatial attention (Lamme et al. 2000).

As for the reported increase in LGN activity with spatial attention (Chapter 15, this volume), it would seem that further experiments are required to determine whether this activity reflects increased feedforward signals, delayed feedback signals, or a shift in baseline activity. The finding from ERP studies that the initial C1 is not affected by attention appears to weigh against a feedforward gain control at the level of the LGN, but differences in experimental design between the ERP and fMRI studies make it difficult to rule out this possibility altogether.

19.4 **Mechanisms of spatial selection**

Reynolds and Fallah (Chapter 18, this volume) describe how the biased-competition model introduced by Desimone and Duncan (1995) can account for a good deal of the attention effects observed in monkey single-unit studies. In particular, when multiple stimuli are presented in close spatial proximity so as to fall within a neuron's receptive field, attention can bias the competitive interactions between the stimuli so that one input exerts a dominant influence over the cell's firing. Kastner (Chapter 15, this volume) has reported analogous competitive interactions between neighboring stimuli using fMRI, and by varying their spatial separation it was possible to estimate the dimensions of receptive fields in different visual areas of humans.

The biased-competition model seems to provide a good account of attentional selections among neighboring stimuli, but its application is less obvious when attended stimuli and distractors are widely separated and hence minimally competitive. How, then, can we explain the substantial modulations of neural activity that have been observed in ERP, fMRI and single-unit studies under conditions where attended and ignored stimuli were presented in opposite visual fields and separated by 8–10° or more (e.g. Hillyard and Anllo-Vento 1998; Tootell *et al.* 1998; Maunsell and McAdams 2000). Kastner (Chapter 15, this volume) has proposed a mechanism for attentional selection under such conditions, which involves a 'push–pull' facilitation of inputs from the attended location together with a suppression of inputs from outside the attended location. This suppressive effect was postulated to decrease systematically as a function of distance from the locus of attention and to be mediated by long-range horizontal and transcallosal connections.

While Kastner's push–pull hypothesis appears to have considerable explanatory power, further work is needed to clarify its neural underpinnings. In particular, the role of the frontal–parietal control network in allocating attention across the visual field needs to be explicated. It may well be the case that top-down projections from this network are primarily responsible for establishing the zones of excitatory and inhibitory bias over widely separated regions of the visual field. The studies reviewed by Driver *et al.* (Chapter 14, this volume) indicate further that spatial attention has an intrinsically multimodal organization, such that directing attention to a stimulus in one modality facilitates the processing of stimuli in multiple modalities at the attended location. This suggests that the top-down control network imposes concurrent patterns of facilitatory and inhibitory influences within the sensory cortices of multiple modalities and that these patterns are kept in spatial register with one another.

19.5 Object-based selection

As noted by several contributors (Chapters 14, 16, and 18, this volume), it is now evident that attention may be directed towards, and guided by, objects as well as locations in the visual field. For example, it has been found that attention can be divided more efficiently between different aspects of the same object than between separate objects, and that attention directed towards one part of an object may extend automatically to its other parts. In a serious of ingenious experiments, Valdes-Sosa and co-workers (Chapter 16, this volume) have investigated the neural mechanisms underlying object-based selection in humans by means of combined psychophysical/ERP experiments. A key feature of their designs has been to trigger ERPs by transforming the shape of the object being attended, rather than by superimposing a transient stimulus upon it that may not be integrated into the object's perceptual representation. In one such experiment it was found that a shape change was judged more accurately when the shape belonged to the same object in which a preceding shape change had occurred, as opposed to a different object.

Recordings of ERPs in this experiment revealed that the early N1 component (at a latency of around 200 ms) was reduced when successive shape changes occurred within different objects (relative to the same object). The scalp topography of this N1 resembled that of the N1 previously found to be modulated by purely spatial attention, which appears to be generated in extrastriate visual areas (Clark and Hillyard 1996; Martinez *et al.* 2001; DiRusso *et al.* 2003). Accordingly, Valdes-Sosa *et al.* suggested that the different-object cost may be associated with an early suppression of sensory information in extrastriate cortex. An important question for future study is whether this object-based selection is achieved by the same mechanisms that enable purely spatial selection—that is, does spatial attention automatically spread from an attended part of an object throughout the entire object? This could be investigated by recording ERPs to shape-change stimuli when attention is explicitly directed to their locations, and comparing the scalp distributions and neural sources of the ERPs modulated by space-based and object-based attention.

In other experiments Valdes-Sosa and colleagues (Chapter 16, this volume) have shown that attention can be directed selectively to one of two spatially superimposed moving dot patterns having the appearance of overlapped transparent surfaces. Remarkably, although the design precludes selection by location, the ERP triggered by displacement of the attended surface showed enlarged P1/N1 components that resembled the ERP modulations reported to occur during spatial attention. Again, it would be of great interest to compare the properties of these ERP modulations under conditions of spatial versus surface-based selection, to determine whether common mechanisms are involved. Further insight into the cortical mechanisms of surface-based attention comes from preliminary studies in monkeys by Reynolds and Fallah (Chapter 18, this volume), who found that neurons in area V4 can be driven selectively by a transparent surface that captures attention. Taken together, these human and animal studies suggest that attentional selections based on locations, features, and objects may all be carried out in low-level extrastriate visual areas, presumably under the influence of task-specific top-down and feedback influences from higher cortical areas.

19.6 Conclusion

From this brief overview, it is apparent that different imaging methods must be combined in common experiments if we are to achieve a full understanding of where, when, and how sensory information is modulated by the brain's attentional systems. While hemodynamic imaging with fMRI can give a detailed anatomical picture of the brain regions participating in attentional selection, methods with better temporal resolution, such as ERP/MEG, are needed to reveal the timing and sequencing of neural events in the different regions. In particular, we cannot assume that neural activity at an anatomically early stage of the sensory pathways necessarily occurs earlier in time than activity at a higher level in the cortical processing hierarchy. There is mounting evidence that feedback from higher to lower anatomical levels plays an essential role in visual perception and attention and acts in concert with feedforward signals to form a bidirectional cascade of sensory information.

References

Clark, V.P. and Hillyard, S.A. (1996). Spatial selective attention affects early extra striate but not striate components of the visual evoked potential. *Journal of Cognitive Neuroscience*, **8**, 387–402.

Desimone, R. and Duncan, J. (1995). Neural mechanisms of selective visual attention. *Annual Review of Neuroscience*, **18**, 193–222.

Di Russo, F., Martinez, A., and Hillyard, S.A. (2003). Source analysis of event-related cortical activity during visuo-spatial attention. *Cerebral Cortex*, **13**, 486–99.

Hillyard, S.A. and Anllo-Vento, L. (1998). Event-related brain potentials in the study of visual selective attention. *Proceedings of the National Academy of Sciences*, **95**, 781–87.

Lamme, V.A.F., Super, H., Landman, R., Roelfsema, P.T., and Spekreijse, H. (2000). The role of primary visual cortex (V1) in visual awareness. *Vision Research*, **40**, 1507–21.

Martinez, A., Di Russo, F., Anllo-Vento, L., Sereno, M.I., Buxton, R.B., and Hillyard, S.A. (2001). Putting spatial attention on the map: timing and localization of stimulus selection processes in striate and extrastriate visual areas. *Vision Research*, **41**, 1437–57.

Maunsell, J.H.R. and McAdams, C.J. (2000). Effects of attention on neuronal response properties in visual cerebral cortex. In M. Gazzaniga (Ed.), *The new cognitive neurosciences*. MIT Press, Cambridge MA.

Noesselt, T., Hillyard, S.A., Woldorff, M.G., Schoenfeld, A., Hagner, T., Jancke, L., Tempelmann, C., Hinrichs, H., and Heinze, H.J. (2002). Delayed striate cortical activation during spatial attention. *Neuron*, **35**, 575–87.

Schroeder, C.E., Mehta, A.D., and Foxe, J.J. (2001). Determinants and mechanisms of attentional modulation of neural processing. *Frontiers in Bioscience*, **6**, 672–84.

Tootell, R.B.H., Hadjikhani, N., Hall, E.K., Marrett, S., Vanduffel, W., Vaughan, J.T., and Dale, A.M. (1998). The retinotopy of visual spatial attention. *Neuron*, **21**, 1409–22.

Vidyasagar, T.R. (1999). A neuronal model of attentional spotlight: parietal guiding the temporal. *Brain Research Reviews*, **30**, 66–76.

Part 4

Sensorimotor integration

Chapter 20

Neural selection and control of action

Jeffrey D. Schall

Abstract

This chapter is concerned with how perception guides action and with how actions are chosen and controlled to achieve goals. Human performance seems unexplainable without referring to unobservable processes such as encoding, categorizing, deciding, selecting, preparing, and monitoring. Neural correlates of each of these processes have been identified through both non-invasive and invasive monitoring methods. Indeed, single neurons can convey signals apparently sufficient to mediate these cognitive processes. A description of neural properties though, does not explain how the behavior and intermediate processes are produced. A complete explanation requires computational theories implemented in mathematical models. However, redundant models based on different assumptions can mimic each other. The signals observed in neurons can provide definite constraints to help exclude alternative models. This requires obtaining neural data under strict behavioral conditions that invoke the processes of interest.

20.1 Introduction

Cognitive neuroscience has witnessed staggering advances in technology, but have the essential questions advanced in proportion? The two pillars on which our science stands were erected over a century ago. Firstly, Fechner (1860/1912) showed how the mental could be grounded in the physical through precise measurement, and thus was born psychophysics. Secondly, Helmholtz's pivotal measurement of the modest speed of conduction of nerve impulses gave birth to the analysis of reaction time (Posner 1978; Luce 1986; Meyer *et al.* 1988).

Psychophysics and the measurement of reaction time have formed the empirical foundation of experimental psychology. However, observing behavior alone cannot provide a satisfying explanation of how the behavior came about. In experimental psychology, the pursuit of a mechanistic understanding has involved mathematical models of covert processes that seek to replicate performance (Townsend and Ashby 1983; Luce 1986). Current versions of these models have become very sophisticated in mechanism and subtle in performance (Nosofsky and Palmeri 1997; Ratcliff and Rouder 1998; Bundesen 1998; Van Zandt *et al.* 2000; Logan 2002). However, alternative models with different assumptions and architectures cannot be discriminated based on behavioral testing (Townsend and Ashby 1983).

Models can be constrained by discovering the neural processes that realize the covert cognitive processes. Non-invasive methods allow monitoring of brain processes during performance of tasks. However, while providing important information, event-related scalp potentials and functional imaging methods suffer from coarse spatial and temporal resolution which challenges obtaining a mechanistic understanding (Uttal 2001; Friston 2002). Monitoring the states of individual neurons within ensembles seems necessary because the neuron is the essential level at which to understand how the brain produces perception and action (Barlow 1995).

Neural recordings have afforded remarkable insights into how the brain represents the world, selects among alternatives, and produces actions. The pioneering work of Mountcastle (1957) and Hubel and Wiesel (1962) determined what kinds of stimuli activated neurons in anesthetized animals. As neural recordings commenced in awake, behaving monkeys, the approach of designing experiments for neurons continued (Evarts 1966; Wurtz 1969). However, as more sophisticated attributes are studied, designing experiments for neurons can result in unusual behavioral conditions that are not necessarily well grounded in experimental psychology. Consequently, many investigators now train macaque monkeys to perform tasks with a firm empirical and theoretical foundation in psychology, so neural activity is monitored under conditions corresponding precisely to those used in human studies. However, descriptions of the properties of neurons do not explain the functions performed. In the first place, it is not yet clear how neurons signal as they do (Marder 1998). In the second place, a higher level of description is necessary; computational theories provide the basis for explaining function (Marr 1982). Hence, an intellectual synergy is developing; neurophysiological data can constrain redundant models while models embody the interpretation of neural function.

This chapter will review these issues in several domains. To achieve the desired scope in the limited space, citations are selective and figures, absent. Also, the extensive evidence for impairments following lesions or experimental inactivation will not be included. The interested reader should consult the cited reviews and primary literature for more comprehensive treatments.

20.2 How does perception guide action?

20.2.1 How are stimuli discriminated?

One of the cornerstones of neuroscience is the observation that neurons in the primary sensory areas of the cerebral cortex represent the attributes of stimuli through selective responses (Hubel and Wiesel 1962). Mounted on this cornerstone is the compelling observation that the stimulus selectivity of neurons in sensory areas of the cerebral cortex relates reliably to psychophysical reports (Parker and Newsome 1998; Romo and Salinas 2001). The work that will be reviewed is commonly construed in terms of neural correlates of decision making because subjects are presented with alternatives, but can we assume that every response produced in the context of alternatives is derived from a deliberate decision process? Philosophical analysis says no (Oldenquist 1967).

A choice is required when an organism is confronted with alternatives for which some action is required to acquire or avoid one of the alternatives because of a desire, goal, or preference. In its most basic sense, a choice is an overt action performed in the context of alternatives for which explanations in terms of purposes can be given. Choices can be good or bad, defined according to attainment of a goal. Choices are not true or false; the idea of choosing falsely is incoherent. Choices take time; a choice process evolves from a state of more or less equipotentiality after the alternatives are presented to a state of commitment before the overt action is performed. Finally, with prior knowledge of the alternatives and preferences, choices can be predicted with certainty, i.e. it is possible to choose in advance.

A decision process precedes a choice that involves perplexity because the alternatives are difficult to distinguish, have uncertain pay-offs, or require prior knowledge to disambiguate. Thus, in contrast to choices, decisions take more time, require more attention and deliberation, and are more error prone. For this reason, unlike choices, decisions cannot be predicted with certainty, even by the agent. If you can say what you will decide, then you will have already decided. A corollary of this is that decisions, like perceptions, seem just to happen; introspection cannot find the source of the decision (Dennett 1984).

We must distinguish two kinds of decisions—*decide that* versus *decide to*. *Decide that* is a categorization among alternative interpretations which can be true or false. *Decide to* is a choice among alternative actions which cannot be true or false but can be good or bad, like a choice. Can we distinguish between *choose* and *decide to*? Choosing can be synonymous with deciding when we speak of choosing in advance, that is when a choice is determined but the action that expresses that choice is delayed. Typically, though, choosing is more closely related to the commitment of an action and deciding is more closely related to the deliberation and formation of an intention.

20.2.1.1 Tactile discrimination

The earliest experiments combining psychophysical and neurophysiological methods were carried out in the somatosensory system (Werner and Mountcastle 1965; Talbot *et al.* 1968). A comprehensive series of experiments has investigated how tactile stimuli are discriminated (reviewed by Romo and Salinas 2001). In one task, monkeys were trained to discriminate the frequencies of two flutter stimuli presented sequentially on a fingertip, and neural activity was monitored in primary somatosensory cortex (Hernández *et al.* 2000; Salinas *et al.* 2000). The response of the neurons was predictive of the psychophysical judgment. To test the causal efficacy of activation in somatosensory cortex in the perceptual judgment, microstimulation was delivered at sites in somatosensory cortex with receptive fields on the fingertip on which the discriminative stimuli were presented. Microstimulation at various frequencies affected the judgment in a predictable manner (Romo et al. 2000).

To perform the two-interval, forced-choice discrimination task, a representation of the sample frequency must be maintained throughout the delay until the comparison stimulus is presented, the comparison stimulus must be encoded and compared to the sample stimulus, and the response must be produced. Neurons recorded in dorsolateral prefrontal cortex exhibited activation during the delay period that varied monotonically with stimulus frequency, thus forming an enduring representation of the sample stimulus (Romo et al. 1999). In the supplementary motor area (SMA) of monkeys performing this task, different kinds of neurons were found (Hernández et al. 2002). Some neurons were active in relation to the motor response, but others signaled the sample stimulus frequency during the delay until the comparison stimulus was presented. Still other neurons seemed to mediate the comparison process; during presentation of a given comparison frequency these neurons exhibited significantly different activity if the sample frequency was higher (or lower) than the comparison frequency.

In another experiment, monkeys were trained to categorize as low or high the speed of a stimulus moving across one finger (Romo et al. 1996). The stimulus moved at one of ten closely spaced speeds from 12 to 30 mm/s. Monkeys were rewarded for signaling with a keypress whether the speed was low (<20 mm/s) or high (>20 mm/s). A series of experiments traced the neural processes mediating performance of this task. On the sensory side, some neurons in somatosensory cortex signaled only the presence of the stimulus, but others exhibited a linear relationship between discharge rate and stimulus speed, which did not vary with the monkeys' performance or when the stimulus was delivered passively (Romo et al. 1996). On the motor side, the activity of most neurons in primary motor cortex (M1) and SMA was related to the limb movement that was the operant response regardless of the stimulus speed. However, a fraction of neurons in both M1 and SMA exhibited an explicit representation of the low versus high categorization of speed; the discharge rate of these neurons varied according to the category (low, high) and not the overt response (Romo et al. 1997; Salinas and Romo 1998). Thus, the representation of the stimuli on which the decision is based is made explicit in the parts of the brain that produce the overt response. These results highlight the intimate relation between forming a decision and planning a movement, between *deciding that* and *deciding to*.

These results show that individual neurons in premotor areas of the frontal lobe signal every step of the decision process, including stimulus encoding, comparison with a sample stimulus, and preparation of the overt response. Therefore, it is incorrect to imagine that sensory processes in the back of the brain send fully formed explicit representations to the front of the brain that produces the overt response (also see Requin *et al.* 1988). Also, the fact that single cortical areas are populated by neurons that convey qualitatively different signals highlights the challenge of interpreting non-invasive event-related potential and functional imaging data.

20.2.1.2 Visual discrimination

Another series of experiments has described the neural processes responsible for performance of macaque monkeys trained to discriminate the net direction of motion in an aperture of randomly moving dots (Newsome 1997). Monkeys reported the direction of motion of the dots by shifting gaze to one of two targets on either side of the display. When most of the dots move in the same direction, monkeys can produce a high fraction of correct responses with short reaction times. But, when a small fraction of dots are moving coherently, monkeys require more time and make more errors (Roitman and Shadlen 2002). Neural recordings have demonstrated that performance on this task is based on the representation of the motion stimulus in area MT (Britten *et al.* 1992; Britten *et al.* 1996). In fact, trial-by-trial psychophysical reports correlate with random fluctuations of neural activity, even when the motion signal is absent (Britten *et al.* 1996; see also Logothetis and Schall 1989; Dodd *et al.* 2001). The causal efficacy of activity in area MT for the motion discrimination was demonstrated by showing that electrical stimulation of area MT can influence the psychophysical performance (Salzman and Newsome 1994). Quantitative modeling shows that this discrimination can be based on the weakly correlated activity of small populations of neurons, not all responding optimally, pooled with some noise to signal one of the alternative directions of motion (Shadlen *et al.* 1996). The probability of choosing the alternative directions was accounted for quantitatively by the difference in activity in the pools of neurons representing the alternative directions of motion.

These observations demonstrate that the behavioral discrimination is based on the activity of neurons in area MT, but such activity is not sufficient to produce the saccade by which the discrimination is reported because neurons in area MT do not innervate the necessary ocular motor structures. To investigate how the stimulus representation in area MT may be read out, activity has been monitored in parts of the brain that MT innervates, such as the superior colliculus (SC) (Horwitz and Newsome 2001); area LIP in posterior parietal cortex (Shadlen and Newsome 2001); and dorso-lateral prefrontal cortex including the frontal eye fields (FEF) (Kim and Shadlen 1999). The SC and FEF, and to lesser extent area LIP and dorso-lateral prefrontal cortex, are involved in producing saccades (Schall 1997; Wurtz *et al.* 2001). Unlike the experiments on area MT in which the motion stimulus was placed in the receptive field, in these studies the motion stimulus was placed at the focus of gaze, with one of the saccade targets in the response field of the neuron and the other target in the opposite hemifield.

The results of these studies are in good agreement. After the motion stimulus appeared, the response of many neurons in these structures became active (concurrently) in a manner that predicted the ultimate overt choice. Neurons with the target of the saccade mapped on to the motion direction in their response field became more active, and neurons with the non-selected target in their response field became less active. The stronger the motion signal, the more rapid and extreme the difference of activity representing the alternatives. These studies also observed neurons that exhibited responses that varied significantly with the direction of motion of the stimulus at the fovea. This was unexpected because neurons in the SC and frontal and parietal lobes of macaque monkeys are typically not direction selective, but the direction-sensitive response was evident even when the motion was irrelevant to another task monkeys performed. In the population of neurons that exhibited direction-biased responses, the magnitude of the activity during the delay period of the discrimination task was positively correlated with the coherence of the motion stimulus that instructed the saccade. Evidently, the process of learning stimulus–response mapping can change the stimulus representation in sensorimotor association structures, conferring stimulus selectivity that otherwise is not observed (Bichot et al. 1996). The neurons that were not direction-sensitive exhibited activity that was more clearly modulated around the saccade and less correlated with the strength of the motion stimulus. These two populations probably subserve rather distinct functions, the latter preparing the saccade and the former participating more immediately in the discrimination process.

The time course of this discrimination has been studied using electrical microstimulation of the FEF to elicit saccadic eye movements (Gold and Shadlen 2000). Stimulation of a particular site in the FEF elicits saccades with a particular direction and amplitude (Schall 1997). At different times after presentation of the motion stimulus, some trials were interrupted by electrical stimulation of the FEF. The direction of the saccade evoked from the FEF was influenced by the direction of the saccade the monkeys were preparing to make to report the direction of motion in a random dot display. The magnitude of this deviation increased with the duration of the motion presentation in a monotonic, decelerating function with an asymptote that was proportional to the strength of the motion signal. The form of this function corresponds to what has been observed in numerous behavioral studies with limited exposure durations (e.g. Usher and McClelland 2001), and other neurophysiological studies have demonstrated that the quality of the neural discrimination between alternatives improves with a similar time course (Bichot et al. 2001a). The variation of the deviation with motion strength and time could be fit by a model in which the saccade command corresponds to the accumulation of the difference in responses between pools of motion-sensitive neurons in area MT that represent the alternative directions of motion, that is, by a quantity that corresponds to the log of the likelihood ratio[1] favoring one alternative over the other (Gold and Shadlen 2001).

This framework is motivated by the fact that likelihood ratio provides the basis for optimal decisions, and any quantity that is monotonically related to likelihood ratio can be used to generate a rule of equivalent effectiveness (Green and Swets 1966). This model corresponds to the general form of sequential sampling models known as random walk or diffusion, in which a single accumulator represents the relative evidence for two alternative stimuli (Ratcliff and Rouder 1998).

This formulation is powerful because it provides a sensible rationale for the modulation of neural activity in sensorimotor association structures. The discharge rate of neurons in these structures cannot be explained entirely or exclusively as representing sensory evidence about stimuli. Activity in these structures varies with the probability of a saccade into the response field (Basso and Wurtz 1998; Dorris and Munoz 1998; Platt and Glimcher 1999; Bichot and Schall 2002) and with the probability or amount of reinforcement (Glimcher 2001). Forming decisions by calculating the log of the likelihood ratio affords the ability to incorporate the influence of prior probability and utility with sensory evidence (Gold and Shadlen 2001).

This interpretation of the neural basis of visual discrimination is based on the premise that errors of categorization arise from noise in the representation of the stimuli. However, such an account is not sufficient. Shadlen *et al.* (1996) found that the pooled responses of the neurons representing the motion stimulus actually was too good, so additional noise was added that was attributed to the decision process. Also, to account for errors, Ratcliff and Rouder (1998, 2000) needed to add more noise through variability in the starting level of the diffusion process. From a strict strong-inference point of view, the failure of the stimulus representation alone to explain performance could be taken as evidence to exclude signal detection theory as a complete account.

An alternative to a random walk or diffusion of the difference between alternatives is a race among accumulators representing each alternative, with the first to reach a threshold dictating the response. In most such models, the racing accumulators increment according to a Poisson process. The formula describing the outcome of this model is at the heart of biased choice theory (Luce 1963), which is the historical and conceptual counterpart to signal detection theory. Both signal detection theory and choice theory provide quantitative descriptors of discriminability and response criterion and are mathematically equivalent under certain reasonable assumptions. But choice theory attributes the unpredictability of the response to the decision process instead of the stimulus representation. Race models can explain stimulus discrimination and categorization as well as diffusion models (Logan 2002). In fact, race and diffusion models can account for common sets of data (Van Zandt *et al.* 2000).

20.2.2 How are stimuli selected?

We have considered how a categorical decision is made about a single stimulus. Usually, though, we are confronted with many stimuli among which we must select one to act on. For vision, this selectivity is necessary because primates have a fovea. The process of visual selection has been investigated using the visual search paradigm (Wolfe 1998). In a visual search task, multiple stimuli are presented among which a target is discriminated. Search is efficient (pop-out) if stimuli differ along basic visual feature dimensions—color, form, direction of motion. In contrast, if the distractors resemble the target or no single feature clearly distinguishes stimuli, then search becomes less efficient. Search results were popularly interpreted as proceeding through an early parallel preattentive stage operating on stimuli represented discretely, so that a target is either detected or not and the representation of the distractors cannot be confused for the representation of the target followed by a late, serial, limited capacity attentional stage (Treisman 1988; Wolfe 1994). However, alternative models based on signal detection theory (Geisler and Chou 1995; Palmer *et al.* 2000) and biased choice theory (Bundesen 1998) have been developed to account for the same observations.

20.2.2.1 Neural basis of visual selection

Numerous studies have described neural correlates of visual selection in primary and extrastriate visual cortex (Colby and Goldberg 1999; Treue 2001). A series of experiments has described how neural activity in FEF selects the target for a saccade during visual search (Schall and Thompson 1999; Thompson *et al.* 2001). The selection observed in FEF resembles, and is surely based on, visual selection processes observed in visual areas in the occipital, temporal, and parietal lobes. But FEF is commonly regarded as an ocular motor area (Bruce and Goldberg 1985), so is the selection observed in FEF nothing more than saccade preparation? Several lines of evidence indicate that neural activity in FEF may be related to covert and not just overt orienting. In monkeys trained to perform visual search, neural selection of a singleton happens even if no eye movement is planned (Thompson *et al.* 1997) or if gaze shifts to a location different from the target (Murthy *et al.* 2001). Hence, the activation of the visual neurons in FEF can be described as signaling the location of an oddball, regardless of whether or which saccade is produced, and may, therefore, correspond to covert but not necessarily overt orienting. These results are consistent with other physiological studies of FEF (Kodaka *et al.* 1997; Moore and Fallah 2001). Also, functional imaging studies have found activation of the FEF and surrounding cortex associated with both overt and covert orienting (Nobre *et al.* 2000; Culham *et al.* 2001; Corbetta and Shulman 2002).

20.2.2.2 Neural chronometry of visual selection

Explaining the duration and variability of response times is a central problem (Posner 1978; Luce 1986; Meyer *et al.* 1988). One explanation supposes that response times are comprised of distinct stages of processing that occupy variable intervals (Donders 1868; Sternberg 2001). Different stages can be defined by the influence of distinct factors. For example, the difficulty of discriminating stimuli should influence encoding but not response preparation. Similarly, interfering with response production should not influence stimulus encoding. Such models suppose that a later stage does not begin until a preceding stage is completed; this is referred to as discrete flow. An alternative explanation supposes that the output of stimulus encoding flows continuously into response preparation; a particular response is produced when the appropriate activation reaches a threshold (Eriksen and Schultz 1979; McClelland 1979). Behavioral evidence has not conclusively ruled out the discrete or the continuous flow alternatives (Luce 1986; Miller 1988).

If it were possible to identify physiological markers for the end of one stage and the beginning of another, then the alternative hypotheses about the existence of stages and the time course of activation could be distinguished more clearly. For instance, the P300 component of event-related potentials (ERP) has been interpreted as indicating the termination of perceptual processing, and the lateralized readiness potential (LRP) has been interpreted as occurring during response preparation (Coles *et al.* 1995). These physiological measures have been used to provide evidence to distinguish between the discrete flow and the continuous flow hypotheses. For example, measurement of the LRP indicates that responses can be partially prepared before stimulus evaluation is completed (Smid *et al.* 1990). However, another study demonstrated selective effects of manipulating both stimulus discriminability and response complexity on the timing of P300 and LRP, and concluded that the changes provide evidence in support of discrete stages (Smulders *et al.* 1995). Thus, further work is needed.

Single-unit activity can also contribute to distinguishing between discrete and continuous models of processing. For example, the target selection process observed through neural recording takes time to transpire (Thompson *et al.* 1996; Bichot *et al.* 2001*a*; Sato *et al.* 2001) comparable to the time of visual selection observed under comparable conditions in extrastriate visual cortex (Motter 1994; Luck *et al.* 1997; Chelazzi *et al.* 1998, 2001). Accordingly, the time of target selection observed in neural activity can be used as a marker of the time when an explicit representation of the image is achieved; in other words, the conclusion of the encoding stage. During an efficient oddball search, the time of target selection by most visually responsive neurons in FEF did not predict when gaze would shift (Thompson *et al.* 1996; Sato *et al.* 2001); thus, a significant fraction of the variability in reaction time in efficient search arises in post-perceptual response preparation. However, during less efficient search, more time was taken by FEF neurons to select the target, and the time of target discrimination accounted for more of the variability of reaction time, with the increase in the time taken to select the target accounting for nearly all of the increase in response time (Bichot *et al.* 2001*a*; Sato *et al.* 2001). A complementary observation has been reported from neural activity in the parietal lobe of monkeys performing the random dot direction of motion discrimination described above (Roitman and Shadlen 2002).

Data from single neurons can also be employed to test whether information about a stimulus influences the response stage, which has been regarded as an important test of discrete versus continuous flow models. Certain neurons in the FEF and other ocular motor structures contribute directly to preparing and producing saccades (see below). A recent study examined movement-related neurons in the FEF of monkeys performing a conjunction visual search or a feature visual search with a singleton distractor

(Bichot et al. 2001b). While movement-related neurons were activated maximally when the target of the search array was in their movement field, the neurons were also activated when monkeys shifted gaze to distractors that appeared in the movement field, even though a saccade was made to the target outside of the movement field. Furthermore, the level of activation depended on the properties of the distractor, with greater activation for distractors that shared a target feature or were the target during the previous session during conjunction search, and for the singleton distractor during feature search. Similar evidence for the influence of sensory factors on unit activity in motor cortex has been obtained (Miller et al. 1992; Requin and Riehle 1995; Riehle et al. 1997). These results are consistent with continuous information processing, but cannot conclusively rule out certain forms of discrete models (Miller 1988).

Nevertheless, the existence of distinct, if not necessarily discrete, stages seems a necessary hypothesis because not all response errors can be explained by confusion in the representation of stimuli. For example, a search task was designed in which, on some trials, the target swaps locations with one distractor (Murthy et al. 2001). Visually responsive neurons selected the new location of the singleton target even when gaze shifted to the old location. If the selection of the target obliged an accurate gaze shift, this could not happen. This error can be explained, though, by a saccade-production stage producing the gaze shift before, or in spite of, visual processing. Such errors of overt but not covert target selection can also be observed during more natural scanning behavior (Hooge and Erkelens 1996). These considerations lead to the issue of how movements are chosen.

20.2.3 How are actions chosen?

The results just reviewed suggest that psychological stages of processing can be identified in the activity of neurons. Sensorimotor structures commonly are comprised of sensory neurons in upper layers and motor neurons in deeper layers. A deceptively simple hypothesis is that sensorimotor transformation occurs through activation spreading from upper sensory to lower motor layers of the cerebral cortex or SC. However, inactivation studies show that normal signals can be observed in the different layers of the cerebral cortex even if other layers are inactivated (Malpeli 1983; Schwark et al. 1986). Also, the motor error signaled by neurons in the deeper layers of the SC can be dissociated from the retinal error signaled by neurons in the upper layers (Mays and Sparks 1980). These observations argue against a one-to-one mapping of stimulus onto response hard-wired in the nervous system. As primates, our survival depends on flexible responses in a complex environment, which is only afforded through a many-to-many mapping soft-wired in the nervous system (Wise et al. 1997). However, the overgrown complexity of a primate brain is not necessary for arbitrary stimulus–response associations (Shaw and Kristan 1997).

The process of mapping an arbitrary response to a particular stimulus has been investigated extensively (Kornblum and Stevens 2002). Certain key properties of the process have been observed. For example, the speed and accuracy of performance suffers under incompatible, as compared to compatible, mappings. Also, some mappings are more automatic than others, and the more automatic can interfere with performance of the less automatic. Event-related potential studies have measured the influence of stimulus–response compatibility on different ERP components (Coles *et al.* 1995). Early work showed that the latency of the P300 is the same in compatible and incompatible conditions, although response time is longer in the incompatible condition (McCarthy and Donchin 1981). A more recent study showed that the onset of the stimulus-locked LRP is delayed in incompatible as compared to compatible trials, but P300 latency was also longer for incompatible than for compatible stimuli (Masaki *et al.* 2000). Other studies have reported partial activation of the incorrect response on incompatible trials that is subsequently replaced by activation for the correct response (Gratton *et al.* 1988; Eimer 1995; DeSoto *et al.* 2001).

Several neurophysiological studies have investigated how particular stimuli become mapped on to arbitrary behaviors (Wise and Murray 2000). The results of these studies demonstrate an evolution of neural activity in frontal cortex during the reaction time from an early representation of the location of the sensory cue stimulus to a later representation of the direction of movement. They also identified neurons in premotor and primary motor cortex exhibiting variations of stimulus-evoked and tonic activity contingent on the stimulus–response mapping rule (Riehle *et al.* 1997; Shen and Alexander 1997; White and Wise 1999; Hoshi *et al.* 2000; Wallis *et al.* 2001). The interference suffered in incompatible trials may result from activation of neurons by stimuli that should be ignored (Chen *et al.* 2001; Lauwereyns *et al.* 2001).

One study examined the timing of neural modulation to account for the increase in response time on incompatible as compared to compatible mapping trials (Mouret and Hasbroucq 2000). The evidence indicated that the latency of the sensory response relative to stimulus presentation and the timing of the movement-related activation relative to response initiation were not affected by the mapping compatibility, but the onset of the movement-related activation was delayed relative to stimulus presentation in incompatible as compared to compatible trials. These observations were interpreted as evidence in support of discrete flow between sensory, mapping, and motor stages. However, the latency of the neural sensory response to the stimulus measures the time when sensory processing begins, not when it concludes. A more telling measure is the time when sensory activation representing the alternatives becomes different (Thompson *et al.* 1996; Sato *et al.* 2001).

20.2.4 How are actions produced?

Several reviews have considered how movements are produced (Kawato 1999; Wolpert and Ghahramani 2000; Scudder *et al.* 2002). This section will focus on the preliminaries to movement.

20.2.4.1 Are movements prepared before execution?

Only overt responses can be observed, but many lines of evidence indicate that processes occurring before the movement is initiated influence, in an orderly manner, the time and quality of the movement. For example, reaction times following a trigger signal are shorter if subjects are given a warning signal, but only if the time that elapses from the warning to the trigger is predictable (Niemi and Näätänen 1981).

Characteristic scalp potentials have been identified with response preparation. As early as 1000 ms before execution of a movement a readiness potential can be recorded over the medial frontal lobe (Deecke 1987). The readiness potential recorded before pre-planned, externally cued movements is qualitatively different from that recorded prior to spontaneous movements (Libet et al. 1982; Keller and Heckhausen 1990). The most dramatic report is that the readiness potential begins before the subject reports being aware of the intention to act (Libet 1985; but see Haggard and Eimer 1999; Trevena and Miller 2002). The LRP is a more specific index of movement preparation than the readiness potential, because it is a measure of the difference of activation between the hemispheres (Coles et al. 1995). Functional imaging studies have also reported activation preceding movements in the frontal lobe (Richter et al. 1997; Deiber et al. 1999; Lee et al. 1999).

Specific criteria for identifying neural modulation with response preparation have been proposed (Riehle and Requin 1993). First, the modulation in neuronal discharge must occur during the warning period before the movement. Secondly, the level of neural activity must be modulated in proportion to the likelihood of a movement being produced into the response field of the neuron. Thirdly, the modulation in discharge rate must predict some attribute of motor performance such as reaction time. Several studies have shown modulation of neural activity that meets these criteria in the cortex (Lecas et al. 1986; Riehle and Requin 1993; Riehle et al. 2000; Crammond and Kalaska, 2000) as well as in the SC (Dorris and Munoz 1998).

Further evidence for response preparation is obtained in subjects producing sequences of movements. The problem of serial order was emphasized by Lashley (1951), who demonstrated that stimulus–response chaining could not explain rapid sequences of movements such as an arpeggio. The latency of the first movement in a sequence increases with the number of movements in the sequence for speech or typing (Sternberg et al. 1978) or for saccades (Zingale and Kowler 1987). Several neurophysiological studies have described particular patterns of neural activity in the frontal lobe of monkeys performing arbitrary movement sequences (Tanji et al. 1996; Hikosaka et al. 1999). Neural correlates of the latency variation with sequence length have not been reported, but a central role of the cerebellum seems clear (Inhoff et al. 1989).

20.2.4.2 Can movements be prepared but not executed?

A critical characteristic of voluntary control is the ability to withhold planned movements. If movements can be prepared, then the preparation process must be controllable, which means that movements can be prepared but not executed.

Many studies have contrasted performance and brain activation in trials with *go* and *no-go* cues, and found that activation in primary motor and premotor areas of the frontal lobe, but not in the parietal lobe, predicted whether a movement would be executed (Schall 1991; Kalaska and Crammond 1995; Paré and Wurtz 2001; Sakagami *et al.* 2001). Complementary results have been observed with event-related potentials (Pfefferbaum and Ford 1985; Falkenstein *et al.* 1999; Bokura *et al.* 2001) and with functional imaging (Menon *et al.* 2001; Durston *et al.* 2002).

The ability of subjects to inhibit partially prepared responses has been investigated fruitfully using the countermanding task that requires subjects to cancel a planned speeded response following an unexpected stop signal (Logan and Cowan 1984). Performance can be accounted for in terms of a stochastic model of a race between a process that produces the response and a process that cancels the movement in response to the stop signal (Logan and Cowan, 1984). This model permits estimation of the time needed to cancel the movement, the stop-signal reaction time. The validity of the estimate of stop-signal reaction time depends on the validity of the premises of the race model, which have stood up to scrutiny even with modest violations of the assumptions (De Jong *et al.* 1990). The LRP has been measured in subjects performing the countermanding task (De Jong *et al.* 1990, 1995; van Boxtel *et al.* 2001). These studies show that the LRP is of lower magnitude and duration when movements are partially prepared but not executed.

The countermanding task has been used with saccadic eye movements (Hanes and Schall 1995; Cabel *et al.* 2000; Logan and Irwin 2000; Asrress and Carpenter 2001; Colonius *et al.* 2001). With this task, neural activity has been recorded in the FEF to determine how neural activity controls whether saccades are produced (Hanes *et al.* 1998). The chief virtue of the countermanding paradigm is that one can determine whether single neurons generate signals that are sufficient to control the production of movements. The logic of the countermanding paradigm establishes two criteria a neuron must meet to play a direct role in the control of movement. First, the neuron must discharge differently when a saccade is initiated versus when a saccade is withheld. Secondly, this difference in activity must occur within the stop-signal reaction time. When saccades were initiated, movement-related activity in FEF exhibited a clear increase preceding the movement, followed by a decrease after the movement. When saccades were canceled, movement-related activity in FEF began to increase but then exhibited a rapid decrease that occurred within the stop-signal reaction time. A complementary pattern of neural activity was produced by FEF fixation neurons. However, visual neurons in FEF were modulated not at all or too late to influence saccade initiation. Therefore, the activity of movement and fixation but not visual neurons in FEF is logically sufficient to specify whether or not a saccade will be produced. Results indicate that the same holds true in the SC (Paré and Hanes 2003). The ability to distinguish between neurons that can control response initiation and neurons that cannot constitutes further evidence that the two kinds of neurons instantiate distinct functions.

20.2.4.3 How are movements initiated?

The stochastic variability of response times must originate in neural processes. Research has discovered that saccadic eye movements are initiated when the activity of certain neurons in a network of structures, including the FEF and SC, reaches a particular level (Sparks 1978; Hanes and Schall 1996; Dorris *et al.* 1997). The same holds true for neural activity in motor cortex leading to movements of the limbs (Lecas *et al.* 1986). This observation has been made with event-related potentials as well; the magnitude of the LRP at movement initiation does not vary with reaction time (Gratton *et al.* 1988). The variability in reaction time can be accounted for mainly by the time taken by the activity of movement-related neurons to reach the threshold. The origin of such variability in the growth of movement-related activity is not known, but may include the state of neuromodulatory systems (Aston-Jones *et al.* 1994).

This observation has two theoretical implications. First, the relation of movement-related neural activity to reaction time corresponds to an accumulator architecture with variable growth to a fixed threshold (Ratcliff 1978; Carpenter and Williams 1995) and directly contradicts an architecture with a fixed growth process and random threshold (Grice *et al.* 1982).

Secondly, the stochastic variability of response preparation challenges the assumption of a fixed efferent delay made by most models (Ratcliff and Rouder 1998, 2000; Usher and McClelland 2001). It is clear that the time taken to decide between unclear alternatives is variable and can account for much of the variability of response time (e.g. Sato *et al.* 2001). However, it is equally clear that response preparation can occupy variable and unpredictable amounts of time. Indeed, the separability of these two processes is the basis of the hypothesis of different stages of processing. However, if both stimulus encoding and response preparation occupy variable intervals, then behavioral response time amounts to a mixture of two random processes. This presents a serious challenge for elucidating the contribution of different stages to response time (Luce 1986; van Zandt and Ratcliff 1995). However, as described above, pivotal information can be obtained by using particular neural signals as markers of the beginning and end of different processing stages. Future work monitoring the activity of ensembles of neurons realizing the different stages is necessary to evaluate these alternatives.

20.3 How do actions achieve goals?

The goal of science is to explain the causes of events. For example, we can explain the event of a finger moving in terms of the causal pathway extending back through muscles contracting and neurons firing. However, many movements produced by humans are distinguished from mere events by having explanations in terms of not just causes but reasons (Davidson 1963). We describe certain movements as actions directed toward a goal for a purpose and not as just events that occur following some chain of causes.

However, *actions* are not always identical to movements of the body (Goldman 1970). An action is what we do. Defining a body movement as an action depends on the context. An action (typing on a keyboard) is distinguished from a mere event (a hand bumping a keyboard) by reference to an intelligible plan. Actions are performed to achieve goals. Often we do one thing (write a chapter) by doing something else (typing on a keyboard). But to type on a keyboard, one must do something else (move a finger). However, although we can say that we type on a keyboard by moving a finger, we cannot say what we do to move a finger. A *basic actio*n, such as moving a finger, is an action we perform without any preliminaries. If an agent's intention corresponds to (is caused by?) a particular brain state, then the intention gains the causal efficacy of neurophysiology. However, such determinism challenges common views of freedom of choice because the events preceding the finger movement—such as muscles contracting and neurons firing—are not things we do. Hence, if voluntary movements are grounded in such involuntary processes, how can we be responsible for the consequences of our actions? One solution is to base responsibility on intention rather than the means by which an action occurs. We are not responsible for unintended consequences, and as agents we usually (but not always) can report whether an action was intended. Some have suggested that a relationship between brain states and actions beyond the causal chain leading to muscle contraction is conceivable. The hypothesis is that a neural representation of an intention corresponds to the reason for the action (Kim 1998; Juarrero 1999). This entails that the brain can monitor the actions it produces to determine whether the intended goal is achieved.

The ability to recognize and correct errors has been investigated intensively. Original research found slowed responses following errors (Rabbitt 1966). The ability to recognize errors and adjust performance has been suggested to be mediated by a supervisory executive control system (Logan 1985; Norman and Shallice 1986; Posner and DiGirolamo 1998; Botvinick *et al.* 2001), and models have been developed to explain how the executive can control subordinate processes (Logan and Gordon 2001). The first physiological evidence for such a monitoring system was the error-related negativity (ERN), a scalp potential observed specifically when subjects produce errors (Falkenstein *et al.* 2000; Coles *et al.* 2001). Current evidence suggests that this signal corresponds to the detection, but not necessarily the correction, of errors. The source generator of the ERN seems centered in the anterior cingulate cortex, but may include the supplementary motor area (Dehaene *et al.* 1994).

Two mechanisms (that are not mutually exclusive) for computing the error signal have been conceived. The first hypothesis explains the error signal as a mismatch between the intended and the performed behavior (Falkenstein *et al.* 2000; Coles *et al.* 2001). However, models of this sort are challenged to explain how intended actions are represented. One solution to this problem supposes that reward-related dopaminergic signals from the midbrain produce the error signal (Holroyd and Coles 2002). This model is based on the observation that dopamine cells exhibit a transient reduction of activity when an unexpected error occurs (Schultz and Dickinson 2000), and the timing of this activity coincides with the timing of error signals observed in the ERN. The second hypothesis explains the error signal as a manifestation of the conflict engendered when mutually incompatible processes are activated simultaneously but cannot both run to completion, so that errors are made (Botvinick et al. 2001). A virtue of this model is that it derives a measure of conflict directly from response preparation signals. The error-based and conflict-based models make different predictions regarding the presence and magnitude of brain activation and the ERN under different conditions, which have been tested with divergent results (Bernstein et al. 1995; Carter et al. 1998; MacLeod and MacDonald 2000; Coles et al. 2001; Gehring and Fencsik 2001).

Recent neurophysiological recordings in the supplementary eye field (SEF) of monkeys performing the countermanding task have identified single neurons that convey signals related to monitoring of performance (Stuphorn et al. 2000). The error-related activity occurs at the same time as the ERN is observed, so these neurons probably contribute to the source generator. A second type of neuron was active on successful trials when monkeys anticipated and received reinforcement. Other studies have also described neural activity associated with reinforcement in SEF (Amador et al. 2000). This type of neuron represents a functional complement of the error-related neurons.

A third type of neuron appeared to signal processing conflict. During the saccade countermanding task, gaze-shifting and gaze-holding neurons are activated concurrently when movements are canceled but not when the movements fail to be canceled (Hanes et al. 1998). Because they are mutually incompatible, coactivation of the gaze-holding and gaze-shifting systems engenders conflict in processing; the magnitude of the conflict should be proportional to the magnitude of the activation of the gaze-holding and gaze-shifting neuron pools. Now, the probability of canceling a planned eye movement is dictated by the balance of activation of gaze-holding and gaze-shifting neurons, because movements are canceled only if the gaze-shifting activation does not reach the threshold to trigger the movement because it is countered by the gaze-holding activation. Thus, the probability of failing to cancel a partially prepared saccade increases as gaze-shifting activation grows. Accordingly, as the probability of failing to cancel the movement increases, the combined magnitude of gaze-shifting and gaze-holding activation sufficient to cancel a planned movement will be higher, thereby engendering more conflict. This measure of processing conflict corresponds to the variation in the magnitude of the neural modulation observed in these neurons in SEF.

The diversity of signals observed in unit recordings may provide some reconciliation of the alternative hypotheses. In particular, the neurons exhibiting error-related activity during countermanding were not active in trials in which conflict between gaze-shifting and gaze-holding was most pronounced (trials when the planned movement was canceled), and error-related activity was observed in trials in which no conflict was present because the gaze-holding neurons were not modulated (trials when the movement was produced in spite of the stop signal). This indicates that the search for an exclusive distinction between error-based, conflict-based, and reward-based models may be misguided.

Regardless of differences of interpretation, the evidence is clear that medial frontal lobe activation occurs when the environment is ambiguous or presents competing demands, or the mapping of stimulus on to response is complex or contrary to habit, making performance prone to errors—these are just the conditions under which deciding and not just choosing is involved. This suggests that medial frontal activation could be an empirical basis for identifying the extent to which an action should be identified as a choice versus a decision. In fact, in humans performing the random dot motion discrimination task, activation in anterior cingulate cortex increases as the motion strength decreases, making the alternatives more ambiguous (Rees *et al.* 2000). Under such perplexing conditions, the assignment of intention can become more relevant but less clear, but if the medial frontal lobe performs self-monitoring, then activation in the medial frontal lobe may be critical for distinguishing *I did* from *it happened*.

20.4 Conclusions

It seems clear Fechner and Helmholtz and the other fathers of experimental psychology would be amazed and impressed with the accomplishments reviewed in this chapter. However, it seems equally likely that they would realize that the essential questions remain unanswered about how neural events produce behavior and cognition. Further progress toward elucidating the physical basis of human cognition can be anticipated from well-controlled behavioral paradigms that challenge subjects at the threshold of performance, coupled with theoretically grounded analyses linking neural processes to behavioral reports and inferred cognitive processes.

Acknowledgements

Support provided by the National Eye Institute, National Institute of Mental Health and the McKnight Endowment Fund for Neuroscience. The author is very grateful to G. Fox, E. Gotcher, and C. Wiley for preparing and M. Chun, I. Gauthier, G. Logan, R. Marois, J. Miller, T. Palmeri, S. Shorter-Jacobi, and V. Stuphorn for commenting on the manuscript.

Notes

1. Likelihood ratio refers to the ratio of the probability of one possibility to the probability of the alternative, given the occurrence of a particular stimulus. Optimal decisions can be made based on the likelihood ratio.

References

Amador, N., Schlag-Rey, M., and Schlag, J. (2000). Reward-predicting and reward-detecting neuronal activity in the primate supplementary eye field. *J Neurophysiol*, **84**, 2166–70.

Asrress, K.N. and Carpenter, R.H.S. (2001). Saccadic countermanding: a comparison of central and peripheral stop signals. *Vision Res*, **41**, 2645–51.

Aston-Jones, G., Rajkowski, J., Kubiak, P., and Alexinsky, T. (1994). Locus coeruleus neurons in monkey are selectively activated by attended cues in a vigilance task. *J Neurosci*, **14**, 4467–80.

Barlow, H.B. (1995). The neuron doctrine in perception. In Gazzaniga M.S. *The Cognitive Neurosciences*, pp. 415–35. MIT Press, Cambridge MA.

Basso, M.A. and Wurtz, R.H. (1998). Modulation of neuronal activity in SC by changes in target probability. *J Neurosci*, **18**, 7519–34.

Bernstein, P.S., Scheffers, M.K., and Coles, M.G. (1995). 'Where did I go wrong?' A psychophysiological analysis of error detection. *J Exp Psychol: Hum Percept Perform*, **21**, 1312–22.

Bichot, N.P. and Schall, J.D. (2002). Priming in macaque frontal cortex during popout visual search:Feature-based facilitation and location-based inhibition of return. *J Neurosci*, **22**, 4675–85.

Bichot, N.P., Schall, J.D., and Thompson, K.G. (1996). Visual feature selectivity in frontal eye field induced by experience in mature macaques. *Nature*, **381**, 697–9.

Bichot, N.P., Thompson, K.G., Rao S.C., and Schall, J.D. (2001*a*). Reliability of macaque frontal eye field neurons signaling saccade targets during visual search. *J Neurosci*, **21**, 713–25.

Bichot, N.P., Rao, S.C., and Schall, J.D. (2001*b*). Continuous processing in macaque frontal cortex during visual search. *Neuropsychologia*, **39**, 972–82.

Bokura, H., Yamaguchi, S., and Kobayashi, S. (2001). Electrophysiological correlates for response inhibition in a Go/NoGo task. *Clin Neurophysiol* **112**, 2224–32.

Botvinick, M.M., Braver, T.S., Barch, D.M., Carter, C.S., and Cohen, J.D. (2001). Conflict monitoring and cognitive control. *Psychol Rev*, **108**, 624–52.

Britten, K.H., Shadlen, M.N., Newsome, W.T., and Movshon, J.A. (1992). The analysis of visual motion: a comparison of neuronal and psychophysical performance. *J Neurosci*, **12**, 4745–65.

Britten, K.H., Newsome, W.T., Shadlen, M.N., Celebrini, S., and Movshon, J.A. (1996). A relationship between behavioral choice and the visual responses of neurons in macaque MT. *Visual Neuro*, **13**, 87–100.

Bruce, C.J. and Goldberg, M.E. (1985). Primate frontal eye fields I: Single neurons discharging before saccades. *J Neurophysiol*, **53**, 603–35.

Bundesen, C. (1998). A computational theory of visual attention. *Phil Trans R Soc Lond B Biol Sci*, **353**, 1271–81.

Cabel, D.W.J., Armstrong, I.T., Reingold, E., and Munoz, D.P. (2000). Control of saccade initiation in a countermanding task using visual and auditory stop signals. *Exp Brain Res*, **133**, 431–41.

Carpenter, R.H.S. and Williams, M.L.L. (1995). Neural computation of log likelihood in the control of saccadic eye movements. *Nature*, **377**, 59–62.

Carter, C.S., Braver, T.S., Barch, D.M., Botvinick, M.M., Noll, D., and Cohen, J.D. (1998). Anterior cingulate cortex, error detection and the online monitoring of performance. *Science*, **280**, 747–9.

Chelazzi, L., Duncan, J., Miller, E.K., and Desimone, R. (1998). Responses of neurons in inferior temporal cortex during memory-guided visual search. *J Neurophys*, **80**, 2918–40.

Chelazzi, L., Miller, E.K., Duncan, J., and Desimone, R. (2001). Responses of neurons in macaque area V4 during memory-guided visual search. *Cereb Cortex*, **11**, 761–72.

Chen, N.H., White, I.M., and Wise, S.P. (2001). Neuronal activity in dorsomedial frontal cortex and prefrontal cortex reflecting irrelevant stimulus dimensions. *Exp Brain Res*, **139**, 116–19.

Colby, C.L. and Goldberg, M.E. (1999). Space and attention in parietal cortex. *Annu Rev Neurosci*, **22**, 319–49.

Coles, M.G.H., Smid, H.G.O.M., Scheffers, M.K., and Otten, L.J. (1995). Mental chronometry and the study of human information processing. In M.D. Rugg and M.G.H. Coles, (Eds) *Electrophysiology of mind: event-related brain potentials and cognition,* pp. 86–131. Oxford University Press, Oxford.

Coles, M.G., Scheffers, M.K., and Holroyd, C.B. (2001). Why is there an ERN/Ne on correct trials? Response representations, stimulus-related components, and the theory of error-processing. *Biol Psychol,* **56,** 173–89.

Colonius, H., Ozyurt, J., and Arndt, P.A. (2001). Countermanding saccades with auditory stop signals: testing the race model. *Vision Res,* **41,** 1951–68.

Corbetta, M. and Shulman, G.L. (2002). Control of goal-directed and stimulus-driven attention in the brain. *Nat Rev Neurosci,* **3,** 201–15.

Crammond, D.J. and Kalaska, J.F. (2000). Prior information in motor and premotor cortex: Activity during the delay period and effect on pre-movement activity. *J Neurophysiol,* **84,** 986–1005.

Culham, J.C., Cavanagh, P., and Kanwisher, N.G. (2001). Attention response functions: characterizing brain areas using fMRI activation during parametric variations of attentional load. *Neuron,* **32,** 737–45.

Davidson, D. (1963). Actions, reasons and causes. *J Philosophy,* **60,** 685–700.

Deecke, L. (1987). Bereitschaftspotential as an indicator of movement preparation in supplementary motor area and motor cortex. *Ciba Found Symp,* **132,** 231–50.

Dehaene, S., Posner, M.I., and Tucker, D.M. (1994). Localization of a neural system for error detection and compensation. *Psych Science,* **5,** 303–5.

Deiber, M.P., Honda, M., Ibanez, V., Sadato, N., and Hallett, M. (1999). Mesial motor areas in self-initiated versus externally triggered movements examined with fMRI: effect of movement type and rate. *J Neurophysiol,* **81,** 3065–77.

De Jong, R., Coles, M.G., Logan, G.D., and Gratton, G. (1990). In search of the point of no return: the control of response processes. *J Exp Psychol: Hum Percept Perform,* **16,** 164–82.

De Jong, R., Coles, M.G., and Logan, G.D. (1995). Strategies and mechanisms in nonselective and selective inhibitory motor control. *J Exp Psychol: Hum Percept Perform,* **21,** 498–511.

Dennett, D.C. (1984). *Elbow Room: The Varieties of Free Will Worth Wanting.* MIT Press, Cambridge MA.

DeSoto, M.C., Fabiani, M., Geary, D.C., and Gratton, G. (2001). When in doubt, do it both ways: brain evidence of the simultaneous activation of conflicting motor responses in a spatial Stroop task. *J Cogn Neurosci,* **13,** 523–36.

Dodd, J.V., Krug, K., Cumming, B.G., and Parker, A.J. (2001). Perceptually bistable three-dimensional figures evoke high choice probabilities in cortical area MT. *J Neurosci,* **21,** 4809–21.

Donders, F.C. (1868). On the speed of mental processes. (Translated by W.G. Koster, 1969) In *Attention and Performance II* pp. 412–31. North-Holland Publishing, Amsterdam.

Dorris, M.C. and Munoz, D.P. (1998). Saccadic probability influences motor preparation signals and time to saccadic initiation. *J Neurosci,* **18,** 7015–26.

Dorris, M.C., Paré, M., and Munoz, D.P. (1997). Neuronal activity in monkey SC related to the initiation of saccadic eye movements. *J Neurosci,* **17,** 8566–79.

Durston, S., Thomas, K.M., Worden, M.S., Yang, Y., and Casey, B.J. (2002). The effect of preceding context on inhibition: An event-related fMRI study. *Neuroimage,* **16,** 449–53.

Eimer, M. (1995). Stimulus-response compatibility and automatic response activation: evidence from psychophysiological studies. *J Exp Psychol: Hum Percept Perform,* **21,** 837–54.

Eriksen, C.W. and Schultz, D.W. (1979). Information processing in visual search: a continuous flow conception and experimental results. *Percept Psychophys,* **25,** 249–63.

Evarts, E.V. (1966). Pyramidal tract activity associated with a conditioned hand movement in the monkey. *J Neurophysiol*, **29**, 1011–27.

Falkenstein, M., Hoormann, J., and Hohnsbein, J. (1999). ERP components in Go/Nogo tasks and their relation to inhibition. *Acta Psychol (Amst)*, **101**, 267–91.

Falkenstein, M., Hoormann, J., Christ, S., and Hohnsbein, J. (2000). ERP components on reaction errors and their functional significance: a tutorial. *Biol Psychol*, **51**, 87–107.

Fechner, G.T. (1860/1912). Elements of psychophysics. In Rand B. (Ed.). *The Classical Psychologists* (pp. 562–572). Houghton Mifflin, Boston MA.

Friston, K. (2002). Beyond phrenology: What can neuroimaging tell us about distributed circuitry? *Annu Rev Neurosci*, **25**, 221–50.

Gehring, W.J. and Fencsik, D.E. (2001). Functions of the medial frontal cortex in the processing of conflict and errors. *J Neurosci*, **21**, 9430–7.

Geisler, W.S. and Chou, K.L. (1995). Separation of low-level and high-level factors in complex tasks: visual search. *Psychol Rev*, **102**, 356–78.

Glimcher, P.W. (2001). Making choices: The neurophysiology of visual-saccadic decision making. *Trends Neurosci*, **24**, 654–9.

Gold, J.I., and Shadlen, M.N. (2000). Representation of a perceptual decision in developing oculomotor commands. *Nature*, **404**, 390–4.

Gold, J.I. and Shadlen, M.N. (2001). Neural computations that underlie decisions about sensory stimuli. *Trends Cogn Sci*, **5**, 10–16.

Goldman, A.I. (1970). *A Theory of Human Action*. Prentice-Hall, Englewood Cliffs NJ.

Gratton, G., Coles, M.G., Sirevaag, E.J., Eriksen, C.W., and Donchin, E. (1988). Pre- and poststimulus activation of response channels: a psychophysiological analysis. *J Exp Psychol: Hum Percept Perform*, **14**, 331–44.

Green, D.M. and Swets, J.A. (1966). *Signal Detection Theory and Psychophysics*. John Wiley & Sons, New York.

Grice, G.R., Nullmeyer, R., and Spiker, V.A. (1982). Human reaction time: Toward a general theory. *J Exp Psych*, **111**, 135–53.

Haggard, P. and Eimer, M. (1999). On the relation between brain potentials and the awareness of voluntary movements. *Exp Brain Res*, **126**, 128–33.

Hanes, D.P. and Schall, J.D. (1995). Countermanding saccades in macaque. *Vis Neurosci*, **12**, 929–37.

Hanes, D.P. and Schall, J.D. (1996). Neural control of voluntary movement initiation. *Science*, **274**, 427–30.

Hanes, D.P., Patterson, W.F., and Schall, J.D. (1998). The role of FEF in countermanding saccades: Visual, movement and fixation activity. *J Neurophysiol*, **79**, 817–34.

Hernández, A., Zainos, A., and Romo, R. (2000). Neuronal correlates of sensory discrimination in the somatosensory cortex. *Proc Natl Acad Sci*, **97**, 6191–6.

Hernández, A., Zainos, A., and Romo, R. (2002). Temporal evolution of a decision-making process in medial premotor cortex. *Neuron*, **33**, 959–72.

Hikosaka, O., Nakahara, H., Rand, M.K., Sakai, K., Lu, X., Nakamura, K., Miyachi, S., and Doya, K. (1999). Parallel neural networks for learning sequential procedures. *Trends Neurosci*, **22**, 464–71.

Holroyd, C.B. and Coles, M.G.H. (2002). The neural basis of human error processing: Reinforcement learning, dopamine, and the error-related negativity. *Psychol Rev*, **109**, 679–709.

Hooge, I.T. and Erkelens, C.J. (1996). Control of fixation duration in a simple search task. *Percept Psychophys*, **58**, 969–76.

Horwitz, G.D. and Newsome, W.T. (2001). Target selection for saccadic eye movements: prelude activity in the superior colliculus during a direction-discrimination task. *J Neurophysiol*, **86**, 2543–58.

Hoshi, E., Shima, K., and Tanji, J. (2000). Neuronal activity in the primate prefrontal cortex in the process of motor selection based on two behavioral rules. *J Neurophysiol*, **83**, 2355–73.

Hubel, D.H. and Wiesel, T.N. (1962). Receptive fields, binocular interactions and functional architecture in the cat's visual cortex. *J Physiol (Lond)*, **160**, 106–54.

Inhoff, A.W., Diener, H.C., Rafal, R.D., and Ivry, R. (1989). The role of cerebellar structures in the programming end execution of movement sequences. *Brain*, **112**, 565–81.

Juarrero, A. (1999). *Dynamics in Action: Intentional Behavior as a Complex System*. MIT Press, Cambridge MA.

Kalaska, J.F. and Crammond, D.J. (1995). Deciding not to GO: neuronal correlates of response selection in a GO/NOGO task in primate premotor and parietal cortex. *Cereb Cortex*, **5**, 410–28.

Kawato, M. (1999). Internal models for motor control and trajectory planning. *Curr Opin Neurobiol*, **9**, 718–27.

Keller, I. and Heckhausen, H. (1990). Readiness potentials preceding spontaneous motor acts: voluntary vs. involuntary control. *Electroencephalogr Clin Neurophysiol*, **76**, 351–61.

Kim, J. (1998). *Mind in a Physical World: An Essay on the Mind-body Problem and Mental Causation*. MIT Press, Cambridge MA.

Kim, J. N. and Shadlen, M.N. (1999). Neural correlates of a decision in the dorsolateral prefrontal cortex of the macaque. *Nat. Neurosci*, **2**, 176–85.

Kodaka, Y., Mikami, A., and Kubota, K. (1997). Neuronal activity in the frontal eye field of the monkey is modulated while attention is focused on to a stimulus in the peripheral visual field, irrespective of eye movement. *Neurosci Res*, **28**, 291–8.

Kornblum, S. and Stevens, G. (2002). Sequential effects of dimensional overlap: Findings and issues. In *Attention and Performance*, pp. 9–54. Oxford University Press, Oxford.

Lashley, K.S. (1951). The problem of serial order in behavior. In: L.A. Jeffress, Ed. *Cerebral Mechanisms in Behavior: The Hixon Symposium*, pp. 112–46. Wiley, New York NY.

Lauwereyns, J., Sakagami, M., Tsutsui, K.I., Kobayashi, S., Koizumi, M., and Hikosaka, O. (2001). Responses to task-irrelevant visual features by primate prefrontal neurons. *J Neurophysiol*, **86**, 2001–10.

Lecas, J.C., Requin, J., Anger, C., and Vitton, N. (1986). Changes in neuronal activity of the monkey precentral cortex during preparation for movement. *J Neurophysiol*, **56**, 1680–702.

Lee, K.M., Chang, K.H., and Roh, J.K. (1999). Subregions within the supplementary motor area activated at different stages of movement preparation and execution. *Neuroimage*, **9**, 117–23.

Libet, B. (1985). Unconscious cerebral initiative and the role of conscious will in voluntary action. *Behav Brain Sci*, **8**, 529–66.

Libet, B., Wright, E.W., and Gleason, C.A. (1982). Readiness potentials preceding unrestricted 'spontaneous' and pre-planned voluntary acts. *Electroenceph Clin Neurophysiol*, **54**, 322–35.

Logan, G.D. (1985). Executive control of thought and action. *Acta Psychol*, **60**, 193–210.

Logan, G.D. (2002). An instance theory of attention and memory. *Psychol Rev*, **109**, 376–400.

Logan, G.D. and Cowan, W.B. (1984). On the ability to inhibit thought and action: A theory of an act of control. *Psychol Rev*, **91**, 295–327.

Logan, G.D. and Gordon, R.D. (2001). Executive control of visual attention in dual-task situations. *Psychol Rev*, **108**, 393–434.

Logan, G.D. and Irwin, D.E. (2000). Don't look! Don't touch! Inhibitory control of eye and hand movements. *Psychon Bull Rev*, **7**, 107–12.

Logothetis, N.K. and Schall, J.D. (1989). Neuronal correlates of subjective visual perception. *Science*, **245**, 761–3.

Luce, R.D. (1963). Detection and recognition. In Luce, R.D., Bush, R.R., and Galanter E. (Eds), *Handbook of Mathematical Psychology*, pp. 105–89. Wiley, New York NY.

Luce, R.D. (1986). *Response Times: Their Role in Inferring Elementary Mental Organization*. Oxford University Press, Oxford.

Luck, S., Chelazzi, J.L., Hillyard, S.A., and Desimone, R. (1997). Neural mechanisms of spatial selective attention in areas V1, V2, and V4 of macaque visual cortex. *J Neurophysiol*, **77**, 24–42.

MacLeod, C.M. and MacDonald, P.A. (2000). Interdimensional interference in the Stroop effect: Uncovering the cognitive and neural anatomy of attention. *Trends Cog Sci*, **4**, 383–91.

Malpeli, J.G. (1983). Activity of cells in area 17 of the cat in absence of input from layer A of lateral geniculate nucleus. *J Neurophysiol*, **49**, 595–610.

Marder, E. (1998). From biophysics to models of network function. *Annu Rev Neurosci*, **21**, 25–45.

Marr, D. (1982). *Vision: A Computational Investigation into the Human Representation and Processing of Visual Information*. W.H. Freeman, San Francisco.

Masaki, H., Takasawa, N., and Yamazaki, K. (2000). An electrophysiological study of the locus of the interference effect in a stimulus-response compatibility paradigm. *Psychophysiology*, **37**, 464–72.

Mays, L.E. and Sparks, D.L. (1980). Dissociation of visual and saccade-related responses in SC neurons. *J Neurophysiol*, **43**, 207–32.

McCarthy, G. and Donchin, E. (1981). A metric for thought: A comparison of P300 latency and reaction time. *Science*, **211**, 77–80.

McClelland, J.L. (1979). On the time relations of mental processes: An examination of systems of processes in cascade. *Psych Rev*, **86**, 287–330.

Menon, V., Adleman, N.E., White, C.D., Glover, G.H., and Reiss, A.L. (2001). Error-related brain activation during a Go/NoGo response inhibition task. *Hum Brain Mapp*, **12**, 131–43.

Meyer, D.E., Osman, A.M., Irwin, D.E., and Yantis, S. (1988). Modern mental chronometry. *Biol Psychol*, **26**, 3–67.

Miller, J.O. (1988). Discrete and continuous models of human information processing: theoretical distinctions and empirical results. *Acta Psychol*, **67**, 191–257.

Miller, J., Riehle, A., and Requin, J. (1992). Effects of preliminary perceptual output on neuronal activity of the primary motor cortex. *J Exp Psychol: Hum Percept Perfor*, **18**, 1121–38.

Moore, T. and Fallah, M. (2001). Control of eye movements and spatial attention. *Proc Natl Acad Sci USA*, **98**, 1273–6.

Motter, B.C. (1994). Neural correlates of attentive selection for color or luminance in extrastriate area V4. *J Neurosci*, **14**, 2178–89.

Mountcastle, V.B. (1957). Modality and topographic properties of single neurons of cat's somatic sensory cortex. *J Neurophysiol*, **20**, 408–34.

Mouret, I. and Hasbroucq, T. (2000). The chronometry of single neuron activity: testing discrete and continuous models of information processing. *J Exp Psychol Hum Percept Perform*, **26**, 1622–38.

Murthy, A., Thompson, K.G., and Schall, J.D. (2001). Dynamic dissociation of visual selection from saccade programming in frontal eye field. *J Neurophysiol*, **86**, 2634–7.

Newsome, W.T. (1997). The King Solomon lectures in neuroethology. Deciding about motion: linking perception to action. *J Comp Physiol*, **18**, 5–12.

Niemi, P., Näätänen, R. (1981). Foreperiod and simple reaction time. *Psych Bull*, **89**, 133–62.

Nobre, A.C., Gitelman, D.R., Dias, E.C., and Mesulam, M.M. (2000). Covert visual spatial orienting and saccades: overlapping neural systems. *Neuroimage*, **11**, 210–6.

Norman, D.A. and Shallice, T. (1986). Attention to action: Willed and automatic control of behavior. In Davidson, R.J., Schwartz, G.E., Shapiro, D. (Eds.), *Consciousness and Self-regulation* (Vol. 4): *Advances in Research and Theory*, pp. 1–18. Plenum Press, New York.

Nosofsky, R.M. and Palmeri, T.J. (1997). An exemplar-based random walk model of speeded classification. *Psychol Rev*, **104**, 266–300.

Oldenquist, A. (1967). Choosing, deciding and doing. In Edwards, P. (Ed.) *Encyclopedia of Philosophy*, pp. 96–104. Macmillian, New York.

Palmer, J., Verghese, P., and Pavel, M. (2000). The psychophysics of visual search. *Vision Res*, **40**, 1227–68.

Paré, M. and Wurtz, R.H. (2001). Progression in neuronal processing for saccadic eye movements from parietal cortex area lip to SC. *J Neurophysiol*, **85**, 2545–62.

Paré, M. and Hanes, D.P. (2003). Controlled movement processing: Superior colliculus activity associated with countermanded sarcades. *J Neurosci*, **23**, 6480–9.

Parker, A.J. and Newsome, W.T. (1998). Sense and the single neuron: Probing the physiology of perception. *Annu Rev Neurosci*, **21**, 227–77.

Pfefferbaum, A. and Ford, J.M. (1985). ERPs to response production and inhibition. *Electroenceph Clin Neurophysiol*, **60**, 423–34.

Platt, M.L. and Glimcher, P.W. (1999). Neural correlates of decision variables in parietal cortex. *Nature*, **400**, 233–8.

Posner, M.I. (1978). *Chronometric Explorations of Mind*. Hillsdale, New Jersey: L Erlbaum Associates.

Posner, M.I. and DiGirolamo, G.J. (1998). Conflict, target detection and cognitive control. In R. Parasuraman, (Ed.) *Executive Attention: The Attentive Brain*, pp. 401–23. MIT Press, Cambridge, MA.

Rabbitt, P.M.A. (1966). Errors and error-correction in choice-response tasks. *J Exp Psychol*, **71**, 264–72.

Ratcliff, R. (1978). A theory of memory retrieval. *Psych Rev*, **85**, 59–108.

Ratcliff, R. and Rouder, J.N. (1998). Modeling response times for two-choice decisions. *Psych Sci*, **9**, 347–56.

Ratcliff, R. and Rouder, J.N. (2000). A diffusion model account of masking in two-choice letter identification. *J Exp Psychol: Hum Percept Perform*, **26**, 127–40.

Rees, G., Friston, K., and Koch, C. (2000). A direct quantitative relationship between the functional properties of human and macaque V5. *Nat Neurosci*, **3**, 716–23.

Requin, J. and Riehle, A. (1995). Neural correlates of partial transmission of sensorimotor information in the cerebral cortex. *Acta Psychol*, **90**, 81–95.

Requin, J., Riehle, A., and Seal, J. (1988). Neuronal activity and information processing in motor control: from stages to continuous flow. *Biol Psychol*, **26**, 179–98.

Richter, W., Andersen, P.M., Georgopoulos, A.P., and Kim, S.G. (1997). Sequential activity in human motor areas during a delayed cued finger movement task studied by time-resolved fMRI. *Neuroreport*, **8**, 1257–61.

Riehle, A. and Requin, J. (1993). The predictive value for performance speed of preparatory changes in neuronal activity of the monkey motor and premotor cortex. *Behav Brain Res*, **53**, 35–49.

Riehle, A., Kornblum, S., and Requin, J. (1997). Neuronal correlates of sensorimotor association in stimulus-response compatibility. *J Exp Psychol: Hum Percept Perform*, **23**, 1708–26.

Riehle, A., Grammont, F., Diesmann, M., and Grun, S. (2000). Dynamical changes and temporal precision of synchronized spiking activity in monkey motor cortex during movement preparation. *J Physiol Paris*, **94**, 569–82.

Roitman, J.D. and Shadlen, M.N. (2002). Response of neurons in the lateral intraparietal area (LIP) during a combined visual discrimination reaction time task. *J Neurosci*, 22, 9475–89.

Romo, R. and Salinas, E. (2001). Touch and go: decision-making mechanisms in somatosensation. *Annu Rev Neurosci*, 24, 107–37.

Romo, R., Merchant, H., Zainos, A., and Hernández, A. (1996). Categorization of somaesthetic stimuli: sensorimotor performance and neuronal activity in primary somatic sensory cortex of awake monkeys. *NeuroReport*, 7, 1273–9.

Romo, R., Merchant, H., Zainos, A., and Hernández, A. (1997). Categorical perception of somesthetic stimuli: psychophysical measurements correlated with neuronal events in primate medial premotor cortex. *Cereb Cortex*, 7, 317–26.

Romo, R., Brody, C.D., Hernández, A., and Lemus, L. (1999). Neuronal correlates of parametric working memory in the prefrontal cortex. *Nature*, 339, 470–3.

Romo, R., Hernández, A., Zainos, A., Brody, C.D., and Lemus, L. (2000). Sensing without touching: psychophysical performance based on cortical microstimulation. *Neuron*, 26, 273–8.

Sakagami, M., Tsutsui, Ki., Lauwereyns, J., Koizumi, M., Kobayashi, S., and Hikosaka, O. (2001). A code for behavioral inhibition on the basis of color, but not motion, in ventrolateral prefrontal cortex of macaque monkey. *J Neurosci*, 21, 4801–8.

Salinas, E. and Romo, R. (1998). Conversion of sensory signals into motor commands in primary motor cortex. *J Neurosci*, 18, 499–511.

Salinas, E., Hernández, H., Zainos, A., and Romo, R. (2000). Periodicity and firing rate as candidate neural codes for the frequency of vibrotactile stimuli. *J Neurosci*, 20, 5503–15.

Salzman, C.D. and Newsome, W.T. (1994). Neural mechanisms for forming a perceptual decision. *Science*, 264, 231–7.

Sato, T., Murthy, A., Thompson, K.G., and Schall, J.D. (2001). Effect of search efficiency but not response interference on visual selection in frontal eye field. *Neuron*, 30, 583–91.

Schall, J.D. (1991). Neuronal activity related to visually guided saccadic eye movements in the supplementary motor area of rhesus monkeys. *J Neurophysiol*, 66, 530–58.

Schall, J.D. (1997). Visuomotor areas of the frontal lobe. In K. Rockland, A. Peters, J. Kaas, (Eds.) *Cerebral Cortex* Volume 12: *Extrastriate Cortex of Primates*, pp. 527–638. Plenum, New York.

Schall, J.D. and Thompson, K.G. (1999). Neural selection and control of visually guided eye movements. *Annu Rev Neurosci*, 22, 241–59.

Schultz, W. and Dickinson, A. (2000). Neuronal coding of prediction errors. *Annu Rev Neurosci*, 23, 473–500.

Schwark, H.D., Malpeli, J.G., Weyand, T.G., and Lee, C. (1986). Cat area 17. II. Response properties of infragranular layer neurons in the absence of supragranular layer activity. *J Neurophysiol*, 56, 1074–87.

Scudder, C.A., Kaneko, C.S., and Fuchs, A.F. (2002). The brainstem burst generator for saccadic eye movements: A modern synthesis. *Exp Brain Res*, 142, 439–62.

Shadlen, M.N. and Newsome, W.T. (2001). Neural basis of a perceptual decision in the parietal cortex (area LIP) of the rhesus monkey. *J Neurophysiol*, 86, 1916–36.

Shadlen, M.N., Britten, K.H., Newsome, W.T., and Movshon, J.A. (1996). A computational analysis of the relationship between neuronal and behavioral responses to visual motion. *J Neurosci*, 16, 1486–510.

Shaw, B.K. and Kristan, W.B. (1997). The neuronal basis of the behavioral choice between swimming and shortening in the leech: control is not selectively exercised at higher circuit levels. *J Neurosci*, 17, 786–95.

Shen, L. and Alexander, G.E. (1997). Neural correlates of a spatial sensory-to-motor transformation in primary motor cortex. *J Neurophysiol*, 77, 1171–94.

Smid, H.G., Mulder, G., and Mulder, L.J. (1990). Selective response activation can begin before stimulus recognition is complete: a psychophysiological and error analysis of continuous flow. *Acta Psychol (Amst)*, **74**, 169–201.

Smulders, F.T., Kok, A., Kenemans, J.L., and Bashore, T.R. (1995). The temporal selectivity of additive factor effects on the reaction process revealed in ERP component latencies. *Acta Psychol*, **90**, 97–109.

Sparks, D.L. (1978). Functional properties of neurons in the monkey superior colliculus: Coupling of neuronal activity and saccade onset. *Brain Res*, **156**, 1–16.

Sternberg, S. (2001). Separate modifiability, mental modules, and the use of pure and composite measures to reveal them. *Acta Psychol*, **106**, 147–246.

Sternberg, S., Monsell, S., Knoll, R.L., and Wright, C.E. (1978). The latency and duration of rapid movement sequences: Comparisons of speech and type writing. In G.E. Stelmach, (Ed.) *Information Processing in Motor Control*. Academic Press, New York NY.

Stuphorn, V., Taylor, T.L., and Schall, J.D. (2000). Performance monitoring by the supplementary eye field. *Nature*, **408**, 857–60.

Talbot, W.H., Darian-Smith, I., Kornhuber, H.H., and Mountcastle, V.B. (1968). The sense of flutter-vibration: comparison of the human capacity response patterns of mechanoreceptive afferents from the monkey hand. *J Neurophysiol*, **31**, 301–34.

Tanji, J., Shima, K., and Mushiake, H. (1996). Multiple cortical motor areas and temporal sequencing of movements. *Brain Res Cogn Brain Res*, **5**, 117–22.

Thompson, K.G., Hanes, D.P., Bichot, N.P., and Schall, J.D. (1996). Perceptual and motor processing stages identified in the activity of macaque frontal eye field neurons during visual search. *J Neurophysiol*, **76**, 4040–55.

Thompson, K.G., Bichot, N.P., and Schall, J.D. (1997). Dissociation of target selection from saccade planning in macaque frontal eye field. *J Neurophysiol*, **77**, 1046–50.

Thompson, K.G., Bichot, N.P., and Schall, J.D. (2001). From attention to action in frontal cortex. In J. Braun, C. Koch, J. Davis (Eds) *Visual Attention and Cortical Circuits* pp. 137–57. MIT Press, Cambridge MA.

Townsend, J. T. and Ashby, F.G. (1983). *The Stochastic Modeling of Elementary Psychological Processes*. Cambridge University Press, New York .

Treisman, A. (1988). Features and objects: The Fourteenth Bartlett Memorial Lecture. *Q J Exp Psychol*, **40**A, 201–37.

Treue, S. (2001). Neural correlates of attention in primate visual cortex. *Trends Neurosci*, **24**, 295–300.

Trevena, J. and Miller, J. (2002). Cortical movement preparation before and after a conscious decision to move. *Conscious Cogn*, **11**, 162.

Usher, M. and McClelland, J.L. (2001). The timecourse of perceptual choice: the leaky, competing accumulator model. *Psychol Rev*, **108**, 550–92.

Uttal, W.R. (2001). *The New Phrenology: The Limits of Localizing Cognitive Processes in the Brain*. MA: MIT Press, Cambridge MA.

van Boxtel, G.J., van der Molen, M.W., Jennings, J.R., and Brunia, C.H. (2001). A psychophysiological analysis of inhibitory motor control in the stop-signal paradigm. *Biol Psychol*, **58**, 229–62.

Van Zandt, T. and Ratcliff, R. (1995). Statistical mimicking of reaction time data: Single-process models, parameter variability and mixtures. *Psychon Bull Rev*, **2**, 20–54.

Van Zandt, T., Colonius, H., and Proctor, R.W. (2000). A comparison of two response time models applied to perceptual matching. *Psychon Bull Rev*, **7**, 208–56.

Wallis, J.D., Anderson, K.C., and Miller, E.K. (2001). Single neurons in prefrontal cortex encode abstract rules. *Nature*, **411**, 953–6.

White, I.M. and Wise, S.P. (1999). Rule-dependent neuronal activity in the prefrontal cortex. *Exp Brain Res*, **126**, 315–35.

Wise, S.P. and Murray, E.A. (2000). Arbitrary associations between antecedents and actions. *Trends Neurosci*, **23**, 271–6.

Wise, S.P., Boussaoud, D., Johnson, P. B., and Caminiti, R. (1997). Premotor and parietal cortex: Corticocortical connectivity and combinatorial computations. *Annu Rev Neurosci*, **20**, 25–42.

Werner, G. and Mountcastle, V.B. (1965). Neural activity in mechanoreceptive cutaneous afferents: stimulus-response relations, Weber functions, and information transmission. *J Neurophysiol*, **28**, 359–97.

Wolfe, J.M. (1994). Guided search 2.0: A revised model of visual search. *Psychon Bull Rev*, **1**, 202–38.

Wolfe, J.M. (1998). Visual search. In Pashler, H. (Ed.) *Attention* (pp. 13–74). Psychological Press, Hove, East Sussex.

Wolpert, D.M. and Ghahramani, Z. (2000). Computational principles of movement neuroscience. *Nat Neurosci*, **3**, 1212–7.

Wurtz, R.H. (1969). Visual receptive fields of striate cortex neurons in awake monkeys. *J Neurophysiol*, **32**, 727–42.

Wurtz, R.H., Sommer, M.A., Pare, M., and Ferraina, S. (2001). Signal transformations from cerebral cortex to SC for the generation of saccades. *Vision Res*, **41**, 3399–412.

Zingale, C.M. and Kowler, E. (1987). Planning sequences of saccades. *Vision Res*, **27**, 1327–41.

Chapter 21

Human brain imaging reveals a parietal area specialized for grasping

Jody Culham

Abstract

We used functional magnetic resonance imaging to reveal a region of the anterior intraparietal (AIP) sulcus that is activated during visually guided grasping (versus reaching) actions for which information about object shape and size is required to preshape the hand. Human AIP is a potential homolog of macaque AIP, an area containing neurons that are activated by the viewing and grasping of specific shapes. Like macaque AIP, human AIP is active during both the vision and action phases of a delayed grasping trial. Human AIP is not activated by the visual presentation of two-dimensional images of objects, and does not seem to be an all-purpose object-processing area. Conversely, the lateral occipital complex, a ventral stream area that responds to two-dimensional object images, shows no enhanced activation for grasping compared to reaching. AIP is activated in a visual form agnosia patient with poor object recognition but intact grasping abilities. Taken together, these results suggest that AIP within both monkey and human parietal cortex is the critical neural substrate for the processing of object shape for actions but not for recognition. These results demonstrate the utility of using information about specialized regions within the macaque to suggest functional modules within human association cortex.

21.1 Mapping human parietal cortex: A call to 'action'

Human brain mapping studies have been particularly valuable in understanding vision and visual cognition. For example, topographic mapping studies have revealed numerous retinotopic regions of the occipital cortex (e.g. Sereno *et al.* 1995), as well as less topographic higher-order areas, such as those involved in processing motion (Tootell *et al.* 1995), objects (Malach *et al.* 1995), faces (Kanwisher *et al.* 1997), places (Epstein and Kanwisher 1998), and bodies (Downing *et al.* 2001). In many of these cases, I contend that the most valuable studies have focused on specific areas—or, in neuroimaging jargon, 'regions of interest' (ROIs)—rather than producing 'laundry lists' of regions activated by a particular stimulus or task, or employing meta-analytic techniques (Cabeza and Nyberg 1997, 2000), which can appear disappointing even in zones known to contain distinct subregions (Farah and Aguirre 1999).

Typically, new areas have first been identified using criteria expected from other species (e.g. Dukelow et al. 2001) or from stages of cognitive processing suggested by behavioral or neuropsychological techniques (e.g. Kanwisher et al. 1997). Following the reliable identification of a particular ROI, interesting theoretical questions can be addressed. In established areas, important questions have been asked about the relationships between activation in the area and:

(1) hypothesized stages of cognitive processing (e.g. Grill-Spector et al. 1998) and the functional properties expected of units in those stages (e.g. Grill-Spector et al. 1999);

(2) neuropsychological deficits (e.g. Marotta et al. 2001);

(3) development and plasticity (e.g. Kanwisher 2000; Tarr and Gauthier 2000);

(4) the organizational principles that could determine specialization (Ishai and Sagi 1995; Levy et al. 2001); and

(5) the properties of areas in other species (e.g. Tootell et al. 1997; Sereno 1998; Logothetis et al. 2001).

Despite the successful mapping of areas in the early visual areas of occipital cortex and visual perceptual areas of temporal cortex, there is a surprising dearth of systematic exploration in the posterior parietal cortex (PPC, the regions of the parietal lobes posterior to somatosensory cortex). PPC forms part of the 'dorsal stream', which is thought to use visual information for the guidance of actions, in contrast to the 'ventral stream', which is thought to use visual information for perceptual tasks such as object recognition. While some dorsal-stream functions, such as motion perception, have been well studied at the early stages (especially the MT^+ complex), little is known about the areas and functioning of human PPC. A recent literature search by Mel Goodale (personal communication) in the PubMed database of scientific papers revealed considerably more papers on brain imaging and perception topics than imaging and action topics, by a factor of about 10:1. This is likely due, in part, to relative ease of studying perceptual tasks compared to action tasks in the neuroimaging environment. In addition, a number of other caveats exist (Culham and Kanwisher 2001). In particular, expected parietal functions such as attention, working memory, eye movement planning, and spatial encoding are common components of many cognitive tasks and may account for much of the overlapping activation found across many functionally disparate tasks. We have even suggested (Culham and Kanwisher 2001) that, in some cases, demonstrations of what does *not* activate parietal cortex can be more surprising and informative than demonstrations of what does (e.g. Wojciulik and Kanwisher 1999). However, this approach requires an examination of regional activity across numerous tasks and control conditions to really determine its role in cognition.

Despite these difficulties, with careful imaging studies it may be possible to identify functional areas within parietal cortex. Certainly, neurophysiological evidence from non-human primates has suggested a mosaic of specialized subregions within parietal cortex. Specific areas of macaque cortex appear specific to functions such as eye movements and attention, reaching, grasping, processing of shape and surface orientation, and motion near the head (Colby and Duhamel 1991; Andersen *et al.* 1997). Our review of neuroimaging of human parietal cortex suggested that similar areas may exist in human parietal cortex, in a somewhat similar neuroanatomical arrangement to the monkey (Culham and Kanwisher 2001). And recently, comparisons of different tasks within the same subjects have demonstrated segregation of functions that are expected from monkey cortex (eye movements, attention, reaching, grasping) as well as functions that are more uniquely human (calculation and language) (Simon *et al.* 2002).

21.2 Visually guided grasping: A case study of interspecies comparisons in parietal cortex

> It is not the hand that is perfect, but the whole nervous mechanism by which movements of the hand are evoked, coordinated, and controlled
>
> Frederick Wood Jones (1942, pp. 298–300)

In this chapter, I will review recent research from my lab to illustrate how brain imaging, guided by knowledge of monkey physiology, can explore the human brain for possible homologies. The results here will focus on a region in the anterior intraparietal sulcus that appears to be specialized for grasping in both the monkey and the human.

21.2.1 Why would there be an area specialized for grasping?

Consider the act of opening a door. Humans easily recognize common opening mechanisms, such as a door knob, a door handle, or a push bar. Once the type of mechanism has been identified, say a pull door with a vertical handle, the hand must be transported to the handle, the hand must be oriented with a vertical opening between the fingers and thumb, and the hand opening must be large enough to surround the handle prior to closure (Jeannerod 1981). Both recognition of the door handle and the preshaping of the hand for grasping require *object processing* to determine the characteristics of the handle. However, the computational requirements of these two functions are considerably different. In the case of recognition, the system must be able to identify the door handle from a variety of potential locations, distances, and viewpoints. However, when the person goes to grasp the handle, viewpoint and distance information are critical to precisely encode the spatial location of the handle relative to the arm, hand, or fingers.

Indeed, the computational requirements of perceptual and action mechanisms appear sufficiently different to be performed in different systems of the brain. For example, a patient, D.F., with visual form agnosia and poor object recognition abilities can none the less grasp an object accurately (Goodale *et al.* 1991). Conversely, a patient with optic ataxia can easily recognize objects but fails to preshape the hand appropriately during grasping (Jakobson *et al.* 1991). Based on such data, as well as behavioral evidence from normal subjects, it has been suggested that object recognition is performed in the ventral stream from occipital to temporal cortex, whereas object grasping is performed in the dorsal stream from occipital to parietal cortex (Goodale and Milner 1992; Milner and Goodale 1995).

Neuroimaging studies have been quite successful in identifying the ventral stream substrates of object perception. One key area that is involved in object perception and recognition is the lateral occipital (LO) complex in the ventral stream (Malach *et al.* 1995; Grill-Spector *et al.* 2001). As would be expected, area LO is invariant to size and position, though not viewpoint (Grill-Spector *et al.* 1999). The homology between LO and monkey areas involved in object perception is not yet clear, though macaque inferotemporal (IT) cortex is a likely homolog (Tanaka, 1996).

The characterization of dorsal stream substrates involved in object processing has been much more limited. Monkey physiology suggests that an area of parietal cortex is specialized for grasping and other forms of hand manipulation that require processing of object features. Based on these results in the monkey, in the research described here, my colleagues and I have used functional brain imaging to investigate whether a similar area exists in the human, and how its properties differ from object areas within the ventral stream.

21.2.2 A 'grasping area' in monkeys

Initial forays into PPC were challenging for physiologists because, unlike basic sensory and motor areas, PPC neurons were unexpectedly complex (Hyvärinen and Poranen 1974; Mountcastle *et al.* 1975). PPC cells demonstrated various combinations of visual, motor, somatosensory, motion, and eye movement responses. Two particularly intriguing populations were the 'arm projection' and 'hand manipulation' cells reported in the posterior bank of the anterior intraparietal sulcus (within Brodmann's area 7). Whereas arm projection neurons responded when the monkey reached out with the arm towards a target, hand manipulation neurons fired when the monkey manipulated a target object with the hand, as in a grasping action. In both cases, activation only occurred when the monkey was actively trying to acquire a target object or location, but not when the arms or hands were passively stimulated. Mountcastle (1975) reported that hand manipulation neurons fired furiously when the monkey manipulated the hand within a small box to obtain food ('winkling' as he called it).

Hand manipulation neurons were studied more systematically by Sakata and his colleagues (Taira *et al.* 1990; Sakata *et al.* 1995; Murata *et al.* 1996, 2000). They reported that neurons in an area they called AIP, in the anterior end of the intraparietal sulcus, fired when the monkey viewed and manipulated different types of objects. The majority of neurons displayed a preference for one particular type of object over the others tested; for example, a particular neuron might fire most when the monkey manipulated a lever at a particular orientation, more so than when it manipulated a lever at another orientation or manipulated a pull-knob. Based on responses to manipulation in the light, manipulation in the dark, and viewing of the objects without manipulation, they reported several subpopulations of AIP neurons. 'Motor dominant' neurons responded only when manipulation occurred (in the light or dark); 'visual dominant' neurons responded only when the object was visible (with or without manipulation); and 'visual–motor' neurons responded best when the object was both seen and manipulated. As a population, AIP cells were generally selective to shape, size, and orientation (Murata *et al.* 2000) but were relatively insensitive to position (Taira *et al.* 1990).

Hand manipulation neurons were also reported in Brodmann's area 5, a somatosensory association area on the anterior bank of the intraparietal sulcus (across the sulcus from AIP) (Mountcastle *et al.* 1975). Recently, the responses in somatosensory cortex (SI), area 5, and area 7 (AIP) have been contrasted (Debowy *et al.* 2001). Whereas, SI neurons were most affected during hand contact, lifting, and holding, PPC neurons (5 and 7) fired most vigorously during hand preshaping prior to contact.

Thus it seems that PPC cortex, particularly in AIP, is involved in the recognition of objects during preparation of hand actions toward them and perhaps also in the evaluation of whether the planned grip matches the goal (based on visual feedback and/or corollary discharge from premotor cortex; Sakata *et al.* 1995). The most compelling evidence for a crucial role of AIP in preshaping the hand to the target object comes from studies of AIP inactivation (Gallese *et al.* 1994). When processing in AIP is chemically disrupted, the monkey loses the ability to preshape the contralateral hand to grasp a particular object, although reaching to the object's location remains intact and, upon contacting the object, the monkey can correct the grasp.

21.2.3 A 'grasping area' in humans?

The strongest suggestion that humans, like monkeys, have a brain region specialized for grasping comes from the neuropsychological literature. Patients with PPC lesions can demonstrate optic ataxia, a deficit in grasping and reaching movements (Balint 1909; Jeannerod 1986). Furthermore, these grasping and reaching deficits appear to be dissociable (Jeannerod *et al.* 1994). In an elegant study, Binkofski and colleagues (1998) demonstrated that human patients with lesions in the anterior intraparietal sulcus displayed deficits in grasping while reaching remained relatively intact (much like the performance of AIP-inactivated monkeys).

Neuroimaging data also suggest a human region specialized for grasping. Binkofski *et al.* (1998) demonstrated greater fMRI activation for grasping compared to reaching in normals in the same area that, when lesioned, had led to grasping deficits. However, other studies that attempted to find such a 'grasping area' have had mixed results. One PET study reported activity in the anterior PPC with visually guided grasping (versus reaching) (Faillenot *et al.* 1997); however, two other PET studies did not (Grafton *et al.* 1996; Matsumura *et al.* 1996). Other imaging studies reported activity in the vicinity with non-visual grasping (e.g. Ehrsson *et al.* 2000), but the possibility remains that this region is simply driven by tactile finger stimulation. A number of reasons could account for the mixed results in the neuroimaging of grasping literature, including: the loss of sensitivity with intersubject averaging in PET, the limited number of objects that are presentable in a confined space, the presence of motion artifacts that are particularly troublesome in motor control experiments, and the lack of control for somatosensory finger stimulation that co-occurs with grasping.

21.3 New investigations of a 'grasping area' in humans

We began our search for a human grasping area, the possible homolog of AIP, by developing a robust paradigm to study grasping. We further examined the contribution of somatosensory, visual, and action components. Once a 'grasping area' had been established across numerous individual subjects, we began to address theoretical questions about its role in object processing for action, as compared to the ventral object-selective area, LO. We also addressed the effect of introducing a delay between the sight of the object and the initiation of the movement. Delayed actions were useful to dissociate the vision- and action-related activation . Finally, we investigated whether the preserved grasping abilities of D.F., the patient with visual form agnosia, may indeed be subserved by the human grasping area.

21.3.1 General methods

We developed a paradigm to study visually guided reaching and grasping in a high-field (4 Tesla) magnetic environment using blood-oxygenation level dependent (BOLD) activation. The subject lay supine within the magnet, with the torso tilted slightly and the head tilted at an angle (~30°) which permitted direct viewing of the target, without mirrors. A whole head coil was used to sample 13 oblique slices at an orientation approximately parallel to the calcarine sulcus to include occipital, parietal, posterior temporal, and superior-posterior frontal cortex. We sampled the brain at a resolution of $3 \times 3 \times 6$ mm with a volume collected every 2 s. A light-emitting diode (LED) was mounted on the ceiling of the magnet bore to provide a fixation point at the subject's natural line of gaze, while graspable objects were presented in the lower visual field. We presented a diverse and unpredictable sequence of objects using custom equipment, which became known as the 'grasparatus'. It consisted of an octagonal rotating drum with four translucent rectangular shapes mounted horizontally on each of the eight faces. Targets were rectangular shapes of varied length and orientation. Beneath each of the four target locations, a superbright LED was positioned to illuminate the target. The grasparatus was connected to a computer-controlled solenoid that could release air into a hose connected to a piston to rotate the drum to the next face.

Experiments were performed in the dark to reduce confounding activation due to visual stimulation, such as the visual motion produced by movement of the hand. On each trial, one of the rectangular target objects would be illuminated briefly (250 ms). Between trials, the drum would rotate to the next face in the dark. The color of the fixation LED prior to target illumination cued the subject regarding the motor response. When the fixation point was green, the subject used the index finger and thumb (precision grip) to grasp the target along the long axis. When the fixation point was red, the subject reached forward with the arm to touch the object without preshaping the hand.

We took great care to minimize confounding effects of head motion (and distortions of the magnetic field due to the changing position of the mass of the hand). An arm brace restrained the subject's upper right arm and foam padding was placed around the head. A single trial design ensured that the activation occurred with the appropriate hemodynamic lag (~4–6 s) and profile, unlike motion-related artifacts which would have occurred immediately.

Data from Experiments 1–3 were collected in the same population of seven subjects. Data were analyzed and visualized using Brain Voyager software (Brain Innovation, Maastricht, The Netherlands). We co-registered each subject's anatomical and functional data from each session to a 'canonical' high-resolution anatomical. Each subject's canonical anatomical was segmented at the gray–white matter boundary, rendered in three dimensions, and inflated to show the cortical surface. Statistical analyses (general linear models) were performed both for time courses extracted from individual subjects and for group-averaged data in Talairach space. All effects illustrated in the figures were significant at $P < 0.001$, uncorrected, or better.

21.3.2 Experiment 1: Identification of a 'grasping area'

21.3.2.1 Rationale

We began by investigating whether any brain area showed significantly greater fMRI activation for grasping versus reaching (Culham *et al.* 2000). Grasping and reaching both share the 'transport' component of moving the hand to the target object's location. However, only grasping also requires a 'grip' component that uses object orientation and size information to preshape the hand (Jeannerod 1981).

21.3.2.2 Methods and results

Subjects viewed one trial every 14 s. To reduce the frequency of task switching, the task alternated between grasping and reaching every four trials. General linear models used two predictors: one for the grasping response and one for the reaching response, both convolved with the hemodynamic response function.

The actions of reaching and grasping considered together (relative to an intertrial interval baseline) activated a large extent of parietal cortex along the intraparietal sulcus in addition to numerous other somatosensory, motor, premotor, and visual areas. However, few areas were significantly more activated by grasping compared to reaching and no areas were significantly more activated by reaching than grasping. As shown in Fig. 21.1A, when reaching (transport) activity was subtracted from grasping (transport + grip) activity, one reliable focus of residual (grip) activation was found at the anterior intraparietal sulcus (IPS) and the postcentral sulcus (PostCS). This region has been found consistently in over 10 subjects and is reliably observed within subjects scanned across multiple sessions.

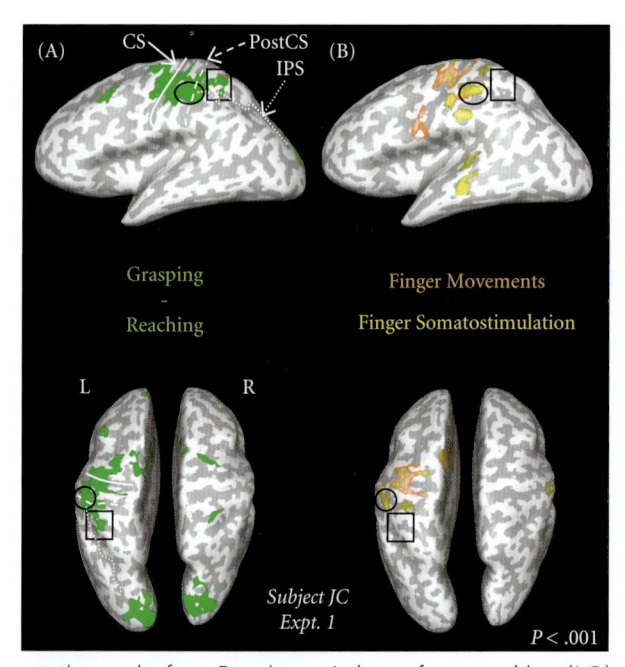

Fig. 21.1 Representative results from Experiment 1 shown for one subject (J.C.) on the inflated cortical surface. The top row shows a lateral view of the left hemisphere and the lower row shows a view of both hemispheres from above (with the frontal lobes at the top of the image and the occipital lobes at the bottom). Sulci appear dark gray and gyri appear light gray. The central (CS), postcentral (PostCS) and intraparietal (IPS) sulci are indicated by solid, dashed, and dotted lines, respectively. (A) Activation for immediate grasping – immediate reaching in Experiment 1. (B) Activation for finger movements - rest (orange) and finger somatostimulation – rest (yellow) in Experiment 1. The circle outlines the region of grasping (versus reaching) activation that was also activated by somatosensory finger stimulation. The square outlines the region of grasping activation that did not overlap with the somatosensory activation.

For our purposes, the critical difference between grasping and reaching is the utilization of object properties to preshape the grip. However, it is possible that other differences between the two tasks, particularly differences in the degree of digit control and somatosensory feedback to the fingers, could account for the activity seen in the anterior IPS. Indeed, we frequently observed activation in the hand area of motor cortex, as defined anatomically (Yousry *et al.* 1997). Clearly, one would not argue that motor cortex processes object information for grip formation. Instead, the activity is likely due to the fine motor control of the fingers necessary during grasp preparation and execution. Similarly, the activation in the anterior IPS/PostCS could reflect somatosensory processing.

21.3.2.3 Comparison with finger movement and tactile finger stimulation

To determine whether our grasping (−reaching) activation was due to grasping *per se*, or was merely a motor or somatosensory confound, we compared it to activation produced by motor control or somatosensory stimulation of the fingers. In a finger movement condition performed with a block design, subjects repeatedly moved the index finger and thumb together without touching. In a finger somatostimulation condition, subjects remained still with the palm facing upward and the finger and thumb approximately 6 cm apart while the experimenter moved a stick repeatedly to stimulate the fingertip and tip of the thumb (the skin areas that would receive tactile stimulation in a precision grip). In both conditions, subjects kept their eyes open but could not see their hands.

As illustrated in Fig. 21.1A, grasping (compared to reaching) produces activation in a contiguous region that includes both the postcentral sulcus and the anterior intraparietal sulcus. Comparisons with finger movement and somatosensory finger stimulation suggest, however, that this continuous region of activation may in fact consist of two subregions. The more anterior region, within the PostCS, is activated by both finger movements and somatosensory finger stimulation (Fig. 21.1B) in addition to grasping (−reaching). However, the more posterior region, in the IPS behind the PostCS, is activated by grasping (−reaching) but not by finger movement or somatostimulation. Furthermore, as we will see in Experiment 3 (below), the anterior IPS region, but not the PostCS region, is strongly activated by the visual presentation of the object to be acted upon in addition to the action itself.

In the monkey, AIP does not show somatosensory responses or motor responses to random hand actions that are not directed towards a goal (Mountcastle *et al.* 1975). However, although the majority of neurophysiological studies of grasping have focused on AIP, hand manipulation neurons have also been reported in areas 5 (Mountcastle *et al.* 1975) and 7b (Gallese *et al.* 2002), two areas with strong somatosensory inputs that have not yet been identified in human cortex. Thus the more anterior region of grasping activation may, like SI, area 5 or area 7b, have a stronger emphasis on the somatosensory aspects of grasping (such as using tactile feedback to control grip force); whereas, the more posterior focus may correspond to monkey AIP and rely on vision to plan hand actions. Indeed, as will be seen in Experiment 3, only the more posterior focus shows a visual response in a delayed action paradigm. We thus suggest that the posterior region is putative candidate for area AIP in the human and tentatively refer to this area as AIP.

21.3.3 Experiment 2: Contrasting object grasping and object recognition

21.3.3.1 Rationale

Our next experiment was designed to address the generality of object-related activation in both area AIP and in area LO, a well-established object-selective area in the ventral stream (Culham et al. 2001, 2003). Specifically, we wondered whether LO would also show a preference for grasping over reaching and whether AIP would respond to two-dimensional images of objects even when no manipulation task was required, or even possible.

21.3.3.2 Methods and results

We first identified AIP and LO using standard localizers. AIP was identified by the subtraction of grasping – reaching in the paradigm described above. LO was identified comparing the activation to two-dimensional images of objects versus scrambled images of the same objects (Kourtzi and Kanwisher 2000). To maintain attention in both viewing conditions of the LO localizer, we employed a '1-back' task, in which the subject was required to hit a button when an image was repeated twice in a row. After identifying each region with its standard localizer comparison, we then examined the activation within each area during the task used to localize the *other* region (AIP during object and scrambled viewing, LO during grasping and reaching).

We observed overlapping activation in LO for both object viewing (versus scrambled objects) and reaching and grasping (versus the intertrial interval) as in Fig. 21.2A; however, no activity was found in LO for grasping when reaching was subtracted out (Fig. 21.2B). These effects can be observed in sample time courses of activation (Fig. 21.2C). As expected, AIP showed higher activity during grasping than reaching and LO showed higher activity during the viewing of objects than scrambled objects. However, neither the posterior grasping region (putative AIP) nor the anterior grasping region (somatosensory-related activation) showed any response to images of intact objects.

These results suggest that indeed AIP and LO are highly specialized with little functional overlap. AIP showed no response whatsoever to objects when shown in two-dimensional images that did not afford an action. LO produced a response when subjects grasped or reached towards visible objects, but showed no differential activity for grasping compared to reaching.

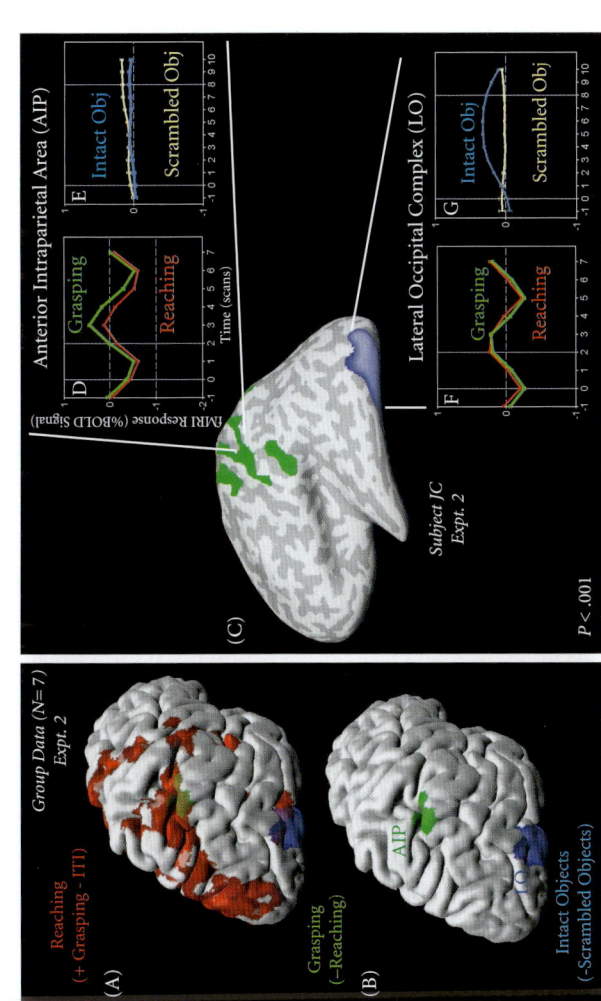

Fig. 21.2 Results of Experiment 2, comparing activation for grasping versus reaching and for images of intact objects versus scrambled objects. (A) Group activation rendered on one subject's outer gray matter surface (not necessarily representative of group anatomy), viewed from a postero-lateral perspective to show parietal and lateral occipital regions. Activation for reaching + grasping − ITI is shown in red. Activation for grasping − reaching is shown in green, but appears yellowish where it overlaps with activation for reaching. Activation for viewing of two-dimensional images of intact objects − scrambled objects appears in blue. Note the overlap between intact − scrambled reaching + grasping and reaching + grasping in purple in lateral occipital cortex. (B) The data in (A) are shown without the red reaching + grasping activation. When the reaching component is subtracted from grasping, there is no residual overlap with the object (−scrambled) activation. (C) Time courses of activation from one typical subject. Error bars show the standard error of the mean. While AIP-proper showed the expected pattern of greater activation for grasping versus reaching, it had no response to intact objects versus scrambled ones. LO showed the expected pattern of greater activation for intact versus scrambled objects. However, the LO response for grasping and reaching towards the real target objects was equal, with no greater activity for grasping.

21.3.4 **Experiment 3: Grasping after a delay**

21.3.4.1 Rationale

Although visual information is typically used to plan a grasp, people can grasp unseen objects, such as one's bedroom door handle, in the dark. How are such memory-guided actions performed? Which brain areas are involved in different phases of a memory-guided action—the viewing of the visual stimulus, the storing of the information through the delay period, and the execution of the action?

We introduced a delay between the viewing of an object and the reaching or grasping movement towards it in order to discriminate between responses to the visual presentation of an object and responses to the motor action and somatosensory feedback from making tactile contact (Culham *et al.* 2002). Recall that neurons in monkey AIP can be tuned to the visual component alone, the action component alone, or a combination of both (Taira *et al.* 1990; Murata *et al.* 2000). Thus in a true homolog of AIP, we would expect the population response measured by fMRI to show both vision and action responses.

Delayed action paradigms are frequently used to dissociate phases of the action; however, behavioral evidence suggests that the introduction of a delay can also profoundly affect the nature of the processing. To date, our analyses have focused on the use of the delay to tease apart the visual and action components. However, ongoing analyses and experiments are also using this paradigm to investigate delay-period activity. Although macaque physiology data (Murata *et al.* 1996) suggest that AIP neurons can maintain object-specific information over the delay period, human data in both patient D.F. (Goodale *et al.* 1994) and normal subjects (Goodale *et al.* 1994; Hu and Goodale 2000; Westwood *et al.* 2001) have led to the suggestion that the control of memory-guided actions depends on perceptual mechanisms within the ventral stream (Goodale *et al.* 1994). Neuroimaging of grasping with and without a delay between vision and the action may help reconcile whether the object information required for a delayed grasp is stored in dorsal (especially AIP) or ventral (such as LO) regions.

21.3.4.2 Methods

We examined subjects' activation patterns in AIP and LO during delayed grasping and reaching. At the start of each trial, the color of the fixation point cued the subject to make either a delayed grasping movement or a delayed reaching movement. Subjects remained in the dark except for a brief illumination (250 ms) of the target object. After target illumination, subjects withheld the action for 10 s until the fixation point blinked, at which point they initiated the movement. After the movement cue, there was a 10 s intertrial interval (ITI) while the subject waited for the next trial to begin.

21.3.4.3 Vision and action responses

The first analysis sought to dissociate vision-related and action-related responses within parietal cortex, particularly the regions activated by grasping (compared to reaching).

A general linear model analysis was performed with four regressors: the visual response for grasping, the visual response for reaching, the action response for grasping, and the action response for reaching. As illustrated for one subject in Fig. 21.3A, delayed grasping (grasp vision + grasp action versus reach vision + reach action) activated similar regions as immediate grasping (compared to immediate reaching). When the vision component (grasp vision + reach vision) and the action component (grasp action + reach action) were displayed (Fig. 21.3B), we typically observed that in the PostCS (circle outline in Figs 21.1 and 21.3A/B), only the action components were significant. By comparison, in the grasp-selective region of the anterior IPS, just posterior to the PostCS (square outline in Figs 21.1 and 21.3A/B), both the vision and action components were significant.

Time courses from group data further illustrate the responses, as shown in Fig. 21.3C. The activation has been color-coded such that areas with only an action response are shown in blue, areas with only a vision response would be shown in pink (but no such regions were observed), and areas with both vision and action responses are shown in purple. In Fig. 21.3C, note that in the left PostCS activation, particularly in the most anterior strip, the action-related response is very strong, whereas the visual response is much weaker and far less significant. For the action phase, the response was greater for grasping than reaching. The extent of PostCS activation was quite large in the left hemisphere, contralateral to the right hand that was used for the actions. Only a small region of activation near the PostCS was seen in the ipsilateral right hemisphere, which showed similar time courses as the left PostCS. Posterior to this action-related activation, the region we have called putative AIP, in the anterior IPS, showed both vision and action components that were highly significant. The grasping response in this posterior area was significantly stronger than the reaching response for the visual component in both hemispheres, and for the motor component in the right hemisphere.

These results demonstrate that the delayed action paradigm is an excellent tool to dissociate regions that may be activated by grasping (versus reaching) for different reasons. PostCS grasping activation was predominantly related to the motor or somatosensory processing that occurred at the execution of the action. This result corroborates our earlier description that only the PostCS responded to finger movements and finger somatostimulation. PostCS activation could be due to the feedback from the proprioceptive information about joint position or the tactile response to finger stimulation which would be stronger during grasping. Alternatively, the activation could be occurring in the human equivalent of areas 5 and/or 7, which, like AIP, respond during grasping (Debowy *et al.* 2001). By contrast, the anterior IPS displayed properties consistent with those expected for a human homolog of AIP—a clear response during both the vision and action phases of the trials. Reassuringly, this area showed a greater grasping (versus reaching) response in *both* the vision and action phases, indicating that the activation is not merely a confound of visual or somatosensory stimulation differences between grasping and reaching.

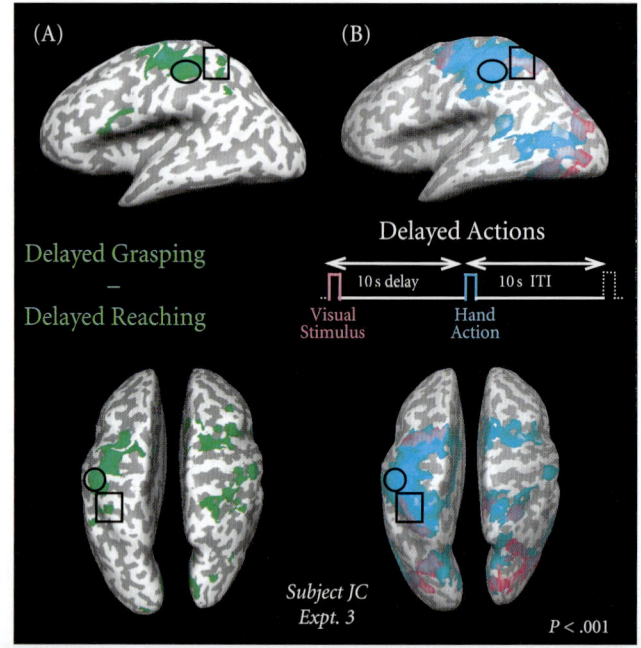

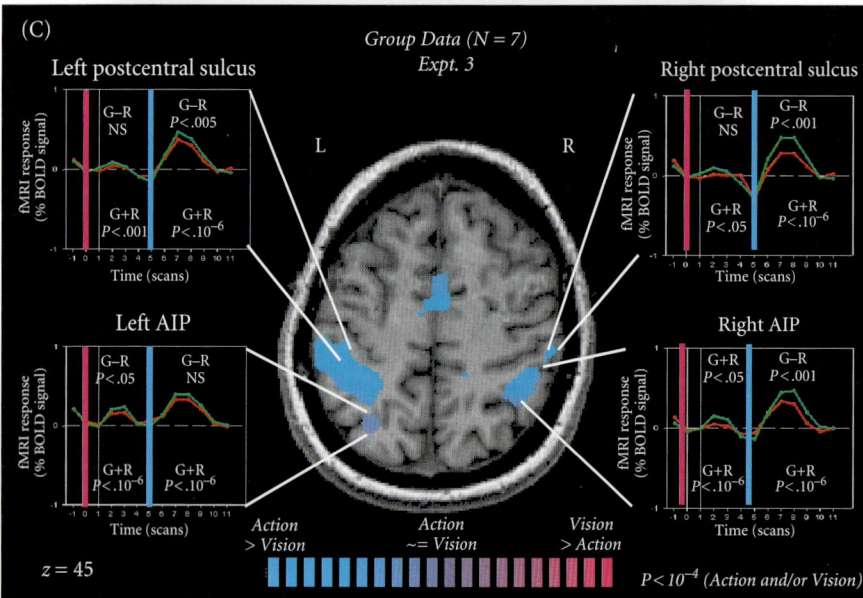

Fig. 21.3 Results of Experiment 3, examining activation in delayed reaching and grasping. In (A) and (B), the representative single-subject results are shown for the same subject seen in Fig. 21.1. (A) Green indicates regions with significantly higher activation for delayed grasping than delayed reaching (including both the vision and action components). (B) Blue indicates regions where the regressors for the visual phase of the trial (combining both reaching and grasping trials) were significant; pink indicates regions where the regressors for the action

21.3.4.4 Effect of the delay on dorsal and ventral activity

We have conducted preliminary analyses of brain activation during the delay period. We are collecting new data with this paradigm using a longer delay period and a larger sampling volume that will include frontal regions such as dorso-lateral prefrontal cortex and the possible homolog of monkey F5 in human Broca's area (Rizzolatti and Arbib 1998). Our preliminary analysis of data collected from seven subjects tested thus far with brief (10 s) delays indicates little delay-related activity in AIP (in a comparison of delay period activity during reaching and grasping as compared with the ITI). Where then is the information stored during the delay? Our preliminary analyses suggest that both area LO and visual areas along the occipital midline show 'reactivation' upon initiation of the action as well as an overall 'baseline shift' for grasping compared to reaching (increased activity throughout all phases of the grasping trials compared to reaching trials) along midline occipital visual areas. Further experiments will investigate the delay-related activity more systematically.

21.3.5 Experiment 4: Grasping without recognition in a visual agnosia patient

21.3.5.1 Rationale

As mentioned earlier, a well-studied patient with visual agnosia, D.F., can grasp objects accurately in real time despite a severe deficit in object recognition (Goodale *et al.* 1991; Milner *et al.* 1991). Her deficits in object recognition have been presumed to result from

---

phase of the trial (combining both reaching and grasping trials) were significant. As in Fig. 21.1, the circle and square outlines highlight the two postulated subregions of grasping activation. Note that the circled PostCS region, which was responsive during somatosensory finger stimulation in Experiment 1, is only activated during the action phase of the delayed action trials. In contrast, the anterior IPS region outlined by the square, which was not activated by somatosensory finger stimulation in Experiment 1, showed not only an action response, but a visual response as well. (C) Group results have been averaged in Talairach space and time courses have been extracted from the group data. The activation was first selected by including activation in which either the vision regressor (for reaching and grasping combined) or the action regressor (for reaching and grasping combined) was significant at $P < 10^{-4}$ (corrected for serial correlations but not multiple comparisons). Then the activation was color-coded by the relative contribution between the vision and action regressors (action-only in blue, vision-only in pink, equal contribution in purple). Time courses are shown for four different brain regions. Green and red lines show the BOLD activation for grasping and reaching trials, respectively. The vertical pink line indicates the real-time occurrence of the visual object illumination (250 ms) and the vertical blue line indicates the real-time initiation of the action. Based on the hemodynamic response function, the activation would be expected to peak approximately 6 seconds (three scans) later. For each time course, statistics were computed for each ROI using the general linear model to determine whether, for both types of action (G + R), the visual or action components were significant and whether the difference between grasping and reaching (G – R) for the vision or action components were significant.

damage to her ventral stream; whereas her spared grasping abilities have been presumed to be subserved by an intact dorsal stream. We used fMRI to examine whether these assumptions were indeed borne out by functional activation patterns (James *et al.* 2003).

21.3.5.2 Methods and results

We had a brief opportunity to collect preliminary neuroimaging data on patient D.F. We scanned her with a high-resolution anatomical ($256 \times 256 \times 256$ 1 mm^3 voxels) and then collected functional images while she performed visually guided grasping and reaching in the same paradigm as was used for normal subjects in Experiment 1. Her anatomical image showed the most extreme damage in the lateral occipital area. In fact, the area of damage almost exactly overlapped with the group ROI for LO in seven normal subjects (Fig. 21.4A). D.F.'s brain also exhibited atrophy and enlarged sulcal spaces elsewhere, particularly in the occipital and parietal lobes and including the intraparietal sulcus (Fig. 21.4B). Despite this atrophy, activation was still observed in dorsal stream areas during grasping (Fig. 21.4C), reaching (Fig. 21.4B) and saccades (not shown). Of particular interest, grasping (compared to reaching) activation was observed at the PostCS/IPS junction, especially in the right hemisphere (Fig. 21.4C), with activation comparable to neurologically intact subjects (Fig. 21.4D).

These results provide encouraging support that:

(1) AIP can subserve the ability to compute grasping actions towards objects even in the absence of ventral stream activity related to objects; and

(2) D.F.'s spared object grasping does indeed arise from intact grasp-related regions in the dorsal stream.

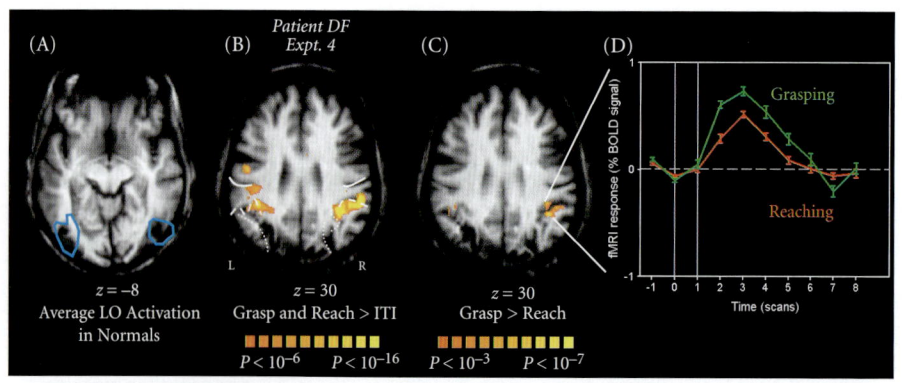

Fig. 21.4 Results from Experiment 4 on a patient with visual form agnosia, D.F. (A) An axial slice showing the average extent of LO activation in normals, outlined in blue, reveals considerable damage in that area in D.F. (B) An axial slice shows activation for reaching + grasping – ITI. (C) Activation for grasping – reaching on the same slice as shown in B. Note the activation shown in the intraparietal sulcus, particularly on the right. (D) The time course from the right hemisphere AIP region in (C).

They further demonstrate that the LO complex, in particular, is essential for intact object recognition capabilities. They also illustrate the limitations of studying neuropsychological patients using only anatomical scans and the utility of including functional scans as well. Specifically, D.F.'s anatomical scans indicated widespread abnormalities in brain structure, but functional scanning revealed that much of the cortex that had enlarged sulci none the less appeared to function normally, as predicted from behavioral studies of her performance.

21.4 General discussion

To summarize, I have described the properties of a monkey area, AIP, specialized for visually guided grasping, and suggested the existence of a comparable human area based on numerous criteria. The degree of homology between the two areas has yet to be firmly established, but some degree of functional equivalence is suggested by several comparisons:

1. Both regions appear at similar locations relative to the sulci, namely at the anterior end of the IPS, behind somatosensory regions.
2. Evidence from monkey AIP inactivation (Gallese *et al.* 1994) and human AIP lesions suggest that disruptions to AIP processing selectively impairs grasping but not reaching (Binkofski *et al.* 1998).
3. In both species, AIP has both visual and motor responses.
4. In both species, hand manipulation responses in AIP occur bilaterally (Mountcastle *et al.* 1975).
5. In humans, AIP appears to be critical in coding grasping compared to reaching (no activation differences between grasping and reaching were observed in ventral stream area LO) but does not participate in general object recognition when no action is afforded (although LO does).
6. AIP appears to subserve the spared grasping abilities of a patient with visual object agnosia.

Now that these experiments have systematically identified and characterized a human parietal area involved in grasping, future experiments can begin to investigate the role of AIP in other cognitive and action functions. Already we have shown a clear dissociation between the way objects are processed in AIP versus LO, and we have begun to reveal how object information may be stored during a delay. Future studies may also examine the AIP response during: object matching tasks with real objects; images of tools, graspable and non-graspable objects (see Chao and Martin 2000); grasping actions made by other individuals (Rizzolatti *et al.* 2001); and other types of unnatural grasping actions (such as pantomimed grasping).

Despite the suggestive comparisons, extrapolation from monkey to human brains must be taken with caution. One cannot necessarily expect the human brain to be that of a 'morphed monkey' (Van Essen *et al.* 2001), particularly in parietal cortex where the surface area in the human is 20× larger than in the monkey (compared to visual cortex where it is 2× as large, or temporal cortex where it's 9× as large) (Van Essen *et al.* 2001). Even where similarities are established, they do not necessarily imply homology in the strict sense (structures in two species that are inherited from a common ancestor), but could rather indicate homoplasies (structures that have arisen independently in separate lineages) or analogies (similar structures with no common origin) (Krubitzer 1995; Butler and Hodos 1996). Furthermore, human dexterity in hand actions and tool use is far more sophisticated than in other non-human primates. Thus it is possible that humans possess additional association areas involved in control of the hand, or a greater range of hand- and tool-related functions within AIP than we have reported thus far.

None the less, given that humans and monkeys have shared common evolutionary challenges as well as common ancestral species, monkey brain organization provides important clues as to the likely functional modules within the human brain. Functional considerations are only one means to identify an 'area' or a 'homolog'. Ideally, the human brain mapping will benefit as well from using other criteria to determine relationships, including anatomical and cytoarchitectonic parcellation, inter-area connectivity, and topology. The determination of homologies should also be aided by the development of MRI in monkeys (e.g. Logothetis *et al.* 2001) and, ideally, in other species as well, to enable comparative analyses.

To summarize, neuroimaging holds the potential to help understand the cognitive functions subserved by 'association cortex' such as that in the posterior parietal lobes. A useful starting point is the identification of the functional regions that comprise PPC, an endeavor that may be motivated in part by our knowledge of PPC organization in other species. Once tentative functional regions have been proposed, more extensive experiments can be conducted to characterize the activation across a wide range of cognitive tasks, preferably within the same set of subjects. Indeed, a review of the neuroimaging literature (Culham and Kanwisher 2001) and experiments in which several tasks have been compared within subjects (Simon *et al.* 2002) both suggest that PPC is indeed comprised of numerous functional regions, some of which are expected from data on other species and some of which may be uniquely human. Although 'the map is not the territory', I believe that in trying to understand complex brain regions, efforts to establish maps would greatly aid in exploring the territory.

Acknowledgements

This work was done in collaboration with Mel Goodale and with the assistance of Stacey Danckert, Joe DeSouza, Joe Gati, and Ravi Menon. It was supported by a grant from the McDonnell-Pew Program in Cognitive Neuroscience (to Jody Culham) and the Canadian Institutes of Health Research (operating grant to Mel Goodale and Multi-User Maintenance Grant to Ravi Menon). We thank Dave Woytowich,

Bob Stuart, Dan Pulham, Leopold van Cleeff, Derek Quinlan Philip Servos, and Zoe Kourtzi for technical assistance.

References

Andersen, R. A., Snyder, L. H., Bradley, D. C., and Xing, J. (1997). Multimodal representation of space in the posterior parietal cortex and its use in planning movements. *Annual Review of Neuroscience,* 20, 303–30.

Balint, R. (1909). Seelenhammung des 'Schauens', optische Ataxie, raümliche Störungen des Aufmersamkeit. *Monastchrift für Psychiatrie und Neurologie,* 25, 51–81.

Binkofski, F., Dohle, C., Posse, S., Stephan, K. M., Hefter, H., Seitz, R. J., and Freund, H. J. (1998). Human anterior intraparietal area subserves prehension. *Neurology,* 50, 1253–9.

Butler, A. B. and Hodos, W. (1996). *Comparative Vertebrate Neuroanatomy: Evolution and Adaptation.* Wiley-Liss, New York NY.

Cabeza, R. and Nyberg, L. (1997). Imaging cognition: An empirical review of PET studies with normal subjects. *Journal of Cognitive Neuroscience,* 9, 1–26.

Cabeza, R. and Nyberg, L. (2000). Imaging cognition II: An empirical review of 275 PET and fMRI studies. *Journal of Cognitive Neuroscience,* 12, 1–47.

Chao, L. L. and Martin, A. (2000). Representation of manipulable man-made objects in the dorsal stream. *Neuroimage,* 12, 478–84.

Colby, C. L. and Duhamel, J. R. (1991). Heterogeneity of extrastriate visual areas and multiple parietal areas in the macaque monkey. *Neuropsychologia,* 29, 517–37.

Culham, J. C. and Kanwisher, N. G. (2001). Neuroimaging of cognitive functions in human parietal cortex. *Current Opinion in Neurobiology,* 11, 157–63.

Culham, J. C., DeSouza, J. F. X., Osu, R., Milner, A. D., Gati, J. S., Menon, R. S., and Goodale, M. A. (2000). Visually-guided grasping produces fMRI activation in human anterior intraparietal sulcus. Paper presented at the Joint Meeting of the Experimental Psychology Society (UK)/Canadian Society for Brain, Behaviour and Cognitive Science, Cambridge, UK.

Culham, J. C., DeSouza, J. F. X., Woodward, S., Kourtzi, Z., Gati, J. S., Menon, R. S., and Goodale, M. A. (2001). Visually-guided grasping produces fMRI activation in dorsal but not ventral stream brain areas. *Journal of Vision,* 1, 194.

Culham, J. C., Danckert, S. L., and Goodale, M. A. (2002). fMRI reveals a dissociation of visual and somatomotor responses in human AIP during delayed grasping. *Journal of Vision,* 2, 701.

Culham, J. C., Danckert, S. L., De Souza, J. F. X., Gati, J. S., Menon, R. S., and Goodale, M. A. (2003: in press). Visually-guided grasping produces activation in dorsal but not ventral stream brain areas. *Experimental Brain Research.*

Debowy, D. J., Ghosh, S., Ro, J. Y., and Gardner, E. P. (2001). Comparison of neuronal firing rates in somatosensory and posterior parietal cortex during prehension. *Experimental Brain Research,* 137, 269–91.

Downing, P. E., Jiang, Y., Shuman, M., and Kanwisher, N. (2001). A cortical area selective for visual processing of the human body. *Science,* 293, 2470–3.

Dukelow, S. P., DeSouza, J. F., Culham, J. C., van Den Berg, A. V., Menon, R. S., and Vilis, T. (2001). Distinguishing subregions of the human MT+ complex using visual fields and pursuit eye movements. *Journal of Neurophysiology,* 86, 1991–2000.

Ehrsson, H. H., Fagergren, A., Jonsson, T., Westling, G., Johansson, R. S., and Forssberg, H. (2000). Cortical activity in precision- versus power-grip tasks: an fMRI study. *Journal of Neurophysiology,* 83, 528–36.

Epstein, R. and Kanwisher, N. (1998). A cortical representation of the local visual environment. *Nature*, **392**, 598–601.

Faillenot, I., Toni, I., Decety, J., Gregoire, M. C., and Jeannerod, M. (1997). Visual pathways for object-oriented action and object recognition: functional anatomy with PET. *Cerebral Cortex*, **7**, 77–85.

Farah, M. J. and Aguirre, G. K. (1999). Imaging visual recognition: PET and fMRI studies of the functional anatomy of human visual recognition. *Trends in Cognitive Sciences*, **3**, 179–86.

Gallese, V., Murata, A., Kaseda, M., Niki, N., and Sakata, H. (1994). Deficit of hand preshaping after muscimol injection in monkey parietal cortex. *Neuroreport*, **5**, 1525–9.

Gallese, V., Fadiga, L., Fogassi, L., and Rizzolatti, G. (2002). Action representation and the inferior parietal lobule. In W. Prinz and B. Hommel (Eds.), *Attention and Performance XIX: Common Mechanisms in Perception and Action* pp. 334–55. Oxford University Press, Oxford.

Goodale, M. A. and Milner, A. D. (1992). Separate visual pathways for perception and action. *Trends in Neurosciences*, **15**, 20–5.

Goodale, M. A., Milner, A. D., Jakobson, L. S., and Carey, D. P. (1991). A neurological dissociation between perceiving objects and grasping them. *Nature*, **349**, 154–6.

Goodale, M. A., Jakobson, L. S., and Keillor, J. M. (1994). Differences in the visual control of pantomimed and natural grasping movements. *Neuropsychologia*, **32**, 1159–78.

Grafton, S. T., Fagg, A. H., Woods, R. P., and Arbib, M. A. (1996). Functional anatomy of pointing and grasping in humans. *Cerebral Cortex*, **6**, 226–37.

Grill-Spector, K., Kushnir, T., Hendler, T., Edelman, S., Itzchak, Y., and Malach, R. (1998). A sequence of object-processing stages revealed by fMRI in the human occipital lobe. *Human Brain Mapping*, **6**, 316–28.

Grill-Spector, K., Kushnir, T., Edelman, S., Avidan, G., Itzchak, Y., and Malach, R. (1999). Differential processing of objects under various viewing conditions in the human lateral occipital complex. *Neuron*, **24**, 187–203.

Grill-Spector, K., Kourtzi, Z., and Kanwisher, N. (2001). The lateral occipital complex and its role in object recognition. *Vision Research*, **41**, 1409–22.

Hu, Y. and Goodale, M. A. (2000). Grasping after a delay shifts size-scaling from absolute to relative metrics. *Journal of Cognitive Neuroscience*, **12**, 856–68.

Hyvärinen, J. and Poranen, A. (1974). Function of the parietal associative area 7 as revealed from cellular discharges in alert monkeys. *Brain*, **97**, 673–92.

Ishai, A. and Sagi, D. (1995). Common mechanisms of visual imagery and perception. *Science*, **268**, 1772–4.

Jakobson, L. S., Archibald, Y. M., Carey, D. P., and Goodale, M. A. (1991). A kinematic analysis of reaching and grasping movements in a patient recovering from optic ataxia. *Neuropsychologia*, **29**, 803–9.

James, T. W., Culham, J. C., Humphrey, G. K., Milner, A. D., and Goodale, M. A. (2003: in press). Ventral occipital lesions impair object recognition but not object-directed grasping: a fMRI study. *Brain*.

Jeannerod, M. (1981). Intersegmental coordination during reaching at natural visual objects. In J. Long and A. Baddeley (Eds.), *Attention and Performance IX*. pp. 153–68. Erlbaum. Hillsdale NJ.

Jeannerod, M. (1986). Mechanisms of visuomotor coordination: a study in normal and brain-damaged subjects. *Neuropsychologia*, **24**, 41–78.

Jeannerod, M., Decety, J., and Michel, F. (1994). Impairment of grasping movements following a bilateral posterior parietal lesion. *Neuropsychologia*, **32**, 369–80.

Jones, F. W. (1942). *The Principles of Anatomy as Seen in the Hand* (2nd ed.). Williams and Wilkins, Baltimore.

Kanwisher, N. (2000). Domain specificity in face perception. *Nature Neuroscience*, **3**, 759–63.

Kanwisher, N., McDermott, J., and Chun, M. M. (1997). The fusiform face area: a module in human extrastriate cortex specialized for face perception. *Journal of Neuroscience*, **17**, 4302–11.

Kourtzi, Z. and Kanwisher, N. (2000). Cortical regions involved in perceiving object shape. *Journal of Neuroscience*, **20**, 3310–18.

Krubitzer, L. (1995). The organization of neocortex in mammals: are species differences really so different? *Trends in Neurosciences*, **18**, 408–17.

Levy, I., Hasson, U., Avidan, G., Hendler, T., and Malach, R. (2001). Center-periphery organization of human object areas. *Nature Neuroscience*, **4**, 533–7.

Logothetis, N. K., Pauls, J., Augath, M., Trinath, T., and Oeltermann, A. (2001). Neurophysiological investigation of the basis of the fMRI signal. *Nature*, **412**, 150–7.

Malach, R., Reppas, J. B., Benson, R. R., Kwong, K. K., Jiang, H., Kennedy, W. A., Ledden, P. J., Brady, T. J., Rosen, B. R., and Tootell, R. B. H. (1995). Object-related activity revealed by functional magnetic resonance imaging in human occipital cortex. *Proceedings of the National Academy of Science, USA*, **92**, 8135–9.

Marotta, J. J., Genovese, C. R., and Behrmann, M. (2001). A functional MRI study of face recognition in patients with prosopagnosia. *Neuroreport*, **12**, 1581–7.

Matsumura, M., Kawashima, R., Naito, E., Takahashi, T., Satoh, K., Yanagisawa, T., and Fukuda, H. (1996). Changes in rCBF during grasping in humans examined by PET. *Neuroreport*, **7**, 749–52.

Milner, A. D., and Goodale, M. A. (1995). *The Visual Brain in Action*. Oxford University Press, Oxford.

Milner, A. D., Perrett, D. I., Johnston, R. S., Benson, P. J., Jordon, T. R., Heeley, D. W., Bettucci, D., Mortara, F., Mutani, R., Terazzi, E., and Davidson, D. L. W. (1991). Perception and action in visual form agnosia. *Brain*, **114**, 405–28.

Mountcastle, V. B., Lynch, J. C., Georgopoulos, A., Sakata, H., and Acuna, C. (1975). Posterior parietal association cortex of the monkey: command functions for operations within extrapersonal space. *Journal of Neurophysiology*, **38**, 871–908.

Murata, A., Gallese, V., Kaseda, M., and Sakata, H. (1996). Parietal neurons related to memory-guided hand manipulation. *Journal of Neurophysiology*, **75**, 2180–6.

Murata, A., Gallese, V., Luppino, G., Kaseda, M., and Sakata, H. (2000). Selectivity for the shape, size, and orientation of objects for grasping in neurons of monkey parietal area AIP. *Journal of Neurophysiology*, **83**, 2580–601.

Rizzolatti, G. and Arbib, M. A. (1998). Language within our grasp. *Trends in Neurosciences*, **21**, 188–94.

Rizzolatti, G., Fogassi, L., and Gallese, V. (2001). Neurophysiological mechanisms underlying the understanding and imitation of action. *Nature Reviews Neuroscience*, **2**, 661–70.

Sakata, H., Taira, M., Murata, A., and Mine, S. (1995). Neural mechanisms of visual guidance of hand action in the parietal cortex of the monkey. *Cerebral Cortex*, **5**, 429–38.

Sereno, M. I. (1998). Brain mapping in animals and humans. *Current Opinion in Neurobiology*, **8**, 188–94.

Sereno, M. I., Dale, A. M., Reppas, J. B., Kwong, K. K., Belliveau, J. W., Brady, T. J., Rosen, B. R., and Tootell, R. B. (1995). Borders of multiple visual areas in humans revealed by functional magnetic resonance imaging. *Science*, **268**, 889–93.

Simon, O., Mangin, J. F., Cohen, L., Le Bihan, D., and Dehaene, S. (2002). Topographical layout of hand, eye, calculation, and language-related areas in the human parietal lobe. *Neuron*, **33**, 475–87.

Taira, M., Mine, S., Georgopoulos, A. P., Murata, A., and Sakata, H. (1990). Parietal cortex neurons of the monkey related to the visual guidance of hand movement. *Experimental Brain Research,* **83**, 29–36.

Tanaka, K. (1996). Inferotemporal cortex and object vision. *Annual Review of Neuroscience,* **19**, 109–39.

Tarr, M. J. and Gauthier, I. (2000). FFA: a flexible fusiform area for subordinate-level visual processing automatized by expertise. *Nature Neuroscience,* **3**, 764–9.

Tootell, R. B. H., Reppas, J. B., Kwong, K. K., Malach, R., Born, R. T., Brady, T. J., Rosen, B. R., and Belliveau, J. W. (1995). Functional analysis of human MT and related visual cortical areas using magnetic resonance imaging. *Journal of Neuroscience,* **15**, 3215–30.

Tootell, R. B. H., Mendola, J. D., Hadjikhani, N. K., Ledden, P. J., Lui, A. K., Reppas, J. B., Sereno, M. I., and Dale, A. M. (1997). Functional analysis of V3A and related areas in human visual cortex. *Journal of Neuroscience,* **17**, 7060–78.

Van Essen, D. C., Lewis, J. W., Drury, H. A., Hadjikhani, N., Tootell, R. B., Bakircioglu, M., and Miller, M. I. (2001). Mapping visual cortex in monkeys and humans using surface-based atlases. *Vision Research,* **41**, 1359–78.

Westwood, D. A., McEachern, T., and Roy, E. A. (2001). Delayed grasping of a Muller–Lyer figure. *Experimental Brain Research,* **141**, 166–73.

Wojciulik, E. and Kanwisher, N. (1999). The generality of parietal involvement in visual attention. *Neuron,* **23**, 747–64.

Yousry, T. A., Schmid, U. D., Alkadhi, H., Schmidt, D., Peraud, A., Buettner, A., and Winkler, P. (1997). Localization of the motor hand area to a knob on the precentral gyrus. A new landmark. *Brain,* **120**, 141–57.

Chapter 22

A framework for the investigation of directed cortical interactions: Theoretical background and application to dynamic sensorimotor mapping

Rainer Goebel, Alard Roebroeck, Dae-Shik Kim, and Elia Formisano

Abstract

We present a general framework for the investigation of directed interactions of activated brain areas using time-resolved fMRI and vector autoregressive (VAR) modeling in the context of Granger causality. This framework can be extended by including information on anatomical connectivity patterns from diffusion tensor imaging (DTI). We apply the proposed approach to a dynamic sensorimotor mapping paradigm. In an event-related fMRI experiment, subjects performed a visuomotor mapping task for which the mapping of two stimuli ('faces' versus 'houses') to two responses ('left' or 'right') alternated periodically between the two possible mappings. Besides expected activity in sensory and motor areas, a fronto-parietal network was found to be active during presentation of a cue indicating a change in the stimulus–response (S–R) mapping. The observed network includes the superior parietal lobule, dorso-lateral prefrontal cortex, and premotor areas. These areas might be involved in setting up and maintaining stimulus–response associations. The Granger causality analysis revealed a directed influence exerted by the left lateral prefrontal cortex and premotor areas on the left posterior parietal cortex.

22.1 Introduction

Functional brain imaging has contributed substantial insights into the neural correlates of human information processing and cognitive operations. Yet the limitations in temporal resolution have led researchers to focus on relevant information about *where* information is processed in the human brain (*functional segregation*). To gain a deeper understanding of *how* the brain processes information, more knowledge about the interaction of activated brain areas (*functional integration*) is needed.

Following the seminal work of several researchers (Horwitz 1991; McIntosh and Gonzales-Lima 1994; Friston *et al.* 1995; Buchel and Friston 1997), functional brain integration has been investigated during various cognitive or sensorimotor tasks using positron emission tomography (PET) and functional magnetic resonance imaging (fMRI). Significant methodological developments, such as the application of covariance structural equation modeling (McIntosh and Gonzales-Lima 1994) and non-linear system identification techniques (Friston and Buchel 2000) to neuroimaging data, have supported the idea that a statistical model of interacting neuronal systems can be obtained from metabolic/hemodynamic measurements of task-related neural activation. These models of the interaction between brain areas are often referred to as models of *effective connectivity*, defined as *the influence one neuronal system exerts over another* (Buchel and Friston, 2000).

Here we present a general framework to investigate effective connectivity (or *directed influences*) between activated brain areas using time-resolved fMRI and vector autoregressive (VAR) modeling in the context of Granger causality (Granger 1969, 1980). In this framework, time-resolved fMRI measurements provide topographical as well as temporal information about the brain areas subserving a cognitive task. VAR modeling of fMRI time-series and computation of Granger causality maps provide the mathematical framework for modeling effective connectivity. Furthermore, we discuss the possibility of extending this framework by including information on anatomical connectivity patterns between activated brain regions. This information might be obtained non-invasively using diffusion tensor imaging (DTI) measurements and subsequent fiber tracking algorithms, and might serve as a constraint for the mathematical modeling of functional interactions (Fig. 22.1A).

22.2 The proposed framework

In the following, the essential components of the proposed framework are outlined (see Fig. 22.1). Each subsequent section describes the contributions provided by a particular technique (fMRI, DTI) or modeling method (autoregressive modeling and Granger causality). In addition, problems and limitations of each part are discussed.

22.2.1 Time-resolved functional MRI

The non-invasive investigation of the timing of human cognitive functions has relied so far almost exclusively on electroencephalography (EEG) and magnetoencephalography (MEG). Because recordings obtained with these methods reflect phenomena that accompany changes in neuronal activity with a millisecond temporal accuracy, EEG

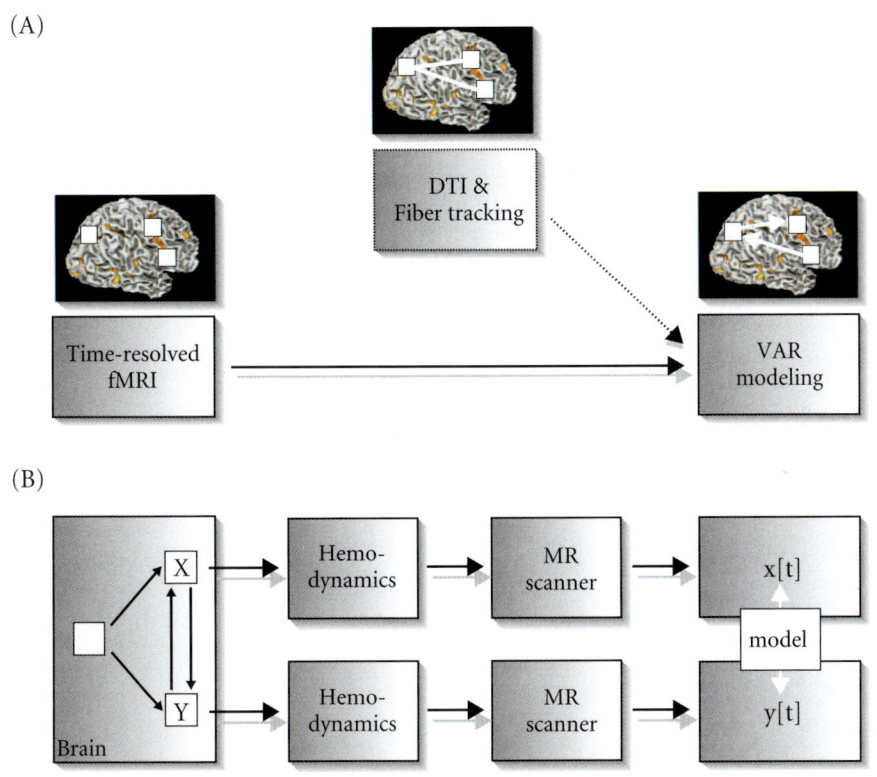

Fig. 22.1 (A) An illustration of the major components of the proposed framework and their relations. Information obtained with time-resolved fMRI is used for the statistical modeling (using vector autoregressive (VAR) models and Granger causality) of directed cortical interactions. DTI and fiber tracking might provide individual anatomical constraints for these models. (B) An illustration of the indirect access to interacting brain regions with fMRI. Hemodynamics and the MR scanner contribute unwanted artifacts to the signals of interest and might confound modeling efforts. Confounding is especially deleterious when the unwanted contributions are different for the brain regions under investigation (e.g. different hemodynamic responses in different regions).

and MEG appear to be particularly suited to track cortical activation dynamics during a cognitive task. As a consequence, EEG and MEG are often used in combination with reaction time (RT) experiments in the attempt to relate sequences of neural events to sequences of processing stages that are presumed to form a cognitive task (mental chronometry) (Renault *et al.* 1982; Hillyard 1993; Hari *et al.* 2000).

Due to the limited spatial resolution, however, EEG and MEG provide only crude information about the spatial layout of the neural activity. Conversely, maps obtained with techniques such as fMRI provide spatially more highly resolved information on cortical activation loci but, conventionally, only crude information about their relative timing.

Together with numerous other researchers, we believe that the most powerful approach to investigate the spatio-temporal patterns of cortical activation will consist in the combination of fast (EEG/MEG) with slow brain imaging methods (PET, fMRI) (Dale and Halgren 2001). However, because many unresolved theoretical, methodological, and technical issues still prevent an optimal use of this combined approach, we consider the possibility of using only one of the techniques to simultaneously gain information about space and time in cortical activation studies.

Here we propose the use of time-resolved fMRI to trace both the topography and the sequence of cortical activation across brain areas during perceptual or cognitive tasks.

In a recent article (Formisano *et al.* 2002) our group used fMRI–mental chronometry to address the question of functional differentiation between the various cortical areas involved in a complex cognitive task, the mental clock task (Paivio 1978; Trojano *et al.* 2000). In this mental imagery paradigm, subjects are asked to imagine pairs of clock faces on the basis of acoustically presented times, to compare the mental images, and to report in which of the two faces the clock hands form the greater angle. This 'mental clock task' thus involves auditory perception, the translation of the auditory information into mental representations that preserve the angular differences, the comparison of the angles, and a behavioral response (Paivio 1978).

The time-resolved analysis revealed a sequence of cortical activation from auditory perception to motor response with an asymmetric (sequential) activation of the left and right posterior parietal cortex. The detected temporal difference between left and right posterior parietal cortex suggested that these regions perform distinct functions in this imagery task. This hypothesis was confirmed by a trial-by-trial analysis of correlations between reaction time and onset, width, and amplitude of the hemodynamic response (HR) and by a subsequent study using repetitive transcranial magnetic stimulation (rTMS) (Sack *et al.* 2002).

Results from these and several other studies (see, for example, Menon *et al.* 1998; Menon and Kim 1999; Richter *et al.* 2000; Thierry *et al.* 1999) suggest that time-resolved fMRI provides the possibility to trace the flow of information in cortical networks at a relevant time scale for various perceptual and cognitive tasks. The achievable time resolution appears to be in the order of few hundreds of milliseconds. However, because latency differences of evoked HRs in different brain areas can have many non-neuronal biophysical causes (e.g. differences between areas in coupling between neural activity and BOLD), the validity of the assumption that timing differences in observed BOLD responses are attributable to underlying neuronal dynamics needs to be assessed separately in each investigation (see Formisano and Goebel 2003).

Based on these arguments, we maintain that, if a sub-second time-scale is an acceptable scale for the investigation, fMRI might have several advantages over EEG/MEG. First, as opposed to simple sensory or motor tasks, complex cognitive tasks involve networks with many activated brain regions, which are difficult to model at the level of the brain sources, even with the help of fMRI data from the same subject. Secondly, EEG/MEG measurements are not very sensitive for long-lasting, sustained processes and are better suited to reveal effects closely time-locked to external stimulus-onsets. If higher temporal resolution is required, EEG or MEG measurements will still be necessary.

In the proposed framework, time-resolved fMRI represents a very relevant aspect since temporal information of sufficient precision constitutes a prerequisite for applying vector autoregressive modeling (see below) or similar methods that aim to characterize not only instantaneous effects between coactivated brain regions but also 'causal' (directed) effects acting over time. Such non-instantaneous effects occur if activity changes in area A affect activity changes in area B at a later point in time.

22.2.2 Autoregressive modeling and Granger causality

In the following, we will describe how effective connectivity is modeled within our proposed multivariate framework. We treat the sequence of fMRI measurements of selected regions of interest x_i (individual voxels and/or averages of multiple contiguous voxels in a region of interest) as the elements of a discrete vector time-series $\mathbf{x}[t] - (x_1[t], \ldots, x_M[t])$, where t represents discrete time. In the simplest scenario M (the number of voxels or regions of interest) will be two and we will be investigating the dependencies between these two time series.

22.2.2.1 Functional connectivity and effective connectivity

The linear functional connectivity between the elements of $\mathbf{x}[t]$ is fully contained in the cross-covariance matrix $\Gamma[k]$, which is a function of lag (or delay) k. The diagonal elements $\gamma_{ii}[0]$ and off-diagonal elements $\gamma_{ij}[0]$ of $\Gamma[0]$ are simply the variances and covariances of the individual elements. For $k > 0$, diagonal elements $\gamma_{ii}[k]$ are scalar auto-covariances of x_i at lag k and off-diagonal elements $\gamma_{ij}[k]$, $i \neq j$, are scalar cross-covariances between $x_i[t]$ and $x_j[t]$ at lag k. The cross-covariance matrix is a model-free characterization of the linear statistical association between the element time-series x_i without any regard for the underlying dependence-structure between the elements (and possible external components z_j), which generated such association. To make inferences about such underlying structure, i.e. about effective connectivity, some additional assumptions are needed and have to be incorporated into a multivariate process-model of the vector time-series $\mathbf{x}[t]$. In fact, models for effective connectivity have two distinct parts: a neuroanatomical model and a mathematical model (Buchel and Friston 2000). Thus, assumptions can be of two sorts. First, *structural assumptions* (incorporated in the neuroanatomical model) determine *which* components x_i can depend directly on or be directly influenced by which other components x_j (or exogenous components z_j). Eventually, DTI tractography (see below) might provide a more direct way of forming these assumptions. Secondly, *functional dependence assumptions* (incorporated in the mathematical model) determine *how*, mathematically, the value $x_i[t]$ can statistically depend on (or be a function of) values $x_j[t, \ldots, t-k]$ of other components (or values $z_j[t, \ldots, t-k]$ of exogenous components). Such functional dependence assumptions are essentially contained in the specific process-model one chooses to employ. For instance, in covariance structural equation modeling, the functional dependence assumptions are that the value $x_i[t]$ can only be a linear function of the instantaneous values of other components, as in $x_1[t] = a_2 x_2[t] + a_3 x_3[t] + b_1 z_1[t] + e[t]$, where $e[t]$ denotes unexplained noise.

We propose to treat **x**[n] as a vector autoregressive (VAR) process, and thus to use vector autoregressive models to make inferences about effective connectivity. We choose autoregressive modeling to assess the degree of dependence between several components for several reasons. First, VAR models are dynamical models that can capture the temporal structure in the variations of individual components and in the interdependence between them. Secondly, the parameters of autoregressive models are relatively easy to estimate by solving a *linear* regression problem. Thirdly, many random processes can be very well approximated by a sufficiently high-order AR model. Finally, as we will show, VAR models form a natural context in which measures of directed influence based on the concept of Granger causality can be defined.

22.2.2.2 Granger causality

Temporal precedence is a feature of data that can inform us about the *direction* of possible influences without further structural assumptions. Thus, capturing the temporal structure of signal time-courses is potentially useful in investigating effective connectivity. Furthermore, our commonsense concept of causality is entangled with considerations of time: causes always precede effects. These considerations have led the econometrist Clive Granger to propose a testable definition of causality for stochastic time-series (Granger 1969, 1980) that has recently been employed in the investigation of neuronal interactions (Bernasconi and Konig 1999; Freiwald *et al.* 1999; Kaminski *et al.* 2001). Conceptually, it amounts to the following: if a time-series *y* causes (or has an influence on) *x*, then knowledge of *y* should help predict future values of *x*. Thus, causality (or influence) is framed in terms of predictability. More in detail, let Ω_t denote all knowledge available in the universe at time *t*, and $\Omega_t - Y_t$ denote all knowledge except values of *y* up to time *t*. Then, if we can predict $x[t+1]$ better using Ω_t than we can using $\Omega_t - Y_t$, we say that *y Granger causes x*, or that there exists Granger causality from *y* to *x*. Analogously, if we can predict $y[t+1]$ better using Ω_t than we can using $\Omega_t - X_t$, we say that *x Granger causes y*, or that there exists Granger causality from *x* to *y*. In this way the temporal structure in the dependency between *x* and *y* is used to decide on the *direction* of possible influences between them. Granger causality from *x* to *y* can exist independently of causality in the other direction. Finally, when we can predict $x[t+1]$ better from $\Omega_t + \{y[t+1]\}$ than we can from Ω_t alone we say that there is *instantaneous causality* between *x* and *y*. In contrast to directed Granger causality, instantaneous causality does not have a direction, since its definition can be seen to be symmetric with respect to *x* and *y*.

To become operational, these definitions need to be specialized in several ways. First, in practice we cannot (and need not) use all knowledge available in the universe, and we will have to replace Ω_t with a smaller information set J_t, containing all relevant variables available. However, any notion of Granger causality then becomes critically dependent on the used set J_t and, as discussed below, when J_t is too small, spurious causality can occur. Secondly, we are generally not able (nor obliged) to use information from the infinite past for our predictions and, thus, we will only use information from the recent past. Thirdly, we need a time-series model that can generate predictions of future values $x[t+1]$ and that can precisely quantify when one prediction is better than another. Here we will use linear autoregressive models to produce predictions of values of x from values in its own past and in the past of other variables. The mean squared error (MSE) that such autoregressive models make in their forecasts can be used as a measure of their ability to predict $x[t+1]$. Moreover, we can compare models that use a different set of past values in their prediction of $x[t+1]$ by comparing their MSE. Note, however, that the definition of Granger causality is, in principle, not limited to the use of linear autoregressive models to generate predictions.

Geweke (1982) has proposed a measure of linear dependence (or feedback, as he calls it) $\mathbf{F}_{x,y}$ between the time-series x and y (themselves possibly vector time-series), which can be regarded as an implementation of the concept of Granger causality in terms of vector autoregressive models (Bernasconi and Konig 1999). The dependence measure $\mathbf{F}_{x,y}$ is the sum of three components: the linear influence from x to y ($\mathbf{F}_{x \to y}$), the linear influence from y to x ($\mathbf{F}_{y \to x}$), and the instantaneous influence between x and y ($\mathbf{F}_{x \cdot y}$):

$$\mathbf{F}_{x,y} = \mathbf{F}_{x \to y} + \mathbf{F}_{y \to x} + \mathbf{F}_{x \cdot y}$$

We can give the following interpretations of the measures. The four measures take their values in the interval $[0, \infty)$, i.e. they are by construction non-negative. $\mathbf{F}_{x,y}$ is a measure of the total linear dependence between the series x and y. If nothing of the value at a given instant of one can be explained by a linear model containing all the values (past, present, and future) of the other, $\mathbf{F}_{x,y}$ will be zero. $\mathbf{F}_{x \to y}$ is a measure of linear directed influence from x to y. If past values of x will not improve the best linear prediction of the current value of y over the prediction obtained from using past values of y alone, then $\mathbf{F}_{x \to y}$ will be zero. A value larger than zero will mean that past values of x do improve the prediction of the current value of y, or that there is Granger causality or directed linear influence from x to y. A similar interpretation holds, of course, for $\mathbf{F}_{y \to x}$.

Thus, the two directed components, $\mathbf{F}_{x \to y}$ and $\mathbf{F}_{y \to x}$, use the arrow of time to decide on the direction of influence. However, the total linear dependence between x and y often does not consist fully of these directed components. Much of the total linear dependence can be contained in the undirected instantaneous influence $\mathbf{F}_{x \cdot y}$ between them. $\mathbf{F}_{x \cdot y}$ quantifies the improvement in the prediction of the current value of x (or y) by including the current value of y (or x) in a linear model already containing the past values of *x and y*. From this symmetry it can be seen that $F_{x \cdot y}$ indeed contains no directional information at all. In practice, non-zero values of $F_{x \cdot y}$ can be caused by directed influence between *x* and *y* at a finer time scale than that at which *x* and *y* are observed. Thus, poor temporal sampling of the processes of interest (at a frequency lower than that required to detect relevant interactions) can obscure the true directed linear influence between them.

As illustrated in Fig. 22.1B, this problem is very prominent in fMRI data. Two stages are interposed between the true neural activity and the time series actually used to model interaction between brain areas. Both stages have adverse effects on the temporal information contained in the signal. The first stage, the hemodynamic response, effectively works as a low-pass filter, and temporally and spatially integrates the original neural/synaptic activity. In the second stage, image acquisition by the MRI scanner, the continuous hemodynamic response is discretely sampled, generally at different instants for different parts of the brain, since only a single slice can be acquired at a time. In this context of 'poor' temporal sampling, true directional influence (as computed with the above measures) might either not be detected or might 'leak' into the instantaneous component, hiding the true direction of influence. Thus, an advantage of computing the instantaneous component arises when the two directed components are both estimated to be (nearly) zero. In this case, the value of the instantaneous component can distinguish between two alternative interpretations of this finding. Temporal sampling too poor to detect the direction of influences (where such influences *are* present) can be diagnosed by positive values for the instantaneous component. In contrast, an absolute lack of evidence for the existence of any interactions would be signaled by a (near) zero value for $F_{x \cdot y}$.

There is an additional issue that troubles the interpretation of the influence measures and, in fact, many other effective connectivity approaches as detecting directed influence between observed processes. This is the problem of spurious causality (Granger 1980) that can appear between two processes when both are influenced by other (external) sources that are not taken into account. The dependence measure $F_{x,y}$ takes the information set J_t to contain *only x* and *y* and, thus, is only valid when *x* is the sole source of influence on *y* and vice versa. Any additional external source of influence will confound the inferences made from these measures.

For instance, consider three groups of voxels **X**, **Y**, and **Z**, generating three time-courses $x[t]$, $y[t]$, and $z[t]$ as measured with fMRI. We consider the situation in which activity in **Z** has a profound influence on the activity in **X** and **Y** (perhaps because neural populations in **Z** interact with those in **X** and Y), but no influence is present between **X** and **Y** themselves. If we calculate $F_{x \to y}$, $F_{y \to x}$, and $F_{x \cdot y}$ from the time courses $x[t]$ and $y[t]$ alone, we will find non-zero values and wrongfully conclude that there exist influences between **X** and **Y**. The exact kind of spurious influence (**X** to **Y**, **Y** to **X**, or instantaneous) depends on the difference in temporal lag with which **Z** interacts with **X** and **Y**. It should be clear from this example that, when trying to infer the degree of influence between two processes, we should take *any* source of influence on either process into account and add it to the information set. Mathematically, this amounts to including any such external process $z_i[t]$ in our analysis as a confound. An important candidate in fMRI data for such confounding external influence is the 'driving' influence exerted by experimental manipulations (stimuli, task demands) as they are characterized in explanatory variables used in standard general linear model (GLM) analysis for regional effects. In creating activation maps, we are assuming that experimental manipulations are accompanied by (or even causing) localized brain activity (the tenet of functional specialization). Therefore, we could consider the same experimental manipulations to be confounding external sources of influence when analyzing for interactions between systems that go beyond mere 'stimulus-locked' localized changes. Thus, a first step in trying to avoid spurious causality inferences could be to enforce the very strict demand that interactions between brain areas underlying cognitive processing should explain features of observed data that cannot already be explained by 'stimulus-locked' localized changes. Therefore, in the study discussed below, we have applied the influence analysis to the remaining residuals of the GLM analysis for experimentally induced BOLD responses.

22.2.3 Diffusion tensor imaging (DTI) and measurements of anatomical connectivity

The performance of a cognitive or a sensory-motor task requires the active interaction and exchange of information between several neural processing units (active brain regions), which are located in different places of the cortex. White matter fibers constitute the structural substrate subserving this exchange and integration of neural information. Knowledge of the patterns of white matter fibers (*anatomical connectivity*) thus appears to be essential for an integrated understanding of the neural processing that takes place in distributed brain networks (Ramnani *et al.* 2002).

So far, anatomical connectivity has been studied mainly through invasive tracer techniques that cannot be used *in vivo*. Only very recently has it been shown that the axonal connections in the living brain can be investigated non-invasively using a range of techniques based on *diffusion MRI*. Diffusion MRI relies on the measurement of Brownian motion of molecules and allows the characterization of water diffusion properties at each volume element (voxel) of an image using the so-called 'gradient' system (Moseley *et al.* 1990; Le Bihan 1995). The use of diffusion MRI to study anatomical connectivity relies on the principle that brain water diffuses with strong directionality (*anisotropy*) where myelination and/or protein in fiber bundles of axons are present. If there is a region where fibers align in a direction, water molecules tend to diffuse preferably along the direction parallel to the neuronal fibers (Sakuma *et al.* 1991; Wimberger *et al.* 1995).

In a recent application of diffusion MRI (*diffusion tensor imaging*, *DTI*), a water *diffusion tensor* is obtained at each voxel through the collection of at least seven diffusion-weighted images. An appropriate combination of magnetic field gradients ensures that each of these images is sensitive to water diffusion along an independent direction (Basser *et al.* 1994; Hsu and Mori 1995; Conturo *et al.* 1996; Pierpaoli *et al.* 1996). Besides the calculation of the degree of diffusion anisotropy (*anisotropy index*), a diffusion tensor allows the estimation of the principal direction of diffusion at each voxel. Starting from a voxel-wise discrete representation of the water-diffusion direction field, a 'tracking algorithm' is used to estimate continuous three-dimensional trajectories of bundles of fibers. Encouraging results have been obtained in the rat (Mori *et al.* 1999), monkey (Parker *et al.* 2002) and human (Conturo *et al.* 1999; Catani *et al.* 2002; Mori *et al.* 2002). An alternative, interesting approach is to compute the probability of anatomical connectivity between two brain areas rather than the exact trajectory of the connecting fibers (Koch *et al.* 2002).

It is important to note that results of DTI-tractography studies are influenced by multiple factors, including signal-to-noise ratio, image acquisition parameters, size of the voxels relative to the size of the fiber's cross-section and post-processing algorithms and that, at present, ultimate limitations and possibilities of this technique are not fully assessed. Notwithstanding these uncertainties, the perspective of using the information obtained with DTI in the context of systems-level modeling of functional neuroimaging data appears to be highly valuable. In particular, DTI data may serve as a constraint for all data analysis methods that involve modeling interactions of brain regions during a cognitive task (e.g. structural equation modeling, Volterra series and vector autoregressive models). Embedding the information on anatomical connectivity in the statistical models of interactions between brain regions would allow an explicit distinction between directly and indirectly connected regions. Direct connectivity would imply the existence of a single anatomical fiber pathway between two regions; indirect connectivity would imply a functional chain of such anatomical links (Horwitz *et al.* 2000).

Several aspects need to be specified carefully for the proposed approach to become operational. In the following, we list three main caveats that might seriously undermine its applicability.

1. As neural activity is measured indirectly with fMRI through the visualization of the BOLD contrast, the trajectories of fiber bundles revealed by DTI are not visualized directly but only estimated indirectly by measuring the diffusion of brain water molecules along them. This can lead to inaccurate estimates, especially at locations in the brain where several fiber tracts are densely packed in a small volume, cross each other, or 'kiss' (i.e. converge, contact and diverge again).

2. At present, DTI tractography has been used to visualize the three-dimensional structure of the major white matter fasciculi within living human brain (Conturo *et al.* 1999; Catani *et al.* 2002; Mori *et al.*, 2002). The capability to resolve dedicated bundles of fibers that all go from a cortical region A to a cortical region B, without fibers passing on to other regions, still needs to be demonstrated.

3. Since DTI does not allow one to distinguish the direction of connections between areas, obtained information can only be used to include constraints on the existence of a structural connection, but not on its directionality. Directionality of a connection, thus, needs to be inferred exclusively based on functional measurements and modeling methods (e.g. using Granger causality).

Promises to overcome these limitations, at least partially, come from recent advancements in diffusion MRI acquisition techniques (King *et al.* 1994; Basser 2002; Tuch *et al.* 2002). In addition, a particularly relevant role is being assumed by studies in animals (Lin *et al.* 2001; D.-S. Kim *et al.* 2002). These studies allow the comparison between patterns of fibers obtained using these non-invasive techniques with those obtained using more invasive methods, and will serve as the gold standard for defining possibilities and limitations of the application of these techniques also in the human brain.

In summary, the knowledge of patterns of fiber connections between different brain regions is a much-needed source of information. It would allow the inclusion of structural constraints in systems-level models of functional neuroimaging data. Ideally, this knowledge might be obtained *in vivo*, non-invasively, and on an individual basis using DTI or related approaches. However, the precision and accuracy of these methods has yet to be quantified in extensive and careful validation studies, requiring a comparison with *post-mortem* anatomical measurements of fiber's trajectories.

22.3 An application: Dynamic sensorimotor mapping

In this part, we describe an application of the outlined framework to a dynamic sensorimotor mapping paradigm. Sensorimotor coordination is dynamic in nature and involves the selection and execution of appropriate actions based on the perceived, changing environment and previous experience. Here we focus on a particularly interesting aspect of sensorimotor coordination, namely how sensorimotor associations are established and dynamically changed over time. In particular, the task we used requires sudden remapping of established stimulus–response couplings. It is expected that such a remapping process takes at least several hundreds of milliseconds until it is 'implemented' in cortical systems mediating the coupling of sensory to motor areas, and therefore can be investigated using our framework. We chose an event-related design to also study transient responses to individual events, such as the presentation of a stimulus or execution of a response.

There are several behavioral studies on related phenomena, most notably in the context of task-switching paradigms (Allport *et al.* 1994). Although previous imaging studies have provided important knowledge of which areas are involved in sensorimotor coordination and task switching in the human brain (Dove *et al.* 2000; Sohn *et al.* 2000; Toni *et al.* 2001*a, b*, 2002; Rushworth *et al.* 2002) the direction of interactions between these areas are largely unknown (but see Toni *et al.* 2002).

We remark that our approach to applying the influence measures has two important features. First, as discussed above, we used the remaining residuals of the GLM analysis for regional activations for the influence analysis. In essence, we used only variance left unexplained by conventional analysis to detect interactions among regions. Secondly, we mapped the computed influence measures for a given region of interest (ROI) over all voxels in the scanning volume. In this way, regions that interact with a given ROI are identified by the maps, instead of prespecified by the experimenter in the form of an anatomical model. This represents a more data-driven, exploratory method of investigation than used in other methods (e.g. covariance structural equation modeling) using as little a priori knowledge or expectations as possible.

22.3.1 Materials and methods

22.3.1.1 Cognitive task

The subjects performed a visuomotor mapping task in the scanner for which the mapping of two stimulus categories ('houses' and 'faces') to two responses ('left' or 'right') alternated periodically between one of two possible mappings (see Fig. 22.2A). All brief changes in the visible display took place *in* or *on* a square bounding box just above the fixation cross and constituted either a stimulus to which the subject could respond or a cue to change the stimulus–response (S–R) mapping. An unchanging white bounding box formed the 'fixation' or baseline condition.

A 1-second change in color of the bounding box cued a change in the required mapping of stimulus to response ('mapping cue'). A yellow bounding box cued *mapping one*: images of faces require a left-hand response and images of houses require a right-hand response. A blue bounding box cued *mapping two*: faces require a right-hand response and houses require a left-hand response. The control stimulus was a 'no-go' stimulus (requiring no response) and consisted of images of objects.

The experiment consisted of eight event-blocks of one mapping cue followed by 15 trials of stimuli: 5 faces, 5 houses, and 5 objects in randomized order. The mapping alternated continually, starting with mapping one. Thus, for each of the S–R mappings there were four event blocks. The stimulus onset asynchrony (SOA) was synchronized to the volume acquisition of the scanner and was therefore always an integer multiple of the volume repetition time (TR) (960 ms, see below). The interval between a mapping cue and the first response stimulus was always 5 TRs, while the SOA for the response stimuli in a block was randomly chosen between 2 and 6 TRs.

22.3.1.2 MRI scanning and experimental set-up

Images were acquired using a 3 Tesla scanner ('Trio', Siemens, Erlangen, Germany). Functional images were acquired for two subjects during a single run of the task using a T_2-weighted echo planar sequence (echo time (TE), 40 ms; volume repetition time (TR), 960 ms; field of view, 224 mm × 224 mm, 64 × 64 matrix, giving 3.5 mm × 3.5 mm in-plane resolution). The images consisted of 12 oblique transverse

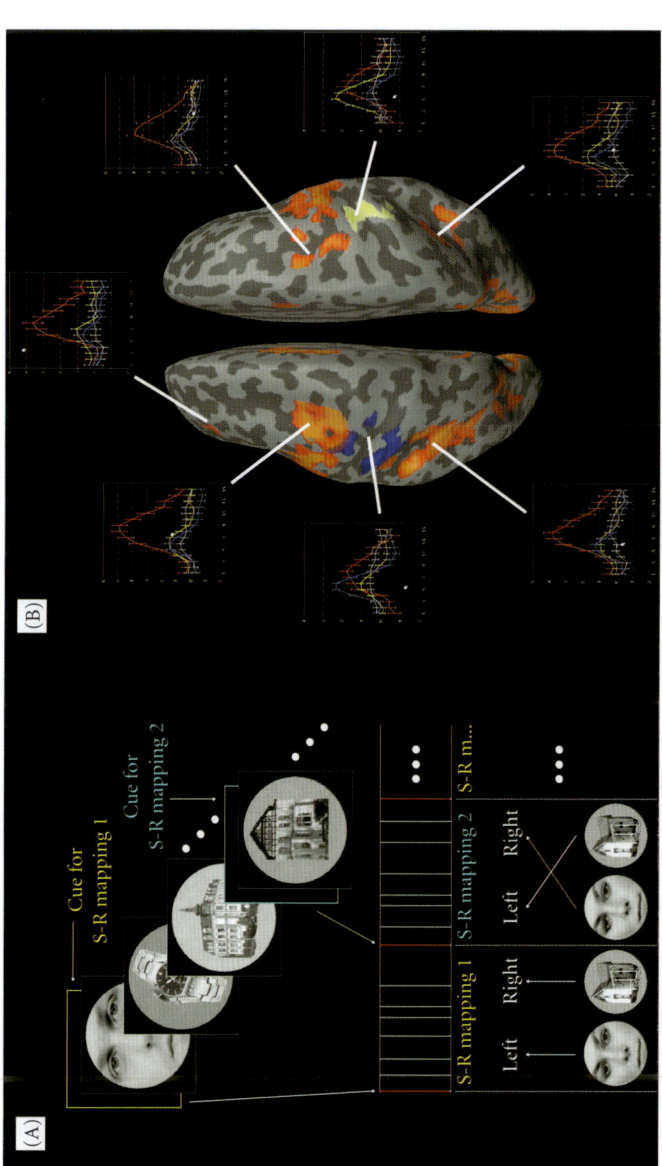

Fig. 22.2 (A) The stimuli (top), temporal organization of events (middle), and required responses (bottom) in the visuomotor mapping task. White dots in the event sequence imply continuation of the task regime. Refer to the text for further explanation. (B) Maps for regional activations and event-related deconvolved BOLD responses for all runs of all subjects. The map for a contrast ($P < 0.001$, corrected) of right responses against left responses is shown in blue. The map for the reverse contrast (left against right) is shown in yellow ($P < 0.001$, corrected). Regional activation for the presentation of the mapping cue is shown in orange/red ($P < 10^{-6}$, corrected). Deconvoled BOLD responses for indicated regions are shown for right response events (in blue), left response events (in yellow), presentation of a control stimulus that required no response (in gray), and for the presentation of the mapping cue (in red).

slices (interleaved acquisition), 6 mm thick with a 2 mm inter-slice gap. During performance of the task, 540 volumes were acquired. Structural images were acquired using a T_1 MPRAGE sequence (TE = 4 ms, 256 × 256 × 192 matrix, 1 × 1 × 1 mm³ voxels). For a third subject, functional images were acquired during two runs of the task, consisting of 16 slices (TE = 27 ms, TR = 960 ms). For this subject, diffusion-weighted images, six non-collinear directions, and one reference image without diffusion weighting, were collected using an echo planar sequence (TE = 80 ms, TR = 8250 ms, field of view 256 × 256 mm², 128 × 128 matrix, giving 2 × 2 mm² in-plane resolution). The images consisted of 55 contiguous transverse slices, 2.5 mm thick. Diffusion-weighted images were averaged over 14 acquisitions. Stimulus presentation, response registration, and synchronization to the scanner acquisition were performed by custom-built software. Two response boxes (Psychology Software Tools Inc., Pittsburgh, USA) were used to record button presses with the left and right index fingers. Based on the recorded data, accuracy and reaction time was determined for each trial.

22.3.1.3 Data analysis

Imaging data were analyzed using BrainVoyager 2000 (Brain Innovation, Maastricht, The Netherlands). The anatomical volume was transformed to the Talairach coordinate system (Talairach and Tournoux 1988). The cortical surface was reconstructed (Kriegeskorte and Goebel 2001) and inflated for visualization of results. The time courses of activation of individual voxels were constructed from the functional images and corrected for the temporal difference in acquisition of different slices (slice scan time correction). Subsequently, linear trends and low-frequency components (up to and including three cycles in the time course) were removed prior to any analysis. Voxel time courses were then coregistered to the structural volume and transformed into Talairach space with a resolution of 3 × 3 × 3 mm using trilinear interpolation. No spatial or temporal smoothing was applied to the functional time courses.

22.3.1.4 Conventional statistical analysis (general linear model)

Regional activations were analyzed using a GLM, testing for the (differential) contribution of several predictor functions to the explanation of variation in individual voxel time-courses. Six predictor functions reflecting the main stimuli and cues in the task were constructed as box-car functions (value one at the single scan where the relevant event took place, value zero otherwise) filtered through a linear model of the BOLD response (Boynton *et al.* 1996). Predictors were created for the mapping cue, the control stimulus, the face stimuli for which a left-handed response was required, the face stimuli requiring a right-handed response, the house stimuli requiring a right handed response, and the house stimuli requiring a left handed response. Individual voxel time courses were regressed on to a model containing these predictors and an additional constant-level predictor to correct for the signal level.

22.3.1.5 Granger causality mapping

The network of interacting regions subserving performance of the visuomotor mapping task was investigated by mapping the influence measures discussed above over the whole brain, giving what we call Granger causality maps (GCMs). Each of the computed GCMs centers on a single region of interest (reference region) that is considered *both* as the source of influence to voxels in the rest of the brain *and* as the target of influence from voxels in the rest of the brain. The reference regions were chosen as the activated regions found with the GLM analysis. In the notation used above, the averaged BOLD response time-course of voxels in a specified reference region was regressed on to the predictors described above and the remaining residual of the regression was then considered as the time course $x[t]$. Subsequently, the BOLD response time-course of each single voxel in the functional volume was regressed on to the same model and the remaining residual was considered as the time course $y[t]$, and the influence measures $F_{x \to y}$, $F_{y \to x}$, and $F_{x \cdot y}$ were computed. For each reference region this procedure resulted in three GCMs:

(1) the map of $F_{x \to y}$ (reference to voxel map), showing voxels that are influenced by the activity in the reference region;

(2) the map of $F_{y \to x}$ (voxel to reference map), showing voxels whose activity influence the activation in the reference ROI; and

(3) the map of $F_{x \cdot y}$ (instantaneous influence map), showing voxels whose activation shows an instantaneous dependency relation with activation of the reference ROI without any clear direction in time.

The GCM analysis was performed only on the data of the third subject. The influence maps were computed using an autoregressive model order of 1. Thus, Granger causality between brain regions was considered looking *one* TR (i.e. 960 ms) into the past. Autoregressive models were estimated using an orthogonalization procedure (Bagarinao and Sato 2002) allowing us to compute pooled estimates over the two runs of the task performed by the subject.

22.3.1.6 DTI tractography

Principal directions of diffusion at each voxel and fractional anisotropy maps were computed from the averaged diffusion weighted images using conventional fitting and diagonalization techniques (Basser *et al.* 1994). Tractography was performed using a novel tracking algorithm (K. H. Kim *et al.* 2002). Starting points for fiber-tracking were selected in activated frontal and parietal regions (Conturo *et al.* 1999).

22.3.2 Results and discussion

22.3.2.1 Activated regions and deconvolved event-related time courses

Maps for regional activation and deconvolved event-related BOLD responses, computed over all runs of all subjects, are shown in Fig. 22.2B. As expected, the maps for left and right responses show activation for the right and left motor and somatosensory cortex, respectively. Surprisingly, these regions also show a rise in BOLD response at presentation of the mapping cue, which required no manual response. Activation for the mapping cue exists prominently in posterior parietal areas (somewhat lateralized to the left) and premotor areas, both on the medial (supplementary motor area and presupplementary motor area) and lateral cortical surface (dorsal and ventral lateral premotor cortex). Furthermore, visual and prefrontal areas are also activated at the mapping cue. The high and wide peak in the deconvolution plots for the mapping cue for these frontoparietal areas show a transient, but temporally extended (over a few seconds) rise of activity when the stimulus-response mapping changed. Therefore, we investigated the interaction between these regions further by choosing highly activated frontal and parietal areas as reference regions for the GCM analysis described above.

22.3.2.2 Granger causality mapping and influence analysis

Two sets of GCMs for reference regions in the posterior parietal cortex (PPC) and lateral prefrontal cortex (LPF) of the left hemisphere are shown in Fig. 22.3A and B, respectively. The GCM analysis was performed in two runs of an individual subject. Two important points have to be made about the interpretation of these maps. First, all inferences about influence between regions of the cortex can only be interpreted with respect to the reference region. Strictly, for a given set of influence maps we *cannot* talk about influence between two regions which are both *not* the reference region. Secondly, direct influence can only be inferred in as far as the reference region and a given voxel are each other's only source of influence (save driving influence from the experimentally induced stimulation accounted for by the GLM predictors). Indirect influence (e.g. mediated by a third region) will wrongly appear as direct influence unless additional mediating sources of influence are taken into account (by adding them to the model). Considering this, the GCMs for the left posterior parietal region show large amounts of instantaneous dependence between this region and other parts of the brain, notably other parietal regions and premotor and prefrontal regions bilaterally without any clear direction. However, the voxel-to-reference map for this region also clearly shows an influence exerted by prefrontal and premotor regions (bilaterally) on the left PPC. This result is confirmed by the left PPC appearing in the reference-to-voxel map for the left LPF region shown in Fig. 22.3B. Furthermore, the reference-to-voxel map for the left PPC region shows a great deal of influence exerted by this region, first, on other parietal areas, both ipsilateral and contralateral, and, secondly, on higher-order visual areas in the inferotemporal cortex. The reference-to-voxel map for the left PFC also shows influence on parietal *and* inferotemporal areas. However, keeping the above cautions in mind, we should note that influence from prefrontal areas on visual areas might very well be mediated by parietal areas. Adding both prefrontal

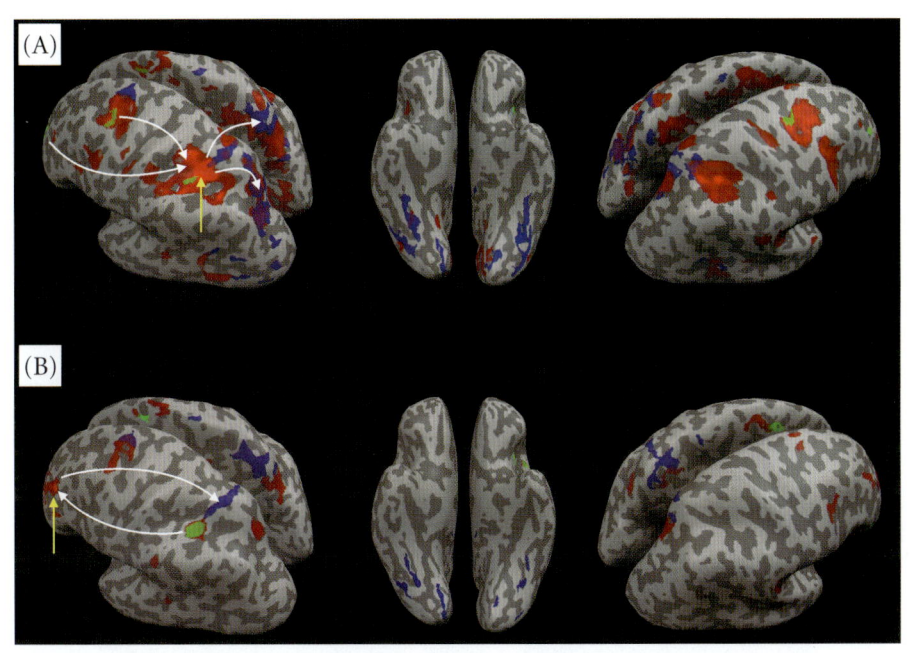

Fig. 22.3 Linear Granger causality maps (GCMs) computed over two runs of subject 3 for reference regions in (A) the left posterior parietal and (B) the left lateral prefrontal cortex, marked by yellow arrows. Maps for instantaneous influence between the reference area and other parts of the brain (instantaneous maps) are shown in red. Maps for influence from the reference region to other parts in the brain (reference to voxel maps) are shown in blue. Maps for influence from other parts of the brain to the reference region (voxel to reference maps) are shown in green. White arrows illustrate the interpretation of some of the maps and mark main discussed results.

and parietal areas to the autoregressive models could allow us to dissociate these two possibilities.

22.3.2.3 DTI-tracking of frontoparietal connections

Figure 22.4 shows the result of fiber tracking superimposed on anatomical and functional MRI data from the same subject. First, tracking was performed within a large part of the white matter in the left hemisphere (Fig. 4A). Fiber tracts potentially reflecting direct anatomical connections between task-related frontal or parietal regions were visualized by selecting only the fiber tracts hitting those regions. Figures 22.4B, C, and D show an example in which a 'seeding' region in the frontal lobe was selected. Some of the fiber tracts originating from this region reached the posterior parietal cortex in the proximity of regions that were also highlighted in the influence analysis (see Fig. 22.3).

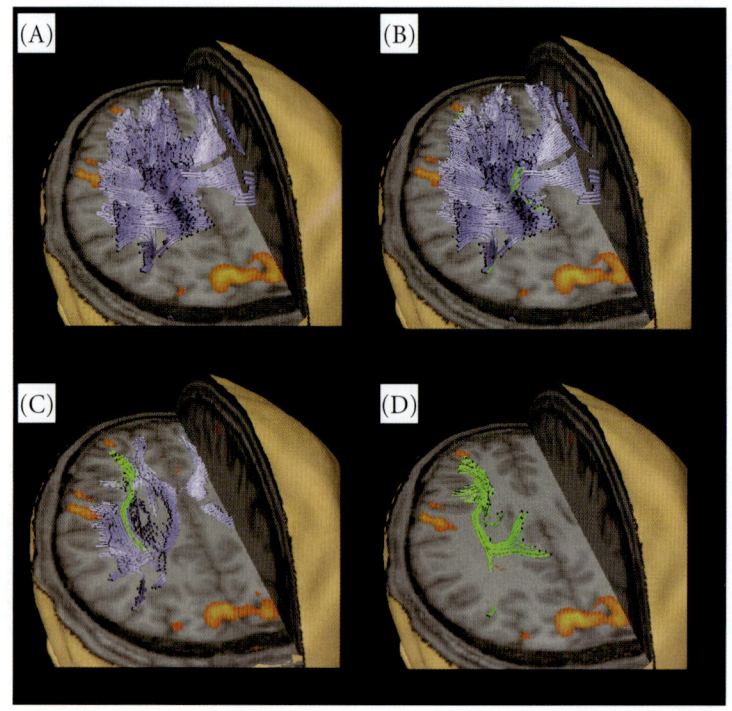

Fig. 22.4 Visualization of reconstructed fiber tracts on anatomical and functional MRI data from the same subject. (A) Large bundles of association and callosal fiber tracts originating from a large region of interest in the white matter in the left hemisphere. (B), (C), and (D) Different visualizations highlighting fiber tracts (green) potentially reflecting direct anatomical connections between frontal or parietal regions as revealed by the GLM and influence analysis.

22.4 General discussion

We have presented a framework for modeling directed cortical interactions based on time-resolved fMRI and a mathematical realization of the concept of Granger causality.

The proposed framework rests on the assumption that time-resolved fMRI contains enough temporal information to determine directed influences based solely on temporal precedence. Recent work suggests that time-resolved fMRI has a temporal resolution in the sub-second range. While such a temporal resolution can not reveal neuronal interactions, it might be sufficient to reveal temporal dependencies among cognitive components of complex cognitive tasks. If, for example, an area A needs 100 ms to compute a representation which is required as input for area B, a 'cognitive' temporal order is imposed on these areas in the 100 ms range. At the neuronal level, areas A and B might exchange some information continuously during these 100 ms, but only if area A has generated an adequate representation for area B can area B start to process the input to generate another useful representation for other brain areas. The increased processing in area B is expected to be accompanied by a rise of activity in that area, which is measurable using BOLD-fMRI.

Although interactions between brain areas might occur more gradually than in the outlined scenario, a temporal order between brain areas emerges if the input-representation for one areas is the resulting output of other areas. We assume that our framework is able to reveal directed interactions at such a psychological level as opposed to the much faster neuronal level (Formisano and Goebel 2003).

In our application, the processing delay between sensory areas in fusiform regions (including FFA and PPA) and parietal and motor areas within a trial of our task appeared to be too short to be detected by the directed influence analysis method. Due to low temporal sampling, noise and limited temporal resolution, we expect that in many cases true directed influences between brain areas cannot be resolved with our approach. None the less, these directed influences will appear in our analysis in the (non-directional) instantaneous maps. From this consideration it also follows that, on the basis of the instantaneous maps, it cannot be inferred that no directed effects among areas exist. If the influence analysis, however, shows areas in the voxel-to-reference and reference-to-voxel influence maps, this would indicate the presence of directed effects at a cognitive level, as outlined above. In our task, we could observe directed effects between frontal and parietal areas, which we attribute to the processes initiated by the (re-)mapping cues. The GLM maps show that the mapping cues lead to increased activation in frontal, premotor, and parietal areas, while the influence analysis reveals that the frontal areas drive or 'Granger cause' activity fluctuations in specific subregions of the activated parietal areas. We interpret this directed influence by assuming that frontal areas are involved in generating and maintaining an appropriate self-instruction for the new sensorimotor mapping, and that the corresponding neural representations act upon parietal areas to implement a respective new motor program. Our finding and its interpretation can only be tentative at present since we have performed the influence analysis only in a single subject. We are currently analyzing the data of other subjects to evaluate whether the directed influence from frontal to parietal areas can be generalized to the population level.

22.4.1 Future improvements

Since exploitation of temporal information in fMRI time courses is critical for the usefulness of our approach, one should optimize all scanning-related aspects influencing temporal resolution. The temporal properties of the hemodynamic response can not be changed, but if MRI techniques sensitive to neuronal currents (Bodurka and Bandettini 2002) become available in the future, our method could be used without change with much more power. At present, the temporal sampling is the only relevant scanning parameter that can be optimized. A shorter recording time per slice can be obtained by using higher magnetic fields (since shorter TEs can be used) and parallel imaging using phased array coils. Since the volume TR determines temporal sampling, a small number of slices is recommended. In the first studies described here, we covered the whole brain to potentially reveal directed effects between any brain areas. With specific hypothesis, one could increase temporal sampling by running the same experiment again using only a few slices to cover the relevant brain regions.

One of the attractive properties of the chosen cognitive task is that the same visual stimuli require a different response, depending on the current mapping context. We would thus expect that influence patterns between brain areas also change with respect to the currently valid mapping rule. These potential mapping-specific influence maps could not be revealed in the current analysis because the maps were computed over the whole time course. As a future improvement, we want to compute influence maps for each mapping type separately, to potentially reveal dynamic remapping effects. Such a 'windowed' extension to our approach might also be useful for other paradigms in which aspects of the task are changing over time within, or across, functional runs.

Another improvement would be to include non-linear terms in the multivariate vector autoregressive modeling approach. This would allow, for example, the revealing of modulatory (multiplicative) effects in which one area influences the coupling strength between two other areas. In our task, for example, we would expect the sensorimotor coupling (mediated probably by parietal regions) to change with respect to the current mapping rule because, in one mapping context, houses (faces) require a left (right) response and in the other, houses (faces) require a right (left) response. The necessary sensorimotor associations could be implemented in the parietal lobe as discussed above. However, it could also be established by modulating the flow of information from the sensory areas (i.e. FFA, PPA) to parietal and motor areas. We would expect such a modulatory sensorimotor effect especially in the first trials after presentation of a (re-)mapping cue. As discussed in a previous section, the use of DTI in the proposed framework is preliminary. In the present study, DTI was only used for the visualization of potential direct anatomical connections between activated frontal and parietal brain regions. The group of fiber tracts that we were able to track likely belongs to the large associative bundle composed of long and short fibers that connect the frontal lobe with the parietal, occipital, and temporal lobes (Catani et al. 2002; Mori et al. 2002). Ideally, individual information on anatomical connectivity from DTI (or other similar non-invasive technique) could be used as a structural constraint for the VAR modeling of functional time-series. Verifying the feasibility of this idea and its methodological implementation will require further, extensive research.

While presenting an fMRI-based framework to exploit temporal information might appear very limited at first sight, we think that our framework provides a viable complement to other neuroimaging approaches, such as combined fMRI–EEG/MEG studies. We hope that the proposed framework will stimulate further investigations of how brain areas communicate with each other.

Acknowledgements

The authors are grateful to David E. J. Linden and Nikolaus Kriegeskorte for useful comments on the manuscript. This work was supported by the Human Frontiers Science Program.

References

Allport, D. A., Styles, E. A., and Hsieh, S. (1994). Shifting intentional set: Exploring the dynamic control of tasks. In C. U. M. Moscovitch (Ed.), *Attention and Performance XV* pp. 421–52. MIT Press, Cambridge, MA.

Bagarinao, E. and Sato, S. (2002). Algorithm for vector autoregressive model parameter estimation using an orthogonalization procedure. *Ann Biomed Eng,* **30**(2), 260–71.

Basser, P. J. (2002). Relationships between diffusion tensor and q-space MRI. *Magn Reson Med,* **47**(2), 392–7.

Basser, P. J., Mattiello, J., and LeBihan, D. (1994). Estimation of the effective self-diffusion tensor from the NMR spin echo. *J Magn Reson B,* **103**(3), 247–54.

Bernasconi, C. and Konig, P. (1999). On the directionality of cortical interactions studied by structural analysis of electrophysiological recordings. *Biol Cybern,* **81**(3), 199–210.

Bodurka, J. and Bandettini, P. A. (2002). Toward direct mapping of neuronal activity: MRI detection of ultraweak, transient magnetic field changes. *Magn Reson Med,* **47**(6), 1052–8.

Boynton, G. M., Engel, S. A., Glover, G. H., and Heeger, D. J. (1996). Linear systems analysis of functional magnetic resonance imaging in human V1. *Journal of Neuroscience,* **16**(13), 4207–21.

Buchel, C. and Friston, K. J. (1997). Modulation of connectivity in visual pathways by attention: cortical interactions evaluated with structural equation modelling and fMRI. *Cereb Cortex,* **7**(8), 768–78.

Buchel, C. and Friston, K. (2000). Assessing interactions among neuronal systems using functional neuroimaging. *Neural Netw,* **13**(8–9), 871–82.

Catani, M., Howard, R. J., Pajevic, S. and Jones, D. (2002). Virtual *in vivo* interactive dissection of white matter fasciculi in the human brain. *Neuroimage,* **17**(17), 77–94.

Conturo, T. E., McKinstry, R. C., Akbudak, E., and Robinson, B. H. (1996). Encoding of anisotropic diffusion with tetrahedral gradients: a general mathematical diffusion formalism and experimental results. *Magn Reson Med,* **35**(3), 399–412.

Conturo, T. E., Lori, N. F., Cull, T. S., Akbudak, E., Snyder, A. Z., Shimony, J. S., McKinstry, R. C., Burton, H., and Raichle, M. E. (1999). Tracking neuronal fiber pathways in the living human brain. *Proc Natl Acad Sci USA,* **96**(18), 10422–27.

Dale, A. M. and Halgren, E. (2001). Spatiotemporal mapping of brain activity by integration of multiple imaging modalities. *Curr Opin Neurobiol,* **11**(2), 202–8.

Dove, A., Pollmann, S., Schubert, T., Wiggins, C. J., and von Cramon, D. Y. (2000). Prefrontal cortex activation in task switching: an event-related fMRI study. *Brain Res Cogn Brain Res,* **9**(1), 103–9.

Formisano, E. and Goebel, R. (2003). Tracking cognitive processes using fMRI mental chronometry. *Curr Opin Neurobiol,* **13**(2), 174–81.

Formisano, E., Linden, D. E., Di Salle, F., Trojano, L., Esposito, F., Sack, A. T., Grossi, D., Zanella, F. E., and Goebel, R. (2002). Tracking the mind's image in the brain I: time-resolved fMRI during visuospatial mental imagery. *Neuron,* **35**(1), 185–94.

Freiwald, W. A., Valdes, P., Bosch, J., Biscay, R., Jimenez, J. C., Rodriguez, L. M., Rodriguez, V., Kreiter, A. K., and Singer, W. (1999). Testing non-linearity and directedness of interactions between neural groups in the macaque inferotemporal cortex. *J Neurosci Methods*, **94**(1), 105–19.

Friston, K. J. and Buchel, C. (2000). Attentional modulation of effective connectivity from V2 to V5/MT in humans. *Proc Natl Acad Sci USA*, **97**(13), 7591–6.

Friston, K. J., Ungerleider, L. G., Jezzard, P., and Turner, R. (1995). Characterizing modulatory interactions between V1 and V2 in human cortex with fMRI. *Human Brain Mapping*, **2**, 211–24.

Geweke, J. (1982). Measurement of linear dependence and feedback between multiple time series. *Journal of the American Statistical Association*, **77**(378), 304–13.

Granger, C. W. J. (1969). Investigating causal relations by econometric models and cross-spectral methods. *Econometrica*, **37**(3), 424–38.

Granger, C. W. J. (1980). Testing for causality: A personal viewpoint. *Journal of Economic Dynamics and Control*, **2**(4), 329–52.

Hari, R., Levanen, S., and Raij, T. (2000). Timing of human cortical functions during cognition: role of MEG. *Trends Cogn Sci*, **4**(12), 455–62.

Hillyard, S. A. (1993). Electrical and magnetic brain recordings: contributions to cognitive neuroscience. *Curr Opin Neurobiol*, **3**(2), 217–24.

Horwitz, B. (1991). Functional interactions in the brain: use of correlations between regional metabolic rates. *J Cereb Blood Flow Metab*, **11**(2), A114–120.

Horwitz, B., Friston, K. J., and Taylor, J. G. (2000). Neural modeling and functional brain imaging: an overview. *Neural Netw*, **13**(8–9), 829–46.

Hsu, E. W. and Mori, S. (1995). Analytical expressions for the NMR apparent diffusion coefficients in an anisotropic system and a simplified method for determining fiber orientation. *Magn Reson Med*, **34**(2), 194–200.

Kaminski, M., Ding, M., Truccolo, W. A., and Bressler, S. L. (2001). Evaluating causal relations in neural systems: granger causality, directed transfer function and statistical assessment of significance. *Biol Cybern*, **85**(2), 145–57.

Kim, D.-S., Kim, M., Ronen, I., Formisano, E., Kim, K. H., Ugurbil, K., Mori, S., and Goebel, R. (2002). Bridging the gap between functional neuroimaging and neuroanatomy. Paper presented at the Society for Neuroscience, Orlando FL (USA).

Kim, K. H., Ronen, I., Formisano, E., Mori, S., Ugurbil, K., Goebel, R., and Kim, D.-S. (2002). A novel fiber-tracking method using vector criterion and predictive directivity in diffusion tensor imaging. Paper presented at the International Society for Magnetic Resonance in Medicine, Honolulu, Hawaii HI (USA).

King, M. D., Houseman, J., Roussel, S. A., van Bruggen, N., Williams, S. R., and Gadian, D. G. (1994). q-Space imaging of the brain. *Magn Reson Med*, **32**(6), 707–13.

Koch, M. A., Norris, D. G., and Hund-Georgiadis, M. (2002). An investigation of functional and anatomical connectivity using magnetic resonance imaging. *Neuroimage*, **16**(1), 241–50.

Kriegeskorte, N. and Goebel, R. (2001). An efficient algorithm for topologically correct segmentation of the cortical sheet in anatomical mr volumes. *Neuroimage*, **14**(2), 329–46.

Le Bihan, D. (Ed.) (1995). *Diffusion and Pefusion Magnetic Resonance Imaging: Application to Functional MRI*. New York: Raven Press.

Lin, C. P., Tseng, W. Y., Cheng, H. C., and Chen, J. H. (2001). Validation of diffusion tensor magnetic resonance axonal fiber imaging with registered manganese-enhanced optic tracts. *Neuroimage*, **14**(5), 1035–47.

McIntosh, A. R. and Gonzales-Lima, F. (1994). Structural equation modelling and its application to network analysis in functional brain imaging. *Human Brain Mapping,* **2**, 2–22.

Menon, R. S. and Kim, S. G. (1999). Spatial and temporal limits in cognitive neuroimaging with fMRI. *Trends Cogn Sci,* **3**(6), 207–16.

Menon, R. S., Luknowsky, D. C., and Gati, J. S. (1998). Mental chronometry using latency-resolved functional MRI. *Proc Natl Acad Sci USA,* **95**(18), 10902–7.

Mori, S., Crain, B. J., Chacko, V. P., and van Zijl, P. C. (1999). Three-dimensional tracking of axonal projections in the brain by magnetic resonance imaging. *Ann Neurol,* **45**(2), 265–9.

Mori, S., Kaufmann, W. E., Davatzikos, C., Stieltjes, B., Amodei, L., Fredericksen, K., Pearlson, G. D., Melhem, E. R., Solaiyappan, M., Raymond, G. V., Moser, H. W., and van Zijl, P. C. (2002). Imaging cortical association tracts in the human brain using diffusion-tensor-based axonal tracking. *Magn Reson Med,* **47**(2), 215–23.

Moseley, M. E., Cohen, Y., Kucharczyk, J., Mintorovitch, J., Asgari, H. S., Wendland, M. F., Tsuruda, J., and Norman, D. (1990). Diffusion-weighted MR imaging of anisotropic water diffusion in cat central nervous system. *Radiology,* **176**(2), 439–45.

Paivio, A. (1978). Comparisons of mental clocks. *J Exp Psychol Hum Percept Perform,* **4**(1), 61–71.

Parker, G. J., Stephan, K. E., Barker, G. J., Rowe, J. B., MacManus, D. G., Wheeler-Kingshott, C. A., Ciccarelli, O., Passingham, R. E., Spinks, R. L., Lemon, R. N., and Turner, R. (2002). Initial demonstration of in vivo tracing of axonal projections in the macaque brain and comparison with the human brain using diffusion tensor imaging and fast marching tractography. *Neuroimage,* **15**(4), 797–809.

Pierpaoli, C., Jezzard, P., Basser, P. J., Barnett, A., and Di Chiro, G. (1996). Diffusion tensor MR imaging of the human brain. *Radiology,* **201**(3), 637–48.

Ramnani, N., Lee, L., Mechelli, A., Phillips, C., Roebroeck, A., and Formisano, E. (2002). Exploring brain connectivity: a new frontier in systems neuroscience. Functional Brain Connectivity, 4–6 April 2002, Dusseldorf, Germany. *Trends Neurosci,* **25**(10), 496–7.

Renault, B., Ragot, R., Lesevre, N., and Remond, A. (1982). Onset and offset of brain events as indices of mental chronometry. *Science,* **215**(4538), 1413–5.

Richter, W., Somorjai, R., Summers, R., Jarmasz, M., Menon, R. S., Gati, J. S., Georgopoulos, A. P., Tegeler, C., Ugurbil, K., and Kim, S. G. (2000). Motor area activity during mental rotation studied by time-resolved single-trial fMRI. *J Cogn Neurosci,* **12**(2), 310–20.

Rushworth, M. F., Hadland, K. A., Paus, T., and Sipila, P. K. (2002). Role of the human medial frontal cortex in task switching: a combined fMRI and TMS study. *J Neurophysiol,* **87**(5), 2577–92.

Sack, A. T., Sperling, J. M., Prvulovic, D., Formisano, E., Goebel, R., Di Salle, F., Dierks, T., and Linden, D. E. (2002). Tracking the mind's image in the brain II: transcranial magnetic stimulation reveals parietal asymmetry in visuospatial imagery. *Neuron,* **35**(1), 195–204.

Sakuma, H., Nomura, Y., Takeda, K., Tagami, T., Nakagawa, T., Tamagawa, Y., Ishii, Y., and Tsukamoto, T. (1991). Adult and neonatal human brain: diffusional anisotropy and myelination with diffusion-weighted MR imaging. *Radiology,* **180**(1), 229–33.

Sohn, M. H., Ursu, S., Anderson, J. R., Stenger, V. A., and Carter, C. S. (2000). Inaugural article: the role of prefrontal cortex and posterior parietal cortex in task switching. *Proc Natl Acad Sci USA,* **97**(24), 13448–53.

Talairach, J. and Tournoux, P. (1988). *Co-planar Stereotaxic Atlas of the Human Brain: 3-Dimensional Proportional System: An Approach to Cerebral Imaging.* Thieme, Stuttgart.

Thierry, G., Boulanouar, K., Kherif, F., Ranjeva, J. P., and Demonet, J. F. (1999). Temporal sorting of neural components underlying phonological processing. *Neuroreport,* **10**(12), 2599–603.

Toni, I., Ramnani, N., Josephs, O., Ashburner, J., and Passingham, R. E. (2001*a*). Learning arbitrary visuomotor associations: temporal dynamic of brain activity. *Neuroimage*, **14**(5), 1048–57.

Toni, I., Rushworth, M. F., and Passingham, R. E. (2001*b*). Neural correlates of visuomotor associations. Spatial rules compared with arbitrary rules. *Exp Brain Res*, **141**(3), 359–69.

Toni, I., Rowe, J., Stephan, K. E., and Passingham, R. E. (2002). Changes of cortico-striatal effective connectivity during visuomotor learning. *Cereb Cortex*, **12**(10), 1040–7.

Trojano, L., Grossi, D., Linden, D. E., Formisano, E., Hacker, H., Zanella, F. E., Goebel, R., and Di Salle, F. (2000). Matching two imagined clocks: the functional anatomy of spatial analysis in the absence of visual stimulation. *Cereb Cortex*, **10**(5), 473–81.

Tuch, D. S., Reese, T. G., Wiegell, M. R., Makris, N., Belliveau, J. W., and Wedeen, V. J. (2002). High angular resolution diffusion imaging reveals intravoxel white matter fiber heterogeneity. *Magn Reson Med*, **48**(4), 577–82.

Wimberger, D. M., Roberts, T. P., Barkovich, A. J., Prayer, L. M., Moseley, M. E., and Kucharczyk, J. (1995). Identification of 'premyelination' by diffusion-weighted MRI. *J Comput Assist Tomogr*, **19**(1), 28–33.

Chapter 23

From viewing of movements to understanding and imitation of other persons' acts: MEG studies of the human mirror-neuron system

Riitta Hari and Nobuyuki Nishitani

Abstract

For successful social interactions, humans continuously monitor, and often imitate, motor acts, postures, and gaze of their co-citizens. Recent data suggest that this type of intention reading and action imitation could be supported by 'mirror neurons', first identified in the monkey frontal lobe. We have characterized, with whole-scalp magnetoencephalographic (MEG) recordings, the human mirror-neuron system (MNS), which seems to consist of several cortical areas activated in a clear temporal sequence. Observation of orofacial gestures activated in both hemispheres and within about 250 ms, first the visual cortex, then the superior temporal sulcus (STS) region, the inferior parietal cortex, Broca's area, and finally the primary motor cortex. Activation of Broca's area was significantly stronger during on-line imitation than execution or observation of finger and mouth movements. Activation of somatosensory cortices was also modified during action viewing. Studies of the human MNS raise intriguing questions about the functional relationship between speech production and orofacial/hand gestures, and also emphasize the importance of motor functions in human cognition. Future studies should address the generality and abnormalities of the human MNS.

23.1 Introduction

Humans monitor other subjects' motor-act-based intentions continuously, automatically, and without effort. This very human behavior, essential for successful social interactions, could be supported by a 'mirror-neuron system' (MNS) that provides shared representations for action execution and representation.

Since the pioneering findings of 'mirror neurons' in the monkey frontal cortex by the Parma group (Di Pellegrino *et al.* 1992; Gallese *et al.* 1996; Rizzolatti *et al.* 1996), attempts have been made to find similar neuronal behavior in the human brain. We have followed the millisecond-range activation sequences of the human MNS with whole-scalp magnetoencephalography (MEG; Hari *et al.* 2000). We will therefore start our chapter with a brief presentation of the basics of the MEG method. We will then discuss our MEG data on the human MNS, ending with some speculative statements of the role of the MNS in human behavior.

23.2 MEG as a tool to study human brain function

Postsynaptic neuronal currents, arising in synchronously activated cortical pyramidal cells, produce extremely tiny magnetic fields, only about 10^{-9} times the steady magnetic field of the Earth. These fields, mainly generated by currents in fissural cortex, can be measured with sensitive superconducting quantum interference devices (SQUIDs) totally non-invasively outside the head. To locate the activated areas in the brain, it is often useful to consider the local neuronal activation as a current dipole. The purpose of MEG recordings is then to identify and locate the dipoles accurately in the brain, and to follow changes in their strengths as a function of time.

The magnetic field outside the head is first picked up with superconducting flux transformers, i.e. wire loops, the shapes of which are crucial for the shape of the measured field patterns. When the source can be well described as a single current dipole, recordings with a magnetometer (a single coil) or an axial gradiometer (two coils wired in opposite directions, one close to and the other a few centimeters above the scalp) show the largest signals a few centimeters away from the dipole, symmetrically on both sides. Instead, recordings with a planar gradiometer, consisting of a figure-of-eight coil wired in the same plane (such as used in studies of this chapter), show the strongest signals just above the current dipole at the site of the steepest field gradient (Hämäläinen, *et al.* 1993; Hari 1999).

The next step is to find the site of the current dipole by means of a least-squares fit to signals measured (preferably simultaneously) at several sites above the scalp. In the first visual evaluation of the data, the strongest signals measured with planar gradiometers point directly towards the most probable source locations, greatly facilitating the subsequent analysis of complex field patterns. Finally, a multi-dipole model can be constructed by adding all local sources to the model, and by taking into account the signals of all sensors at all time points.

The current-dipole model is well suited for studies of local discrete activation areas, whereas distributed source models may be preferred for proper description of very complex current distributions (Uutela *et al.* 1999). However, it should be emphasized

MEG STUDIES OF THE HUMAN MIRROR-NEURON SYSTEM | 465

that the analysis method affects the results: applying a distributed model to a point-like source will indicate distributed activity, and vice versa! The localization accuracy of the source models naturally depends on the signal-to-noise ratio.

The requirement of measuring the magnetic field pattern at the same time all over the whole head has led to rapid development of multichannel neuromagnetometers during the past two decades. Figure 23.1 shows our helmet-shaped MEG device, comprising 306 sensors; 204 sensors are planar gradiometers and the remaining 102 are magnetometers (VectorView™, Neuromag Ltd). With the emergence of whole-scalp neuromagnetometers, MEG recordings have been increasingly applied for exploration of human cognitive functions, including 'motor cognition', as the MNS studies below will demonstrate.

Skull and scalp distort the electric potential distributions (and thus EEG) but are transparent for magnetic signals, so that the MEG pattern is restricted to a more local area than the corresponding EEG distribution. MEG measurements are often advantageous when multiple sources are to be resolved. For example, the distinction between the primary and the secondary somatosensory cortices is easy with MEG but, due to the overlapping potential distributions, tedious with EEG (Hari *et al.* 1984; Kaukoranta *et al.* 1986).

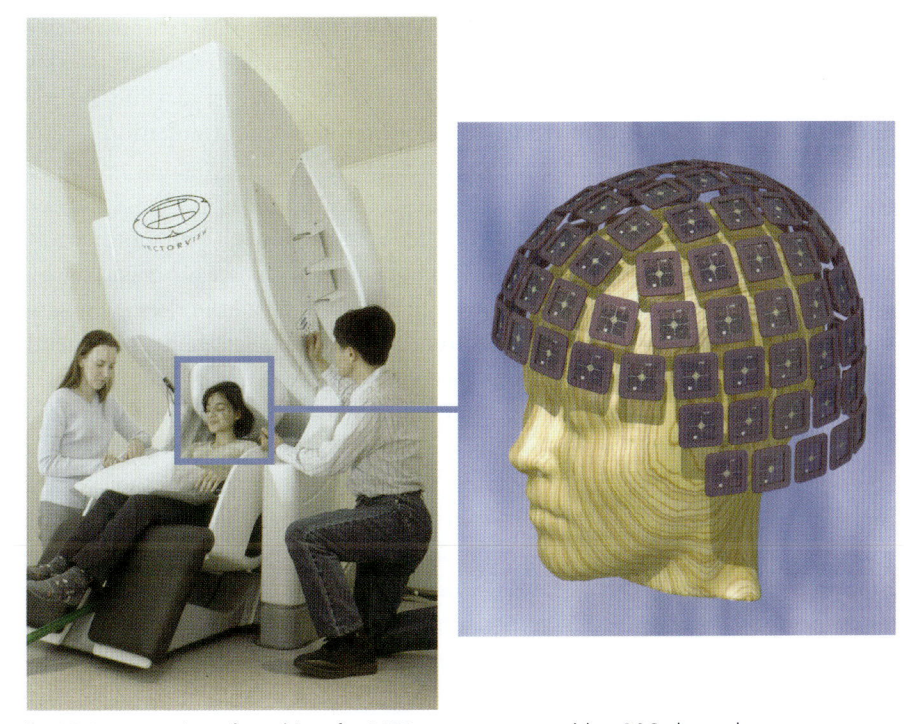

Fig. 23.1 Preparation of a subject for MEG measurement with a 306-channel neuromagnetometer. The sensor array of the device is shown on the right (courtesy of Mika Seppä).

In a sphere, radial currents affect only the potential distribution. For accurate EEG modeling, the electric conductivities of skull and scalp should be known. Both MEG and EEG signals reflect neural activation directly, and not blood flow or metabolism.

The advantages of MEG in human brain research (Hari *et al.* 2000) include—besides non-invasiveness and the excellent temporal resolution—the very good replicability of the signals, even over months, and the possibility of obtaining quantitative information on activation strengths (net intracellular currents) in local neuronal populations. The selectivity of MEG to activation of fissural cortex can be considered an advantage as well, because these areas are difficult to reach with other means, including intracranial recordings. Moreover, the complete source configuration can be more easily resolved if the tangential (fissural) parts of the currents are identified first by MEG.

Because of the non-uniqueness of the inverse problem, the MEG and EEG source analyses require human interaction and are rather demanding for a beginner. Another limitation (common with fMRI and PET but not with EEG) is that the subject has to keep his or her head immobile during the recording.

23.3 Social brains and mirror neurons

Much of the previous and highly important animal and human electrophysiological work on sensory functions has been carried out with rather artificial stimuli, such as tone pips, flashing checkerboard patterns, or electric pulses delivered on the skin. However, such elementary stimuli are rarely encountered in everyday life. Instead, other humans provide us with strong and continuous 'social stimulation' and we certainly have 'social brains' that have been tuned to help us to get along among our co-citizens.

Seeing other people's movements may influence the observer's own motor system and even result in unintended imitation. This phenomenon can be experienced while just viewing athletic performance, and noticing that one's own knee rises at the same time as does the athlete's knee. Similarly, students of body language have frequently noted that humans easily imitate another person's postures. One candidate for a brain system supporting this type of copying and imitation of movements and postures is the neuronal network forming the MNS.

Several years ago, Rizzolatti and coworkers found neurons in area F5 of the monkey frontal cortex that were activated both when the monkey acted himself, taking a raisin from a tray, and when he viewed another monkey or the human experimenter making the same act (Di Pellegrino *et al.* 1992; Gallese *et al.* 1996; Rizzolatti *et al.* 1996). These 'mirror neurons', more jovially 'monkey see, monkey do neurons' (Carey 1996) seem to match action representation with action execution. They thereby could form a link between the sender and the receiver of a motor-act-based message: two persons can communicate effectively only if they share a common code or possess similar mechanisms to perceive, and act in, the world.

We have been interested in finding, by means of MEG, evidence about the existence of a human MNS which, at its simplest level, should fulfill the following requirements.

1. The system's functional state should change when the subject herself makes a movement.
2. The functional state should be modulated to the same direction as in (1), but most likely less vigorously, when the subject just views another person making a similar movement. In this situation the subject has no requirement for subsequent reproduction of the movement and her own muscles should be relaxed.
3. Modulation, probably even stronger than during (2), would be expected to occur when the subject imitates on-line (as simultaneously as possible) the motor acts of the other person.

A transcranial magnetic stimulation (TMS) study was the first to show modulation of pathways from the human primary motor cortex to the spinal level while the subject viewed another person's motor acts (Fadiga *et al.* 1995). However, this study did not pinpoint the exact level (cortical versus subcortical/spinal) where the effect of viewing took place. Thus our MEG study, reported below (Hari *et al.* 1998), was the first one to demonstrate that action viewing affects the primary motor cortex. A more recent TMS study has confirmed this proposal (Strafella and Paus 2000). The early human imaging studies by means of positron emission tomography (PET) demonstrated involvement of Broca's area in the human MNS, activated when the subject observed the examiner using a precision grasp to enclose an object (Grafton *et al.* 1996).

23.4 Motor cortex and the MNS

In our first MNS-related MEG study (Hari *et al.* 1998), we tried to find out whether the human primary motor cortex, M1, would be activated when subjects just view another person's motor acts. To probe the functional state of M1, we monitored modulation of the motor cortex 20-Hz MEG rhythm (Hari and Salmelin 1997). Movements of either the left or the right hand suppress this rhythm, with slightly contralateral dominance; the movements are followed by 'rebounds'—increases in the level of the MEG rhythm (Nagamine *et al.* 1996). Electric stimuli applied to median nerves at the wrists are also followed by clear rebounds of the 20-Hz rhythm, and this increase likely reflects an inhibitory state of the motor cortex (Salmelin and Hari 1994). The relationship between the rebound and motor cortex inhibition is also supported by TMS results of decreased excitability in the motor cortex with a time course similar to that of the 20-Hz rebound (Chen *et al.* 1998).

In addition to source analysis that indicates that the main origin of the 20-Hz rolandic activity is in the precentral motor cortex (Salmelin and Hari 1994; Hari *et al.* 1998; see insert of Fig. 23.2), strong further evidence about the generation of the rhythm in M1 derives from studies demonstrating oscillatory cortico-muscular coupling at about 20-Hz during isometric contraction of different muscles (Conway *et al.* 1995; Salenius *et al.* 1997; Hari and Salenius 1999). The generators of the maximally coherent

468 | FUNCTIONAL NEUROIMAGING OF VISUAL COGNITION

MEG signals agree with the location of the individual M1 cortex, evaluated anatomically and functionally in each individual (Salenius *et al.* 1997) and by stimulations during neurosurgical operations (Mäkelä *et al.* 2001).

Figure 23.2 (left) shows our three experimental conditions: resting, acting, and viewing. During all conditions, the left and right median nerves of the subject were stimulated alternately to vary the level of the motor-cortex rhythm in a well-controlled manner (see above). During acting, the subject manipulated a small object with the fingers of

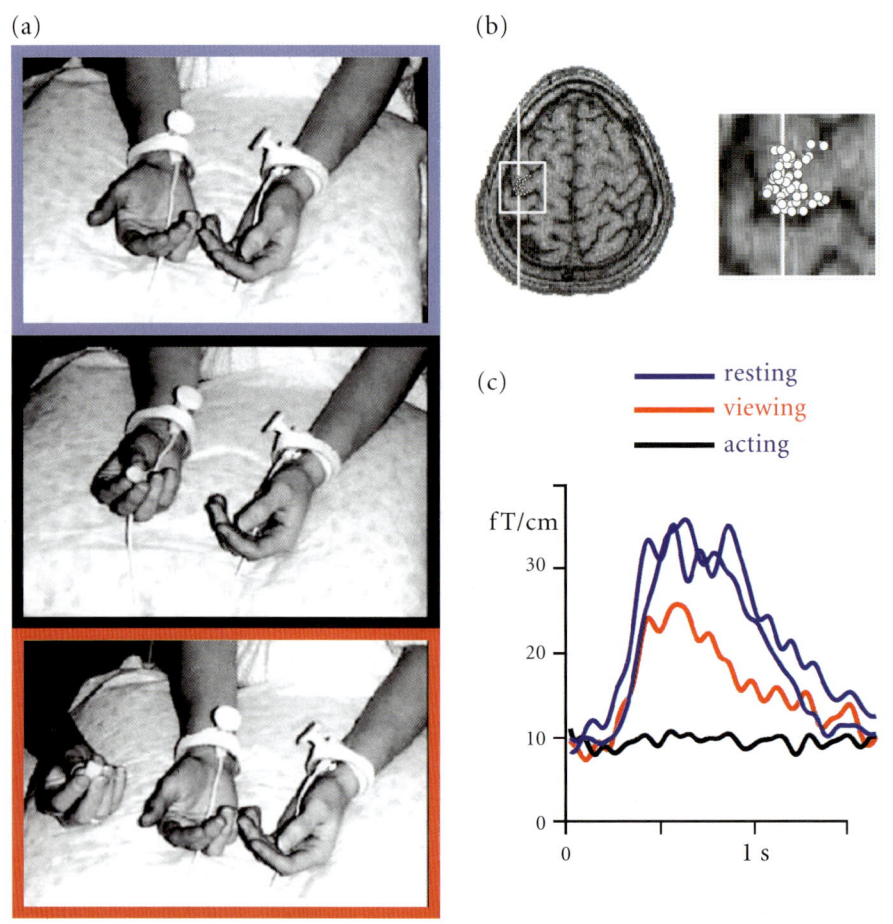

Fig. 23.2 (a) Experimental conditions during an action viewing experiment (from top to bottom: resting, acting, and viewing). (b) Sources of the unaveraged 20 Hz oscillations in the left rolandic area. The locations of the dipoles (circles) agree with activation of the 'hand knob' (Yousry *et al.* 1997) of the precentral primary motor cortex. (c) Level of the left-hemisphere 20-Hz motor-cortex rhythm after right median nerve stimuli (presented at time zero) during all conditions; two repetitions of the same stimulation are shown for the rest condition. (Adapted from Hari *et al.* 1998.)

her right hand, and in the viewing condition she viewed the experimenter to perform similar movements. An important new experimental approach was to have a real person performing the movements in the subject's field of view, instead of some video presentation of the stimuli. In fact, a later study showed the effects of interest to be 15–20% stronger for motor acts presented live than those seen on a video (Järveläinen et al. 2001).

Figure 23.2c shows that stimulation of the right median nerve was followed by a typical 20-Hz rebound in the left hemisphere: the 20-Hz level increased to its maximum within 500 ms after the median nerve stimulus (presented at time zero). This rebound reflects changes in the M1 cortex, as discussed above. When the subject was herself manipulating the small object, the rebound (inhibition of M1) was totally abolished, as was expected because the movements required activation of the M1 cortex. Interestingly, the rebound was also significantly suppressed when the subject just viewed another person's manipulation movements without moving herself. It is important to note that surface EMG did not show any increase of muscle activity during action viewing.

These data indicate that the human primary motor cortex is activated when the subject just views another person's actions. The motor cortex thus fulfills our requirements (1) and (2) for a MNS, and we can suggest it to be a part of the human MNS.

23.5 Temporal sequence during pinching

Because modulation of the 20-Hz motor-cortex rhythm reflects changes in M1 only, it is not of much use for monitoring inter-areal cortical dynamics during action viewing. In the next MNS study we therefore recorded brain activity time-locked to hand movements. The subject either himself stretched his right arm and hand towards a manipulandum, ending the movement with a pinch of the tip (see insert of Fig. 23.3), or he imitated on-line similar movements made by the experimenter (who was again performing live in the measurement chamber in the subject's field of view), or only observed the experimenter's movements.

Figure 23.3 shows examples of responses of a single subject in the imitation condition (Nishitani and Hari 2000). The first activation occurred in the posterior visual areas about 400 ms before the subject pinched the top of the manipulandum (and thereby released a trigger pulse for signal averaging). The next main activation occurred in Broca's area and was followed by signals in the left motor cortex, and finally in the right motor cortex some 200 ms later.

These brain regions were activated both during action execution and observation, and they can thus be considered parts of the human MNS. The relative timing of the visual cortex, but not of the other brain areas, differed between the conditions because the visual stimulus, the experimenter's or the subject's hand, appeared into the subject's visual field at different stages of the movement sequence, depending on the task. A similar sequence of activation was seen in other subjects, and the

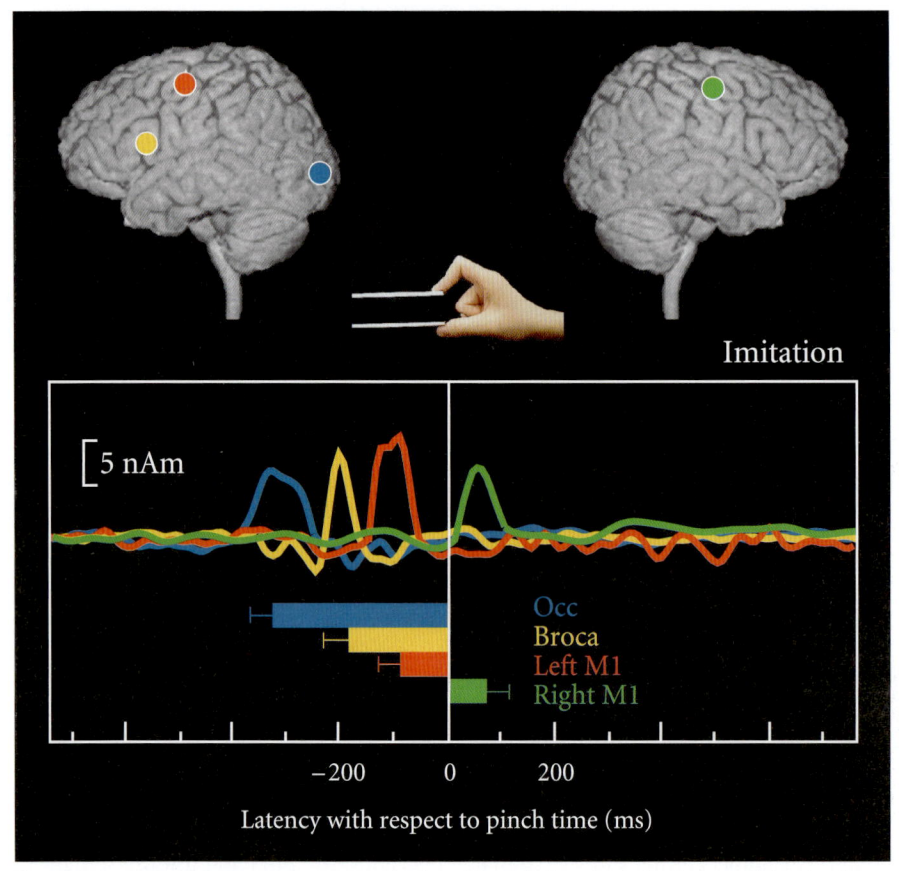

Fig. 23.3 Averaged MEG responses of a single subject during the imitation condition in the pinching experiment. Different colors in the traces, bars, and brain locations refer to visual cortex (Occ), Broca's area (Broca), and the left and right primary motor cortices (M1). The mean ± SEM latencies of eight subjects are given below the traces. The time zero refers to the time when the reaching movement ended with a pinch at the tip of the manipulandum. (Adapted from Nishitani and Hari, 2000.)

sequence was similar in the imitation and observation conditions. Activation in Broca's area and in the left primary motor cortex was significantly stronger during imitation than during the other conditions (Nishitani and Hari 2000). Thus either the same cells fired more vigorously during imitation than during viewing and execution, or the neurons activated by viewing and execution were closely intermingled in a small area.

23.6 Frozen orofacial gestures

The above data indicate that many types of hand actions can activate the human MNS. In monkeys, the mirror neurons of the F5 region are also activated by orofacial gestures.

MEG STUDIES OF THE HUMAN MIRROR-NEURON SYSTEM | 471

We therefore wondered whether still pictures of lip forms, 'frozen orofacial gestures', only implying motion, could also activate the human MNS. In one session, the subject was asked to imitate the lip forms immediately after having seen them in the picture, and in the other session the subject had to view the pictures without any requirement for subsequent production of the movements.

Figure 23.4 shows brain responses from a single subject while he imitated the lip forms (red traces). The signals illustrate a clear prolongation of the response latencies

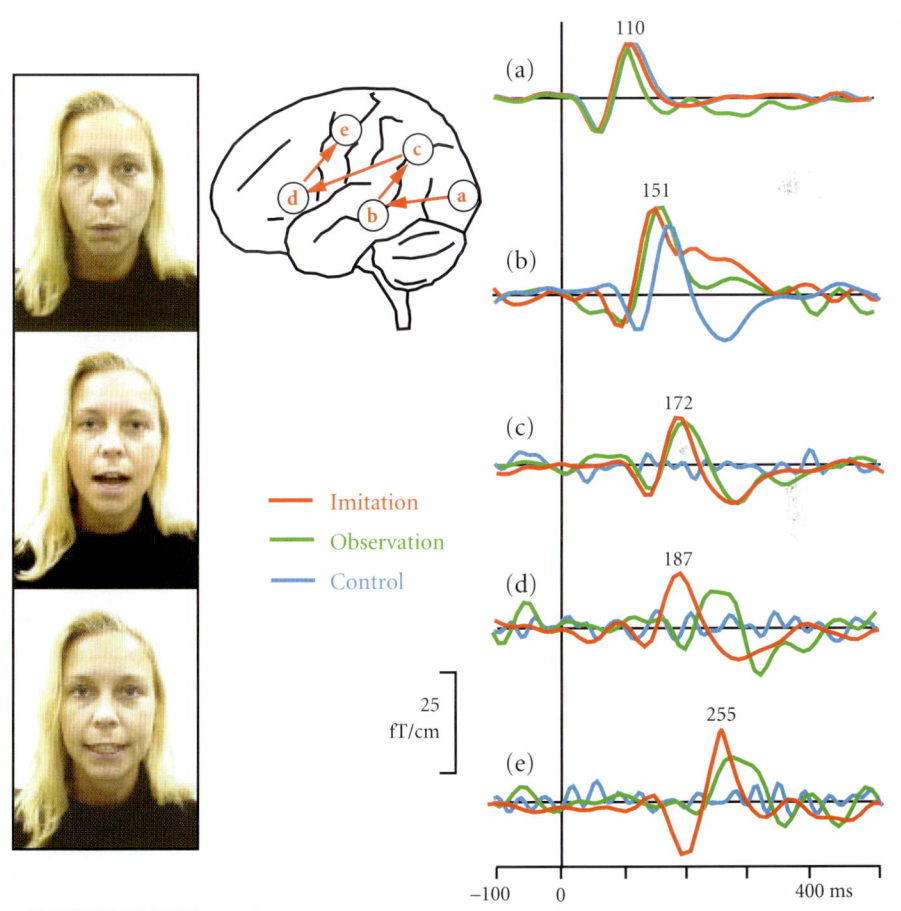

Fig. 23.4 Averaged MEG responses of a single subject to observation and imitation of lip forms, shown on the left, and to viewing of landscapes (control). The signals were picked up from five locations, corresponding approximately to the visual cortex, the superior temporal sulcus, the inferior parietal lobe, Broca's area, and the primary motor cortex, respectively. The numbers give the peak latencies of the responses. The schematic brain picture in the middle indicates the assumed sequence of activation based on the response latencies. (Nishitani and Hari, unpublished data from a Finnish subject in a study very similar to that of Nishitani and Hari 2002.)

from sensor location a to location e. Because the signals were recorded with planar gradiometers, this temporal order could correspond to progression of activation from the occipital visual cortex to the superior temporal sulcus (STS), then to inferior parietal area, to Broca's region, and finally to the left primary motor cortex. This order of activation, occurring in 200–250 ms, has been confirmed by source analysis in 10 Finnish (Nishitani and Hari, unpublished) and in 10 Japanese subjects (Nishitani and Hari 2002).

A very similar sequence of activation was also seen when the subject just observed the lip forms (green traces). Interestingly, the control stimuli of landscapes activated the occipital and temporal-lobe regions (areas a and b), similarly to the lip form stimuli, but thereafter these blue traces did not show any significant activation, suggesting clear stimulus-specificity of the activation pattern.

It is again important to note that a surface electromyogram from mouth muscles was silent during the observation condition, whereas during imitation and execution (not shown in Fig. 23.4 but reported by Nishitani and Hari 2002) it reflected real mouth movements, as expected.

The STS region activated in our subjects (mean Talaraich coordinates: x from ±49 to ±55, y from –43 to –48, and z from 10 to 13) agrees with locations related to social perception (Allison et al. 2000) and with areas activated during reciprocal imitation (Decety et al. 2002). The STS area activated in our study was in the left hemisphere, about 15 mm more anterior to landscapes than to faces.

Monkey studies have remained indecisive about the activation routes from the STS, where neurons with mirror properties have also been observed, to the inferior frontal cortex, as there are no direct pathways between these areas. Our timing data (see also the mean data in Fig. 23.5) suggest that the human STS area involved in the MNS is connected to the inferior frontal cortex (Broca's area) via the inferior parietal lobe. However, at present the correspondence between the human and monkey STS areas is still poorly understood. Whereas the monkey mirror neurons in STS region are located in the anterior part of the STS region (STPa; Perrett et al. 1990), the human MEG activation was observed in the middle or posterior STS. Of course it is possible, and even likely, that the mirror-neuron systems differ between monkeys and humans.

We would also like to be cautious about the conclusiveness of the exact activation sequence derived from the timing of peak activations at different brain areas. One reason is that microelectrode recordings in monkeys have demonstrated considerable temporal overlap in activations of different visual cortices that are anatomically considered to represent hierarchical stages of visual processing (Schmolesky et al. 1998).

Before making a strong claim about activation of Broca's region by just viewing lip forms, we should rule out the possibility that the activation was elicited by the inferred verbal content of the lip forms. If that were the case, activation in Broca's area could be considered speech-related rather than addressed to the human MNS. We therefore

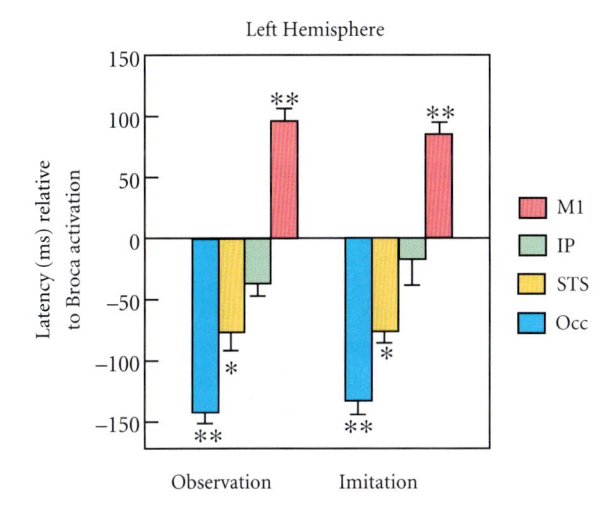

Fig. 23.5 Mean ± SEM latencies of 10 subjects who were observing and imitating verbal-type lip forms. The latencies are given with respect to activation times at Broca's region. All data are from the left hemisphere. Abbreviations: M1, primary motor cortex; IP, inferior parietal cortex; STS, superior temporal sulcus; Occ, occipital visual cortex. (Adapted from Nishitani and Hari 2002.)

also compared activations triggered by neutral faces and by faces with verbal and non-verbal lip forms (Nishitani and Hari 2002).

Neutral faces activated the primary motor cortex in none of our subjects and weak activation was seen Broca's region in only 3 out of 10 subjects. Instead, both verbal and non-verbal lip forms activated Broca's region and the M1 cortex in a very similar manner, with slightly stronger sources to verbal than non-verbal lip forms in the left hemisphere, and vice versa in the right hemisphere. We are thus confident that the observed activation in Broca's region is not explained by linguistic or verbal content of the stimuli.

23.7 Problem of agency

The existence of the human MNS, which all the above data suggest, means that overlapping brain areas may be activated during action execution and observation. This leads to the problem of agency: how does the subject know that she made a certain motor act and not only saw it, provided that the corresponding brain activation patterns are very similar? This may sound a ridiculous question, but is not trivial at all because misattribution of one's own acts is possible. For example, patients with some psychiatric disorders may address their own actions to external agents and experience that they are under 'alien control'.

In the healthy human brain, the problem of agency may be solved by first sending an efference copy (corollary discharge) from the movement preparation areas to other brain areas to inform them about the forthcoming consequences of the acts. Then the feedback provided by proprioceptive and other somatosensory afferents during own movements can be compared with the expectations tuned by the efference copy.

We have observed that parts of the somatosensory cortical network demonstrate behavior that may contribute to the sense of agency. When median nerve stimuli were used to probe the functional state of the primary and secondary somatosensory cortices, SI and SII, responses of SI were increased and responses of SII decreased both when the subject manipulated a small object with the fingers of her right hand and when she saw another person performing similar movements (Avikainen *et al.* 2002); however, as the only exception, the responses of the left SI cortex were suppressed during the subject's own manipulation movements.

These data imply that the cortical somatosensory network, and especially the primary somatosensory cortex contralateral to the moving limb, could help in resolving the agent of a motor act. A recent study on a haptically deafferented patient suggests that subjects cannot become aware of their actions on the basis of efference copies only, but that they might be aware of the result of the comparison process (Fourneret *et al.* 2002). Of course, the execution and observation conditions differ also in the sense that activation of the primary motor cortex is under threshold during observation but over threshold during execution, and this difference certainly affects the awareness of own action as well.

23.8 **Role of the MNS in human behavior?**

23.8.1 **Imitation**

The strengthening of activation during imitation in Broca's area (Talairach coordinates: x from ± 42 to ± 50, y from 15 to 22, and z from 12 to 14 in our studies) is interesting and would support the role of the MNS in imitation behavior. Imitation is essential for the learning of many motor skills. Children learn by watching adults and other children, developing under continuous social feedback. Fortunately for the child's development, imitation seems highly rewarding: healthy children greatly enjoy both imitating and being imitated themselves.

One can copy movements and motor acts without understanding their meaning, as happens when a flock of birds escapes from a lake following the first frightened bird. However, true imitation goes beyond copying of action patterns and is more flexible, resisting, for example, changes in the physical environment (Gattis *et al.* 2002). Imitation is likely to be guided by a set of hierarchically organized goals which compete with each other if the processing capacity of the actor is limited (Gattis *et al.* 2002). Therefore imitation cannot be based on a direct sensory–motor mapping only, but rather requires some degree of action recognition and understanding. The MNS might provide such a sophisticated mechanism that helps to recognize and understand motor acts of other persons (Gallese 2001).

Imitation skills are abnormal in many autistic subjects, who imitate less and in a different manner than healthy subjects (Williams *et al.* 2001). If a healthy person is asked to imitate another person face-to-face, she typically prefers to imitate as in mirror; this preference is seen already in small children. In our recent behavioral study with a pen–cup task (Wohlschläger and Bekkering 2002), Asperger and high-functioning autistic subjects did not benefit from the mirror-image pose of the person to be imitated (Avikainen *et al.* 2003). The autistic group made considerably more errors than the control group in selecting the correct grip and correct hand for the imitation in the mirror-image position but were equally accurate as the control subjects in imitating in crossed-over fashion (with anatomical rather than spatial correspondence with the person to be imitated). The relationship of the deficit in mirror-image imitation to possible dysfunctions of the MNS remains to be shown. It is interesting that the autistic subjects seemed to be normally reading the goals of other subjects' acts, as they achieved the end-point of the movement (the cup) as accurately as the control subjects. The end-points of actions are considered to be at the top of the set of goals guiding the imitation behavior (Gattis *et al.* 2002) and thus could be the last ones to be disturbed in case of deteriorating imitation behavior.

23.8.2 Mind-reading skills

Besides imitation, another possible role for the human MNS could be involvement in 'mind reading' skills, i.e. computation of other people's intentions continuously and automatically, based on observed and understood movements, postures, and gaze. These skills are essential for social communication and seem to be defective in, for example, autism and schizophrenia. Studies of the MNS might therefore provide novel insights into the pathophysiology of these brain disorders.

23.8.3 Representation of hand and orofacial gestures in Broca's area

One fascinating puzzle arising from the available MNS data is the representation of hand and orofacial gestures in the speech production area. In this context it is of interest to note that Brodmann's area 44, i.e. a part of Broca's area, is considered the human homologue of the monkey mirror-neuron area F5 (Passingham 1993; Petrides and Pandya 1994; Matelli and Luppino 1997). Moreover, Broca's area is not only a speech-production area but can be activated in association of various hand actions; for example, when stroke patients try to use their paralyzed hand or when people mentally imagine hand grasping (for references, see Gallese *et al.* 1996). Thirdly, there are close connections between speech-related gestures and speech production, which are often even considered as outlets of the same thought process (Goldin-Meadow 1999).

Speech-related gesturing seems to be independent of whether others are able to see the gestures or not. Thus humans gesture while speaking in phone, and even congenitally blind persons gesture—and while speaking with persons whom they know to be blind as well (Iverson and Goldin-Meadow 1998)! In stutterers, speech-related hand gestures may freeze at the time of stuttering whereas speech-unrelated hand movements continue (Mayberry and Jaques 2000). Interestingly, hearing babies born to deaf parents may not babble aloud with their mouths but, instead, silently with their hands, so that the hand movements then contain the proper speech rhythm (Petitto *et al.* 2001).

Thus abundant evidence points towards rather intimate connections between speech-related gestures and speech production, and the co-representation of these two functions in Broca's region is in full agreement with suggestions about common evolutionary roots of gestures and speech. For example, Rizzolatti and Arbib (1998), in the framework of the MNS, speculated that hand and orofacial gestures—rather than sounds—might have served as the precursors of human language. This proposal was not without precedents, but is now grounded, on a feasible neurophysiological basis.

The evolution of Broca's region first for gestural communication and only later for speech is supported by findings that this area is larger in the left than the right brain half of great apes, who do not speak (Cantalupo and Hopkins 2001).

23.9 Future studies

One interesting question for future studies is whether the human MNS functions extend beyond pure motor behavior. Modulation of the SI and SII cortices (Avikainen *et al.* 2002), discussed above, supports such a possibility. Interestingly, recent intracranial recordings have shown that the human anterior cingulate cortex can react to thermal pain perceived by the subject and also to mere viewing of another subject receiving similar stimuli (Hutchison *et al.* 1999). It is thus tempting to speculate that the human MNS might comprise a general and socially important mechanism to allow intentions, emotions, and even the intensity of pain to be matched and communicated between individuals.

Studies of the human MNS could provide insights into brain disorders that manifest themselves as abnormal imitation or mind-reading skills, such as autism and schizophrenia. Studies of the 'social brain' will certainly provide exciting challenges for the whole neuroscience community.

Acknowledgements

Supported by the Academy of Finland, the Sigrid Jusélius Foundation, the Human Frontier Science Program Organization Grant RG 39–98 (P.I. Vittorio Gallese), and Grant-in-Aid for Scientific Research (B) 15 300111 from JSPS.

References

Allison, T., Puce, A., and McCarthy, G. (2000). Social perception from visual cues: role of the STS region. *Trends in Cognitive Sciences*, **4**, 267–78.

Avikainen, S., Forss, N., and Hari, R. (2002). Modulated activation of the human SI and SII cortices during observation of hand actions. *Neuroimage*, **15**, 640–6.

Avikainen, S., Wohlschläger, A., Liuhanen, S., Hänninen, R., and Hari, R. (2003). Impaired mirror-image imitation in Asperger and high-functioning autistic subjects. *Current Biology*, **13**, 339–41.

Cantalupo, C. and Hopkins, W. (2001). Asymmetric Broca's area in great apes. *Nature*, **414**, 505.

Carey, D.P. (1996). 'Monkey see monkey do' cells. *Current Biology*, **6**, 1087–8.

Chen, R., Corwell, B., Cohen, L., and Hallett, M. (1998). Reduction of motor cortex excitability after median nerve stimulation. *Muscle and Nerve*, **21**, 1585.

Conway, B., Halliday, D., Farmer, S., Shahani, U., Maas, P., Weir, A., and Rosenberg, J. (1995). Synchronization between motor cortex and spinal motoneuronal pool during the performance of a maintained motor task in man. *Journal of Physiology*, **489**, 917–24.

Decety, J., Chaminade, T., Grezes, J., and Meltzoff, A.N. (2002). A PET exploration of the neural mechanisms involved in reciprocal imitation. *Neuroimage*, **15**, 265–72.

Di Pellegrino, G., Fadiga, L., Fogassi, L., Gallese, V., and Rizzolatti, G. (1992). Understanding motor events: a neurophysiological study. *Experimental Brain Research*, **91**, 176–80.

Fadiga, L., Fogassi, L., Pavesi, G., and Rizzolatti, G. (1995). Motor facilitation during action observation: A magnetic stimulation study. *Journal of Neurophysiology*, **73**, 2608–11.

Fourneret, P., Paillard, J., Lamarre, Y., Cole, J., and Jeannerod, M. (2002). Lack of conscious recognition of one's own actions in a haptically deafferented patient. *Neuroreport*, **13**, 541–7.

Gallese, V. (2001). The 'shared manifold' hypothesis—From mirror neurons to empathy. *Journal of Consciousness Studies*, **5–7**, 33–50.

Gallese, V., Fadiga, L., Fogassi, L., and Rizzolatti, G. (1996). Action recognition in the premotor cortex. *Brain*, **119**, 593–609.

Gattis, M., Bekkering, H., and Wohlschläger, A. (2002). Goal-directed imitation. In A. L Meltzoff and W. Prinz (Eds.). *The imitative mind: Development, evolution, and brain bases*, pp. 183–203. Cambridge University Press, Cambridge.

Goldin-Meadow, S. (1999). The role of gesture in communication and thinking. *Trends in Cognitive Sciences*, **3**, 419–29.

Grafton, S.T., Arbib, M.A., Fadiga, L., and Rizzolatti, G. (1996). Localization of grasp representation in humans by positron emission tomography. 2. Observation compared with imagination. *Experimental Brain Research*, **112**, 103–11.

Hämäläinen, M., Hari, R., Ilmoniemi, R., Knuutila, J., and Lounasmaa, O.V. (1993). Magnetoencephalography – theory, instrumentation, and applications to noninvasive studies of the working human brain. *Reviews of Modern Physics*, **65**, 413–97.

Hari, R. (1999). Magnetoencephalography as a tool of clinical neurophysiology. In E. Niedermeyer and F. Lopes da Silva (Eds.): *Electroencephalography. Basic principles, clinical applications and related fields*. 4th ed., pp. 1107–34. Williams and Wilkins, Baltimore MD.

Hari, R. and Salmelin, R. (1997). Human cortical rhythms: a neuromagnetic view through the skull. *Trends in Neurosciences*, **20**, 44–9.

Hari, R. and Salenius, S. (1999). Rhythmical corticomotoneuronal communication. *Neuroreport*, **10**, R1–R10.

Hari, R., Reinikainen, K., Kaukoranta, E., Hämäläinen, M., Ilmoniemi, R., Penttinen, A., Salminen, J., and Teszner, D. (1984). Somatosensory evoked cerebral magnetic fields from SI and SII in man. *Electroencephalography and Clinical Neurophysiology*, **57**, 254–63.

Hari, R., Forss, N., Avikainen, S., Kirveskari, E., Salenius, S., and Rizzolatti, G. (1998). Activation of human primary motor cortex during action observation: A neuromagnetic study. *Proceedings of the National Academy of Sciences USA*, **95**, 15061–5.

Hari, R., Levänen, S., and Raij, T. (2000). Timing of human cortical activation sequences during cognition: role of MEG. *Trends in Cognitive Sciences*, **4**, 455–62.

Hutchison, W.D., Davis, K.D., Lozano, A.M., Tasker, R.R., and Dostrovsky, J.O. (1999). Pain-related neurons in the human cingulate cortex. *Nature Neuroscience*, **2**, 403–5.

Iverson, J.M. and Goldin-Meadow, S. (1998). Why people gesture when they speak. *Nature*, **396**, 228.

Järveläinen, J., Schürmann, M., Avikainen, S., and Hari, R. (2001). Stronger reactivity of the human primary motor cortex during observation of live rather than video motor acts. *Neuroreport*, **12**, 3493–5.

Kaukoranta, E., Hari, R., Hämäläinen, M., and Huttunen, J. (1986). Cerebral magnetic fields evoked by peroneal nerve stimulation. *Somatosensory Research*, **3**, 309–21.

Mäkelä, J., Kirveskari, E., Seppä, M., Hämäläinen, M., Forss, N., Avikainen, S., Salonen, O., Salenius, S., Kovala, T., Randell, T., Jääskeläinen, J., and Hari, R. (2001). Three-dimensional integration of brain anatomy and function to facilitate intraoperative navigation around the sensorimotor strip. *Human Brain Mapping*, **12**, 180–92.

Matelli, M. and Luppino, G. (1997). Functional anatomy of human motor cortical areas. In F. Boller and J. Grafman (Eds.): *Handbook of neuropsychology*, Vol. **11**, pp. 9–26. Elsevier Science, Amsterdam.

Mayberry, R. and Jaques, J. (2000). Gesture production during stuttered speech: insights into the nature of gesture–speech integration. In D. McNeill (Ed.): *Language and gesture*, 199–214. Cambridge University Press, Cambridge,

Nagamine, T., Kajola, M., Salmelin, R., Shibasaki, H., and Hari, R. (1996). Movement-related slow magnetic fields and changes of spontaneous MEG and EEG brain rhythms. *Electroencephalogrraphy and Clinical Neurophysiology*, **99**, 274–86.

Nishitani, N. and Hari, R. (2000). Temporal dynamics of cortical representation for action. *Proceedings of the National Academy of Sciences USA*, **97**, 913–18.

Nishitani, N. and Hari, R. (2002). Viewing lip forms: Cortical dynamics. *Neuron*, **36**, 1211–20.

Passingham, R. (1993). *The frontal lobes and voluntary action.* Oxford University Press, Oxford.

Perrett, D., Harries, M., Mistlin, A., and Chitty, A. (1990). Three stages in the classification of body movements by visual neurons. In H. Barlow, C. Blakemore, and M. Weston-Smith (Eds.): *Images and understanding: Thought about images: ideas about understanding*, pp. 94–107. Cambridge University Press, Cambridge.

Petitto, L.A., Holowka, S., Sergio, L.E., and Ostry, D. (2001). Language rhythms in baby hand movements. *Nature*, **413**, 35–6.

Petrides, M. and Pandya, D.N. (1994). Comparative architectonic analysis of the human and the macaque frontal cortex. In F. Boller and J. Grafman, (Eds.): *Handbook of neuropsychology*, Vol. **11**, pp. 17–58. Elsevier Science, Amsterdam.

Rizzolatti, G. and Arbib, M.A. (1998). Language within our grasp. *Trends Neuroscience*, **21**, 188–94.

Rizzolatti, G., Fadiga, L., Gallese, V., and Fogassi, L. (1996). Premotor cortex and recognition of motor actions. *Cognitive Brain Research*, **3**, 131–41.

Salenius, S., Portin, K., Kajola, M., Salmelin, R., and Hari, R. (1997). Cortical control of human motoneuron firing during isometric contraction. *Journal of Neurophysiology*, **77**, 3401–5.

Salmelin, R. and Hari, R. (1994). Spatiotemporal characteristics of rhythmic neuromagnetic activity related to thumb movement. *Neuroscience*, **60**, 537–50.

Schmolesky, M., Wang, Y., Hanes, D., Thompson, K., Leutgeb, S., Schall, J., and Leventhal, A. (1998). Signal timing across the macaque visual system. *Journal of Neurophysiology*, **79**, 3272–8.

Strafella, A.P. and Paus, T. (2000). Modulation of cortical excitability during action observation: a transcranial magnetic stimulation study. *Neuroreport*, **11**, 2289–92.

Uutela, K., Hämäläinen, M., and Somersalo, E. (1999). Visualization of magnetoencephalographic data using minimum current estimates. *NeuroImage*, **10**, 173–80.

Williams, J., Whiten, A., Suddendorf, T., and Perrett, D.I. (2001). Imitation, mirror neurons and autism. *Neuroscience and Biobehavioral Reviews*, **25**, 287–95.

Wohlschläger, A. and Bekkering, H. (2002). The role of objects in imitation. In M. Stamenov and V. Gallese (Eds.). *Mirror neurons and the evolution of brain and language*. John Benjamins Publishing Company, Amsterdam PA.

Yousry, T., Schmid, U., Alkadhi, H., Schmidt, D., Peraud, A., Buettner, A., and Winkler, P. (1997). Localization of the motor hand area to a knob on the precentral gyrus. A new landmark. *Brain*, **120**, 141–57.

Chapter 24

Neuropsychological perspectives on sensorimotor integration: Eye-hand coordination and visually-guided reaching

David P. Carey

Abstract

Neuroimaging work has done much to confirm hypotheses of a role for the posterior parietal cortex (PPC) in sensorimotor integration. Nevertheless, in the future such studies will need to test more specific models of the processes in which the PPC participates. A recurrent theme cutting across neuropsychology, physiology, and computational modeling is the idea of transformations in reference frames from retinal- to arm- (or hand-) centered coordinates. Understanding sensorimotor control processes at this level with neuroimaging will always be difficult. Evidence from neurophysiology suggests that any populations of cells related to encoding target location, preparing, and executing saccadic eye movements, and preparing, executing, and monitoring arm movements overlap considerably. If similar circuits exist in the human brain, they may be difficult to distinguish with neuroimaging, given the spatial resolution of fMRI and similar techniques. These are better suited for investigation of hemispheric contributions to sensorimotor control of each hand in each hemispace. Additionally, some of the feedforward/feedback processes that occur before, during, and after visually guided actions may be difficult to resolve, given the temporal resolution of neuroimaging. Finally, some of the neuropsychological data on eye–hand coordination may not fit easily into the hierarchical feedforward models based on computational considerations.

24.1 The posterior parietal cortex and input to output transformations

Neuropsychological work in the later part of the nineteenth century on disorders such as 'optic ataxia' did much to anticipate the discovery of cortical fields in the posterior parietal cortex related to eye and hand movements (for reviews, see Grusser and Landis 1991; Carey *et al.* 2003). At the same time, lesion studies in monkey were telling a similar story. For example, Munk described deficits in making visually-guided eye movements after area 7 lesions as early as 1888. Arm movement deficits after parietal lesions which largely spared primary somatosensory and motor cortex were described by Ferrier in 1876 (although his interpretations were all wrong, see Glickstein 1985). Even in these early days of exploration of the effects of lesions of the posterior parietal cortex, there was considerable debate about the mechanisms behind the obtained deficits. In non-human primates in particular, pinpointing a precise reason for a post-lesion deficit (or the response pattern of a single neuron for that matter) is often no easy task (Culham and Kanwisher 2001).

The purpose of this chapter is to review some of the processes that underlie arm movements to visual targets from a neuropsychological perspective. The importance of direction of gaze will be examined from single-unit neurophysiology, neuroimaging, and single-case neuropsychology. Finally, the chapter will re-examine the bias towards hierarchical, feedforward models of sensorimotor control for visually guided reaching.

The brain systems that control sensory-guided movements such as reaching and grasping clearly must process a number of sources of non-visual information, in addition to retinal location, in order to acquire visual targets. For grasping, information about object size and orientation in space must be extracted using computations involving depth, probably provided by stereopsis, vergence, and differential motion cues on the two retinas. For reaching, position of the target on the retinas must be integrated with information about position of the eyes in the head, the head on the trunk and the arm/hand/shoulder in order to guide the finger or hand successfully to the target.

In primates in particular, highly mobile, frontally-directed eyes have created enormous computational challenges for these systems. One of the earlier processes in such transformations may depend on integrating position of the eyes in the orbit with retinal locus of a visual target of interest. In the 1990s, Andersen, Galletti, and colleagues introduced the neuroscience community to the notion of gaze coding in single neurons of the monkey neocortex.

24.2 Single-unit investigations of eye–hand transformations

In its strongest form, a neuron exhibiting 'gaze-coding' would respond to a visual target only when the eyes of the animal were in a certain position in the orbit. A more common type of gaze signal is found when a retinotopic neuron (i.e. responds best when a target falls on a certain retinal locus) is 'modulated' by eye position—the retinotopic response grows or weakens depending on the position of the eyes within the head, even though the visual stimulus, from a retinal point of view, is kept constant. Gaze modulation has also been found in other types of single-cell activity, including in neurons with activity related to arm movements in a particular direction (e.g. Ferraina *et al.* 1997). Such neurons are obviously strong candidates for playing a major role in sensorimotor transformations of visual inputs. This type of neuronal response is easily incorporated into serial models of the control of limb movement to visual targets. The serial nature of such transformations is strongly implied by the logical ordering of processes from retina to arm. For example, in structures that respond rapidly to visual stimuli, such as the superior colliculus and V1, retinotopic coding of visual targets should be the norm. In intermediate regions in the sensorimotor pathways, such as subdivisions of early occipitoparietal cortex, gaze position coding should be more common than in regions further 'downstream' (in an anatomical rather than functional sense), where head-centered and subsequently body-centered, hand-centered, and/or arm-centered codes should predominate. There are a few examples in the neurophysiological literature which seem to fit into such a scheme, such as the fascinating neurons with arm-centered visual receptive fields in prefrontal cortex and the basal ganglia, described by Graziano and Gross (Graziano and Gross 1993; Graziano *et al.* 1994; see also Fogassi *et al.* 1996). Nevertheless, gaze modulation has now been claimed for a number of different cortical and subcortical brain regions that are not easily mapped on to early visual areas versus later visuomotor and premotor regions (Desmurget *et al.* 1998). Similarly, although 'head-centered' coding has been found in downstream areas such as the supplementary eye field, it has also been found in regions much closer to primary visual cortex, such as the ventral intraparietal area (VIP; Duhamel *et al.* 1997; Bremmer *et al.* 2001).

Jouffrais and Boussaoud (1999) argue that if gaze modulation is 'used to build limb motor commands, limb-related activity should depend on whether or not the monkey made a saccade to the target prior to the reaching movement' (p. 205). However, from a computational point of view (so often embraced by various coordinate transformation models; for reviews, see Xing and Andersen 2000; Salinas and Their 2000) the presence of a saccade towards a visual target should be largely irrelevant. In fact, foveation of the target in any way is irrelevant, as long as its retinotopic locus can be registered accurately (some of the arguments about the function of foveation assume the poor resolving power of the peripheral retina is the main issue—probably an incorrect assumption for reaching and grasping; for critiques in human and animal work, see Goodale and Murphy 1997 and Goodale and Carey 1988, respectively). Eye position (in head) coupled with retinotopic position of a non-foveated target is computationally no different than eye position (in head) coupled with retinotopic position of a foveated target. Many of these models also presuppose that, even though on-line feedback may be used to improve final movement accuracy, hand-centered codes are constructed before the movements of the eyes or the hands begin.

Even if gaze-modulated cells are playing some sort of intermediate role in computing target position in an arm- or hand-centered frame of reference, the data from physiological and psychological experiments of actual reaching suggest a more dynamic eye–hand synergy than that implied by a static signal providing eye position in the head. Data from Fisk and Goodale (1985), Neggers and Bekkering (2000), and others suggest special linkages in time between *saccades* and moving hands during visually guided reaching (Carey *et al.* 2003). Another example is a study by Honda (1984), who found that acquiring a target by a saccade produced more accurate hand movements in darkness than when pursuit eye movement was used to acquire the same target. A special role for foveation in visually-guided reaching has also been suggested in the comprehensive two-channel model of feedback loops for error correction, suggested by Paillard and colleagues (reviewed in Paillard 1996).

24.3 Overlapping saccade and arm movement activity in primate posterior parietal cortex

Gaze modulation of single cells in primate neocortex suggests some role in sensorimotor transformations in the primate. Another obvious candidate link between visual targets and eye–hand coordination is suggested by neurons that show both saccadic and arm-movement-driven activity, and regions that are dominated by populations of cells that code for both saccades and arm movements. Several regions in the parietal lobe have been described which are active when monkeys make eye and hand movements (Battaglia-Mayer *et al.* 2001; Galletti *et al.* 2001; Marconi *et al.* 2001; Buneo *et al.* 2002; Cohen and Andersen 2002). Andersen and colleagues have compared single-unit activity in areas specialized for the control of saccadic eye movements (e.g. the lateral intraparietal area, LIP) and control of arm movements (e.g. the parietal reach region, PRR). The early studies emphasized the relative independence of eye- and hand-related coding of these two regions (and in single cells within the PRR and LIP). More recent accounts have uncovered the importance of eye-movement- and eye-position-related activity, even in 'reaching' cells in PRR. Batista *et al.* (1999) have found that the responses of reaching neurons were modulated by the initial position of eyes prior to the arm movement, whereas changes in the initial position of the arm had no effect on the subsequent arm-movement-related activity.

Snyder *et al.* (2000) subsequently discovered that initial eye position is not the only eye-related property that influences the activity of cells supposedly restricted to coding arm movements. They reported that, despite a clear relationship between arm movements and firing patterns of PRR cells, 29% of the cells tested were influenced by saccadic eye movements in a task where no arm movement was required. The neurons were tuned for the same preferred directions for both delayed saccades made without reaches and delayed reaches made without saccades. An important property of these cells is that the activity was not related to *preparing* a saccade *per se*. In their delayed saccade task, monkeys were required to make saccadic eye movements to targets after a delay period; therefore activity related to preparing the saccade should have been elicited *during* the delay. In fact, most PRR arm movement neurons increased firing during or immediately after the saccade began, and not during the delay. Such units may play a role in circuits that utilize signals about saccading to a target for subsequent foveation, or in post-saccade control of arm movement.

These gaze- and saccade-related effects on single-cell activity in the macaque brain have rekindled interest in models of coordinate transformations as a central issue in sensorimotor control for primates. Recent neuropsychological work has also made use of these same types of model to help understand disorders of sensory-guided reaching seen in the clinic. The coordinate transformation models have also inspired the first wave of neuroimaging studies in human primates.

24.4 **Neuroimaging of gaze coding**

Eye movements can be studied easily within scanners, without any of the space-related difficulties posed by neuroimaging studies of whole-arm movements. Gaze modulation is no exception to this general rule. One of the first studies designed explicitly to look at gaze direction and hand movement was performed by Baker *et al.* (1999). They had participants make sequential finger movements with their right hand, while gazing straight ahead, to the left, or to the right (the hand itself was not visible). They found that rightwards and central gaze during right-hand finger movements produced significant increases relative to a baseline condition with no movement, while leftward gaze produced equivalent activation to the baseline (that is no increase in activation relative to baseline).

DeSouza *et al.* (2000) utilized a similar methodology, but required participants to make actual pointing movements to visual targets (experiments that require even more care than usual, in that arm movements made within a magnet may affect properties of the magnetic field). In one condition, participants pointed to one of three remembered target positions while gaze was directed 14° left or right of center. Both hands of most of the participants were tested, unlike in Baker *et al.* (1999). Like Baker *et al.* (1999), DeSouza *et al.* (2000) found significant modulation of arm-movement-related activation by the direction of gaze. When participants were reaching with their right hand, rightwards deviation of gaze increased activity in the left hemisphere rostral intraparietal sulcus (rIPS). Similarly, when the left hand was used, the greatest modulation of right hemisphere rips followed from with leftward eye deviation. These modulatory effects were only seen in the rIPS of the hemisphere contralateral to the reaching limb.

Of course, as noted above, there is evidence that the presence of an actual saccade is important for hand movements to visual targets, and some of the PRR single-unit activity in the monkey is indeed perisaccadic. It is more difficult to imagine how neuroimaging can speak to issues related to active foveation during visually-guided hand movements. In many eye-movement studies, foveation of a stimulus would be the control condition to be actively subtracted from a pursuit or a saccade condition in order to identify cortical fields related to eye-movement control—little is known about the cortical control of fixation *per se*. The literature on antisaccades might be of some interest in this context, as that task initially requires inhibition of eye movements to visual transients, a process shared in some instances with foveation of a specified target at the expense of other targets. Neurophysiologists studying the superior colliculus have been interested in the activity of fixation neurons in that structure for some time, and have built models of saccade generation and fronto-striatal inhibition which incorporate the fixation neurons as part of the circuitry (Munoz and Fecteau 2003; see also Ben Hamed *et al.* 2002 for similar work in LIP).

The Baker *et al.* (1999) and DeSouza *et al.* (2000) studies represent the closest neuroimaging has yet come to finding a functional equivalent to the eye-position modulation of single-unit activity seen in monkey cortex. They also suggest an important direction forward for neuroimaging studies of eye–hand coordination and sensorimotor control.

24.5 Asymmetries in sensorimotor processing: stimulus–response compatibility?

The effects found in the Baker *et al.* (1999) study have some resemblance to the literature on stimulus–response compatibility—gaze in the same (or a similar) direction as the tapping hand seemed to facilitate the movement-related cortical activation. The DeSouza *et al.* (2000) experiment went one step further and required participants to make movements to the targets with and without gaze directed towards the side of the target relative to the body midline. Earlier demonstrations of facilitated movement into the same side of space as the reaching limb were typically interpreted as resulting from within- versus between-hemisphere processing. That is, a target in left hemispace will be initially processed in the visual cortex of the right hemisphere; the same hemisphere which controls the reaching limb when the left hand is being used (although for a different, more biomechanical interpretation, see Carey *et al.* 1996; Carey and Otto-de Haart 2001).

What is of interest for the current discussion is that sensorimotor control systems of the central nervous system seem well-acquainted with these directional/hemispatial anisotropies of arm movement and build them into internal models important for motor control (cf Flanagan and Lolley 2001). Eye movements which are made to targets that are followed by a contralateral arm movement are slowed in a similar fashion (relative to an eye movement to the same target coupled with an ipsilateral arm movement). Of course, there is no obvious reason why the eyes could not arrive well ahead of the forthcoming reaching hand (Fisk and Goodale, 1985). Attentional systems may be influenced by these hemispatial asymmetries effects as well as eye-hand coordinative systems. For example, Meegan and Tipper (1998) have shown that the effect of visual distractors on reaching produces larger interference in ipsilateral space than in contralateral space. The functional significance of such asymmetries remains unknown, but surely they have an attentional and/or intentional flavor that deserves further investigation, within and outwith the scanner. Of interest in the DeSouza *et al.* (2000) study was the finding that any activity in the rostral intraparietal sulcus on the ipsilateral side as the reaching hand was not modulated by eye position, as it was in the contralateral rostral intraparietal sulcus. These data, coupled with the larger activation increases with gaze in the same direction as the reaching hand, suggest something unique about sensorimotor transformations for movements of the contralateral hand and/or movements into the contralateral space. Indeed, the gaze modulation effects on single units in cortex are largest for eye gaze direction away from the hemisphere being recorded from (Brotchie *et al.* 1995).

24.6 Asymmetries in sensorimotor processing: Handedness and the hemispheres

In the Baker *et al.* (1999) experiments, the increased activation when gaze was directed towards the tapping fingers is of interest for a related, but somewhat different reason; the participants were all right-handed, and made sequential finger movements using their right hands only. A branch of neuropsychology quite distinct from sensorimotor control has as its subject matter human right sidedness, which includes hand preference, foot preference, and even eye preference (when tasks force monocular sighting). Although the spatial resolution of current imaging techniques (still improving of course) may be insufficient for resolving subtle differences in overlapping networks for eye and hand movements in parietal cortex, processes related to attention and to preparation of right and left hand movements could profitably be examined in the scanner (cf. Rushworth *et al.* 2001). Presumably many of the effects at input and output would be restricted to the contralateral cerebral hemisphere. In his review of neuroimaging studies on handedness, Peters (2000) refers to the idea of 'substantial asymmetries'—instances where asymmetrical brain activation (e.g. during a hand movement) is not the same for both hands/sides of the brain. If that participant happens to be a self-professed right-hander and the degree of contralateral versus ipsilateral activation is greater when the right hand is used, a substantial asymmetry is present. The first wave of neuroimaging studies which compare handedness groups have tended to focus on simple motor tasks such as finger-to-thumb opposition. Using such procedures, most labs find that right handers show greater left hemispheric activation when they move their dominant hand than right hemispheric activation when they move their non-dominant hand. Left handers usually show some evidence of bilateral activation when they move their dominant hand, *or* look somewhat like the mirror-image of right-handers (i.e. greater right hemisphere activity with left hand movement than left hemisphere activity with right hand movement). For example, Kim *et al.* (1993) had left- and right-handed participants make finger–thumb opposition movements in an fMRI scanner. They found strong activation of right hemisphere motor cortex with movements of the left hand in both groups. They also saw increased left-hemisphere activation by the same movements. In other words, left motor cortex was activated by movements of either hand, while right motor cortex was not. Nevertheless, some recent work occasionally suggests that left-handers are indeed as asymmetrical as right handers. For example, Volkmamn *et al.* (1998) used magnetoencephalography to estimate the size of the hand motor cortex in both hemispheres of left- and right-handed participants. Both groups had larger representations of the dominant hand in the contralateral hemisphere. However, Kawashima *et al.* (1997) used positron emission tomography (PET) scans on six left-handed participants while they made index finger flexion and extension movements. They claim that, opposite to right-handed participants, the right hemisphere premotor area was activated by both ipsilateral (right, non-dominant hand) and contralateral (left, dominant hand) movements. This intriguing finding suggests that Kawashima *et al.* (1997) had managed to find six left handers who had mirror-image motor lateralization to right handers assessed in the same way.

Of course, the *sensorimotor* components (if indeed there are any) of left hemisphere motor control processes in participants with conventional dominance are completely unknown. Of the few imaging experiments on reaching, virtually none are concerned with asymmetries in control of the hands or in eye–hand coordination. Apraxia following left-hemisphere damage has been linked to hemispheric asymmetries of motor control (e.g. Goodale 1990; Kimura 1993). Nevertheless, most of the arguments from neuropsychological studies of apraxia tend to emphasize high-level semantic representations of movement in the left parietal cortex of most right handers (Rothi and Heilman 1997). The less cognitive arguments made by theorists interested in the left hemisphere in motor control tend to be related to motor output *per se* rather than sensory control of movement (e.g. motor sequencing superiority of the left hemisphere, reviewed in Kimura 1993). Theorists interested in sensorimotor control are rarely interested in handedness or hemispheric asymmetry. Milner and Goodale (1995) have tentatively suggested that some occipitotemporal functions may be handled differentially by each hemisphere, while occipitoparietal sensorimotor functions are more symmetrically organized. Other authors have attempted to link asymmetrical disorders such as apraxia after left-brain damage to the 'dorsal stream' or to optic ataxia (e.g. Marsden 1998; Passingham and Toni 2001), although the evidence for these links (beyond anatomical proximity of superior and inferior parietal cortex) is not overwhelming. Instead, it is the neuropsychological syndromes usually associated with the *superior* parietal lobe, including optic ataxia, that include disturbances that have an obvious 'sensorimotor' control flavor.

24.7 **Neuropsychological disturbances of reaching**

To date, the study of impaired reaching in patients with posterior brain damage has been concerned almost exclusively with the visual guidance of aiming. This bias is related to the popularity of misreaching disorders as a component of 'Holmes–Balint' syndrome, a triad of visuomotor/attentional disturbances that has been described in patients with bilateral parietal lobe dysfunction. These include gaze dysfunction and attentional disorders such as simultanagnosia (an inability to perceive more than one visual target at a time). The third component of Holmes–Balint syndrome is the symptom referred to as 'optic ataxia' (Harvey and Milner 1995; De Renzi 1996). Optic ataxia is a great difficulty in reaching to a visual target, which cannot be explained by visual difficulties such as field defects or amblyopia. In some cases of Holmes–Balint syndrome, gaze dysfunction and attentional disorders largely resolve over time, whereas optic ataxia for targets in peripheral vision remains ('extrafoveal' optic ataxia).

Optic ataxia seems an obvious candidate for a disturbance in sensorimotor integration, because the misreaching difficulties are restricted to visual targets. If a patient could not reach accurately under other conditions (e.g. to named parts of his/her body) the disorder would be 'motoric' in nature and of less interest to the neuropsychologists who study brain function at the level of perceptual, attentional, and spatial processes.

Many of the same coordinate transformations necessary for reaching to visual targets will share some of their circuitry with movements made in responses to non-visual targets. And yet auditory localization, with the exception of a few clinical reports (e.g. Tzavaras and Masure 1975; Kase *et al.* 1977; De Renzi 1988), has rarely been assessed in any rigorous way in patients showing disordered reaching. In cases of optic ataxia, intact reaching to proprioceptive targets is all that is usually reported (typically named body parts, e.g. Damasio and Benton 1979, or the nose as part of finger-to-nose assessment for cerebellar ataxia, cf. Kase *et al.* 1977, case 2). The assessment of movements to non-visual targets are usually informal (i.e. as part of a bedside neurological exam) or just implied by the diagnostic labeling (cf. Hausser *et al.* 1980; Jakobson *et al.* 1991; Jeannerod *et al.* 1994; Milner *et al.* 2001). In some cases, proprioceptive guidance of movement has been assessed more rigorously by having the blindfolded participant's thumb or finger (positioned by the examiner) as the target of a grasping movement made with the other hand (Boller *et al.* 1975; Perenin and Vighetto 1988). Unfortunately, the accuracy and efficiency of such movements have never been recorded and quantified (although for an early attempt see Levine *et al.* 1978).

In spite of these considerations, the theoretical focus in assessing patients with poor reaching has remained on visual input and analysis. For example, theorists debate whether or not deficits in the perception of space can account for the disordered reaching in patients with full Holmes–Balint's syndrome or optic ataxia (Holmes 1918; Rondot *et al.* 1977; Perenin and Vighetto 1988; Milner and Goodale 1995; Baylis and Baylis 2001). Other studies have addressed issues such as whether or not the disturbances in visual attention in such patients could account for their poor spatial behavior in general (Hecaen and De Ajuriaguerra 1954; Kim and Robertson 2001). Recently, scientists have documented disturbances in the visual guidance of grasping (Jakobson *et al.* 1991; Goodale *et al.* 1994; Jeannerod *et al.* 1994; Milner *et al.* 2001) in patients with full Holmes–Balint syndrome or optic ataxia. Other reports have been concerned with the time course of aiming movement responses to visual targets or changes in target position (Milner *et al.* 1999, 2001; Pisella *et al.* 2000).

This bias towards visual inputs may be responsible for the lack of definitive theories accounting for optic ataxia in its various forms. Some authors have argued for a 'disconnection' syndrome account of the disorder, whereby motor systems are deafferented from the visual inputs that specify target position (Balint favored such an interpretation, see De Renzi 1996). In their recent analysis of optic ataxia, Buxbaum and Coslett (1997, 1998) are critical of disconnection accounts (e.g. Rondot *et al.* 1977; for a review of earlier arguments, see De Renzi 1982). They argue that disconnection of frontal lobe structures from visual cortex should result in 'complete' optic ataxia; therefore, demonstrations of 'subtotal' syndromes, where patients only misreach to extrafoveal targets, make such accounts untenable.

Such a criticism is most appropriate when authors have drawn direct comparisons to the frontal lobe deafferentation experiments of Kuypers and his colleagues (Haaxma and Kuypers 1975) because they are suggesting visual deafferentation of the neurons controlling arm movements as the mechanism behind misreaching. Those who have not made such comparisons could just as easily have been referring to disconnection phenomena at some intermediate level of control. For example, parietal lobe lesions could disconnect the visual cortices from the frontal eye fields, or the cerebellum (cf. Glickstein 1998). Such lesions could also damage parietal lobe efferents to oculomotor control centers in the superior colliculus and/or brainstem. These types of disconnection could disrupt eye movements to visual targets in some fashion, for example, which might contribute to misreaching of the hand. Indeed, since many patients with optic ataxia have difficulties with oculomotor control, the relative impact of gaze disorders on misreaching has also been a focus of some of the discussion. However, such analyses have not examined gaze during reaching *per se*; instead they claimed that optic ataxia and the degree of gaze dysfunction were dissociable (Rondot *et al.* 1977; Perenin and Vighetto 1988). In other words, optic ataxia can be seen in patients whose eye movements to visual targets are relatively spared, in a few cases at least.

Rather less effort has been directed to understanding the complex interactions between position sense of eye, head, and limb, efference copy of motor commands, etc., and how disruptions in these interactions can lead to different disturbances in visually-guided behavior. The potential points of breakdown in cases of optic ataxia are numerous, given what is known about the complex control processes involved in visually-guided reaching and grasping in normals.

24.8 Feedforward and feedback processes during visually guided reaching

A detailed review of the vast literature on controlled processing in visually-guided reaching and grasping is beyond the scope of this chapter (see Wing *et al.* 1996; Jeannerod 1997; Desmurget *et al.*, 1998). Briefly, the popular idea is that visual feedback of the reaching limb is normally used as part of a feedback loop that ensures movement accuracy and smoothness of trajectory. Nevertheless, relatively accurate reaching is possible in the absence of visual feedback of the reaching limb, particularly if the target being reached to remains visible *or* if the participants get a view of their static hand in between actual reaching trials. Visibility of the static hand between trials seems to allow for some sort of between-trial calibration of hand and eye, although the exact mechanisms remain unknown (Prablanc *et al.* 1979; Desmurget *et al.* 1995). Target visibility during hand-invisible reaching increases terminal accuracy because fixation- or saccade-related signals about target position from the eye movement can be compared in some way to the position of the limb, which is still on its way (see Desmurget *et al.* 1998; Desmurget and Grafton 2000; Carey *et al.* 2003). Arguments about the utility of arm proprioceptive/kinesthetic feedback during a ballistic movement have led many theorists to posit that a copy of the motor commands to limb muscles is used as part of the comparison. This 'efference copy' can be compared with feedback at the end of a movement and used to fine tune how it is compared to eye position on later trials. Recently these ideas have been extended to include some sort of 'feedforward model', which incorporates knowledge of limb dynamic properties and can estimate where the limb should be at a given stage of the movement, such as when the eye arrives at the target (cf Desmurget and Grafton 2000; Gaveau *et al.* 2003). Nevertheless, such models usually presuppose that the arm movement was planned in a hand- or arm-centered frame of reference before the movement was initiated—or they are agnostic when it comes to this issue. Eye–hand localization systems adapt extremely well to sudden shifts in target positions after the primary saccadic eye movement has been initiated. These transformations may depend on circuits of the posterior parietal cortex, as suggested by some of the neuroimaging work on reaching and grasping.

Neuroimaging of reaching and grasping is explored more fully in Chapter 21 (this volume), although a few comments are relevant here. Most efforts in this domain to date are interested in identifying locations of regions in cortex specialized for aiming, looking, or grasping, such as the PET experiment of Grafton *et al.* (1996). More recently, the increased resolution of fMRI has allowed for better visualization of the intraparietal sulcus, and a few groups have begun to identify a posterior to anterior gradient of areas with saccadic, pointing, and grasping activity, respectively (e.g. DeSouza *et al.* 2000; see also Simon *et al.* 2002).

Only a few studies have attempted to go beyond questions of localization of regions involved in reaching and grasping. Desmurget *et al.* (2001) compared and contrasted pointing movements made with and without visual feedback using PET. The critical contrast in that study was between trials where targets were jumped during the saccade, which required more error correction than trials in which the targets did not jump. Intraparietal sulcus activation in the hemisphere contralateral to the reaching limb was found when non-jump trials were subtracted from jump trials (Desmurget *et al.*, 2001).

One of the obvious limitations for neuroimaging studies of sensorimotor control processes are the type and size of actual movements that can be made in the scanner. For scientists working on oculomotor control, technological advances in presentation of visual stimuli within the magnet and non-metallic devices for measuring eye movements have eliminated virtually all of the difficulties. For arm movement control, studies of pointing, such as DeSouza *et al.* (2000), and grasping (for example see Chapter 21, this volume) within fMRI scanners in particular, will always be limited by the rather small workspace available for participants to move in. The aiming movements examined by DeSouza *et al.* (2000) were limited in extent and largely restricted to movements around the wrist. The upper arm and torso were restrained to prevent unwanted movement of the head. Examining differences in movements to lateralized targets in such circumstances will be very distinct from the inertial anisotropies (and related higher-level processes which integrate the saccadic system with the hand) when participants make large, whole-arm, pointing movements involving proximal and distal musculature, and rotations around shoulder, elbow, and wrist. The inertial consequences of unrestricted ballistic arm movements are dramatic, and require postural compensation in trunk musculature on both sides of the body. Clearly, such requirements are virtually absent when lying supine in a scanner.

The temporal resolution of fMRI and related techniques may also pose some limitations on the utility of neuroimaging for studies of sensorimotor control. For example, some of the single-unit activity related to eye movement in PRR seems to be largely perisaccadic and is absent during a delay period before initiating a saccade to a remembered target position. Even event-related fMRI will be limited in its usefulness at distinguishing saccadic planning versus processes related to the saccade itself, which seem to play some special role in visually-guided arm movement, as noted earlier.

Blocked fMRI designs usually require presenting stimuli in blocks in order to be able to identify any signal changes. For the participants in the DeSouza *et al.* (2000) study this meant that several trials required identical fixation and pointing movements to the same target position several times in a row. Presumably such a procedure does not engage the same attentional/oculomotor mechanisms, which are required for localizing targets when they appear randomly.

A final limitation of the DeSouza *et al.* (2000) study for our understanding of sensorimotor transformations is that all of the reaching movements made by the participants were done in hand-invisible conditions and *without any foveation* of the actual target. In the literature on optic ataxia, this distinction is important. Further evidence for a unique role of the fovea and processes related to foveation for reaching comes from patients with 'magnetic misreaching'.

24.9 Magnetic misreaching

My colleagues and I have described a patient, Ms D, who presented with a slowly progressive bilateral ideomotor apraxia, coupled with apraxic gait and constructional disturbance.

MRI scanning revealed severe bilateral parietal atrophy, particularly evident in the left superior parietal region. The most striking feature of Ms D's case was a peculiar reaching disorder that we called 'magnetic misreaching'. Magnetic misreaching refers to a pathological 'yoking' of the reaching hand to the direction of gaze, such that attempts to reach to extrafoveal targets inevitably fail, but *these failures are very specific*. Every time Ms D. attempted to reach towards an extrafoveal target, she slavishly reached to the place where her eyes were pointing (Carey *et al.* 1997). Buxbaum and Coslett (1997, 1998) refer to what appears to be the same symptom in at least one (D.P.) of their two patients, although their interpretation is largely restricted to the distinction between foveal and extrafoveal optic ataxia. However, there is a key distinction between magnetic misreaching and extrafoveal optic ataxia: a patient with magnetic misreaching *must* reach to the place that she or he is looking at, while the errors of extrafoveal optic ataxics are randomly spread or simply biased towards the side of the brain lesion (cf Revol *et al.* 2003).

Magnetic misreaching may be related to vision of the reaching limb interacting in some complex way with eye position signals. Foveation of a target to be reached to may provide a directional error signal for on-line control of the reaching limb, which is monitored by fast visual feedback loops of the extrafoveal retinas. These fast loops track the hand as it moves towards the foveated target (Paillard 1996). Use of such motion cues in extrafoveal vision appears to be disrupted in the patient of Rizzo *et al.* (1992). This patient had severe deficits in the perception of visual motion, as well as difficulties reaching ballistically to visual targets.

If the sensorimotor systems controlling reaching operate in such a fashion, signals about target position might be compared to the *visual* position of the limb. Thus, once Ms D initiated a ballistic movement under visual guidance towards a non-foveated target, her hand was 'captured' by an on-line *visual* control system, which normally minimizes target-hand distance. This perspective implies that Ms D's misreaching might be conceptualized as a variant of optic ataxia, because the problem would be related to the visual guidance of movement. Such a view of optic ataxia is somewhat broader than the traditional interpretation, where the problem is thought to be with encoding the *position of the visual target*, or getting information about target position to the motor and premotor cortices (see Perenin and Vighetto 1988; Buxbaum and Coslett 1997, 1998).

My colleagues and I have proposed a different interpretation of magnetic misreaching. Ms D was not just *capable* of using eye position to specify targets, she was *compelled* to reach towards the place she was foveating. This pathological yoking of eye and hand suggests that eye proprioception (or efference copy of oculomotor commands) was the only way that target position could be specified for the purposes of preparation and execution of a ballistic aiming movement. This hypothesis was supported by demonstrations of poor reaching to proprioceptively specified targets as well as poor reaching to auditory targets (Carey *et al.* 2003). And yet her misreaching difficulties were not obviously related to control of the arm *per se*, because Ms D could make perfectly accurate reaches to targets that she was allowed to foveate.

24.10 Can a coordinate transformation disorder explain extrafoveal optic ataxia?

In the Buxbaum and Coslett (1997, 1998) scheme, a head-centered representation of the target is computed by combining retinal target position with a feedforward or feedback signal of eye position. This signal is then combined with feedback from neck muscles to produce a body-centered representation of the target (usually conceived as relative to the torso or body midline). Failures in transforming retinal or oculocentric (i.e. relative to the geometric center of the eye) coordinate schemes into arm-centered coordinates lead to various subtypes of optic ataxia.

If the model presupposes breakdown at these early stages from eye to hand, then patients with certain variants of optic ataxia would fail at reaching because, for example, the retinocentric or oculocentric codes could not be constructed. Some of the eye movement abnormalities seen in patients with full-blown Holmes–Balint syndrome could theoretically have disruptions of this sort. The consequence of such impairments would be poor reaching accuracy because of a failure to construct and/or utilize the retinocentric or oculocentric code. (Of course, as noted above, many patients with optic ataxia can make reasonably accurate eye movements to the very targets they cannot reach, so clearly some retinocentric and oculocentric processing is intact in such cases.)

Foveation (or lack thereof) in such patients should be largely irrelevant, so a breakdown at this stage in the Buxbaum and Coslett model could explain 'full-blown' optic ataxia. The Buxbaum and Coslett coordinate transformation failure account (or any particular model of how such transformations take place) emphasizes how retinal information is combined with eye-in-head information. The position of any possible target on the retina is irrelevant as long as it can be specified in a retinocentric code. Therefore, it is not clear how the model accounts for extrafoveal optic ataxia—where reaching to foveal targets is relatively spared but reaching to non-foveal targets is not. To explain extrafoveal optic ataxia, Buxbaum and Coslett (1997) speculate that 'the actions of the eye and hand may be linked in a common system of spatio-motor coordinates' (p. 164). Such a system is difficult to place in their model, although is reminiscent of the Andersen group account of a common spatial coding scheme in posterior parietal cortex.

A somewhat different interpretation is this: foveation is the 'default option' for primates. That is, orienting movements to auditory (and proprioceptive) targets allow primates to utilize the extremely well-adapted resolving power of the fovea and/or reduce the computational requirements of representing all of the retina in particular sensorimotor transformations. If there is any common spatial coding, it makes perfect evolutionary sense for it to be centered on the eye. In fact, a recent report by

Ben Hamed *et al.* (2002) showed that active fixation by an animal increases the proportion of neurons in LIP sensitive to central regions of the retina at the expense of the periphery. Whatever the functional interpretation of these intriguing data, our data suggest foveation signals provide the only remaining route for guiding hand movement by patients with magnetic misreaching. Labeling such cases as further instances of extrafoveal optic ataxia is probably inappropriate because these rarer patients reach to where they foveate.

24.11 Conclusions

Neuroimaging has confirmed hypotheses about roles for parietal (and frontal) cortex in sensorimotor control. Nevertheless, some of the outstanding questions related to the sensory control of movement may not be easily resolved, given the measurement limitations of current neuroimaging technology. Of course, improvements in these technologies will go some way towards overcoming many of the difficulties. Nevertheless, attacking outstanding questions using single-unit investigations, case studies of brain-damaged patients, as well as studies of sensorimotor control in neurologically intact participants, remain the obvious route forward, given the differing strengths and weaknesses of each of these fields in isolation. Eye–hand coordination in visually-guided reaching is a good case in point, currently establishing the utility and necessity of multi-disciplinary methods to uncover the principles of control.

One of the most attractive features of visual control of hand movement as a research topic, is that many of the necessary steps seem specified by the ambiguity of retinal and head-centered signals, and the logical ordering of the building blocks of a hand-centered representation of a visual target. However, the data summarized above suggest that strictly feedforward models may be oversimplified, and that even more sophisticated approaches that explicitly acknowledge the importance of post-movement onset feedback may need revision. One key question that remains unanswered is whether or not a truly hand-centered representation of the target position is available for control of the initial arm movement before the arm (and perhaps even the eye) have begun their respective movements.

In neuropsychology, criticism of oversimplistic disconnection accounts of optic ataxia (Buxbaum and Coslett 1997, 1998) is appropriate. The Buxbaum and Coslett model does note the potential contributions of non-visual sources of information to reaching to visual targets, and that some of the computations from eye to hand may not take place in a completely serial, ordered way. Nevertheless, it does not take into account any of the sensorimotor processes that happen after the target has appeared and the eyes begin to move, or the evidence reviewed above suggesting a special role for foveation and saccadic eye movements in visually-guided reaching.

It may be the case that foveation as the default option saves considerable processing space for sensorimotor systems. Rather than represent the entire retina (or both retinas) as well as all possible eye positions in order to compute head-centered codes, active foveation of targets for manual movements means that eye (and head) movements could be initiated without any further processing beyond a retinotopic error signal. Once initiated, the feedforward and feedback processes (which include vergence, accomodation, and stereoscopic signals from other non-target objects) related to eye–arm coordination *dynamically* produce the required arm movement without any explicit representation of target location in space, constructed before the hand begins to move.

Effectively, much controversy remains regarding what attributes of target position influence movement planning before eyes or hand begin their movements. The data of Fisk and Goodale (1985) suggest that arm movement direction is known to eye–hand coordinative structures before hand onset. It may be that extent is also known in a similar fashion (Gribble *et al.* 2002), in which case hierarchical coordinate transformation accounts make more sense. In other words, if target position in direction *and distance* is known to sensorimotor control systems before movement onset, then apparently, serial transformations to an arm- or hand-centered frame have to have taken place before movement onset. On-line computations in this view have evolved to deal with slight adjustments in hand and eye position due to noise in output specifications, or indeed possible changes in target position (insect and rodent prey of primates often move).

The plasticity of eye–hand coordination systems has been documented in prism adaptation studies, as well as in experiments that change eye movement system gain. In most laboratory experiments, including that of Gribble *et al.* (2002), a restricted number of targets are located in one plane, which required only versional eye movements across that plane. In such a set-up, retinal error is completely predictive of subsequent arm movement amplitude as well as direction, which may be learned by an internal model that relates arm movement gain to eye movement gain. (It may be worth re-examining older published experiments on extrafoveal reaching in this regard as well—motor memory might play a role in reaching to repeated targets in extrafoveal vision, particularly if the hand was visible at any time during testing.) In the real world, retinal position alone is not so predictive of arm movement amplitude, unless feedforward signals from both eyes are integrated in some fashion to predict a head-centered target location. If not, foveation of the target after the appropriate vergence and version would suffice to drive the arm in the usual fashion (i.e. as in the case of two-dimensional studies).

If foveation is the default option, given that head movements usually follow saccadic eye movements, a head-on-body signal might be sufficient to drive the arm in a direction appropriate enough to get the arm on the retina, at which time retinocentric coding (using both eyes) could be sufficient for guiding the limb to the target. In fact, in unrestrained circumstances, movements of the head also contribute to final gaze position and may further reduce the computational space required for computing head-centered representations (because final head position would code for gaze position and location of the eye in orbit would be aligned with the head). Little research has been done on unrestrained head and eye movements to targets for manual movement in cluttered, natural environments, although recently a few groups have tackled this kind of situation. The consensus seems to be that the saccadic system is a step or two ahead of arm movement, in that during complex tasks the eye is saccading to the next target while the arm is approaching the last saccadic target (cf. Johannsson et al. 2001; Hayhoe et al. 2003). Such data tell a quite different story from Neggers and Bekkering (2000). They found that it was very difficult for their participants to make a saccade to a second target while the hand was still approaching the target of an earlier saccade and hand movement. Task demands on speeded responses and the simplicity of the displays in most experiments on eye–hand coordination may explain much of this difference.

Remarkably little is known about how neck muscle afference and efference contributes to the sensorimotor control of reaching. In fact over 40 years ago Cohen (1961) showed that neck muscle deafferentation disrupted arm movement accuracy in the monkey, even under conditions when the animal was actively foveating the target. These experiments, as well as tendon vibration studies, produce undeniable evidence that neck position sense is a crucial part of any sensorimotor chain involving vision (and probably audition as well). Of course, head-on-torso signals do play a part in the coordinate transformation accounts of reaching and misreaching (although typically as a feedforward signal before movement onset, not as a dynamic signal utilized after movement initiation).

An understanding of processes from input to output is necessary for developing a comprehensive neuropsychological model of misreaching and the contributions of disordered sensorimotor processes to optic ataxia, Holmes–Balint syndrome, and other disturbances of action guided by vision, sound, and position sense. Magnetic misreaching is a good case in point: models of normal eye–hand control go some way in understanding the pathological yoking of hand to eye in Ms D. and D.P., although the absence of dynamic, post-movement processes in some models may be too restrictive. More spatially complex target arrays should be used with neurologically intact participants in studies of eye–hand coordination, to tease apart some of the feedforward and feedback processes discussed above, and to resolve the debate about what is explicitly coded before movement begins. Asymmetries in eye–hand coordination remain relatively unexplored, yet can reveal what is known about target position before movement initiation (cf. Fisk and Goodale 1985). Neuroimaging technologies will aid in these efforts because of their incredible power for examining attentional and cognitive contributions to sensorimotor control, as well as the contributions of gaze and foveation to aiming, grasping, and representations of target position in space.

References

Baker, J.T., Donoghue, J.P., and Sanes, J.N. (1999). Gaze direction modulates finger movement activation patterns in human cerebral cortex. *Journal of Neuroscience*, **19**, 10044–52.

Batista, A.P., Buneo, C.A., Snyder, L.H., and Andersen, R.A. (1999). Reach plans in eye-centred coordinates. *Science*, **285**, 257–60.

Battaglia-Mayer, A., Ferraina, S., Genovesio, A., Marconi, B., Squatrio, S., Molinari, M., Lacquaniti, F., and Caminiti, R. (2001). Eye–hand coordination during reaching. II. An analysis of the relationships between visuomanual signals in parietal cortex and parieto-frontal association projections. *Cerebral Cortex*, **11**, 528–44.

Baylis, G.C. and Baylis, L.L. (2001). Visually misguided reaching in Balint's syndrome. *Neuropsychologia*, **39**, 865–75.

Ben Hamed, S., Duhamel. J.-R., Bremmer, F., and Graf, W. (2002). Visual receptive field modulation in the lateral intraparietal area during attentive fixation and free gaze. *Cerebral Cortex*, **12**, 234–45.

Boller, F., Cole, M., Kim, Y., Mack, J.L., and Patawaran, C. (1975). Optic ataxia: clinical–radiological correlations with EMIscan. *Journal of Neurology, Neurosurgery and Psychiatry*, **38**, 954–8.

Bremmer, F., Schlack, A., Duhamel, J.-R., Graf, W., and Fink, G.R. (2001). Space coding in primate posterior parietal cortex. *Neuroimage*, **14**, S46–S51.

Brotchie, P.R., Andersen, R.A., Snyder, L.H., and Goodman, S.J. (1995). Head position signals used by parietal neurons to encode locations of visual stimuli. *Nature*, **375**, 232–5.

Buneo, C.A., Jarvis, M.R., Batista, A.P., and Andersen, R.A. (2002). Direct visuomotor transformations for reaching. *Nature*, **416**, 632–6.

Buxbaum, L.J. and Coslett, H.B. (1997). Subtypes of optic ataxia: reframing the disconnectionist account. *Neurocase*, **3**, 159–66.

Buxbaum, L.J. and Coslett, H.B. (1998). Spatio-motor representations in reaching: evidence for subtypes of optic ataxia. *Cognitive Neuropsychology*, **15**, 279–312.

Carey, D.P. and Otto-de Haart, E.G. (2001). Hemispatial differences in visually guided aiming are neither hemispatial nor visual. *Neuropsychologia*, **39**, 885–61.

Carey, D.P., Hargreaves, E.L., and Goodale, M.A. (1996). Reaching to ipsilateral or contralateral targets: Within-hemisphere visuomotor processing cannot explain hemispatial differences in motor control. *Experimental Brain Research*, **112**, 496–504.

Carey, D.P., Coleman, R.J., and Della Sala, S. (1997). Magnetic misreaching. *Cortex*, **33**, 639–52.

Carey, D.P., Ietswaart, M., and Della Sala, D. (2003). Neuropsychological perspectives on eye–hand coordination in visually-guided reaching. In J. Hyönä, D. Munoz, W. Heide, and R. Radach (Eds.), *Progress in Brain Research: Vol. 140. The brain's eyes: neurobiological and clinical aspects of oculomotor research*, pp. 311–27. Elsevier Science, Oxford.

Cohen, L.A. (1961). Role of eye and neck proprioceptive mechanisms in body orientation and motor coordination. *Journal of Neurophysiology*, **24**, 1–11.

Cohen, Y.E. and Andersen, R.A. (2002). A common reference frame for movement plans in the posterior parietal cortex. *Nature Neuroscience*, **3**, 553–62.

Culham, J.C. and Kanwisher, N.G. (2001). Neuroimaging of cognitive functions in human parietal cortex. *Current Opinion in Neurobiology*, **11**, 157–63.

Damasio, A.R. and Benton, A.L. (1979). Impairments of hand movements under visual guidance. *Neurology*, **29**, 170–4.

De Renzi, E. (1982). *Disorders of space exploration and cognition*. Wiley, Chichester.

De Renzi, E. (1988). Visuo-spatial disorders. In C. Kennard and F. Clifford Rose (Eds.), *Physiological aspects of clinical neuro-ophthalmology* (pp. 155–71). Chapman and Hall, London.

De Renzi, E. (1996). Holmes–Balint syndrome. In C. Code (Ed.) *Classic cases in neuropsychology* (pp. 123–143). Psychology Press, Hove, East Sussex.

Desmurget, M. and Grafton, S. (2000). Forward modelling allows feedback control for fast feedback loops. *Trends in Cognitive Science*, **4**, 423–31.

Desmurget M., Rossetti, Y., Prablanc, C., Stelmach, G.E., and Jeannerod, M. (1995). Representation of hand position prior to movement and motor variability. *Canadian Journal of Physiology and Pharmacology*, **73**, 262–72.

Desmurget, M., Péllisson, D., Rossetti, Y., and Prablanc, C. (1998). From eye to hand: planning goal-directed movements. *Neuroscence and Biobehavioral Reviews*, **22**, 761–88.

Desmurget, M., Grea, H., Grethe, J.S., Prablanc, C., Alexander, G.E., and Grafton, S.T. (2001). Functional anatomy of nonvisual feedback loops during reaching: A positron emission tomography study. *Journal of Neuroscience*, **21**, 2919–28.

DeSouza, J.F.X., Dukelow, S.P., Gati, J.S., Menon, R.S., Andersen, R.A., and Villis, T. (2000). Eye position signal modulates a human parietal pointing region during memory-guided movements. *Journal of Neuroscience*, **20**, 5835–40.

Duhamel, J.R., Bremmer, F., Ben Hamed, S., and Graf, W. (1997). Spatial invariance of visual receptive fields in parietal cortex neurons. *Nature*, **389**, 845–8.

Ferraina, S., Garasto, M.R., Battaglia-Mayer, A., Ferraresi, P., Johnson, P.B., Lacquaniti, F., and Caminiti, R. (1997). Visual control of hand reaching movement: activity in parietal area 7m. *European Journal of Neuroscience*, **9**, 1090–5.

Ferrier, D. (1876). *The Functions of the Brain*. Smith, Elder & Co., London.

Fisk, J.D. and Goodale, M.A. (1985). The organization of eye and limb movements during unrestricted reaching to targets in contralateral and ipsilateral space. *Experimental Brain Research*, **60**, 159–78.

Flanagan, J.R. and Lolley, S. (2001). The inertial anisotropy of the arm is accurately predicted during movement planning. *Journal of Neuroscience*, **21**, 1361–9.

Fogassi, L.T., Gallese, V., Fadiga, L., Luppino, G., Matelli, M., and Rizzolatti, G. (1996). Coding of peripersonal space in inferior premotor cortex (area F4). *Journal of Neurophysiology*, **76**, 141–57.

Galletti, C., Gamberini, M., Kutz, D.F., Fattori, P., Luppino, G., and Matelli, M. (2001). The cortical connections of area V6: an occipito-parietal network processing visual information. *European Journal of Neuroscience*, **13**, 1572–88.

Gaveau, V., Vindras, P., Prablanc, C., Pelisson, D., and Desmurget, M. (2003). Eye–hand coordination in reaching movements. In M. Arbib (Ed.) *The handbook of brain theory and neural networks* (2nd ed.) pp. 431–4. MIT Press, Cambridge MA.

Glickstein, M. (1985). Ferrier's mistake. *Trends in Neuroscience*, **8**, 341–4.

Glickstein, M. (1998). Cerebellum and the sensory guidance of movement. In G.R. Bock and J.A. Goode (Eds.) *The sensory guidance of movement* (pp. 252–66). John Wiley and Sons, Chichester.

Goodale, M.A. (1990). Brain asymmetries in the control of reaching. In M.A. Goodale (Ed.) *Vision and action: the control of grasping* (pp. 14–32). Ablex, Norwood NJ.

Goodale, M.A. and Carey, D.P. (1988). The role of cerebral cortex in visuomotor control. In B. Kolb and R.C. Tees, (Eds.) *Cerebral cortex of the rat* (pp. 309–40). MIT Press, Cambridge MA.

Goodale, M.A. and Murphy, K.J. (1997). Action and perception in the visual periphery. In P. Their and H.-O. Karnath (Eds.) *Parietal lobe contributions to orientation in 3D space* (pp. 447–61). Springer-Verlag, Heidelberg.

Goodale, M.A., Meenan, J.P., Bülthoff, H.H., Nicolle, D.A., Murphy, K.J., and Racicot, C.I. (1994). Separate neural pathways for the visual analysis of object shape in perception and prehension. *Current Biology*, **4**, 604–10.

Grafton, S.T., Fagg, A.H., Woods, R.P., and Arbib, M.A. (1996). Functional anatomy of pointing and grasping in humans. *Cerebral Cortex*, **6**, 226–37.

Graziano, M.S.A. and Gross, C.G. (1993). A bimodal map of space: somatosensory receptive fields in the macaque putamen with corresponding visual receptive fields. *Experimental Brain Research*, **97**, 96–109.

Graziano, M.S.A., Yap, G.S., and Gross, C.G. (1994). Coding of visual space by premotor neurons. *Science*, **266**, 1054–7.

Gribble, P.L., Everling, S., Ford, K., and Mattar, A. (2002). Hand–eye coordination for rapid pointing movements: arm movement direction and distance are specified prior to saccade onset. *Experimental Brain Research*, **145**, 372–82.

Grusser, O.-J. and Landis, T. (1991). *The visual agnosias and other disturbances of visual perception and cognition*. CRC Press, London.

Haaxma, H. and Kuypers, H.G.J.M. (1975). Intrahemispheric cortical connections and visual guidance of hand and finger movements in the rhesus monkey. *Brain*, **98**, 239–60.

Harvey, M. and Milner, A.D. (1995). Balints patient. *Cognitive Neuropsychology*, **12**, 261–4.

Hausser, C.O., Robert, F., and Giard, N. (1980). Balint's syndrome. *Canadian Journal of Neurological Science*, **7**, 157–61.

Hayhoe, M., Aivar, P., Shrivastavah, A., and Mruczek, R. (2003). Visual short-term memory and motor planning. In J. Hyönä, D. Munoz, W. Heide, and R. Radach (Eds.) *Progress in Brain Research: Vol. 140. The brain's eyes: neurobiological and clinical aspects of oculomotor research*, pp. 349–63. Elsevier Science, Oxford.

Hecaen, H. and De Ajuriaguerra, J. (1954). Balint's syndrome (psychic paralysis of visual fixation) and its minor forms. *Brain*, **77**, 373–400.

Holmes, G. (1918). Disturbances of visual orientation. *British Journal of Ophthalmology*, **2**, 449–468, 506–16.

Honda, H. (1984). Functional between-hand differences in outflow eye position information. *Quarterly Journal of Experimental Psychology*, **36A**, 75–88.

Jakobson, L.S., Archibald, Y.M., Carey, D.P., and Goodale, M.A. (1991). A kinematic analysis of reaching and grasping movements in a patient recovering from optic ataxia. *Neuropsychologia*, **29**, 803–9.

Jeannerod, M. (1997). *The cognitive neuroscience of action*. Cambridge MA: Blackwell Publishers.

Jeannerod, M., Decety, J., and Michel, F. (1994). Impairment of grasping movements following a bilateral posterior parietal lesion. *Neuropsychologia*, **32**, 369–80.

Johannsson, R.S., Westling, G., Bäckström, A., and Flanagan, J.R. (2001). Eye–hand coordination in object manipulation. *Journal of Neuroscience*, **21**, 6917–32.

Jouffrais, C. and Boussaoud, D. (1999). Neuronal activity related to eye–hand coordination in the primate premotor cortex. *Experimental Brain Research*, **128**, 205–9.

Kase, C.S., Troncoso, J.F., Court, J.E., Tapia, J.F., and Mohr, J.P. (1977). Global spatial disorientation: clinico-pathologic correlations. *Journal of the Neurological Sciences*, **34**, 267–78.

Kawashima, R., Inoue, K., Sato, K., and Fukada, H. (1997). Functional symmetry of cortical motor control in left-handed subjects. *Neuroreport*, **8**, 1729–32.

Kim, M.-S. and Robertson, L.C. (2001). Implicit representations of space after bilateral parietal lobe damage. *Journal of Cognitive Neuroscience*, **13**, 1080–7.

Kim, S.-G., Ashe, J., Hendrich, K., Ellerman, J.M., Merkle, H., Ugurbil, K., and Georgopoulos, A.P. (1993). Functional magnetic resonance imaging of motor cortex: hemispheric asymmetry and handedness. *Science*, **261**, 615–6.

Kimura, D. (1993). *Neuromotor mechanisms in human communication*. Oxford University Press, Oxford.

Levine, D.N., Kaufman, K.J., and Mohr, J.P. (1978). Inaccurate reaching associated with a superior parietal lobe tumor. *Neurology*, **28**, 556–61.

Marconi, B., Genovesio, A., Battaglia-Mayer, A., Ferraina, S., Squatrio, S., Molinari, M., Lacquaniti, F., and Caminiti, R. (2001). Eye–hand coordination during reaching. I. Anatomical relationships between parietal and frontal cortex. *Cerebral Cortex*, **11**, 513–27.

Marsden, C.D. (1998). The apraxias are higher-order deficits of sensorimotor integration. In G.R. Bock and J.A. Goode (Eds.) *The sensory guidance of movement* (pp. 308–331). John Wiley and Sons, Chichester.

Meegan, D.V. and Tipper, S.P. (1998). Reaching into cluttered visual environments: spatial and temporal influences of distracting objects. *Quarterly Journal of Experimental Psychology*, **51A**, 225–49.

Milner, A.D. and Goodale, M.A. (1995). *The visual brain in action*. Oxford Univeristy Press, Oxford.

Milner, A.D., Paulignan, Y., Dijkerman, H.C., Michel, F., and Jeannerod, M. (1999). A paradoxical improvement of misreaching in optic ataxia: new evidence for two separate neural systems for visual localization. *Proceedings of the Royal Society of London*, **266B**, 2225–9.

Milner, A.D., Dijkerman, H.-C., Pisella, L., McIntosh, R.D., Tilikete, C., Vighetto, A., and Rossetti, Y. (2001). Grasping the past: delay can improve visuomotor performance. *Current Biology*, **11**, 1–20.

Munoz, D.P. and Fecteau, J.H. (2003). Vying for dominance: Dynamic interactions control visual fixation and saccadic initiation in the superior colliculus. In J. Hyönä, D. Munoz, W. Heide, and R. Radach (Eds.) *Progress in Brain Research: Vol. 140. The brain's eyes: neurobiological and clinical aspects of oculomotor research*, pp. 3–19. Elsevier Science, Oxford.

Neggers, S.F.W. and Bekkering, H. (2000). Ocular gaze is anchored to the target of an ongoing pointing movement. *Journal of Neurophysiology*, **83**, 639–51.

Paillard, J. (1996). Fast and slow feedback loops for the visual correction of spatial errors in a pointing task: a reappraisal. *Canadian Journal of Physiology and Pharmacology*, **74**, 401–17.

Passingham, R.E. and Toni, I. (2001). Contrasting the dorsal and ventral visual systems: guidance of movement versus decision making. *Neuroimage*, **14**, S125–S131.

Perenin, M.T. and Vighetto, A. (1988). Optic ataxia: a specific disturbance in visuomotor mechanisms. *Brain*, **111**, 643–74.

Peters, M. (2000). Contributions of imaging techniques to our understanding of handedness. In K. Manas, M. Mandal, B. Bulman-Fleming and G. Tiwari (Eds.) *Side bias: A neuropsychological perspective*. Kluwer Academic Publishing, Dordrecht.

Pisella, L., Gréa, H., Tilikete, C., Vighetto, A., Desmurget, M., Rode, G., Boisoon, D., and Rosetti, Y. (2000). An 'automatic pilot' for the hand in the human posterior parietal cortex. Towards a reinterpretation of optic ataxia. *Nature Neuroscience*, **3**, 729–36.

Prablanc, C., Echalier, J.F., Jeannerod, M., and Komilis, E. (1979). Optimal response of eye and hand motor systems in pointing at a visual target-I. Spatio-temporal charcteristics of eye and hand movements and their relationship when varying the amount of visual information. *Biological Cybernetics*, **35**, 113–24.

Revol, P., Rossetti, Y., Vighetoo, A., Rode, G., Boison, D., and Pisella, L. (2003). Pointing errors in immediate and delayed conditions in unilateral optic ataxia. *Spatial Vision*, **16**, 347–64.

Rizzo, M., Rotella, D., and Darling, W. (1992). Troubled reaching after right occipito-temporal damage. *Neuropsychologia*, **30**, 711–22.

Rondot, P., DeRecondo, J., and Ribadeau Dumas, J.L. (1977). Visuomotor ataxia. *Brain*, **100**, 355–76.

Rothi L.J. and Heilman, K.M. (Eds.) (1997). *Apraxia: the neuropsychology of action*. Psychology Press, Hove, East Sussex.

Rushworth, M.F.S., Krams, M., and Passingham, R.E. (2001). The attentional role of the left parietal cortex: The distinct lateralization and localization of motor attention in the human brain. *Journal of Cognitive Neuroscience,* **13**, 698–710.

Salinas, E. and Their, P. (2000). Gain modulation: a major computational principle of the central nervous system. *Neuron,* **27**, 15–21.

Simon, O., Mangin, J.F., Cohen, L., Le Bihan, D., and Dehaene, S. (2002). Topographical layout of hand, eye, calculation and language-related areas in the human parietal lobe. *Neuron,* **33**, 475–87.

Snyder, L.H., Batista A.O., and Andersen, R.A. (2000). Saccade-related activity in the parietal reach region. *Journal of Neurophysiology,* **83**, 1099–102.

Tzavaras, A. and Masure, M.C. (1975). Aspects différents de l'ataxie optique selon la latéralisation hémisphérique de la lésion. *Lyon Médical,* **236**, 673–83.

Volkmann, J., Schnitzler, A., Witte, O.W., and Freund, H.-J. (1998). Handedness and asymmetry of hand representation in human motor cortex. *Journal of Neurophysiology,* **79**, 2149–54.

Wing, A.M., Haggard, P., and Flanagan, J.R. (1996). *Hand and brain: the neurophysiology and psychology of hand movements.* Academic Press, San Diego CA.

Xing, J. and Andersen, R.A. (2000). Models of the posterior parietal cortex which perform multimodal integration and represent space in several coordinate frames. *Journal of Cognitive Neuroscience,* **12**, 601–14.

Finale

Chapter 25

The achievement of brain imaging: Past and future[1]

Michael I. Posner

Abstract

Brain imaging has forged an impressive link between psychology and neuroscience. Many studies show consistent activation of a small number of widely separated brain areas that differ among tasks. Do these results imply a modular organization of the human brain? One view of modularity requires that there be no top down or cognitive input from other systems. This view has almost no support even for primary sensory cortex, which can be influenced by attention. A less restrictive version of modularity requires the ability to manipulate one area of activation independently from another. The studies in the sections of this volume on visual pathways and on orienting of attention, provide support for this form of modularity, but require specification of the time course and the connections involved. The section on plasticity indicates that brain circuits change with experience, development, and brain injury. Images of the brain will have more lasting impact if we can convert the insights they make possible into understanding how genes and experience influence neural networks and can apply this knowledge to issues of normal and abnormal development. Sensory-motor integration raises the difficult issue of how we move from brain activity to behavior. Progress on this question has been slower because it involves questions of motivation, and voluntary choice. The ability to view attention as an organ system with its own anatomy, circuitry, and individual differences could lead us toward an understanding of the mechanisms of volition.

25.1 Introduction

At Attention and Performance XIV, which was the 25th anniversary of the Association, Donald Broadbent (Broadbent 1991) delivered a talk at the end of the meeting, which he titled 'A word before leaving'. He warned that the task he had been given to summarize the meeting was not possible. Rather than attempting to summarize, I try to provide a personal perspective about the general topics discussed at this meeting.

Although the headings in the present meeting might seem familiar to those who attended previous or have read the volumes from past Attention and Performance meetings, few, if any, of the papers given at this meeting could have been given at the meeting 12 years ago. That in itself, is rather amazing. Techniques such as fMRI, diffusion tensor imaging, transcranial magnetic stimulation, high-density electrical recording related to specific generators (although for related material see Mangun *et al.* 1990), and recording from many implanted electrodes in freely performing animals were simply unknown at the 1990 meeting.

However, we are urged to go beyond the methodological progress to say how these methods illuminate the underlying cognitive questions. To carry out this task I have broken this discussion into four sections that roughly correspond with the sections of the meeting. First, I deal with modularity, starting with the issue of its definition applied to the study of faces and attention and then considering the objections that have been raised to localization of function (e.g. Uttal 2001). The form of modularity supported by brain imaging requires communication between anatomically defined modules. It is to this topic I turn next, in a discussion of efforts to uncover the circuitry for performing cognitive tasks. In the third section, I consider the area of plasticity, particularly as it arises in the study of differences among individual in development of the nervous system. I believe that there needs to be a special effort to make the study of individuality as well as development central topics of Attention and Performance, in part, because the new imaging methods allow for such studies, and because individual variation represents an important way of relating brain networks to genes. In the final section, I consider the theme of sensory motor integration, in particular how we can deal with the issue of volition, sometimes called decision making (Chapter 20, this volume), in understanding how to go from brain activity to behavior. This effort may allow understanding of the regulatory processes developed by the brain that are so central to control of our thoughts, feelings, and behavior.

25.2 **Modularity**

25.2.1 **Definition**

It is clear from the literature and the many papers presented in this volume that psychological tasks usually activate several circumscribed areas that are often well separated in the brain. These areas are sufficiently common among subjects to allow averaging of data at least within the millimeter range available to fMRI, and are often sufficiently different between tasks so they can be identified with particular functions. Should these brain areas be considered as modules?

This clearly depends upon the definition of module. According to Sternberg (Chapter 5, this volume) activated brain areas would be a module if functionally distinct processes are carried on that could be manipulated independently in experiments. In my view, brain areas related to aspects of processing faces, numbers, words, motion, and attention are all quite likely to meet this definition, even if there are reasons given by Sternberg (Chapter 5, this volume) to think that in many cases the most appropriate fMRI experiments remain to be done.

A quite different definition of modularity comes from Fodor (1983). According to this view, modularity requires that candidate modules be sufficiently encapsulated that there is no penetrability by attention or other top-down brain systems. Even V1, and probably not even the lateral geniculate, could not serve as a module by this definition (Posner and Gilbert 1999; Chapter 15, this volume).

Spelke (Chapter 2, this volume) says that modules should be domain specific. This idea is related to her concept of domain-specific core knowledge. However, attentional orienting to stimuli is an example of a network employed in a wide range of domains (Chapter 14, this volume) and yet orienting of attention shows the same type of activation of widely separated brain areas as is found for domain-specific tasks such as number and language. For this reason, I see no need to restrict modularity to domain-specific systems.

Although I prefer the Sternberg definition, and thus like to think of brain organization as modular, there is really very little importance to the decision of whether to apply this particular term. The important thing is that consistent empirical results show that many psychological tasks can be identified with a specific set of brain activations.

These consistent findings from imaging do provide a brain-based result to support the more abstract idea arising from cognitive science, that tasks are decomposable into elementary operations. Stemming from the work of Lashley (1931), and from his interpretation of the frontal lesion data that was probably oversimplified (for a better view see Duncan 1995), the idea came to be accepted that higher mental processes were not subject to the kind of localization found for simpler sensory and motor tasks. However, the distributed nature of the nodes in even simple cognitive tasks helps to reconcile a strictly localizationist view (i.e. one task one brain area) with the view of the brain operating as a whole that arose from Lashley's work. Since the assembly of a visual percept can involve many visual areas, it seems fitting that reading or mental arithmetic would be no less complex. Even if each operation were strictly localized, since psychological tasks involve many operations, the task would involve different parts of the brain. In addition, localization of mental operations allows use of clues from anatomical connectivity and evolution, and provides information to animal studies about the brain areas from which it might be desirable to record neuronal activity.

25.2.2 Faces

There is no general agreement about what is localized. The first day of the meeting was heavily devoted to the idea that the fusiform face area localized representations of the face (Chapters 3 and 7, this volume). This led to considerable dispute, because there was evidence that other kinds of stimuli (e.g. dogs for those who were dog experts) could activate the same general area. Later in the meeting, when attention and motor control were discussed, areas of activation were most often related to mental operations that might be performed in that area (e.g. Chapters 14, 17, and 20, this volume). It has often proved difficult to discriminate between operations and the representations on which these operations are performed. This is certainly true in neuroimaging data. In order to know whether the BOLD signal is sensitive to the representation or to operations upon that representation we would have to have a detailed account of what goes on within the area of activation, or a strong theory of how different brain areas work together to carry out a task. In general, we don't yet have the information to decide. However, in the case of faces, I argue below that the issues involved become less controversial if one believes in the localization of operations rather than representations.

The data from the study of faces presented in this meeting and outside have been quite consistent in showing that an area of the fusiform gyrus is activated by faces and that when one attends to the face, activation in this area is amplified (Chapters 3 and 7, this volume). This is a remarkable level of empirical agreement. However, if representations are localized, evidence that experts in dogs or cars activate this area when they process material within their expertise (e.g. dogs) calls into question localization of function. If, however, operations of the type discussed by Ullman (Chapter 6, this volume) and Malach *et al.* (Chapter 8, this volume) are localized, the controversy seems to evaporate. Faces are part of core knowledge and most often require careful individuation. As such, it seems very likely that the features or fragments used in their identification might differ in scope or spatial frequency from those required by other objects. Expertise brings with it the need to carry out operations that lead to identification of individuals, so at that point the area of activation would be expected to overlap that used by faces. Indeed, the finding by Puce *et al.* (1999) that more anterior temporal areas, difficult to study by fMRI, might be recruited in order to recognize the particular face is very consistent with the general view of a network of operations involved in cognitive tasks.

25.2.3 Attention

In many ways the strongest evidence for localization of mental operations stems from the area of orienting of attention toward sensory stimuli (Chapters 14, 15, 17, and 19, this volume). In this area of research a brilliant combination of method and experimental sophistication has shown how exquisitely separate brain areas can be brought together to orchestrate a very simple shift in the orienting of attention.

Hillyard *et al.* (Chapter 19, this volume), in their discussion, outlined a consensus view in which orienting of attention to a visual stimulus produced an amplification in prestriate regions, which affected processing in all subsequent regions and which was fed back to influence processing in V1 and perhaps in the lateral geniculate (Kastner, Chapter 15, this volume). The source of this attention effect was in a network of parietal and frontal regions. These results were based upon fMRI work and EEG studies that attempted to explore the time course of activation of specific generators. The extent of convergence between the activations in fMRI and the modeling data attempting source localization with high-density event-related potential (ERP) was impressive (Chapter 19, this volume; also see Dale *et al.* 2000).

What does orienting to a visual object do? It seems clear from the papers in this volume that one role that attention plays in the prestriate areas is to counteract the suppression of unit activity that occurs when two stimuli occupy the same receptive field. Support for the biased-competition model (Desimone and Duncan 1995) was presented both from fMRI studies (Chapter 15, this volume) and from cellular studies (Chapter 18, this volume). However, there are many hints that orienting of attention may play additional roles as well, for example, in amplifying input when only a single item is present in the visual field (see Posner 1980).

The superior parietal lobe appears to be crucial in orchestrating a voluntary shift of attention by a cue, which precedes a target. This brain area is closely tied with the eye movement system, as would be predicted by the premotor theory of attention shifts. However, a quite different area of the parietal lobe, the temporal parietal junction (TPJ) is the area through which a stimulus at an unexpected location acts to summon orienting. The TPJ effect seems to be strongly lateralized to the right side, while the voluntary shift appears to be more bilateral (Chapter 17, this volume). This is not the separation of operations that I had originally expected, based on what was known in the late 1980s (Posner *et al.* 1987). I had proposed that the superior parietal lobe was related to the 'disengage' operation, but it now appears that it relates more to voluntary shifts of attention irrespective of whether the person is engaged visually or not. However, the emerging evidence does provide a more satisfactory resolution of mysteries, such as why lesions producing unilateral neglect do not generally involve the superior parietal lobe, but instead areas in or near the temporal parietal junction. The lesson I have learned from these findings is that it is difficult to know a priori exactly how tasks should be broken down. Too much of my reasoning in 1987 was based upon peripheral cues, which can both act like targets or be the basis for a voluntary shift. The general brain areas now shown to be involved in orienting appear to be similar in literally dozens of visual experiments and also appear to be general for cues and targets in other modalities (Chapters 14 and 15, this volume).

These studies have shown that it is important not to treat attention as a unified field as Uttal (2001) does in his critique, but rather to separate functions such as orienting to sensory stimuli from quite different functions such as resolving conflict among responses. The imaging approach to cognitive function will require us to think quite hard about how cognitive functions should be divided and what we have learned from behavioral studies about the complex nature of what may seem very simple tasks.

I believe that this general story of localization of quite specific mental operations will prove true for more anterior regions as well, but here it is much more disputed (Duncan and Owen 2000). However, as event-related fMRI experiments are applied to this area, there appear to be separate operations occurring in the lateral and medial frontal areas (MacDonald *et al.* 2000), that are so often activated in common that they have been thought of as a single global workspace (Dehaene *et al.* 1998*b*).

25.2.4 Summary

The area of orienting of attention toward sensory stimuli, as represented in this meeting, provides strong support for the idea of localization of mental operations. A set of dorsal brain areas, including the superior parietal lobe and temporal parietal junction, serve as a common source of attention to sensory stimuli. They produce effects within a network of areas that depend upon modality (e.g. ventral visual areas in the case of visual input). What is still missing is direct evidence that activated areas in the dorsal stream, such as the superior parietal cortex and TOJ, hypothesized to be the source of the attentional effect, are causing the increased activation found in prestriate visual cortex. The need to activate separate areas of the parietal and lobes and trace their influence upon the ventral object recognition pathway will require careful study of these networks in real time. In the next section we turn to evidence that existing methods can be used to trace this kind of circuitry.

25.3 Circuitry

In 1978 I defined mental chronometry as the time course of information flow in the human nervous system (Posner 1978). I had in mind functional models of information flow, that followed the ideas of Broadbent (1973). With low-density ERPs, a few monkey studies, and the human lesion work, we did not have the tools to carry out studies of the human brain in a very systematic way.

25.3.1 Scalp electrical recording

The situation has changed dramatically. Methods that separate cue and target over sufficient intervals can use event-related fMRI effectively to separate preparatory activity from computations specific to the target (Chapter 17, this volume). It is also possible to use event-related fMRI together with correlational techniques to work out the time and direction of information flow (Chapter 22, this volume). However, the relatively slow hemodynamic response raises special problems for real-time measurement.

By combining fMRI studies with electrical recordings, it is possible to trace the time course of brain activity and to index communication between neural areas. The methods used for linking electrical or magnetic recording to hemodynamic imaging have been reviewed by experts in the field (Dale *et al.* 2000; Chapter 19, this volume;) and they have found both EEG and MEG helpful in providing specific time relations for the active brain areas (Liu *et al.* 2002).

The visual system, including visual attention as reviewed by Hillyard *et al.* (Chapter 19, this volume), has been the best area for the close integration of hemodynamic, lesion, and EEG work. In my view, the results have been very impressive. Mapping studies have provided convincing evidence that generators in primary and extrastriate visual systems can be studied from scalp recordings, using algorithms for fitting dipoles to the scalp distribution (Chapter 19, this volume) and careful analysis of the time course of their activation has suggested that the V1 activity is due to feedback from higher visual areas (Martinez *et al.*, 2001). These findings have shown the importance of very detailed temporal information in interpreting brain activation and have demonstrated that scalp recordings can provide the needed information.

However, there has been less acceptance of the extension of these ideas to more complex cognitive processes where brain networks involve many distributed sites. Goebel *et al.* (Chapter 22, this volume) provide a good example of using fMRI to trace the network involved in imagery tasks. However, the problem of the slowness of the hemodynamic response suggests that this method will need to be supplemented by relating scalp-recorded electrical activity to the hemodynamic generators. One of the major problems in using algorithms for relating scalp electrical distribution to underlying generators is when a number of generators are close together in space and active at the same time. Tasks such as generating the use of a written word, or deciding whether a number is above or below five, work well because they involve distant generators that are distributed over the relatively long time needed to carry out the task.

25.3.2 Word reading

Utall ends his book criticizing imaging methods with the task of reading visual words. His arguments are instructive because they represent such a strong a priori philosophical position. He says that the argument we took in *Images of Mind* (Posner and Raichle 1994), that cognitive tasks are performed by networks of widely distributed neural systems, and that computations within a network interact by re-entrant processes, are clearly inconsistent with modularity. He seems to be accepting the idea of modularity that comes from Fodor (1983). If for localization to have any utility it had to rely upon little or no interaction among brain areas, it would be clearly inconsistent with the imaging data. However, as I pointed out in the previous section, it is quite consistent to hold a form of modularity (Sternberg, Chapter 5, this volume) and believe in a network of interacting nodes. If such views are more in line with the empirical philosophy outlined by Broadbent (1973) than the philosophical a priori views of Fodor, perhaps that is not such bad company.

The study of visual words provides a good example of how much we can learn from examining the time course of generators found in hemodynamic studies (Posner and McCandliss 1999). Five consistent brain areas have been found to be active in the study of visual words. They are: a left lateralized occipital–temporal area that can be associated with visual word form (Dehaene, Chapter 9, this volume), left ventral frontal and medial frontal areas related to attention, a left lateralized temporal–parietal area (Wernicke's area) and, finally, the right cerebellum. These results are relatively consistent, but my conclusion that they represent a reasonable consensus about how words are processed, seems so utterly different than the conclusion one might reach from Utall's book, namely that the field is a mess, that it may be worth some deep consideration of the reasons for this discrepancy.

To understand this issue it is important to trace the time course of activity in the various generators involved when reading a word aloud is subtracted from generating a word use. This subtraction eliminates the posterior word form area, which is involved in both tasks. In our ERP studies (Abdullaev and Posner 1998; Posner and McCandliss 1999) the electrical components that relate to the anterior cingulate activation occur quite early, after about 150 ms. However, in many studies of the Stroop effect, the cingulate activity seems to occur later and has been related to the monitoring of conflict (Botvinick *et al.* 2001). It appears certain that the cingulate is often related to aspects of conflict, but the time-course information is going to be critical to understanding its exact role. Lateral frontal activity starts its activation around 200 ms and only much later, at about 500 ms, do we find activity in Wernicke's area.

Our PET work of generating the use of a word identified the medial frontal area with attention and the left frontal area with semantic processing (Petersen *et al.* 1987). This was because we only found the left frontal area active in semantic tasks and not in reading words aloud. However, subsequent data seem to suggest that the specifically semantic processing involves the left posterior (Wernicke's) area. The two frontal areas are more related to attention although, as discussed before, I believe they perform somewhat different mental operations, neither of which is related to a semantic store. The lateral area may be involved in temporarily storing the input word. Duncan *et al.* (2000) showed that this area was active in a wide variety of tasks, which loaded on the g factor from intelligence tests. It seems reasonable that what all these tests have in common is the need to represent information in some temporary store while the brain provides information on what is known about the item. This fits with findings that lateral frontal areas are involved in working memory. It is also our finding that this frontal area is active after 200 ms in the generate uses task, which requires over a second. This leaves time for Wernicke's area, which has been shown by many investigators to be related to semantic operations, to retrieve the word uses needed to carry out the generate uses task.

A somewhat similar story may account for the controversy concerning the visual word form system. This was first identified by a PET study (Petersen *et al.* 1987) and has been found in several recent fMRI studies (Chapter 9, this volume; Shaywitz *et al.*, 2002). One reason for the controversy is that the strength of its activation may depend upon the depth of the orthography used in the language (Paulesu *et al.* 2001), by the degree of attention needed for processing this information in a particular task, and perhaps also by the teaching method by which reading was originally acquired.

In addition, there may be several sub-areas that respond selectively to different aspects of the organization of letters into words.

The visual word form system is of great importance because its relatively late development in conjunction with the acquisition of literacy represents an opportunity to trace how different methods of acquiring the reading skill might influence its development. Studies of 10-year-olds and adults, using fMRI, have shown a relatively larger area of activation in children than in adults (Schlaggar *et al.* 2002). Adult studies, using high-density electrical recording, have shown scalp signatures of a left lateralized posterior activation, which is present for words and orthographically regular non-words, but not consonant strings (Compton *et al.* 1991; McCandliss *et al.* 1997). This scalp signature is active by about 170 ms after input and seems to fit the characteristics of the visual word form system. This method has been used by McCandliss *et al.* (1997) to show that, in adults, the visual word form area is not responsive to familiarity induced by high levels of training (50 hours of learning a new set of lexical items in an artificial language). However, in ten-year-old children this scalp signature seems to occur only for words for which the child is familiar. These studies suggest that the visual system first recognizes words through familiarity, but as the density of items becomes high, only orthography and not familiarity matters (Posner and McCandliss 1999). Results of this type may make imaging studies of significance for tracing the influence of different educational methods.

25.3.3 Information transfer

Time-course information can be very useful in understanding the role of particular hemodynamic generators. However, knowing when something happens does not, by itself, tell us how brain areas are connected. One step that has been taken to examine connections between remote brain sites is to use correlational methods. This can be done using fMRI (Chapter 22, this volume), but of course suffers from the imprecise time courses involved. EEG coherence and other correlational measures provide better time resolution, but raise the problem of contamination through volume conduction between sites. When the sites are remote, volume conduction is less of a problem. However, many studies using these techniques (Gevins 1996) have found evidence for such complex networks of connections that it has been difficult to know whether they are all genuine, despite statistical verification. Recently, we hypothesized that the left lateralized frontal information found during our word-processing task would have to be in contact with posterior regions to connect the presented word being represented in frontal areas, with posterior brain areas that know about the word uses. The time-course information we had recorded suggested that there was plenty of time for this to take place, since the frontal activity began at about 200 ms and the posterior activity not until about 500 ms. Singer (Chapter 4, this volume) discussed the importance of high-frequency bands in the integration of local information and slower frequencies for large-scale integration of information. We (Nikolaev *et al.* 2001) found evidence of correlation within the β-frequency band that began prior to the activation of the posterior area, at about 450 ms. Because there are often many electrodes that show strong correlation, I believe we need specific predictions of what areas need to be related to test whether a correlation between remote sites is likely to serve as a good index of the transfer of information between these brain areas.

New methods are now available for examining the development of white matter pathways in the human brain by use of diffusion tensor MRI. Dehaene and colleagues (Chapter 9, this volume) have shown that this method can be used to confirm ideas about connectivity between brain areas. They studied a patient who was unable to read fluently when words were presented to the left of fixation, but who could do so for right-lateralized words. Diffusion tensor imaging showed a disconnection between brain areas of the left and right hemisphere with the location and properties of the visual word form system. Lesion data also suggest that this visual word form system works automatically, because patients with lesions of the right parietal lobe, who neglect the parts of consonant strings to the left of fixation, are spared from extinction when the string is a word (Sieroff *et al.* 1988). Apparently words are chunked into a whole, by the word form system, and then no separate attentional orienting is required.

Diffusion tensor imaging could also open up the prospect of using measures of the development of coherence between distant electrode sites from EEG as a means of probing the earliest functional use of particular white matter pathways.

25.3.4 Other methods

An entirely different combination of techniques for tracing circuitry combines fMRI with transcranial magnetic stimulation (TMS) (Pascal-Leone and Hamilton 2001). The fMRI studies revealed that both somatosensory and primary visual cortex was active during the process of learning to read Braille. TMS was used to show that performance was reduced when a pulse was given over the somatosensory system at the time of stimulus presentation, while stimulation of the visual system by TMS reduced performance at a delay for about 60 ms. This rapid time course suggested that the somatosensory stimulation activated visual cortex in the process of understanding the character. This network finding was coupled with evidence that extensive visual deprivation could be used to improve the rate of learning Braille, presumably by removing conflicting visual input (Kaufman *et al.* 2002).

25.3.5 Summary

Large-scale neural circuits used in the performance of many complex human skills involve connections between several brain areas. Although much effort in cognitive neuroscience has been devoted to improvements in spatial resolution, the nature of these complex circuits allow them to be studied with a combination of existing methods.

By combining hemodynamic techniques with scalp recordings or with stimulation by TMS, it is possible to examine these networks in real time. Correlational methods and diffusion tensor analysis give promise of allowing us to study the transfer of information between nodes of the network during the performance of skills. These results show how tracing the flow of information during tasks such as reading print (Posner and McCandliss 1999); learning Braille or learning ASL can help us understand how tasks are acquired and performed. These circuits should not be thought of as purely static but have to change with the direction of attention, new learning, and development. In the next section of this chapter I turn to the study of plasticity in these networks.

25.4 Plasticity

Plasticity can occur at many different time scales. At this meeting we have mainly discussed relatively slow time scales involved in the maturation of networks during child development. Spelke (Chapter 2, this volume) discussed the initial state in the newborn for the understanding of places, objects, number, and people as part of core knowledge. Each of these domains only reaches the adult level after many years of maturation and experience. Neville and Bavelier (2001) showed how many years of visual or auditory deprivation could induce changes in the organization of sensory systems in blind and deaf individuals. They also showed a role for the specific learning of American Sign Language in understanding laterality in the deaf. Learning these skills may require many months or years of practice. Development, recovery from brain injury, and the acquisition of complex skills are all relatively slow sources of plasticity.

25.4.1 Attentional networks

At the other end of the time scale, attention plays an important role in the plasticity of human thought processes, allowing rapid shifts from one scene to another or between streams of thought. Although not usually thought of in this way, this attention is a primary means of introducing plasticity among the networks that dominate behavior from moment to moment. Momentary changes are certainly apparent in visual orienting, where head and eye movements reflect very intimately the direction of our moment-to-moment interest. There is a similar control by central networks over the organization of our thought processes. The very same network of brain areas, for example one involved in understanding written words, may organize the order of computations in different ways depending on how attention is directed (Posner and Pavese 1998).

Attention can be studied in terms of at least three networks of neural areas. Studies using functional neuroimaging to examine the network involved in sensory orienting have been well represented at this meeting. In our work we have also examined networks related to obtaining the alert state and resolving conflict among responses (executive control) (Posner and Fan in press).

Recently we developed an integrated attention network test (ANT) (Fan et al. 2002) that builds on several chronometric tasks (Posner and Boies 1972; Ericksen and Ericksen 1974; Posner 1980) often used individually in studies of normal individuals and patients. The ANT requires the person to respond to a central arrow by pressing a left key when the arrow points left and a right key when it points right. The arrow is surrounded either by congruent (same direction) or incongruent flankers. The ANT also uses cues to direct attention. The four cue conditions are: no cue, spatial cue, double cue, and central cue. By making three subtractions, it is possible to isolate performance in the alerting, orienting, and executive network. For alerting, we subtract the double cue from the no cue RTs to get the effect of a warning prior to the target. Since the target is presented above or below fixation, a subtraction of RTs to a spatial cue at the target location from a central cue provides a measure of orienting. Finally, the flanker effect (incongruent flanker RTs – congruent flanker RTs) provides a measure of the efficiency of resolving conflict, one of the primary functions of the executive attention network (Botvinick *et al.* 2001).

All of these tasks had been studied in various forms of neuroimaging that provide a consensus on the neural networks involved. We have also conducted a neuroimaging study of the ANT that activates each of these networks (Fan *et al.* 2001*a*). Each of the three subtractions, discussed in the previous paragraph, activated a separate network that corresponded reasonably well to what had been found in previous studies of alerting, orienting, and conflict using separate tasks.

In addition, each of these networks appears to have a dominant neurotransmitter. Alerting appears to be noradrenergic (Marrocco and Davidson 1998); orienting, cholinergic (Davidson and Marrocco in press) and the executive network, dopaminergic (Fossella *et al.*, 2002*a*, *b*; Posner and Fan in press).

The ANT has a number of useful properties as a measure of the efficiency of attentional networks. It does not use linguistic stimuli, so can be used with children, animals, or speakers of any language. While individual tests of each network might provide more precise scores, using the ANT, sufficient trials can be acquired in 20 min to obtain a measure of the efficiency of each network with reasonable reliability. This allows a rapid assay of attention to be used in conjunction with diagnosis or therapy.

25.4.2 **Individual differences**

A future goal of imaging will be to understand differences among individuals. The availability of MRI makes it possible to specify individual brains anatomically and, by superimposing hemodynamic and electrical responses, specify functions related to individual anatomy. However, the study of attention has not been much concerned with individual differences; although the topic of group differences in attention in cases of brain injury or other neurological or psychiatric disorders has been a major topic, the study of normal variability in attentional networks has not. With the new information available from the human genome project, and with some evidence already of the important role genes can play in performance (Greenwood *et al.* 2000). It does seem to be the time to develop such an effort.

Our goal was to try to provide a phenotypic characterization of attention in individuals that would allow links both to imaging and genetic studies. In addition to overall reaction time and error rates, subtractions can be used to obtain measures of the efficiency of the alerting, orienting, and conflict networks. Rather surprisingly, the scores of these three networks are relatively independent from each other and from the scores of overall RT and errors (Fan *et al.* 2002; Rueda *et al.* 2002).

We have used the ANT with fMRI to produce images of the attentional networks involved in processing both cues and targets. The brain areas found active in orienting correspond quite closely to the parietal and frontal network discussed at this meeting (Chapters 15, 17, and 19, this volume). In addition, we found a strong thalamic node for alerting and an activation of the cingulate in response conflict (Fan *et al.* 2001).

25.4.3 Genetics

We conducted a small-scale twin study (Fan *et al.* 2001*b*), which showed correlations among the twin pairs in overall reaction time, conflict and alerting, and a substantial heritability estimate for the conflict network. These results show that the network scores are valid, in the sense that they correlate among pairs of subjects with similar genomes.

The use of genetic analysis with chronometric studies has been applied to the neuropsychology of reading disorder (Olson *et al.* 1999), early Alzheimer dementia (Greenwood *et al.* 2000) and attention deficit disorder (Swanson *et al.* 2000). For example, Parasuraman and his colleagues (Greenwood *et al.* 2000) were able to show a deficit in orienting networks in early Alzheimer's disease. Patients had difficulty in using central cues to improve their handling of subsequent targets. This chronometric analysis was supported by a finding of reduction in blood flow specific to the superior parietal lobe, an area known to be important in orchestrating voluntary shifts of attention toward targets. Patients with early Alzheimer's disease frequently have a characteristic genetic allele related to the tendency to produce deposits of amyloid. In the study of asymptomatic subjects who had this genetic variant, they found a small deficit in orienting to central cues that closely resembled what they had reported for early Alzheimer's (Greenwood *et al.* 2000).

We have been using the links between specific neural networks of attention and particular chemical modulators to investigate the genetic basis of normal attention (Fossella *et al.* 2002*a, b*; Fan *et al.* 2003*b*) and pathologies (Swanson *et al.* 2000). While this work is in an early stage, we believe it will be very useful in helping to understand the range of individual response to pharmacological and other forms of therapy.

25.4.4 Development

Changes in RT for eye movements or key press responses can be used to trace changes in performance from a few weeks of age. In general, one finds continual reductions in speed from early infancy to adulthood, and increases later in life.

However, this does not mean that all neural networks that might be studied by chronometric techniques are likely to show the same function. Recently we developed a child version of the ANT in which the arrows are replaced by fish swimming to the left or right. We compared children and adult performance in the child version of the ANT (Rueda *et al.* 2002). We found evidence of clear and surprising differences between attentional networks in the rate of their development. Covert orienting of attention in our task was already developed to adult levels at 4 years of age. Our task required only rather gross distinctions between targets above and below fixation, so that its early development is not too surprising. Alerting continued to show development at least to adolescence. The alerting affect appeared to be mainly because children have trouble maintaining alertness in the absence of a warning. This could reflect difficulty in maintaining the task set over a brief delay. Most interesting, the executive network, involved with monitoring conflict between the flanker and the target, developed from 4–7 but not after that. This surprising effect might be due to the relatively simple nature of the child ANT. When we ran older children (10 years of age) and adults in the arrow version of the ANT, we found much longer RTs than for the child ANT and the children were slower than the adults, but the children and adults were not different in conflict scores, supporting the relatively early development of this network.

Modern studies of how the human genome differs among individuals have given new emphasis to studies of brain networks underlying the individual functioning of cognition and emotion. This approach is of obvious utility to neuropsychology. If we understood the range of normal variation and the genetic and experiential events that produce it, we would be in a better position to understand pathologies. This goal is somewhat remote, but even the early studies of genetics (e.g. Swanson *et al.* 2000; Fossella *et al.* 2002*a, b*) are already providing some useful illumination to our understanding of neuropsychology.

25.4.5 Summary

The brain is not fully formed at birth and studies of child development are revealing how maturation and experience shape its early growth (Posner *et al.* 2001; Spelke, Chapter 2, this volume). Of great importance are control networks involved in sensory orienting, achieving and maintaining the alert state, and in executive control, which can take many years to develop to adult levels. While all humans undergo a generally similar developmental course, brain networks are also influenced by differences among individuals in their genes and experiences. The study of individual differences in cognition, personality, and attention includes studies of structural and functional imaging. These efforts, together with other methods, are aimed at understanding how genes and experience work to shape networks, both during child development and following maturity.

25.5 Volition

The chapters on sensory motor integration raise the difficult question of going from brain images to behavior. It has proven possible to image areas of the brain that are driven by sensory input or respond to motor output, but what is involved in determining if the presentation of a word or picture will lead the person to name the word, free associate, or race out of the room? Volition, under the title *Control of Cognitive*

Processes, was the primary topic of *Attention and Performance XVIII* (Monsell and Driver 2000). In the present volume, Schall (Chapter 20) argues that automatic or reflexive responses to stimuli, what he calls choice, involve different mechanisms from the more reflective or effortful responses, which he labels 'deciding'.

The study of effortful control, based on deciding what response to make to a stimulus, raises the issue of the mechanisms involved in volition. If imaging is to live up to its promise, it will have to tackle such questions directly. In this section I consider three closely related issues concerning the nature of volition. First are techniques that have already begun to be used to relate images of the brain to outcomes. I think this is an important area. A second part of this section discusses the links between the executive attention network and volition. The final section reviews efforts to relate the anatomy and circuitry of executive attention to the complex world of natural behavior of children and adults through correlations between questionnaires and experiments.

25.5.1 Imaging behavior

While almost all of the chapter in this volume image the brain during cognitive tasks, we also need to be able to ask, how directly does a particular brain state, which can be imaged, lead to a particular behavior? In other words, how predictive is the state of the brain as measured by current methods of the behavior that might emerge?

Dehaene *et al.* (1998*a*) put the issue as follows:

> most neuroimaging experiments aim to identify the brain areas whose activation correlates tightly with an aspect of the subject's behavioral task. If the logic of neuroimaging is correct, however, it should also be possible to reverse this sequence of operations. Once we understand the function of a given brain area or network of areas, it should be possible to infer what kind of task the subject was performing.

To illustrate this idea Dehaene *et al.* (1998*a*) used data from a task of deciding whether a digit was above or below 5 by having subjects press a key with the left or right hand. It was found that an average of 91% of motor responses could be predicted correctly from the BOLD activity in the motor area. Other investigators have used the lateralized readiness potential from the EEG or MEG recordings to carry out similar, but somewhat less accurate, predictions.

Deheane and his associates (Piazza *et al.* 2003) have recently extended this idea to determine whether or not a subject employs attention in enumerating the number of stimuli present in a display. The idea is that if people are able to calculate the number preattentively, there should be no activation of the orienting network, which is involved in shifting attention between items. He finds that most individuals can process 2–3 items (dots in this case) without evidence of changes in the orienting network, but above that number activation of the orienting network increases with the number presented. Usually we control the stimulus presentation, for example, by a mask, to keep subjects unaware, but perhaps now we could simply use activation of an attentional network related to awareness or volition to determine, on a trial-by-trial basis, whether the response was based on 'deciding' (Schall, Cahpter 20, this volume) or was an automatic choice.

25.5.2 Conflict

A major theme of this meeting is the importance of competition as a determinant of the organization of neural circuitry. In the attention section we discussed the biased-competition model (Desimone and Duncan 1995) that emphasizes the role of attention in the resolution of local conflict within the receptive fields of neurons in prestriate areas. On a longer time scale, competition between sensory inputs can be reduced by deprivation, as in blindness or deafness, leading to a reorganization of normal sensory systems. Similarly, Kaufman *et al.* (2002) showed that reducing sensory input by blindfolding could lead to easier recruitment of the visual system in the service of somatosensory input during learning to read Braille. Work by Taub *et al.* (2002) has shown that recovery from brain injury can be enhanced by preventing use of the good limb, during exercise of the affected limb, thus improving reorganization.

When subjects obey an instruction to select between simultaneously presented dimensions in conflict tasks, they activate a common network of prefrontal and midline frontal brain areas that we have labeled the executive attention network. A central node of this network appears to be located in the anterior cingulate gyrus. Studies of tasks designed to involve conflict have invariably found activity in this area (Bush *et al.* 1998, 1999; Fan *et al.* 2003*a*). In these tasks, the source of the conflict is contained within the stimulus display, a situation that is relatively uncommon in real life. However, a network including the anterior cingulate, but also lateral prefrontal, cortex is activated in a wide variety of tasks that draw upon the general intelligence factor (g), as reported by Duncan and his colleagues (Duncan *et al.* 2000). Thus even when the conflict is implicitly elicited by the difficult nature of the thought process, the executive attention network appears to be needed. We might then consider a reflexive response as one that is so clearly driven by input that no activity within this network is required. We would then have a candidate for distinguishing volitional from involuntary responses to input and at least a start in understanding the nature of the decision process that is involved when tasks involve reflection. Schall (Chapter 20, this volume) cites an interesting fMRI study (Rees *et al.* 2000) in which displays of moving dots were presented to subjects who had to report the direction of movement. When the movement was induced by coherent dots all moving in the same direction, the task was easy and mostly occipital areas were active, these areas increased in activity the more coherent the motion. However, the anterior cingulate and insula increased systematically as the coherence of the dots decreased and hence the difficulty of the judgment increased.

While conflict seems to be a central way of activating the executive attentional network, there is evidence of other situations in which it has been activated; for example, when subjects are required to respond rapidly to stimuli (Naito *et al.* 2000) for error detection (Dehaene *et al.* 1994), even when the error is indicated by feedback, and in emotional responding (Bush *et al.* 2000). It may be that the network could better be seen as relating to many forms of effortful or voluntary control, even when it involves regulation of positive and negative affect.

Imaging studies have begun to provide a basis for exploring frontal networks involved in effortful control (Dehaene *et al.* 1998*b*). These studies suggest that both anterior cingulate and areas of the lateral prefrontal cortex operate together during tasks involving high levels of mental effort. Dehaene *et al.* (1998*b*) have proposed a global workspace, suggesting that both medial and lateral frontal areas exert a common influence during mental processing, However, the neuroimaging studies of orienting to sensory stimuli, discussed previously, show that within the large-scale distributed network found to orchestrate shifts of attention, each specific area plays a different role. We expect that a similar story will eventually emerge with respect to the lateral and medial frontal structures involved in executive attention. Some proposals have been made as to what the roles for different frontal areas might be, but as we pointed out (Bush *et al.* 2000) none have clear support in the current data.

One possibility is that the more lateral areas are involved as holding circuits in which domain-specific activity is represented while acted upon by midline circuits. This would fit with the role of lateral frontal cortex in working memory (Smith Jonides *et al.* 1998).

A different idea about how lateral and medial frontal areas might be involved developed from the idea that there are both automated and attended routes to output (Raichle *et al.* 1994). Raichle presented subjects with the same list of nouns many times, each time requiring them to generate a use. While a novel list activated a brain network including the cingulate and lateral prefrontal cortex, when the list had been repeated several times these activations disappeared and instead a different area, the anterior insula, increased in activity. According to this idea, the anterior insula represents a more automated pathway to word output, which is active in reading aloud and in generating uses when the association has been recently practiced. The cingulate is part of a pathway used when effortful control is required, such as when generating the use of a noun for the first time.

Another view involves the distinction between monitoring and control (Botvinick *et al.* 2001). According to this view, the cingulate function is largely on the sensory end. It is involved in the detection of potential cross-talk or confusions between separable modules. Lateral areas of the cortex are then activated to provide control operations, which might include increased activation or inhibition that will eliminate the confusion between modules.

These three views all propose to break down executive attention into components and preserve the general approach to brain activity in which specific anatomical locations are related to mental operations (Posner and Raichle 1994). Other ways of organizing frontal function have also been proposed (Frith 2000; Petrides 2000). Future research will have to determine whether high-level attention can be identified with distributed components, each of which carries out a specific function, or whether the frontal lobe activity follows new principles of brain organization.

25.5.3 Self-regulation

A somewhat overlapping perspective on these mechanisms involves the study of control of emotional and cognitive responses to stimulation. There is substantial animal work suggesting that the frontal midline area connects with and regulates brain systems related to emotion (Davidson *et al.* 2000). While much of the fMRI work has involved negative emotion (Whalen *et al.* 1998), there is now evidence the efforts to control positive emotional states, such as the erotic responses to explicitly sexual movies, activates a cingulate generator (Beauregard *et al.* 2001).

The idea that the cingulate region is involved in self-regulation also comes from our studies of the development of the executive attention network. We found strong evidence of development of this network in 2- to 5-year-olds, using a spatial conflict task which induces conflict between identity and location, two of the earliest developing visual system operations (Gerardi-Caulton 2000; Rothbart *et al.* in press). In adults the spatial conflict task activates the cingulate in an area similar to the color Stroop and flanker tasks (Fan *et al.* 2003).

As mentioned previously, our work with the child version of the ANT found that development in the executive attention network continues up to age 7 but not after that age. In order to test this with a slightly harder task, we compared 10-year-olds and adults in the adult version of the ANT and found no difference in the conflict network, although other aspects of the task continued to develop. Recent studies of adolescents seem to indicate about the same level of conflict as found in children above 7 and adults. Because a striking feature of children between 2 and 7 years of age is the increase in their ability to regulate their emotions and actions, we have thought of the network as related to self-regulation. In support of this idea, we have found significant correlations between parental reports of the ability of the child to control themselves in the practical situations of real life (effortful control) and performance in the spatial conflict task (Gerardi-Caulton 2000; Rothbart *et al.* in press). Recently we have found evidence of a similar correlation in adults between a self-report measure of effortful control and performance on the conflict network of the ANT.

Studies of the executive attention network and effortful control have both found substantial heritability (Goldsmith *et al.* 1999). Moreover, effortful control related to empathy is stronger, with children high in effortful control showing greater empathy (Rothbart *et al.* 1994). Effortful control also appears to play a role in the development of conscience. Studies of the development of conscience show it to be facilitated in children high in effortful control (Kochanska *et al.* 1996). Lesions of the medial frontal area that includes the cingulate gyrus produces the tendency toward psychopathy, both in children (Anderson *et al.* 2000) and adults (Damasio 1994). Recently, Ellis (2002) found evidence that both effortful control and conflict measured by the ANT aided in the prediction of antisocial outcomes in adolescence. These findings suggest that the executive attention network is a part of the mechanisms by which children learn to regulate their behavior in accordance with social norms.

The importance of imitation of others on our behavior and mental state is strongly emphasized in this volume in the association lecture (Rizzolatti, Chapter 1, this volume) and in the work of Hari and Nishitani in humans (Chapter 23, this volume). Their studies provide one basis for viewing the social world as part of the infants' core knowledge (Spelke, Chapter 2, this volume). Like all aspects of core knowledge, this function also must have a long developmental history in reaching adult levels. The ability to examine the mechanisms by which these influences control pro-social behavior should prove a significant contribution of brain imaging.

25.5.4 **Summary**

The difference between automatic choice and reflective 'deciding' can be explored by understanding the circumstances where high-level attentional networks are brought to bear upon control of thoughts, actions, and emotions. Studies of the executive attention system in conflict-related tasks in children, have shown the close relation between its development and parental reports of the ability of their children to control their behavior and emotions. In turn, effortful control places an important role in successful outcomes for children. These findings suggest the start of efforts to relate the results of imaging studies to central issues of fostering the growth and development of children.

25.6 **Personal perspective**

It has been 36 years since the first Attention and Performance meeting in The Netherlands in the summer of 1966. I sat in that audience in hopes that our collective work might make a contribution to understanding human nature, to bettering the lot of people, and to explaining the human brain. For me, out of North America for the first time, it was a stunning experience. The presentations represented the work of people who shared the goal of understanding the functions and limitations of the human nervous system. However, the range of measures was somewhat limited. Large leaps were required for us to pass from the number of items reported following a cue, to the inference of an iconic trace present in the visual system, or from delayed reaction time to the second of two stimuli to inferences about a central refractory mechanism.

Today, 36 years later, we have a powerful array of new methods. I believe that a large number of difficult inferences from RT and error methods have been validated by these new techniques. Thirty years ago we speculated that subjects could react to the visual form of input based on attention to a visual code. Now we know they can do so, because attention amplifies neural activity within narrowly defined visual system areas. Twenty years ago we were defining the objective characteristics of visual images created internally without stimulation, while today we can observe changes in visual cortex due to such computations. These findings ensure a permanent role for both behavioral studies and imaging studies in understanding links between brain and mind.

Donald Broadbent was not at the first meeting, but his research represented another important value strong in the field at that time. He supported the application of human psychological studies to real-world problems. In this regard human performance as a field had been greatly influenced by World War II applications. It seems to me that cognitive neuroscience is now in a similar position to contribute to, and gain from, studies applying our new knowledge of how neural networks are built under the influence of genetics and experience to issues such as child socialization, psychopathology, and the design of educational systems. Imaging is not just a new tool for studying the human brain, it is an appropriate tool for understanding how our brains control our behavior.

Notes

1. This chapter was written at the invitation of the editors to discuss material presented at the meeting in relation to future developments in brain imaging. Research reported here was supported by a James S. McDonnell 21st Century grant.

References

Abdullaev, Y.G. and Posner, M.I. (1998). Event-related brain potential imaging of semantic encoding during processing single words. *Neuroimage*, **7**, 1–13.

Anderson, S.W., Damasio, H., Tranel, D., and Damasio, A.R. (2000). Long-term sequelae of prefrontal cortex damage acquired in early childhood. *Developmental Neuropsychology*, **18** (3), 281–96.

Beauregard, M., Levesque, J., and Bourgouin, P. (2001). Neural correlates of the conscious self-regulation of emotion. *Journal of Neuroscience*, **21**, RC165.

Blair, R.J.R., Morris, J.S., Frith, C.D., Perrett, D.I., and Dolan, R.J. (1999). Dissociable neural responses to facial expression of sadness and anger. *Brain*, **122**, 883–93.

Botvinick, M.M., Braver, T.S., Barch, D.M., Carter, C.S., and Cohen, J.D. (2001). Conflict monitoring and cognitive control. *Psychological Review*, **108**, 624–52.

Broadbent, D.E. (1973). *Decision and Stress*. Academic Press, London.

Broadbent, D.E. (1991). A word before leaving. In D.E. Meyer and S. Kornblum (Eds.) *Attention and Performance XIV* pp. 863–79. Bradford Books MIT Press, Cambridge MA.

Bush, G., Whalen, P.J., Rosen, B.R., Jenike, M.A., McInerey, S.C., and Rauch, S.L. (1998). The counting Stroop: An interference task specialized for functional neuroimaging—validation study with functional MRI. *Human Brain Mapping*, **6**(4), 270–82.

Bush, G., Luu, P., and Posner, M.I. (2000). Cognitive and emotional influences in the anterior cingulate cortex. *Trends in Cognitive Science*, **4/6**, 215–22.

Compton, P., Grossenbacher, P., Posner, M.I., and Tucker, D.M. (1991). A cognitive-anatomical approach to attention in lexical access. *Journal of Cognitive Neuroscience*, **3**(4), 304–12.

Dale, A.M., Liu, A.K., Fischi, B.R., Ruckner, R., Beliveau, J.W., Lewine, J.D., and Halgren, E. (2000). Dynamic statistical parameter mapping: combining fMRI and MEG for high-resolution cortical activity. *Neuron*, **26**, 55–67.

Damasio, A.R. (1994). *Descarte's Error: Emotion, Reason and the Human Brain*. G.P. Putnam, New York.

Davidson, M.C. and Marrocco, R.T. (2000). Local infusion of scopoplamine into intraparietal cortex slows cover orienting in rhesus monkeys. *Journal of Neurophysiology,* **83**, 1536–49.

Davidson, R.J., Putnam, K.M., and Larson, C.L. (2000). Dysfunction in the neural circuitry of emotion regulation – a possible prelude to violence. *Science,* **289**, 591–5.

Dehaene, S., Gurvan, L.C., Cohen, L., Poline, J.P., van de Moortele, F., and LeBihan, D. (1998a). Inferring behavior from functional brain images. *Nature Neuroscience,* **1**(7), 549–50.

Dehaene, S., Posner, M.I., and Tucker, D.M. (1994). Localization of a neural system for error detection and compensation. *Psychological Science,* **5**, 303–5.

Dehaene, S., Kerszberg, M., and Changeux, J.-P. (1998b). A neuronal model of a global workspace in effortful cognitive tasks. *Proceedings National Academy of Science USA,* **95**, 14529–34.

Desimone, R. and Duncan, J. (1995). Neural mechanisms of selective visual attention. *Annual Review of Neuroscience,* **18**, 193–222.

Duncan, J. (1995). Attention intelligence and frontal lobes. In M.S. Gazzaniga (Ed). *The Cognitive Neurosciences* pp. 721–733. MIT Press, Cambridge MA.

Duncan, J. and Owen, A.M. (2000). Common regions of the human frontal lobe recruited by diverse cognitive demands. *Trends in Neurosciences,* **23**, 475–83.

Duncan, J., Seitz, R.J., Kolodny, J., Bor, D., Herzog, H., Ahmed, A., Newell, F.N., and Emslie, H. (2000). A neural basis for general intelligence. *Science,* **289**, 457–60.

Ellis, L. (2002). Individual differences and adolescent. psychosocial development. Unpublished doctoral dissertation, University of Oregon.

Eriksen, B.A. and Eriksen, C.W. (1974). Effects of noise letters upon the identification of target letter in a nonsearch task. *Perception and Psychophysics,* **16**, 143–9.

Fan, J., McCandliss, B.D., Flombaum, J.I., and Posner, M.I. (2001a). *Imaging Attentional Networks.* Society for Neuroscience Abstracts.

Fan, J., Wu, Y., Fossella, J., and Posner, M.I. (2001b). Assessing the heritability of attentional networks. *BioMed Central Neuroscience,* **2**, 14.

Fan, J., McCandliss, B.D., Sommer, T., Raz, M., and Posner, M.I. (2002). Testing the efficiency and independence of attentional networks. *Journal of Cognitive Neuroscience,* **3**(14), 340–7.

Fan, J., Flombaum J.I., McCandliss, B.D., Thomas, K.M., and Posner, M.I. (2003a). Cognitive and brain consequences of conflict. *Neuroimage,* **18**, 42–57.

Fan, J., Fossella, J.A., Summer, T., and Posner, M.I. (2003b). Mapping the genetic variation of executive attention onto brain activity. *Proc of the Nat'l Acad of Sci USA* **100**, 7406–11.

Fodor. J. (1983). *Modularity of Mind.* Cambridge, MA, MIT Press, Bradford Books.

Fossella, J., Posner, M.I., Fan, J., Swanson, J.M., and Pfaff, D.M. (2002a). Attentional phenotypes for the analysis of higher mental function. *The Scientific World Journal,* **2**, 217–23.

Fossella, J., Sommer, T., Fan, J., Wu , Y., Swanson, J.M., Pfaff, D.W., and Posner, M.I. (2002b). Assessing the molecular genetics of attention networks. *BMC Neuroscience,* **3**, 14.

Frith, C. (2000). The role of dosolateral prefrontal cortex in the selection of action as revealed by functional imaging. In S. Monsell and J. Driver eds. *Control of Cognitive Processes Attention and Performance XVIII.* Cambridge MA, MIT Press, 549–66.

Gerardi-Caulton, G. (2000). Sensitivity to spatial conflict and the development of self-regulation in children 24–36 months of age. *Developmental Science,* **3/4**, 397–404.

Gevins, A. (1996). Electrophysiological imaging of brain function. In A.W. Toga and J.C. Mazziotta eds. *Brain Mapping: The Methods.* New York, Academic Press, 259–73.

Goldsmith, H.H., Lemery, K.S., Buss, K.A., and Campos, J.J. (1999). Genetic analysis of focal aspects of infant temperament. *Developmental Psychology,* **35**, 972–85.

Greenwood, P.M., Sunderland, T., Friz, J.L., and Parasuraman, R. (2000). Genetics of visual attention: Selective deficits in healthy adults carriers of the epsilon 4 allele of the apolipoprotein E gene. *Proceedings of the National Academy of Sciences USA*, **97**, 11661–6.

Kaufman, T., Theoret, H., and Pascual-Leone, A. (2002). Braille character discrimination in blindfolded human subjects. *Neuroreport*, **13**, 571–4.

Kochanska, G., Murray, K., Jacques, T.Y., Koenig, A.L., and Vandegeest, K.A. (1996). Inhibitory control in young children and its role in emerging internalization. *Child Development*, **67**, 490–507.

Lashley, K.S. (1931). Mass action in cerebral function. *Science*, **73**, 245–54.

Liu, A.K., Dales, A.M., and Belliveau, J.W. (2002). Monte carlo simulation studies of EEG and MEG localization accuracy. *Human Brain Mapping*, **16**, 47–62.

Mangun, G.R., Hillyard, S.A., and Luck S.J. (1990). Electrocortical substrates of visual selective attention In D.E. Meyer and S. Kornblum eds. *Attention and Performance XIV* pp. 219–44. Academic Press, New York.

Marrocco, R.T. and Davidson, M.C. (1998). Neurochemistry of attention. In R. Parasuraman (ed.) *The Attention Brain* pp. 35–50. MIT Press, Cambridge MA.

Martinez, A., DiRusso, F., Anllo-Vento, L. Sereno, M ., Buxton, R., and Hillyard, S. (2001). Putting spatial attention on the map: timing and localization of stimulus selection processing in striate and extrastriate visual areas. *Vision Reseach*, **41**, 1437–57.

McCandliss, B.D., Posner, M.I., and Givon, T. (1997). Brain plasticity in learning visual words. *Cognitive Psychology*, **33**, 88–110.

McCandliss, B.D., Cohen, L., and Dehaene, S. (2003). The visual word form area: expertise for reading in the fusiform gyrus. *Trends in Cognitive Science*, **7**(7), 293–9.

MacDonald, A.W., Cohen, J.D., Stenger, V.A., and Carter, C.S. (2000). Dissociating the role of the dorsolateral prefrontal and anterior cingulate cortex in cognitive control. *Science*, **288**, 1835–8.

Monsell, S. and Driver, J. (2000). *Control of Cognitive Processes: Attention and Performance XVIII.* MIT Press, Cambridge MA.

Naito, E., Kinomura, S., Geyer, S., Kawashima, R., Roland, P.E., and Zilles, K. (2000). Fast reaction to different sensory modalities activates common fields in the motor areas, but the anterior cingulate cortex is involved in the speed of reaction. *Journal of Neurophysiology*, **83**(3), 1701–9.

Neville, H.J. and Bavelier, D. (2001). Effects of auditory and visual deprivation on human brain development. *Clinical Neuroscience Research*, **1**, 248–57.

Nikolaev, A.R., Ivanitsky, G.A., Ivanitsky, A.M., Abdullaev, Y.G., and Posner, M.I. (2001). Short-term correlation between frontal and Wernicke's areas in word association. *Neuroscience Letters*, **298**, 107–10.

Olson, R.K., Datta, H., Gayan, J., and DeFries (1999). A behavioral-genetic analysis of reading disabilities and component processes. In R. Klein and P. McMullen (eds.) *Converging Methods for Understanding Reading and Dyslexia* pp. 133–51. MIT Press, Cambridge MA.

Pascal-Leone, A. and Hamilton, R. (2001). The metamodal organization of the brain. *Vision: from neurons to cognition progress in brain research*, **134**, 427–45.

Paulesu, E., Demonet, J.-F., Fazio, F. *et al.* (2001). Dyslexia–cultural diversity and biological unity. *Science*, **291**, 2165–7.

Petersen, S.E., Fox, P.T., Posner, M.I., Mintun, M., and Raichle, M.E. (1987). Positron emission tomographic studies of the cortical anatomy of single word processing. *Nature*, **331**, 585–9.

Petrides, M. (2000). Middorsolateral and Midventrolateral prefrontal cortex: Two levels of executive control for the processing of mnemonic information. In S. Monsell and J. Driver eds. *Control of Cognitive Processes Attention and Performance XVIII* pp. 535–48. MIT Press, Cambridge MA.

Piazza, M., Giacomini, E., LeBihan, D., and Dehaene, S. (2003). Single trial detection of the parietal attention system during subitizing and counting. *Proceedings of the Royal Society of London Series B-Biological Science*, **270**(1521), 1237–45.

Posner, M.I. (1978). *Chronometric Explorations of Mind*. Lawrence Erlbaum Associates, Hillsdale NJ.

Posner, M.I. (1980). Orienting of attention. The 7th Sir F.C. Bartlett Lecture. *Quarterly Journal of Experimental Psychology*, **32**, 3–25.

Posner, M.I. and Boies, S.J. (1972). Components of attention. *Psychological Review*, **78**, 391–408.

Posner, M.I. and Fan, J. (2003: in press). Attention as an organ system. In J. Pomerantz ed. *Neurobiology of Perception and Communication: From Synapse to Society, the IVth De Lange Conference*. Cambridge University Press, Cambridge.

Posner, M.I. and Gilbert, C.D. (1999). Attention and primary visual cortex: *Proceedings of the National Academy of Sciences, USA*, **96/6**, 2585–7.

Posner, M.I. and McCandliss, B.D. (1999). Brain circuitry during reading. In R. Klein and P. McMullen, (Eds.) *Converging Methods for Understanding Reading and Dyslexia* pp. 305–26. MIT Press, Cambridge MA.

Posner, M.I. and Pavese, A. (1998). Anatomy of word and sentence meaning. *Proceedings of the National Academy of Sciences, USA*, **95**, 899–905.

Posner, M.I. and Raichle, M.E. (1994). *Images of Mind*. Scientific American Books, New York NY.

Posner, M.I., Inhoff, A., Friedrich, F.J., and Cohen, A. (1987). Isolating attentional systems: A cognitive–anatomical analysis. *Psychobiology*, **15**, 107–21.

Posner, M.I., Rothbart, M.K., Farah, M., and Bruer, J. (2001). Human brain development. *Developmental Science*, **4/3**, 253–384.

Price, C.J. and Devlin, J.T. (2003). The myth of the visual word form area. *Neuroimage*, **19**, 473–81.

Puce, A., Allison, T., and McCarthy, G. (1999). Electrophysiological studies of human face perception. III: Effects of top-down processing on face-specific potentials. *Cerebral Cortex*, **9**(5), 445–58.

Raichle, M.E., Fiez, J.A., Videen, T.O., McCleod, A.M.K., Pardo, J.V., Fox, P.T., and Petersen, S.E. (1994). Practice-related changes in the human brain: functional anatomy during nonmotor learning. *Cerebral Cortex*, **4**, 8–26.

Rees, G., Friston, K., and Koch, C. (2000). A direct quantitative relationship between functional properties of human and macque V5. *Nature Neuroscience*, **3/7**, 717–23.

Rothbart, M.K., Ahadi, S.A., and Hershey, K. (1994). Temperament and social behavior in children. *Merrill–Palmer Quarterly*, **40**, 21–39.

Rothbart, M.K., Ellis, L., Reuda, M.R., and Posner, M.I. (in press). Developing mechanisms of temperamental self regulation. *J. of Personality*.

Rueda, M.R., Fan, J., McCandliss, B.D., Haprin, J., Gruber, D., Pappert, L. and Posner, M.I. (2002). Assaying the development of attentional networks in six to ten-year-old children. *Cognitive Neuroscience Society Abstracts*, p. 21.

Schlaggar, B.L., Brown, T.T., Lugar, H.M., Visscher, K.M., Miezin, F.M., and Petersen, S.E. (2002). Functional neuroanatomical differences between adults and school-age children in processing single words. *Science*, **296**, 1476–9.

Shaywitz, B.A., Shaywitz, S.E., Pugh, K.R., Mencl, W.E., Fulbright, R.K., Skudlarski, P., Constable, R.T., Marchione, K.E., Fletcher, J.M., Lyon, G.R., and Gord, J.C. (2002). Disruption of posterior brain systems for reaching in children with developmental dyslexia. *Biological Psychiatry*, **52**, 101–10.

Sieroff, E., Pollatsek, A., and Posner, M.I. (1988). Recognition of visual letter strings following injury to the posterior visual spatial attention system. *Cognitive Neuropsychology*, **5**, 427–49.

Smith, E.E., Jonides, J., Marshuetz, C., and Koeppe, R.A. (1998). Components of verbal working memory. Evidence from neuroimaging. *Proceedings of the National Academy of Sciences, USA*, **95**, 876–82.

Swanson, J.M., Floodman, P., Kennedy, J.M., Spence, A.M., Moyzes, M., Schruck, S., Murias, M., Moriarty, J., Barr, C., Smith, M., and Posner, M.I. (2000). Dopamine genes and ADHD. *Neuroscience and Biobehavioral Reviews*, **24**(1), 21–5.

Taub, E., Uswatte, G., and Elbert, T. (2002). New treatments in neurorehabilitation founded on basic research. *Nature Reviews Neuroscience*, **3**(3), 228–36.

Uttal, W.R. (2001). *The New Phrenology*. Bradford Books MIT Press, Cambridge MA.

Whalen, P.J., Bush, G., McNally, R.J., Wilhelm, S., McInerney, S.C., Jenike, M.A., and Rauch, S.L. (1998). The emotional counting Stroop paradigm: A functional magnetic resonance imaging probe of the anterior cingulate affective division. *Biological Psychiatry*, **44**(12), 1219–28.

Author Index

Aakalu, G. 75, 76
Abdi, H. 94
Abdullaev, Y.G. 512, 513
Abeles, M. 106, 107
Abrams, R.I. 205, 212, 214
Abu-Mostafa, Y.S. 145
Acampora, D. 64
Acosta, Y. 333, 335, 336, 337
Acuna, C. 20, 420, 421, 425, 433
Adelson, E.H. 110
Adini, Y. 145, 151
Adleman, N.E. 403
Aertsen, A. 106, 109
Aggleton, J.P. 22
Agid, Y. 13
Agnetta, B. 40
Aguirre, G.K. 84, 170, 349, 352, 418
Ahadi, S.A. 522
Ahmed, A. 512, 520
Aihara, T. 234
Akbudak, E. 348, 353, 448, 453
Alain, C. 329
Alais, D. 107
Alborzian, S. 373, 374, 375, 376
Albright, T.D. 110, 241, 252
Alexander, G.E. 401, 491
Alexinsky, T. 404
Alkadhi, H. 425, 468
Allison, T. 48, 84, 95, 170, 198, 226, 472, 508
Allman, J.M. 61
Allport, A. 349
Allport, D.A. 449
Alonso, J.-M. 107
Amador, N. 406
Amaral, D.G. 11
Amedi, A. 198, 226, 236
Amir, Y. 113, 197
Amit, D.J. 150, 156
Amodei, L. 458
Amunts, K. 13
Andersen, P.M. 402
Andersen, R.A. 346, 419, 483, 484, 485, 486, 487, 491, 492
Anderson, A.W. 84, 159
Anderson, C.H. 162
Anderson, J.M. 236
Anderson, J.R. 349, 352, 449
Anderson, K.C. 401
Anderson, S. 65
Anderson, S.W. 522
Andrews, J. 129
Anger, C. 402, 404

Anllo-Vento, L. 270, 322, 323, 337, 382, 383, 384, 385, 386, 387, 511
Arbib, M.A. 13, 22, 376, 422, 431, 467, 491
Archibald, Y.M. 420, 489
Arieli, A. 243
Armstrong, I.T. 403
Arndt, P.A. 403
Arnell, K.M. 323, 374
Arrington, C.M. 323, 338, 357
Asgari, H.S. 447
Asgari, M. 170, 226
Ashburner, J. 449
Ashby, F.G. 391, 392
Ashe, J. 487
Asrress, K.N. 403
Assad, J.A. 348
Astafiev, S.V. 358
Aston-Jones, G. 404
Aubert-Vazquez, E. 329
Augath, M. 116, 186, 418, 434
Avidan, G. 95, 159, 160, 170, 189, 198, 199, 201, 214, 418, 420
Avikainen, S. 21, 467, 468, 469, 474, 475, 476

Bach, M. 328
Bachevalier, J. 317
Bagarinao, E. 453
Bailey, P. 13
Baillargeon, R. 38
Bair, W. 107
Baizer, J.S. 235
Baker, C.I. 10
Baker, J.T. 485, 486, 487
Bakircioglu, M. 434
Baldwin, D.A. 40
Balint, R. 421
Ball, K. 346, 353
Bandettini, P.A. 457
Bar, M. 210
Barbe, M.F. 65, 225
Barcelo, F. 272, 275
Barch, D.M. 357, 358, 405, 406, 512, 516, 521
Barker, G.J. 448
Barkovich, A.J. 447
Barlow, H.B. 392
Barnes, C.L. 235
Barnett, A. 448
Baron, G. 59
Barr, C. 517, 518
Barrett, T.W. 226
Barth, H.C. 43
Bashore, T.R. 127, 128, 135, 399

Basser, P.J. 448, 449, 453
Basso, M.A. 397
Bateson, P.G. 24
Batista, A.P. 346, 484
Battaglia-Mayer, A. 483, 484
Bauer, R. 110
Baumgartner, G. 193
Bavelier, D. 515
Baylis, G.C. 171, 235, 252, 322, 371, 489
Baylis, L.L. 489
Beachy, P.A. 65
Beauchamp, M.S. 198
Beauregard, M. 522
Beck, D.M. 221, 357
Becklen, R. 371
Behar, K.L. 221
Behrmann, M. 84, 95, 189, 199, 322, 330, 332, 418
Bekkering, H. 13, 19, 20, 21, 40, 474, 475, 484
Belger, A. 226
Belliveau, J.W. 92, 304, 417, 418, 449, 509, 510
Ben Bashat, D. 170, 197
Ben Hamed, S. 483, 486, 495
Bender, D.B. 241
Benson, P.J. 10, 11, 241, 252, 431
Benson, R. 158, 159, 197, 198, 241
Benson, R.R. 417, 420
Bental, E. 226
Benton, A.L. 489
Benuzzi, F. 14
Berger, R.L. 136
Berglan, L.R. 325
Berman, R.A. 352
Bernasconi, C. 444, 445
Bernstein, P.S. 406
Bertenthal, B.I. 37, 46
Bertrand, O. 115, 119
Bettinardi, V. 13
Bettucci, D. 431
Bevan, R. 10, 11, 241, 252
Bhatt, R.S. 35
Bichot, N.P. 396, 397, 398, 399, 400, 401
Biederman, I. 149, 161, 162
Binkofski, F. 13, 14, 15, 421, 422, 433
Bisazza, A. 32
Biscay, R. 444
Bishop, K.M. 65
Bittner, R.A. 115, 117
Blake, R. 107, 117
Blakemore, C. 328, 338
Blakemore, S.-J. 19
Blandin, E. 211
Blanz, V. 94
Blaser, E. 327, 338, 371
Bliss, D. 226
Block, E. 213
Bloom, P. 35, 37, 39
Blumenfeld, H. 221
Bobes, M.A. 328, 329, 330, 333, 335, 336, 337
Bock, G. 79

Bodurka, J. 457
Boies, S.J. 515
Boise, L.H. 62
Boison, D. 489, 493
Bokura, H. 403
Boller, F. 25, 489
Bonatti, L. 39
Boncinelli, E. 64
Bonhoeffer, T. 248
Bor, D. 512, 520
Borenstein, E. 163
Borghesani, P.R. 76
Born, R.T. 417
Bosch, J. 444
Bosch-Bayard, J. 329
Boser, B. 151
Bottou, L. 151
Botvinick, M.M. 358, 405, 406, 512, 516, 521
Boulanouar, K. 442
Bourgouin, P. 522
Boussaoud, D. 400, 483
Bowers, J.S. 205
Boyes-Braem, P. 174
Boynton, G.M. 270, 322, 452
Boysen, S.T. 45
Bradley, D.C. 419
Brady, T.J. 92, 158, 159, 197, 198, 241, 304, 417, 418, 420
Brandt, S.A. 48
Brannon, E.M. 42, 43
Brass, M. 13, 20, 21
Braun, J. 117, 157, 170, 370
Braver, T.S. 357, 358, 405, 406, 512, 516, 521
Brecht, M. 107
Brefczynski, J.A. 270, 322
Bremmer, F. 483, 486, 495
Bressler, S.L. 444
Britten, K.H. 395, 397
Broadbent, D.E. 268, 269, 510, 511
Broccoli, V. 64
Brody, C.D. 394
Bronchti, G. 74
Brotchie, P.R. 487
Brothers, L. 11
Brown, T.T. 513
Bruce, C. 235, 241, 252, 398
Bruer, J. 518
Brunia, C.H. 403
Brunot, A. 151
Brunswick, N. 48, 209
Brysbaert, M. 213
Buccino, G. 13, 14, 15
Büchel, C. 226, 227, 351, 442
Buchel, G.M. 440
Buckner, R.L. 214
Buergel, U. 13
Buettner, A. 425, 468
Buhl, E.H. 119
Bulfone, A. 65
Bülthoff, H.H. 157, 161, 163, 198, 489

Bundesen, C. 364, 398
Buneman, O.P. 150
Buneo, C.A. 484
Buonocore, M.H. 117, 270, 271, 273
Buracas, G. 119
Burchert, W. 270, 323
Burgess, N. 31
Burton, H. 21, 448, 453
Bush, G. 520, 521, 522
Bushnell, M.C. 270, 367
Buss, K.A. 522
Butler, A.B. 66, 434
Butterworth, G.E. 40
Buxbaum, L.J. 490, 493, 494, 495
Buxton, R.B. 270, 303, 306, 323, 382, 383, 384, 385, 387, 511
Buzsaki, G. 109
Byrne, R.W. 3, 4, 19, 21

Cabel, D.W.J. 403
Cabeza, R. 226, 233, 418
Caessens, B. 213
Calford, M. 21
Call, J. 19, 40
Camarda, R. 25
Camberini, M. 484
Cameron, E.L. 302
Caminiti, R. 400, 483, 484
Campos, J.J. 522
Canessa, N. 14
Cantalupo, C. 376
Cappa, S.F. 209
Caramazza, A. 30
Cardew, G. 79
Carey, D.P. 420, 431, 466, 482, 483, 484, 486, 489, 493
Carey, S. 35, 36, 37, 38, 39, 40, 41, 43, 46, 47
Carmichael, S.T. 11
Carpenter, R.H.S. 403, 404
Carpenter, T.A. 236
Carr, T.H. 212, 323, 338, 357
Carrasco, M. 302, 310
Carter, A.R. 76
Carter, C.S. 349, 352, 358, 405, 406, 449, 510, 512, 516, 521
Casey, B.J. 403
Castelo-Branco, M. 107, 110
Catani, M. 448, 458
Catania, K.C. 69, 70
Cavada, C. 317
Cavanagh, P. 48, 398
Cave, K.R. 372
Celebrini, S. 395
Chack, V.P. 448
Chaminade, T. 472
Chang, H.S. 163, 367
Chang, K.H. 402
Changeux, J.P. 220, 510, 521
Chanoine, V. 209

Chao, L.L. 84, 170, 433
Chawla, D. 151, 271
Chelazzi, J.L. 399
Chelazzi, L. 303, 307, 309, 316, 364, 365, 367
Chen, C. 79
Chen, J.H. 449
Chen, N.H. 401
Chen, R. 467
Chen, W.G. 75
Cheng, H.C. 449
Cheng, K. 31, 32, 33, 35, 157, 170, 201, 243
Chenn, A. 63
Chiang, C. 65
Chiang, W.C. 35
Chistiakova, M. 114
Chitty, A.J. 10, 11, 472
Chochon, F. 207, 208
Chorev, Z. 349
Chou, K.L. 398
Christ, S. 405
Chrobak, J.J. 109
Chun, M.M. 48, 84, 158, 170, 183, 197, 226, 230, 323, 339, 417, 418
Church, R.M. 42
Ciccarelli, O. 448
Clarey, J. 21
Clark, D.A. 61
Clark, V.P. 84, 352, 387
Clarke, K. 316
Cobo, A. 322, 326, 327, 328, 338, 371, 372, 373, 376
Cohen, A. 509
Cohen, J.D. 358, 405, 406, 510, 512, 516, 521
Cohen, K. 519
Cohen, L. 13, 45, 206, 207, 208, 221, 419, 434, 467, 491
Cohen, L.A. 497
Cohen, Y. 322, 447
Cohen, Y.E. 484
Colby, C.L. 282, 348, 398, 419
Cole, J. 474
Cole, M. 489
Coleman, R.J. 493
Coles, M.G.H. 127, 399, 401, 402, 403, 404, 405, 406
Collett, M. 30, 46
Collett, T.S. 30, 46
Colombo, M. 234
Colonius, H. 392, 397, 403
Compton, P. 513
Condry, K. 46
Constable, R.T. 209, 512
Conturo, T.E. 348, 353, 448, 453
Conway, B. 467
Cook, E.P. 306
Cooper, E.E. 162
Cooper, H.M. 74
Corbetta, M. 270, 272, 273, 274, 292, 293, 294, 303, 314, 346, 348, 353, 354, 355, 357, 358, 398
Corden, J.L. 65
Corfas, G. 76
Cornette, L. 352

Cortes, C. 151
Corwell, B. 467
Coslett, H.B. 490, 493, 494, 495
Cosmides, L. 46
Cossu, G. 209
Costes, N. 13, 20
Court, J.E. 489
Courtney, S.M. 84, 91
Cowan, M.J. 358
Cowan, W.B. 403
Cowan, W.M. 80
Craft, S. 259
Crain, B.J. 448
Crammond, D.J. 402, 403
Crawley, A.P. 355, 359
Crick, F. 306
Crossley, P.H. 65
Csibra, G. 259, 261
Culham, J.C. 48, 330, 398, 418, 419, 423, 426, 428, 432, 434, 482
Cull, T.S. 448, 453
Cumming, B.G. 395

Dafny, N. 226
Dagenbach, D. 212
Dale, A.M. 48, 91, 176, 197, 198, 200, 210, 270, 303, 382, 386, 417, 418, 442, 509, 510
Dalva, M.B. 75
Damasio, A.R. 170, 225, 489, 522
Damasio, H. 170, 225, 522
Damian, M.F. 213, 214
Danckert, S.L. 426, 428
Darian-Smith, I. 393
Darling, W. 493
Darlington, R.B. 61
Datta, H. 517
Dautenhahn, K. 26
Davatzikos, C. 458
Davidson, D. 404
Davidson, D.L.W. 431
Davidson, M.C. 516
Davidson, R.J. 522
Davies, D.R. 52
Davis, C. 205, 207
Davis, G. 372
Davis, K.D. 355, 359, 476
d'Avossa, G. 346, 358, 359
De Ajuriaguerra, J. 489
de Gelder, B. 276
de Haan, M. 259, 260
De Jong, R. 403
De Recondo, J. 490
De Renzi, E. 489, 490
De Weerd, P. 116, 230, 234, 236, 271, 272, 273, 291, 303, 304, 306, 307, 308, 309, 310, 315, 370
Debowy, D.J. 421, 429
Decety, J. 13, 19, 20, 422, 472, 489
Deecke, L. 402

Deffenbacher, K.A. 94
Dehaene, S. 13, 42, 43, 44, 45, 48, 132, 135, 138, 205, 206, 207, 208, 209, 210, 211, 212, 213, 220, 221, 405, 419, 434, 491, 510, 519, 520, 521
Dehaene-Lambertz, G. 207, 210, 211
Deiber, M.P. 13, 226, 402
Della Sala, D. 482, 484, 493
Della Sala, S. 493
Demonet, J.F. 209, 442, 512
Denker, J.S. 151
Dennett, D.C. 393
DeRecondo, J. 489
Desimone, R. 114, 117, 230, 234, 235, 236, 241, 252, 270, 271, 272, 273, 274, 291, 303, 304, 306, 307, 308, 309, 310, 311, 313, 315, 316, 322, 330, 348, 364, 365, 367, 368, 369, 370, 374, 385, 399, 509, 520
Desmurget, M. 483, 489, 490, 491
DeSoto, M.C. 401
DeSouza, J.F.X. 418, 423, 426, 485, 486, 491, 492
D'Esposito, M. 84, 170, 349, 352
Dettmers, C. 351
Deutsch, D. 268
Deutsch, J.A. 268
Deweese, M. 119
DeYoe, E.A. 270, 322
Di Chiro, G. 448
Di Pellegrino, G. 3, 464, 466
Di Russo, F. 323, 337, 382, 383, 384, 385, 387
Di Salle, F. 442
Di Salle, R. 117
Diamond, A. 258
Dias, E.C. 398
Dickinson, A. 406
Diener, H.C. 402
Dierks, T. 117, 357, 442
Diesmann, M. 106, 109, 402
DiGirolamo, G.J. 405
Dijkerman, H.-C. 489
Dill, M. 162
Ding, L. 62
Ding, M. 444
Dinse, H.R. 58
DiRusso, F. 511
Disbrow, E. 21
Dobmeyer, S. 270, 303
Dodd, J.V. 395
Dohle, C. 421, 422, 433
Dolan, R.J. 214, 215, 221, 227, 236
Dold, G. 226
Dolmetsch, R.E. 75
Donald, M. 23
Donchin, E. 127, 401, 404
Donders, E.C. 131, 398
Donnelly, N. 107
Donoghue, J.P. 485, 486, 487
Donovan, T. 236
Dorris, M.C. 397, 402, 404
Dosher, B.A. 302
Dostrovsky, J.O. 476

Dove, A. 449
Downar, J. 355, 359
Downing, P.E. 30, 48, 84, 136, 137, 159, 170, 275, 277, 330, 371, 417
Doya, K. 402
Driver, J. 37, 40, 171, 214, 226, 257, 268, 270, 271, 272, 273, 274, 275, 276, 277, 278, 279, 280, 282, 283, 284, 285, 286, 287, 288, 289, 290, 292, 294, 316, 322, 338, 371, 372, 520
Driver, M. 279
Drucker, H. 151
Drury, H.A. 434
D'Sa-Eipper, C. 62
Dubeau, M.-C. 21
Dubowitz, D.J. 270, 303, 306, 323
Dudchenko, P.A. 32
Duhamel, J.R. 282, 348, 419, 483, 486, 495
Dukelow, S.P. 330, 418, 485, 486, 491, 492
Duncan, J. 268, 273, 274, 275, 277, 302, 303, 309, 322, 325, 326, 337, 352, 364, 367, 371, 373, 385, 399, 507, 509, 512, 520
Dunn, C.A. 72
Dupont, P. 352
Durston, S. 403
Duty, T.L. 357

Eagleson, K.L. 65, 225
Eason, R. 270
Echalier, J.F. 491
Eckhorn, R. 110
Edelman, S. 159, 160, 161, 170, 180, 188, 197, 198, 214, 418, 420
Egeth, H. 323, 325
Egly, R. 322, 357, 371
Ehrsson, H.H. 13, 14, 422
Eimas, P.D. 35, 37
Eimer, M. 211, 271, 275, 277, 278, 279, 280, 281, 285, 287, 288, 289, 290, 292, 401, 402
Elbert, T. 520
Elizondo, M.I. 304, 307, 308, 310
Ellerman, J.M. 487
Elliffe, M.C.M. 150
Ellis, L. 522
Ellison, A. 357
Elston, G. 21
Emslie, H. 512, 520
Engel, A.K. 107, 108, 109, 110, 111, 112, 113, 114, 115, 117
Engel, S.A. 92, 452
Enns, J.T. 372
Epstein, C.M. 226
Epstein, R. 30, 48, 84, 136, 137, 158, 170, 191, 198, 417
Erb, M. 198
Eriksen, B.A. 515
Eriksen, C.W. 322, 399, 401, 404, 515
Erkelens, C.J. 400
Ermentrout, B. 119
Eskandar, E.N. 348

Esposito, F. 117, 442
Evarts, E.V. 392
Everling, S. 496
Eysel, U.T. 248

Fabiani, M. 401
Fadiga, L. 5, 6, 7, 10, 11, 13, 14, 15, 425, 433, 464, 466, 467, 475, 483
Fagergren, A. 13, 14, 422
Fagg, A.H. 422, 491
Fahle, M. 107, 162
Faillenot, I. 422
Falkenstein, M. 403, 405
Fallah, M. 275, 376, 398
Fan, J. 515, 516, 517, 518, 520, 522
Farah, M. 418, 518
Farah, M.F. 322, 338
Farah, M.J. 160, 170, 371
Farmer, S. 467
Farroni, T. 261
Fattori, P. 484
Fazio, E. 209
Fazio, F. 13, 512
Fechner, G.T. 391
Fecteau, J.H. 486
Feig, S.L. 306
Feldman, J. 30
Feldman, S. 226
Felleman, D.J. 164, 197
Fencsik, D.E. 406
Fernandes, D.M. 42
Feron, J. 42
Ferraina, S. 395, 483, 484
Ferrand, I. 205
Ferraresi, P. 483
Ferreira, F. 326
Ferrera, V.P. 352
Ferrier, D. 482
Field, D.J. 146
Fiez, J.A. 215, 236, 521
Fink, G.R. 14, 15, 483
Finkbeiner, S. 75
Finlay, B.L. 61
Fischer, C. 115
Fischl, B. 176, 210, 509, 510
Fisk, J.D. 484, 486, 496, 497
Fitzgerald, P.J. 114
Flanagan, J.R. 486, 490
Flavell, R.A. 62
Fletcher, E.M. 323
Fletcher, J.M. 209, 512
Fletcher, P.C. 48, 236
Flombaum, J.I. 516, 517, 520, 522
Floodman, P. 517, 518
Fodiak, P. 248
Fodor, J. 31, 46, 507, 511
Fogassi, L. 5, 6, 7, 9, 10, 11, 12, 14, 15, 22, 40, 425, 433, 464, 466, 467, 475, 483
Foldiak, P. 162

Folk, C.L, 352
Ford, J.M. 403
Ford, K. 496
Formisano, E. 117, 357, 442, 447, 449, 453, 457
Forss, N. 21, 467, 468, 474, 476
Forssberg, H. 13, 14, 422
Forster, B.B. 357
Forster, K.I. 205, 207
Fossella, J. 515, 516, 517, 518, 522
Fourneret, P. 474
Fox, P.T. 131, 215, 236, 512, 521
Foxe, J.J. 385
Frackowiak, R.S. 13, 84, 91, 226, 228, 271, 272, 273
Fragaszy, D. 19
Frahm, H. 61
Frank, L.R. 270, 303, 306, 323
Fredericksen, K. 458
Freedman, D.J. 42
Freeman, E. 271
Freiwald, W. 109, 326, 444
Freund, H.J. 13, 14, 23, 421, 422, 433, 488
Friedman, H.S. 171
Friedman-Hill, S. 110, 372
Friedrich, F.J. 357, 360, 509
Frien, A. 110
Fries, P. 107, 110, 111, 112, 113, 114, 117
Friston, K. 84, 91, 151, 226, 227, 228, 271, 392, 407, 440, 442, 448, 520
Frith, C.D. 48, 209, 221, 226, 270, 271, 272, 273, 282, 283, 284, 285, 286, 291, 292, 294, 303, 311, 316, 357, 521
Frith, U. 48, 209
Friz, J.L. 516, 517
Frot, E. 39
Frotscher, M. 107
Fuchs, A.F. 401
Fujii, S. 62
Fujita, I. 157, 170, 243, 247
Fujita, T. 247
Fukada, H. 488
Fukada, Y. 241
Fukuchi-Shimogori, T. 65, 71
Fukuda, H. 422
Fukui, M.M. 303, 304, 305
Fulbright, R.K. 209, 512
Furey, M.L. 83, 85, 87, 89, 91, 159, 170, 188, 189
Furuta, Y. 65
Fuster, J.M. 367

Gabrieli, J.D.E. 312
Gadian, D.G. 449
Gaffan, D. 317
Galef, B.G. 47
Gallagher, H.L. 48
Gallant, J.I. 157, 171, 310
Gallese, V. 5, 6, 7, 9, 10, 11, 12, 14, 15, 19, 22, 40, 421, 425, 428, 433, 464, 466, 474, 475, 483
Galletti, C. 484

Gallistel, C.R. 30, 42, 43, 46
Gandhi, S.P. 270, 322
Garasto, M.R. 483
Gardner, E.P. 421, 429
Gascia, P. 42
Gatenby, C. 209
Gati, J.S. 170, 234, 423, 426, 442, 485, 486, 491, 492
Gattis, M. 21, 474, 475
Gauthier, I. 30, 84, 159, 161, 183, 187, 258, 259, 418
Gaveau, V. 491
Gayan, J. 517
Gaymard, B. 13
Gazzaniga, M.S. 25 50, 53, 81, 323
Geary, D.C. 401
Gehring, W.J. 406
Geisler, W.S. 398
Gelman, R. 43, 46
Gelman, S. 50, 52
Geman, D. 156
Genovese, C.R. 352, 418
Genovesio, A. 484
Gentaz, E. 42
Gentilucci, M. 25
Georgopoulos, A.P. 20, 402, 420, 421, 425, 428, 433, 442, 487
Gerardi-Caulton, G. 522
Gerardin, E. 13
Gergely, G. 40
Gerhardstein, P.C. 161
Gerstein, G.L. 243
Gerstner, W. 107
Gevins, A. 513
Gewaltig, M.-O. 106
Geweke, J. 445
Geyer, S. 520
Ghahramani, Z. 401
Ghosh, S. 421, 429
Giacomini, E. 519
Giard, N. 489
Gibbon, J. 43
Gibbs, B.J. 37, 325, 339
Gibson, B.S. 163
Gibson, J.R. 234
Gilbert, C.D. 113, 303, 311, 507
Gillespie, N. 276
Girelli, M. 270, 277, 287
Gitelman, D.R. 272, 294, 398
Givon, T. 513
Glaser, D. 243
Glauche, V. 351
Gleason, C.A. 402
Glickstein, M. 482, 490
Glimcher, P.W. 397
Glover, G.H. 92, 403, 452
Gobbini, M.I. 83, 85, 87, 89, 91, 159, 170, 188, 189
Gochin, P.M. 243, 307
Goebel, R. 107, 110, 113, 117, 357, 442, 449, 452, 453, 457
Gold, J.I. 396, 397
Gold, L. 358

Goldberg, M.E. 270, 282, 348, 365, 367, 398
Goldin-Meadow, S. 475
Goldman, A.I. 19, 405
Goldman-Rakic, P.S. 258, 317
Goldsmith, H.H. 522
Gonzalez-Garcia, M. 62
Gonzalez-Lima, F. 440
Goodale, M.A. 25, 170, 234, 420, 423, 426, 428, 431, 432, 483, 484, 486, 488, 489, 496, 497
Goodman, S.J. 487
Goodridge, J.P. 32
Goodyear, B.G. 201
Gopnik, A. 38
Gord, J.C. 512
Gordon, R.D. 405
Gore, J.C. 84, 159, 170, 198, 209, 226
Gorno-Tempini, M.-L. 221
Gos, A. 270, 323
Gottlieb, J. 367
Goudreau, G. 65
Gouteux, S. 32, 35
Grady, C.L. 226
Graf, W. 483, 486, 495
Grafman, J. 25, 226
Grafton, S.T. 13, 226, 422, 467, 491
Grainger, J. 205
Grajski, K.A. 58
Grammont, F. 402
Granger, C.W.J. 440, 444, 446
Grasby, P.J. 84, 91, 228
Grassi, F. 13
Gratton, G. 401, 403, 404
Gray, C.M. 101, 102, 108, 110
Gray, J.R. 357, 358
Gray, M.D. 174
Gray, R. 276
Graziano, M.S.A. 483
Gréa, H. 489, 491
Green, D.M. 396
Greenberg, M.E. 75
Greenwald, A.G. 20, 205, 212, 213, 214
Greenwood, P.M. 516, 517
Gregoire, M.C. 422
Grethe, J.S. 491
Greve, D.N. 210
Grèzes, J. 13, 20, 472
Gribble, P.L. 496
Grice, G.R. 404
Grill-Spector, K. 159, 160, 170, 173, 180, 190, 197, 198, 201, 210, 214, 418, 420
Grinvald, A. 113, 116, 243
Grodd, W. 198
Gross, C.G. 234, 235, 241, 243, 252, 307, 483
Grossenbacher, P. 513
Grossi, D. 117, 442
Grove, E.A. 65, 71
Gruber, D. 517
Grun, S. 109, 402
Grusser, O.-J. 482
Gu, Y. 62

Guillery, R.W. 303, 306
Gulisano, M. 64
Gurvan, L.C. 519
Gutman, D. 371
Guyon, I. 151

Haan, R. 205
Haaxma, H. 490
Habib, M. 209
Hacker, H. 117
Hadjikhani, J.D. 418
Hadjikhani, N. 382, 386, 434
Hadland, K.A. 449
Haggard, P. 402, 490
Hagner, T. 271, 383, 384, 385
Haidt, J. 47
Halgren, E. 95, 198, 200, 442, 509, 510
Hall, E.K. 382, 386
Hallett, M. 402, 467
Halliday, D. 467
Hämäläinen, M. 464, 465, 468
Hamilton, R. 514
Hamzel, F. 351
Handy, T.C. 322
Hanes, D. 472
Hanes, D.P. 399, 401, 403, 404, 406
Hänninen, R. 475
Happe, F. 48
Haprin, J. 517
Hare, B. 40
Harel, M. 113, 170, 197
Hargreaves, E.L. 486
Hari, R. 20, 21, 95, 441, 464, 465, 466, 467, 468, 469, 470, 471, 472, 473, 474, 475, 476
Harle, K. van 54
Harmony, T. 329
Harner, A.M. 330
Harries, M. 10, 11, 241, 252, 472
Harter, M. 270, 287
Harvey, M. 489
Hasbroucq, T. 401
Hasegawa, I. 274
Hasegawa, T. 42
Hashikawa, T. 234
Hasson, U. 84, 95, 158, 170, 189, 198, 199, 201, 311, 418
Hauser, H.D. 42
Hauser, M.D. 37, 38, 40, 46, 47
Hausser, C.O. 489
Haviland-Jones, J.M. 53
Haxby, J.V. 83, 84, 85, 87, 89, 91, 159, 170, 188, 189, 198, 226, 352
Haydar, T.F. 62
Hayward, W.G. 159, 161
Hazeltine, E. 312
He, S. 330
He, Z.J. 171, 322, 371
Head, A.S. 252
Hecaen, H. 489

Hecke, P. 352
Heckhausen, H. 402
Heeger, D.J. 117, 201, 270, 322, 452
Heeley, D.W. 431
Hefter, H. 421, 422, 433
Heil, P. 79
Heilman, K.M. 488
Heimann, M. 52
Heinze, H.J. 270, 271, 303, 306, 323, 383, 384, 385
Hénaff, M.A. 207
Henderson, D. 151
Henderson, J.M. 326
Hendler, T. 84, 95, 159, 170, 180, 189, 190, 197, 198, 199, 201, 210, 226, 236, 418
Hendrich, K. 487
Henson, R.N. 214, 215, 221
Herbin, M. 74
Herculano-Housel, S. 109, 111, 116
Hering, E. 93
Hermer, L. 32, 33
Hermer-Vasquez, L. 33
Hernandez, A. 394
Hernandez, H. 393
Hershey, K. 522
Herzog, H. 512, 520
Hesselmann, G. 328, 338
Heva, R. 80
Hevner, R. 65, 67
Hichiels, J. 352
Hietanen, J.K. 10, 11, 241, 252
Hikosaka, I. 402
Hikosaka, K. 234
Hikosaka, O. 401, 403
Hildesheim, R. 243
Hilgetag, C.C. 272, 275
Hillstrom, A.P. 326
Hillyard, S.A. 270, 271, 276, 277, 279, 303, 306, 316, 322, 323, 329, 337, 365, 382, 383, 384, 385, 386, 387, 399, 441, 506, 511
Hinde, R.A. 24
Hinrichs, H. 270, 271, 303, 306, 323, 383, 384, 385
Hiraki, K. 42
Hirschfeld, L. 50, 52
Hodos, W. 434
Hoffman, E.A. 84
Hoffmann, J. 213, 322
Hofsten, C. von 36
Hogan, B.L.M. 65
Hohnsbein, J. 403, 405
Holcomb, H. 259
Holcombe, A. 327, 338, 371
Holmes, A.P. 84, 91, 227
Holmes, G. 489
Holowka, S. 376
Holroyd, C.B. 405, 406
Hommel, B. 212
Honda, H. 484
Honda, M. 402

Honey, R. 236
Hood, B.M. 40
Hooge, I.T. 400
Hoormann, J. 403, 405
Hopf, J.M. 270, 277, 287
Hopfield, J.J. 150
Hopfinger, J.B. 117, 270, 271, 273, 277, 287, 322, 323
Hopkins, W. 376
Horwitz, B. 226, 440, 448
Horwitz, G.D. 395
Hoshi, E. 401
Houseman, J. 449
Howard, K.I. 136
Howard, R.E. 151
Howard, R.J. 448, 458
Hsiao, S.S. 114, 129
Hsieh, S. 349, 449
Hsu, E.W. 448
Hsu, J.C. 136
Hu, Y. 428
Hubbard, W. 151
Hubel, D.H. 171, 392
Huffman, K.J. 59
Huijzen, V. 236
Huk, A.C. 201
Humphrey, G.K. 170, 432
Humphrey, N. 3
Hund-Georgiadis, M. 448
Hundeshagen, H. 270, 323
Huntley-Fenner, G. 35
Hurvich, L.M. 93
Husain, M. 316
Hutchinson, W.D. 476
Huttenlocher, J. 42
Hutton, C. 170, 192
Huttunen, J. 465
Hyder, F. 221
Hyvärinen, J. 11, 420

Iacoboni, M. 13, 20, 21, 158
Ibanez, V. 226, 402
Ietswaart, M. 482, 484, 493
Iglesias, J. 333, 335, 336, 337
Ilmoniemi, R. 464, 465
Ingeholm, J. 352
Inhoff, A. 402, 509
Inoue, K. 488
Irwin, D.E. 391, 398, 403
Ishai, A. 83, 84, 85, 87, 89, 91, 159, 170, 188, 189, 418
Ishii, Y. 447
Itanitsky, A.M. 513
Ito, M. 157, 170, 243, 303, 311
Itzchak, Y. 159, 160, 170, 180, 197, 198, 214, 418, 420
Ivanitsky, G.A. 513
Iverson, J.M. 475
Ivry, R. 402
Iwai, E. 234

Jääskeläinen, J. 468
Jackel, I.D. 151
Jacques, T.Y. 522
Jakobson, L.S. 420, 428, 431, 489
James, T.W. 234, 432
James, W. 20, 161, 170
Jameson, D. 93
Jancke, L. 271, 383, 384, 385
Jandl, M. 117
Jaques, J. 475
Jarmasz, M. 442
Järvelainen, J. 469
Jarvis, M.R. 484
Jaskowski, P. 211
Jeannerod, M. 13, 419, 421, 422, 423, 474, 489, 490, 491
Jeeves, M.A. 252
Jellema, T. 10, 12, 13
Jenike, M.A. 520, 522
Jenkins, W.M. 58
Jennings, J.R. 403
Jenson, O. 109
Jervey, J.P. 367
Jezzard, P. 440, 448
Jha, A.P. 270, 277, 287, 322
Jiang, C. 75, 76
Jiang, H. 158, 159, 197, 198, 241, 417, 420
Jiang, J. 84
Jiang, Y. 48, 159, 170
Jimenez, J.C. 444
Johannes, S. 270, 322
Johansson, R.S. 13, 14, 422
John, E.R. 329
John, M.S. 329
Johnson, D.M. 174
Johnson, J.C. 335
Johnson, J.E. 65
Johnson, K.O. 114
Johnson, M.H. 40, 46, 259, 260, 261
Johnson, P.B. 400, 483
Johnson, S. 46
Johnston, J.C. 352
Johnston, R.S. 431
Jolly, A. 3
Jones, D. 448, 458
Jones, S.S. 38
Jonides, J. 322, 326, 335, 374, 521
Jonsson, T. 13, 14, 422
Jordon, T.R. 431
Josephs, O. 449
Jouffrais, C. 483
Jousmaki, V. 95
Juarrero, A. 405

Kaas, J.H. 69, 70
Kahan, T.A. 205
Kahn, D.M. 59, 72, 73, 74
Kahneman, D. 37, 325, 339
Kaji, S. 252

Kajola, M. 467, 468
Kalaska, E. 402
Kalaska, J.F. 403
Kaminski, M. 444
Kanade, T. 147
Kaneko, C.S. 401
Kanwisher, N.G. 30, 36, 37, 43, 48, 84, 91, 136, 137, 158, 159, 170, 173, 183, 191, 198, 201, 226, 230, 236, 270, 272, 273, 275, 277, 294, 314, 330, 340, 371, 398, 417, 418, 419, 420, 426, 434, 482
Kapadia, M.K. 311
Kaplan, D.R. 76
Karasuyama, H. 62
Kase, C.S. 489
Kaseda, M. 421, 428
Kastner, S. 117, 230, 234, 236, 270, 271, 272, 273, 291, 303, 304, 305, 306, 307, 309, 310, 311, 314, 315
Katsnelson, A. 33
Katz, L.C. 75, 209
Kaufman, K.J. 489
Kaufman, T. 514, 520
Kaufmann, W.E. 458
Kaukoranta, E. 465
Kawano, K. 252
Kawashima, R. 422, 488, 520
Kawato, M. 401
Kehr, H. 110
Keil, F. 41
Keillor, J.M. 428
Keller, I. 402
Kenemans, J.L. 128, 135, 399
Kennard, C. 226
Kennedy, J.M. 517, 518
Kennedy, W.A. 158, 159, 197, 198, 241, 417, 420
Kennett, S. 280
Kerszberg, M. 510, 521
Kerzberg, M. 220
Kestenbaum, R. 36, 39
Keyes, A.L. 287
Keysers, C. 9, 10, 22
Kherif, F. 442
Khorram-Sefat, D. 117
Kiehl, K.A. 357
Kiesel, A. 213
Kim, D.S. 248, 449, 453
Kim, J. 405
Kim, J.N. 395
Kim, K.H. 449, 453
Kim, M. 449
Kim, S.G. 402, 442, 487, 489
Kim, Y. 489
Kim, Y.H. 294
Kimberg, D.Y. 349, 352
Kimura, D. 488
Kincade, J.M. 273, 292, 293, 353, 355, 358, 359
King, M.D. 449
Kinomura, S. 520
Kinsbourne, M. 284
Kiraly, I. 40

Kirveskari, E. 467, 468
Kisvarday, Z.F. 248
Klein, R. 76
Kleinschmidt, A. 170
Kling, A. 11
Klotz, W. 211, 212
Knab, R. 351
Knight, R.T. 272, 275, 357
Knoll, R.L. 402
Knuutila, J. 464
Kobatake, E. 241, 242, 255
Kobayashi, R. 42
Kobayashi, S. 401, 403
Kobotake, E. 234
Koch, C. 107, 407
Koch, K. 520
Koch, M.A. 448
Kochanska, G. 522
Kodaka, Y. 398
Koechlin, E. 206, 207, 210, 211, 213
Koenig, A.L. 522
Koeppe, R.A. 521
Koester, H.J. 117
Kohler, E. 9, 10, 22
Koizumi, M. 401, 403
Kok, A. 128, 135, 399
Kolodny, J. 512, 520
Komilis, E. 491
Konig, P. 107, 108, 109, 110, 111, 112, 113, 116, 444, 445
Kopell, N. 119
Kornack, D.R. 61, 62
Kornblum, S. 400, 401
Kornhauser, J.M. 75
Kornhuber, H.H. 393
Koski, L.M. 21
Kosslyn, S.M. 207
Kounios, J. 128
Kourtzi, Z. 30, 136, 137, 158, 198, 201, 214, 420, 426
Koutstaal, W. 214
Kovacs, G. 221
Kovala, T. 468
Kowler, E. 402
Kramer, S.J. 46
Krams, M. 13, 487
Kreiter, A.K. 109, 110, 113, 444
Kriegeskorte, N. 115, 117, 452
Kristan, W.B. 400
Krubitzer, L. 21, 59, 70, 72, 73, 74, 434
Krug, K. 395
Kuan, C.-Y. 62
Kuba, M. 328
Kubiak, P. 404
Kubora, Z. 328
Kubota, K. 398
Kucharczyk, J. 447
Kuhlmeier, V.A. 39
Kuida, K. 62
Kunde, W. 213

Kushnir, T. 159, 160, 170, 180, 190, 197, 210, 214, 418
Kussmaul, C.L. 323
Kusunoki, M. 367
Kutz, D.F. 484
Kuypers, H.G.J.M. 490
Kwong, K.K. 92, 158, 159, 197, 198, 241, 304, 330, 417, 418, 420

La Mont, K. 43
LaBar, K.S. 294
LaBerge, D. 322
Lacadie, C. 209
Lacerda, F. 52
Lacquaniti, F. 483, 484
Lagravineze, G. 14
Lai, E. 81
LaLonde, M.E. 287
Lamarre, Y. 474
Lamme, V.A.F. 171, 221, 285, 311, 339, 385
Landau, B. 38
Landis, T. 482
Landman, R. 385
Lanfermann, H. 117
Larkum, M.E. 121
Larson, C.L. 522
Lashley, K.S. 402, 507
Laurens, K.R. 357
Lauritzen, M. 358
Lauwereyns, J. 401, 403
Lavie, N. 221, 303, 311, 322, 323, 338
Lawson, D. 260
Le Clec', H.G. 206, 207, 210, 211
Lea, S.E.G. 36
LeBihan, D. 45, 48, 135, 138, 206, 207, 208, 210, 211, 221, 419, 434, 447, 448, 453, 491, 519
Lecas, J.C. 402, 404
LeCun, Y. 151
Ledden, P.J. 158, 159, 197, 198, 241, 417, 418, 420
Lee, C. 400
Lee, E. 65
Lee, K.E. 198
Lee, K.M. 402
Lee, L. 447
Lee, S.H. 107
Legerstee, M. 40
Lehericy, S. 13, 207, 208
Leinonen, L. 11
Lemer, C. 45, 207, 208
Lemery, K.S. 522
Lemon, R.N. 448
Lemus, L. 394
Leonard, C.M. 235, 252
Leonards, U. 107
Leopold, D. 111
Lerner, Y. 170, 197
Lesevre, N. 441
Leslie, A.M. 39
Leutgeb, S. 472

Levanen, S. 441
Levänen, S. 464, 466
Leventhal, A. 472
Levesque, J. 522
Levi-Montalcini, R. 76
Levine, D.N. 489
Levine, S.C. 42
Levitt, P. 65, 225
Levy, I. 84, 95, 158, 170, 189, 198, 199, 311, 418
Lewine, J.D. 509, 510
Lewis, J.W. 434
Lewis, M. 53
Liberman, A.M. 209
Libet, B.L. 402
Liddle, P.F. 357
Limon, J. 65
Lin, C.P. 449
Lin, C.S. 69
Linden, D.E. 115, 117, 357, 442
Lindsten, T. 62
Linenweber, M. 353, 354
Linnankoski, I. 11
Lipsitt, L.P. 53
Lipton, J.S. 42, 46
Lisman, J.E. 109
Litingtung, Y. 65
Littlestone, N. 147
Liu, A. 176
Liu, A.K. 509, 510
Liuhanen, S. 475
Livingstone, M. 171
Lo, D.C. 75
Lobaugh, N.J. 226, 233
Logan, G.D. 397, 403, 405
Logothetis, N.K. 111, 116, 157, 163, 188, 221, 242, 255, 395, 418, 434
Loh, D.Y. 62
Lolley, S. 486
Longuet-Higgins, H.C. 150
Loos, H. van der 69, 71
Lori, N.F. 448, 453
Lotze, R. 20
Lounasmaa, O.V. 464
Lowel, S. 109, 113
Lozano, A.M. 476
Lozsadi, D.A. 306
Lu, X. 402
Lu, Z.L. 302
Lubke, J. 108
Luce, R.D. 391, 397, 398, 399, 404
Luck, S.J. 221, 303, 316, 322, 323, 325, 338, 365, 399, 506
Lueck, C.J. 226
Lugar, H.M. 513
Lui, A.K. 418
Lui, F. 14
Luna, B. 352
Luppino, G. 5, 6, 421, 428, 475, 483, 484
Luster, A. 76
Luu, P. 520, 521

Lynch, J.C. 20, 420, 421, 425, 433
Lyon, G.R. 209, 512
Maas, P. 467
McAdams, C.J. 367, 382, 386
McAllister, A.K. 75
Macaluso, E. 226, 270, 273, 274, 275, 282, 283, 284, 285, 286, 291, 292, 294, 357
McAvoy, M.P. 273, 292, 293, 353, 355, 358, 359
McCandliss, B.D. 511, 512, 513, 514, 516, 517, 520, 522
McCarthy, G. 48, 84, 95, 170, 198, 226, 401, 472, 508
McCauley, C.R. 46
McClelland, J.L. 396, 399, 404
McCleod, A.M.K. 521
McCrory, E. 209
McCurtain, B.J. 352
McDermott, J. 48, 84, 158, 170, 183, 197, 226, 230, 417, 418
McDonald, A.W. 510
McDonald, J.J. 276, 280
MacDonald, P.A. 406
McDonough, L. 36
McEachern, T. 428
McGoldrick, J. 332
McInerney, S.C. 520, 522
McIntosh, A.R. 226, 233, 440
McIntosh, R.D. 489
Mack, J.L. 489
Mackeben, M. 325
McKinstry, R.C. 448, 453
MacLeod, A.K. 215, 236
MacLeod, C.M. 406
MacManus, D.G. 448
McNally, R.J. 522
Macquistan, A.D. 342
Maisog, L.M. 84, 91, 226
Major, G. 117
Mäkelä, J. 468
Makris, N. 449
Malach, R. 84, 95, 113, 158, 159, 160, 170, 180, 189, 190, 197, 198, 199, 201, 210, 214, 226, 236, 241, 311, 417, 418, 420
Maldonado, P.E. 110
Malonek, D. 116
Malpeli, J.G. 400
Malsberg, C. von der 102
Mandler, G. 44
Mandler, J.M. 36
Mangin, J.F. 13, 206, 208, 221, 419, 434, 491
Mangun, G.R. 117, 270, 271, 273, 277, 279, 287, 322, 323, 506
Mao, X. 62
Marcel, A.J. 205
Marchione, K.E. 209, 512
Marconi, B. 484
Marder, E. 392
Margules, J. 31
Maril, A. 214
Marin, O. 65

Marinkovic, K. 95, 198, 200
Markman, E.M. 37
Markram, H. 108
Marotta, J.J. 418
Marr, D. 136, 149, 392
Marrett, S. 382, 386
Marrocco, T.R. 516
Marsault, C. 13
Marsden, C.D. 488
Marshuetz, c. 521
Marsolek, C.J. 207
Martin, A. 84, 91, 159, 170, 198, 433
Martinez, A. 270, 303, 306, 323, 337, 382, 383, 384, 385, 387, 511
Martinez, L.M. 107
Martinez, S. 65, 67
Masaki, H. 401
Masure, M.C. 489
Matelli, M. 13, 475, 483, 484
Matsumara, M. 422
Matsumoto, M. 250
Matsuzawa, T. 45
Mattar, A. 496
Mattiello, J. 448, 453
Mattingly, J.B. 372
Maunsell, J.H.R. 234, 270, 306, 348, 352, 365, 367, 382, 386
Mayberry, R. 475
Mayer, A.R. 323, 338, 357
Mays, L.E. 400
Mazer, J.A. 310
Mazziotta, J.C. 13, 20, 21, 158
Mechelli, A. 447
Mechner, F. 53
Meegan, D.V. 486
Meenan, J.P. 489
Mehler, J. 39
Mehta, A.D. 303, 306, 385
Meiran, N. 349
Mel, W.B. 146, 156
Melhem, E.R. 458
Meltzoff, A.N. 19, 51, 472
Mencl, W.E. 209, 512
Mendola, J.D. 210, 418
Menon, R.S. 170, 201, 234, 418, 423, 426, 442, 485, 486, 491, 492
Menon, V. 403
Merabet, L. 129
Merchant, H. 394
Merkle, H. 487
Merlau-Ponty, M. 5
Mervis, C.B. 174
Merzenich, M.M. 58, 69
Mesulam, M.M. 272, 294, 398
Meyer, D.E. 127, 391, 398
Meyer, J.R. 294
Michel, F. 207, 489
Miezin, F.M. 270, 272, 294, 303, 513
Mikami, A. 398
Mikulis, D.J. 355, 359

Milders, M. 328, 338
Miller, E. 274
Miller, E.K. 42, 243, 259, 307, 367, 399, 401
Miller, F.D. 76
Miller, J. 400, 402
Miller, J.O. 399
Miller, M.J. 434
Miller, S.L. 287
Mills, D. 260
Milner, A.D. 252, 420, 423, 426, 431, 432, 488, 489
Mine, S. 421, 428
Miniussi, C. 287
Mintorovitch, J. 447
Mintun, M. 131, 512
Mishkin, M. 164, 311, 313
Mistlin, A.J. 10, 11, 252, 472
Mitra, P.P. 61
Mitroff, S.R. 37
Mittelstaedt, H. 31
Mittelstaedt, M.L. 31
Mix, K.S. 42
Miyachi, S. 402
Miyashita, Y. 163, 274, 367
Miyashita-Lin, E.M. 65, 67
Miyauchi, S. 330
Moffett, A. 51
Mohlberg, H. 13
Mohr, J.P. 489
Molfese, D.L. 357, 358
Molinari, M. 484
Molnar, Z 66
Monsell, S. 272, 349, 402, 519
Moore, C.M. 325
Moore, T. 275, 398
Moortele, P.-F. van de 45
Moran, J. 270, 303, 310, 311, 313, 365, 367
Mori, S. 448, 449, 453, 458
Moriarty, J. 517, 518
Moriya, M. 241
Morris, J. 227
Morris, R.K. 326
Mortara, F. 431
Mortelmans, L. 352
Morton, J. 46, 261
Moseley, M.E. 447
Moser, H.W. 458
Moses, Y. 145, 151
Motoyama, N. 62
Motter, B.C. 303, 365, 367, 399
Mountcastle, V.B. 20, 367, 392, 393, 420, 421, 425, 433
Mouret, I. 401
Movshon, J.A. 110, 395, 397
Moyzes, M. 517, 518
Mozer, M.C. 322, 330
Muckli, L. 117
Mueller, M. 206, 207, 210, 211
Mulder, G. 399
Mulder, L.J. 399
Müller, K.R. 148

Muller, M.M. 329
Muller, U.A. 151
Munakata, Y. 32, 37, 38, 46
Munk, M.H.J. 109, 110, 111, 115, 116
Munkholm, P. 51
Munoz, D.P. 397, 402, 403, 404, 486
Münte, T.F. 270, 271, 277, 279, 323
Murata, A. 6, 421, 428
Murias, M. 517, 518
Murphy, K.J. 483, 489
Murray, B. 129
Murray, E.A. 401
Murray, K. 522
Murthy, A. 398, 399, 400, 401, 404
Mushiake, H. 402
Mutani, R. 431

Naccache, I. 205, 206, 207, 208, 210, 211, 212, 213, 220, 221
Nagamine, T. 467
Naito, E. 422, 520
Nakagawa, T. 447
Nakagawa, Y. 65
Nakahara, H. 402
Nakahara, K. 274
Nakamura, K. 402
Nakayama, K. 62, 170, 171, 322, 325, 371
Nakayama, K.I. 62
Nätänen, R. 402
Nazir, T. 162
Needham, A. 38
Neely, J.H. 205, 212
Neggers, S.F.W. 484
Negishi, I. 62
Nehaniv, C. 26
Neisser, U. 371
Nelson, C.A. 258, 259
Nelson, R.J. 69
Neuenschwander, S. 107, 110, 113, 116
Neumann, O. 211, 212
Neville, H.J. 260, 515
Nevo, E. 74
Newell, F.N. 512, 520
Newsome, W.T. 365, 393, 395, 397
Nguyen, N. 75, 76
Nicholas, 370
Nicholls, M.E. 276
Nicolle, D.A. 489
Niebur, E. 107, 114
Niedeggen, M. 328, 338
Nieder, A. 42
Nielsen, M. 330
Niemi, P. 402
Nieuwenhuys, R. 236
Niki, N. 421
Nikolaev, A.R. 513
Nishitani, N. 20, 469, 470, 471, 472, 473
Nishizaki, M. 243, 244, 245, 246, 247
Nobre, A. 226, 272, 287, 294, 314, 398

Noesselt, T. 271, 383, 384, 385
Noll, D. 406
Nomura, Y. 447
Norman, D.A. 405, 447
Norris, D.G. 448
Nosofsky, R.M. 392
Nothdurft, H.C. 308
Nullmeyer, R. 404
Nunez, G. 62
Nyberg, L. 418
Nyman, G. 11

O'Connor, D.H. 303, 304, 305
O'Craven, K.M. 236, 275, 277, 330, 371
Oeffelen, M.P. van 43
Oeltermann, A. 116, 188, 418, 434
Oepold, D.A. 121
Ohbayashi, M. 274
Ohkubo, Y. 65
O'Kane, G.C. 45
O'Keefe, J. 31
Oldenquist, A. 393
O'Leary, D.D.M. 65
Ollinger, J.M. 273, 292, 293, 348, 353, 354, 355, 358
Olshausen, B.A. 162
Olson, R.K. 517
O'Neill, S.G. 234
Oram, M.W. 10, 241, 252
Orban, G.A. 221, 352
O'Regan, J.K. 162
O'Reilly, R.C. 32, 37, 38, 46
Ortega, J.E. 10, 11
Osman, A. 127, 391, 398
Ostry, D. 376
Osu, R. 423, 426
O'Toole, A.J. 94
Otten, L.J. 399, 401, 402
Otto-de Haart, E.G. 486
Owen, A.M. 509
Ozyurt, J. 403

Paillard, J. 474, 484, 493
Paivio, A. 442
Pajevic, S. 448, 458
Palmer, J. 398
Palmeri, T.J. 392
Palomo, D. 330
Pandya, D.N. 10, 197, 235, 475
Pappert, L. 517
Parasuraman, R. 52, 516, 517
Pardo, J.V. 215, 236, 521
Paré, M. 395, 403, 404
Parga, N. 150, 162
Parker, A.J. 393, 395
Parker, G.J. 448
Parrish, T.B. 294
Pascalis, O. 259, 260

Pascual-Leone, A. 129, 226, 272, 275, 514, 520
Pashler, H. 302
Passarotti, 261
Passingham, R.E. 13, 448, 449, 475, 487, 488
Pasternak, T. 274, 367, 368, 369
Pasti, G. 31
Patawaran, C. 489
Patteri, I. 14
Patterson, W.F. 403, 406
Paulescu, E. 13, 209
Paulesu, E. 512
Pauls, J. 116, 157, 163, 188, 242, 255, 418, 434
Paus, T. 449, 467
Pavani, E. 279
Pavel, M. 398
Pavese, A. 515
Pavesi, G. 467
Pearlson, G.D. 458
Peifer, M. 62
Pelak, V.S. 317
Peled, S. 198, 226, 236
Péllisson, D. 483, 490
Pentland, A. 147
Penttinen, A. 465
Pepperberg, I.M. 45
Peralta, M.R. 307, 310, 370
Perani, D. 13
Peraud, A. 425, 468
Perenin, M.T. 489, 490, 493
Perrett, D.I. 10, 11, 12, 241, 252, 431, 472, 474
Peters, M. 487
Petersen, S.E. 131, 215, 272, 294, 303, 348, 353, 354, 359, 512, 513, 521
Peterson, M.A. 163, 171
Petit, L. 352
Petitto, L.A. 376
Petrides, M. 13, 475, 521
Peyrin, J.M. 76
Pfaff, D.M. 516, 518
Pfaff, D.W. 517
Pfefferbaum, A. 403
Phillips, A.T. 39, 250
Phillips, C. 447
Piaget, J. 258
Piazza, M. 519
Picton, T.W. 323, 329
Pierpaoli, C. 448
Pietrini, P. 83, 85, 87, 89, 91, 159, 170, 188, 189, 226
Pigarev, I.N. 308
Pinel, P. 44, 45, 48, 132, 135, 138, 213
Pinilla, T. 322, 326, 327, 328, 335, 338, 371, 372, 373, 376
Pinsk, M.A. 117, 234, 236, 271, 272, 273, 291, 303, 304, 305, 307, 310, 311, 315
Pisella, L. 489, 493
Piston, D.W. 65
Pitzalis, S. 323, 337
Platt, M.L. 397
Poggio, T. 155, 156, 157, 161, 163, 242, 255

Polakis, P. 62
Poldrack, R.A. 312
Poline, J.B. 13, 84, 91, 207, 208, 221, 228
Poline, J.P. 519
Pollatsek, A. 326, 514
Pollmann, S. 449
Polonsky, A. 117
Pomeroy, S.L. 76
Poo, M-M. 75
Poranen, A. 420
Porro, C.A. 14
Portin, K. 467, 468
Posner, M.I. 44, 131, 212, 272, 293, 302, 322, 360, 391, 398, 405, 507, 509, 510, 511, 512, 513, 514, 515, 516, 517, 518, 520, 521, 522
Posse, S. 13, 421, 422, 433
Postema, C.E. 62
Potter, D.D. 252
Prablanc, C. 483, 490, 491
Prayer, L.M. 447
Price, J.L. 11
Price, N.J. 287
Prinz, W. 19, 20
Proctor, R.W. 392, 397
Procyk, E. 13
Proffitt, D.R. 46
Prvulovic, D. 115, 357, 442
Psaltis, D. 145
Puce, A. 48, 84, 95, 170, 198, 226, 472, 508
Pugh, K.R. 209, 512
Putnam, K.M. 522
Pylyshyn, Z.W. 36, 37, 44, 326, 327, 338, 371
Pytkanen, A. 11

Quinn, P.C. 35, 37

Rabbitt, P.M.A. 405
Racicot, C.I. 489
Rado, R. 74
Rafal, R.D. 312, 322, 357, 360, 371, 402
Rager, G. 107
Ragot, R. 441
Ragsdale, C.W. 65
Raichle, M.E. 131, 215, 236, 448, 453, 511, 512, 521
Raij, T. 95, 441, 464, 466
Rajkowski, J. 404
Rakic, P. 61, 62, 225
Ramachandran, V.S. 110
Rambaldi, M. 64
Ramnani, N. 447, 449
Rand, M.K. 402
Randell, T. 468
Ranjeva, J.P. 442
Ranson, J. 276
Rao, S.C. 396, 399, 400
Rao, S.M. 323, 338, 357
Raos, V. 6
Ratcliff, R. 392, 397, 404

Rauch, S.I. 520, 522
Raymond, G.V. 458
Raymond, J. 323, 332, 374
Rayner, K. 326
Recanzone, G.H. 58, 307, 309, 365, 367
Rees, G. 151, 221, 271, 273, 303, 312, 316, 407, 520
Reese, T.G. 449
Regolin, L. 36
Reid, R.C. 107
Reingold, E. 403
Reinikainen, K. 465
Reisenhuber 156
Reiss, A.L. 403
Remington, R.W. 335, 352
Remond, A. 441
Renart, A. 150
Renault, B. 441
Rensink, R.A. 372
Reppas, J.B. 92, 158, 159, 197, 198, 241, 304, 417, 418, 420
Requin, J. 395, 400, 401, 402, 404
Reuda, M.R. 522
Revol, P. 493
Reynolds, J.H. 114, 117, 298, 306, 307, 309, 310, 364, 365, 367, 368, 369, 370, 373, 374, 375, 376
Reynvoet, B. 213
Ribadeau Dumas, J.L. 489, 490
Richter, W. 402, 442
Riehle, A. 109, 395, 400, 401, 402
Riera, J. 329
Riera-Diaz, J. 329
Riesenhuber, M. 155
Ring, B. 11
Ringo, J.L. 234
Rivaud, S. 207, 208
Rivière, D. 48, 135, 138, 206, 208, 221
Rizzo, M. 493
Rizzolatti, G. 5, 6, 7, 9, 10, 11, 12, 13, 14, 15, 20, 21, 22, 40, 376, 425, 431, 433, 464, 466, 467, 468, 475, 483
Ro, J.Y. 421, 429
Robert, F. 489
Roberts, T. 21, 447
Robertson, L.C. 489
Robinson, B.H. 448
Robinson, C.J. 21
Robinson, D.L. 270, 365
Rocha-Miranda, C.E. 241
Rock, I. 371
Rockland, K.S. 197, 248
Rode, G. 489, 493
Rodriguez, L.M. 444
Rodriguez, V. 326, 328, 329, 330, 333, 335, 336, 337, 444
Roebroeck, A. 447
Roelfsema, P.R. 107, 109, 110, 111, 112, 113, 116, 339, 385
Rogers, J.L. 136, 349

Roh, J.K. 402
Roitman, J.D. 395, 399
Roland, P.E. 270, 520
Rolls, E.T. 150, 162, 221, 235, 252
Romo, R. 393, 394
Rondot, P. 489, 490
Ronen, I. 453
Ronen, L. 449
Rorie, A.E. 114, 117
Rosch, E.M. 171, 174
Rosen, B.R. 92, 158, 159, 197, 198, 200, 210, 214, 241, 304, 330, 417, 418, 420, 520
Rosenbaum, D.A. 346
Rosenberg, J. 467
Rosenblatt, F. 147
Rosier, A. 352
Rossant, J. 80
Rossetti, Y. 483, 489, 490, 491, 493
Rossion, B. 259
Rotella, D. 493
Roth, K.A. 62
Rothbart, M.K. 518, 522
Rothi, L.J. 488
Rothman, D.I. 221
Rouder, J.N. 392, 397, 404
Roussel, S.A. 449
Rovee-Collier, C. 53
Rowe, J. 449
Rowe, J.B. 448
Roy, A. 114
Roy, E.A. 428
Rozin, P. 47
Rubenstein, J.L.R. 65, 67
Rubin, E. 171
Rubin, J. 76
Ruckner, R. 509, 510
Rudolph, K.K. 352
Rueda, M.R. 517
Rushworth, M.F. 13, 259, 357, 449, 487
Ryan, C.M.E. 36

Sack, A.T. 117, 442
Sackinger, E. 151
Sadato, N. 226, 402
Sagami, M. 403
Sagi, D. 271, 418
Sagiv, N. 312
Sahraie, A. 328, 338
Saito, H. 241
Sakagami, M. 401
Sakai, K. 402
Sakata, H. 20, 420, 421, 425, 428, 433
Sakmann, B. 108
Sakuma, H. 447
Saleem, K.S. 234, 248
Salenius, S. 467, 468
Sali, S. 154, 156, 157, 158, 200
Salinas, E. 393, 394, 483
Salmelin, R. 467, 468

Salminen, J. 465
Salonen, O. 468
Salzman, C.D. 395
Sanders, A.F. 138
Sanes, J.N. 485, 486, 487
Santos, L.R. 37, 38, 40, 46, 47
Sapir, A. 349
Sasaki, Y. 330
Sathian, K. 226
Sato, K. 488
Sato, S. 453
Sato, T. 399, 401, 404
Satoh, K. 422
Savoy, R.L. 330
Sawa, H. 62
Scabini, D. 357
Schacter, D.L. 210, 214
Schall, J.D. 111, 221, 395, 396, 397, 398, 399, 400, 401, 403, 404, 406, 472
Schatz, J. 259
Scheffers, M.K. 401, 402, 405, 406
Scheich, H. 79
Schein, S.J. 311, 313
Scherg, M. 270, 323, 329
Schiller, J. 107, 117
Schiller, P.H. 370
Schiller, Y. 117
Schlack, A. 483
Schlag, J. 406
Schlag-Rey, M. 406
Schlaggar, B.L. 513
Schlaghecken, F. 211
Schleicher, A. 13
Schlotterbeck, E. 211
Schmid, U. 425, 468
Schmidt, D. 425, 468
Schmidt, K.E. 113
Schmidt, T. 211
Schmolesky, M. 472
Schneider, W. 212
Schneiderman, H. 147
Schnitzler, A. 488
Schoenfeld, A. 271, 383, 384, 385
Scholl, B.J. 36, 37
Scholz, M. 270, 323
Schouten, J.L. 83, 84, 85, 87, 89, 91, 159, 170, 188, 189
Schräger, E. 277, 278, 279
Schreiner, C.E. 58
Schroder, J.-H. 107, 111, 117
Schroeder, C.E. 303, 306, 385
Schruck, S. 517, 518
Schubert, T. 449
Schülkopf, B. 148
Schult, C.A. 41
Schultz, D.W. 399
Schultz, W. 406
Schuman, E.M. 75, 76
Schürmann, M. 469
Schwark, H.D. 400

Schwartz, P.M. 76
Schwarz, U. 307, 365
Scudder, C.A. 401
Seal, J. 395
Sears, C.R. 326
Sebestyen, G.N. 272, 287
Segal, R.A. 76
Seidemann E. 365
Seiffert, A.E. 270
Seiterle, D.A. 32
Seitz, R.J. 13, 14, 15, 421, 422, 433, 512, 520
Sekuler, R. 346, 353
Seltzer, B. 11, 235
Senju, S. 62
Seppä, M. 468
Sereno, M.I. 92, 97, 176, 197, 198, 200, 270, 303, 304, 306, 323, 337, 382, 383, 384, 385, 387, 417, 418, 511
Sergio, L.E. 376
Sestokas, A.K. 367
Shadlen, M.N. 395, 396, 397, 399
Shadmehr, R. 259
Shahani, U. 467
Shallice, T. 126, 207, 215, 221, 236, 405
Shanks, D.R. 236
Shankweiler, D.P. 209
Shapiro, K.L. 221, 323, 325, 338, 373, 374
Shaw, B.K. 400
Shaw, M. 357
Shaw, P. 357
Shaywitz, A.J. 75, 209
Shaywitz, B.A. 512
Shaywitz, S.E. 209, 512
Shebo, B.J. 44
Sheffers, M.K. 399
Sheinberg, D.L. 111
Shen, L. 401
Shepard, R.N. 12
Sherman, S.M. 303
Shi, L. 65
Shibasaki, H. 467
Shiffrin, R.M. 212
Shikata, E. 351
Shima, K. 401, 402
Shimojo, S. 171
Shimony, J.S. 448, 453
Shindler, K.S. 62
Shiu, L.P. 302
Shmuel, A. 243
Shofield, W.N. 21
Shoham, D. 243
Shorter, A.D. 372
Shoup, R.E. 310
Shtoyerman, E. 243
Shulman, G.L. 270, 271, 272, 273, 274, 292, 293, 294, 303, 314, 346, 348, 353, 354, 355, 357, 358, 359, 398
Shulman, R.G. 221
Shuman, M. 48, 84, 159, 170, 417
Sieroff, E. 514

Simard, P. 151
Simeone, A. 64
Simion, F. 261
Simon, H.A. 136
Simon, O. 13, 419, 434, 491
Simons, D. 39
Simpson, G.V. 271, 277
Singer, W. 101, 102, 107, 108, 109, 110, 111, 112, 113, 114, 115, 116, 117, 444
Sipila, P.K. 449
Sireteanu, R. 107
Sirevaag, E.J. 401, 404
Sirigu, A. 13
Skrebitsky, V.G. 226
Skudlarski, P. 84, 159, 209, 512
Slater, A.M. 36
Slaughter, V. 40, 46
Slovin, H. 243
Smid, H.G. 399, 401, 402
Smith, A.J. 221
Smith, E.E. 521
Smith, H.E. 226
Smith, L.B. 38
Smith, M. 517, 518
Smith, P.A.J. 252
Smith, W.B. 75, 76
Smola, A. 148
Smulders, F.T. 128, 135, 399
Snyder, A.Z. 348, 353, 357, 358, 448, 453
Snyder, C.R.R. 212
Snyder, L.H. 346, 419, 484, 487
Sobel, D. 37
Sohn, M.H. 349, 352
Sohn, M.J. 449
Soja, N. 38
Solaiyappan, M. 458
Solimando, A. 35
Soloviev, S. 162
Somers, D.C. 270
Somersalo, E. 464
Sommer, M.A. 395
Somorjai, R. 442
Sorrentino, C. 39
Sovrano, V.A. 32
Spada, F. 64
Sparks, D.L. 400, 404
Spekreijse, H. 221, 385
Spelke, E.S. 32, 33, 35, 36, 37, 38, 39, 42, 43, 44, 45, 46, 47, 213
Spence, A.M. 517, 518
Spence, C. 271, 275, 276, 279, 280
Spencer, D.D. 84, 95
Sperlich, E.-J. 277, 279
Sperling, J.M. 442
Spiker, V.A. 404
Spinelli, D.N. 226
Spinks, R.I. 448
Spiridon, M. 91
Spitzer, H. 367
Squatrio, S. 484

Squire, L.R. 207
Stanescu, R. 45, 213
Stanescu-Cosson, R. 45
Starr, A. 226
Stedron, J.M. 32
Steinmetz, M.A. 367
Steinmetz, P.N. 114
Stelmach, G.E. 491
Stenger, V.A. 349, 352, 449, 510
Stephan, H. 61
Stephan, K.E. 448, 449
Stephan, K.M. 421, 422, 433
Sternberg, S. 126, 136, 137, 398, 402
Stevens, C.F. 107
Stevens, G. 400
Stewart, M.I. 372
Stieltjes, B. 458
Stiles, C.D. 76
Stoner, G.R. 110, 373, 374, 375, 376
Storm, R.W. 37
Stornaiuolo, A. 64
Strafella, A.P. 467
Streri, A. 42
Strojwas, M.H. 352
Stuphorn, V. 406
Styles, E.A. 349, 449
Su, M.S.-S. 62
Suddendorf, T. 474
Summer, T. 515, 516, 517, 522
Summers, R. 442
Sunaert, S. 352
Sunderland, T. 516, 517
Super, H. 221, 385
Sur, M. 69
Suwazono S. 272, 275
Suzuki, W. 234
Swanson, J.M. 516, 517, 518
Sweeney, J.A. 352
Swets, J.A. 396
Swick, D. 312

Tagami, T. 447
Tai, J.C. 302
Taira, M. 421, 428
Takahashi, T. 422
Takasawa, N. 401
Takasu, M.A. 75
Takeda, K. 447
Talairach, J. 452
Talbot, W.H. 393
Tallon-Baudry, C. 115, 119
Tamagawa, Y. 447
Tan, H.Z. 276
Tanaka, J. 175
Tanaka, K. 157, 170, 201, 234, 241, 242, 243, 248, 252, 255, 420
Tanifuji, M. 243, 244, 245, 246, 247, 250, 252
Tanji, J. 401, 402
Tansy, A.P. 346, 358, 359

Tao, W. 81
Tao, X. 75
Tapia, J.F. 489
Tarr, J.J. 84
Tarr, M.J. 30, 159, 161, 183, 187, 418
Tasker, R.R. 476
Taub, E. 520
Taube, J.S. 32
Taya, C. 62
Taylor, J.B. 226
Taylor, J.G. 448
Taylor, T.L. 406
Teder-Salejarvi, W.A. 276, 277, 279, 329
Tegeler, C. 442
Teller, D. 135
Tempelmann, C. 271, 383, 384, 385
Temple, E. 44
Tennigkeit, F. 108
Terazzi, E. 431
Terkel, J. 74
Terrace, H.S. 42
Teszner, D. 465
Their, P. 483
Theoret, H. 272, 275, 514, 520
Thierry, G. 442
Thinus-Blanc, C. 32, 35
Thomas, K.M. 403, 520, 522
Thomas, S. 10, 11, 241, 252
Thompson, C.B. 62
Thompson, K.G. 221, 396, 398, 399, 400, 401, 404, 472
Thompson, R.F. 226
Thulborn, K.R. 352
Thut, G. 129
Tilikete, C. 489
Tipper, S.P. 371, 486
Tole, S. 65
Tomasello, M. 40
Tomita, H. 274
Tong, F. 117, 170
Toni, I. 422, 449, 488
Tooby, J. 46
Tootell, R.B.H. 48, 92, 158, 159, 197, 198, 200, 210, 226, 241, 270, 304, 382, 386, 417, 418, 420, 434
Torres, K. 327, 328
Tournoux, P. 452
Tovee, M.J. 221
Townsend, J.T. 391, 392
Tranel, D. 170, 522
Traub, R.D. 119
Treisman, A. 37, 325, 330, 339, 340, 398
Treue, S. 365, 398
Trevena, J. 402
Trick, L. 44
Triesman, A. 269
Trinath, T. 116, 188, 418, 434
Trojano, 117
Trojano, L. 442
Troncoso, J.F. 489

Truccolo, W.A. 444
Trujillo, N.J. 329
Trujillo-Barreto, N. 329
Tsao, F. 42
Tseng, W.Y. 449
Tsivkin, S. 45, 213
Ts'o, D.Y. 113
Tsukamoto, T. 447
Tsunoda, K. 170, 243, 244, 245, 246, 247, 250
Tsuruda, J. 447
Tsutsui, K.I. 401, 403
Tuch, D.S. 449
Tucker, D.M. 405, 513, 520
Tucker, L.A. 259
Turk, M. 147
Turka, L.A. 62
Turner, R. 440, 448
Tweedale, R. 21
Tzavaras, A. 489

Ugurbil, K. 442, 449, 453, 487
Ulbert, I. 303, 306
Ullman, S. 144, 145, 151, 154, 156, 157, 158, 160, 161, 162, 163, 164, 200
Ullrich, D. 328
Umilta, M.A. 9, 10, 22
Ungerleider, L.G. 84, 91, 117, 159, 164, 170, 226, 230, 234, 235, 236, 270, 271, 272, 273, 291, 303, 304, 306, 307, 308, 309, 310, 314, 315, 317, 330, 348, 370, 440
Ursu, S. 349, 352, 449
Usher, M. 107, 396, 404
Usrey, W.M. 107
Uswatte, G. 520
Uttal, S.J. 509
Uttal, W.R. 392, 506
Uutela, K. 464
Uylings, H.B.M. 13

Valdes, P. 444
Valdes-Sosa, M. 322, 326, 327, 328, 329, 330, 332, 335, 336, 337, 338, 371, 372, 373, 376
Valentin, D. 94
Valenza, E. 260
Vallortigara, G. 30, 32, 36
van Boxtel, G.J. 403
van Bruggen, N. 449
van de Moortele, F. 519
van de Moortele, P.F. 207, 210, 211
van Den Berg, A.V. 418
van der Lubbe, R.H. 211
van der Molen, M.W. 403
van Essen, D.C. 157, 162, 164, 171, 197, 348, 434
Van Hoesen, G.W. 225
Van Velzen, J. 279, 287, 288, 289, 290
Van Voorhis, S. 270, 271, 277
Van Zandt, T. 392, 397, 404
van Zijl, P.C. 448, 458

Vandegeest, K.A. 522
Vanduffel, W. 382, 386
Vanzetta, I. 243
Vapnik, V. 147, 151
Vauclair, J. 32, 33
Vaughan, J.T. 170, 382, 386
Vecera, S.P. 322, 332, 338, 371
Vento, L.A. 303, 306
Verghese, P. 398
Verleger, R. 211
Verstarten, F.A.J. 330
Vessey, J.T. 136
Vetter, T. 94
Viana Di Prisco, G. 110
Vidal-Naquet, M. 154, 156, 157, 158, 200
Videen, T.O. 215, 236, 521
Vidyasagar, T.R. 285
Vighetto, A. 489, 490, 493
Vigliocco, G. 205
Vilis, T. 418
Villis, T. 485, 486, 491, 492
Vindras, P. 491
Virues-Alba, T. 329
Visalberghi, E. 19
Visscher, K.M. 513
Vitton, N. 402, 404
Vogel, E.K. 221, 322, 338
Vogels, R. 221
Volgushev, M. 114
Volkmann, J. 488
Vollinger, M. 357
Von Bonin, G. 13
von Cramon, D.Y. 449
von der Heydt, R. 171
Voogd, J. 236
Voronin, L.L. 226
Vos, P.G. 43
Vroomen, J. 276
Vuilleumier, P. 214, 272, 312

Waggoner, R.A. 201
Wagner, A.D. 214
Walker, J.A. 360
Wallis, J.D. 401
Walsh, C.A. 63
Walsh, V. 357
Waltz, J.A. 115, 117
Wandell, B.A. 92
Wang, F. 80
Wang, G. 243, 252, 255
Wang, R.F. 32, 33
Wang, S.S.-H. 61
Wang, Y. 472
Ward, L.M. 280
Ward, R. 323, 325, 373
Warrington, E.K. 45, 161, 207
Wassarman, K.M. 65, 67
Watanabe, T. 330
Watson, J.D. 226

Weaver, B. 371
Webster, M.J. 317
Wedeen, V.J. 449
Wehner, R. 31
Wehner, S. 31
Weiller, C. 351
Wein, D. 39
Weir, A. 467
Weiskrantz, I. 161
Welch, J. 36, 37
Welker, E. 69, 71
Wellman, H.M. 41
Wendland, M.F. 447
Werner, G. 393
Wespatat, V. 108
West, A.E. 75
Westheimer, G. 311
Westling, G. 13, 14, 422
Westphal, H. 65
Westwood, D.A. 428
Weyand, T.G. 400
Whalen, J. 43
Whalen, P.J. 520, 522
Wheller-Kingshott, C.A. 448
White, C. 270
White, C.D. 403
White, I.M. 401
White, L.M. 401
Whiten, A. 3, 474
Whittington, M.A. 119
Wiegell, M.R. 449
Wiesel, T.N. 113, 392
Wiggs, C.L. 159, 170
Wijnbergen, C. 243
Wild, H.A. 94
Wilder, K. 156
Wilhelm, S. 522
Wilhelmsen, A. 212
Wilkinson, F. 234
Willen, J.D. 40
Williams, C.R. 84, 91
Williams, J. 474
Williams, M.L.I. 404
Williams, P. 161
Williams, S.C.R. 228
Williams, S.R. 449
Williams, T.D. 36, 38, 46
Willshaw, D.J. 150
Wilson, H.R. 234
Wimberger, D.M. 447
Wing, A.M. 490
Winkler, P. 425, 468
Wise, S.P. 400
Witte, O.W. 488
Wiuggins, C.J. 449
Woelbern, T. 110
Wohlschläger, A. 19, 20, 21, 474, 475
Wojciulik, E. 270, 272, 273, 294, 314, 316, 419
Woldorff, M.G. 271, 383, 384, 385

Wolfe, J.M. 339, 372, 397, 398
Wollberg, Z. 74
Wolpert, D.M. 401
Wong, E.C. 270, 303, 306, 323
Woods, D.L. 271, 277, 329
Woods, R.P. 13, 20, 22, 158, 422, 491
Woodward, A.L. 39, 46
Worden, M.S. 403
Wright, C.E. 402
Wright, E.W. 402
Wu, Y. 516, 517, 518
Wurtz, R.H. 307, 309, 365, 367, 392, 395, 397, 403
Wusthoff, C.J. 43
Wynn, K. 35, 37, 39, 43, 46

Xing, J. 419, 483
Xu, F. 36, 37, 39, 42

Yamaguchi, S. 403
Yamane, Y. 243, 244, 245, 246, 247, 250, 252
Yamazaki, K. 401
Yanagisawa, T. 422
Yang, Y. 403
Yantis, S. 323, 326, 335, 374, 391, 398

Yap, G.S. 483
Yeshurun, Y. 302, 310
Yip, L.-W. 79
Young, K.E. 65
Young, M.P. 248, 252
Yousry, T. 468
Yousry, T.A. 425
Yu, K.P. 372

Zador, A. 107
Zainos, A. 393, 394
Zanella, F. 117, 357, 442
Zanforlin, M. 31, 36
Zangaladze, A. 226
Zangl, R. 39
Zarahn, E. 84, 170
Zeki, S. 226
Zemel, R.S. 322, 330
Zentall, T. 51
Zhang, L.I. 75
Zhang, Q. 62
Zheng, T.S. 62
Zhou, H. 171
Zilles, K. 13, 14, 15, 520
Zingale, C.M. 402
Zohary, E. 198, 201, 226, 236

Subject Index

action representation 466
actions 4–22
activity-dependent changes 75–6
activity-dependent contributions 72
activity-dependent mechanisms 59, 71, 74, 76–8, 162
ADAN 287–91
adaptation paradigm 201
agent 6, 16, 29, 31, 38–46, 393, 405, 473, 474
alerting 516–18
anagrams 208–10
animal navigation 31, 46
anterior cingulate cortex 358, 405, 476, 512, 520–21 *see also* cingulate
anterior intraparietal (AIP) 417, 419, 420, 421, 424, 425, 427
apoptosis 62
area F5 4–10
area LP 10
area MT 107, 225, 330, 395, 396
area TE 241–54
area TEO 248
arithmetic operations 43
arm movement 484–6, 491, 496, 497
assembly coding 102, 103, 118
associative learning 225–39
asymmetries 486–8
attention 301–19, 321–44, 381–8
attention networks test (ANT) 515
attention-related baseline increases 303, 304, 312
attentional blink (AB) 221, 321, 323–6, 376
attentional competition 322–3
attentional deficits 272, 313, 372
attentional load 312–13, 314
attentional networks 515–16
attentional response enhancement 304, 312, 314
attentional response modulation 303, 305, 306
attentional-load dependent suppression 304, 312
audio-visual associations 226
audio-visual learning 225–39
audition 233, 270, 278–9, 281, 287, 288, 497
autism 476
  autistic subjects 474–5
automatic spreading activations theory 212

baseline-shifts 273, 274
basic action 405
basic level category 173–5, 177, 181
basket cells 248
Bcl-2 62, 64
BDNF *see* brain-derived neurotrophic factor
behaving monkeys 111, 392

bias 261, 273, 274, 352, 364, 367, 376, 482, 488, 490
biased choice theory 397, 398
biased-competition 363, 364
  model 364–70, 386, 509
bilateral enucleated animals 74
binding 102, 103, 105, 114, 115, 118, 339
BMP *see* bone morphogenic protein
body language 466
bone morphogenic protein 65, 75, 77
bootstrapping 261
brain development 257–64
brain-derived neurotrophic factor 75, 76
Broca's area 13, 14, 16, 69, 463–80
burst firing 117

Ca^{2+} entry 117
canonical neurons 6, 228, 423
capacity limit 268, 302
beta-catenin 62–4, 76
cell assemblies 118
central states 111–14
cetaceans 58, 64
choice 393, 519
cholinergic 516
cingulate 221, 358, 405, 476, 512, 517, 520–22
  see also anterior cingulate cortex
circuitry 510–15
classical RF 303, 307–16
co-evolution of the motor system 75
cognitive architecture 29, 30, 47, 339
column 95, 109, 113, 243–9, 253
combination rule 127, 128, 131, 137
comparative approach 59
  studies 40, 57, 70, 72, 74
competition 273–5, 307–9, 321–3, 364–70
competitive circuits 363–80
competitive interactions 274, 307–11, 315, 365, 370
composite measure 127, 128, 131
conflict 406–7, 516–17, 520–22
conjunction visual search 399
conjunction-specific neurons 99–102, 115, 118–19
consciousness 206, 210, 220, 322
contextual factors 311
contextual modulation 311
continuous flow 399
contrast response functions 367
control processes 272–3, 287–91, 488, 490, 492
control structures 272, 274, 275, 294
coordinate transformation 483, 485, 489, 494–5
core knowledge 29–55

core representations 29, 31, 41, 47, 48
    of numbers 44
    of objects 44–5
    of persons 41
cortical domain 67, 70, 72, 74
cortical field emergence 65
cortical field magnification 70
cortical magnification 67, 69, 70
cortical plasticity 58
cortical sheet expansion 64
cortical sheet size 61, 62, 63
cortico-cortical feedback projections 314
corticogenesis 62
corticothalamic feedback 306
countermanding task 403, 406
covert orienting 398, 518
cross-modal interactions 265, 275, 277, 287, 294
cross-modal learning 225, 232, 236
cross-modal links in spatial attention 275–82
current dipole 464
current-dipole model 464

decision process 393, 394, 397, 520
delayed grasping 428–30
development 29–55, 57–81, 257–63, 505–28
diffusion MRI 447, 449
diffusion tensor imaging (DTI) 447–9
    diffusion tensor MRI 514
discrete flow 398, 399, 401
distracter stimuli 398, 399, 400
distributed network 302, 306, 314, 316, 317, 521
distributed representation 93, 95, 248–9
divided-attention 276
dolphins 64
domain-general systems 29, 30, 34, 35, 47
domain-specific cognitive mechanisms 30
domain-specific modules 47, 507
dopaminergic 405, 516
dorsal pathways 306
dorsal stream 195, 330, 367, 418, 420, 432, 488, 510
dorso-lateral prefrontal cortex 236, 258, 345, 354, 395, 431, 439
double dissociation of subprocesses 126
double dissociation of tasks 126, 129
dual-task 33, 276
dynamic grouping 106
dyslexia 209

early selection 267, 302
eccentricity biases 189, 199
EEG 111, 115, 116, 440–3, 465–6
effective connectivity 285, 440, 443–4, 446
efferent copy 473–4
effortful control 519, 521–23
electrical stimulation 108, 170, 395, 396
elephants 64
Emx1 64
Emx2 64, 65, 66

encapsulation 29, 30, 32, 33, 35, 36, 40, 46, 126, 135, 507
epigenetic 57
ERPs *see* event-related potentials
error-related negativity 405
event-related fMRI 226, 439, 492, 509, 510
event-related potentials 321–44, 386, 387
evolutionary psychologists 46
executive 272, 405, 515–23
executive attention network 516
expectancy 35, 114, 293, 355
expectancy violation 35
experience 16, 38, 505–28
expertise 159, 184, 198, 209, 258–9, 508
extinction 372, 514
extrastriate cortex 303, 312, 321–44, 363–79
eye-hand coordination 481–502

face neurons 252–3
face perception 84, 94
false alarms 156, 181
feedback 301–20, 364–5, 384–5, 481–504
FFA *see* fusiform face area
FGF8 *see* fibroblast growth factor-8
fibroblast growth factor-8 65, 70, 71, 77
filtering out unwanted information 301
fMRI *see* functional magnetic resonance imaging
fMRI adaptation 201, 214
frontal eye fields (FEF) 221, 314, 315, 345, 395, 490
frozen orofacial gestures 470–3
frugivory 61
functional architecture 83, 84, 94, 114, 197
functional connectivity 443–4
functional dependence assumptions 443, 444
functional magnetic resonance imaging 13, 83–97, 301–20, 330, 382–7, 427–32, 440–8, 456–8, 491–2, 510–14
functional map organization 59
functional specialization 225, 226, 447
fusiform face area (FFA) 92, 95, 183–90, 230, 232, 234–6, 457, 458
fusiform gyrus 48, 158, 170, 175, 180, 184, 185, 196, 198–200, 205, 207–8, 225–6, 230–4, 283–4, 352, 382, 508

gamma frequency 105
gaze 284, 286, 287, 395, 398–400, 406, 482–90, 493
gaze-coding 482
GCMs *see* Granger causality maps
general attention network 315
genes 57–81, 257, 505, 506, 516, 518
genetic specification of cortical areas 67
genetics 517
genotypic 57
gestural communication 476
gesture 475–6
goals 11, 30, 38–41, 129, 150, 322, 345, 364, 365, 367, 370, 404–7, 474, 475

SUBJECT INDEX | 551

Granger causality 440, 443–7
Granger causality maps (GCMs) 440, 453, 454–5
grasparatus 422, 423
grasping 4–6, 417–38, 482, 483, 489–92

habituation 35, 39, 42, 218
hallucinations 117
hand gestures 463, 475
hearing 30, 226, 234, 275, 276, 475
Hebbian learning 106, 234
hemoglobin 243
heritability 517, 522
hierarchy 163, 181, 188, 195, 197, 314, 387
high-density electrical recording 506, 513
hits 175, 176, 179–81, 183, 187
Holmes-Balint syndrome 488–9, 494, 497
homeobox genes 64
homology 419, 420, 433, 434
human ventral stream 165, 175, 177, 179, 187–9, 195
human visual object areas 195–204
20-Hz rhythm 467

imagery 117, 151, 236, 271, 442, 511
imitation 19, 21, 463–79
inferior parietal 11–14, 16, 20, 285, 293, 463, 471, 472, 473, 488
inferotemporal cortex 158, 196, 216, 221, 234, 241–56, 454
information transfer 513–14
inhibition 105, 247–8, 306, 308, 339, 340, 371, 381, 386, 467, 469, 486, 521
instantaneous causality 444
integration 93, 94, 267–99, 307, 339, 364, 370, 372, 386, 447, 482, 496, 515
intentions 19, 212, 463, 475, 476
inter-regional interactions 259, 260
interactive specialization 259–61
interspecies comparisons 419–22
intra- and extra-RF mechanisms 307, 311, 313, 314
intracortical connections 113
intraparietal sulcus 48, 196, 205, 213, 236, 282–3, 291–4, 356, 419–25, 432, 485, 487, 491
intrinsic signal imaging 243–6, 248, 250, 252
intrinsic signals 243–6
invalid targets 357–9
invariance 127, 132, 162–3, 197, 209
inverse problem 466

labeled line coding 101
language 22, 30, 33–8, 39, 41, 44–6, 47
late selection 268, 269, 302
lateral geniculate nucleus (LGN) 66, 72, 74, 301–3, 305, 317, 318, 382
lateral intraparietal area (area LIP) 317, 348, 484
lateral object selective foci (LO) 180

lateral occipital see lateral occipital complex; lateral occipital cortex
lateral occipital complex 417, 427
lateral occipital cortex (LOC) 129, 158, 175, 241, 311, 427
lateral prefrontal cortex 236–7, 258, 354, 355, 394–5, 431, 439, 454, 455, 520–21
lateralized readiness potential (LRP) 136, 210, 399, 519
LDAP 287–91
left handers 488
LFP fluctuations 116, 117
likelihood ratio 152, 155, 396–7, 407
lip forms 471–3
local features 250–2, 254–5
localization of mental operations 507, 508, 510
long-term potentiation 108
longevity 61

M1 cortex 468, 469, 473
macaque monkeys 60, 62, 158, 241–56, 392, 395, 396, 417, 419, 420, 428, 485
magnetic brain stimulation 129–31
magnetic fields 457, 464
magnetic misreaching 492–4
magnetoencephalography (MEG) 12, 267, 268, 381, 382, 383, 441, 464, 488
masking 205, 214, 221, 230, 323, 324, 332, 335
maturation 257–61, 515, 518
MEG see magnetoencephalography
mental chronometry 441, 442, 510
mental processor 126
microstimulation 396
mind-reading 475, 476
mirror neurons 5–10, 463–79
mirror-neuron system (MNS) 12–22
misses 175, 176, 180, 181
modularity 29–55, 125–6, 506–10
modules 125–39, 273, 275, 507, 521
molecular neurobiology 64
monitoring 325, 358, 391, 392, 405–7
motion processing 330, 348
motion selectivity 348, 349, 351, 353
motor act 463, 466
motor cognition 465
motor cortex 127, 128, 205, 212, 400, 401, 404, 425, 454, 467–74, 488
motor cortex rhythm 468, 469
multimodal cortex 236
multiple-object tracking 36–7, 44, 48
mutimodal processing 73, 77, 236, 282–6, 291–4, 386

N1 277–81, 323, 336–9, 338, 384, 387
N1 component 277, 279, 387
natural language 35, 45, 46
natural number concepts 43, 44
natural number representations 45

navigation 31, 33, 35, 46, 48
neglect 215, 312, 357, 509, 514
nerve growth factor (NGF) 76
neural code 214, 217, 220
neurogenesis 61–3
neuromagnetometers 465
neurotrophins 75–6
neutral faces 227, 473
newborns 260, 261
NGF *see* nerve growth factor
NMDA-receptor 114
non-human primates 31, 40, 46, 58, 69, 419, 434
noradrenergic 516
novelty detection 42, 225, 237
NT3 76
NT4/5 76
number comparison 212
number concepts 31, 43, 44, 46
number line 213
number sense 42–6
numbers 31, 42–6, 47, 507
numerical discrimination 42
numerosity 42–6

object categorization 171
object detection 169, 171, 173, 177, 180, 181, 187
object identification 160, 177, 180, 181
object parsing 36, 37, 38
object recognition 141–68, 169–93, 426–7, 431–3
object representation 36, 38, 43, 44, 91, 149, 157, 188, 241, 242, 249, 340
object vision 83, 84, 235
object-based selection 386, 387
object-file 321, 335, 339
object-tokens 324, 326, 327, 330
observation 3–26, 463–79
occipitotemporal cortex 84, 158, 170, 171, 189, 195–204
occipitotemporal sulcus 175, 186
odontoceti 64
opossum (*Monodelphis domestica*) 60, 68, 69, 72–5
optic ataxia 420, 421, 482, 488–95, 497
orienting of attention 505–28
oscillations 107–9, 113–16, 468
oscillatory activity 108–9, 113, 115
oscillatory cortico-muscular coupling 467
OTS *see* occipitotemporal sulcus
Otx1 64
Otx2 64

P1 277–81, 289–91, 323, 328, 338, 383, 384, 387
P300 221, 399, 401
pain 476
parietal lobe 12, 13, 16, 20, 21, 267–99, 272, 301–20, 345–62, 372, 395, 396, 398, 399, 403, 417–38, 439–62, 471, 472, 484, 488, 490, 509, 510, 514

parahippocampal place area (PPA) 158, 186, 189, 191, 457, 458
partial reinforcement 227
perceptual expertise 259
perceptual grouping 109–10, 322
perceptual sensitivity 276, 302
perceptual task 177
peripheral innervation 69, 71
peripheral morphology 67, 70, 72, 77, 78
phenotypic variability 57, 58, 76
plaid stimuli 110
plasticity 108, 257–63, 418, 496, 505, 506, 515
platypus 60, 69, 70, 74, 77
population response 83, 85, 93–5, 104, 105, 428
positron electron tomography (PET) 12, 440, 467, 488
Posner task 347, 355, 356, 358, 359
posterior parietal cortex 317, 395, 418, 439, 442, 454, 455, 481, 482, 484–5, 491, 494
PPA *see* parahippocampal place area
PPC *see* posterior parietal cortex
preattentive 114, 118, 272, 398, 519
precursors of human language 476
prefrontal cortex 78, 225, 236, 258, 259, 272, 274, 345, 346, 349, 355, 394, 395, 431, 439, 454, 455, 520, 521
preparatory activity 267, 271, 291, 292, 346, 352, 510
preparatory signals 349, 351, 352
primary motor cortex 394, 401, 463, 467–74
primary visual cortex 107, 147, 216, 241, 248, 270, 382, 483, 514
priming 205–24
problem of agency 473–4
proboscidea 64
process decomposition method 126–31
process decomposition use fMRI 131–3
processes and processors 133–5
processing hierarchy 314, 387
pulvinar 261, 306
pure measures 126–7, 128, 131, 132
purpose 150, 151, 154, 161, 393, 404, 417, 493
push-pull mechanism 301, 312, 313–14

race 397, 403
random walk 397
rapid serial object transformation (RSOT) 321, 325
rapid serial visual presentation (RSVP) 321, 323, 324
rate codes 119
reaching 417–38, 481–504
reaction time 113, 173, 175, 207–10, 228–9, 232, 357, 391, 395, 398, 399, 401–4, 441, 442, 452, 471, 516, 517, 523
readiness potential 136, 210, 399, 402, 519
receptive field 105, 109, 147, 303, 309, 311, 317, 363, 368, 375, 385, 395, 509
reciprocal imitation 472
recognition 83–97, 143–67, 169–93, 417–38

SUBJECT INDEX

regions of interest 180, 418, 443
reinforcement 227, 234, 397, 406
relational code 109
reorientation 31–3
repetition priming 206, 207, 209, 212, 219–20
repetition suppression 207–9, 213–15, 217–19
representation of agents 31, 38–41, 46, 48
representation of objects 38–41
representations of human agents 39, 40
representations of space 29
representing objects 35–8
response preparation 398, 399, 402, 404, 406
response synchronization 106, 108, 109, 110, 111, 114, 115–16
retinal sampling 200
retinotopic coding 483
retinotopic cortex 170, 196, 197, 198
RF *see* receptive field
right handers 487, 488
right sidedness 487
rivalry
ROIs *see* regions of interest
RSOT *see* rapid serial object transformation
RSVP *see* rapid serial visual presentation

saccades 395–8, 400, 403, 404, 406, 481, 483–6, 491, 492, 495–6, 497
scalp signature 513
scene organization 321–44
schizophrenia 475, 476
secondary somatosensory cortices 465, 474
SEF *see* supplementary eye field
selective attention 267–99, 301–20, 321–44
selective influence 127, 129, 137
selective processing 260, 364
self-organization 103, 114
semantic processing 212, 512
sensorimotor integration 481–504
sensory domain shifts 59
sensory receptor array 67, 72, 77
sensory-specific ERP 270, 279, 294
separate modifiability 125–7, 131, 132, 135
serial modules 128
Shh 65, 74, 77
short-term memory 115–16
signal detection theory 397, 398
signaling centers 64, 65, 66, 71, 74
simulation theory 19
single-unit recordings 241, 243, 245, 247, 248, 250, 363, 365, 385
skill learning 77, 257, 260
social brain 466–7
social interactions 463
sociality 61
somatosensory cortex 66, 68, 69, 114, 129, 275, 282, 285, 393, 394, 418, 421, 474
somatosensory system 393, 514
somatostimulation 424, 425, 429
sonic hedgehog 65

spatial attention 310, 313, 315, 357, 363, 364, 370, 371, 372, 381–7
spatial language 33
spatial layout 31–5, 441
spatial representation 31, 33, 34, 46
spatial selective attention 309
spatio-temporal representations 35–8
specialized learning mechanisms 4
speech 13, 14, 22, 402, 463, 475, 476
speech production 463, 475, 476
speech-related gestures 475
SPL *see* superior parietal lobule
spontaneous (baseline) firing rates 316
spurious causality 445, 446, 447
SQUIDs 464
star-nosed mole 69, 70
stimulus salience 268, 302, 322, 352, 355, 358
stimulus-driven control 352–9
stimulus-driven mechanisms 345–62
stimulus-response associations 400, 439
structural assumptions 443, 444
subcortical structures 65, 257, 261, 272, 303
subliminal priming 205–24
subordinate category 173, 175, 176, 179, 181, 183, 190
subordinate recognition 183–7
subtractions method 516
superior colliculus 107, 306, 395, 483, 486, 490
superior parietal 13, 20, 21, 314, 315, 439, 488, 509, 510, 517
superior temporal sulcus 10–11, 20, 48, 95, 234, 235, 243, 471–3
superposition problem 104, 105
supplementary eye field 314, 315, 406, 483
supplementary motor area 394, 405, 454
suppressive interactions 308–10, 330, 339, 377
surfaces 110, 321–44, 372–6
symbolic systems 29
synaptic plasticity 108, 117
synchronization 99–123
synchronization codes 119

tactile discrimination 114, 289, 393–5
Talaraich coordinates 472
target stimuli 216, 217, 273, 302, 311–13, 346, 352, 354, 355, 358, 359, 360
task preparation 345, 346, 348, 351, 354, 359, 360
task-comparison method 125, 126, 128, 129, 131
task-general processing modules 133
temporal attention 211, 212
temporal codes 99, 116
temporal cortex 83–97, 418, 420, 434
temporal patterning 106, 108, 116, 117
temporo-parietal junction 272, 274, 293
thalamic reticular nucleus 303
thalamocortical connections 65, 72, 74, 75

time-resolved fMRI 439–62
TMS *see* transcranial magnetic stimulation
top-down processing 114, 211, 212, 221, 269–74, 301–6, 314, 316, 317, 339, 352, 355, 359, 386, 387, 507
topgraphic pattern analysis 83–97
topography 92, 93, 94, 170, 188, 189, 200, 387, 442
touch 20, 267–99, 423
transcranial magnetic stimulation 5, 12, 125, 126, 275, 289, 357, 442, 467, 506, 514
transparent motion 326–30, 337, 338
TRN *see* thalamic reticular nucleus
true imitation 474

unconscious processes 206
unilateral neglect 509

V4 157, 225, 230, 304, 306–14, 330, 352, 367–70, 376, 382, 385, 387
valid targets 356, 357
valid trials 293, 355, 357
VARETA 329, 336, 337
vector autoregressive (VAR) process 439–62
ventral occipitotemporal cortex 235, 352

ventral stream 169, 169–93, 170, 175, 177, 181, 187–9, 195, 197, 234, 317, 365, 417, 418, 420, 426
ventral visual pathway 169–93
verbal interference 33, 35
verbal and non-verbal 473
visual agnosia 431–3
visual attention 301–20, 321–44, 381–8
visual discrimination 395–7
visual features 93, 94, 106, 157, 163, 207, 209, 241, 248–50, 253, 254
visual search 325, 352, 353, 371, 397–9
visual short term memory 339
visual word form system 207–10, 512–14
visually-guided eye movements 482
visually-guided reaching 481–504
VOT *see* ventral occipitotemporal cortex
VWFA *see* visual word form area

whales 64
Wnt 65, 74, 77
Wnt3 64
word classification 128
word reading 209
word recognition 207, 209, 260
workspace model 220